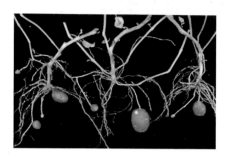

1. Potatoes reproduce vegetatively by stem tubers.

2. Crape myrtle tissue culture.

3. Strawberry plants reproducing by runners.

4. Rooting conifer cuttings under mist.

5. Small quantities of seeds stratified in plastic bags.

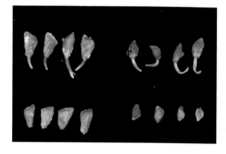

6. Viability tests on conifer seeds.

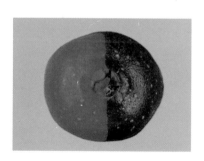

7. A mericlinal chimera in an orange.

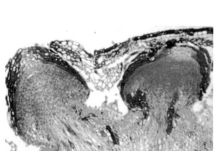

8. Adventitious roots from a stem cutting × 100.

9. Quonset-type greenhouses.

10. Potting machine in wholesale nursery.

11. Some variegated foliage plants are natural chimeras.

12. Germinated avocado seed.

13. Biodegradable pots.

14. Air layering India rubber plants.

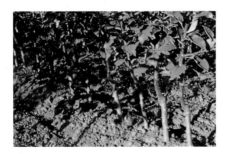

15. Nursery trees propagated by T-budding.

16. Tissue-cultured microplants.

17. Top-grafted plum tree.

18. Technician in a tissue-culture laboratory.

PLANT PROPAGATION
Principles and Practices

FIFTH EDITION

Hudson T. Hartmann, Ph.D.
University of California, Davis

Dale E. Kester, Ph.D.
University of California, Davis

Fred T. Davies, Jr., Ph.D.
Texas A & M University
College Station

Prentice-Hall International, Inc.

The authors and publisher are not responsible, nor do we accept any liability for, loss, damage, or injury resulting from the application of any product, concept, or procedure discussed in or derived from any part of this book. It is expected that propagators would initially conduct small empirical trials to test products or techniques to fit their particular propagation system and environmental conditions before initiating large scale production. In the interest of factual reporting, trade names and proprietary products may be mentioned. No endorsement of names or products is intended, nor is criticism implied of products that are not mentioned.

This edition may be sold only in those countries to which it is consigned by Prentice-Hall International. It is not to be re-exported and it is not for sale in the U.S.A., Mexico, or Canada.

Prentice-Hall International (UK) Limited, *London*
Prentice-Hall of Australia Pty. Limited, *Sydney*
Prentice-Hall Canada Inc., *Toronto*
Prentice-Hall Hispanoamericana, S.A., *Mexico*
Prentice-Hall of India Private Limited, *New Delhi*
Prentice-Hall of Japan, Inc., *Tokyo*
Simon & Schuster Asia Pte. Ltd., *Singapore*
Editora Prentice-Hall do Brasil, Ltda., *Rio de Janeiro*
Prentice-Hall, Inc., *Englewood Cliffs, New Jersey*

Contents

iii

Preface

The fifth edition revision of *Plant Propagation* was necessary due to the considerable amount of new information that has been forthcoming in the field of plant propagation since the fourth edition was published in 1983. A general textbook on plant propagation covers many areas. A great amount of new technical and scientific knowledge has recently been developed in most of these areas by university researchers, private industry, and commercial plant propagators. In this new edition we have attempted to include all the new significant information so as to keep abreast of the changing times while retaining the essential material that has been included in previous editions. The sections on propagation facilities, cutting propagation, clonal selection and purity, as well as tissue culture micropropagation, have, in particular, been considerably enlarged and updated. The chapters on propagation of ornamentals have also been enlarged.

This book was written primarily as a text for university-level plant propagation courses. It provides information concerning the fundamental principles involved in plant propagation and, in addition, serves as a manual that describes the many useful techniques for propagating plants. It is presumed that the students using this book will have had some background training, either in high school or college biology or botany courses, although the chapters covering the techniques of plant propagation can usually be handled successfully without such background.

This book covers all aspects of the propagation of higher plants, both sexual and asexual, especially in reference to *human* efforts to increase plant numbers, as contrasted to plant reproduction *in nature*. The text is organized into five major units:

The first, consisting of two chapters, covers general information pertaining to the various types of propagation, such as the facilities, equipment, and supplies needed, nomenclature, and organizations supporting propagation activities.

The second unit, consisting of five chapters,

deals with the production of seeds and the methods of using them in sexually producing new plants.

The third unit, consisting of seven chapters, considers all the types of asexual or vegetative propagation—the use of cuttings; grafting and budding; layering, suckers, and runners; as well as the use of specialized structures, such as bulbs, corms, rhizomes, and tubers, to produce new plants.

The fourth unit is comprised of two chapters that deal with micropropagation and tissue culture, using aseptic culture methodology to grow tiny plant pieces for sexual or asexual regeneration under confined, controlled conditions for producing vast numbers of new plants.

The fifth unit consists of three chapters giving, in encyclopedic format, the accepted propagation methods for the important fruit and nut crops, the principal ornamental trees, shrubs, and vines, as well as the common annual and herbaceous perennial plants used as ornamentals.

In considering the various ways by which plants can be propagated—as seeds, cuttings, grafting, etc.—this book has separated the theoretical implications from the techniques involved into different, consecutive chapters. For example, the basic botanical principles underlying grafting and budding are stressed in one chapter, while the techniques of the various grafting and budding methods are described in detail in the two following chapters. Although some readers are very much interested in the fundamental principles of plant propagation, others may be concerned primarily in the actual techniques involved.

An extensive bibliography of the important research literature is included for each subject. In addition, suggested supplementary readings are listed for most of the subjects. These specialized works deal with the topic more extensively than is possible in this book. These references should be valuable for those wishing to study the subject in greater detail.

The term *cultivar* (*culti*vated *va*riety)—now widely used throughout the world in the horticultural literature—is used in this book rather than the earlier and less precise term *variety,* which does not distinguish between botanical varieties and the commonly cultivated varieties of horticultural interest. The Latin plant names used generally conform to those given in *Hortus Third* (Macmillan, 1976).

In preparing the fifth edition of this book, we have depended upon the assistance of authorities in the various fields of propagation and related subjects. They gave their time most generously in reading sections of the manuscript and offering suggestions. We especially wish to thank Bill Barr, Gene Blythe, Donald Durzan, Sharon Duray, Wesley Hackett, Brian Howard, Anton Kofranek, Jaime Lazarte, Keith Loach, George Martin, Art Nightingale, Peter Van de Pol, Douglas Shaw, Dan Struve, Ellen Sutter, Yin-Tung Wang, Rick Wells, and Richard Zimmerman. The responsibility, however, for the final version of the text is that of the authors.

We also thank our wives, Hazel M. Hartmann, Daphne Kester, and Maritza Davies, for their assistance in typing and proofreading throughout the various stages of preparing the fifth edition of this book. The typing of Leslie Wercham was greatly appreciated. We wish to thank, too, both Marilyn Hartmann and Janet Williams, who made many of the drawings used in this book.

Hudson T. Hartmann
Dale E. Kester
Fred T. Davies, Jr.

1

Introduction

Plant propagation is the multiplication of plants by both sexual and asexual means. A study of plant propagation has three different aspects. First of all, successful propagation of plants requires a knowledge of mechanical, environmental, and chemical manipulations and technical skills that take a certain amount of practice and experience to master, such as how to bud or graft, how to make cuttings, or how to use tissue culture procedures. These may be considered as the **art of propagation**.

Second, successful propagation requires a knowledge of plant growth, development, and morphology. This may be said to be the **science of propagation**. The propagator can obtain some of this information empirically by working with the plants themselves, but it should be supplemented with formal courses in chemistry, botany, horticulture, genetics, and plant physiology. Such knowledge aids propagators in understanding why they do the things they do, in doing them

better, and in coping with unexpected problems (7).

A third aspect of successful plant propagation is a knowledge of the different **kinds of plants** and the various possible methods by which certain plants can be propagated. To a large extent the method selected must be related to the responses of the kind of plant being propagated and to the situation at hand.

The propagation of plants has been a fundamental occupation of humankind since civilization began. Agriculture started some 10,000 years ago when ancient peoples learned to plant and grow kinds of plants that fulfilled nutritional needs for themselves and their animals (11). As civilization advanced, people added to the variety of plants, cultivating not only additional food crops but also those that provided fibers, medicine, recreational opportunities, and beauty (10, 16). From the great diversity and variation in plant life it has been possible to select kinds of

1

plants particularly useful to the welfare of humans and their animals and to develop new kinds of plants through the science of genetics (*3, 5, 14*).

Progress in agricultural development has involved an interplay of two separate activities. One is the selection of specific kinds of desirable plants; the second is the reproduction of these kinds in such a manner as to retain their valuable characteristics in large scale production.

Much progress in plant improvement was made long before the modern period of plant breeding (*5, 6, 8*). Some of our present cultivated plants originated mainly by three general methods:

First, some kinds of plants were selected directly from wild species, but under the selective hand of humans, evolved into types that differ radically from their wild relatives. Examples of this group are lima bean, tomato, barley, and rice.

Second, other kinds of plants arose as hybrids between species, accompanied by changes in chromosome number. These plants are completely unique to cultivation and have no single wild relative. Examples of this group include maize (corn), wheat, tobacco, and strawberry.

Third, another group of plants occurs naturally as rare monstrosities. Although unadapted to a native environment they are useful to humans. Among these are heading cabbage, broccoli, and Brussels sprouts.

These important but empirical methods of plant improvement have gradually given way to modern plant breeding efforts that can be said to have started with the Austrian monk, Gregor Mendel, in the 1860s. The explosion of genetic knowledge of principles and application has now provided new and major cultivars.

This effort is being expanded with the development of genetic engineering and biochemical engineering that now promises even greater advancement in crop improvement.

This progress in plant improvement would have been of little significance, however, without methods whereby improved forms could be maintained in cultivation. Consequently, there has been a process of invention and development of techniques for plant propagation. Most cultivated plants either would be lost or would revert to less desirable forms if they were not propagated under certain conditions that preserve the unique characteristics that make them desirable. Throughout history, as new kinds of plants became available, the knowledge and techniques to preserve them had to be learned; conversely, as advances in propagation techniques developed, the number of plants that became available for cultivation increased.

Listed below are the general methods of propagating plants. Most of them antedate recorded history. It is probably no mere coincidence that some of the oldest fruit crops—grape, olive, mulberry, quince, pomegranate, and fig—are also the easiest to be propagated by means of simple techniques using hardwood cuttings. To grow most other tree fruits, budding and grafting procedures had to be developed. Invention of glasshouses in the nineteenth century made the rooting of leafy cuttings possible. The discovery of root-inducing chemicals and the development of mist and fog propagation and tissue culture–produced liners have greatly enhanced nursery procedures. Production of seed crops has been revolutionized by the discovery of genetic principles leading to, among other advances, the production of hybrid seed. A significant step forward occurred with the development of the new biotechnology and genetic engineering, including aseptic micropropagation and in vitro tissue culture techniques described in Chaps. 16 and 17.

Methods of Propagating Plants with Typical Examples

I. Seed (sexual)

    A. Propagation by seed—annuals, biennials, and many perennial plants

    B. In vitro culture systems

        1. Microspore and pollen culture—tobacco, *Datura*

        2. Ovule culture—carnation, tobacco, petunia

        3. Somatic embryogenesis—conifers

        4. Seed culture—orchid

        5. Spores—ferns

II. Apomictic (asexual)

    A. Seed

        1. Nucellar embryos—citrus, mango

2. Adventitious embryony—
   Kentucky bluegrass

III. Vegetative (asexual)

  A. Propagation by cuttings

    1. Stem cuttings

      a. Hardwood—fig, grape,
crape myrtle, rose, willow,
poplar

      b. Semihardwood—lemon,
camellia, holly,
rhododendron

      c. Softwood—lilac, forsythia,
weigela, crape myrtle

      d. Herbaceous—geranium,
coleus, chrysanthemum

    2. Leaf cuttings—*Begonia rex,
Bryophyllum, Sansevieria,* African
violet

    3. Leaf-bud cuttings—blackberry,
hydrangea

    4. Root cuttings—red raspberry,
horseradish, phlox

  B. Propagation by grafting

    1. Root grafting

      a. Whip graft—apple and pear

    2. Crown grafting

      a. Whip graft—Persian walnut

      b. Cleft graft—camellia

      c. Side graft—narrow-leaved
evergreens

    3. Top grafting

      a. Cleft graft—various fruit
trees

      b. Saw-kerf or notch graft—
various fruit trees

      c. Bark graft—various fruit
trees

      d. Side graft—various fruit
trees and conifers

      e. Whip graft—various fruit
trees

    4. Approach grafting—mango

  C. Propagation by budding

    1. T-budding—stone and pome
fruit trees, rose

    2. Patch budding—walnut and
pecan

    3. Ring budding—walnut and pecan

    4. I-budding—walnut and pecan

    5. Chip budding—grape, mango,
fruit and ornamental trees

  D. Propagation by layering

    1. Tip—trailing blackberry, black
raspberry

    2. Simple—honeysuckle, spirea,
filbert

    3. Trench—apple, pear, cherry

    4. Mound or stool—gooseberry,
apple

    5. Air (pot or Chinese)—India
rubber plant, lychee

    6. Compound or serpentine—
grape, honeysuckle

  E. Propagation by runners—strawberry,
Chlorophytum comosum

  F. Propagation by suckers—red raspberry, blackberry

  G. Separation

    1. Bulbs—hyacinth, lily, narcissus,
tulip

    2. Corms—gladiolus, crocus

  H. Division

    1. Rhizomes—canna, iris, banana

    2. Offsets—houseleek, pineapple,
date palm

    3. Tubers—Irish potato

    4. Tuberous roots—sweet potato,
dahlia

    5. Crowns—everbearing
strawberry, phlox

  I. In vitro culture systems

    1. Shoot-tip culture—orchid, carnation, asparagus, chrysanthemum,
fern, strawberry, and many other
kinds of herbaceous and woody
plants

    2. Adventitious shoot formation—
rhododendron, African violet, lily

    3. Micrografting—citrus, apple,
plum

4. Tissue and protoplast culture—tobacco, carrot
   a. Organogenesis
   b. Somatic embryogenesis

CELLULAR BASIS FOR PROPAGATION

Plant propagation involves the control of two basically different types of developmental life cycles—**sexual** (p. 55) and **asexual** (p. 165). Preservation of the unique characteristics of a plant or group of plants depends on the transmission from one generation to the next of a particular combination of genes present on the chromosomes in the cells. The sum total of these genes make up the **genotype** of the plant. The genotype, in combination with the environment, produces a plant of a given outward appearance (the **phenotype**). The function of any plant propagation technique is *to preserve a particular genotype or population of genotypes* that will reproduce the particular kind of plant being propagated.

Meiosis and Sexual Reproduction

Sexual reproduction involves the union of male and female sex cells, the formation of seeds, and the creation of a population of seedling individuals with new and similar or differing genotypes (Figs. 1–1 and 1–2). The cell division **(meiosis)** which produces the sex cells involves reduction division of the chromosomes, in which their number is reduced by half. The original chromosome number is restored during fertilization, resulting in new individuals containing chromosomes from both the male and the female parents. Offspring may resemble either, neither, or both of the parents, depending upon their genetic similarities. Among the progeny from a particular combination of parents, considerable variation may occur.

The outward appearance (phenotype) of a plant and the way characteristics are inherited from generation to generation are controlled through the action of genes present on the chromosomes. Some traits are controlled by a single gene as shown for pea height in Fig. 1–3. Figure 1–4 shows the more complex case where two independent genes affect the appearance of peach fruit. The analysis of inheritance for traits controlled by many genes, which is the usual case, is much more complex and requires statistical analysis to show how closely the offspring resembles the parents. The kind of genes present and how much the environment influences their effect also must be taken into account (*5, 6*).

The two terms **homozygous** and **heterozygous** are useful to describe the genotype of a particular plant. If a high proportion of the genes present on one chromosome are the same as those

---

DEFINITION OF TERMS

Meiosis. Two successive nuclear divisions, in the course of which the diploid chromosome number is reduced to the haploid and genetic segregation occurs.

Mitosis. A form of nuclear cell division in which the chromosomes duplicate and divide to yield two nuclei that are identical with the original nucleus. Usually mitosis includes cellular division (cytokinensis).

Heterozygous. Having different genes of a Mendelian pair present in the same cell or organism; for instance, a tall pea plant with genes for both tallness (*T*) and dwarfness (*t*).

Homozygous. Having similar genes of a Mendelian pair present in the same cell or organism; for instance, a dwarf pea with genes for dwarfness (*tt*) only.

Homologous Chromosomes. The members of a chromosome pair; they may be heterozygous or homozygous.

Genotype. The genetic makeup of a nucleus or of an individual.

Phenotype. The external physical appearance of an organism. It can be expressed as the genotype interacting with the environment.

Fertilization. The union of an egg and a sperm to form a zygote.

Inheritance. The acquisition of characters or qualities by transmission from parent to offspring.

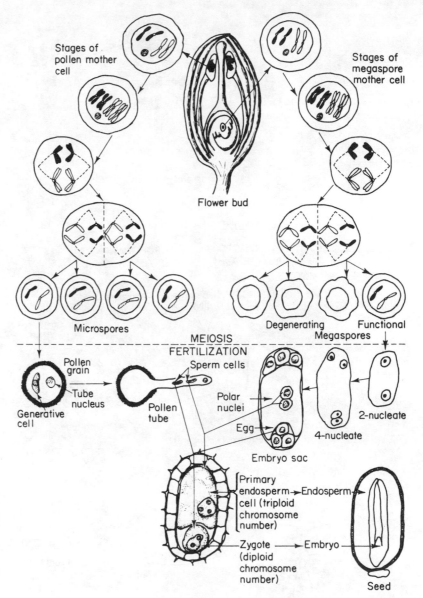

FIGURE 1–1 Diagrammatical representation of the sexual cycle in angiosperms. Meiosis occurs in the flower bud in the anther (male) and the pistil (female) during the bud stage. During this process the pollen mother cells and the megaspore mother cells, both diploid, undergo a reduction division in which homologous chromosomes segregate to different cells. This is followed immediately by a mitotic division which produces four daughter cells, each with half the chromosomes of the mother cells. In fertilization a male gamete unites with the egg to produce a zygote, in which the diploid chromosome number is restored. A second male gamete unites with the polar nuclei to produce the endosperm.

on the opposite member of the chromosome pair **(homologous chromosomes)**, the plant is homozygous and will "breed true" if self-pollinated or if the other parent is genetically similar. This means that the particular traits or characteristics that the plant possesses will be transmitted to its offspring, and the offspring will resemble the parent. On the other hand, if a sufficient number of genes on one chromosome differ from those on the other member of the chromosome pair, then the plant is said to be heterozygous. In such a case,

important phenotypic traits of the parent may not be transmitted to its offspring, and the seedlings may differ in appearance not only from the parent but also from each other. The amount of seedling variation may be very great in some kinds of plants.

To minimize variation and to be sure that seedling offspring will possess the particular characteristics for which they are to be grown, certain procedures must be followed during seed production. These are discussed in Chap. 4.

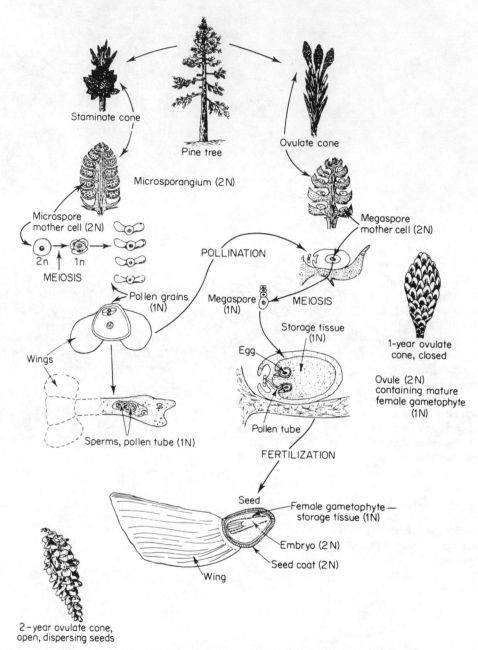

Staminate cone

Pine tree

Ovulate cone

Microsporangium (2N)

Microspore mother cell (2N)

2n   1n

MEIOSIS

POLLINATION

Megaspore mother cell (2N)

Pollen grains (1N)

Megaspore (1N)

MEIOSIS

Storage tissue (1N)

1-year ovulate cone, closed

Wings

Egg

Ovule (2N) containing mature female gametophyte (1N)

Sperms, pollen tube (1N)

Pollen tube

FERTILIZATION

Seed

Female gametophyte — storage tissue (1N)

Embryo (2N)

Seed coat (2N)

Wing

2-year ovulate cone, open, dispersing seeds

FIGURE 1-2 Diagrammatical representation of the sexual cycle in a gymnosperm (pine) showing meiosis and fertilization. It differs from that in the angiosperms by the formation of a ''naked'' seed, and the development of a haploid female gametophyte that comprises the storage tissue of the seed.

Mitosis and Asexual Reproduction

Asexual propagation is possible because each cell of the plant contains all the genes necessary for growth and development and, during the cell division (**mitosis**) that occurs during growth and regeneration, the genes are replicated in the daughter cells. Regeneration of a new organism by asexual methods occurs readily in higher plants but not in higher animals. In some forms of lower

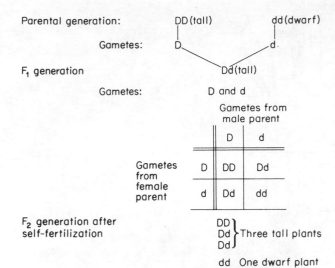

Parental generation: DD (tall)   dd (dwarf)

Gametes: D   d

F_1 generation   Dd (tall)

Gametes: D and d

	Gametes from male parent	
	D	d
Gametes from female parent — D	DD	Dd
d	Dd	dd

F_2 generation after self-fertilization

DD
Dd } Three tall plants
Dd

dd   One dwarf plant

FIGURE 1–3 Inheritance involving a single gene pair in a monohybrid cross. In garden peas tallness (D) is dominant over dwarfness (d). A tall pea plant is either homozygous (DD) or heterozygous (Dd). Segregation occurs in the F_2 generation to produce three genotypes (DD, Dd, dd) and two phenotypes (tall and dwarf).

animal life, however, for example the flatworm, *Planaria,* in the phylum Platyhelminthes, asexual multiplication can take place. A flatworm cut in half will develop into two worms, each half regenerating the missing part.

The details of mitosis are shown in Fig. 1–5, its principal feature being that individual chromosomes split longitudinally, the two identical parts going to two daughter cells. As a result, the complete chromosome system of an individual cell is duplicated in each of its two daughter cells (with certain exceptions). The chromosomes produced will be the same as in the cell from which they came. Consequently, the characteristics of the new plant that grows will be the same as that from which it originated.

Mitosis occurs in specific growing points or areas of the plant to produce growth (see Fig. 1–5). These are the **shoot apex,** the **root apex,** the **vascular cambium,** and the **intercalary zones** (internode bases of monocotyledonous plants). Mitosis also occurs when **callus** forms on a wounded plant part and when new growing points are initiated on root and stem pieces. Callus

Parental generation: YYgg (white nectarine)   yyGG (yellow peach)

Gametes: Yg   yG

F_1 generation after cross–fertilization: Yy Gg (white peach)

Gametes:
YG
Yg
yG
yg

	Gametes from male parent			
	YG	Yg	yG	yg
Gametes from female parent — YG	YYGG	YYGg	YyGG	YyGg
Yg	YYGg	YYgg	YyGg	Yygg
yG	YyGG	YyGg	yyGG	yyGg
yg	YyGg	Yygg	yyGg	yygg

F_2 generation after self–fertilization:

1 YYGG
2 Yy GG
2 YYGg } Nine white peaches
4 Yy Gg

1 YYgg
2 Yy gg } Three white nectarines

1 yy GG
2 yy Gg } Three yellow peaches

1 yy gg } One yellow nectarine

FIGURE 1–4 Inheritance in a dihybrid cross involving peach (*Prunus persica*). Fuzzy skin (G) of a peach is dominant over the glabrous (i.e., smooth) skin of the nectarine (g). White flesh color (Y) is dominant over yellow flesh color (y). In the example shown, the phenotype of the F_1 generation is different from either parent. Segregation in the F_2 generation produces nine genotypes and four phenotypes.

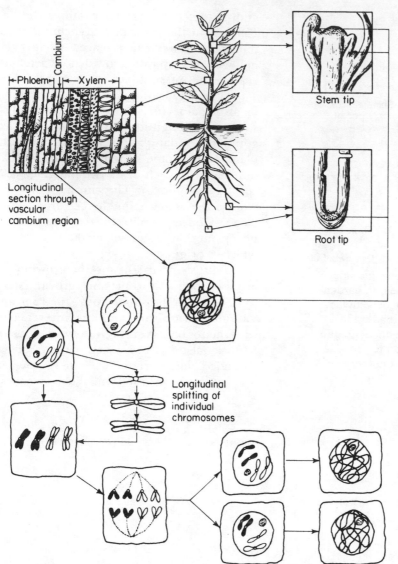

Stem tip

Root tip

Phloem — Xylem

Cambium

Longitudinal section through vascular cambium region

Longitudinal splitting of individual chromosomes

FIGURE 1–5 Diagrammatical representation of the process by which growth and asexual reproduction take place in a dicotyledonous plant. Mitosis occurs in three principal growing regions of the plant: the stem tip, the root tip of primary and secondary roots, and the cambium. A meristematic cell is shown dividing to produce two daughter cells whose chromosomes will (usually) be identical with those of the original cell.

parenchyma consists of new cells proliferating from cut tissues in response to wounding. When new growing points are initiated on a vegetative structure, such as root, stem, or leaf, they are referred to as **adventitious roots** or **adventitious shoots** (see Fig. 1–6 and Chap. 9).

Adventitious roots are those that arise from aerial plant parts, from underground stems, or from relatively old roots. All roots other than those arising from the embryo axis and all their branches formed in normal sequence can be considered adventitious roots. Adventitious shoots are those appearing on roots or internodally on stems after the terminal and lateral growing points are produced. New shoots **(watersprouts)** sometimes arise from latent growing points or buds that are not adventitious but originated along with the branch on which they occur. These are common on old branches of woody plants and can be stimulated into active growth if the part terminal to it is removed.

Mitosis is the basic process of normal vegetative growth, regeneration, and wound healing which makes possible such vegetative propagation techniques as cuttage, graftage, layerage, separation, and division. These methods of propagation

FIGURE 1-6 Types of regeneration occurring in asexual propagation. *Left:* Adventitious shoots growing from a root cutting. *Center:* Adventitious roots developing from the base of a stem cutting. *Right:* Callus tissue produced to give healing of a graft union.

are important because they permit large-scale multiplication of an **individual plant** into as many separate plants as the amount of parent material will permit. Each separate plant produced by such means is (in most cases) genetically identical with the plant from which it came. The primary reason for using these vegetative propagation techniques is to reproduce exactly the genetic characteristics of any individual plant, although there may be additional advantages from the standpoint of culture.

PLANT NOMENCLATURE

Since plant propagation involves the preservation of genotypes that are important to humans, it is essential to have some means of labeling them (*15*). A system of nomenclature that has gradually evolved through the efforts of botanists and horticulturists provides the basis for uniform world-wide plant identification. The system is embodied in the *International Code of Botanical Nomenclature* (*13*).

Botanical Classification

Classification of plants in nature is the function of taxonomists. The system of classification in use today is based upon increasing specialization and complexity in structure and organization resulting from evolutionary processes. For example, the plant **kingdom** is divided into **divisions**—first, with one-celled plants—i.e., Schizophyta (bacteria and blue-green algae)—through somewhat more complex plants, such as the Bryophyta (mosses, liverworts), to specialized higher plants, Pterophyta.

This text is principally concerned with the Pterophyta, which includes three **classes:** Filicinae (ferns), Gymnospermae (gymnosperms, as *Ginkgo* and conifers), and Angiospermae (flowering plants). Classification within these classes is based principally on flower structure. The gymnosperms produce seeds that are not enclosed. The angiosperms include plants in which the seed is produced within an enclosed structure—the ovary. The angiosperms are divided into two **subclasses**—the monocotyledons (e.g., grasses, palms, orchids) and the dicotyledons (e.g., beans, roses, peaches). These subclasses are divided into **orders,** the orders into **families,** families into **genera,** and genera into **species.**

The **species** is the fundamental unit customarily used by taxonomists to designate groups of plants that can be recognized as distinct kinds. In nature, individuals within one species normally interbreed freely but do not interbreed with a separate species because they are separated either by distance or by some other physiological, morphological, or genetic barrier that prevents the interchange of genes between the two species (*18*). Consequently, it is possible to reproduce a species by seed and to maintain it through propagation. However, if one reproduces a species from individual plants or from different parts of the natural range of that species, there may be a great deal of natural variability in appearance and adaptability among individuals of that species (*9, 12, 17, 19*). To get a total picture of species variability, one should examine individuals from all parts of the range rather than relying on individuals from only a narrow selection.

A distinct morphological subgrouping within a species (usually resulting from geographical separation) may be recognized taxonomically as a **botanical variety.** The botanical code also makes provision for certain other naturally occurring subdivisions within the species such as **subspecies, subvariety, group, form,** and **individual.**

Natural variation among native plants of a species can also be described with the two terms, **cline** and **ecotype** (*12, 17*). A cline refers to the continuous differences in genetically controlled physiological and morphological characteristics that occur within a species in different parts of its range. The differences are related to continuous variations in the environment and result from the evolution of populations of plants adapted to these variations. Where the differences are distinct and discontinuous, the term *ecotype* is used. Distinct ecotypes of many native forest species are known (see p. 86 and Fig. 4–3).

Classification of Cultivated Plants (Cultivars)

The classification and naming of those special kinds of cultivated plants has a different basis from that for plants growing wild in nature. The basis for separating one kind from another in cultivation is not because of naturally occurring variation but because each kind has some practical significance in agriculture and horticulture. The group of plants representing each kind has invariably arisen as a variant, either within a species or as a hybrid between one or more species. Very often it is derived from a single plant that is repro-

duced asexually. Such a group of plants representing a single type whose unique characteristics are reproduced during propagation is known as a **cultivar,** which is synonymous with such older terms as *variety* (English), *variete* (French), *variedad* (Spanish), *sorte* (German), *sort* (Scandinavian), *ras* (Russian), or *razza* (Italian). Such groups of plants are labeled by a non-Latin name, usually bestowed by its originator.

Naming of cultivated plants can be handled best if it conforms to a widely agreed upon system of nomenclature. The *International Code of Nomenclature for Cultivated Plants* (*4*) covers the special categories and names of cultivated plants. First, it recognizes the **cultivar** as a special taxonomic category for cultivated plants which is defined as "an assemblage of cultivated plants which is clearly distinguished by any characters (morphological, physiological, cytological, chemical, or others) and which when reproduced, sexually or asexually, retains its distinguishing characters." Second, it establishes guidelines by which cultivar names can be applied, and third, it establishes registration procedures by which cultivar names can be recorded.

The word **cultivar** is a contraction of the words "**culti**vated **vari**ety," and should be distinguished from the analogous, naturally occurring category—botanical variety—which is a population of plants within a species that has unique characteristics and occurs naturally. A cultivar is also a unique population of plants, but it is artificially maintained by human efforts and is named. Many cultivars would not continue to exist without human efforts.

The complete scientific name of any plant developed or maintained under cultivation includes the: (a) genus, (b) species, and (c) cultivar name, the first two in the usual Latin form and the latter in words of a common language. The cultivar name can be set off by the abbreviation **cv.,** or by single quotation marks, but not both. It can also just be attached to the common name.

> *Syringa vulgaris* cv. Mont Blanc
> *Syringa vulgaris* 'Mont Blanc'
> Lilac 'Mont Blanc'

A cultivar should have a proper name that can be recognized by horticulturists everywhere, one that will not confuse its identification with other cultivars. Multiple names for the same cultivar, or one name for several similar cultivars, arising either by accident or from deliberate changes in name, can only lead to confusion and misrepresentation. The most important principle in naming cultivars is that normally the earliest name applied should have priority. A second principle is that once a name is correctly applied, it should be changed only for exceptional reasons. There are additional rules which can be of assistance in choosing a plant name (*1, 2, 15*).

Categories of Cultivars

Sexually reproduced cultivars are propagated by seed. The basic category is the **line,** which is a population of seed-propagated plants in which genetic variability is controlled and the uniformity is maintained to a standard appropriate for that cultivar. An example is *Triticum aestivum* 'Marquis' wheat, in which the uniformity is readily maintained in propagation because it is homozygous and self-pollinated.

As knowledge of genetic principles developed, plant breeders have formulated various other types of seed-propagated lines that require special methods of parent selection and use of seed production procedures to maintain. These categories of cultivars include **inbred lines, composite** or **synthetic lines,** and **hybrid lines.** In addition, certain mixtures, as well as natural variants, known as **provenances** (see p. 86), can be classed as cultivars. Definitions of these special categories and their underlying genetic basis are discussed in Chap. 4.

Asexually reproduced cultivars are propagated by various vegetative methods, as by tissue culture, cuttings, division, or grafting. The basic category of this type of cultivar is the **clone.** It is defined in the Code (*4*) as ''a genetically uniform assemblage of individuals (which may be chimeral in nature), derived originally from a single individual by asexual propagation, for example, by cuttings, division, grafts, or obligate apomixis. Individuals propagated from a distinguishable bud mutation form a cultivar distinct from the parent plant.'' (Chimeras are discussed on page 178 and apomixis on page 71.) Examples of clones are 'Redhaven' peach, 'King Alfred' daffodil, and 'Burbank Russet' potato. Concepts of the

clone and maintenance of its uniformity during propagation are discussed in Chap. 8.

Another type of asexual cultivar is comprised of those groups of seedling plant populations reproduced by **obligate apomixis.** This concept is discussed in Chap. 3 (p. 71).

Plants showing special or unique growth phases (see p. 176) within clones and which can be reproduced by vegetative means may also be classed as separate cultivars. Examples of these are *Sequoia sempervirens* 'Prostrata', a prostrate form of the redwood tree, and *Picea abies* 'Pygmaea', a witches' broom type of spruce.

PLANT PROPAGATION ORGANIZATIONS

There are several kinds of organizations that promote sexual or asexual propagation of plants. Some of the principal ones are described below:

The International Plant Propagators' Society (IPPS)[1] was organized in 1951 and now has regional sections in the eastern, western, and southern United States, Great Britain and Ireland, Australia, and New Zealand. Each region holds annual meetings and the papers presented at these meetings are published in an annual *Combined Proceedings.* This society has over 2000 members; membership is confined to individuals active in some phase of plant propagation. A requirement for continued membership is participation in the society's activities, for example, by attending a regional meeting at least once in four years or writing an article for publication in the society's newsletters. Publications are available only to members and libraries.

The International Dwarf Fruit Tree Association[2] has over 1500 members worldwide, and holds an annual meeting, usually in Michigan. Papers are presented on various aspects of fruit tree rootstocks and propagation as well as general orchard culture. These papers are published in the annual proceedings, *Compact Fruit Tree.* The association sponsors an annual orchard tour in different fruit-

[1]Urban Horticulture GF-15, University of Washington, Seattle, Wash. 98195 U.S.A.

[2]Department of Horticulture, Mich. State University, East Lansing, Mich. 48824 U.S.A.

growing areas of the world, and it financially supports rootstock research.

International Association for Plant Tissue Culture[3] has about 2500 members representing 63 countries, including 300 members in the United States. Any person interested in in vitro plant cell, tissue, or organ culture is eligible for membership. The association promotes the interests of plant tissue culture workers by publishing a newsletter and by sponsoring an International Congress of Plant Tissue Culture every four years. Distributed three times a year, the newsletter contains feature articles on plant tissue culture and work underway in various laboratories, book reviews, bibliography listings of articles on plant tissue culture, and names of new members. Members represent industry as well as academic institutions and their interest includes commercial propagation; freeing stock plants from virus, bacterial, and fungal infestation; plant improvement by new genetic technology; and secondary product production in vitro.

Professional Plant Growers Association, Inc.[4] (*Bedding Plants, Inc.*—B.P.I.) is an organization of more than 3000 members growing or interested primarily in bedding plants. Its annual meetings are held in various parts of the United States; papers are published in an annual proceedings. A *Buyers' Guide* is published every two years to help bedding plant growers obtain their supplies. A newsletter is published several times a year.

The American Association of Nurserymen (AAN),[5] organized in 1875, is a national trade organization of the U.S. nursery and landscape industry. It serves about 3200 member firms involved in the nursery business—wholesale growers, garden center retailers, landscape firms, mail-order nurseries, and allied suppliers to the horticultural community.

Through information, educational meetings, and other services, AAN helps its members to manage their business more effectively. Its twice-a-month newsletter, *Up-date,* keeps members informed of legislative and regulatory decisions with which they must comply. Through active committees, the association maintains an up-to-date system for sizing and describing plants for ease in trading, and information on plant nomenclature, quarantines, pesticides, and the like.

AAN publishes the *American Standard for Nursery Stock* (*1*) prepared by a committee of nurserymen, landscape architects, landscape contractors, and others trading in or specifying nursery plants. These standards are reviewed and endorsed by interested national and regional societies, associations, and governmental agencies, and approved by the American National Standards Institute. Categories of nursery stock for which standards are established are: shade and flowering trees; deciduous shrubs; coniferous evergreens; broadleaf evergreens; roses; vines and ground covers; fruit trees; small fruits; lining out stock; seedling trees and shrubs; bulbs, corms, and tubers; and Christmas trees. The American Association of Nurserymen also holds annual meetings in various U.S. cities. AAN presents annually the Norman J. Colman Award (named after the first U.S. Secretary of Agriculture) to an individual for outstanding horticultural research.

Several organizations deal with plant reproduction and growing by the use of seeds. Some of the principal ones are:

The International Seed Testing Association (ISTA)[6] is an intergovernmental association with worldwide membership accredited by the governments of 59 countries. There are 137 official seed-testing stations that belong to the association. The chief purposes of the association are to develop, adopt, and publish standard procedures for sampling and testing seeds, and to promote uniform application of these procedures for evaluation of seeds moving in international trade.

Secondary purposes of ISTA are to promote research in all areas of seed science and technology, to encourage cultivar certification, and to participate in conferences and training courses aimed at promoting these objectives.

At a general meeting held every three years,

---

[3]Soil and Crop Sciences Dept., Texas A & M University, College Station, Tex. 77843 U.S.A.

[4]P.O. Box 27517, Lansing, Mich. 48909 U.S.A.

[5]Suite 500, 1250 I Street, N.W., Washington, D.C. 20005 U.S.A.

[6]P.O. Box 412, CH-8046, Zurich, Switzerland.

there is a symposium for scientific and technical papers presented by specialists working in seed technology.

ISTA publishes the journal, *Seed Science and Technology;* a quarterly newsletter, *ISTA News Bulletin;* and a number of technical handbooks such as the *Handbook for Seed Evaluation.*

Association of Official Seed Analysts[7] is an organization of member laboratories, serving both private seed analysts and workers on seed research problems in recognized government seed laboratories. Most members are in continental North America. The objectives of the organization are to improve all branches of seed testing and to make it more useful to agriculture and society.

The association holds an annual meeting and publishes the proceedings in the *Journal of Seed Technology.* It also publishes a quarterly newsletter along with several other publications, such as *Rules for Testing Seeds.*

American Seed Trade Association, Inc. (ASTA),[8] an organization of seed companies, has been serving the industry since 1883. ASTA holds a general meeting each year, as well as conferences on farm seed, lawn seed, soybean seed, the garden seed industry, and hybrid corn and sorghum industry research. The association publishes a semimonthly newsletter, an annual yearbook, and proceedings of the farm seed, soybean seed, and hybrid corn and sorghum industry conferences. ASTA becomes involved in regulatory activities, for example, by participating in the drafting of amendments to the Recommended Uniform State Seed Law, by evaluating proposed changes in tariffs or amendments to the Federal Seed Act, studying and evaluating benefits of international seed certification, and formulating trade rules for domestic and foreign commerce in seeds. ASTA also maintains close liaison with seed-testing authorities and evaluates the benefits of international seed certification. It serves as the industry's center of information on current business facts and problems and it actively promotes industry sales in both U.S. and overseas markets.

The Canadian Seed Trade Association,[9] similar to ASTA, promotes the seed industry of Canada.

The Association of Official Seed Certifying Agencies (AOSCA)[10] is an organization whose members are U.S. and Canadian agencies responsible for seed certification in their respective areas. AOSCA was organized in 1919 as the International Crop Improvement Association. These agencies maintain a close working relationship with the seed industry, seed regulatory agencies, governmental agencies involved in international seed market development and movement, and agricultural research and extension services.

REFERENCES

1. American Association of Nurserymen. 1986. *American standard for nursery stock.* Washington, D.C.: Amer. Assn. Nurs.

2. American Association of Nurserymen. 1988. *How to use, select, and register cultivar names.* Washington, D.C.: Amer. Assn. Nurs.

3. Baker, H. G. 1978. *Plants and civilization* (3rd ed.). Belmont, Calif.: Wadsworth.

4. Brickell, C. D., ed. 1980. International code of nomenclature for cultivated plants—1980. *Regnum Vegetabile* 104:7–32. (Obtainable from Crop Science Society of America, 677 South Segoe Road, Madison, Wis. 53771.)

5. Briggs, F. N., and P. F. Knowles. 1967. *Introduction to plant breeding.* New York: Reinhold.

6. Fehr, W. R. 1987. *Principles of cultivar development,* Vol. 1, *Theory and technique.* New York: Macmillan.

7. Flemer, W., III. 1988. Careers in nursery stock growing. *Amer. Nurs.* 167(10):73–82.

[7]P.O. Box 4906, State Fairgrounds, Springfield, Ill. 62708-4906 U.S.A.
[8]1030 15th Street, N.W., Washington, D.C. 20005 U.S.A.

[9]2948 Baseline Road, #207, Ottawa, Ontario K2H 8T5, Canada.
[10]3709 Hillsborough Street, Raleigh, N.C. 27607 U.S.A.

8. Frey, K. J., ed. 1981. *Plant breeding II*. Ames, Iowa: Iowa State Univ. Press.

9. Harlan, J. R. 1956. Distribution and utilization of natural variability in cultivated plants. In *Genetics in plant breeding,* Brookhaven Symposia in Biol., Vol. 9, pp. 191–206.

10. ———. 1976. The plants and animals that nourish man. *Scient. Amer.* 235(3):88–97.

11. Hartmann, H. T., A. M. Kofranek, V. E. Rubatzky, and W. J. Flocker. 1988. *Plant science: Growth, development, and utilization of cultivated plants* (2nd ed.). Englewood Cliffs, N.J.: Prentice-Hall.

12. Langlet, O. 1962. Ecological variability and taxonomy of forest trees. In *Tree growth,* T. T. Kozlowski, ed. New York: Ronald Press, pp. 357–69.

13. Lanjouw, J., ed. 1966. International code of botanical nomenclature. *Regnum Vegetabile* 46:402.

14. Sauer, C. O. 1969. *Agricultural origins and dispersals.* Cambridge, Mass.: Massachusetts Institute of Technology Press.

15. Sheldon, A. 1988. What's in a name? *Amer. Hort.* 67(4):34–36.

16. Simmonds, N. W., ed. 1976. *Evolution of crop plants.* London: Longman.

17. Stebbins, G. L. 1950. *Variation and evolution in plants.* New York: Columbia Univ. Press.

18. ———. 1971. *Processes of organic evolution* (2nd ed.). Englewood Cliffs, N.J.: Prentice-Hall.

19. Vavilov, N. I. 1930. Wild progenitors of the fruit trees of Turkestan and the Caucasus and the problem of the origin of fruit trees. *Rpt. and Proc. 9th Inter. Hort. Cong.,* London, pp. 271–86.

SUPPLEMENTARY READING

AYALA, F. J., and J. A. KIGER, JR. 1984. *Modern genetics* (2nd ed.). Menlo Park, Calif.: Benjamin-Cummings Publ. Co.

BAILEY, L. H., E. Z. BAILEY, and STAFF OF L. H. BAILEY HORTORIUM. 1976. *Hortus third.* New York: Macmillan.

DIRR, M., and C. W. HEUSER, JR. 1987. *The reference manual of woody plant propagation: From seed to tissue culture.* Athens, Ga.: Varsity Press.

DUNMIRE, J. R., ed. 1988. *Sunset western garden book* (5th ed.). Menlo Park, Calif.: Sunset Publ. Company.

FRANKEL, O. H., and E. BENNETT. 1970. *Genetic resources in plants: Their exploration and conservation.* Oxford: Blackwell Scientific Company.

HARTMANN, H. T., and J. BEUTEL. 1981. Propagation of temperate zone fruit plants. *Leaflet 21103.* Berkeley: Univ. Calif. Div. Agr. Sci.

HARTMANN, H. T., A. M. KOFRANEK, V. E. RUBATSKY, and W. J. FLOCKER. 1988. *Plant science: Growth, development, and utilization of cultivated plants* (2nd ed.). Englewood Cliffs, N.J.: Prentice-Hall.

HEISER, C. B., JR. 1981. *Seed to civilization* (2nd ed.). San Francisco: W. H. Freeman & Company Publishers.

HOWARD, B. 1987. Propagation. In *Rootstocks for fruit crops,* R. C. Rom and R. F. Carlson, eds. New York: John Wiley.

LORETI, F. ed. 1987. International symposium on vegative propagation of woody species. *Acta Hort.* 227:1–516.

MACDONALD, B. 1987. *Practical woody plant propagation for nursery growers.* Portland, Oreg.: Timber Press.

SCHWANITZ, F. 1966. *The origin of cultivated plants* (English translation from German edition of 1957). Cambridge, Mass.: Harvard Univ. Press.

WEIER, T. E., C. R. STOCKING, M. G. BARBOUR, and T. L. ROST. 1982. *Botany: An introduction to plant biology* (6th ed.). New York: John Wiley.

2

Propagation Structures, Media, Fertilizers, Sanitation, and Containers

In propagating and growing young nursery plants, facilities and procedures are best arranged to optimize the response of plants to the five fundamental environmental factors influencing their growth and development: **light, water, temperature, gases,** and **mineral nutrients.** In addition, young nursery plants require protection from pathogens and other pests, as well as control of salinity levels in the growing media. The propagation structures, equipment, and procedures described in this chapter, if handled properly, maximize the plants' growth and development by controlling their environment.

Facilities required for propagating plants by seed, cuttings, and grafting, and other methods include two basic units. One is a structure with temperature control and ample light, such as a greenhouse, modified quonset house, or hotbed, where seeds can be germinated or cuttings rooted or rooted and unrooted microplants acclimatized. The second unit is a structure into which the young, tender plants (liners) can be moved for

hardening, preparatory to transplanting out-of-doors. Cold frames or lathhouses are useful for this purpose. Any of these structures may, at certain times of the year and for certain species, serve both purposes.

PROPAGATING STRUCTURES

Aseptic Micropropagation Facilities

These are described in Chap. 17 (p. 496).

Greenhouses (Glasshouses)

Greenhouses have a long history of use by horticulturists as a means of forcing more rapid growth of plants (40, 88). Most of the greenhouse area in the United States is used for the wholesale production of floricultural crops, such as pot plants, foliage plants, bedding plants, and cut flowers. Lesser amounts are used for vegetable crops, and

15

nursery stock (*118*). A substantial amount of greenhouse space is used in Europe for production of vegetable crops.

There are several types of greenhouses ranging from homemade structures to elaborate commercial installations. A simple one is a shed-roof, lean-to structure utilizing one side, preferably the south or east, of another building as one wall. In fact, as a solar heat utilization measure, greenhouses may be constructed as part or all of the south wall of a dwelling, the heat accumulated in the greenhouse being used to heat the house (*83, 119, 121*).

Small, inexpensive greenhouses are available commercially, or can be constructed from a number of standard 3-by 6-ft hotbed sashes fastened to a 2-by-4 wood framework. Such a framework can also be covered with rigid plastic panels or polyethylene sheeting (*154*).

Commercial greenhouses are usually independent structures of even-span, gable-roof construction, proportioned so that the space is well utilized for convenient walkways and propagating benches (*56*). In large operations, several single greenhouse units are often attached side by side, eliminating the cost of glassing-in the adjoining walls (see Figs. 2–1 and 2–2). Gutter-connected houses are more fuel-efficient than a range of separate houses.

Quonset-type construction is very popular. Such houses are inexpensive to build, with a framework of piping, and are easily covered with one or two layers of polyethylene (see Figs. 2–3 and 2–21).

Arrangement of benches in greenhouses vary considerably. Some propagation installations do not have permanently attached benches, their placement varying according to the type of equipment, such as lift trucks or electric carts, used to move flats and plants (*154*). An innovation that can reduce aisle space and increase the usable space in a greenhouse is the use of rolling benches, which are pushed together until one needs to get between them, and then rolled apart (Fig. 2–4).

Greenhouse construction begins with a metal framework, to which are fastened metal sash bars to support either panes of glass embedded in putty or some type of plastic material (see p. 22). All-metal prefabricated greenhouses also are widely used and are available from several manufacturers.[1]

In any type of greenhouse or bench construction using wood, the wood should be pressure-treated with a preservative such as copper naphthenate, which will add a great many years to its life (*15*).

Ventilation to provide air movement and air exchange with the outside is necessary in all greenhouses to aid in controlling temperature and humidity. A mechanism for manual opening of panels at the ridge and sides can be used in smaller greenhouses, but most larger installations use forced-air fan ventilation controlled by thermostats.

Traditionally, greenhouses have been heated by steam or hot water from a central boiler through banks of pipes (some finned to increase radiation surface) suitably located in the greenhouse. Sometimes unit heaters for each house, with fans for improved air circulation, also are used. If oil, gas, or wood heaters (*33*) are used, they must be vented to the outside because the combustion products are toxic to plants or the ethylene generated can adversely affect plant growth. In large greenhouses heated air is often blown into large—30 to 60 cm (12 to 24 in.)—4-mil polyethylene tubes hung overhead and running the length of the greenhouse. Small—5 to 7.5 cm (2 to 3 in.)—holes spaced throughout the length of these tubes allow the hot air to escape, thus giving uniform heating throughout the house (see Fig. 2–5). These same tubes can be used for forced air ventilation in summer, eliminating the need for manual side and top vents.

Gas-fired infrared heaters are sometimes installed in the ridges of greenhouses with the concept of heating the plants but not the air mass. Infrared heaters consist of several lines of radiant tubing running the length of the house with reflective shielding above the tubes and installed at a height of 6 to 12 ft above the plants. The principal advantage of the relatively new infrared heating systems in greenhouses is lower energy use (*162*). Cultural practices may need to be changed with

---

[1] For sources of commercial greenhouses, contact National Greenhouse Manufacturers Association, P.O. Box 567, Pana, Ill. 62557.

FIGURE 2-1 *Above:* Air-supported plastic greenhouse (*right*) in comparison with a range of conventional glass-covered greenhouses. *Below:* Interior of air-supported greenhouse where a crop of tomatoes is being grown. (Courtesy W. L. Bauerle and T. H. Short, Ohio Agricultural Research and Development Center.)

infrared heating due to heating the plant but not the soil underneath.

The reverse of infrared heating in the greenhouse is to place pipes on or below the soil surface in the floor of the greenhouse, or on the benches, with hot water, controlled by a thermostat, circulating through the pipes. This places the heat below the plants, which hastens the germination of seeds, rooting of cuttings, or growth of plants. This popular system has been very satisfactory in many installations, heating the plant's roots and tops, but not the entire air mass in the greenhouse, giving substantial fuel savings.

Solar heating of greenhouses occurs naturally. The cost of fossil fuels has evoked considerable interest in methods of conserving daytime solar heat for night heating (*8, 53, 72, 94, 121, 143*). Otherwise, high heating costs may eventually make winter use of greenhouses in the colder regions economically unfeasible, relegating green-

FIGURE 2–2   Interior of a large, well-kept greenhouse used to grow foliage plants for sale. Efficient use of expensive space is accomplished by hanging ferns above lower benches. Note sanitary practice of keeping hose nozzle off the floor where it might pick up disease pathogens.

house operations to areas with relatively warm winters (*39, 147*). Consequently, better conservation of energy in the greenhouse is of considerable interest (*96*).

Most heat loss in greenhouses takes place through the roof. One method of reducing heat loss in winter is to install double-layered, sealed polyethylene sheeting outside over the glass, or just two layers of polyethylene sheeting, as in a quonset house (see Fig. 2–3). This form of insulation is very effective. The two layers are kept separate by an air cushion from a low-pressure blower. Energy savings from the use of this system are substantial—more than 50 percent reduction in fuel compared to conventional greenhouses—but the greatly lowered light intensity with the double-layer plastic cover lowers yields of the greenhouse crops (*18, 19, 50*).

Another device that reduces heat loss dramatically is a movable thermal curtain (Fig. 2–6), which at night comes between the crop and the greenhouse roof and walls (*69, 132, 142, 148*). Insulating the north wall reduces heat loss without appreciably lowering the available light.

A low-cost horizontal airflow system has been used effectively to maintain a constantly circulating air movement in the greenhouse. A series of blade-style fans (16-in. diameter) spaced 35 to 50 ft apart down one side of the greenhouse and up the other sets up a circular air movement that uniformly distributes the heated or cooled air very well. It is also helpful for distributing CO_2-enriched air (*16, 27*).

Greenhouses can be cooled mechanically in the summer by the use of large evaporative cooling units, as shown in Fig. 2–5. The "pad and fan" system, in which a wet pad of some material, such as a special honeycombed cellulose, aluminum mesh, or plastic fibers, is installed at one side (or end) of a greenhouse and large exhaust fans at the other, has proved to be the best method of cooling greenhouses, especially in low-humidity climates (*7*).

Greenhouses are often sprayed on the outside at the onset of warm spring weather with a thin layer of whitewash or a white, cold-water paint (see p. 383). This coating reflects much of the heat from the sun, thus preventing excessively though varying with the species, a minimum night temperature of 13 to 15.5°C (55 to 60°F) is suggested. The evaporator cooling thermostats should be set to start the blowers at about 24°C (75°F).

Computerized Environmental Controls[2]

In the early days of greenhouse operation, light, temperature, and humidity were about the only environmental controls attempted, spraying the greenhouse with whitewash in summer, opening and closing side and ridge vents with a crank to control temperatures, along with turning on steam valves at night to prevent freezing. Humidity was increased by spraying the walks and benches by hand at least once a day. Later it was found that thermostats, operating solenoid valves, would turn electric motors on and off, to raise and lower vents, and would turn steam and water valves on and off, thus giving some degree of automatic control.

But with the advent of computer technology, controls can be much more precise, although more costly. Not only temperature, ventilation, and humidity can be controlled but many other factors, such as propagating bed temperatures, application of liquid fertilizers through the irrigation system (also controlled), daylength lighting, light-intensity regulation with mechanically operated shade cloth, operation of a mist or fog system, CO_2 enrichment, and so on, all of which can be varied for different times of the day and night and for different banks of propagation units (*20, 48, 55, 135, 152*). Alarms can be sounded, triggered by the computer, if deviations from preset levels occur, such as a heating failure on a cold winter night, or a mist system failure on cuttings on a hot summer day. Some of these operations are shown in Fig. 2–7. The computer can provide data sheets on all factors being controlled for review by the greenhouse manager and possible changes that may be needed. Computer-con-

FIGURE 2–3 Versatility of a quonset-polyethylene house. *Above:* Rooting media floor beds are prepared and sterilized with methyl bromide. *Center:* Cuttings inserted and rooted under mist. *Below:* Rooted liners are pruned and maintained in beds until transplanting the following spring.

high temperatures in the greenhouse during summer. The whitewash is removed in the fall. Too heavy a coating of whitewash, however, can reduce the light intensity to undesirably low levels.

Controls are recommended for greenhouse heating and evaporative cooling systems. Al-

[2]Available from Priva Computers, Inc., P.O. Box 110, Vineland Station, Ontario, Canada LOR 2EO; Oglevee, 151 Oglevee Lane, Connellsville, Pa. 15425; Wadsworth Control Systems, 5541 Marshall Street, Arvada, Colo. 80002.

FIGURE 2–4 For more efficient use of costly greenhouse space, movable benches on rollers have been installed to reduce aisle space.

trolled greenhouses are widely used in the Netherlands and Denmark.

Greenhouse Covering Materials (*54, 117, 140*)

Glass

Glass-covered greenhouses are expensive, but for a permanent long-term installation may be more satisfactory than the popular low-cost plastic-covered houses, which require renewal of the plastic every few years. Glass continues to be widely used due, in part, to its superior light-transmitting properties.

Plastic

Lightweight frames covered with plastic materials are popular for small home-garden structures as well as for large commercial installations (*105, 130*). Worldwide, the area covered by plastic greenhouses is about three times that under glass-covered greenhouses (*149*). Several types of plastic are available and in general use—polyethylene, either single or double layered, fiberglass, acrylic, polycarbonate, polyvinyl chloride, and polyvinyl fluoride. All are lightweight and relatively inexpensive compared to glass. Their light weight permits a less expensive supporting framework than is required for glass.

Plastic-covered greenhouses tend to be much tighter than glass-covered ones, with a consequent increase in humidity and, especially in winter, give undesirable water drip on the plants. This problem can be overcome, however, by maintaining adequate ventilation (*34*).

Flexible Covering Materials

Polyethylene (polythene).   About 50 percent of the greenhouse area in the United States is covered with low-cost polyethylene, much with inflated double layers, giving good insulating properties. Polyethylene has a relatively short life. It breaks down in summer and must be replaced after one or two years, generally in the fall in preparation for winter. Ultraviolet (UV)-resisting polyethylene lasts longer (14 to 30 months) but is more costly. A thickness of 4 to 6 mils (1 mil = 0.001 in.) is recommended. For better insulation and lowered winter heating costs, a double layer of UV-inhibited copolymer material is available

FIGURE 2–5 Completely automated heating and cooling systems installed in fiberglass-covered greenhouse. *Upper left:* Hot air from hot-water heaters (at top) is blown into polyethylene distribution tubes. *Upper right:* Distribution tubes, which have outlet holes spaced to disperse heated air uniformly, extend the length of the house. *Lower left and right:* The opposite end wall of greenhouse has an insert of a wettable pad (*right*) through which air is pulled by exhaust fans for cooling. Automatic closure panels (*left*) shut off outside air movement into the house through this pad when heating is required. All components of both heating and cooling systems are thermostatically controlled.

with a 2.5-cm (1-in.) air gap between layers, kept separated by air pressure from a small blower. Laid over conventional glass greenhouses, this material substantially reduces winter heat loss (*8*).

A single-layer polyethylene-covered greenhouse loses more heat at night or in winter than a glass-covered house since polyethylene allows passage of heat energy from the soil and plants inside the greenhouse much more readily than glass. Glass stops most infrared radiation, whereas polyethylene is transparent to it. Polyethylene is available in widths up to 12 m (40 ft).

Only materials especially prepared for greenhouse covering should be used. Many installations, especially in windy areas, use a supporting material, usually welded wire mesh, for the polyethylene film. Occasionally, other supporting materials, such as Saran cloth, are used.

Polyethylene transmits about 85 percent of the sun's light, which is low compared to glass, but it passes all wavelengths of light required for the growth of plants (*137*). A tough, white, opaque film consisting of a mixture of polyethylene and vinyl plastic is available. This film stays

FIGURE 2–6 Movable insulating blankets above the plants in a greenhouse (see arrow). On cold winter nights they automatically close to conserve heat. Double- or single-layer thermal white polyethylene sheets can also be used to conserve heat. In addition, such sheets can be used for shading in summer.

more flexible under low winter temperatures than does polyethylene but is more expensive. Because temperature fluctuates less under such film than under clear plastic, it is suitable for winter storage of container-grown plants. Polyethylene permits the passage of oxygen and carbon dioxide, necessary for the growth processes of plants, while reducing the passage of water vapor.

There are a number of materials especially prepared for the horticultural industry that are proving to be satisfactory, such as an UV-treated cross-woven polyethylene fabric that resists ripping and tearing, and a knitted high-density UV polyethylene shade cloth that is strong and has a long life.

Polyvinyl fluoride (Tedlar).   This is a long-lasting (10 years or more) flexible covering that has excellent light transmission properties. Its initial cost is about five times that of polyethylene but since it lasts so long, the annual cost may be lower.

Polyvinyl chloride film.   This is not the same as the rigid PVC sheets. It will outlast polyethylene but is more expensive. It will become brittle and tear in cold weather, but in winter will retain more heat than polyethylene and, in summer, results in a hotter house for the same reason. In Japan it has become a popular greenhouse covering material.

Rigid Covering Materials

Fiberglass.   Rigid panels, corrugated or flat, of polyester resin reinforced with fiberglass have been widely used for greenhouse construction. This material is strong, long-lasting, lightweight, and easily applied, and comes in a variety of widths, lengths, and thicknesses but is not as permanent as glass. Only the clear material—especially made for greenhouses and in a thickness of 0.096 cm (0.038 in.) or more and weighing 4 to 5 oz per square foot—should be used. When new this material transmits about 80 to 90 percent of the available light, but light transmission decreases over the years, which can be a serious problem. Since fiberglass burns rapidly, an entire greenhouse may quickly be consumed by fire, so insurance costs can be higher. Fiberglass is more expensive than polyethylene.

FIGURE 2–7   *Upper and lower left*, computer-controlled environmental manipulation of propagation facilities, including: *upper right*, a mechanized traveling mist boom under high-pressure sodium vapor lights; *lower right*, automated shade material programmed to close along the top of the house when preset radiant energy levels are reached. This system works well with contact polyethelene propagation, *lower center*, in commercial English nurseries. *Lower left*, automated metering system for monitoring CO_2 injection.

Acrylic (Plexiglass, Lucite, Exolite).   This is highly weather resistant, does not yellow with age, has excellent light transmission properties, retains twice the heat of glass, and is very resistant to impact, but is brittle. It is somewhat more expensive and nearly as combustible as fiberglass. It is available in twin-wall construction which gives good insulation properties and eliminates condensation problems.

Polycarbonate (Polygal, Lexan, Cyroflex, Dynaglas).   This material is similar to acrylic in heat retention properties, with about 90 percent the light transmission of glass. It has high impact strength, about 200 times that of glass. It is lightweight, about one-sixth that of glass, making it easy to install. Polycarbonate's textured surface diffuses light and reduces condensation drip. It is available in twin-wall construction which gives good insulation properties. Polycarbonate can be cut, sawn, drilled, or nailed. It is UV stabilized so will resist long outdoor exposure but will eventually yellow with age (*101*).

The economics of using these greenhouse covering materials must be considered carefully before a decision is made. New materials are continually coming into the market.

Hotbeds

The hotbed is a small, low structure, used for the same purposes as a greenhouse. Seedlings can be started and leafy cuttings rooted in hotbeds early in the season. Heat is provided below the propagating medium by electric heating cables, hot water, steam pipes, hot air flues, or fermenting manure. As in the greenhouse, close attention must be paid in hotbeds to shading and ventilation, as well as to temperature and humidity control. For small propagation operations, hotbed structures are suitable for producing many thousands of nursery plants without the expenditure of large sums for construction of walk-in greenhouses (*67*).

The hotbed is a large wooden box or frame with a sloping, tight-fitting lid made of window sash or, preferably, regular hotbed sash. It should be placed in a sunny but protected and well-drained location. The size of the frame conforms to the size of the glass sash available. A standard size is 0.9 by 1.8 m (3 by 6 ft). If polyethylene or fiberglass is used as the covering, any convenient dimensions can be used. The frame can be built easily of 1-in. or 2-in. lumber nailed to 4-by-4 corner posts set in the ground. Decay-resistant wood such as redwood, cypress, or cedar should be used, and preferably pressure-treated with a wood preservative, such as copper naphthenate. This compound retards decay for many years and does not give off fumes toxic to plants. Creosote must not be used on wood structures in which plants will be grown, since the fumes released, particularly on hot days, are toxic to plants. Publications are available giving in detail the construction of

such equipment (*131*). Hotbeds can be used throughout the year, except in areas with severe winters, where their use may be restricted to spring, summer, and fall.

Lead- or plastic-covered electric soil-heating cables are quite satisfactory for providing bottom heat in hotbeds. Automatic temperature control can be obtained with inexpensive thermostats. For a hotbed 1.8 by 1.8 m (6 by 6 ft) about 18 m (60 ft) of heating cable is required. The details of a typical installation are shown in Fig. 2–8. Low-voltage soil-heating systems are sometimes used. A transformer reduces the regular line voltage to about 30 V, lessening the danger of electrical shock. The heating element consists of low-cost No. 8 or No. 10 bare, galvanized wire (*38, 106*).

The hotbed is filled with 10 to 15 cm (4 to 6 in.) of a rooting or seed-germinating medium over the heating cables. Alternatively, flats containing the medium can be used; these are placed directly on a thin layer of sand covering wire netting, placed over the heating cables to protect them from tools.

Cold Frames (*1, 131*)

Low-cost cold-frame (sun frame) construction (Figs. 2–9 and 2–10) is the same as for hotbeds, except that no provision is made for supplying bottom heat. The standard glass 0.9 by 1.8 m (3 by 6 ft) hotbed sash is often used as a covering for the frame, although lightweight, less expensive frames can be constructed with polyethylene or fiberglass covering. The covered frames should fit tightly in order to retain heat and obtain high humidity. Cold frames should be placed in locations protected from winds, with the sash cover sloping down from north to south (south to north in the southern hemisphere).

A primary use of cold frames is in conditioning or hardening rooted cuttings or young seedlings (liners) preceding field, nursery-row, or container planting. They may be used also for starting new plants in late spring, summer, or fall when no external supply of heat is necessary (*157*). In cold frames only the heat of the sun, retained by the transparent covering, is utilized.

Close attention to ventilation, shading, watering, and winter protection is necessary for success with cold frames. When young, tender plants

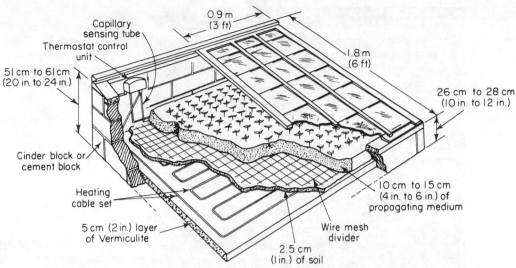

FIGURE 2-8 Construction of a hotbed showing the installation of an electric heating cable and thermostat. (Courtesy General Electric Co.)

are first placed in a cold frame, the coverings are generally kept closed tightly to maintain a high humidity, but as the plants become adjusted, the sash frames are gradually raised to permit more ventilation and dryer conditions. The installation of a mist line or frequent sprinkling of the plants in a cold frame is essential to maintain humid conditions. During sunny days temperatures can build up to excessively high levels in closed frames unless ventilation and shading are provided. Spaced lath, cloth-covered frames, or reed mats

are useful to lay over the sash to provide protection from the sun. In areas where extremely low temperatures occur, plants being overwintered in cold frames may require additional protective coverings.

Lathhouses

Lathhouses (Fig. 2-11) provide outdoor shade and protect container-grown plants from high

FIGURE 2-9 Banks of cold frames that were used for starting tender plants. Frames are opened after protection is no longer required. Kew Gardens, Richmond, England.

FIGURE 2-10 Older commercial use of glass-covered cold frames in propagating ground cover plants by cuttings. Glass coverings are rarely used due to the high labor costs in moving the heavy sash. Plastic coverings are more suitable.

FIGURE 2–11 Lathhouses are often covered by Saran or polypropylene shade cloth supported by wires stretched between metal upright poles. These may extend to cover many acres.

summer temperatures and high light intensities. They reduce moisture stress and decrease the water requirements of plants. Lathhouses have many uses in propagation, particularly in conjunction with transplanting, and with maintenance of shade-requiring or tender plants. At times a lathhouse, in which watering needs are relatively low, is used by nurseries simply to hold plants for sale. In mild climates they are used for propagation, along with a mist facility.

Lathhouse construction varies widely. Aluminum prefabricated lathhouses are available but may be more costly than wood structures. More commonly, pipe supports are used, set in concrete with the necessary supporting cross-members. Sometimes shade is provided by thin wood strips about 5 cm (2 in.) wide, placed to give one-third to two-thirds cover, depending on the need. Both sides and the top are usually covered. Rolls of snow fencing attached to a supporting framework can be utilized for inexpensive construction.

Woven plastic material—Saran or polypropylene fabric—is widely used in covering structures to provide shade. These materials are available in different densities, thus allowing lower intensities of light, such as 50 percent sunlight, on the plants. They are lightweight and can be attached to heavy wire fastened to supporting posts.

In regions having naturally cool summers, as an alternative to shading, applying water from overhead sprinklers on the rare hot days may be more economical.

Miscellaneous Propagating Structures

Fluorescent Light Boxes

Young plants of many species grow satisfactorily under the artificial light from fluorescent lamps or other light sources. These units may be used to start young seedlings and to root cuttings (*82*). By using the lamps in boxes or stacked in small rooms it is possible to maintain high humidity. The "cool white" fluorescent tubes are generally preferable and more economical than those types claiming to have both red and far-red light characteristics (*80*). With such equipment it is often helpful to provide bottom heat, from either thermostatically controlled soil-heating cables or a heater or a lamp in the air space below the rooting medium. Although adequate growth of many plant species may be obtained under fluorescent lamps, the results rarely equal or surpass those obtained under good greenhouse conditions.

Propagating Frames

Even in a greenhouse, humidity is not always high enough to permit satisfactory rooting of certain kinds of leafy cuttings. Enclosed frames covered with glass or one of the plastic materials may be necessary for successful rooting (see Fig. 2-12). There are many variations of such devices (small ones were called Wardian cases in earlier days). Such enclosed frames are also useful for completed grafts of small potted nursery stock, since they retain high humidity during the healing process.

Sometimes in cool summer climates, or as far south as Virginia in the United States, when fall semihardwood cuttings are taken, a layer of very thin (1 or 2 mils) polyethylene laid directly on top of a bed of newly prepared leafy cuttings in a greenhouse or lathhouse will provide a sufficient increase in relative humidity to give good rooting.

Bell jars (large inverted glass jars) can be set over a container of cuttings to be rooted. Humidity can be kept high in such devices, but shading and ventilation are necessary as soon as rooting starts. As shown in Figure 2-13, polyethylene plastic bags can be placed over a simple wire framework set in the rooting container to provide an inexpensive protective cover that maintains high humidity when rooting a few cuttings.

In using all such structures, care is necessary to avoid the buildup of pathogenic organisms. The warm, humid conditions, combined with lack of air movement and relatively low light intensity,

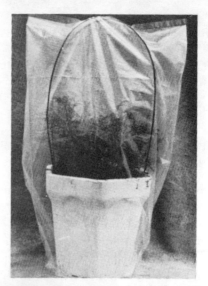

FIGURE 2-13 A small unit for rooting a few cuttings. A polyethylene bag has been placed over a framework made from two wire coat hangers. The bag is folded under the pot to give a tight seal. This unit should be set in a fairly light place but never in the direct sunlight, which would cause overheating within the bag.

provide excellent conditions for the growth of various fungi and bacteria. Cleanliness of all materials placed in such units is important but, in addition, use of fungicides is sometimes necessary (see section on sanitation, p. 37).

FIGURE 2-12 Polyethylene-covered beds used in a greenhouse to maintain high humidity surrounding the cuttings during rooting.

MEDIA[3] FOR PROPAGATING AND GROWING NURSERY PLANTS

Various materials and mixtures of materials are used for germinating seeds and rooting cuttings. For good results the following characteristics of the medium are required (110):

1. The medium must be sufficiently firm and dense to hold the cuttings or seeds in place during rooting or germination. Its volume must be fairly constant when either wet or dry; that is, excessive shrinkage after drying is undesirable.
2. It must retain enough moisture so that watering does not have to be too frequent.
3. It must be sufficiently porous so that excess water drains away, permitting adequate penetration of oxygen to the roots.
4. It must be free from weed seeds, nematodes, and various pathogens.
5. It must not have a high salinity level.
6. It should be capable of being pasteurized with steam or chemicals without harmful effects.
7. It should provide adequate nutrients in situations where plants are to remain for a long period—although supplementary slow-release fertilizers are frequently recommended.

Soil

A soil is composed of materials in the solid, liquid, and gaseous states. For satisfactory plant growth these materials must exist in the proper proportions. The solid portion of a soil is comprised of both inorganic and organic forms. The inorganic part consists of the residue from parent rock after decomposition resulting from the chemical and physical process of weathering. Such inorganic components vary in size from gravel down to extremely minute colloidal particles of clay, the texture of the soil being determined by the relative porportions of particles of different size. The coarser particles serve mainly as a supporting framework for the remainder of the soil, whereas

the colloidal clay fractions of the soil serve as storehouses for nutrients that are absorbed by plants. The organic portion of the soil consists of both living and dead organisms. Insects, worms, fungi, bacteria, and plant roots generally constitute the living organic matter, whereas the remains of such animal and plant life in various stages of decay make up the dead organic material. The residue from such decay (termed **humus**) is largely colloidal and assists in holding the water and plant nutrients.

The liquid part of the soil, the soil solution, is made up of water containing dissolved minerals in various quantities, as well as dissolved oxygen and carbon dioxide. Mineral elements, water, and possibly some carbon dioxide enter the plant from the soil solution.

The gaseous portion of the soil is important to good plant growth. In poorly drained, water-logged soils, water replaces the air, thus depriving plant roots, as well as certain desirable aerobic microorganisms, of the oxygen necessary for their existence.

The **texture** of a soil depends upon the relative proportions of *sand* (0.05 to 2 mm particle diameter), *silt* (0.05 to 0.002 mm particle diameter), and *clay* (less than 0.002 mm particle diameter). The principal texture classes are sand, loamy sand, sandy loam, loam, silt loam, clay loam, and clay. A typical sandy loam might consist of 70 percent sand, 20 percent silt, and 10 percent clay; whereas a clay loam might have 35 percent sand, 35 percent silt, and 30 percent clay.

In contrast to soil *texture,* which refers to the proportions of individual soil particles, soil **structure** refers to the arrangement of those particles in the entire soil mass. These individual soil grains are held together in aggregates of various sizes and shapes. Maintenance of a favorable granular and crumb soil structure is very important. For example, working heavy clay soils when they are too wet can so change the soil structure that the heavy clods formed may remain for years.

Sand

Sand consists of small rock particles, 0.05 to 2.0 mm in diameter, formed as the result of the weathering of various rocks, its mineral composition depending upon the type of rock. Quartz

[3]*Media* is plural—*medium* is singular.

sand, consisting chiefly of a silica complex, is generally used for propagation purposes. The type used in plastering is the grade ordinarily the most satisfactory for rooting cuttings. Sand is the heaviest of all rooting media used, a cubic foot of dry sand weighing about 45 kg (100 lb). It should preferably be fumigated or steam pasteurized before use, as it may contain weed seeds and various harmful pathogens. Sand contains virtually no mineral nutrients and has no buffering capacity, or cation exchange capacity (C.E.C.). It is used mostly in combination with organic materials.

Peat

Peat consists of the remains of aquatic, marsh, bog, or swamp vegetation which has been preserved under water in a partially decomposed state. The lack of oxygen in the bog slows bacterial and chemical decomposition of the plant material. Composition of different peat deposits varies widely, depending upon the vegetation from which it originated, state of decomposition, mineral content, and degree of acidity (*28, 84, 95, 150*).

There are three types of peat as classified by the U.S. Bureau of Mines: moss peat, reed sedge, and peat humus. *Moss peat* (usually referred to in the market as "peat moss") is the least decomposed of the three types and is derived from sphagnum or other mosses. It varies in color from light tan to dark brown. It has a high moisture-holding capacity (15 times its dry weight), has a high acidity (pH of 3.2 to 4.5), and contains a small amount of nitrogen (about 1 percent) but little or no phosphorus or potassium. This type of peat generally comes from Canada, Ireland, or Germany, although some is produced in the northern United States. This is the type most used in horticulture, the coarse grade being the best.

Reed sedge peat consists of the remains of grasses, reeds, sedges, and other swamp plants (e.g., Florida peat). This type of peat varies considerably in composition and in color, ranging from reddish brown to almost black. The pH ranges from about 4.0 to 7.5 and its water-holding capacity is about 10 times its own dry weight. This is not used for horticultural purposes.

Peat humus is in such an advanced state of decomposition that the original plant remains cannot be identified; it can originate from either hypnum moss or reed sedge peat. It is dark brown to black in color with a low moisture-holding capacity but with 2.0 to 3.5 percent nitrogen.

When peat moss is to be used in mixtures, it should be broken apart and moistened before adding to the mixture. Continued addition of coarse organic materials such as peat moss or sphagnum moss to greenhouse soil mixtures can cause a decrease in wettability. Water will not penetrate easily, and many of the soil particles will remain dry even after watering. No good method for preventing this nonwettability is known, although the repeated use of commercial wetting agents, such as Aqua-Gro, may improve water penetration (*87*).

Peat, as used in propagation, is not a uniform product and can be a source of weed seed, insects, and disease inoculum. Peat should be pasteurized along with the other media components (*22, 35*). Peat moss is relatively expensive so is used less and less in plant-growing media, being replaced by other components, such as pulverized or shredded bark.

Sphagnum Moss

Commercial sphagnum moss is the dehydrated young residue or living portions of acid-bog plants in the genus *Sphagnum,* such as *S. papillosum, S. capillaceum,* and *S. palustre.* It is relatively sterile, light in weight, and has a very high water-holding capacity, being able to absorb 10 to 20 times its weight of water. The stem and leaf tissues of sphagnum moss consist largely of groups of water-holding cells. This material is generally shredded, either by hand or mechanically, before it is used in a propagating or growing medium. It contains such small amounts of minerals that plants grown in it for any length of time require added nutrients. Sphagnum moss has a pH of about 3.5 to 4.0. It contains a specific fungistatic substance, or substances, which accounts for its ability to inhibit damping-off of seedlings germinated in it (*36, 43*).

Vermiculite

Vermiculite is a micaceous mineral that expands markedly when heated. Extensive deposits are found in Montana and North Carolina. Chemically, it is a hydrated magnesium–aluminum–iron silicate. When expanded, vermiculite is very light in weight, 90 to 150 kg per cubic meter (6 to 10 lb per cubic foot), neutral in reaction with good buffering properties, and insoluble in water; it is able to absorb large quantities of water—3 to 4 gal per cubic foot. Vermiculite has a relatively high cation exchange capacity and thus can hold nutrients in reserve and later release them. It contains enough magnesium and potassium to supply most plants.

In the crude vermiculite ore, the particles consist of a great many very thin, separate layers with microscopic quantities of water trapped between them. When run through furnaces at temperatures near 1090°C (2000°F), the water turns to steam, popping the layers apart, forming small, porous, spongelike kernels. Heating to this temperature provides complete sterilization. Horticultural vermiculite is graded to four sizes: No. 1 has particles from 5 to 8 mm in diameter; No. 2, the regular horticultural grade, from 2 to 3 mm; No. 3, from 1 to 2 mm; No. 4, which is most useful as a seed-germinating medium, from 0.75 to 1 mm. Expanded vermiculite should not be compacted when wet, as pressing destroys its desirable porous structure. Do not use nonhorticultural (construction grade) vermiculite, as it is treated with chemicals toxic to plant tissues.

Perlite

Perlite, a gray-white silicaceous material, is of volcanic origin, mined from lava flows. The crude ore is crushed and screened, then heated in furnaces to about 760°C (1400°F), at which temperature the small amount of moisture in the particles changes to steam, expanding the particles to small, sponge-like kernels that are very light, weighing only 5 to 8 lb per cubic foot. The high processing temperature provides a sterile product. Usually, a particle size of 1.6 to 3.0 mm ($\frac{1}{16}$ to $\frac{1}{8}$ in.) in diameter is used in horticultural applications. Perlite holds three to four times its weight of water. It is essentially neutral with a pH of 6.0 to 8.0 but with no buffering capacity; unlike vermiculite, it has no cation exchange capacity and contains no mineral nutrients. Perlite presents some problems with fluoride-sensitive plants. It is

most useful in increasing aeration in a mixture. Perlite, in combination with peat moss, is a very popular rooting medium for cuttings (*32, 98*).

Pumice

Chemically, pumice is mostly silicon dioxide and aluminum oxide, with small amounts of iron, calcium, magnesium, and sodium in the oxide form. It is mined in several regions in the western states of the United States, one source being in the Sierra Nevada mountains near Bishop, California. Pumice is screened to different size grades but is not heat-treated. It increases aeration and drainage in a rooting mix and can be used alone or mixed with peat moss (*71*).

Rockwool

This material is used as a rooting and growing medium in Europe, Australia, and the United States. It is prepared from various rock sources, melted at a temperature of about 1600°C, then, as it cools, is spun into fibers, then pressed into blocks with a binder added. Horticultural rockwool is available in several forms—shredded, prills (pellets), slabs, cubes, or combined with peat moss as a mixture. Rockwool will hold a considerable amount of water, yet retains good oxygen levels. With the addition of fertilizers it can be used in place of the Peat-Lite mixes (see p. 33). Since rockwool has not been as widely used as some of the other materials, small-scale trials should be done initially (*51*).

Shredded Bark
and Wood Shavings

Shredded or pulverized bark from redwood, cedar, fir, pine, hemlock, or various hardwood species can be used as a component in growing and propagating mixes, serving much the same purposes as peat moss and at a lower cost (*89, 91, 102, 107, 115, 127, 158*). Additional nitrogen may be needed in an amount sufficient for the decomposition requirements of the material, plus an additional amount for use by the plants (*161*). Rate of decomposition varies with the wood species. Because of their relatively low cost, light weight, and availability, these materials are very popular and widely used in soil mixes for container-grown plants, but supplementary nutrients must be added. Some types, when fresh, may contain materials toxic to plants, such as phenols, resins, terpenes, and tannins, so they require composting for 10 to 14 weeks before using (*25*).

Synthetic Plastic Aggregates

These materials are used, especially in Europe and some parts of the United States, as substitutes for sand or perlite. *Expanded polystyrene flakes* improve drainage and aeration and decrease bulk density. They are chemically neutral, do not absorb water, and do not decay, but they can be difficult to incorporate uniformly in the media. *Urea-formaldehyde foam* consists of sponge-like particles that have a high water-holding capacity and a 30 percent nitrogen content, which is slowly released over a period of several years. This material should not be used in plant growing media until the odor of formaldehyde has disappeared.

Compost

Composting can be defined as the biological decomposition of bulk organic wastes under controlled conditions, which takes place in piles or bins. The process occurs in three steps: (a) an initial stage lasting a few days in which decomposition of easily degradable soluble materials occurs; (b) a second stage, lasting several months, during which high temperatures occur and cellulose compounds are broken down; (c) a final stabilization stage when decomposition decreases, temperatures lower, and microorganisms recolonize the material. Microorganisms include bacteria, fungi, and nematodes; larger organisms, such as millipedes, soil mites, beetles, springtails, earthworms, earwigs, slugs, sowbugs, and fruit flies, can often be found in compost piles in great numbers. Compost prepared largely from leaves may have a high soluble salt content, which will inhibit plant growth, but can be lowered by leaching with water before use (*113*).

In the home garden a compost mixture may be useful as a moisture-holding humus material. Mixed with soil, compost adds organic matter. To start a compost, leaves and garden refuse are accumulated and allowed to decompose, preferably

in a bin 1.2 by 2.4 m (4 by 6 ft) with slatted sides to give good aeration. Moisture should be added from time to time during the dry seasons; decomposition is hastened if some nitrogenous fertilizer is sprinkled through each batch of newly added material. The mass of material should be stirred once a week to ensure even decomposition. Several bins are preferable—one for newly started material, one for material undergoing decomposition, and one for completely decomposed compost, ready to use. Twelve to 24 months may be necessary for complete decomposition to humus. Since compost may contain weed seeds and nematodes, as well as noxious insects and pathogens, preferably it should be pasteurized (*5, 52, 68, 74, 113*).

In the United States compost mainly refers to a gradually decomposing mixture of bulk organic materials—leaves, grass clippings, soil, and any garden refuse, as described above. In Great Britain and some parts of the United States compost refers to plant-growing media containing such material as soil, peat moss, perlite, bark, and vermiculite, ready to be used for germinating seeds, rooting cuttings, or growing plants (*28*).

MIXTURES FOR CONTAINER GROWING

In propagation procedures, young seedlings or rooted cuttings (liners) are sometimes planted directly in the field, but frequently they are started in a blended mix in some type of container. Container growing of young seedlings and rooted cuttings has become an important alternative for field growers. For this purpose special growing mixes are needed (*28, 57, 89, 112, 139, 155, 160*).

To provide uniform potting mixtures of better textures, sand and some organic matter, such as peat moss, or shredded bark, may be used alone or added to a loam soil. In preparing these mixtures, the soil should be screened to make it uniform and to eliminate large particles. If the materials are very dry, they should be moistened slightly; this applies particularly to peat, which if mixed when dry, absorbs moisture very slowly. The soil should not be wet and sticky, however. In mixing, the various ingredients may be arranged in layers in a pile and turned with a shovel. A power-driven cement mixer, soil shredder, or front-end loader is used in large-scale operations. Many nurseries omit the soil and just use organic materials in their mixes.

Preparation of the mixture should preferably take place at least a day prior to use. During the ensuing 24 hours the moisture tends to become equalized throughout the mixture. The mixture should be just slightly moist at the time of use so that it does not crumble; on the other hand, it should not be sufficiently wet to form a ball when squeezed in the hand (44).

Suggested potting mixtures containing soil and mixed by volume are:

1. Heavy soils, such as clay loams or clay

    1 part soil

    2 parts perlite or sand

    2 parts peat moss (or composted shredded bark, or possibly leaf mold)

2. Medium soils, such as silt loams

    1 part soil

    1 part perlite or sand

    1 part peat moss (or composted shredded bark, or possibly leaf mold)

3. Light soils, such as sandy loams

    1 part soil

    1 part peat moss (or composted shredded bark, or possibly leaf mold)

    For each bushel (35 liters) of the foregoing mixes add:

      224 g (8 oz) dolomitic limestone (for calcium and magnesium)

      280 g (10 oz) 20 percent superphosphate (for phosphorus and sulfur)

      (For plants requiring acid soils, substitute calcium sulfate for the limestone.)

After starting new plants by rooting cuttings, germinating seeds, or by tissue culture, commercial producers of nursery stock grow many plants to a salable size in containers, using growing media that usually do not contain soil. These mixes vary widely throughout the nursery industry, but generally include a fine sand mixed in varying proportions with such materials as peat moss, shredded fir, pine, or hardwood bark. Such mixes require fertilizer supplements and continued feeding of the plants until they become established in their permanent locations (160). For example, one successful mix for small seedlings, rooted cuttings, and bedding plants consists of one part each of shredded fir or pine bark, peat moss, perlite, and sand. To this mixture is added gypsum, superphosphate, dolomite lime, and potash. Nitrogen and potassium are added subsequently in the irrigation water or as a top dressing of slow-release fertilizer, such as Osmocote.

In summary, nurseries have generally been changing from loam-based growing media, as exemplified by the John Innes composts (4) developed in England in the 1930s, to mixes incorporating such materials as sand, peat, perlite, vermiculite, pumice, and finely shredded bark in varying proportions. The trend away from loam-based mixes is due to a lack of suitable uniform soils, the added costs of having to pasteurize soil mixes, and the costs of handling and shipping the heavier soils compared to the lighter materials. Much experimentation takes place in trying to develop other low-cost, readily available bulk material to be used as a component of growing mixes, such as spent mushroom compost, papermill sludge (30, 37), rice hulls, or sugarcane bagasse.

The U.C. Potting Mixes

The well-known U.C. potting mixtures were developed in the mid-1950s by plant pathologists and others at the University of California, Los Angeles, to provide growing media that could be readily prepared in large quantities for commercial nurseries as an integral part of a pathogen-free propagating and cultural program (90). Since the U.C. mixes are based upon materials that are uniform, generally available, and require no previous preparation, they can easily be duplicated. The basic components are:

1. An inert type of fine sand, plus
2. Finely shredded peat moss—mixed with each other in varying proportions
3. Fertilizer mixtures as described below

The sand consists of round, wind-blown particles, uniform in size and relatively small (0.05

to 0.5 mm in diameter), thus having a rather high moisture-holding capacity. Such sand does not tend to compact even though the particles are small, because of their round shape and uniformity. The absence of colloidal clay particles in sand tends to prevent compaction or shrinkage. Disadvantages of the U.C. mixes are their heavy weight and poor aeration and drainage.

The chief purpose of the peat moss in this soil mixture is to increase its moisture and nutrient-holding capacities. In a mixture of equal parts of sand and peat moss, the maximum moisture-holding capacity is about 48 percent.

Basic fertilizer additives recommended (*90*) for a U.C. mix of 50 percent fine sand and 50 percent peat moss are as follows:

1. *If the mix is to be stored for an indefinite period before using:* This supplement furnishes a moderate supply of available nitrogen, but the plants will soon require further feeding. To each cubic yard (0.76m³) of the mix, add

112 g (4 oz) potassium nitrate	3.38 kg (7½ lb) dolomite lime
112 g (4 oz) potassium sulfate	1.13 kg (2½ lb) calcium carbonate lime
1.13 kg (2½ lb) single superphosphate	

2. *If the mix is to be planted within one week of preparation:* This supplement furnishes available nitrogen as well as a moderate nitrogen reserve. To each cubic yard (0.76 m³ of the mix, add:

2.25 kg (5 lb) of an organic nitrogen form, such as cottonseed meal (7% nitrogen)	1.13 kg (2½ lb) single superphosphate
	3.38 kg (7½ lb) dolomite lime
112 g (4 oz) potassium nitrate	1.13 kg (2½ lb) calcium carbonate lime
112 g (4 oz) potassium sulfate	

The organic nitrogen is omitted if the mix is to be stored before using, since such organic forms break down during storage, releasing a high content of water-soluble nitrogen, which may cause plant injury.

In preparing the U.C. mix, the fine sand, shredded peat moss, and fertilizer must be mixed together thoroughly. The peat moss should be moistened before mixing. If the mixing is done well, the peat moss will not separate and float to the top when the mixture is saturated with water. This mixture, including the fertilizer, can be safely steam pasteurized or chemically sterilized without the subsequent harmful effects to the plants that often occur when other soil-containing mixes are pasteurized.

If large amounts of superphosphate are added to a growing mix, it may contain toxic levels of fluoride as an impurity. The fluoride can be "fixed" in an insoluble form, however, by the addition of lime (*116*).

The Cornell Peat-Lite Mixes

The Cornell Peat-Lite mixes are artificial "soils," first developed in the mid-1960s, used primarily for seed germination and for container-growing of bedding plants and annuals (*23*). The components are lightweight, uniform, and readily available, and have chemical and physical characteristics suitable for the growth of plants. Excellent results have been obtained with these mixes. Mixes of this type have an important advantage of usually not requiring any decontamination before use. It may be desirable, however, to pasteurize the peat moss before use to eliminate any disease inoculum or other plant pests. Finely shredded bark is sometimes substituted for the peat moss.

Peat-Lite Mix A: To Make 0.76 m³ (1 cubic yard)

0.39 m³ (11 bu) shredded German or Canadian sphagnum peat moss

0.39 m³ (11 bu) horticultural grade vermiculite (No. 2 or 4)

2.25 kg (5 lb) ground limestone (preferably dolomitic), finely ground

0.45 to 0.9 kg (1 to 2 lb) single superphosphate (20 percent), preferably powdered

0.45 kg (1 lb) calcium nitrate

84 g (3 oz) fritted trace elements (FTE 555)

56 g (2 oz) iron sequestrene (330)

84 g (3 oz) wetting agent

Peat-Lite Mix B (same as A, except that horticultural perlite is substituted for the vermiculite)

Peat-Lite Mix C (for germinating seeds)

0.035 m³ (1 bu) shredded German or Canadian sphagnum peat moss

0.035 m³ (1 bu) horticultural grade vermiculite No. 4 (fine)

42 g (1½ oz) (4 level tbsp) ammonium nitrate

42 g (1½ oz) (2 level tbsp) superphosphate (20 percent), powdered

210 g (7½ oz) (10 level tbsp) finely ground dolomitic limestone

The materials should be mixed thoroughly, with special attention to wetting the peat moss during mixing. Adding a nonionic wetting agent, such as Aqua-Gro [28 g (1 oz) per 23 l (6 gal) of water] to the initial wetting usually aids in wetting the peat moss.

Many commercial ready-mixed preparations[4] are available in bulk or bags and are widely used by nurseries and home gardeners. Some mixes, already filled into cell packs, seed trays, or pots, are available, ready to be planted. Some soilless proprietary mixes are very sophisticated, containing peat moss, vermiculite, and perlite, plus a nutrient charge of nitrogen, potassium, phosphorus, dolomitic limestone, micronutrients, and a wetting agent, with the pH adjusted to about 6.0.

Proprietary micronutrient materials, such as Esmigran, FTE 555, or Micromax, consisting of combinations of minor elements, are available for adding to growing media. Adding a slow-release fertilizer such as Osmocote, MagAmp, or Nutriform, to the basic Peat-Lite mix is useful if the

[4]Examples of suppliers in the United States are: Vaughn Seed Co., 5300 Katrine Avenue, Downers Grove, Ill. 60515; Burpee Seed Co., Burpee Building, Warminster, Pa. 18974; Premier Brands, Inc., 145 Huguenot Street, New Rochelle, N.Y. 10801; Ball Seed Co., P.O. Box 335, West Chicago, Ill. 60185.

plants are to be grown in it for an extended period of time. Osmocote pellets containing micronutrient materials (Micromax) are available.

PREPLANTING TREATMENTS OF SOIL AND SOIL MIXES

Soils themselves often contain weed seeds, nematodes, and various fungi and bacteria harmful to plant tissue. The so-called "damping-off" commonly encountered in seedbeds is caused by soil fungi, such as species of *Pythium, Phytophthora, Rhizoctonia,* and *Fusarium.* To avoid loss from these pathogens, it is desirable to treat the soil, or mixtures with soil or leaf mold in them, before using.

Soil can be heated or fumigated with chemicals to eliminate weeds, insects, nematodes, and disease organisms. Heating soil mixes high in leaf mold or compost hastens decomposition of the organic matter, especially if it is already partially rotted, and leads to the formation of toxic compounds necessitating leaching with water or a three- to six-week delay in planting. Undecomposed materials, like the brown types of peat moss, are relatively unaffected.

Certain of the complex chemical compounds in the soil are broken down by excessive heat—above 85°C (185°F)—increasing the amounts of soluble salts of nitrogen, manganese, phosphorus, potassium, and others. Some, particularly nitrogen in the form of ammonia, may be present during the first few weeks after steaming in such quantities as to be toxic to the plants, particularly in mixes high in organic nitrogen forms. Later the ammonia is converted to nitrate nitrogen, which reaches a peak in about six weeks. Added superphosphate in the soil mix is useful for tying up excessive manganese released during heating, particularly of acid soils—thus preventing injury from manganese toxicity.

Heat Treatment

Although the term soil "sterilization" has been commonly used, a more accurate word is **pasteurization,** since the recommended heating processes do not kill all organisms (*9–13*). Pasteurization of growing media with aerated steam is generally preferable to fumigation with chemicals.

After treatment with steam, the medium can be used much sooner. Steam is nonselective for pests, whereas chemicals may be selective. Aerated steam, when properly used, is much less dangerous to use than fumigant chemicals, to both plants and the operator. Chemicals do not vaporize well at low temperatures, but steam pasteurization can be used for cold, wet media.

The moist heat is advantageous; it can be injected directly into the soil in covered bins or benches from perforated pipes placed 15 to 20 cm (6 to 8 in.) below the surface. In heating the soil, which should be moist but not wet, a temperature of 82°C (180°F) for 30 minutes has been a standard recommendation since this procedure kills most harmful bacteria and fungi as well as nematodes, insects, and most weed seeds, as indicated in Fig. 2–14. However, a lower temperature, such as 60°C (140°F) for 30 minutes, is more desirable since it kills pathogens but leaves many beneficial organisms that prevent explosive growth of harmful organisms if recontamination occurs. The lower temperature also tends to avoid toxicity problems, such as the release of excess ammonia

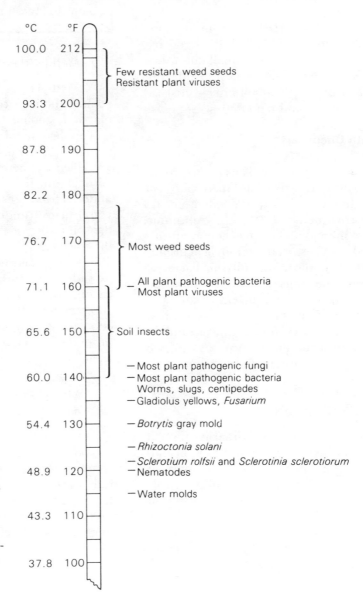

FIGURE 2–14 Soil temperatures required to kill weed seeds, insects, and various plant pathogens. Temperatures given are for 30 minutes under moist conditions. (From University of California Division of Agricultural Sciences, *Manual 23* (*9*).)

and nitrite, as well as manganese injury (*21, 146, 151*), which is often encountered at higher steaming temperatures.

Air mixed with steam (**aerated steam**) in the ratio of 4.1 to 1 by volume gives a temperature of 60°C (140°F) (*13*). This temperature will not kill many weed seeds but, if held for 30 minutes, and if the soil is moist, it will kill most pathogenic bacteria and fungi but does not kill beneficial microorganisms. Designing equipment to mix air and steam in the proper proportions to give this temperature presents some difficulties, but successful equipment has been developed (*3, 26, 60*) and is being widely used.

Electric heat pasteurizers are in use for amounts of soil up to 0.4 m³ (½ yd³). Microwave ovens can be used effectively for small quantities of soil. They do not have the undesirable drying effect of oven heating and will kill insects, disease organisms, weed seed, and nematodes.

Fumigation with Chemicals

Chemical fumigation kills organisms in the propagating mixes without disrupting their physical and chemical characteristics to the extent occurring with heat treatments (*100, 133*).[5] Ammonia production may increase following chemical fumigation, however, because of the removal of organisms antagonistic to the ammonifying bacteria. The mixes should be moist (between 40 and 80 percent of field capacity) and at temperatures of 18 to 24°C (65 to 75°F) for satisfactory results. Before using the mixture after chemical fumigation, allow a waiting period of two days to two weeks, depending upon the material, for dissipation of the fumes.

Chloropicrin (Tear Gas)

Chloropicrin is a liquid ordinarily applied with an injector, which should put 2 to 4 ml into holes 3 to 6 in. deep, spaced 9 to 12 in. apart. It may also be applied at the rate of 5 ml per cubic foot of soil. Chloropicrin changes to a gas that penetrates through the soil. The gas should be confined by sprinkling the soil surface with water and then covering it with an airtight material, which is then left for three days. Seven to 10 days

[5]Recommendations on the pesticide labels must be followed to conform to the permitted usages.

are required for thorough aeration of the soil before it can be planted. Chloropicrin is effective against nematodes, insects, some weed seeds, verticillium, and most other resistant fungi. Chloropicrin fumes are very toxic to living plant tissue.

> Chloropicrin and methyl bromide are hazardous materials to use, especially in confined areas. They should be applied only by persons trained in their use, who must take the necessary precautions as stated in the instructions on the containers or in the accompanying literature.

Methyl Bromide

Methyl bromide is an odorless material, very volatile and **very toxic to animals, including humans.** It should be mixed with other materials and applied only by those trained in its use. Most nematodes, insects, weed seeds, and some fungi are killed by methyl bromide, but it will not kill verticillium. It is most often used by injecting the material from pressurized containers into an open vessel placed under a plastic sheet which covers the soil to be treated. The cover is sealed around the edges with soil, and should be kept in place for 48 hours. Penetration is very good, and its effect extends to a depth of about 30 cm (12 in.). For treating bulk soil, methyl bromide at 333 ml per cubic meter (10 ml per cubic foot) or 0.6 kg per cubic meter (4 lb per 100 ft³) can be used. Methyl bromide is a very lethal material and its future legal use is uncertain.

Methyl Bromide and Chloropicrin Mixtures

Proprietary materials are available containing both methyl bromide and chloropicrin. Such combinations are more effective than either material alone in controlling weeds, insects, nematodes, and soil-borne pathogens. Aeration for 10 to 14 days is required following applications of methyl bromide–chloropicrin mixtures.

Vapam

Vapam (sodium *N*-methyl dithiocarbamate dihydrate) is a water-soluble soil fumigant that kills weeds, germinating weed seeds, most soil fungi, and under the proper conditions, nematodes. It undergoes rapid decomposition to produce a very penetrating gas. Vapam is applied by sprinkling on the soil surface, through irrigation systems, or with standard injection equipment. For seed-bed fumigation, 0.95 liter (1 qt) of the liquid formulation of Vapam in 7.6 to 11.4 liters (2 to 3 gal) of water is used, sprinkled uniformly over 9 m^2 (100 ft^2) of area. After application, the Vapam is sealed with additional water or with a roller. Three weeks after application the soil can be planted. Although Vapam has a relatively low toxicity to humans, care should be taken to avoid inhaling fumes or splashing the solution on the skin.

Fungicidal Soil Drenches

Fungicidal soil drenches can be applied to soil in which young plants are growing or are to be grown to inhibit growth of many soil-borne fungi. These materials may be applied either to the soil or to the plants. Preferably, a wetting agent should be added to the chemicals before application. It is very important in using such chemicals to read and follow the manufacturer's directions and dilutions carefully, and to try the chemicals on a limited number of plants first before going to large-scale applications. Examples of these materials are:

Quintozene (*PCNB, Terraclor*) controls *Rhizoctonia, Sclerotina,* and *Sclerotium.*

Benomyl (*Benlate, Tersan 1991*) is a systemic fungicide that inhibits growth of such soil pathogens as *Rhizoctonia, Cylindrocladium, Fusarium,* and *Verticillium.* It is ineffective against the water molds, *Pythium* and *Phytophthora.* A 50 percent wettable powder is available for use on ornamentals. Rubber gloves should be worn when handling this material.

Captan (*Orthocide*) added to the rooting or potting medium is effective against *Pythium* and *Fusarium* but gives only slight inhibition of *Rhizoctonia.* Wash excess material from leaves after drenching with the fungicide.

Etridiazole (*Terrazole, Truban*) incorporated into the propagating medium, gives good control of *Pythium* and *Phytophthora* and some control of *Fusarium* and *Rhizoctonia.*

SANITATION IN PROPAGATION

During propagation procedures losses of young seedlings, rooted cuttings, including tissue-cultured rooted plants, and grafted nursery plants to various pathogens and insect pests can sometimes be devastating, especially under the warm, humid conditions found in greenhouses (*47, 81, 92, 97, 103*). It is preferable, by far, to operate in a pathogen-free and insect-free environment, rather than continually attempting to suppress such pests on infected plants and propagating and growing plants by the continual use of pesticides. Ideally, sanitation problems should be considered even in the construction phase of propagation structures (*104*).

In recent years the importance of sanitation during propagation and growing procedures has become widely accepted and recognized as an essential part of nursery operations. Some large nurseries have detailed pest management programs with crews supervised by trained plant pathologists and entomologists (*31*). Such programs involve primarily the prevention of plant diseases and the avoidance of insects, mites, and weed problems. (Virus control is considered in Chap. 8.)

Harmful pathogens and other pests are best eliminated by dealing with the three situations where they can enter and become a problem during propagation procedures:

1. The physical propagation facilities—propagating room, pots, flats, knives, shears, working surfaces, hoses, greenhouse benches, and the like.
2. The propagation media—rooting and growing mixes for cuttings and seedlings.
3. The stock plant material itself as used for propagation—seeds, cutting material, scion and stock material for grafting.

If pathogens and other pests are eliminated in each of these areas it is likely that the young plants can be propagated and grown to a salable size with no disease, insect, or mite infestations. The pathogens most likely to cause disease development during propagation are species of *Pythium, Phytophthora, Fusarium, Cylindrocladium, Thelaviopsis, Sclerotinia,* and *Rhizoctonia solani.* These are all soil-borne organisms that attack plant roots. They can best be controlled by using nonsoil mixes, pasteurization of the propagating and growing mixes, general hygiene of the plants and facilities, avoidance of overwatering, good drainage of excess water, and the use of the proper fungicides (93, 138).

Physical Propagation Facilities

The space where the actual propagation (making cuttings, planting seeds, grafting, and so on) takes place should be a light, very clean, cool room, completely separated from areas where the soil mixing, pot and flat storage, growing, and other operations take place. Traffic and visitors in this room should be kept to a minimum. At the end of each working day all plant debris and soil should be cleaned out, the floors hosed down, and working surfaces washed with a sodium hypochlorite solution (Clorox, diluted 1 part to 9 parts water), or a solution of Consan CTA 20 or Physan 20 disinfectants, diluted according to directions.

Flats and pots coming into this room should have been washed thoroughly and, if used previously, sterilized with steam or chemicals, for example, a 30-minute soak in sodium hypochlorite (Clorox) diluted 1 to 9. No dirty flats or pots should be allowed in the propagating area. Knives, shears, and other equipment used in propagation should be sterilized periodically during the propagation day by dipping in a disinfectant.

Mist propagating and growing areas in greenhouses, cold frames, and lathhouses should be kept clean and dead plant debris should be removed daily. The bench tops and frames should be painted annually with 2 percent copper naphthenate, thus providing self-disinfecting surfaces that help control algae, fungi, and bacteria. Water to be used for misting should be free of pathogens. Water from ponds or reservoirs to be used for propagation purposes should be chlorinated to kill algae and pathogens.

Propagation Media

Since certain components of propagation mixes, particularly leaf mold, soil, sand, and peat moss (22, 35), can contain harmful pathogens, they should be pasteurized, preferably by aerated steam (13, 60), otherwise, by chemicals before being brought into the "clean" propagation area. The containers (bins, flats, pots) for such pasteurized mixes should, of course, have been treated to eliminate pathogens. Never put pasteurized mixes into dirty containers. New materials, such as vermiculite, perlite, pumice, and rockwool, which have been heat-treated during their manufacture, need not be pasteurized until reuse.

Plant Material

In selecting propagating material, use only those source plants that are disease and insect-free. Some nurseries maintain stock plant blocks, which they keep meticulously "clean." However, stock plants of particularly disease-prone plants, such as euonymus, might well be sprayed with a suitable fungicide several days before cuttings are taken.

It is best to select cutting material from the upper portion of stock plants rather than from near the ground where the material could possibly be covered with soil pathogens. As cutting material is being collected it should be placed in new plastic bags [or in used ones that have been washed and rinsed in a weak (30 mg/liter) chlorine solution].

After the cuttings have been made and before they are stuck in the flats they can be dipped in a weak chlorine solution, followed by a dip in a broad spectrum fungicidal solution (such as captan plus Benlate), before any hormonal treatment.

SUPPLEMENTARY FERTILIZERS

Even with a good soil mixture, complete with added mineral nutrients, continued growth of plants in containers necessitates the addition at intervals of supplementary minerals, especially nitrogen and potassium. Artificial, nonsoil mixes especially must have added fertilizers (125, 160).

A satisfactory feeding program for growing container plants is to combine a slowly available dry fertilizer in the original mix with a liquid fertilizer applied at frequent intervals during the growing season or with controlled-release fertilizers added as top dressings at intervals, as needed (*49*).

Of the three major elements—nitrogen, phosphorus, and potassium—nitrogen has the most control on the amount of vegetative growth, but phosphorus is very important, too, for root development and plant energy reactions.

To supply nitrogen, phosphorus, and potassium in dry form, 2 heaping teaspoonfuls of the following mixture is suggested for plants growing in 3.8-liter (1-gal) containers:

0.8 kg (1.8 lb) IBDU[6] or 3.6 kg (8 lb) cotton-seed meal

1.8 kg (4 lb) single superphosphate

0.45 kg (1 lb) potassium sulfate

As a supplement to this, a dilute inorganic nutrient feeding at weekly intervals throughout the growing season is desirable. A simple solution may be prepared by dissolving 1 tsp of potassium nitrate and 1 tsp of ammonium nitrate in 3.8 liters (1 gal) of water. Adding 2 tsp of a mixed fertilizer, such as 10-6-4, to 3.8 liters (1 gal) of water will also make a satisfactory solution. A complete nutrient solution for fertilizing container-grown plants is:

water	380 liters (100 gal)
ammonium nitrate	168 g (6 oz)
monoammonium phosphate	168 g (6 oz)
potassium nitrate	168 g (6 oz)

For large-scale operations, it is more practical to prepare a liquid concentrate and inject it into the regular watering or irrigating system by the use of a proportioner (*17, 24*)—"fertigation." For this, a nutrient concentrate formula such as the following can be used but, since soluble forms of phosphorus are expensive, this chemical may be added as superphosphate in the soil mix and only potassium nitrate and ammonium nitrate used in the liquid concentrate. Indicator dyes are sometimes added to serve as a visual guide to the fertilizer flow. The following nutrient concentrate formula can be used in large-scale operations:

water	38 liters (10 gal)
ammonium nitrate	6.75 kg (15 lb)
monoammonium phosphate[7]	1.8 kg (4 lb)
potassium chloride	2.7 kg (6 lb)

The chemicals above should be thoroughly dissolved. *This concentrate should then be diluted—1 part to 200 parts water—before being applied to the plants.* Some nurseries use a combination of top-dressing with dry fertilizers, plus liquid-feeding through the irrigation system. Large nurseries usually have periodic analyses made of their soil mixes and irrigation water to ensure that the proper nutrient levels are being maintained. Satisfactory nutrient levels in the irrigation water are:

nitrogen	100–200 ppm
phosphorus	20 ppm
potassium	100 ppm

Controlled-Release Fertilizers

Controlled-release fertilizers provide nutrients to the plants gradually over a long period and reduce the possibility of injury from excessive applications (*75, 85, 86, 145, 153*). They are expensive, however, and are used chiefly on high-value container-grown plants. This type of fertilizer has become very popular especially for container-grown nursery stock. Some fertilizers of this type, as Osmocote, are available with micronutrients added in the pellets. For both cutting and seed propagation a mixture of macro and micro slow-release fertilizers can be included in the propagating mix, so the newly formed roots can have nutrients available for absorption. This is particularly important with mist propagation where nutrients can be leached out from both the plant and the medium.

There are three types of slow-release fertilizers: (a) coated water-soluble pellets or granules; (b) inorganic materials that are slowly soluble; and (c) organic materials of low solubility that gradually decompose by biological breakdown or by chemical hydrolysis.

---

[6]Isobutylidene diurea (an organic slow-release fertilizer).

[7]Phosphoric acid is sometimes used instead since it aids in controlling the pH of the irrigation water. A final pH of about 6.7 is satisfactory.

Examples of the resin-coated-type pellets are: (a) Osmocote (111), whose release rate depends on the thickness of the coating, and (b) Nutricote (120), whose release rate depends on a release agent in the coating. After a period of time the fertilizer will have completely diffused out of the pellets (159).

Another kind of controlled-release fertilizer is the sulfur-coated urea granules, consisting of urea coated with a sulfur wax mixture so that the final product is made up of about 82 percent urea, 13 percent sulfur, 2 percent wax, 2 percent diatomaceous earth, and 1 percent clay conditioner.

An example of the slowly soluble, inorganic type is MagAmp (magnesium ammonium phosphate), an inorganic material of low water solubility (86). Added to the soil mix, it supplies nutrients slowly for up to two years. MagAmp may be incorporated into media prior to steam pasteurization without toxic effects (86). On the other hand, steam sterilization and sand abrasion in the preparation of mixes containing resin-coated, slow-release fertilizers, such as Osmocote, can lead to premature breakdown of the pellets. Potassium glass frit is a relatively soft glass that slowly supplies potassium at adequate levels for up to 18 months.

An example of the organic, low-solubility type is urea-formaldehyde (UF), which will supply nitrogen slowly over a long period of time (76). Another organic slow-release fertilizer is isobutylidene diurea (IBDU), which is a condensation product between urea and isobutylaldehyde, having 31 percent nitrogen. The diffusion and/or breakdown mechanisms for all these products are not well understood.

SALINITY IN SOIL MIXES

Excessive salts in the propagating or growing mixes or in irrigation water [EC (electrical conductivity) over 2 millimho per cm][8] can reduce plant growth, burn the foliage, or even kill the plants (41, 42, 45, 75, 122). The required fertilization programs also contribute to salt accumulations. Overfertilization causes rapid and severe salinity symptoms, starting with foliage wilting and tip and marginal leaf burning. These symptoms may be accompanied by a white surface accumulation of salts on the medium. To prevent salt buildup, the containers or flats should be leached with water periodically. If the irrigation water contains 250 ppm of salts, the leaching should be done every 12 weeks; for 500 ppm, every six weeks; and for 1000 ppm every three weeks. In addition, fertilizers should not be used that tend to contribute to excess salinity; for example, use potassium nitrate rather than potassium chloride.

There is considerable variability among the various plant species in their tolerance to salts in the growing medium (45, 123). Some, such as araucaria, bougainvillea, callistemon, carnation, gladiolus, olive, petunia, hibiscus, poinsettia, and portulaca, have high tolerance, while others, such as azalea, blackberry, gardenia, mahonia, photinia, pittosporum, and strawberry, have low tolerance. If only water with high salinity is available, only plants from the first group, or others of similar tolerance, should be selected for propagation and growing.

WATER QUALITY

Water quality is an important factor in rooting cuttings, germinating seeds, and growing-on the young plants (42, 73, 136). For good results the available water supply should not contain total soluble salts in excess of 1400 ppm (approximately 2 millimhos/cm)—ocean water averages about 35,000 ppm. The salts are combinations of such cations as sodium, calcium, and magnesium with such anions as sulfate, chloride, and bicarbonate. Water containing a high proportion of sodium to calcium and magnesium can adversely affect the physical properties and water-absorption rates of soils and should not be used for irrigation purposes. It is well to have nursery irrigation water tested at least twice a year by a reputable laboratory that is prepared to evaluate all the elements in the water affecting plant growth.

When using the popular soilless media, such as Peat-Lite, a problem can arise in maintaining

---

[8]Salinity levels in water extracts of the soil (saturation-extract method) can be measured by electrical conductivity (E.C.) using a "Solubridge." Such readings are expressed as millimhos per centimeter. Readings of less than 2 (1400 ppm) indicate no salinity problem; readings of 4 or over indicate a level at which most plants are likely to be affected. At readings over 8, only salt-tolerant plants will grow.

the proper calcium/magnesium levels, both being important elements for plant growth. Usually, the irrigation water will carry adequate amounts of these elements, but some "purified" water sources may not. When the calcium level in the water sources is 25 ppm or less and the magnesium level is 15 ppm or less, calcium and magnesium supplementary fertilizers should be added (*136*).

Although not itself detrimental to plant tissue, so-called "hard" water contains relatively high amounts of calcium and magnesium (as bicarbonates and sulfates) and can be a problem in mist-propagating units or in evaporative water cooling systems as deposits build up wherever evaporation occurs. When water is much over 6 grains per gal (100 ppm) in hardness it is often run through a water softener for household use. Some types of equipment are based upon the replacement of the calcium and magnesium in the water by sodium ions. Such "soft," high-sodium water is toxic to plant tissue and should never be used for watering plants.

A better, but more costly, method of improving water quality is the **deionization** process. Here calcium, magnesium, and sodium are removed by substituting hydrogen ions for them. Water passes over an absorptive medium charged with hydrogen ions, which absorbs calcium and other ions in exchange for hydrogen. For further deionization the water is passed through a second filter charged with hydroxyl (OH) ions, which replace carbonates, sulfates, and chlorides. The proper nutritive ions may then be added back in suitable amounts (*14*).

Boron salts are not removed by deionization units and if present in water in excess of 1 ppm can cause plant injury. There is no satisfactory method for removing excess boron from water. The best solution is to acquire another water source.

Another good but expensive method for improving water quality is **reverse osmosis,** a process in which pressure is applied to a solvent to force it through a semipermeable membrane from a more concentrated solution to a less concentrated solution eliminating unwanted salts from an otherwise good water source (*14*).

Municipal treatment of water supplies with chlorine (0.1 to 0.6 ppm) is not sufficiently high to cause plant injury. However, the addition of fluoride to water supplies at 1 ppm can cause leaf damage to a few tropical foliage plants (*65*).

When the water source is a pond, well, lake, or river, contamination by weed seeds, mosses, or algae can be a problem. Chemical contamination from drainage into the water source from herbicides applied to adjoining fields or from excess fertilizers on crop fields can also be damaging to nursery plants (*134*).

Nurseries using such water for their plants should treat the water before use. A good procedure is to:

1. Use strainers to remove large debris.
2. Run the water through sand filters with automatic back flushing. This removes coarse particles and weed seed.
3. Add chlorine to suppress algae and disease pathogens at about ½ ppm (0.5 mg/liter). This water can then be used for field watering of container nursery plants after soluble fertilizers have been injected into the system.
4. If the water has a high salt content, it can be improved by running it through a deionization or reverse osmosis unit, but this is very expensive.

SOIL pH

Soil reaction (or pH) is a measure of the concentration of hydrogen ions in the soil. Although not directly influencing plant growth, it has a number of indirect effects, such as the availability of various nutrients and the activity of beneficial microorganisms. A pH range of 5.5 to 7.0 is best for growth of most plants (7.0 is neutral—below this level is acid and above is alkaline). To lower pH of an alkaline soil, use ammonium sulfate fertilizer; to raise the pH of acid soils, use calcium nitrate. A faster and more precise method of lowering the pH and controlling carbonate problems in the nursery is to inject sulfuric or phosphoric acid into the irrigation water supplies.

SUPPLEMENTAL WINTER LIGHTING IN THE GREENHOUSE

Plant growth in the winter in greenhouses can be abnormally slow due to the lack of sufficient light for photosynthesis, especially so in the higher latitudes (*29, 129*). This is due to several reasons:

1. Low number of daily light hours
2. Low angle of the sun, resulting in more of the earth's atmosphere the sun's rays must penetrate
3. Many cloudy and overcast days in the winter
4. Shading by the greenhouse structure itself and dirt accumulating on the glass or other covering materials

To overcome the problem of low natural winter light and reduced plant growth, supplemental artificial light can be used over the plants. The best light source for greenhouse lighting is high-pressure sodium vapor lamps. Most of the radiation from these lamps is in the red and yellow wavelengths, being very deficient in the blue, but if used in conjunction with the natural daylight radiation they are quite satisfactory.

The high-pressure sodium vapor lamps emit more photosynthetically active radiation (PAR) for each input watt of electricity than any other lamp that is commercially available. Sodium vapor lamps are long lasting, degrading very slowly. They emit a considerable amount of heat that can be a benefit in the greenhouse in winter. They use a smaller fixture than fluorescent lamps, thus avoiding the substantial shading effect from the fluorescent lamp fixture itself.

The installation should provide a minimum of about 500 footcandles (10 W m^{-2} PAR)[9] at the plant level with lighting taking place about 16 hours a day. For large greenhouses the services of a lighting consultant should be used in designing the installation.

CARBON DIOXIDE (CO₂) ENRICHMENT IN THE GREENHOUSE

Carbon dioxide is one of the required ingredients for the basic photosynthetic process that accounts for the dry-weight materials produced by the plant (*66, 77, 78, 79, 99, 108, 156*).

[9]Watts per square meter of photosynthetically active radiation.

$$6CO_2 + 12H_2O \xrightarrow[\text{green plant cell}]{\text{light energy}} C_6H_{12}O_6 + 6O_2 + 6H_2O$$

Carbon dioxide exists normally in the atmosphere at about 300 to 350 ppm. Sometimes the concentration in winter in closed greenhouses may drop to 200 ppm, or lower, during the sunlight hours, owing to its use by the plants. Under conditions where CO_2 is limiting the photosynthetic rate [at adequate light intensities and relatively high temperatures—about 29.5°C (85°F)] an increase in the CO_2 concentration, to between 1000 and 2400 ppm, can be expected to result in an increase in photosynthesis, as much as 200 percent over the rate found at 300 ppm. To take full advantage of this potential increase in dry-weight production, plant spacing must be adequate to prevent shading of overcrowded leaves. When supplementary CO_2 is used during periods of sunny weather, the temperature in the greenhouse should be kept relatively high. Added CO_2 would be of little benefit whenever the light intensity drops to very low levels. Adding CO_2 at night is of no value. Good horizontal circulation of air in the greenhouse will prevent undesirable lowering of the CO_2 level just at the leaf surface (see p. 18). A tightly closed greenhouse is necessary to be able to increase the ambient CO_2.

Sources of CO_2 for greenhouses are either burners using kerosene, propane, or natural gas, or by direct injection with liquid CO_2 (*2*). Liquid CO_2 is expensive but almost risk free. With kerosene burners, high-quality, low-sulfur kerosene must be used or SO_2 pollution can occur. With propane or natural gas, incomplete combustion is possible. The flames should be a solid blue color. Control of the CO_2 level in the greenhouse is very important, but accurate sensors are available and should be used. With the newer computer technology, sensors in different parts of the greenhouse can give excellent control of the CO_2 levels. Excessively high levels of CO_2 in the greenhouse (over 5000 ppm) can be dangerous to humans.

CONTAINERS FOR PROPAGATING AND GROWING YOUNG PLANTS

New types of containers for propagating and growing plants are continually being developed,

FIGURE 2–15 Plastic containers (Rootrainers) made of preformed, hinged plastic sheets (*left*). These fold together and lock to form a set of four containers that fits into a special plastic tray. The vertical grooves on the sides of the containers reduce the likelihood of undesirable root spiralling. The containers can be opened to permit inspection of the roots or removal of the plants. This type of container is widely used in reforestation for propagating and growing seedlings.

usually with a goal of reducing handling costs. Direct sticking of cuttings into small liner containers, as opposed to sticking into conventional propagation trays, saves production steps and avoids root disturbance of cuttings, which can lead to transplant shock.

Flats

Flats are shallow plastic, wooden, or metal trays, with drainage holes in the bottom. They are useful for germinating seeds or rooting cuttings, since they permit young plants to be moved around easily when necessary. Durable kinds of wood, such as cypress, cedar, or redwood, are preferable for flats. Galvanized-iron and plastic flats are available in various sizes, but zinc released from the galvanized flats can cause toxicity to plants. Both of the latter types will nest, and thus require relatively little storage space.

Clay Pots

The familiar red clay flower pots, long used for growing young plants, are heavy and porous and lose moisture readily. They are easily broken, and their round shape is not economical of space. After continued use, toxic salt accumulations build up, requiring soaking in water before reuse.

Clay pots are rarely used today in large-scale, commercial propagation.

Plastic Pots

Plastic pots, round and square, have numerous advantages; they are nonporous, reusable, lightweight, and use little storage space since they will "nest." Some types are fragile, however, and require careful handling, although other types, made from polyethylene, are flexible and quite sturdy. Square pots are also made up into "packs" of 8 or 12 for easier handling. Plastic pots (and flats) cannot be steam sterilized, but some of the more common plant pathogens can be eliminated by a hot water dip, 70°C (158°F), for three minutes followed by a rinse in a dilute sodium hypochlorite solution (Clorox, Purex, etc.). Figure 2–15 shows a folding type of plastic container widely used in reforestation projects.[10]

Fiber Pots

Containers of various sizes, round or square, are pressed into shape from peat plus wood fiber, with fertilizer added. They are dry and will keep indef-

[10]Available from Spencer-Lemaire Industries, Edmonton, Alberta, Canada.

initely. Since these pots are biodegradable, they are set in the soil along with the plants. Peat pots find their best use where plants are to be held for a relatively short time and then put in a larger container or in the field. Small peat pots with plants growing in them eventually deteriorate because of constant moisture, and may fall apart when moved. On the other hand, unless the pots are kept moist, roots will fail to penetrate the walls of the pot and will grow into an undesirable spiral pattern. Units of 6 or 12 square peat pots fastened together are available. When large numbers of plants are involved, time and labor are saved in handling by the use of these units.

Peat, Fiber, and Expanded Foam Blocks

Blocks of solid material, sometimes with a prepunched hole (Fig. 2–16) have become popular as a germinating medium for seeds or as a rooting medium for cuttings, especially for such plants as chrysanthemums and poinsettias. Fertilizers are sometimes incorporated into the material. One type (Fig. 2–17)[11] is made of highly compressed peat and, when water is added, swells to its usable size and is soft enough for the cutting or seed to be inserted. Such blocks become a part of the plant unit and are set in the soil along with the plant. These blocks replace not only the pot but the propagating mix also.

Synthetic rooting blocks are becoming more widely used in the nursery industry, being well adapted to automation. Other advantages are their light weight, reproducibility, and sterile condition. Watering must be carefully controlled to provide constant moisture, while maintaining adequate aeration.

Plastic and Metal Growing Containers

Many millions of nursery plants are grown and marketed each year in 3.8-liter (1-gal) and—to a lesser extent—in 11-liter (3-gal) and 19-liter (5-gal) containers. They are tapered for nesting and all have drainage holes. However, heavy-wall, injection-molded plastic containers have largely replaced the older-type metal containers. Machine

[11]Available from Jiffy Products of America, West Chicago, Ill. 60185.

FIGURE 2–16   Solid blocks of compressed fibers with a prepunched hole are sometimes used for cuttings of easily rooted plants. After rooting, the block plus the rooted cutting is planted into a growing container or the field nursery.

planters have been developed utilizing containers, in which rooted cuttings or seedlings can be planted as rapidly as 10,000 or more a day. Plants are easily removed from tapered containers by inverting and tapping. Untapered metal containers must be cut down each side with can shears or tin snips to permit removal of the plant.

In areas having high summer temperatures, use of light-colored (white or aluminum) con-

FIGURE 2–17   Use of solid block rooting medium. *Left:* Compressed sphagnum peat discs, encased in a plastic netting and containing some added mineral nutrients. *Right:* Adding water causes peat to swell to size shown. Chrysanthemum cuttings inserted into full-sized pellets rooted rapidly and are ready to be planted in soil medium.

tainers may improve root growth by avoiding heat damage to the roots, often encountered with dark-colored containers that absorb considerable heat if exposed to the sun. In addition, soil temperatures in metal cans tend to be higher than in plastic containers. However, light-colored containers show dirt marks (as opposed to black or dark green containers) and are not as attractive to the consumer.

Polyethylene Bags

Polyethylene bags are widely used in Europe, Australia, New Zealand, and the tropics, but hardly at all in North America for growing-on rooted cuttings or seedlings (liners) to a salable size. They are considerably less expensive than rigid metal or plastic containers and seem to be satisfactory (see Fig. 2–18) but some types deteriorate rapidly. They are usually black, but one kind is black on the inside and light-colored on the outside. The lighter color reflects heat and lowers the root temperature (*144*). After planting, however, they cannot be stacked as easily as the rigid containers for truck transportation and often the poly bag breaks and the plant is damaged.

FIGURE 2–19 Wood containers are used for large nursery shrubs and trees. Note the use of plastic tubing for watering the trees.

Wood Containers

Large wood containers are used for growing large trees and shrubs to provide "instant" landscaping for the customer. The plant material may be kept in such containers for several years before being sold. Heavy moving equipment is required for handling such large nursery stock (Fig. 2–19).

HANDLING CONTAINER-GROWN PLANTS

Watering of container-grown nursery stock is a major expense. Hand watering of individual containers with a large-volume, low-pressure applicator on a hose several times a week is expensive, and is used only for small-scale operations. In large operations, overhead sprinklers (i.e., Rainbird-type impact sprinklers) are often used, although much runoff waste occurs. Watering of container plants by trickle or drip irrigation (Fig. 2–20), also widely used, results in less waste (*141*). Setting the containers on damp mats or beds of fine sand over plastic with water moving upward into the soil by capillarity is another method of supplying water to plants (*6, 109*). This system, however, may spread pathogens such as *Phytophthora* from container to container.

Fertilizer solutions are usually injected into the irrigation system in large commercial nurseries. After the container stock leaves the wholesale nursery the retailer should maintain the stock by

FIGURE 2–18 Polyethylene bags used as plant-growing containers for rhododendrons in New Zealand.

FIGURE 2–20 Automatic watering system for container-grown plants. A small "spaghetti" tube carries water from the plastic feeder tubes to each plant. Automatic control of the main lines can water hundreds of thousands of plants on a given schedule.

FIGURE 2–21 Winter protection of broad- and narrow-leaved evergreen nursery stock in severe winter areas. Plants are placed close together in beds and covered in late fall with 6-mil white polyethylene on pipe framing. *Bottom:* A basin made from polyethylene is filled with water, which releases heat upon freezing and absorbs heat upon thawing, acting as a buffer against sudden temperature changes. (Courtesy The Conard-Pyle Co., West Grove, Pa.)

adequate irrigation and fertilization until the plants have been purchased by the ultimate grower (*46*). Slow-release fertilizers added to the cans keep the plants growing well until they are planted in the ground.

In areas with severe winters attention must be given to the problems of winter injury (*62, 114, 124*). The amount of injury varies with the species. The chances of cold injury are lessened if the plants are well established in the containers before the onset of winter. In addition, setting the plants close together in large groups (jamming or bunching) tends to prevent damage from rapid temperature fluctuation. Wrapping heavy brown paper or Styrofoam-type sheets around the outer rows of cans is helpful. In cold winter regions, however, some additional form of winter protection must be used, for example, placing mulch covering, such as straw or hay, over the tops of the containers or placing the containers inside a protective structure, such as a cold frame or plastic-covered greenhouse. The most reliable type of winter protection, as shown in Fig. 2–21, is a temporary frame constructed over the plants and covered with opaque polyethylene sheeting, 4 to 6 mils in thickness (*63*). Often two layers of polyethylene, separated by a 2.5 to 5 cm (1 to 2 in.) air space, are used. Soaking the soil in the structure with water before the plants are covered enhances the protection. Condensation of moisture inside the cover as the temperature drops reduces radiant heat loss.

Another method of winter protection is to cover the block of container plants, either upright or on their sides, with soft, pliable microfoam sheets, which are then covered with white polyethylene, then sealed airtight to the ground. But rodents can be a problem with this system (*70*).

Overhead sprinkling is also an effective method of protecting tender plants from subfreezing temperatures. Water must be present continually, changing to ice as long as the subfreezing temperatures occur. When liquid water changes to ice approximately 144 BTU of heat is released per pound of water (*64, 126*).

In most woody plant species the roots do not develop as much winter hardiness as the tops, hence low-temperature damage of container stock occurs primarily to the root system. Winterhardiness of roots varies with the species (*61, 128*).

As shown in Figure 2–22, plants kept in containers too long will form an undesirable constricted root system from which they may never recover when planted in their permanent location (*58*). The plants should be shifted to larger containers before such "root spiralling" occurs. Judicious root pruning, early transplanting, and careful potting during the early transplanting stages can do much to develop a good root system by the time the young plant is ready for setting in its permanent location (*59*). Some types of plastic containers have vertical grooves along the sides which tend to prevent horizontal spiralling of the roots (Figure 2–15).

Often when a container plant grown in one of the synthetic, lightweight mixes in the nursery is transplanted into the home garden—into a

FIGURE 2–22 One disadvantage of growing trees in containers is the possibility of producing poorly shaped root systems. Here a defective, twisted root system resulted from holding the young nursery tree too long in a container before transplanting. Such spiraling roots retain this shape after planting and are unable to anchor the tree firmly in the ground.

much heavier loam or clay soil—a problem is encountered in maintaining sufficient moisture to the plant. If the root ball is covered with the heavier native soil, and water is then applied to the plant, possibly in a shallow basin at ground level, the water tends to stay in the heavier soil with its smaller, more absorptive, pores, while the coarser-textured mix containing the roots remains completely dry. To avoid this, the soil mix in the root ball should remain exposed at the top so that applied water must pass through it and so wet the roots.

REFERENCES

1. Aichele, J. H. 1988. High humidity propagation using sweat box methods. *Proc. Inter. Plant Prop. Soc.* 37:497–99.

2. Aimone, T. 1986. Carbon dioxide injection (report of a talk given by J. O. Donovan and J. Tsujita at Grower Expo 66). *Grower Talks* 49(12):88–97.

3. Aldrich, R. A., and P. E. Nelson. 1969. Equipment for aerated steam treatment of small quantities of soil and soil mixes. *Plant Dis. Rpt.* 53(10):784–88.

4. Alvey, N. G. 1961. Soil for John Innes composts. *Jour. Hort. Sci.* 36:228–40.

5. Ammon, G. L. 1978. Composting and use of hardwood bark media for container growing. *Proc. Inter. Plant Prop. Soc.* 28:368–70.

6. Auger, E., C. Zafonte, and J. J. McGuire. 1977. Capillary irrigation of container plants. *Proc. Inter. Plant Prop. Soc.* 27:467–73.

7. Augsburger, N. D., H. R. Bohanon, and J. L. Calhoun. 1978. *The greenhouse climate control handbook.* Muskogee, Okla.: Acme Eng. & Mfg. Co.

8. Baird, C. D., and W. E. Waters. 1979. Solar energy and greenhouse heating. *HortScience* 14(2):147–51.

9. Baker, K. F., and C. N. Roistacher. 1957. Heat treatment of soil. Section 8 in *Calif. Agr. Exp. Sta. Man. 23.*

10. ———. 1957. Principles of heat treatment of soil. Section 9 in *Calif. Agr. Exp. Sta. Man. 23.*

11. ———. 1957. Equipment for heat treatment of soil. Section 10 in *Calif. Agr. Exp. Sta. Man. 23.*

12. ———. 1962. Principles of heat treatment of soil and planting material. *Jour. Australian Inst. Agr. Sci.* 28(2):118–26.

13. ———. 1976. Aerated steam treatment of nursery soils. *Proc. Inter. Plant Prop. Soc.* 26:52–59.

14. Barnstead Co. 1971. *The Barnstead basic book on water.* Boston: Barnstead Co.

15. Bartok, J. W., Jr. 1987. Chemical pressure treatment helps wood last longer. *Greenhouse Manager* 6(2):151–52.

16. ———. 1988. Horizontal air flow. *Greenhouse Manager* 6(10):197–212.

17. ———. 1988. Feed your plants with a fertilizer injector. *Greenhouse Manager* 6(12):117–20.

18. Bauerle, W. L., and T. H. Short. 1977. Conserving heat in glass greenhouses with surface-mounted air-inflated plastic. *Ohio Agr. Res. and Develop. Center Spec. Circ. 101.*

19. ———. 1978. Greenhouse energy conservation and effects on plant response. *Ohio Rpt. on Res. and Develop.* 63(5):74–76.

20. Bayles, E. 1988. Computers. *Greenhouse Manager* 7(2):90–96.

21. Birch, P. D. W., and D. J. Eagle. 1969. Toxicity to seedlings of nitrite in sterilized composts. *Jour. Hort. Sci.* 44:321–30.

22. Bluhm, W. L. 1978. Peat, pests, and propagation. *Proc. Inter. Plant Prop. Soc.* 28:66–70.

23. Boodley, J. W., and R. Sheldrake, Jr. 1964. Cornell "Peat-Lite" mixes for container growing. Ithaca, N.Y.: Cornell Univ. Dept. Flor. and Orn. Hort., mimeo. rpt.

24. Boodley, J. W., C. F. Gortzig, R. W. Langhans, and J. W. Layer. 1966. Fertilizer propor-

tioners for floriculture and nursery crop production management. *Cornell Ext. Bul. 1175.*

25. Branson, R. L., J. P. Martin, and W. A. Dost. 1977. Decomposition rate of various organic materials in soil. *Proc. Inter. Plant Prop. Soc.* 27:94–96.

26. Brazelton, R. W. 1968. Sterilizing soil mixes with aerated steam. *Agric. Eng.* 49(7):400–401.

27. Brugger, M. F. 1987. Horizontal air flow improves crop growth. *Amer. Veg. Grower* 35(10):32–34.

28. Bunt, A. C. 1976. *Modern potting composts.* London: G. Allen & Unwin.

29. Cathey, H. M., and L. E. Campbell. 1980. Light and lighting systems for horticultural plants. *Hort. Rev.* 2:491–537.

30. Chong, C., R. A. Cline, and D. L. Rinker. 1988. Spent mushroom compost and papermill sludge as soil amendments for containerized nursery crops. *Proc. Inter. Plant Prop. Soc.* 37:347–53.

31. Connor, D. 1977. Propagation at Monrovia Nursery Company: Sanitation. *Proc. Inter. Plant Prop. Soc.* 27:102–6.

32. Cooke, C. D., and B. L. Dunsby. 1978. Perlite for propagation. *Proc. Inter. Plant Prop. Soc.* 28:224–28.

33. Copeland, R. W. 1986. A wood burning furnace for heating propagation houses. *Proc. Inter. Plant Prop. Soc.* 35:684–87.

34. Cotter, D. J., and J. N. Walker. 1966. Climate-humidity relationships in plastic greenhouses. *Proc. Amer. Soc. Hort. Sci.* 89:584–93.

35. Coyier, D. L. 1978. Pathogens associated with peat moss used for propagation. *Proc. Inter. Plant Prop. Soc.* 28:70–72.

36. Creech, J. L., R. F. Dowdle, and W. O. Hawley. 1955. Sphagnum moss for plant propagation. *USDA Farmers' Bul. 2058.*

37. Dallon, J., Jr. 1988. Effects of spent mushroom compost on the production of greenhouse-grown crops. *Proc. Inter. Plant Prop. Soc.* 37:323–29.

38. De Lance, P. 1976. Electric soil and hot house heating. *Proc. Inter. Plant Prop. Soc.* 26:369–72.

39. Duncan, G. A., and J. N. Walker. 1980. How to save energy in the greenhouse. *Amer. Nurs.* 152(10):13, 90–112.

40. Falland, J. 1987. Greenhouses in plant propagation: An historical perspective. *Proc. Inter. Plant Prop. Soc.* 36:158–64.

41. Fireman, M., and R. L. Branson. 1963. Salin-

ity in greenhouse soils. *Calif. Agr. Ext. Ser. OSA 68* (rev.).

42. Fireman, M., and H. E. Hayward. 1955. Irrigation water and saline and alkali soils. In *USDA yearbook of agriculture—water.* Washington, D.C.: U.S. Govt. Printing Office, pp. 321–27.

43. Fleming, G., and C. E. Hess. 1965. The isolation of a damping-off inhibitor from sphagnum moss. *Proc. Inter. Plant Prop. Soc.* 14:153–54.

44. Fonteno, W. C. 1988. Know your media: The air, water, and container connection. *Grower Talks* 51(11):110–11.

45. Francois, L. E. 1980. Salt injury to ornamental shrubs and ground covers. *USDA SEA Home and Garden Bul. 231.*

46. Furuta, T. 1971. Fertilizer and irrigation for plants on retail display. *Univ. Calif. Agr. Ext. Ser. AXT—361.*

47. Geard, I. D. 1979. Fungal diseases in plant propagation. *Proc. Inter. Plant Prop. Soc.* 29:589–94.

48. Germing, G. H., ed. 1986. Symposium on greenhouse climate and its control. *Acta. Hort.* 174:1–563.

49. Gilliam, C. H., and E. M. Smith. 1980. How and when to fertilize container nursery stock. *Amer. Nurs.* 151(2):7, 117–27.

50. Goldsberry, K. L. 1979. Greenhouse heat conservation and the effect of wind on heat loss. *HortScience* 14(2):152–55.

51. ———. 1986. Future looks bright for rockwool use. *Greenhouse Manager* 5(6):103–7.

52. Goluek, C. G. 1972. *Composting: A study of the process and its principles.* Emmaus, Pa.: Rodale Press.

53. Gordon, I. 1988. Structures used in Australia for plant propagation. *Proc. Inter. Plant Prop. Soc.* 37:482–89.

54. Hamrick, D. 1988. The covering choice. *Grower Talks* 51(12):64–74.

55. Hanan, J. 1987. A climate control system for greenhouse research. *HortScience* 22(5):704–8.

56. Hanan, J., W. D. Holley, and K. L. Goldsberry. 1978. Greenhouse construction. Chapter 3 in *Greenhouse management.* New York: Springer-Verlag.

57. Handreck, K. A. 1985. Potting mixes, and the care of plants growing in them. East Melbourne, Australia: CSIRO.

58. Harris, R. W., D. Long, and W. B. Davis.

1967. Root problems in nursery liner production. *Calif. Agr. Ext. Ser. AXT-244.*

59. Harris, R. W., W. B. Davis, N. W. Stice, and D. Long. 1971. Effects of root pruning and time of transplanting in nursery liner production. *Calif. Agr.* 25(12):8–10.

60. Hartmann, H. T., and J. E. Whisler. 1979. Mobile aerated steam soil pasteurizer unit. *Proc. Inter. Plant Prop. Soc.* 29:36–41.

61. Havis, J. T. 1974. Tolerance of plant roots in winter storage. *Amer. Nurs.* 139(1):10.

62. Havis, J. T., and R. D. Fitzgerald. 1976. Winter storage of nursery plants. *Mass. Coop. Ext. Ser. Publ. 125.*

63. Hawley, M. A., and D. E. Hamilton. 1980. How to build a poly house for overwintering nursery stock. *Amer. Nurs.* 152(8):7–9, 111–15.

64. Hendershott, C. H. 1979. Cold protection of low growing plants. *Proc. Inter. Plant Prop. Soc.* 29:533–36.

65. Henley, R. W., R. T. Poole, and C. A. Conover. 1976. Injury to selected plants due to fluoride toxicity. *Proc. Inter. Plant Prop. Soc.* 26:185–89.

66. Hicklenton, P. R., and A. M. Armitage. 1988. *CO_2 enrichment in the greenhouse.* Portland, Oreg.: Timber Press.

67. Hildebrandt, C. A. 1987. Economical propagation structures for the small grower. *Proc. Inter. Plant Prop. Soc.* 36:506–10.

68. Hoitink, H. A. J., and H. A. Poole. 1980. Factors affecting quality of composts for utilization in container media. *HortScience* 15(2):171–73.

69. Huang, K. T., and J. J. Hanan. 1976. Theoretical analysis of internal and external covers for greenhouse heat conservation. *HortScience* 11(6):582–83.

70. Hutt, G. M. 1984. Microfoam use for winter protection—your fifth option. *Proc. Inter. Plant Prop. Soc.* 34:418–24.

71. Inose, K. 1971. Pumice as a rooting medium. *Proc. Inter. Plant Prop. Soc.* 21:82–83.

72. Jensen, M. H. 1977. Energy alternative and conservation for greenhouses. *HortScience* 12(1):14–24.

73. Johnson, C. R. 1977. Some water quality problems faced by horticulturists. *Proc. Inter. Plant Prop. Soc.* 27:202–6.

74. Johnson, C. E. 1980. The wild world of composts. *National Geographic* 158(2):273–84.

75. Kelley, J. D. 1960. Effects of over-fertilization

on container-grown plants. *Proc. Plant Prop. Soc.* 10:58–63.

76. ——. 1962. Response of container-grown woody ornamentals to fertilization with urea-formaldehyde and potassium frit. *Proc. Amer. Soc. Hort. Sci.* 81:544–51.

77. Kohl, H. C. 1966. Carbon dioxide fertilization. *Proc. Inter. Plant Prop. Soc.* 15:300–306.

78. Krizek, D. T. 1970. Controlled atmospheres for plant growth. *Trans. Amer. Soc. Agr. Eng.* 13(3):237–68.

79. Krizek, D. T., W. A. Bailey, H. H. Klueter, and H. M. Cathey. 1968. Controlled environments for seedling production. *Proc. Inter. Plant Prop. Soc.* 18:273–80.

80. LaCroix, L. J., D. T. Canvin, and J. Walker. 1966. An evaluation of three fluorescent lamps as sources of light for plant growth. *Proc. Amer. Soc. Hort. Sci.* 89:714–22.

81. Lambe, R. C., and W. H. Wills. 1979. The major diseases of holly in the nursery. *Proc. Inter. Plant Prop. Soc.* 29:536–44.

82. Lechner, A., D. Sprague, E. Schaufler, B. Schalucha, R. Langhans, and P. Hammer. 1972. The Cornell automated plant grower. *Cornell Agr. Ext. Ser. Bul. 40.*

83. Likums, A. 1980. Greenhouse-residence: A new place in the sun. *Agr. Res.* 29(1):4–7.

84. Lucas, R. E., P. E. Riecke, and R. S. Farnham. 1971. Peats for soil improvement and soil mixes. *Mich. Coop. Ext. Ser. Bul. E-516.*

85. Lunt, O. R. 1965. Controlled availability fertilizers. *Farm Tech.* 21(4):11, 26.

86. Lunt, O. R., A. M. Kofranek, and S. B. Clark. 1964. Nutrient availability in soil: Availability of minerals from magnesium-ammonium-phosphates. *Agr. and Food Chem.* 12:497–504.

87. Lunt, O. R., R. H. Sciaroni, and W. Enomoto. 1963. Organic matter and wettability for greenhouse soils. *Calif. Agr.* 17(4):6.

88. Macdonald, A. B. 1986. Propagation facilities—past and present. *Proc. Inter. Plant Prop. Soc.* 35:170–75.

89. Mastalerz, J. W. 1977. Growing media. Chapter 6 in *The greenhouse environment.* New York: John Wiley.

90. Matkin, O. A., and P. A. Chandler. 1957. The U.C. type soil mixes. Section 5 in *Calif. Agr. Exp. Sta. Man. 23.*

91. Matkin, O. A. 1971. Soil mixes today. *Proc. Inter. Plant Prop. Soc.* 21:162–63.

92. McCain, A. H. 1977. Sanitation in plant propagation. *Proc. Inter. Plant Prop. Soc.* 27:91–93.

93. McCully, A. J., and M. B. Thomas. 1977. Soilborne diseases and their role in propagation. *Proc. Inter. Plant Prop. Soc.* 27:339–50.

94. Mears, D. R., ed. 1979. *Proc. 4th ann. conf. on solar energy for heating of greenhouses.* New Brunswick, N.J.: Rutgers Univ. Dept. Biol. and Agr. Eng.

95. Miller, N. 1981. Bogs, bales, and BTU's: A primer on peat. *Horticulture* 49(4):38–45.

96. Monk, G. J., and J. M. Molnar. 1987. Energy efficient greenhouses. In *Horticultural reviews,* Vol. 9, J. Janick, ed. Westport, Conn.: AVI Publ. Co., pp. 1–52.

97. Moody, E. H., Sr. 1983. Sanitation: A deliberate, essential exercise in plant disease control. *Proc. Inter. Plant Prop. Soc.* 33:608–13.

98. Moore, G. 1988. Perlite: Start to finish. *Proc. Inter. Plant Prop. Soc.* 37:48–52.

99. Mortensen, L. M. 1987. Review: CO_2 enrichment in greenhouses—crop responses. *Scient. Hort.* 33:1–25.

100. Munnecke, D. E. 1957. Chemical treatment of nursery soils. Section 11 in *Calif. Agr. Exp. Sta. Man. 23.*

101. O'Donnell, K. 1988. Polycarbonates gain as growers seek high performance coverings. *Grower Talks* 51(12):60–62.

102. Ogdon, R. J., F. A. Pokorny, and M. G. Dunavent. 1987. Elemental status of pine bark-based potting media. In *Horticultural reviews,* Vol. 9, J. Janick, ed. Westport, Conn.: AVI Publ. Co., pp. 103–31.

103. Ormrod, D. J. 1975. Fungicides and their spectra. *Proc. Inter. Plant Prop. Soc.* 25:112–15.

104. Orndorff, C. 1983. Constructing and maintaining disease-free propagation structures. *Proc. Inter. Plant Prop. Soc.* 32:599–605.

105. Parsons, R. A. 1971. Small plastic greenhouses. *Univ. Calif. Agr. Ext. Ser. AXT-328.*

106. Peterson, H. 1955. Low voltage heating. *N. Y. State Flower Growers' Bul. 115.*

107. Platt, G. C. 1984. Use of *Pinus radiata* bark: A four year experience. *Proc. Inter. Plant Prop. Soc.* 33:320–23.

108. Porter, M. A., and B. Grodzinski. 1985. CO_2 enrichment of protected crops. In *Horticultural reviews,* Vol. 7, J. Janick, ed. Westport, Conn.: AVI Publ. Co.

109. Richards, M. 1978. Capillary watering of con-

tainer plants. *Proc. Inter. Plant Prop. Soc.* 28:411–13.

110. Richards, S. J., J. E. Warneke, and F. K. Aljibury. 1964. Physical properties of soil mixes used by nurseries. *Calif. Agr.* 18(5):12–13.

111. Rutten, Th. 1980. Osmocote® controlled release fertilizer. *Acta. Hort.* 99:187–88.

112. Sabalka, D. 1987. Propagation media for flats and for direct sticking: What works? *Proc. Inter. Plant Prop. Soc.* 36:409–13.

113. Sawhney, B. L. 1976. Leaf compost for container-grown plants. *HortScience* 11(1):34–35.

114. Self, R. 1977. Winter protection of nursery plants. *Proc. Inter. Plant Prop. Soc.* 77:303–7.

115. ———. 1978. Pine bark in potting mixes: Grades and age, disease and fertility problems. *Proc. Inter. Plant Prop. Soc.* 28:363–68.

116. Sheldrake, R., G. E. Doss, L. E. St. John, Jr., and D. J. Lisk. 1978. Lime and charcoal amendments reduce fluoride absorption by plants cultured in a peat-perlite medium. *Jour. Amer. Soc. Hort. Sci.* 103(2):268–70.

117. Sherry, W. J. 1986. Greenhouse covering materials: Optical, thermal, and physical properties. *Grower Talks* 49(12):53–58.

118. ———. 1988. Technology in the greenhouse: State of the industry, a survey report. *Grower Talks* 51(9):48–57.

119. Sherwood, G. 1981. The many faces of the modern greenhouse. *Horticulture* 59(11):43–48.

120. Shibata, A., T. Fujita, and S. Maeda. 1980. Nutricote® coated fertilizers processed with polyolefin resins. *Acta Hort.* 99:179–86.

121. Short, T. H., and Bauerle, W. L. 1980. Greenhouse production with lower fuel costs. In *USDA 1980 yearbook of agriculture,* J. Hayes, ed. Washington, D.C.: U.S. Govt. Printing Office.

122. Schoonover, W. R., and R. H. Sciaroni. 1957. The salinity problem in nurseries. Section 4 in *Calif. Agr. Exp. Sta. Man. 23.*

123. Skimina, C. A. 1980. Salt tolerance of ornamentals. *Proc. Inter. Plant Prop. Soc.* 30:113–18.

124. Smith, E. M., ed. 1977. *Proc. woody ornamentals winter storage symposium.* Columbus, Ohio: Ohio State Univ., Coop. Ext.

125. Smith, E. M. 1980. How and when to fertilize container nursery stock. *Amer. Nurs.,* Jan. 15, pp. 365–68.

126. Steavenson, H. 1976. Using water to provide cold weather protection and conserve energy. *Amer. Nurs.* 143(8):10, 65–67.

127. Stewart, N. 1986. Production of bark for composts. *Proc. Inter. Plant Prop. Soc.* 35:454–58.

128. Studer, E. J., P. L. Steponkus, G. L. Good, and S. C. Wiest. 1978. Root hardiness of container-grown ornamentals. *HortScience* 11(1):34–35.

129. Tibbitts, T. W. 1988. Supplementing sunlight in winter. *Amer. Veg. Grower* 36(4):56–58.

130. Trickett, E. S., and J. D. S. Goulden. 1958. The radiation transmission and heat conserving properties of some plastic films. *Jour. Agr. Eng. Res.* 3(4):281–87.

131. U.S. Dept. of Agriculture. 1965. Hotbed and propagating frames. *USDA Misc. Publ. 986.*

132. U.S. Dept. of Energy. 1978. *Energy audit for growers: A self-inspection guide to reduce energy costs.* Alexandria, Va.: Soc. Amer. Florists.

133. Vaartaja, O. 1964. Chemical treatment of seedbeds to control nursery diseases. *Bot. Rev.* 30:1–91.

134. Vance, B. F. 1975. Water quality and plant growth. *Proc. Inter. Plant Prop. Soc.* 25:136–41.

135. Van der Borg, H. H., ed. 1979. Symposium on computers in greenhouse climate control. *Acta Hort.* 106:1–209.

136. Vetanovetz, R. P., and J. F. Knauss. 1988. Water quality. *Greenhouse Manager* 6(12):64–72.

137. Walker, J. N., and D. C. Stack. 1970. Properties of greenhouse covering materials. *Trans. Agr. Soc. Amer. Eng.* 13(5):682–84.

138. Ward, J. 1980. An approach to the control of *Phytophthora cinnamomi. Proc. Inter. Plant Prop. Soc.* 30:230–37.

139. Ward, J., N. C. Bragg, and B. J. Chambers. 1987. Peat-based composts: Their properties defined and modified to your needs. *Proc. Inter. Plant Prop. Soc.* 36:288–92.

140. Ware, R., and I. S. Frankhauser. 1988. What should we cover the greenhouse with? *Proc. Inter. Plant Prop. Soc.* 37:161–65.

141. Weatherspoon, D. M., and C. C. Harrell. 1980. Evaluation of drip irrigation for container production of woody landscape plants. *HortScience* 15(4):488–89.

142. Weiler, T. C. 1977. Survival strategies of Northern Europe's greenhouse industry. *HortScience* 12(1):30–32.

143. Whitcomb, C. E. 1978. A self-contained solar-heated greenhouse. *HortScience* 13(1):30–32.

144. ———. 1979. Growing plants in poly bags. *Amer. Nurs.* 149(12):10–11, 97–98.

145. White, D. P., and B. G. Ellis. 1965. Nature and action of slow release fertilizers as nutrient sources for forest tree seedlings. *Mich. Quart. Bul.* 47(4):606–14.

146. White, J. W. 1971. Interaction of nitrogenous fertilizers and steam on soil chemicals and carnation growth. *Jour. Amer. Soc. Hort. Sci.* 96(2):134–37.

147. ———. 1979. Energy efficient growing structures for controlled environment agriculture. In *Horticultural reviews,* Vol. 1, J. Janick, ed. Westport, Conn: AVI Publ. Co.

148. White, J. B., and R. A. Aldrich. 1980. *Greenhouse energy conservation.* University Park, Pa.: Pennsylvania State Univ.

149. White, R. A. J. 1978. Some comparisons between plastic and glass-covered greenhouses. *Proc. Inter. Plant Prop. Soc.* 28:273–79.

150. Whittle, J. 1987. Physical and chemical properties of peat. *Proc. Inter. Plant Prop. Soc.* 36:284–87.

151. Wiebe, J. 1958. Phytotoxicity as a result of heat treatment of soil. *Proc. Amer. Soc. Hort. Sci.* 72:331–38.

152. Wildschut, H. C. 1984. Computer controls for greenhouse environments. *Proc. Inter. Plant Prop. Soc.* 33:72–78.

153. Williams, D. J. 1980. How slow-release fertilizers work. *Amer. Nurs.* 151(6):90–97.

154. Williams, T. J. 1978. *How to build and use greenhouses.* San Francisco: Ortho Books, Chevron Chemical Co.

155. Wilson, G. C. S., ed. 1980. Symposium on substrates in horticulture other than soils in situ. *Acta Hort.* 99:1–248.

156. Wittwer, S. H., and W. Robb. 1964. Carbon dioxide enrichment of greenhouse atmospheres for food crop production. *Econ. Bot.* 18:34–56.

157. Wood, J. S. 1985. Sun frame propagation. *Proc. Inter. Plant Prop. Soc.* 34:306–11.

158. Worrall, R. 1976. The use of sawdust in potting mixes. *Proc. Inter. Plant Prop. Soc.* 26:379–81.

159. ———. 1982. High temperature release characteristics of resin-coated slow release fertilizers. *Proc. Inter. Plant Prop. Soc.* 31:176–81.

160. Wright, R. D., and A. X. Niemiera. 1987. Nutrition of container-grown woody nursery crops. In *Horticultural reviews,* Vol. 9, J. Janick, ed. Westport, Conn., AVI Publ. Co., pp. 75–101.

161. Yates, N. L., and M. N. Rogers. 1981. Effects of time, temperature, and nitrogen source on the composting of hardwood bark for use as a plant growing medium. *Jour. Amer. Soc. Hort. Sci.* 106(5):589–93.

162. Youngsman, J. E. 1983. Ten points for infrared. *Florists' Review,* May.

SUPPLEMENTARY READING

ALDHOUS, J. R. 1972. *Nursery practice.* For. Comm. Bul. 43. London: Her Majesty's Stationery Office.

AUGSBURGER, N. D., H. R. BOHANON, and J. L. CALHOUN. 1978. *The greenhouse climate control handbook.* Muskogee, Okla.: Acme Eng. and Mfg. Co.

BAKER, K. F. ed. 1957. The U.C. system for producing healthy container-grown plants. *Calif. Agr. Exp. Sta. Man. 23.*

BALL, V., ed. 1985. *The Ball red book* (14th ed.). Reston, Va.: Reston Publ. Co.

BOODLEY, J. W. 1981. *The commercial greenhouse handbook.* New York: Van Nostrand Reinhold.

BUNT, A. C. 1976. *Modern potting composts.* London: G. Allen & Unwin.

CATHEY, H. M., and L. E. CAMPBELL. 1980. Light and lighting systems for horticultural plants. *Hort. Rev.* 2:491–537.

CORRELL, P. G., and J. G. PEPPER, eds. 1977. *Energy conservation in greenhouses.* Newark, Del.: Univ. of Delaware.

DAVIDSON, H., R. MECKLENBURG, and C. PETERSON. 1988. *Nursery management: Administration and culture* (2nd ed.). Englewood Cliffs, N.J.: Prentice-Hall.

Greenhouse Manager, published monthly. Fort Worth, Tex.: Branch-Smith Publishing.

Grower Talks, published monthly. West Chicago, Ill.: Geo. J. Ball, Inc.

HANAN, J. J., W. D. HOLLEY, and K. L. GOLDSBERRY. 1978. *Greenhouse management.* New York: Springer-Verlag.

LAMB, J. G. D., J. C. KELLEY, and P. BOWBRICK. 1975. *Nursery stock manual.* London: Grower Books.

LANGHANS, R. W. 1980. *Greenhouse management.* Ithaca, N.Y.: Halcyon Press.

LARSON, R. A., ed. 1980. *Introduction to floriculture.* New York: Academic Press.

MASTALERZ, J. W., ed. 1985. *Bedding plants* (3rd ed.). University Park, Pa.: Pennsylvania Flower Growers.

MASTALERZ, J. W. 1977. *The greenhouse environment.* New York: John Wiley.

McGUIRE, J. J. 1972. Growing ornamental plants in containers: A handbook for the nurseryman. *Univ. of Rhode Island Coop. Ext. Ser. Bul. 197.*

NELSON, P. V. 1985. *Greenhouse operation and management* (3rd ed.). Reston, Va.: Reston Publ. Co.

TINUS, P. W., and S. E. McDONALD. 1979. *How to grow tree seedlings in containers in greenhouses.* Gen. Tech. Rpt. RM-60. Ft. Collins, Colo.: Rocky Mt. For. and Range Exp. Sta., For. Ser., USDA.

VERMA, B. P., ed. 1983. *Greenhouse and nursery mechanization: A compilation of published papers.* St. Joseph, Mich.: Amer. Soc. Agr. Eng.

3

The Development of Seeds and Spores

Propagation by seeds is the major method by which plants reproduce in nature and one of the most efficient and widely used propagation methods for cultivated crops. The plants produced are referred to as **seedlings.** This term is significant in horticulture in that in this text it refers to the life cycle of a plant grown from a seed in contrast to a plant started vegetatively from a cutting or grafted plant.

The planting of the seed is the physical beginning of seedling propagation. The seed itself, however, is the end product of a process of growth and development within the parent plant that is described in this chapter. It may arise either from the fusion of male and female gametes to form a single cell (the **zygote**) within the ovule of the flower or from vegetative cells within the ovule. The zygote has the property of **totipotency;** that is, it has all of the genetic information needed to reproduce the full-sized plant and to initiate the seedling cycle of the next generation.

SEEDLING LIFE CYCLE

The life cycle of a seedling plant (Chap. 1) consists of consecutive periods of vegetative growth (Fig. 1–5) and reproductive development [i.e., the formation of flowers and fruits (Figs. 1–1 and 1–2)].

The seasonal pattern of growth and development and the ability to grow new vegetative shoots annually determines the length of the life cycle and the use of seedling plants in propagation. Variations in these cycles are adaptations to natural seasonal or climatic cycles.

Annual plants go through the entire sequence from germination of seeds to flowering and production and dissemination of seeds in one growing season and then they die. **Biennial** plants have a two-year cycle. These plants are vegetative and grow as a low clump or rosette of leaves the first season. During the second season the plants produce flower stalks, flowers, and seeds; then the plants die. The transition from a vegetative to a reproductive stage is often a response to some environmental trigger, such as a cold cycle (**vernalization**) or a certain length of day (**photoperiod**).

Perennial plants live for more than two years and, once past the juvenile phase, repeat the vegetative-reproductive cycle annually. Consecutive growth and dormancy cycles are related either to warm–cold or to wet–dry environmental changes.

Herbaceous perennials are those in which shoots die during the winter or during dry periods. The plants survive such dormant periods as specialized underground structures, such as bulbs, rhizomes, or crowns (see Chaps. 14 and 15). **Woody perennial** plants continue to increase in size each year by growth of the shoot and root tips or by lateral cambium growth, or both. Only a portion of the shoot growing points become reproductive while the others remain vegetative to continue shoot growth of the plant.

Phases of the Seedling Cycle

Figure 3–1 shows further aspects of the life cycle of seedling plants and introduces the concept of **phase change** (*9, 33*). A plant life cycle begins with the formation of a **zygote** within the ovule and ovary of a flower. This single-celled structure divides, grows, and differentiates to produce an embryo enclosed within the seed and fruit. During its development period the embryo is parasitic on the mother plant and undergoes characteristic morphological and physiological stages of development.

With the germination of the seed, the embryo develops into a seedling plant to begin the **juvenile** phase. Vegetative growth predominates as the seedling plant enlarges in size through elongation in stems and roots and increase in cross-sectional area.

Juvenile plants continue to be vegetative for a period of time during which they are unable to

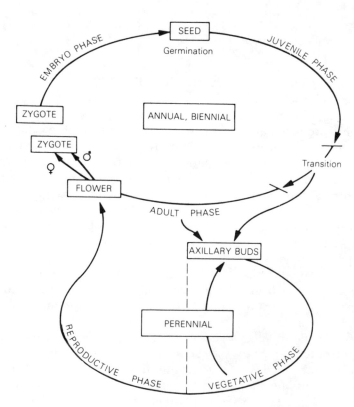

FIGURE 3–1 Model illustrating phase changes that occur in a seedling life cycle. The upper part of the cycle illustrates the progression of a meristem (growing point) in an annual or biennial plant through one or two growing seasons where there is a more or less continuous transition from juvenile-vegetative to flowering to plant senescence. The lower part represents a herbaceous or woody perennial in which the vegetative mature phase is renewed continuously by seasonal cycles of growth and reproduction.

respond to flower-inducing stimuli. Eventually, a **transitional** phase begins in which the plant loses its characteristic juvenility (Fig. 3–2). In the succeeding adult or **mature** phase, reproduction by seed predominates. In herbaceous annuals and biennials the plant reaches its ultimate size, flowers, and then dies. Once the perennial (herbaceous or woody) plant reaches the adult phase, it regularly develops flowers from then on and continues to regenerate itself from vegetative shoots.

The shift from juvenile to adult (or mature) has been called **phase change** (*9*) or **maturation** (*33*). Such a phenomenon occurs in all plants but may not be shown by distinct morphological differences and may not, therefore, be readily recognized; in other plants, particularly in woody and herbaceous perennials, the juvenile phase may differ markedly in appearance from the adult phase and/or last for many years. Juvenility is significant in various aspects of propagation and is described further in Chap. 8 and referred to throughout the text.

PRODUCTION OF THE FLOWER

In the vegetative stage the plant grows by elongation of terminal and lateral shoots producing a series of nodes and internodes. As the shoots shift to the reproductive stage, vegetative growing points develop into flowers. In annual plants this process may be continuous through the life of the plant. In biennial and perennial plants, the difference between vegetative and reproductive stages may be more distinct. The reproductive stage begins with **flower bud induction,** which is an internal physiological change in the individual growing point or meristem that precedes any morphological change. The second stage is the visible **initiation** of the flower parts followed by **differentiation** of the flower structure, leading to flowering (also known as **anthesis**).

In many plants induction and progress through these phases is geared to environmental seasonal cycles. For instance, in biennials and some woody perennials the induction of flowering

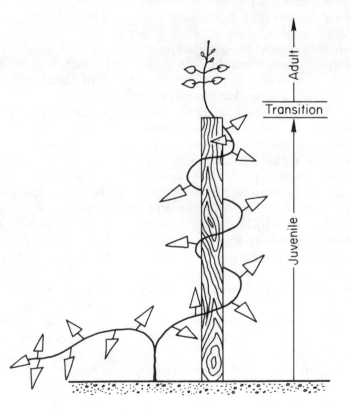

FIGURE 3–2 Phase changes in English ivy (*Hedera helix*) in which the **juvenile** (nonflowering) phase is a vine which, as it grows into a vertical form, can undergo a **transition** into the **mature** phase in which flowers and fruits are produced.

takes place following an exposure to an inductive cold period or other environmental signal (**vernalization**). In many deciduous woody perennials, induction and differentiation begins during the warm summer growing period, but subsequent differentiation of the flower buds requires chilling temperatures for normal progression of flower development and blooming.

Control of flowering in herbaceous and woody perennials is complicated by differences in the responses to flower induction stimuli given by the seedling plant (a) in the **juvenile** phase of its life cycle and (b) in the **mature** phase after it attains reproductive maturity. This difference is illustrated in the upper cycle in Fig. 3–1, contrasted to the lower cycle.

The juvenile phase is extensive in some woody and herbaceous perennials and the length of time before seed crops are produced is very long. Plants of some bamboo species remain vegetative for decades, then suddenly produce flowers and seeds and die. The so-called century plant (*Agave americana*) follows a similar pattern. Trees of many forest species do not produce seed until they are 15 to 20 years old (*53*). Some fruit tree species, such as apple or pear, often require five to seven years before they flower. Seedlings of many orchid and bulb species do not start blooming until they are five to seven years old.

Early induction of maturation and flowering may occur if the plant grows rapidly through the required juvenile phase. Exposure of the plant to optimum growing conditions to maximize vegetative growth during this phase will bring on the mature stage more quickly, particularly among woody perennials. As an example, in controlled experiments apple seedlings have been caused to flower after only 16 months by keeping the plants growing continuously in a greenhouse, compared to three to eight years for similar seedlings grown out-of-doors. Similar results have been reported for pear, crabapple, birch, and spruce seedlings (*59*).

Once past the juvenile phase, the mature plant begins the flowering process by responding to various flower induction stimuli. These may be environmental signals, such as photoperiods (either long or short days), or temperature regimes (cool or warm). Conditions that slow or re-

duce vegetative growth, such as girdling or decreasing nitrogen, may cause flower initiation, whereas excessively vigorous growth depresses it. These effects are opposite to those which favor transition to mature phase from the juvenile phase. Because conditions required for flower induction vary with individual species, one must be familiar with their growth cycles to be able to induce early flowering (*33*).

Many woody perennial plants have a biennial cycle in which shoots are vegetative one season and reproductive the following. Annual cropping is produced because other shoots on the same plant alternate in their cycle. Erratic seed production often seen in forest trees may result from a variety of adverse conditions that inhibit flower induction (*53*). These conditions include: (a) competition from an excessively large seed crop the preceding season, (b) stress produced by inadequate nutrition or water, or excessively low or high temperatures, or (c) defoliation from insects or disease. If the mature plant is growing satisfactorily, a check in vegetative growth at an appropriate stage by girdling or root pruning often induces flowering.

FORMATION OF THE FRUIT, SEED, AND EMBRYO

The miniature flower parts subsequently develop into complete flowers. The sexual cycle includes development of the male (pollen) and female (embryo sac) structures of the flower, as shown for angiospermous flowers in Fig. 1–1. In this part of the cycle reduction division of chromosomes occurs to produce the haploid (*n*) chromosome number. Figure 3–3 shows the basic structure of the flower and Fig. 3–4 shows more details of the pollen, ovary, and ovule structure. Pollen grains and an eight-celled embryo sac contain haploid (*1n*) male and female gametes, respectively. During flowering, pollen is transferred from the anther to the stigma (**pollination**), where it germinates. A pollen tube grows down the style into the ovary until it reaches the embryo sac within the ovule (Fig. 3–4). Two male gametes from the pollen tube are discharged into the embryo sac—one to unite with a female gamete (**fertilization**) to pro-

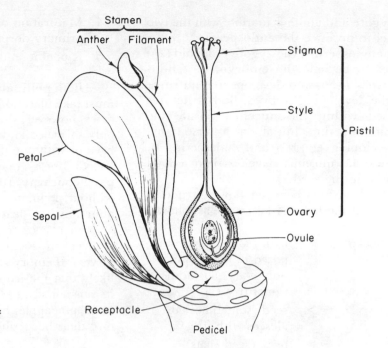

FIGURE 3-3 Flower structure of an angiospermous plant.

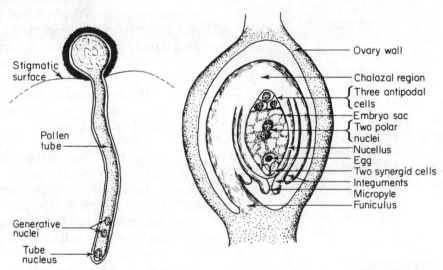

FIGURE 3-4 Sexual reproduction in an angiospermous plant. *Left:* Haploid pollen grain (male) germinates on the stigma, a pollen tube grows into the style. Two generative nuclei (male sperm cells) are produced. *Right:* The ovary is the female structure (pericarp) that encloses the ovule, which consists of an embryo sac surrounded by nucellus tissue, all of which are genetically the same as the maternal plant. The mature embryo sac develops eight haploid nuclei, including an egg nucleus (female sex cell) and two haploid polar nuclei, all of which take part in fertilization.

duce the **zygote** and another to unite with the two polar nuclei to produce the **endosperm.**

With angiosperms, the **zygote** is diploid ($2n$) and divides to become the **embryo;** the **endosperm** is triploid ($3n$) and develops into nutritive tissue for the developing embryo. Both structures are enclosed within the **nucellus** (inside the ovule), which functions initially as a nurse tissue for the developing embryo and endosperm. In some species the nucellus develops into a food storage tissue in the seed.

The relationships between flower structure and the parts of the fruit and seed are as follows:

Ovary (pericarp) ····> fruit (sometimes composed of more than one ovary plus additional tissues)

Ovule ···············> seed (sometimes coalesces with fruit)

  integuments ·······> testa (seed coats)

  nucellus···········> perisperm (usually absent or reduced; sometimes storage tissue)

2 polar nuclei +
  sperm nucleus··> endosperm ($3n$)

egg nucleus +
  sperm nucleus··>
  zygote···········> embryo ($2n$)

Seed development in the gymnosperms is somewhat different (Fig. 1–2). The ovule is not enclosed but is exposed within the ovulate cone. The developing embryo is nourished by the haploid ($1n$) **female gametophyte,** which develops into the storage organ of the mature seed (*55*).

FRUIT, SEED, AND EMBRYO DEVELOPMENT

Fruit and seed development involves five separate processes:

1. Morphological development
2. Acquisition of the ability of the embryo to germinate
3. Accumulation of reserve storage compounds
4. Maturation of the seed plus development of primary dormancy
5. Ripening and dissemination

Both **pollination** and **fertilization** normally must take place to produce a viable seed. In some cases, however, the fruit may mature and contain only shriveled and empty seed coats with no embryo or with one that is thin and shrunken. Such ''seedlessness'' may result from several causes: (a) **parthenocarpy** (the development of the fruit without pollination or fertilization); (b) **embryo abortion** (the death of the embryo during its development); or (c) **nonfilling** (the failure of the embryo to accumulate the required food reserves). If embryo abortion occurs early, it is most likely that the fruit will drop or will not grow to its normal size (*43*). Many important fruit crops (e.g., pineapple, banana, and ''seedless'' grapes) owe their horticultural value to seedlessness.

Morphological Development

Figures 3–5 and 3–6 illustrate the growth patterns of different parts of the fruit and seed as illustrated in four different angiosperm species with different fruit structures. This basic sequence of seed development applies more or less to all species, but unique patterns of fruit development may characterize different plant families (*2*).

There are three morphological stages in fruit and seed development:

Stage 1 is characterized by growth of the fruit (ovary) and the seed (ovule). The developing fruit is also known as the **pericarp.** In the dry lettuce and *Datura* seeds, the final size of both the fruit and seed is reached during this stage. The **nucellus** grows by cell division and enlargement and appears as a clear watery mass enclosed within the ovule walls (integuments). The **endosperm** develops a cellular consistency but remains microscopic during stage 1 enclosing the even smaller embryo. The single-celled **zygote** divides in a characteristic pattern to develop into a polarized mass of cells (**proembryo**) attached to a special embryonic structure known as a **suspensor** in both angiosperms (*4*) and gymnosperms (*41*). This entire structure has been called an **embryo-**

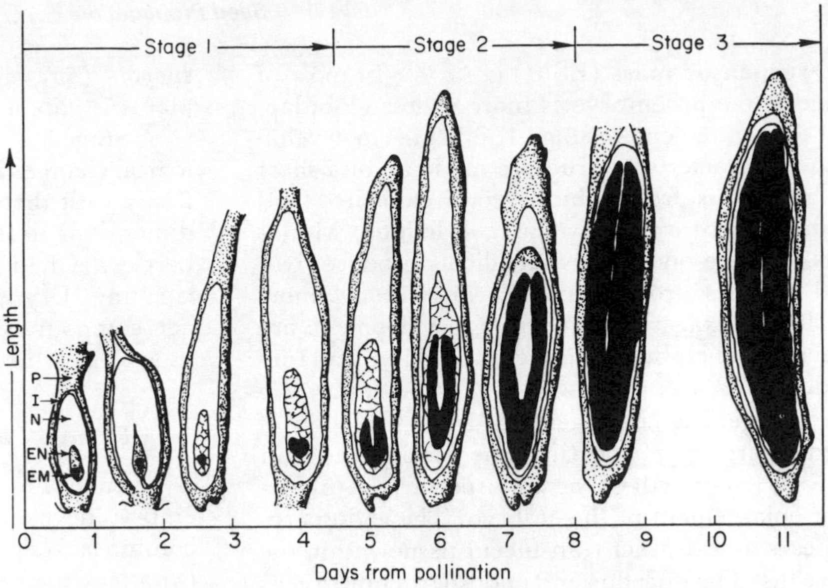

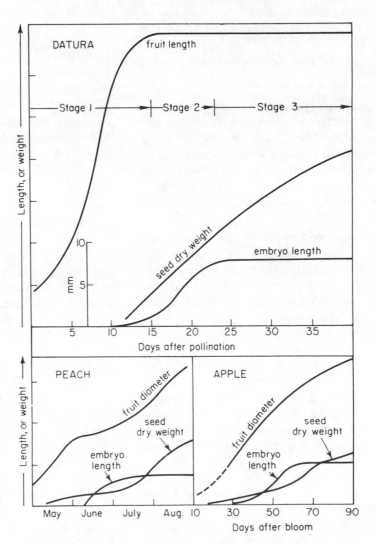

FIGURE 3–5 Growth and development of the fruit and seed structure in lettuce. P, pericarp; I, integruments; N, nucellus; EN, endosperm; EM, embryo. (Redrawn from Jones (*35*).)

FIGURE 3–6 Comparative development of three different fruit types: *Datura* (dry capsule), peach (fleshy drupe), and apple (fleshy pome). Changes in the seed dry weight and embryo are the same among the species represented. Differences occur in the fruit growth pattern. (*Datura* data from Rietsema et al. (*48*), apple data in part from Luckwill (*Jour. Hort. Sci.* 24:32–44).)

nal-suspensor mass (**ESM**) (*24, 31*). In most of stage 1, the proembryo is more or less **globular,** but toward the end of stage 1, it begins to develop into a characteristic structure made up of a short **embryo axis** from which grows elongated and fleshy structures, known as **cotyledons.** Monocots produce one cotyledon, dicots produce two, and conifers produce many. Transitional morphological stages of cotyledon development are described as **globular, heart** (early, late), and **torpedo** (*1, 48, 52*).

Stage 2 is characterized by (a) cessation of seed enlargement and (in some species) the fruit as well, (b) growth of the endosperm, and (c) further enlargement of the embryo. The endosperm appears as a distinct translucent tissue within the nucellus. The nucellus and endosperm function as nurse and conducting tissues for the endosperm and embryo, respectively. Both are digested in the process (*5, 10*).

Embryo development follows either of two basic patterns:

Endospermic type. Cotyledon growth is arrested at different stages of development, such that the embryo may be only one-third to one-half full size at the time of seed ripening. The remainder of the seed cavity contains large amounts of endosperm, nucellus, or gametophyte tissue, depending on the species.

Nonendospermic type. Rapid growth by cell division and enlargement occurs in the endosperm, digesting the enclosing nucellus. This is followed by expansion of the embryo through cell division at the periphery of the cotyledons which digests the endosperm. In these seeds, the endosperm and/or the nucellus is reduced to a remnant between the embryo and the integuments.

Failure of the endosperm to develop properly results in retardation or arrest of embryo development, and embryo abortion can result. This phenomenon is called **somatoplastic sterility** and commonly occurs when two genetically different individuals are hybridized, either from different species (*14–16*) or from two individuals of different ploidy constitution. It can be a barrier to hybridization in angiosperms but not in gymno-

sperms (*55*), since the "endosperm" in these plants is haploid female gametophytic tissue.

Stage 3 is characterized primarily by physiological changes in the fruit and seed except that in fleshy fruit there may be further enlargement in dimensions in the pericarp (e.g., pomes, drupes, berries). Changes in the pericarp walls lead to ripening. Dry weight increases and specific storage compounds accumulate. Changes also occur in seed moisture relationships.

Embryo Damage

Environmental agents can prevent normal embryo development, even though the fruit itself continues to develop. Numerous kinds of insects (*8*) attack the developing seed and fruit, particularly in forest trees (*53*). The developing seed of carrot and other umbelliferous plants is attacked by *Lygus* bugs, which can penetrate the fruit and feed on the embryo. Adverse weather, such as frosts during early fruit development, can kill the embryo, but the fruit itself continues to develop (*26, 54*). Growth tensions, causing the hardening endocarp to split or crack, can result in abortion of stone fruit seeds (*18*). This condition is associated with excessive vigor, reduced crop density, root restriction, or tree injuries resulting in girdling.

Acquisition of the Ability to Germinate

The development from zygote to embryo is called **embryogenesis**. Morphological changes involve cell division and expansion to form a polarized **embryo axis** and **cotyledons.** Physiological and epigenetic changes (see Chap. 6) involve a "switch" from the embryogenic potential of the whole embryo to the localization of cell expansion and division in growing points at each end of the polarized axis, that is, the root and shoot tips. The latter phenomenon is described as germination and seedling growth (see Chap. 6). These changes can be demonstrated by excising the embryo from the seed at various stages and grown in artificial culture in vitro (see Chap. 16). Separated from the enclosing ovule tissues, embryogenesis ceases and the immature embryo shows "precocious germination." The shift is gradual, however (*27, 34, 36, 38, 43, 57*), and a changing mixture of embryonic and germination responses occurs with increasing developmental age.

Figure 3–7A illustrates the general pattern of embryogenic development in rape seed (*Brassica napus*) (*27*). A "germination" response, as defined by radicle (root) emergence, occurs as early as the "heart" stage (19 days). However, during the next 21 to about 40 days, when the cotyledons in the seed are elongating, the embryo continues to retain an **embryogenic potential** (i.e., an inherent tendency to produce an embryo). For example, new **secondary embryos** may develop on the surface of the embryo and at a slightly more advanced age new **secondary cotyledons** can occur. Only at a further embryo age (36 days) is the hypocotyl (see Chap. 6 for description) capable of elongation. Leaves do not appear on the seedling structure until even later (after about 50 days).

Control of embryogenesis is complex and involves the innate potentiality (epigenetic) of the cells at different ages, hormonal controls (both within the embryo and in the surrounding tissue), and physical factors surrounding the embryo, such as osmotic potential (*1*). Epigenetic control of the cell potentiality appears to be mediated through specific kinds of mRNA (**messenger ribonucleic acid**) that translate genetic information from specific genes (see Fig. 3–7C). For example, the mRNA for the storage protein cruciferin can be identified only between days 19 and 40. Endogenous hormones, particularly cytokinins and gibberellins (*5*), are high during the initial stages of embryogenesis. The suspensor tissue may play an essential role in gibberellin synthesis (*5*). High sugar concentrations and/or other osmotic agents are needed in the culture medium to maintain

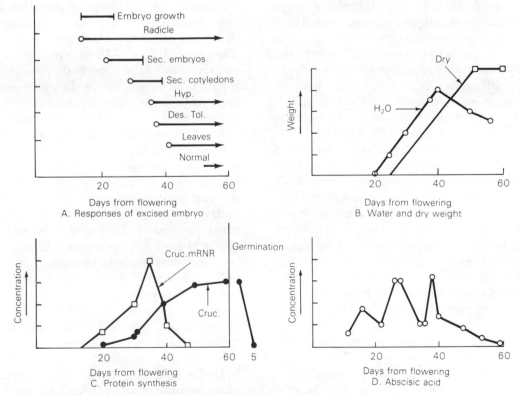

FIGURE 3–7   Changes in specific parameters of development during embryogenesis of rape (*Brassica napus*) seed. (A) Morphological responses in embryo, root emergence, secondary embryo formation, secondary cotyledon formation, hypocotyl elongation, desiccation tolerance (DT), leaf formation, and normal seedling germination. (B) Weight changes: water content, dry weight. (C) Protein synthesis: cruciferin, cruciferin mRNA. (D) Abscisic acid concentration. (Redrawn from Finkelstein and Crouch (*27*).)

embryogenesis in vitro (*5, 14, 36*) and appear to mimic high osmotic pressure within the endosperm and nucellus.

Accumulation of Storage Compounds

Accumulation of complex storage products—carbohydrates, fats, oils, proteins, and other biochemical substances—into the storage organs of the seed (i.e., cotyledons, endosperm, nucellus, and/or gametophyte tissue) is an essential part of seed development. Such substances not only provide essential energy substrates to ensure survival of the germinating seedling but also provide essential food for human beings and animals.

The process of assimilation requires the translocation of small molecular weight compounds, such as sucrose, asparagine, glutamine, and minerals, into the seed. The attachment of the seed to the ovule is by the **funiculus** through which vascular connections (phloem, xylem) extend into the integuments (seed coverings). No vascular connections to the nucellus, endosperm, and embryo exist and assimilates must reach the embryo by **diffusion.** At the same time, most viruses and large complex molecules are effectively screened from the embryo in this process but may accumulate in the outer layers of the seed.

Synthesis of storage compounds requires a battery of specific enzymes with associated mRNAs to direct them (*21–23*). Figure 3–7B shows that the increase in dry weight of the embryo increases with time during the last half of the fruit growth period. In the rape seed this increase in the amount of a unique storage protein—cruciferin—is found only in this species and only at this stage in the life cycle of the plant. Preceding the appearance of cruciferin is the increase in the concentration of cruciferin-mRNA that programs the metabolic machine of the cells of the rape cotyledon to synthesize cruciferin increase. Note that cruciferin-mRNA then disappears. This storage compound is metabolized during germination.

The synthesis and accumulation of reserve compounds must proceed without interference and reach a minimum level if high-quality seeds are to be produced (*17, 19, 45*). Heavier seeds result in better germination and produce more vigorous seedlings. If conditions interfere with the process, the seeds may be thin and light in weight. The more severe the condition, the less the seeds survive storage, the poorer the germination, and the weaker the seedlings that are produced. Such conditions include adverse growing conditions, including poor nutrition, moisture stress, disease and insect damage (*8*), and excessively high or low temperatures. Lack of reserves at harvest is due chiefly to premature seed immaturity (*3, 19, 25*) in which the accumulation process is interrupted.

Maturation of the Seed and Primary Dormancy

Maturation of the Embryo

Studies have suggested that shortly after the embryo (and/or seed) has attained its morphological maturity, a specific-stage **physiological maturity** involving a change in its moisture relationship occurs (*36*). This phenomenon is also illustrated in Fig. 3–7B for rape seed. The rapid enlargement of the rape embryo within the seed is paralleled by an increase in water content. During this time the percent moisture remains relatively constant. At a point near or after the attainment of full morphological maturity, the moisture content begins to decrease. The dry weight continues to increase due to synthesis of protein. During the early development phase, the embryo is **desiccation-sensitive** and is injured if subjected to drying. About the time that the moisture curve turns downward the embryo becomes **desiccation-resistant** and is no longer injured by drying.

An endogenous hormone, **abscisic acid (ABA),** appears to play a major role in maintaining the embryonic phase preventing precocious germination (*5, 22, 46*) and inducing maturation of the embryo, possibly through its effect on moisture relationships (*36*). Changes in ABA content during embryo development are illustrated in Fig. 3–7D for rape seed. During the expansion period of the embryo (stage 2), ABA increases to a relatively high level but decreases sharply during the last period of seed development. In seeds of this species, ABA application to excised embryos promotes continued embryo development and the accumulation of cruciferin.

Primary Dormancy

In seeds of most species, internal controls exist or develop during ripening and for a period of time after harvest that prevent germination. These controls have evolved in particular environments to regulate the time and place of germination to best ensure survival of the species. These are described in Chaps. 6 and 7.

The two major mechanisms that develop during seed maturation and ripening and impose dormancy on the seed include (a) the accumulation of chemical growth inhibitors into different tissues of the fruits and seeds, and (b) the modification of seed coverings. These control the uptake of water, the permeability to gases, and the leaching of inhibitors.

Inhibitors. Many naturally occurring chemicals that can inhibit germination accumulate in different parts of the seed but not necessarily in the embryo. Demonstrating a role in natural dormancy control does not always follow, however. Inhibitors are present in fleshy fruits, such as tomato, lemon, and strawberry. Seeds of some desert plants have a high salt concentration, which has been suggested as an inhibitor.

Abscisic acid has been described as an important compound in controlling embryo development and preventing germination [7]. Although rape seed in Fig. 3–7 shows a concentration decrease at ripening, seeds in many species continue to have a high concentration of ABA at ripening [5, 22, 27, 46]. Later (Chap. 6), ABA is shown to play a role in the initial primary dormancy in seeds of some species.

Seed coverings. Modification of seed coverings primarily affects the outer integument layer of the seed, which may become hard, fibrous, or mucilaginous during dehydration and ripening (Fig. 3–8). Coverings arise primarily from the outer integument layer, which becomes hard, fibrous, or mucilaginous during ripening and dehydration. In addition, layers of the fleshy fruit may dry and become part of the seed covering as in *Cotoneaster* or hawthorn (*Crataegus*). In some drupe fruits, as the olive or *Prunus* species, these layers become the hardened endocarp (pit or stone). In other seeds, such as walnut, they become the surrounding shell. In others, such as caryopsis or achenes in grains or grasses, the fruit covering becomes fibrous and coalesces with the seed.

In various plant families, such as Leguminoseae, the outer seed coats harden and become suberized and impervious to water (Fig. 3–8A). Cells of the outer integument become rearranged, coalesce, incorporate suberin deposits, and develop external cutin coverings. These cells are called **macrosclereid** cells [50].

In other species, such as white mustard and spinach, mucilaginous layers inside and outside the seed coats are produced particularly under high moisture conditions, which also function to restrict gaseous exchange (Fig. 3–8B).

In most families the inner seed coat becomes membranous but remains alive and semipermeable. In the Compositae, for instance, this layer coalesces with the remnant layers of the endosperm. These layers of integument and remnants

---

VIVIPARY

The phenomenon in which seeds germinate in the fruit while still attached to the plant is known as **vivipary** and the germinating embryos are called **viviparous embryos.** This characteristic is controlled both genetically and environmentally. For example, germinating seeds are occasionally found in mature citrus fruits. In mangrove (*Rhizophora mangle*), a swamp-growing tree species, vivipary is an adaptation to its environment. Embryos germinate directly on the tree to produce seedlings with a long javelin-shaped root. These eventually fall and become embedded in the mud below [56].

For most plant species, however, vivipary is undesirable. Premature seed sprouting may occur in some grain crops and certain nut crops, such as pecan, in periods of wet weather during or just before harvest [5, 29, 46]. The tendency toward vivipary is a varietal characteristic in grain crops [5, 58]; such a tendency is automatically selected against as a defective character [20].

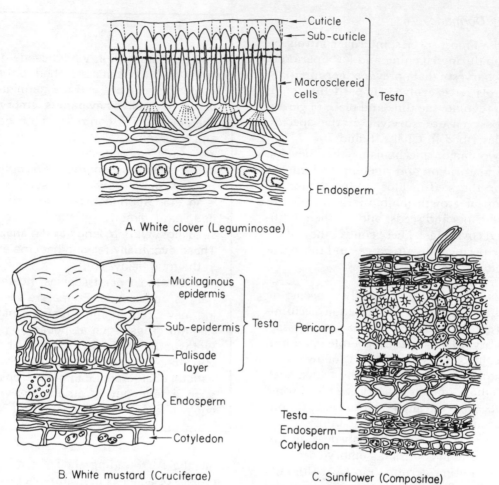

A. White clover (Leguminosae)

B. White mustard (Cruciferae)

C. Sunflower (Compositae)

FIGURE 3-8   Types of seed coats that affect seed dormancy. (A) White clover (*Melilotus alba*) LEGUMINOSEAE: the outer layer of cells becomes impervious to water uptake because of vertically oriented macrosclereid cells that are covered by a layer of cuticle. (B) White mustard (*Brassica hirta*) CRUCIFERAE: outer seedcoats develop a mucilaginous layer with inhibitors when soaked in water. (C) Sunflower (*Helianthus annuus*) COMPOSITAE: pericarp hardens upon drying into a fibrous layer coalescing with the seed coats. Endosperm layer is thin, more or less membranous, and functions in dormancy control. (Courtesy Bewley and Black (*5*).)

of the endosperm and nucellus remain physiologically active during ripening and for a period of time after the seed is separated from the plant (see Fig. 3-8C). Such physiologically active layers play a role in maintaining primary dormancy, mainly because this semipermeable nature restricts aeration and inhibitor movement.

Ripening and Dissemination

Specific physical and chemical changes that take place during maturation and ripening of the fruit lead to fruit senescence and dissemination of the seed. One of the most obvious changes is the drying of the pericarp tissues. In certain species, this leads to dehiscence and the discharge of the seeds from the fruit. Changes may take place in the color of the fruit and the seed coats, and softening of the fruit may occur.

Seeds of most species dehydrate at ripening and prior to dissemination. Moisture content drops to 30 percent or less on the plant. The seed dries further during harvest, usually to about 4 to 6 percent for storage (see Chap. 5). Germination cannot take place at this level of dryness and is an important basis for maintaining viability and controlling germination.

In certain other species, seeds must not dry below about 30 to 50 percent or they will lose their ability to germinate. These plants include (a) species whose fruits ripen early in summer, drop to the ground, and contain seeds that germinate immediately (some maples, poplar, elm); (b) species whose seeds mature in autumn and remain in moist soil over winter (oak); and (c) species from warm, humid tropics (citrus). These are called **recalcitrant seeds** (5) and produce special problems in handling (see Chap. 5).

Seeds of species with fleshy fruits may become dry but are enclosed with soft flesh which can decay and cause injury. In most species this fleshy tissue should be removed to prevent damage from spontaneous heating or inhibiting substance. In some species, however (e.g., *Mahonia* and *Berberis*), the fruits and seeds may be dried together (40).

Seed dispersal is accomplished by many agents. Fish, birds, rodents, and bats consume and carry seeds in their digestive tracts (28). Fruits with spines or hooks become attached to the fur of animals and are often moved considerable distances. Wind dispersal of seed is facilitated in many plant groups by "wings" on dry fruits; tumbleweeds can move long distances by rolling in the wind. Seeds carried by moving water— streams or irrigation canals—can be taken great distances and often become a source of weeds in cultivated fields. Some plants themselves have mechanisms for short-distance dispersal, such as explosive liberation of spores from fern sporangia. Human activities in purposeful shipment of seed lots all over the world are, of course, effective in seed dispersal.

THE MATURE SEED

Botanically, in the angiosperms the seed is a matured ovule enclosed within the mature ovary, or fruit. Seeds and fruits of different species vary greatly in appearance, size, shape, and location and structure of the embryo in relation to storage tissues (Fig. 3-9). These points are not only useful for identification (30, 42) but affect germination requirements. From the standpoint of seed handling, it is not always possible to separate the fruit and seed, since they are sometimes joined in a single unit. In such cases, the fruit itself is treated as the "seed," as in wheat or corn.

Parts of the Seed

The seed has three basic parts: (a) embryo, (b) food storage tissues, and (c) seed coverings.

Embryo

The embryo is a new plant resulting from the union of a male and female gamete during fertilization. Its basic structure is an **embryo axis** with growing points at each end, one for the shoot and one for the root, and one or more seed leaves (**cotyledons**) attached to the embryo axis. Plants are classified by the number of cotyledons. Monocotyledonous plants (such as the coconut palm or grasses) have a single cotyledon, dicotyledonous plants (such as the bean or peach) have two, and gymnosperms (such as the pine or ginkgo) may have as many as 15.

Embryos differ in size, reflecting the extent to which the embryo has developed within the seed. Seeds can be separated into basic types: **nonendospermic** in which the embryo is dominant, and **endospermic** in which the embryo is reduced in size in proportion to the rest of the seed. In the latter, embryos are classified into four basic types characteristic of specific plant families.

Storage Tissues

The nonendospermic seeds store foods in the cotyledons, which are the dominant seed parts. In the endospermic seeds the storage material is in

The following classification of seeds is based upon morphology of embryo and seed coverings. It includes, as examples, families of herbaceous plants [after Atwater (2)].

I.   Seeds with dominant endosperm (or perisperm) as seed storage organs (endospermic).

   A.   Rudimentary embryo. Embryo is very small and undeveloped but undergoes further increase at germination (see Fig. 3–9A, Magnolia).

      RANUNCULACEAE (*Aquilegia, Delphinium*), PAPAVERACEAE (*Eschscholtzia, Papaver*), FUMARIACEAE (*Dicentra*), ARALIACEAE (*Fatsia*), MAGNOLIACEAE (*Magnolia*), AQUIFOLI-ACEAE (*Ilex*)

   B.   Linear embryo. Embryo is more developed than those in A and enlarges further at germination.

      UMBELLIFERAE (*Daucus*), ERICACEAE (*Calluna, Rhododendron*), PRIMULACEAE (*Cyclamen, Primula*), GENTIANACEAE (*Gentiana*), SOLANACEAE (*Datura, Solanum*), OLEACEAE (*Fraxinus*)

   C.   Miniature embryo. Embryo fills more than half the seed.

      CRASSULACEAE (*Sedum, Heuchera, Hypericum*), BEGONIACEAE (*Begonia*), SOLANACEAE (*Nicotiana, Petunia, Salpiglossis*), SCROPHULARIACEAE (*Antirrhinum, Linaria, Mimulus, Nemesia, Penstemon*), LOBELIACEAE (*Lobelia*)

   D.   Peripheral embryo. Embryo encloses endosperm or perisperm tissue.

      POLYGONACEAE (*Eriogonum*), CHENOPODIACEAE (*Kochia*), AMARANTHACEAE (*Amaranthus, Celosia, Gomphrena*), NYCTAGINACEAE (*Abronia, Mirabilis*)

II.   Seeds with embryo dominant (nonendospermic); classified according to the type of seed covering.

   A.   Hard seed coats restricting water entry.

      LEGUMINOSAE, GERANIACEAE (*Pelargonium*), ANACARDIACEAE (*Rhus*), RHAMNACEAE (*Ceanothus*), MALVACEAE (*Abutilon, Altea*), CONVOLVULACEAE (*Convolvulus*)

   B.   Thin seed coats with mucilaginous layer.

      CRUCIFERAE (*Arabis, Iberis, Lobularia, Mathiola*), LINACEAE (*Linum*), VIOLACEAE (*Viola*), LABIATEAE (*Lavandula*)

   C.   Woody outer seed coats with inner semipermeable layer (see Fig. 3–9E).

      ROSACEAE (*Geum, Potentilla*), ZYGOPHYLLACEAE (*Larrea*), BALSAMINACEAE (*Impatiens*), CISTACEAE (*Cistus, Helianthemum*), ONAGRACEAE (*Clarkia, Oenothera*), PLUMBAGINACEAE (*Armeria*), APOCYNACEAE, POLEMONIACEAE (*Phlox*), HYDROPHYLLACEAE (*Nemophila, Phacelia*), BORAGINACEAE (*Anchusa*), VERBENACEAE (*Lantana, Verbena*), LABIATEAE (*Coleus, Moluccela*), DIPSACAEAE (*Dipsacus, Scabiosa*)

   D.   Fibrous outer seed coat with more or less semipermeable membranous layer, including endosperm remnant (see Fig. 3–9F).

      COMPOSITEAE (many species)

III.   Unclassified

   A.   Rudimentary embryo with no food storage.

      ORCHIDACEAE (orchids, in general)

   B.   Modified miniature embryo located on periphery of seed.

      GRAMINEAE (grasses) (see Fig. 3–9G)

   C.   Axillary miniature embryo surrounded by gametophyte tissue (see Fig. 3–9H)

      Gymnosperms, in particular, conifers.

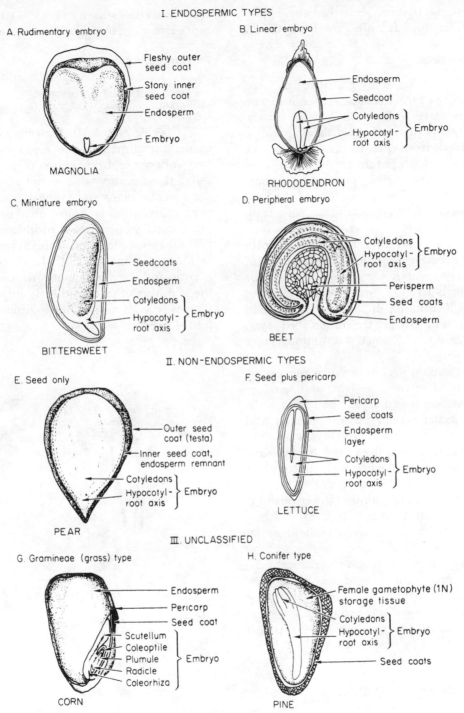

I. ENDOSPERMIC TYPES

A. Rudimentary embryo

— Fleshy outer seed coat
— Stony inner seed coat
— Endosperm
— Embryo

MAGNOLIA

B. Linear embryo

— Endosperm
— Seedcoat
— Cotyledons
— Hypocotyl-root axis } Embryo

RHODODENDRON

C. Miniature embryo

— Seedcoats
— Endosperm
— Cotyledons
— Hypocotyl-root axis } Embryo

BITTERSWEET

D. Peripheral embryo

— Cotyledons
— Hypocotyl-root axis } Embryo
— Perisperm
— Seed coats
— Endosperm

BEET

II. NON-ENDOSPERMIC TYPES

E. Seed only

— Outer seed coat (testa)
— Inner seed coat, endosperm remnant
— Cotyledons
— Hypocotyl-root axis } Embryo

PEAR

F. Seed plus pericarp

— Pericarp
— Seed coats
— Endosperm layer
— Cotyledons
— Hypocotyl-root axis } Embryo

LETTUCE

III. UNCLASSIFIED

G. Gramineae (grass) type

— Endosperm
— Pericarp
— Seed coat
— Scutellum
— Coleoptile
— Plumule
— Radicle
— Coleorhiza } Embryo

CORN

H. Conifer type

— Female gametophyte (1N) storage tissue
— Cotyledons
— Hypocotyl-root axis } Embryo
— Seed coats

PINE

FIGURE 3–9 Morphological types of seeds as described in the text. (*Rhododendron* and bittersweet redrawn from (*53*).)

the **endosperm, perisperm,** or in the case of gymnosperms, the **haploid female gametophyte.**

Seed Coverings

The seed coverings may consist of the seed coats, the remains of the nucellus and endosperm, and sometimes parts of the fruit. The seed coats, or *testa,* usually one or two (rarely three) in number, are derived from the integuments of the ovule. During development the seed coats become modified so that at maturity they present a characteristic appearance. The properties of the seed coat may be highly characteristic of the plant family. Usually, the outer seed coat becomes dry, somewhat hardened and thickened, and brownish in color. In particular families it becomes hard and impervious to water. On the other hand, the inner seed coat is usually thin, transparent, and membranous. Remnants of the endosperm and nucellus are found within the inner seed coat, sometimes making a distinct, continuous layer around the embryo.

In some plants, parts of the fruit remain attached to the seed so that the fruit and seed are commonly handled together as the "seed." In fruits such as achenes, caryopsis, samaras, and schizocarps, the pericarp and seed layers are contiguous. In others, such as the acorn, the pericarp and seed coverings separate but the fruit covering is indehiscent. In still others, such as the "pit" of stone fruits or the shell of walnuts, the covering is a hardened portion of the pericarp, but it is dehiscent and can usually be removed without much difficulty.

The seed coverings provide mechanical protection for the embryo, making it possible to handle seeds without injury, and thus permitting transportation for long distances and storage for long periods of time. The seed coverings can also control germination, as discussed in Chap. 6.

POLYEMBRYONY AND APOMIXIS

These two phenomena represent variations from the normal pattern of zygote formation and embryogenesis. Although related they are not the same phenomenon. **Polyembryony** means that more than one embryo develops within a single seed, sometimes many (Fig. 3–10). Their occurrence can develop from several distinct causes.

Polyembryony

Adventitious Embryony (Nucellar Embryony or Nucellar Budding)

Specific cells in the nucellus or (sometimes) in the integuments have embryogenic potential and undergo embryogenesis (Fig. 3–11). Genetically, these embryos have the same genotype as the parental plant and are also apomictic (see below). Adventitious embryony occurs in many plant species but is most prominent in some important subtropical and tropical tree species, such as *Citrus* and mango (*39*). In these species, both zygotic and apomictic embryos are produced and the stimulus of fertilization is required. In other species (e.g., *Opuntia*) no pollination or fertilization is needed.

FIGURE 3–10 Polyembryony in trifoliate orange (*Poncirus trifoliata*) seeds as shown by the several seedlings arising from each seed. One seedling, usually the weakest, may be sexual, the others nucellar.

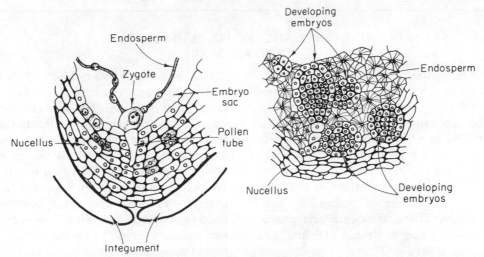

FIGURE 3-11 The development of nucellar embryos in *Citrus. Left:* Stage of development just after fertilization showing zygote and remains of pollen tube. Note individual active cells (shaded) of the nucellus, which are in the initial stages of nucellar embryony. *Right:* A later stage showing developing nucellar embryos. The large one may be the sexual embryo. (Redrawn from Gustafsson (*32*).)

Polyembryogenesis

This phenomenon results with the division of the proembryo at a very early stage of development involving the initial embryonal-suspensor mass (ESM), which is highly embryogenic (*24*). Multiplication results from "cleavage," "splitting," or "budding." Each embryo is a duplicate of the genotype of the others and represents the new sexual generation. Polyembryogenesis is most common in gymnosperms, particularly conifers (*41, 55*).

False Polyembryogenesis

In this case, multiple embryos arise through the fertilization of separate egg nuclei within the same embryo sac apparatus. These have separate genotypes but are related since each has the same seed parent but may have different pollen parents.

Apomixis

Apomixis (*44, 47*) results from the production of a zygote which bypasses the usual process of meiosis and fertilization. The genotype of the zygote and resulting embryo will be the same as the seed parent. Seed production is asexual. Such clonal seedling plants are known as **apomicts.** Some species or individuals produce only apomictic embryos and are known as **obligate apomicts;** others that produce both apomictic and sexual embryos are known as *facultative apomicts* (*6, 32, 41*).

Adventitious Embryony

See above.

Recurrent Apomixis

An **embryo sac** (female gametophyte) develops from the egg mother cell (or from some adjoining cell, the egg mother cell disintegrating), but complete meiosis does not occur. Consequently, the egg has the normal diploid number of chromosomes, the same as the mother plant. The embryo subsequently develops directly from the egg nucleus without fertilization.

This series of events is known to occur in some species of *Crepis, Taraxacum* (dandelion), *Poa* (bluegrass), and *Allium* (onion) without the stimulus of pollination; in others [e.g., species of *Parthenium* (guayule), *Rubus* (raspberry), *Malus* (apple), some *Poa* species, and *Rudbeckia*], pollina-

SIGNIFICANCE OF POLYEMBRYONY AND APOMIXIS

Polyembryony is significant because it demonstrates that embryogenic potential is not limited to the zygote but is possessed by various other somatic (nonreproductive) cells. This is the basis of in vitro somatic embryogenesis (1, 24, 51) which under proper conditions of culture and management provides a potential propagation method of multiplying millions of asexually produced "synthetic" seeds. These concepts and procedures are described in Chaps. 16 and 17.

Apomixis is significant in agriculture and horticulture because the seedling plants have the same genotype. This asexual process automatically eliminates variability and "fixes" the characteristics of any cultivar immediately. Uses in cultivar selection are described in Chap. 4. This **apomictic** life cycle as described in Fig. 3–1 has the same juvenile, transitional, and mature phases as does a seedling cycle. It differs in that no reduction division and fertilization occurred to produce the embryo.

The value of apomictic cultivars depends on the use of the particular cultivar.

Citrus is a noted example of apomixis. For cultivar propagation, apomictic seeds are not used because juvenile apomictic seedlings have many undesirable characteristics for fruit production, such as excessive vigor, slowness to come into bearing, thorniness, and poor quality. On the other hand, viruses are screened out during the process of embryogenesis and apomictic seedlings are used to develop new "clean" versions of citrus cultivars as described in Chap. 8. The most important use of apomictic seedlings is for rootstocks (12, 13) because of their vigor, lack of viruses, and uniformity.

Apomixis is beneficial in plants where vegetative growth predominates. For example, a number of apomictic species and cultivars of grasses exist. Kentucky bluegrass (*Poa pratensis*) plants are **facultative apomicts.** Certain apomictic pasture grass cultivars have been developed, including 'King Ranch' bluestem (*Andropogon*), 'Argentine' Bahiagrass (*Paspalum notatum*), and 'Tucson' side oats grama (*Bouteloua curtipendula*) (11).

tion appears to be necessary, either to stimulate embryo development or to produce a viable endosperm.

Nonrecurrent Apomixis

In **nonrecurrent apomixis** an embryo arises directly from the **egg nucleus** without fertilization. Since the egg is haploid, the resulting embryo will also be haploid. This case is rare and primarily of genetic interest. It does not consistently occur in any particular kind of plant as do recurrent apomixis and adventitious embryony.

Vegetative Apomixis

In some cases vegetative buds or **bulbils** are produced in the inflorescence in place of flowers.

This occurs in *Poa bulbosa* and some *Allium, Agave,* and grass species.

SPORE DEVELOPMENT

Reproduction by spores occurs in ferns (49). Two separate stages or generations are involved as shown in Fig. 3–12. One is the asexual, or **sporophyte** generation, in which the plant has conspicuous roots, stems, and leaves (fronds). The other is the sexual, or **gametophyte,** generation, in which the plant is small; inconspicuous; without roots, stems, or leaves; and is called a **prothallium.** Spores are produced on the underside of fronds in clusters of **sporangia,** or spore cases, that appear as brownish "dots." Groups of spo-

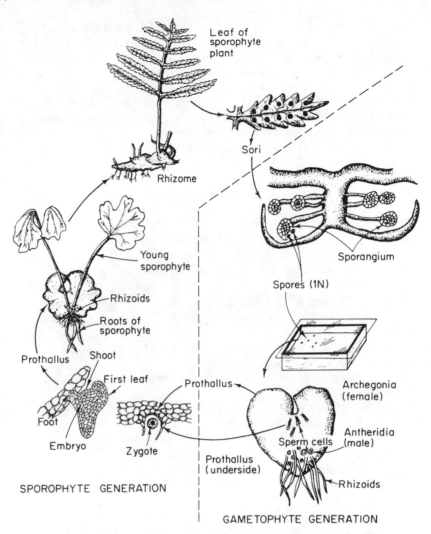

FIGURE 3–12 Reproductive cycle in ferns showing spore development.

rangia are called **sori** and sometimes have a cover known as the **indusium.** Within each sporangium 16 **spore mother cells** form, each of which undergoes meiosis to produce four **spores.** Each spore is haploid, with half the normal chromosome number of the species.

Spores are discharged and under favorable temperature and moisture conditions "germinate" to produce the **prothallus,** a flat green plate of cells with small rootlike structures (**rhizoids**). This grows to about ¼ in. in diameter in about three months. Male (**antheridia**) and female (**archegonia**) structures are produced on the underside of the same prothallus (except in two economically unimportant families). Sperm cells are

discharged and, in the presence of water, are attracted into the archegonium to fuse with the egg and subsequently produce a zygote.

The **zygote** develops into an embryo, which grows to produce the fern plant (sporophyte), which has the diploid chromosome number for the species. In its initial development, the embryo develops a **foot,** through which it absorbs moisture and nutrients from the prothallus. A root is produced which grows downward into the soil. Also produced is the primary leaf, which acts as a temporary photosynthetic organ, and the stem, which develops into the rhizome from which fronds and permanent roots arise. Propagation procedures for ferns are given in Chap. 20.

REFERENCES

1. Ammirato, P. V. 1987. Organizational events during somatic embryogenesis. In *Plant tissue and cell culture,* C. E. Green, D. A. Somers, W. P. Hackett, and D. D. Biesboer, eds. New York: Alan R. Liss, pp. 57–81.

2. Atwater, B. R. 1980. Germination, dormancy and morphology of the seeds of herbaceous ornamental plants. *Seed Sci. and Tech.* 8:523–73.

3. Austin, R. B. 1972. Effects of environment before harvesting on viability. In *Viability of seeds,* E. H. Roberts, ed. New York: Syracuse Univ. Press, pp. 114–49.

4. Bewley, J. D., and M. Black. 1978. *Physiology and biochemistry of seeds in relation to germination,* Vol. 1, *Development, germination, and growth.* Berlin: Springer-Verlag.

5. Bewley, J. D., and M. Black. 1985. *Seeds: Physiology of development and germination.* New York: Plenum Press.

6. Bhatnager, S. P., and B. M. Johri. 1972. Development of angiosperm seeds. In *Seed biology,* Vol. 1, T. T. Kozlowski, ed. New York: Academic Press, pp. 77–149.

7. Black, M. 1980/1981. The role of endogenous hormones in germination and dormancy. *Israel Jour. Bot.* 29:181–92.

8. Bohart, G. E., and T. W. Koerber. 1972. Insects and seed production. In *Seed biology,* Vol. 3, T. T. Kozlowski, ed. New York: Academic Press, pp. 1–54.

9. Brink, R. A. 1962. Phase change in higher plants and somatic cell heredity. *Quart. Rev. Bio.* 37(1):1–22.

10. Brink, R. A., and D. C. Cooper. 1947. The endosperm in seed development. *Bot. Rev.* 13:423–541.

11. Burton, G. W. 1983. Utilization of hybrid vigor. In *Crop breeding,* D. R. Wood, K. M. Rawal, and M. N. Wood, eds. Madison, Wis.: Amer. Soc. Agron. and Crop Sci. Soc. Amer., pp. 89–107.

12. Cameron, J. W., R. K. Soost, and H. B. Frost. 1959. The horticultural significance of nucellar embryony in citrus. In *Citrus virus diseases,* J. Wallace, ed. Berkeley: Univ. Calif. Div. Agr. Sci., pp. 191–96.

13. Campbell, A. J., and D. Wilson. 1962. Apomictic seedling rootstocks for apples: Progress report, III. *Ann. Rpt. Long Ashton Hort. Res. Sta. (1961):* 68–70.

14. Collins, G. B., and J. W. Grosser. 1984. Culture of embryos. In *Cell culture and somatic cell genetics of plants.* Vol. 1, I. K. Vasil, ed. New York: Academic Press, pp. 241–57.

15. Cooper, D. C., and R. A. Brink. 1940. Somatoplastic sterility as a cause of seed failure after interspecific hybridization. *Genetics* 25:593–617.

16. ———. 1945. Seed collapse following matings between diploid and tetraploid races of *Lycopersicon pimpinellifolium. Genetics* 30:375–401.

17. Craig, R. 1976. Flower seed industry. In *Bedding plants: A manual on the culture of bedding plants as a greenhouse crop,* J. W. Mastalerz, ed. University Park, Pa.: Pennsylvania Flower Growers, pp. 25–46.

18. Davis, L. D. 1939. Size of aborted embryos in the Phillips Cling peach. *Proc. Amer. Soc. Hort. Sci.* 37:198–202.

19. Delouche, J. C. 1980. Environmental effects on seed development and seed quality. *HortScience* 15(6):13–18.

20. Dickson, M. H. 1980. Genetic aspects of seed quality. *HortScience* 15(6):771–74.

21. Dure, L. S., III. 1975. Seed formation. *Ann. Rev. Plant Phys.* 26:259–78.

22. ———. 1979. Role of stored messenger RNA in late embryo development and germination. In *The plant seed: Development, preservation, and germination,* I. Rubenstein et al., eds. New York: Academic Press, pp. 113–28.

23. Dure, L. S., III, G. A. Galau, and S. Greenway. 1980/1981. Changing protein patterns during cotton cotyledon embryogenesis and germination as shown by *in vivo* and *in vitro* synthesis. *Israel Jour. Bot.* 29:293–306.

24. Durzan, D. J. 1988. Process control in somatic polyembryogenesis. In *Molecular genetics of forest trees,* J. Hällgren, ed. Umea: Swedish Agri. Univ. (in press).

25. Edwards, D. G. W. 1980. Maturity and quality of tree seeds: A state-of-the-art review. *Seed Sci. and Tech.* 8:625–57.

26. Ehrenberg, C., A. Gustafsson, C. P. Forshell, and M. Simak. 1955. Seed quality and the principles of forest genetics. *Hereditas* 41:291–366.

27. Finkelstein, R. R., and M. L. Crouch. 1987. Hormonal and osmotic effects on developmental potential of maturing rapeseed. *HortScience* 22(5):797–800.

28. Fordham, A. J. 1984. Seed dispersal as it concerns the propagator. *Proc. Inter. Plant Prop. Soc.* 34:531–34.

29. Goldbach, H., and C. Michael. 1976. Abscisic acid content of barley grains during ripening as affected by temperature and variety. *Crop Sci.* 16:797–99.

30. Gunn, C. R. 1972. Seed collecting and identification. In *Seed biology,* Vol. 3, T. T. Kozlowski, ed. New York: Academic Press, pp. 55–144.

31. Gupta, P. K., and D. J. Durzan. 1986. Somatic polyembryogenesis from callus of mature sugar pine embryos. *Biotechnology* 4:643–45.

32. Gustafsson, A. 1946–1947. Apomixis in higher plants, Parts I–III. *Lunds Univ. Arsskrift,* N.F. Avid. 2 Bd 42, Nr. 3:42(2); 43(2); 43(12).

33. Hackett, W. P. 1983. Phase change and intraclonal variability. *HortScience:* 18(6):840–44.

34. Hesse, C. O., and D. E. Kester. 1955. Germination of embryos of *Prunus* related to degree of embryo development and method of handling. *Proc. Amer. Soc. Hort. Sci.* 65:251–64.

35. Jones, H. A. 1927. Pollination and life history studies of the lettuce (*Lactuca sativa* L.). *Hilgardia* 2:425–79.

36. Kermode, A. R., J. D. Bewley, J. Dasgupta, and S. Misra. 1986. The transition from seed development to germination: A key role for desiccation? *HortScience* 21(5):1113–18.

37. Kester, D. E. 1982. The clone in horticulture. *HortScience* 18(6):831–35.

38. Kester, D. E., and C. O. Hesse. 1955. Embryo culture of peach varieties in relation to season of ripening. *Proc. Amer. Soc. Hort. Sci.* 65:265–73.

39. Litz, R. E., R. L. Jarret, and M. P. Asokan. 1986. Tropical and subtropical fruits and vegetables. In *Tissue culture as a plant production system for horticultural crops,* R. H. Zimmerman, R. J. Griesbach, F. A. Hammerschlag, and R. H. Lawson, eds. Dordrecht: Martinus Nijhoff Publishers, pp. 237–52.

40. Macdonald, B. 1986. *Practical woody plant propagation for nursery growers,* Vol. 1. Portland, Oreg.: Timber Press.

41. Maheshwari, P., and R. C. Sachar. 1963. Polyembryony. In *Recent advances in the embryology of angiosperms,* P. Maheshwari, ed. Delhi, India: Univ. of Delhi, Inter. Soc. of Plant Morph., pp. 265–96.

42. Martin, A. C. 1946. The comparative internal morphology of seeds. *Amer. Midland Nat.* 36:512–660.

43. Norstog, K. 1979. Embryo culture as a tool in the study of comparative and developmental morphology. In *Plant cell and tissue culture: Principles and applications,* W. R. Sharp et al., eds. Columbus: Ohio State Univ. Press, pp. 179–202.

44. Nygren, A. 1954. Apomixis in the angiosperms II. *Bot. Rev.* 20:577–649.

45. Pollock, B. M., and E. E. Roos. 1972. Seed and seedling vigor. In *Seed biology,* Vol. 1, T. T. Kozlowski, ed. New York: Academic Press, pp. 314–88.

46. Radley, M. 1979. The role of gibberellin, abscisic acid, and auxin in the regulation of developing wheat grains. *Jour. Exp. Bot.* 30(116):381–89.

47. Raghaven, V. 1986. *Embryogenesis in angiosperms.* Cambridge: Cambridge Univ. Press.

48. Rietsema, J., S. Satina, and A. F. Blakeslee. 1955. Studies on ovule and embryo growth in *Datura.* I. Growth analysis. *Amer. Jour. Bot.* 42:449–54.

49. Roberts, D. J. 1965. Modern propagation of ferns. *Proc. Inter. Plant Prop. Soc.* 15:317–22.

50. Rolston, M. P. 1978. Water impermeable seed dormancy. *Bot. Rev.* 44:365–96.

51. Rowe, W. J. 1986. New technologies in plant tissue culture. In *Tissue culture as a plant production system for horticultural crops.* R. H. Zimmerman, R. J. Griesbach, F. A. Hammerschlag, and R. H. Lawson, eds. Dordrecht: Martinus Nijhoff Publishers, pp. 35–51.

52. Sanders, M. E. 1948. Embryo development in four *Datura* species following self and hybrid pollinations. *Amer. Jour. Bot.* 35:525–32.

53. Schopmeyer, C. S., ed. 1974. *Seeds of woody plants in the United States.* U.S. Dept. Agr. Handbook 450. Washington, D.C.: U.S. Govt. Printing Office.

54. Shepherd, P. H. 1955. The kernel told the tale. *Amer. Fruit Grower* 75:37.

55. Singh, H., and B. M. Johri. 1972. Development of gymnosperm seeds. In *Seed biology,* Vol. 1, T. T. Kozlowski, ed. New York: Academic Press, pp. 22–77.

56. Stephens, W. 1969. The mangrove. *Oceans* 2(5):51–55.

57. Van Overbeek, J., R. Siu, and A. J. Haagen-Smit. 1944. Factors affecting the growth of *Datura* embryos in vitro. *Amer. Jour. Bot.* 31:219–24.

58. Wellington, P. S., and V. W. Durham. 1958. Varietal differences in the tendency of wheat to sprout in the ear. *Empire Jour. Exp. Agr.* 26:47–54.

59. Zimmerman, R. H., ed. 1976. Symposium on juvenility in woody perennials. *Acta Hort.* 56:1–317.

SUPPLEMENTARY READING

BEWLEY, J. D., and M. BLACK. 1985. *Seeds: Physiology of development and germination.* New York: Plenum Press.

KOZLOWSKI, T. T., ed. 1972. *Seed biology,* Vol. 1, *Importance, development and germination.* New York: Academic Press.

MARTIN, A. C., and W. D. BARKLEY. 1961. *Seed identification manual.* Berkeley: Univ. of Calif. Press.

MUSIL, A. F. 1963. *Identification of crop and weed seeds.* U.S. Dept. Agr. Handbook 219. Washington, D.C.: U.S. Govt. Printing Office.

RAGHAVAN, V. 1986. *Embryogenesis in angiosperms.* Cambridge: Cambridge Univ. Press.

RUBENSTEIN, I., R. L. PHILLIPS, C. E. GREEN, and G. GENGENBACH, eds. 1979. *The plant seed: Development, preservation and germination.* New York: Academic Press.

U.S. DEPT. OF AGRICULTURE. 1961. *Seeds: Yearbook of agriculture.* Washington, D.C.: U.S. Govt. Printing Office.

4

Principles
of Seed Selection

In nature higher plants reproduce primarily by seeds. The characteristic genetic variability that occurs among groups of seedlings is important to allow continued adaptation of a particular species to possible changes in the environment. Those individuals within each generation that are best adapted to their environment tend to survive, produce the next generation, and reproduce the species.

On the other hand, propagation of cultivated plants requires that genetic variability during seed reproduction be controlled or the value of a cultivar may be lost. Characteristics selected as important in agriculture, horticulture, and forestry may not be consistently perpetuated into the next seedling generation unless specific principles and procedures are followed.

This chapter deals with the management of genetic variability in seedling populations and cultivars of plants. It parallels Chap. 8, which deals with the control of variation in vegetatively propagated cultivars.

USES OF SEEDLINGS IN PROPAGATION

Mass Propagation

Seedling propagation is an efficient and economical method of propagation as long as genetic variability can be controlled within acceptable limits. Many millions of seeds can be produced in bulk, stored for long periods, transported all over the world, and mass propagated in a highly efficient manner. Viruses are generally not a problem in most seed-propagated cultivars, as they are screened out during seed formation. This is in contrast to vegetatively propagated cultivars where viruses can be perpetuated (see Chap. 8, p. 183).

Annuals and biennials that must be grown from seed account for essentially all the agronomic crops (cereals, forages, grasses, fiber, and oil plants), most vegetable crops, and many garden and florist crops (24). The bedding plant industry depends on large volumes of seeds for the production of flower and vegetable plants (3, 33).

Seed propagation can be used for some ornamental herbaceous perennial species, such as lilies, tulips, cyclamen, and tuberous begonia, although most cultivars are propagated vegetatively (Chap. 20). Many woody landscape tree species are propagated by seed (32, 35) (Chap. 19), but vegetative methods may be needed because of the individual use of the plant (13). Forest trees and shrubs used for reforestation, woodlots, biomass production, screens, and other mass plantings utilize populations of seedling plants (40, 41), although vegetative methods are becoming available (29, 30).

Rootstocks

Seedlings are used extensively to provide rootstocks of vegetatively propagated cultivars of both fruit and nut species (39) (see Chap. 18) and landscape plants (13) (see Chap. 19). Properly selected seed materials can produce uniform, vigorous, virus-free seedling plants at a lower cost than comparable vegetative methods.

Plant Breeding

Growing seedlings is the most important means of developing new cultivars because of the variability that results from segregation of chromosomes during sexual reproduction.

CONTROL OF GENETIC VARIABILITY WITHIN SEEDLING CULTIVARS

Control of genetic variability within seed-propagated species and cultivars depends upon (a) the nature of the breeding system, that is, whether the plants are **self-pollinated, cross-pollinated,** or **apomictic** (7, 18); and (b) the management conditions under which the seed population is grown. The breeding system also influences whether the

genotype of the plant is primarily **homozygous** or **heterozygous.** (See Chap. 1 for definition of terms.)

Self-Pollinated Species and Cultivars

Self-pollination occurs when the pollen tube grows down the style of the same flower or a different flower on the same plant and produces fertilization. The same definition applies to different plants of the same clone (see Chap. 8). Natural self-pollination results from a particular flower structure. In practice, the percentage of self-pollinations may vary (2).

> *Crop plants that are highly self-pollinated,* i.e., less than 4 percent cross-pollinated: barley, oats, wheat, rice, peanut, soybean, lespedeza, field pea, garden bean, cowpea, flax, and some grasses
>
> *Crop plants that are somewhat less self-pollinated:* cotton (upland and Egyptian), pepper, and tomato

Self-pollinated cultivars can be maintained by seed propagation because they are largely homozygous and have been selected through mass selection to a specific standard by eliminating "off-type" individuals and "fixing" the characteristics through consecutive generations of self-pollinations (7, 17). The process of "fixing" or **inbreeding** is as follows: homozygous traits in the parent will also have homozygous offspring unless unexpected chance mutations occur, in which case the trait becomes heterozygous. Heterozygous traits in the parent segregate in the next generation to produce both homozygous and heterozygous pairs of the same allele (see Chap. 1). With continued generations, the proportion of homozygous plants will increase while the proportion of heterozygous· plants will decrease by a factor of one-half each generation. After 6 to 10 generations, the group of descendants of the original parents have segregated into a mixture of more or less true-breeding "lines." In self-pollinated plants there is no loss in vigor with inbreeding.

The end product of the process described represents one of the major categories of cultivars and is known as a **line. A pure line** results when the entire population of seedling plants are all descendants of a single (homozygous) plant (*17*). A **multiline** is a combination of genetically different lines (referred to as **isolines**) designed to incorporate specific traits (*17*).

The fact that wheat, barley, and rice are naturally self-pollinated species in which different "lines" could be maintained genetically true undoubtedly made their cultivation a major factor in the establishment of agriculture some 10,000 years ago. The development of heterogeneous collection of such lines also made possible their widespread adaptation to different locations (*42*).

Cross-Pollinated Species and Cultivars

Most species in nature and cultivars in cultivation are naturally **cross-pollinated** and tend to be more or less heterozygous, however. This seems to be a desirable condition in that it maintains a level of variability and heterozygosity that provides the opportunity for evolutionary adaptation within the population if environments change. In fact, even though flowers in some species may be **perfect** (i.e., have both male and female parts in the same flower), many species have specific mechanisms that prevent self-pollination. These mechanisms include the following types:

Dioecy. Pistillate (female) and staminate (male) flowers are borne on separate plants (buffalo grass, holly, pistachio, date palm, asparagus).

Monoecy. Pistils (female) and stamens (male) develop in separate flowers of the same plant (corn, cucurbits, walnut, conifers).

Dichogamy. Pollen is shed at a different time than when the pistil is receptive (includes many walnut and pecan cultivars).

Self-sterility. Specific genes prevent normal formation of either male (pollen) or female (pistil) reproductive structure. Male (pollen) sterile lines of corn (Fig. 4–1) or onions have been used to produce hybrid plants.

Incompatibility. Inability of pollen to grow into pistils of the same flower or flowers of the same plant even though the pollen is it-self viable. Incompatibility also occurs between flowers on different plants of the same vegetatively propagated cultivar (sweet cherry, almond, lily, petunia, cabbage).

Cross-pollination of most cultivars is carried out either by wind or insects.

Insect pollination is the rule with plants with white, or brightly colored, fragrant, and otherwise conspicuous flowers that attract insects. The honeybee is one of the most important pollinating insects, although wild bees, butterflies, moths, and flies also obtain pollen and nectar from the flower. Generally, pollen is heavy, sticky, and adheres to the body of the insect.

Some important seed crops in this category are the following (*2*): alfalfa, birdsfoot trefoil, red clover, white clover, sweet clover, millet, onion, and watermelon. In addition, many flower and vegetable crops are insect pollinated. **Wind pollination** is the rule with many plants having inconspicuous flowers. Examples are grasses, corn, conifers, olive, and catkin-bearing trees such as the walnut, oak, alder, and cottonwood. The pollen produced from such plants is generally light and dry and in some cases is carried long distances in wind currents. Important seed crops in this category include corn (field or sweet) and grasses.

Categories of Cross-Pollinated Cultivars

Cross-pollinated cultivars are classed as **outcrossers** and have more heterozygosity than those that are naturally self-pollinated. Self-pollination tends to result in reduced vigor (**inbreeding depression**). Consequently, cultivar development utilizes more complex breeding strategies and seed production involves controlled levels of genetic variation (*7, 17*).

Some categories of seed propagated cultivars include the following:

Lines are maintained by mass selection to a standard similar to self-pollinated cultivars but under isolated conditions to allow uncontrolled open **pollination** among the plants.

Inbred lines are true breeding cultivars that result from enforced self-pollination of individual plants through consecutive generations. In most cross-pollinated species an increase in uniformity results, but is accompanied by a decline in vigor and size (**inbreeding depression**). Some lines die out; others stabilize. These selected lines are generally not useful in themselves but are important as parents in hybridization crosses.

Hybrid lines are the F_1 hybrid seedling populations of two or more inbred lines. The result is a population of uniform, vigorous, heterozygous plants which visually exhibit greater vigor than the parents, a phenomenon called **hybrid vigor.** Such hybrids were first developed for corn but the concept has since been applied to many other agronomic, vegetable, and flower crops (*8*).

Hybridization may be made between two inbred lines (**single-cross**), two single-crosses (**double-cross**), an inbred line and an open pollinated cultivar (**top-cross**), or between a single-cross and an inbred line (**three-way cross**) (*43*). Seeds from the F_1 hybrid plants normally are not used for seed production because the seedling plants in the next F_2 generation segregate to produce a wide range of variable progeny plants and the desirable uniformity and vigor of the F_1 generation is lost.

Some cross-pollinated cultivars are seedling populations which involve different genotypes but selected for a single trait, such as flower type or color. These are referred to in the Nomenclature Code (see p. 10) as **nonuniform assemblages.**

Mixtures are made up of combinations of separate lines which are similar in characteristics except in some one trait, such as flower color (*11*).

A **synthetic cultivar** (*7, 8, 17*) is produced by growing together parent plants of a number of genetically distinct but phenotypically similar lines or vegetatively propagated clones. These are allowed to cross-pollinate at random in open pollination and the seeds bulked. 'Ranger' alfalfa and other forages are examples of synthetics (*17*). The first generation of seeds is called *Syn-1*, the second generation, *Syn-2*, and so on. The original stocks are maintained and allowed to increase only a limited number of seed generations from the original, *Syn-0*. The cultivars are also known as **composites** (*47*).

F_2 strains are produced for some flower crops, such as petunia, pansy, and snapdragon (*11*). Relatively uniform populations have been produced by plant breeding in which the range of variability is acceptable for the purpose intended. Since the seed is grown open-pollinated, or self-pollinated, from F_1 hybrids, such seed is less costly to produce than the true F_1 hybrid seed.

SEED PRODUCTION OF TRUE-TO-TYPE HERBACEOUS CULTIVARS

Historically, seed-propagated cultivars of agronomic, vegetable, and flower crops were maintained by retaining selected portions of each propagation cycle to be used to produce the next cycle. This procedure, still practiced in parts of the world, resulted not only in the preservation of specific genetic lines (**landraces** or **local varieties**) adapted to local use but also in the preservation of a great deal of genetic diversity.

The discovery of genetic principles and the parallel development of plant breeding strategies ushered in a new era of plant cultivar improvement and release of new materials to the public. As a consequence of this activity, a whole new technology of seed production and supervision has evolved to protect the genetic identity of these new cultivars and to maintain seed quality (*9, 10, 47*). A further consequence of this activity has been the need to develop a program of preservation of **germplasm resources** (*27*).

For many agricultural crops, genetic identity during production is controlled through a **seed certification** agency (*2, 10, 34*). Seeds of other crops, particularly flowers and vegetables, have been largely produced by commercial firms that follow similar principles and procedures (*6, 20, 25, 38*).

Control of Genetic Identity and Variability

Isolation

Isolation is necessary (a) to prevent mechanical mixing of the seed during harvest, and (b) to prevent contamination by cross-pollination with a different but related cultivar. Isolation is primarily achieved through distance, but it can also be attained by enclosing plants or groups of plants in cages, enclosing individual flowers, or removing male flower parts and then employing artificial pollination.

For naturally self-pollinated cultivars, separation is needed in production only to prevent mechanical mixing of seed of different cultivars during harvest. The minimum distance usually specified between plots is 3 m (10 ft). Careful cleaning of the harvesting equipment is required when a change is made from one cultivar to another. Sacks and other containers used to hold the seed must be cleaned carefully to remove any seed that has remained from previous lots.

More isolation is needed to separate cultivars cross-pollinated by wind or insects. The minimum distance depends upon a number of factors, such as the degree of natural cross-pollination, the pollination agent, the number of insects present, and the direction of prevailing winds. The minimum distance for insect-pollinated plants is 0.4 km (¼ mi) to 1.6 km (1 mi) (*37*). The distance for wind-pollinated plants is 0.2 km (⅛ mi) to 3.2 km (2 mi), depending on species.

Cross-pollination can usually take place between cultivars of the same species; it may also occur between cultivars of a different species but in the same genus; and rarely will it occur between cultivars belonging to another genus. Since the horticultural classification may not indicate taxonomic relationships, seed producers should be familiar with the botanical relationships among the cultivars they grow.

Roguing

Off-type plants, plants of other cultivars, and weeds in the seed production field should be eliminated by removal (**roguing**). A low percentage of such plants may not seriously affect the performance of any one lot of seed, but their continued presence will lead to deterioration of the cultivar over a period of time.

Off-type plants may arise because recessive genes are present in a heterozygous condition even in homozygous cultivars. Recessive genes may arise by mutation (see 176). The effect of a mutant recessive gene controlling a given plant characteristic may not be immediately observed in the plant in which it arises. The plant becomes heterozygous for that gene, and in a later generation the gene segregates and the character appears

in the offspring. Some cultivars have mutable genes that continuously produce specific off-type individuals (*36*). Off-type individual plants should be rogued out of the seed production fields before pollination occurs. Regular inspection of the seed-producing fields by trained personnel is required.

Volunteer plants arising from accidentally planted seed or from seed produced by earlier crops is another source of contamination. Fields for producing seed of a particular cultivar should not have grown a potentially contaminating cultivar for a number of preceding years.

Nuclear Seed Production and Distribution Systems

The production and distribution process utilizing the **nuclear seed system** involves three phases (Fig. 4–1).

Phase 1—development phase. The end product of cultivar development program is a limited amount of seeds referred to as **breeder's seed** that sets the standard for the cultivar. A nucleus of this seed may be maintained by the originating organization.

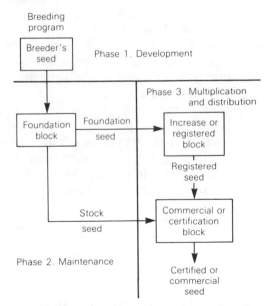

FIGURE 4–1 Generalized scheme for distribution of seeds in a nuclear seed production system with or without certification.

Phase 2—maintenance phase. The primary source of seed is a **foundation planting** which is maintained either as *seedling stocks* (most species) or *vegetatively propagated clones* in other species (asparagus, brussels sprouts) (see Chap. 8). Foundation plants are maintained under highest levels of genetic control utilizing (a) *isolation,* (b) *inspections,* and (c) *roguing* of off-type plants. Foundation seed is supplied for increase by commercial producers (phase 3). Foundation plants can only originate from breeder's seeds or from foundation seed.

Phase 3—multiplication and distribution. Foundation seed is supplied for commercial production either in (a) a two-generation sequence including an **increase** or **registered** block and a **commercial** or **certification** block, or (b) multiplied directly in a single-generation sequence without an intervening increase block. The end product will be the commercial seed sold to the public. If produced under a system of seed certification the seed may be sold as **certified seed.**

Genetic Shifts during Seed Production

During the process of seed increase there is the potentiality for **genetic shifts** to occur in the frequency of particular genes or gene combinations (*17, 47*). These may result from the selection pressure (or lack of) applied to the seedling population during the roguing process or through differential environmental exposure in the growing area. Certain seedlings (genotypes) may survive better than others and contribute more to the next generation. If sufficiently extensive, genetic shift could result in populations of progeny plants that are somewhat different phenotypically from those of the same cultivar grown by other producers or from the original breeders seed. For this reason, establishing **test gardens** for **progeny testing** may be desirable. A problem can result if seed crops of particular perennial cultivars are grown in one environment (such as a mild winter area) to produce seed to be used in a different and more severe environment (such as an area requiring cold-hardiness). This situation has occurred, for example, with alfalfa (*19*) and is a particular problem with synthetic cultivars (*7, 17*). For this reason seed production of forage crop seed produced

in a mild winter area is allowed only one seedling generation of increase.

Production of Hybrid Seed

Hybrid cultivars are the F_1 progeny produced by repetitive crossing of two or more parental lines (Fig. 4–2). These parents are maintained either as (a) seedling *inbred lines,* or (b) vegetatively propagated **clones.** The parental lines are maintained under the control of the parent commercial company, which can then control the production and distribution of the seed.

Mass production of hybrid seed requires a system to prevent self-pollination and to enforce cross-pollination (*7, 21*). Historically, this was first accomplished in such crops as corn by physically removing the male flowers (detasseling). Since then biological methods have been utilized to prevent self-pollination in many crops and to allow natural cross-pollination by insects or wind.

Pollen Sterility	*Dioecy*
Vegetables: onion, carrot, leek, radish, tomato	*Vegetables:* spinach, asparagus
Flowers: petunia, marigold,* dianthus,* geranium,* snapdragon, zinnia*	*Field crops:* some grasses
Field crops: corn, sugar beets, sorghum, pearl millet, wheat, barley	*Monoecy*
	Vegetables: cucurbits (special genotypes)
	Field crops: field and sweet corn (manual removal of tassels)
Self-Incompatibility	*Chemical Control*
Vegetables: cabbage, broccoli, brussels sprouts	*Induction of male sterility:* cotton
Flowers: ageratum, *Bellis*	*Female flower production by gibberellic acid:* cucurbits
Field crops: some grasses	

*Pollen applied manually.

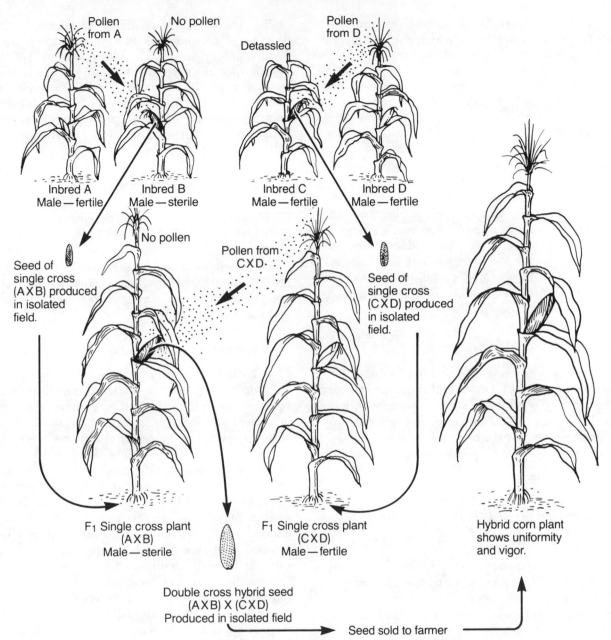

FIGURE 4–2 Hybrid seed production illustrated by two sequential F_1 hybridization crosses involving four inbred lines of corn. Both male-sterile lines and mechanical detasseling are utilized to prevent self-pollination. The final result will be a hybrid line with greater vigor and production than any of the parents. Its genotype is more or less ''fixed'' by repetitive crossing of specific parents. (Redrawn from USDA Yearbook of Agriculture, 1937, Washington, D.C.: U.S. Government Printing Office.)

Hand pollination is sometimes practiced with crops or situations in which the production of seed per flower is very high and the high value of the seed justifies the expense. Hand pollination is used in producing some hybrid flower seeds and in breeding new cultivars.

Apomixis

Apomixis is significant because it allows asexual reproduction in seeds (see p. 71). Various pasture grasses exhibit apomixis, including 'King Ranch' bluestem, 'Argentine' bahiagrass, and 'Tucson' sideoats grama (8). Kentucky bluegrass (*Poa pratense*) is facultative and produces both sexual and apomictic seeds. Apomictic seedlings are completely uniform, but variability may arise through the mixture of apomictic and sexual seedlings in the same population.

Somatic Embryogenesis

This phenomenon is the induction of somatic embryos from certain kinds of embryogenic cell suspensions, as in carrot and some grasses (see Chap. 16). It has the potential for producing not only asexual embryos artificially, but also "seeds" with a combination of artificial seed coverings (see p. 477) combined with fluid drilling (see 149).

SELECTING SEED SOURCES OF WOODY PERENNIALS

Selecting the seed source may be the most important aspect of nursery seedling production (13, 35). Although principles of source selection are the same, strategies for seed selection in landscape nursery production, fruit and ornamental rootstocks, reforestation, woodlots, or energy (biomass production) may differ considerably. Most trees and shrubs are both heterozygous and cross-pollinated, such that there is considerable potential for genetic variability among the seedling progeny.

Wind pollination is the characteristic of many plant species with inconspicuous flowers and includes such important groups as conifers (pines, spruce, etc.), olive, and catkin-bearing families (oak, walnut, alders, etc.) (40). Insect pollination is the rule primarily with families that have large conspicuous flowers, including most fruits (apple, pear, peach, etc.) and some nut (almond) species (49). In addition, biological controls against self-pollination (see p. 79) are common, including **dioecy** (holly, pistachio, date palm), **monoecy** (walnut, pecan, conifers), **dichogamy** (walnut, pecan), and **self-incompatibility** (many fruit trees).

Concepts of Genetic Variability

Problems in seed source selection include (a) variability within individual seed lots, (b) variation among different seed lots of the same species from separate origins, and (c) potential for seed source improvement, including the correlation between the phenotypic appearance of the source plants and the seedling progeny.

Variability within Seed Sources

Variability within seed sources depends not only upon the characteristics of the individual seed tree but of the surrounding plants of the same or related species. Single trees (or separate trees of vegetatively propagated) clones, as might be represented in an arboretum or as a specimen plant in a landscape, may result in seedling progeny with poor quality and considerable genetic variability (15, 44, 48, 50). On the one hand, such plants may be subject to enforced self-pollination, which in itself often results in poorly developed seed (see Chap. 5), inbreeding depression, and weak seedlings. On the other hand, cross-pollination may have come from a separate but related species which may result in vigorous useful hybrid individuals (46, 50) but quite unlike the true species of the seed parent. If too distantly related, the offspring may be weak and inferior.

Before about 1915, "Old French" pear seedlings were grown in the United States from seed produced in Europe by *Pyrus communis* trees. These seedlings were used as rootstocks for pear cultivars. Much variability in the seedlings resulted from seed produced by hybridization of the *P. communis* trees with nearby *P. nivalis* trees. Similarly, "Oriental pear" rootstock seedlings used in California about the same time originated from seeds obtained from Asia with little knowledge

HAND POLLINATION TECHNIQUES

Hands and instruments are washed with alcohol before handling pollen or flowers to eliminate foreign pollen. Pollen should be collected from flower buds immediately preceding bloom and before the anthers open (dehisce). Anthers are extracted by pulling them from the flowers with tweezers, by squeezing the flower between the fingers, or by rubbing the buds across a wire screen. Anthers are dried on a sheet of paper until one can observe their opening under a magnifying glass. Pollen and anthers can be further screened through a fine sieve to remove extraneous material; they are then stored in a glass vial or bottle. Catkins (walnut, birch, aspen) or staminate cones (pine) will open on drying and will shed large quantities of pure, dry pollen if placed on a sheet of paper in a warm room.

Most kinds of pollen will remain viable for only a few days or weeks at warm temperatures, but many kinds can be preserved for several months to several years if stored at low temperatures and relatively low humidity. Effective storage conditions are a combination of 10 to 50 percent relative humidity and a temperature of 0 to 10°C (32 to 50°F). Moisture content of the pollen can be controlled by storing over a desiccant, such as calcium chloride or sulfuric acid. Some pollen—that of grasses, for instance—is best stored at 90 to 100 percent relative humidity. Pollen can be stored effectively at about −18°C (0°F), as in a home freezer.

The stigma at the tip of the pistil must be exposed to apply the pollen. If the plant is self-fertile, the stamens must be removed (emasculated). Petals and stamens can be removed at the bud stage either with tweezers or by cutting the base of the flower between the fingernails. If the flower is to be self-fertilized or is a self-sterile cultivar, the flower need not be emasculated but only enclosed to keep out pollinating insects or wind-blown pollen.

Pollen is applied to the sticky, receptive stigma of insect-pollinated flowers with a fine-hair brush, a glass rod, a pencil eraser, or the tip of the finger. Removal of all the flower parts with the exception of the pistil reduces or eliminates the likelihood of further chance pollination by insects. Complete protection can be obtained by covering the individual flower, the pistil, or the branch with cellophane, plastic, paper, or a cloth bag.

For wind-pollinated plants, such as grasses, pines, and walnut, female flowers (or cones) must be carefully sealed throughout the flowering period. They can be covered with cellophane, paper, or a tightly woven cloth bag. Pollen can be transferred by inverting a paper bag containing pollen over the exposed flower; or dry pollen can be blown into the bag with an atomizer or inserted with a hypodermic needle.

The flower, or the branch on which it is borne, should be carefully labeled to maintain a record of the pollen used. The usual method is to identify the cross as follows: *"seed parent × pollen parent."*

about the characteristics of either the seed tree or the pollen source. Plants grown from these sources have turned out to be a heterogeneous group with mixed parentage of many species and hybrids (*23*).

Differences among flame eucalyptus (*Eucalyptus ficifolia*) trees arising from separate seed collections taken from a particular seed source in Australia have been explained by variable seed collection. Seeds collected from the center of the block of trees reproduced the true species with upright, uniform trees, while seeds from trees at the edge of the block were hybrids with surrounding *E. calophylla* trees, and produced many dwarfed and off-type plants (*44*).

Sometimes desired individuals in variable seedling populations can be identified through **nursery selection.** Variability involves identifiable characters of vigor, appearance, or both. For example, Paradox hybrid walnut seedlings can be identified in nursery plantings of *Juglans hindsii* seedlings (see Fig. 4–5). "Blue" seedlings of the Colorado spruce (*Picea pungens*) appear among others having the usual green form. Differences in fall coloring among seedlings of *Liquidambar* and *Pistacia vera* trees necessitate fall selection of individual trees for landscaping or vegetative propagation as clones. Variation among seedlings grown for rootstocks may be reduced by grading to a size and roguing the weak, small seedlings, a usual practice in U.S. nurseries.

The most desirable procedure is to collect seeds from groups of the same kind of plants with desired characteristics growing in a **pure stand,** where cross-pollination can come from a number of plants of the same type. Seedling plants grown from such sources, although not necessarily homozygous, would be most likely to reproduce the characteristics observed among the parent seed trees.

Variability among Seed Sources (Seed Origin)

In nature, most species can be recognized as a more or less phenotypically (and genotypically) uniform seedling population that has evolved to be adapted to a particular environment (Fig. 4–3). If the species covers a wide area, local variation in environment can result in different populations adapted to that area even though the plants may be phenotypically similar. If there are morphologically recognizable differences, they may be designated as **botanical varieties.** On the other hand, subgroups of a particular species that are morphologically similar but specifically adapted to a particular environmental niche are known as **ecotypes.** Variations that occur continuously between locations are known as **clines** (*28*).

The climatic and geographical locality where tree seed is produced is referred to as **seed origin** or **provenance** (*22, 28, 29, 40*). Variation can occur among groups of plants associated with latitude, longitude, and elevation. Differences may be shown by morphology, physiology, adaptation to climate and to soil, and in resistance to diseases and insects. If seeds of a given species are collected in one locality, plants may be produced that are completely unadapted to another locality. Seed collected from trees in warm climates or at low altitudes is likely to produce seedlings that will not stop growing sufficiently early in the fall to escape freezing when grown in colder regions. Although the reverse situation—collecting seed from

FIGURE 4–3 Differences among seed sources shown by nursery progeny tests of two-year Douglas fir (*Pseudotsuga menziesii*). *Left:* Two fast-growing green strains from Washington and British Columbia. *Center:* Four slow-growing strains from Montana and Wyoming, mostly gray-green in color. *Right:* Fast-growing strain from Arizona, deep blue in color. Seedlings from different sources vary in height from 5 to 10 cm (2 to 4 in.) up to 25 to 43 cm (10 to 16 in.). (Courtesy C. E. Heit.)

colder areas for growth in warmer regions—would be more satisfactory, it might result in a net reduction in growth resulting from the inability of the trees to utilize the growing season fully because of differences in response to photoperiod (*40*).

One application of the seed origin concept is to use only "local" seed whenever the adaptation of a seed source is not known for a given location. "Local seed" means seed from an area subject to similar climatic influences. This is usually considered to mean within 1600 km (100 mi) of the planting and within 333 m (1000 ft) of its elevation. A second application is to use seed from a region having as nearly as possible the same climatic characteristics. Important environmental characteristics to consider are length of growing season, mean temperature during the growing season, frequency of summer droughts, maximum and minimum temperatures, and latitude.

Distinct ecotypes have been identified in many native forest tree species, including Douglas fir, ponderosa pine, lodgepole pine, red pine, eastern white pine, slash pine, loblolly pine, short-leaf pine, and white spruce (*40*). Some ecotypes have been shown to be superior seed sources and, in fact, may be preferable to local sources—for example, the East Baltic race of Scotch pine, the Hartz Mountain source of Norway spruce, the Sudeten (Germany) strain of European larch, the Burmese race of teak, Douglas fir from the Palmer area in Oregon, ponderosa pine from the Lolo Mountains in Montana, and white spruce from the Pembroke, Ontario (Canada) area.

Specific seed sources may be important in growing conifers and hardwood species of trees in the nursery. Plants are grown for individual use in the landscape or for special purposes, such as Christmas trees, in which specific tree characteristics are important (*13, 16, 26, 35*). Douglas fir has at least three recognized races, *viridis, caesia,* and *glauca,* with various geographical strains within them (*22*). The *viridis* strain from the U.S. west coast was not winter-hardy in New York tests but it is well suited to western Europe. Strains collected further inland were winter-hardy and vigorous; those from Montana and Wyoming were very slow growing. Trees of the *glauca* (blue) strain from the Rocky Mountain region were winter-hardy but varied in growth rate and ap-

pearance. Similarly, tree differences due to source occurred in Scotch pine, mugho pine, Norway spruce, and others.

Specific Source Selection

Selection of special seed sources to reproduce desired seedling characteristics for horticultural and forestry use has made much progress with future gains in prospect. Much effort has gone into improvements in fruit and nut tree rootstocks (*39*). Considerable effort has been used to "domesticate" and to upgrade forest tree production over that of "local" seed (*29, 30, 51*). This includes use of "exotic" sources, individual tree selections, and hybridization (*52*).

Correlation between Source and Progeny

There is evidence that for many tree traits the phenotype of the individual seed tree(s) is a good indicator of the phenotype of the offspring that will be produced. Consequently, parent tree selection (**phenotype selection**) can be effective for determining many characteristics, such as stem form, branching habit, growth rate, resistance to diseases and insects, presence of surface defects, and other qualities.

When individual trees in native stands show a superior phenotype, foresters call them "plus" trees. Such trees are significant in natural reseeding and are often used as seed sources. In nature such dominant seed trees may contribute the bulk of natural reseeding in any given area (*40*).

The true value of a particular seed source must be verified by **genotypic selection** in a **seedling progeny** test (*4*). A representative sample of seeds is planted, and the resulting progeny are grown and observed under test conditions that will identify characteristics.

The genetic traits that show a high correlation between parent(s) and average response of the offspring constitute **additive variation** and justify using the "best" trees for seed sources of the next generation (*31*).

In contrast, some seedling variation cannot be predicted from the parents (**nonadditive variation**) and must be determined by testing particular combinations of source trees. Progeny testing is the basis of plant improvement in forestry. To avoid inbreeding depression in the progeny, close relatives need to be avoided as parents of seed source collections.

Foresters refer to seed-source trees with a superior genotype, as demonstrated by a progeny test, as "elite" trees (4). Nursery tests can identify and characterize specific seed sources for landscape and Christmas tree uses (26).

An individual seed tree or shrub source should be selected by the following steps:

1. Evaluate the phenotypic characteristics of the seed tree or trees.
2. Evaluate the characteristics of surrounding trees that could provide cross-pollination.
3. Conduct a seedling progeny test of the trees selected.

Seedling Rootstocks of Fruit and Nut Trees

Historically, seedling rootstocks of many fruit trees have come from wild plants grown locally (peach, apple, pear, cherry) or imported from various parts of the world (39). Seed collections are often made from specific vegetatively propagated cultivars (clones) as parents. For example, to grow *Pyrus communis* seedlings, the so-called "French pear" seeds are obtained from

'Winter Nelis' pear trees grown for cross-pollination purposes in 'Bartlett' ('Williams Bon Cretien') pear orchards. All seedlings are 'Winter Nelis' × 'Bartlett' hybrids. Likewise, peach seeds or apricot seeds collected from fruit-drying yards come from commercial fruit orchards where, for example, the cultivars 'Lovell' peach and 'Royal' apricot are clones grown in a solid block of trees in which self-pollination occurs (see Chap. 18). More recently, specific rootstock seed producing cultivars (clones) have been developed with particular useful characteristics. For example, 'Nemaguard' peach is important for nematode resistance but the fruits are not suitable for eating (39).

Hybrids

Some F_1 interspecific hybrids have been produced for forest and fruit and nut crops, for example, the paradox walnut hybrid (Fig. 4–4). Almond and peach seedling hybrids have been produced by interplanting the parents in a seed orchard (39). In both cases, the hybrid progeny must be sufficiently different from the parents to identify them in the nursery row.

Some forest tree hybrids have been produced as *P. rigida* × *P. taeda* in Korea, and *Larix decidua* × L. *leptolepis* in Europe (52). Because of expense and uncertainty of production, seeds of F_1 hybrids have sometimes been used to produce F_2

FIGURE 4–4 Hybrid seed production in trees illustrated by the paradox walnut (*Juglans hindsii* × *J. regia*). Hybrid seedlings appear spontaneously in the nursery of seedlings from specific seed sources of black walnut (*J. hindsii*). These valuable plants are identified in the nursery row by leaf morphology, lighter bark color, and greater vigor. *Left:* Paradox hybrid: *center:* Black walnut seedlings; *Right top: J. regia; right middle:* Paradox hybrid; *right bottom:* J. hindsii.

seedling populations within which the more vigorous plants dominate.

Procedures in Source Selection and Production of Woody Plant Seed

Seed-Collection Zone

A **seed-collection zone** (*40*) is an area with defined boundaries and altitudinal limits in which soil and climate are sufficiently uniform to indicate high probability of reproducing a single ecotype. Such zones have been mapped and indicated by numbered codes that are recognized by seed certifying agencies for production of **source-identified seed.**

Seed-collection zones, designating particular climatic and geographical areas, have been established in most of the forest tree growing areas in the world (*1, 40*). California, for example, is divided into six physiographic and climatic regions, 32 subregions, and 85 seed-collection zones (Fig. 4–5) (*41*). Similar zones are established in Washington and Oregon and in the central region of the United States.

Seed-Production Areas

A **seed-production area** (*32, 40*) is an area containing a group of trees that has been identified or set aside specifically as a seed source. Such seed trees within the area are selected for their desirable characteristics. The value of the area can be improved by removing off-type trees, those

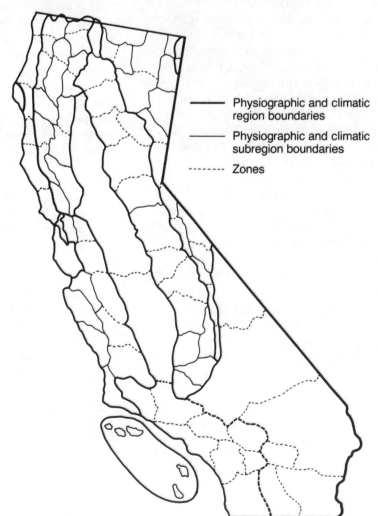

——— Physiographic and climatic region boundaries

——— Physiographic and climatic subregion boundaries

------- Zones

FIGURE 4–5 Distribution of seed collection zones in California. Each zone is identified by a specific number (not shown) and delineated by broken or solid lines representing an area of similar climatic and physiographic characteristics. (Redrawn from G. H. Schubert and R. S. Adams. 1971. Reforestation practices for conifers in California. Sacramento: Division of Forestry.)

that do not meet minimum standards, and other trees or shrub species that would interfere with the operations. It is desirable to eliminate additional trees to provide adequate space for tree development and seed production. An isolation zone at least 120 m (400 ft) wide from which off-type trees are removed should be established around the area. Trees may or may not have been progeny tested prior to establishment of the area.

Seed Orchards

Seed orchards or plantations are established specifically for seed production, propagated from seed trees of particular origin and quality and from progeny-tested seed trees.

Rootstocks for fruit and nuts. Special nematode-resistant rootstock-producing cultivars, such as 'Nemaguard' peach, are grown in seed orchards by grafting onto rootstocks. These are grown in isolation to avoid chance cross-pollination by virus-infected commercial cultivars. Various fruit-tree rootstock clones are self-pollinated and are planted in solid blocks. An isolation zone at least 120 m (400 ft) wide should be established around the orchard to reduce pollen contamination from other sources. The size of this zone can be reduced if a buffer area of the same kind of tree is present around the orchard.

Hybrid seed production of almond (self-incompatible) × peach has to be produced by combining the parents, allowing natural cross-pollination, and then roguing for hybrids in the nursery row.

Some seeds have been obtained by collecting from certain commercial cultivars.

Forest trees (40). There are three general types of seed orchards (*31, 40*): (a) seedling trees produced from selected parents through natural or controlled pollination; (b) clonal seed orchards in which selected clones are propagated by grafting, budding, or rooting cuttings; and (c) seedling-clonal seed orchards in which certain clones are grafted into branches of some of the trees. The choice of the type that should be used will depend upon the particular strategy needed to produce the desired outcome of the seed improvement program.

A site should be selected for good seed production. For forest trees and most other native species enough different selections should be included in a suitable arrangement to ensure cross-pollination and to decrease effects of inbreeding depression. Seven to thirty genotypes have been recommended to avoid this problem.

Apomixis

Some tree species, notably citrus, produce apomictic embryos (see p. 72). The seedlings are vigorous, uniform, usually virus free, and highly desirable as rootstocks (see Chap. 18). Some apple species are also apomictic but, although potentially useful, have not been used much in practice (*39*).

SEED CERTIFICATION

Voluntary **seed certification** (*2, 10, 34*) programs exist in most states of the United States through the cooperative efforts of public research, extension, regulatory agencies in agriculture, and a state seed-certifying agency known as a Crop Improvement Association. The agency is usually designated by law to certify seeds. Individual state organizations are coordinated through the Association of Official Seed Certifying Agencies (AOSCA) (*2*) in the United States and Canada. International certification is regulated through the Organization for Economic Cooperation and Development (OECD).

The principal objective of seed certification is to protect the genetic qualities of a cultivar, but requirements of seed quality may also be enforced, as well as the eligibility of particular cultivars. The seed-certifying agency may determine production standards for isolation, presence of off-type plants, and quality of harvested seed; and make regular inspections of the production fields to see that the standards are maintained. Supervision of seed processing is also involved.

Agricultural Crop Classes

Breeder's seed is that which originates with the sponsoring plant breeder or institution and provides the initial source of all the certified classes.

Foundation seed is the progeny of breeder's seed and is so handled as to maintain the highest standard of genetic identity and purity. It is the source of all other certified seed classes but can also be used to produce additional foundation seed plants. **Select seed** is a comparable seed class used in Canada. Foundation seed is labeled with a white tag or a certified seed tag with the word "Foundation."

Registered seed is the progeny of foundation seed (or sometimes breeder's seed or other registered seed) that is produced under specified standards approved and certified by the certifying agency and designed to maintain satisfactory genetic identity and purity. Bags of registered seed are labeled with a purple tag or with a blue tag marked with the word "Registered."

Certified seed is the progeny of registered seed (or sometimes breeder's, foundation, or other certified seed) and is that which is produced in the largest volume and sold to growers. It is produced under specified standards designed to maintain a satisfactory level of genetic identity and purity and is approved and certified by the certifying agency. Bags of certified seed have a blue tag distributed by the seed-certifying agency. The International OECD scheme includes **basic** (equivalent to either foundation or registered) seed. It also allows for **certified first-generation** (blue tag) and **second-generation** (red tag) seed.

Tree Certification Classes

Certification of forest tree seeds is available in some states and European countries similar to that for crop seed (*10, 14, 40*). Recommended minimum standards are given by the Association of Official Seed Certifying Agencies (*2*).

Source-identified tree seed is collected from natural stands where the geographic origin (source and elevation) is known and specified or from seed orchards or plantations of known provenance, specified by seed-certifying agencies. These seeds carry a *yellow* tag.

Selected tree seed is collected from trees that have been selected for promising phenotypic characteristics but have not been progeny tested (phenotype selection). The source and elevation must be stated. These seeds are given a *green* label.

Certified tree seed is collected from trees of proven genetic superiority, as defined by a certifying agency, and produced under conditions that assure genetic identity (genotypic selection). These could come from trees in a seed orchard, or from superior ("plus") trees in natural stands with controlled pollination. These seeds carry a *blue* tag.

PLANT VARIETY PROTECTION ACT

In the United States, breeders of a new seed-reproduced plant variety (cultivar) may retain exclusive propagation rights through the protection given by the Plant Variety Protection Act, which became effective in 1970 (*5*). The breeder applies to the U.S. Department of Agriculture for a Plant Variety Protection Certificate. For one to be granted, the cultivar must be "novel." It must differ from all known cultivars by one or more morphological, physiological, or other characteristics. It must be uniform; any variation must be describable, predictable, and acceptable. It must be stable; i.e., essential characteristics must remain unchanged during propagation. A certificate is good for 17 years. The applicant may designate that the cultivar be certified and that reproduction continue only for a given number of seed generations from the Breeder or Foundation stock. If designated that the cultivar be certified, it becomes unlawful under the Federal Seed Act to market seed by cultivar name unless it is certified. The passage of this law has greatly stimulated commercial cultivar development in which seed production supervision is controlled by the commercial seed company (*5, 12, 34*).

REFERENCES

1. Aldous, J. R. 1975. *Nursery practice. Forestry Commission Bull. 43,* London.
2. Assoc. Off. Seed Cert. Agencies. 1971. *AOSCA certification handbook.* Publ. 23.
3. Ball, V., ed. 1985. *Ball red book: Greenhouse growing* (14th ed.). Reston, Va.: Reston Publ. Co.
4. Barker, S. C. 1964. Progeny testing forest trees for seed certification programs. *Ann. Rpt. Inter. Crop Imp. Assn.* 46:83–87.
5. Barton, J. H. 1982. The international breeder's rights system and crop plant innovation. *Science* 216:1071–75.

6. Bassett, M. S., ed. 1986. *Breeding vegetable crops.* Westport, Conn.: AVI Publ. Co.

7. Briggs, F. N., and P. Knowles. 1967. *Introduction to plant breeding.* New York: Reinhold.

8. Burton, G. W. 1983. Utilization of hybrid vigor. In *Crop breeding,* D. R. Wood, K. M. Rawal, and M. N. Wood, eds. Madison, Wis.: Amer. Soc. Agron. and Crop Sci. Soc. Amer., pp. 89–107.

9. Copeland, L. O., and M. B. McDonald. 1985. *Principles of seed science and technology* (2nd ed.). New York: Macmillan.

10. Cowan, J. R. 1972. Seed certification. In *Seed biology,* Vol. 3, T. T. Kozlowski, ed. New York: Academic Press, pp. 371–97.

11. Craig, R. 1976. Flower seed industry. In J. W. Mastalerz, *Bedding plants: A manual on the culture of bedding plants as a greenhouse crop.* University Park, Pa.: Pennsylvania Flower Growers, pp. 25–46.

12. Davis, D. W., and J. J. Luby. 1988. Some current options in the use of plant variety protection in horticulture. *HortScience* 23:15–18.

13. Dirr, M. A., and C. W. Heuser, Jr. 1987. *The reference manual of woody plant propagation.* Athens, Ga.: Varsity Press.

14. Edwards, D. G. W., and F. T. Portlock. 1986. Expansion of Canadian tree seed certification. *For. Chron.* 62:461–66.

15. Ehrenberg, C., A. Gustafsson, G. P. Forshell, and M. Simak. 1955. Seed quality and the principles of forest genetics. *Heredity* 41:291–366.

16. Flint, H. 1970. Importance of seed source to propagation. *Proc. Inter. Plant Prop. Soc.* 20:171–78.

17. Frey, K. J. 1983. Plant population management and breeding. In *Crop breeding,* D. R. Wood, K. M. Rawal, and M. N. Wood, eds. Madison, Wis.: Amer. Soc. Agron. and Crop Sci. Soc. Amer., pp. 55–88.

18. Fryxell, P. A. 1957. Mode of reproduction of higher plants. *Bot. Rev.* 23:135–233.

19. Garrison, C. S., and R. J. Bula. 1961. Growing seeds of forages outside their regions of use. In *Seed: Yearbook of agriculture.* Washington, D.C.: U.S. Govt. Printing Office, pp. 401–6.

20. George, R. A. T. 1985. *Vegetable seed production.* London: Longman.

21. Goldsmith, G. A. 1976. The creative search for new F_1 hybrid flowers. *Proc. Inter. Plant Prop. Soc.* 26:100–103.

22. Haddock, P. G. 1968. The importance of provenance in forestry. *Proc. Inter. Plant Prop. Soc.* 17:91–98.

23. Hartman, H. 1961. Historical facts pertaining to root and trunkstocks for pear trees, *Oreg. State Univ. Agr. Exp. Sta. Misc. Paper 109,* 1–38.

24. Hartmann, H. T., A. M. Kofranek, V. Rubatzky, and W. J. Flocker. 1988. *Plant science: Growth, development, and utilization of cultivated plants* (2nd ed.). Englewood Cliffs, N.J.: Prentice-Hall.

25. Hawthorn, L. R., and L. H. Pollard. 1954. *Vegetable and flower seed production.* New York: Blakiston Co.

26. Heit, C. E. 1964. The importance of quality, germinative characteristics and source for successful seed propagation and plant production. *Proc. Inter. Plant Prop. Soc.* 14:74–85.

27. Holden, J. H. W., and J. T. Williams, eds. 1984. *Crop genetic resources: Conservation and evaluation.* London: G. Allen & Unwin.

28. Langlet, O. 1962. Ecological variability and taxonomy of forest trees. In *Tree growth,* T. T. Kozlowski, ed. New York: Ronald Press, pp. 357–69.

29. Libby, W. J. 1987. Genetic resources and variation in forest trees. In *Improving vegetatively propagated plants,* A. J. Abbott and R. K. Atkin, eds. New York: Academic Press, pp. 199–209.

30. ———. 1987. Testing and deployment of genetically engineered trees. In *Tissue culture of forest trees,* J. M. Bonga, and D. J. Durzan, eds. Amsterdam: Elsevier, pp. 167–97.

31. Libby, W. J., and R. M. Rauter. 1984. Advantages of clonal forestry. *For. Chron.,* pp. 145–49.

32. Macdonald, B. 1986. *Practical woody plant propagation for nursery growers,* Vol. 1. Portland, Oreg.: Timber Press.

33. Mastalerz, J. W. 1976. *Bedding plants: A manual on the culture of bedding plants as a greenhouse crop.* University Park, Pa.: Pennsylvania Flower Growers.

34. McDonald, M. B., Jr., and W. D. Pardee, eds. 1985. *The role of seed certification in the seed industry.* CSSA Spec. pub. 10. Madison, Wis.: Crop Sci. Soc. Amer., ASA.

35. McMillan-Browse, P. D. A. 1979. *Hardy, woody plants from seed.* London: Growers Books.

36. Pearson, O. H. 1968. Unstable gene systems in vegetable crops and implications for selection. *HortScience* 3(4):271–74.

37. Pendleton, R. A., H. R. Fennell, and F. C. Reimer. 1950. Sugar beet seed production in Oregon. *Oreg. Agr. Exp. Sta. Bul. 437.*

38. Poehlman, J. M. 1987. *Breeding field crops* (3rd ed.). Westport, Conn.: AVI Publ. Co.

39. Rom, R. C., and R. F. Carlson, eds. 1987. *Rootstocks for fruit crops*. New York: John Wiley.

40. Schopmeyer, C. S., ed. 1974. *Seeds of woody plants in the United States*. U.S. Dept. Agr. Handbook 450. Washington, D.C.: U.S. Govt. Printing Office.

41. Schubert, G. H., and R. S. Adams, 1971. *Reforestation practices for conifers in California*. Sacramento: Calif. State Div. of Forestry.

42. Simmonds, N. W. 1979. *Principles of crop improvement*. New York: John Wiley.

43. Sprague, G. F. 1950. Production of hybrid corn. *Iowa Agr. Exp. Sta. Bul. P48,* pp. 556–82.

44. Stoutemyer, V. T. 1960. Seed propagation as a nursery technique. *Proc. Plant Prop. Soc.* 10:251–55.

45. Struve, D. K., J. B. Jett, D. L. Bramlett. 1987. Production and harvest influences on woody plant seed germination. *Acta Hort.* 202:9–21.

46. Stuke, W. 1960. Seed and seed handling techniques in production of walnut seedlings. *Proc. Plant Prop. Soc.* 10:274–77.

47. Thomson, J. R. 1979. *An introduction to seed technology*. New York: John Wiley.

48. Westwood, M. N. 1966. Arboretums—a note of caution on their use in agriculture. *HortScience* 1:85–86.

49. ———. 1988. *Temperate zone pomology* (2nd ed.). Portland, Oreg.: Timber Press.

50. Wyman, D. 1953. Seeds of woody plants. *Arnoldia* 13:41–60.

51. Zobel, B. J., and J. Talbert. 1984. *Applied tree improvement*. New York: John Wiley.

52. Zobel, B. J., G. van Wyk, and P. Stahl. 1987. *Growing exotic forests*. New York: John Wiley.

SUPPLEMENTARY READING

BASSETT, M. S., ed. 1986. *Breeding vegetable crops*. Westport, Conn.: AVI Publ. Co.

COPELAND, L. O., and M. B. MCDONALD. 1985. *Principles of seed science and technology* (2nd ed.). New York: Macmillan.

DAVIS, D. W., and J. J. LUBY. 1988. Some current options in the use of plant variety protection in horticulture. *HortScience* 23(1):15–18.

POEHLMAN, J. M. 1987. *Breeding field crops* (3rd ed.). Westport, Conn.: AVI Publ. Co.

SCHOPMEYER, C. S., ed. 1974. *Seeds of woody plants in the United States*. U.S. Dept. Agr. Handbook 450. Washington, D.C.: U.S. Govt. Printing Office.

WOOD, D. R., K. M. RAWAL, and M. N. WOOD, eds. 1983. *Crop breeding*. Madison, Wis.: Amer. Soc. Agron. and Crop Sci. Soc. Amer.

U.S. DEPT. OF AGRICULTURE. 1961. *Seeds: Yearbook of agriculture*. Washington, D.C.: U.S. Govt. Printing Office.

ZOBEL, B. J., and J. TALBERT. 1984. *Applied tree improvement*. New York: John Wiley.

5

Techniques of Seed Production and Handling

The production of high-quality seed is of prime importance to propagators, whether they collect or produce the seed themselves or obtain the seed from others. In the production of any crop, the cost of the seed is usually minor compared to other production costs. Yet no single factor is as important in determining the success of the operation.

SOURCES OF SEED

Commercial Seed Sources

Agricultural, Vegetable, and Flower Seed

Commercial seed production is a specialized intensive industry with its own technology geared to the requirements of the individual species. Some agricultural seeds, such as corn, wheat, small grains, and grasses, are produced in the area where the crops are grown. However, very large amounts of seed are produced in specialized areas, particularly those with less summer rainfall and lower humidity, which decreases the danger of disease and rain during harvest (2, 4, 15). Since too low a humidity may cause premature shattering, much flower seed production is in a limited area along the U.S. west coast where moist air and night and morning fog from the ocean reduce the problem.

Seeds can be obtained in bulk but more often are available in smaller lots in bags, sealed cans, or moisture-tight packets which maintain the seeds at a low moisture content. Seed lots need to be properly labeled for identification and seed quality (see Chap. 7). Seeds may be obtained as certified seed (see Chap. 4).

Woody Plant Seed

A large number of commercial and professional seed-collecting firms exist which collect and sell seed of certain species. Lists of such producers are available (*18, 30*). Such seed should be properly labeled as to their origin or provenance (see Chap. 4). Some tree seed can be obtained as certified.

Seed Exchanges

Many arboreta and plant societies have seed exchanges or will provide small amounts of specialty seed.

Seed Collecting

Tree, shrub, and herbaceous plant seed may be collected by the individual nursery or propagator (*36, 40, 46*). These may be collected from specific seed collection zones or from seed production areas (see Chap. 4). Seed may be collected from standing trees, trees felled for logging, or from squirrel caches. They might be collected from parks, roadways, streets, or woodlots. Seed collecting has the advantage of being under the control of the propagator but requires intimate knowledge of each species and the proper method of handling. Most important, the collector should be aware of the importance of the selection principles described in Chap. 4.

Seed Orchards

Seed orchards or plantations are used to maintain seed source trees of particularly valuable species. They are extensively used by nurseries in the production of rootstock seeds of certain species or cultivars and for forest tree improvement. Such seed orchards are described in Chap. 4.

Fruit-Processing Industries

Historically, much fruit-tree rootstock seed has been obtained as by-products of fruit-processing industries such as canneries, cider presses, and dry-yards. Examples include 'Lovell' or 'Halford' peach or 'Royal' apricot in California (see p. 548) as well as apples and pears in the Pacific Northwest. The procedure is satisfactory if the correct cultivar is used (see p. 88). Some apple cultivars are triploid and produce poor seed. Some seed-borne viruses might be present in certain seed sources.

HARVESTING AND PROCESSING SEEDS

Maturity and Ripening

Each crop and plant species undergoes characteristic changes leading to seed ripening which must be known to establish the best time to harvest (*15, 30, 36, 45*). A seed is mature when it can be removed from the plant without impairing the seed's germination. Usually, it has reached a stage on the plant when no further increase in dry weight will occur (*2*). If the fruit ripens too soon or the seed is harvested when the embryo is insufficiently developed, the seed is apt to be thin, light in weight, shriveled, poor in quality, and short-lived (*17, 30*).

Early seed harvest may be desirable for seeds of some species of woody plants that produce a hard seed covering in addition to a dormant embryo. If seeds become dry and the seed coats harden, the seeds may not germinate until the second spring (*44*), whereas otherwise they would have germinated the first spring. On the other hand, if harvesting is delayed too long, the fruit may dehisce ("split open") or "shatter," drop to the ground, or be eaten or carried off by birds or animals. Thus, a balance must be made between late and early harvest to obtain the maximum number of high-quality seed.

Plants can be divided into three types according to the way their fruit ripens (*33*):

1. *Plants with dry fruits that combine seed and fruit covers.* These do not dehisce and are not disseminated immediately upon maturity. Type 1 includes most of the agricultural crops, such as corn, bean, wheat, and other grains. Many of these have undergone considerable selection during domestication for ease of harvest and handling.

2. *Plants producing dry seed from fruits that dehisce readily at maturity.* This type includes seeds in follicles, pods, capsules, siliques, and cones of conifers. Many species of type 2 are either native plants or are used for purposes other than food production.

3. *Plants with fleshy fruits.* Type 3 includes important fruit and vegetable species used for food (such as berries, pomes, and drupes) as

well as many related tree and shrub species used in landscaping or forestry.

Harvesting and Handling Procedures

Type 1.   Field-grown crops (cereals, grasses, corn) can be harvested by a combine, a machine that cuts and threshes the standing plant in a single operation. Plants that tend to fall over or "lodge" are cut, piled, or windrowed for drying and curing. Low humidity is important during harvest. Rain damage results in seeds that show low vigor. The force required to dislodge the seeds may result in mechanical damage and can reduce viability and result in abnormal seedlings. Some of these injuries are internal and not noticeable, but they result in low viability after storage (*1, 24, 33*). Damage is most likely to occur if the machinery is not properly adjusted. Usually less injury occurs if the seeds are somewhat moist at harvest (around 12 to 15 percent).

Type 2.   Fruits of this type dehisce to release the dry seeds. Some crops of this group include many annual or biennial flowers (delphinium, pansy, petunia), various vegetables (onion, cabbage, other cruciferous crops, and okra) which must be harvested before they are fully mature, then dried or cured before extraction. Consequently, some seeds will be undeveloped and immature. Many tree and shrub plants also have fruits which fall into this group and are handled with similar procedures. The steps are as follows:

1. *Drying.* Cut plants (sometimes by hand) or collect dry fruits and place on a canvas, tray, or screen, or lay on the floor to dry for one to three weeks. If only a few plants, cut and hang upside down in a paper bag to dry.
2. *Extraction.* Commercial seeds may be harvested by special machines that both extract by beating, flailing, or rolling the dry fruit and separate the seeds from the fruit parts, dirt, and other debris. Tree and shrub seed may be extracted with special macerators (*36*) also suitable for fleshy fruits. Or seeds may be beaten, flailed, or screened by hand. Lightweight seed may be removed during this screening operation.
3. *Cleaning.* Further cleaning may be required

to eliminate all dirt, debris, and weed and other crop seed. Commercial seed cleaning and processing utilizes various kinds of specialized equipment, such as screen of different sizes, air blasts, and gravity separators (*30, 42, 43*).

Type 3.   Vegetable crops include tomato, pepper, eggplant, and various cucurbits. These are produced in commercial fields and may utilize special macerating machinery. Cucumber and other vine crops are handled with specially developed machines (*39*).

Fleshy fruits include berries (grape), drupes (peach, plum), pomes (apple, pear), aggregate fruits (raspberry, strawberry), and multiple fruits (mulberry). In general, fleshy fruits are easiest to handle if ripe or overripe. However, fruits in the wild are subject to predation by birds (*20*).

1. *Extraction.* For small lots, fruits may be cut open and the seeds scooped out, treaded in tubs, rubbed through screens, or washed with water from a high-pressure spray machine in a wire basket. For larger lots separation is by fermentation, mechanical means, or washing through screens. A macerator can be used that crushes the fruits and mixes the pulverized mass with water that is diverted into a tank.
2. *Fermentation.* The macerated fruits are placed in large barrels or vats and allowed to ferment for about four days at about 21°C (70°F), with occasional stirring. If the process is continued too long, sprouting of the seeds may result. Higher temperatures during fermentation shorten the required time. As the pulp releases the seeds, the heavy, sound seeds sink to the bottom of the vat, and the pulp remains at the surface. Following extraction, the seeds are washed and dried either in the sun or in dehydrators. Additional cleaning is sometimes necessary to remove dried pieces of pulp and other materials. Extraction by fermentation is particularly desirable for tomato seed, because it controls bacterial canker.
3. *Flotation.* A flotation process involves placing the seeds and pulp in water so that the

heavy, sound seeds will sink to the bottom and the lighter pulp, empty seeds, and other extraneous materials will float off the top. This procedure can also be used for removing lightweight, unfilled seeds and other materials from dry fruits, such as acorn fruits infested with weevils, but sometimes both good and bad seed will float.

In the production of rootstock seedlings in oranges, separation of the seed from the surrounding fruit pulp is facilitated by the addition of a commercial pectinase enzyme (5). The small berries of some species, such as *Cotoneaster, Juniperus,* and *Viburnum,* are somewhat difficult to process because of small size and the difficulty in separating the seeds from the pulp. One way to handle such seeds is to crush the berries with a rolling pin, soak them in water for several days, and then remove the pulp by flotation.

4. *Blender.* Another device that removes seeds from small-seeded fleshy fruits is an **electric mixer** or **blender** (*38*). To avoid injuring the seeds, the metal blade of the latter machine can be replaced with a piece of rubber, 1½ in. square, cut from a tire casing. It is fastened at right angles to the revolving axis of the machine (*45*). A mixture of fruits and water is placed in the mixer and stirred for about two minutes. When the pulp has separated from the seed, the pulp is removed by flotation. This procedure is satisfactory for fruits of *Amelanchier* (serviceberry), *Berberis* (barberry), *Crataegus* (hawthorn), *Fragaria* (strawberry), *Gaylussacia* (huckleberry), *Juniperus* (juniper), *Rosa* (rose), and others (*38*).

5. *Drying.* Seeds are thoroughly washed to remove any fleshy remnants and dried except that seeds of recalcitrant species must not be allowed to dry out (see p. 98). If left in bulk for even a few hours, seeds that have more than 20 percent moisture will heat; this impairs viability. Drying may occur either naturally in open air if the humidity is low, or artificially with heat or other devices (*23*). Drying temperatures should not exceed 43°C (110°F); if the seeds are quite wet, 32°C (90°F) is better. Too-rapid drying can cause shrinkage and cracking and can some-

---

CUTTING TESTS

Seed viability of tree and shrub species varies considerably from year to year, from locality to locality, and from plant to plant. Before seeds are collected from any particular source, it is desirable to cut open a number of fruits and examine the seed contents to determine the percentage with sound, well-matured embryos. Although not a reliable viability test, a cutting test avoids taking seed from a source that produces only empty, unsound seed. X-ray examination also can accomplish this purpose (see Chap. 7).

---

times produce hard seed coats. The minimum safe moisture content for most seeds is in the range 8 to 15 percent.

Cones

Conifer cones require special procedures (*36*).

1. *Drying.* Cones of some species will open if they are dried in the open air for 2 to 12 weeks. Others must be force-dried at higher temperatures in special heating kilns. Under such conditions, the cones will open within several hours or at most two days. The temperature of artificial drying should be 46 to 60°C (115 to 140°F) depending upon the species, although a few require even higher temperatures. Jack pine (*Pinus banksiana*) and red pine (*P. resinosa*), for example, need temperatures of 77°C (170°F) for five to six hours. Caution must be used with high temperatures; overexposure will damage seeds. After the cones have been dried, the scales open, exposing the seeds.

2. *Extraction.* The cones are shaken by tumbling or raking to remove the seeds. A revolving wire tumbler or a metal drum is used when large numbers of seeds are to be extracted. The seeds should be removed immediately upon drying since the cones may close.

3. *Dewinging.* Conifer seeds have wings, which

are removed except in species whose seed coats are easily injured, such as incense cedar (*Calocedrus*). Fir (*Abies*) seed is easily injured, but the wings can be removed if the operation is done gently. Redwood (*Sequoia* and *Sequoiadendron*) seed have wings that are an inseparable part of the seed. For small lots of seed, dewinging can be done by rubbing the seeds between moistened hands or trampling or beating the seeds packed loosely in sacks. For larger lots of seeds, special dewinging machines are used.

4. *Cleaning.* The seeds are cleaned after extraction to remove the wings and other light chaff. As a final step, separation of heavy, filled seed from light seed is accomplished by gravity or pneumatic separators.

SEED STORAGE

Seeds are usually stored for varying lengths of time after harvest. Viability at the end of storage depends on (a) the initial viability at harvest, as determined by factors of production and methods of handling; and (b) the rate at which deterioration takes place. This rate of physiological change, or aging (*34*), varies with the kind of seed and the environmental conditions of storage, primarily temperature and humidity.

Longevity

Recalcitrant or Short-Lived Seed

This group is represented by species whose seed normally retain viability for as little as a few days, months, or at most a year. However, with proper handling and storage, seed longevity may be maintained for significant periods. The group includes:

1. Certain spring-ripening, temperate-zone trees such as poplar (*Populus*), some maple (*Acer*) species, willow (*Salix*), and elm (*Ulmus*). Their seeds drop to the ground and normally germinate immediately.

2. Many tropical plants grown under conditions of high temperature and humidity; includes such plants as sugarcane, rubber, jackfruit, macadamia, avocado, loquat, citrus, many palms, litchi, mango, tea, choyote, cocoa, coffee, tung, and kola.

3. Many aquatic plants of the temperate zones, such as wild rice (*Zizania aquatica*), pondweeds, arrowheads, and rushes.

4. Many tree nut and similar species with large fleshy cotyledons—hickories and pecan (*Carya*), birch (*Betula*), hornbeam (*Carpinus*), hazel and filbert (*Corylus*), chestnut (*Castanea*), beech (*Fagus*), oak (*Quercus*), walnut (*Juglans*), and buckeye (*Aesculus*).

Orthodox or Medium-Lived Seed

These remain viable for periods of 2 or 3 up to perhaps 15 years, providing that seeds are stored at low humidity and, preferably, at low temperatures. Seeds of most conifers, fruit trees, and commercially grown vegetables, flowers, and grains fall in this group.

Long-Lived Seeds

These seeds generally have hard seed coats that are impermeable to water. If the hard seed coat remains undamaged, such seeds should remain viable for at least 15 to 20 years. The maximum life can be as long as 75 to 100 years and perhaps more. Records exist of seeds being kept in museum cupboards for 150 to 200 years still retaining viability. Indian lotus (*Nelumbo nucifera*) seeds that had been buried in a Manchurian peat bog for an estimated 1000 years germinated perfectly when the impermeable seed coats were cracked (*13*).

Some weed seeds retain viability for many years (50 to 70 years or more) while buried in the soil, even though they have imbibed moisture (*35*). Longevity seems related to dormancy induced in the seeds by environmental conditions deep in the soil.

Seed Storage Factors Affecting Viability

The storage conditions that maintain seed viability are those that slow respiration and other metabolic processes without injuring the embryo. The most important conditions are low moisture content of the seed, low storage temperature, and modification of the storage atmosphere. Of these, the moisture–temperature relationships have the most practical significance.

Moisture Content

Control of the moisture content of the seed is probably the most important factor in seed longevity and storage. Most species have orthodox seeds (medium- to long-lived) where dehydration is not only their natural state at maturity but a nonfluctuating low moisture content must also be maintained for long-term storage (*23*).

Seeds of orthodox species are *desiccation tolerant* and not only can withstand dryness but must have a low moisture content for long-term storage. A 4 to 6 percent moisture content is favorable for prolonged storage (*16, 23*), although a somewhat higher moisture level is allowable if the temperature is reduced (*41*). For example, for tomato seed stored at 4.5 to 10°C (40 to 50°F), percent moisture content should be no more than 13 percent; if 21°C (70°F), 11 percent, and if 26.5°C (80°F), 9 percent.

Various storage problems arise with increasing seed moisture (*23*). At 8 or 9 percent or more, insects are active and reproduce; above 12 to 14 percent (65 percent relative humidity or more), fungi are active; above 18 to 20 percent, heating may occur; and above 40 to 60 percent, germination occurs.

If the moisture content of the seed is too low (1 to 2 percent), loss in viability and reduced germination rate can occur in some kinds of seeds (*12*). For seeds stored at these low moisture levels, it would be best to rehydrate with saturated water vapor to avoid injury to seed (*32*).

Moisture in seeds is in equilibrium with the ambient relative humidity of the storage container and increases if the relative humidity increases and decreases if it is reduced (*23*). Thus moisture percentage varies with the kind of seed, which is affected by the kind of storage reserves within the seed (*6*). Longevity of seed is maximum if stored at a relative humidity range of 20 to 25 percent (*23*).

Since fluctuations in seed moisture during storage reduce seed longevity (*7*), the ability to store seeds exposed to the open atmosphere varies greatly in different climatic areas. Dry climates are conducive to increased longevity; areas with high relative humidity result in shorter seed life. Seed viability is particularly difficult to maintain in open storage in tropical areas.

Storage in hermetically sealed, moisture-resistant containers is advantageous for long storage, but the moisture content of the seed must be low at the time of sealing (*9*). Seed moisture content of 10 to 12 percent (in contrast to 4 to 6 percent) in a sealed container is worse than storage in an unsealed container (*16*).

Recalcitrant seeds owe their short life primarily to their sensitivity to low moisture content (desiccation sensitivity). For instance, in silver maple (*Acer saccharinum*) seeds, the moisture content was 58 percent in the spring when the seeds matured. Viability was lost when the moisture content dropped below 30 to 34 percent (*27*). Citrus seeds can withstand only slight drying (*8, 14*) without loss of viability. The same is true for seeds of some water plants, such as wild rice, which can be stored directly in water at low temperatures (*31*). The large fleshy seeds of oaks (*Quercus*), hickories (*Carya*), and walnut (*Juglans*) lose viability if allowed to dry after ripening (*36*).

Viability of recalcitrant seeds of the temperate zone can be preserved for a period of time if kept in a moist environment and the temperature is lowered to just above freezing (*13*). Under these conditions many kinds of seeds can be kept for a year or more. Seeds of some tropical species (e.g., cacao, coffee), however, show chilling injury below 10°C (50°F).

Temperature

Reduced temperature invariably lengthens the storage life of seeds and, in general, can offset the adverse effect of a high moisture content. Harrington has given two "rules of thumb" (*23*): (a) for seeds not adversely affected by low moisture conditions, each 1 percent decrease in seed moisture, between 5 and 14 percent, doubles the life of the seed; and (b) each decrease of 5°C (9°F), between 0 and 44.5°C (32 and 112°F) in storage temperature, also doubles seed storage life. On the other hand, seeds stored at low temperature but at a high relative humidity may lose viability rapidly when moved to a higher temperature (*9*).

Subfreezing temperatures, at least down to

−18°C (0°F), will increase storage life of most kinds of seeds but moisture content should be in equilibrium with 70 percent relative humidity or lower, or the free water in the seeds may freeze and cause injury (*23*). Such storage is particularly useful for conifer seeds (*36,37*). Refrigerated storage should be combined with dehumidification or with sealing dried seeds in moisture-proof containers.

Ultrafreezing

Cryopreservation of seeds immersed in liquid nitrogen (LN$_2$) at a very low temperature, −196°C, can be used for many seeds provided that the moisture content is relatively low, about 8 to 15 percent, and the seeds are in sealed containers (*3, 10, 19*). Regulating freezing and thawing rates may be important in the operation.

Special types of equipment are required in order to reduce the temperature to −196°C in liquid nitrogen (*19*). This procedure is useful for very long storage of valuable seed stocks as germplasm (*3, 10, 19, 29*).

Types of Seed Storage

Open Storage without Moisture or Temperature Control

Many kinds of orthodox seeds only need to be stored from harvest until the next planting season. Under these conditions, seed longevity depends on the relative humidity and temperature of the storage atmosphere, the kind of seed, and their condition at the beginning of storage. Basic features (*28*) of the storage structures include (a) protection from water, (b) avoidance of mixture with other seeds, and (c) protection from rodents, insects, fungi, and fire. Retention of viability varies with the climatic factors of the area in which storage occurs. Poorest conditions are found in warm, humid climates; best storage conditions occur in dry, cold regions. Fumigation or insecticidal treatments may be necessary to control insect infestations.

Open storage can be used for many kinds of commercial seeds for at least a year (i.e., to hold seeds from one season to the next). Seeds of many

species, including most agricultural, vegetable, and flower seeds, will retain viability for longer periods up to four to five years (*11, 13, 21*), except under the most adverse conditions.

Seeds that have a water-impervious seed coat will retain viability in open storage for many years (10 to 20 years or more) once they have been dried. Open storage is adequate. Some woody plants whose seeds are handled in this manner are:

Acacia spp.	*Eucalyptus* spp.
Albizia spp. (albizzia)	*Koelreuteria paniculata* (golden rain tree)
Amorpha fruticosa (indigo bush)	*Rhus ovata* (sumac)
Caragana arborescens (Siberian pea shrub)	*Robina pseudoacacia* (locust)
Elaeagnus angustifolia (Russian olive)	*Tilia* (linden)

Sealed Containers

Packaging dry seeds in hermetically sealed, moisture-proof containers is an important method of handling and/or merchandizing seeds. Containers made of different materials vary in durability and strength, cost, protective capacity against rodents and insects, and ability to retain or transmit moisture (*15, 22*). Those completely resistant to moisture transmission include tin or aluminum cans (if properly sealed), hermetically sealed glass jars, and aluminum pouches. Those almost as good (80 to 90 percent effective) are polyethylene (3 mil or thicker) and various types of aluminum-paper laminated bags. Somewhat less desirable, in regard to moisture transmission, are asphalt and polyethylene laminated paper bags and friction-top tin cans. Paper and cloth bags give no protection against moisture change.

Seeds may be protected against moisture uptake by mixing with a desiccant (*15, 23, 28*). A useful desiccant is silica gel treated with cobalt chloride. Silica gel (1 part to 10 parts seed, by weight) can absorb water up to 40 percent of its weight. The CoCl$_2$ turns from blue to pink at 45 percent RH and can act as a useful indicator of excess moisture.

Seeds in sealed containers are more sensitive to excess moisture than when subjected to fluctuating moisture content in open storage. A seed moisture content of 5 to 8 percent or less is desirable, depending upon species.

Conditioned Storage (15, 23, 28)

Conditioned storage includes use of dehumidified and/or refrigerated facilities to reduce temperature and relative humidity. Such facilities are expensive but are justified where particularly valuable seeds are stored, as for research, breeding stocks, and germplasm. Also in some climatic areas, such as in the highly humid tropics, orthodox seeds cannot be maintained from one harvest season to the next planting season.

Cold storage of tree and shrub seed used in nursery production is generally advisable if the seeds are to be held for longer than one year (*18, 25, 26, 36, 37*) except for hard-coated seeds listed previously. Seed storage is useful in forestry because of the uncertainty of good seed-crop years. Seeds of many species are best stored under cold, dry conditions (*45*). Ambient relative humidity in conditioned storage should not be higher than 65 to 70 percent RH (for fungus control) and no lower than 20 to 25 percent.

It is important to control humidity in refrigerated storage since the relative humidity increases with decrease in temperature and moisture will condense on the seed. At 15°C (59°F) this equilibrium moisture may be too high for proper seed storage. Although the moisture content may not be harmful at those low temperatures, rapid deterioration will occur when the seeds are removed from storage and returned to ambient uncontrolled temperatures. Consequently, refrigeration should be combined with dehumidification or sealing in moisture-proof containers (*23*).

Low humidity in storage can be obtained by judicious ventilation, moisture proofing, and dehumidification as well as use of sealed moisture containers, or use of desiccants, as described above. Dehumidifiers utilize desiccants (silica gel) or saturated salt solutions.

The most effective storage is to dry seeds to 3 to 8 percent moisture, place in sealed con-

---

U.S. NATIONAL SEED STORAGE LABORATORY

The National Seed Storage Laboratory (*10*) was built in 1958 on the Colorado State University campus, Fort Collins, Colorado, to preserve seed stocks of valuable germplasm. Seeds are acquired from public agencies, seed companies, and individuals engaged in plant breeding or seed research. Descriptive material is received at the same time. Seed samples are tested for viability and, if suitable for long-term storage, are dried to 5 to 7 percent moisture and stored at −10 to −12°C in hermetically sealed metal cans. Seed lots are tested every five years for germinability. If viability drops, a new generation of seed is produced and stored. Seeds are made available to research workers and plant breeders on request. This facility also conducts seed storage research.

---

tainers, and store at temperatures of 1 to 5°C (41°F). Below-freezing temperatures can be even more effective if the value of the seed justifies the cost.

Moist, Cool Storage

Many recalcitrant seeds that cannot be dried can be mixed with a moisture-retaining medium, placed in a polyethylene bag or other container, and refrigerated at 0 to 10°C (32 to 50°F). The relative humidity in storage should be 80 to 90 percent. Examples of species whose seeds require this storage treatment are: *Acer saccharinum* (silver maple), *Aesculus* spp. (buckeye), *Carpinus caroliniana* (American hornbeam), *Carya* spp. (hickory), *Castanea* spp. (chestnut), *Corylus* spp. (filbert), *Citrus* spp. (citrus), *Eriobotrya japonica* (loquat), *Fagus* spp. (beech), *Juglans* spp. (walnut), *Litchi, Nyssa silvatica* (tupelo), *Persea* spp. (avocado), and *Quercus* spp. (oak). The procedure is similar to moist-chilling (stratification). Acorns and large nuts can be dipped in paraffin or sprayed with latex paint before storage to preserve their moisture content (*26*).

REFERENCES

1. Asgrow. 1959. *A study of mechanical injury to seed beans.* Asgrow Monograph 1. New Haven, Conn.: Associated Seed Growers, 1949.

2. Austin, R. B. 1972. Effects of environment before harvesting on viability. In *Viability of seeds,* E. H. Roberts, ed. Syracuse. N.Y.: Syracuse Univ. Press.

3. Bajaj, Y.P.S. 1979. Establishment of germplasm banks through freeze storage of plant tissue culture and their implications in agriculture. In *Plant cell and tissue culture principles and applications,* W. R. Sharp et al., eds. Columbus: Ohio State Univ. Press, pp. 745–74.

4. Baker, K. F. 1980. Pathology of flower seeds. *Seed Sci. and Tech.* 8:575–89.

5. Barmore, C. R., and W. S. Castle. 1979. Separation of citrus seed from fruit pulp for rootstock propagation using a pectolytic enzyme. *HortScience* 14(4):526–27.

6. Barton, L. V. 1941. Relation of certain air temperatures and humidities to viability of seeds. *Contrib. Boyce Thomp. Inst.* 12:85–102.

7. ———. 1943. Effect of moisture fluctuations on the viability of seeds in storage. *Contrib. Boyce Thomp. Inst.* 13:35–45.

8. ———. 1943. The storage of some citrus seeds. *Contrib. Boyce Thomp. Inst.* 13:47–55.

9. ———. 1953. Seed storage and viability. *Contrib. Boyce Thomp. Inst.* 17:87–103.

10. Bass, L. N. 1979. Physiological and other aspects of seed preservation. In *The plant seed: Development, preservation, and germination,* I. Rubenstein et al., eds. New York: Academic Press, pp. 145–70.

11. ———. 1980. Flower seed storage. *Seed Sci. and Tech.* 8:591–99.

12. ———. 1980. Seed viability during long term storage. *Hort. Rev.* 2:117–41.

13. Bewley, J. D., and M. Black. 1985. *Seeds: Physiology of development and germination.* New York: Plenum Press.

14. Childs, J. F. L., and G. Hrnciar. 1948. A method of maintaining viability of citrus seeds in storage. *Proc. Fla. State Hort. Soc.* 64:69.

15. Copeland, L. O., and M. B. McDonald. 1985. *Principles of seed science and technology* (2nd ed.). New York: Macmillan.

16. Crocker, W., and L. V. Barton. 1953. *Physiology of seeds.* Waltham, Mass.: Chronica Botanica.

17. Delouche, J. C. 1980. Environmental effects on seed development and seed quality. *HortScience* 15:775–80.

18. Dirr, M. A., and C. W. Heuser, Jr. 1987. *The reference manual of woody plant propagation.* Athens, Ga.: Varsity Press.

19. Dougall, D. K. 1978. Preservation of germ plasm. In *Propagation of higher plants through tissue culture: A bridge between research and applications,* K. W. Hughes, R. Henke, and M. Constantin, eds. Springfield, Va.: U.S. Dept. Energy Tech. Inform. Center.

20. Fordham, A. J. 1984. Seed dispersal as it concerns the propagator. *Proc. Inter. Plant Prop. Soc.* 34:531–34.

21. Goss, W. L. 1937. Germination of flower seeds stored for ten years in the California state seed laboratory. *Calif. Dept. Agr. Bul.* 26:326–33.

22. Harrington, J. F. 1963. The value of moisture-resistant containers in vegetable seed packaging. *Calif. Agr. Exp. Sta. Bul.* 792, pp. 1–23.

23. ———. 1972. Seed storage and longevity. In *Seed biology,* T. T. Kozlowski, ed. New York: Academic Press, pp. 145–245.

24. Hawthorn, L. R., and L. H. Pollard. 1954. *Vegetable and flower seed production.* New York: Blakiston Co.

25. Heit, C. E. 1967. Propagation from seed. 10. Storage methods for conifer seed. *Amer. Nurs.* 126(20): 14–15.

26. ———. 1967. Propagation from seed. 11. Storage of deciduous tree and shrub seed. *Amer. Nurs.* 126(21): 12–13, 86–94.

27. Jones, H. A. 1920. Physiological study of maple seeds. *Bot. Gaz.* 69:127–52.

28. Justice, O. L., and L. N. Bass. 1978. *Principles and practices of seed storage.* U.S. Dept. Agr. Handbook 506. Washington, D.C.: U.S. Govt. Printing Office.

29. Kartha, K. K. 1985. *Cryopreservation of plant cells and organs.* Boca Raton, Fla.: CRC Press.

30. Macdonald, B. 1986. *Practical woody plant propagation for nursery growers,* Vol. 1. Portland, Oreg.: Timber Press.

31. Muenscher, W. C. 1936. Storage and germination of seeds of aquatic plants. *New York (Cornell Univ.) Agr. Exp. Sta. Bul. 652,* pp. 1–17.

32. Nutile, G. E. 1964. Effect of desiccation on viability of seeds. *Crop Sci.* 4:325–28.

33. Pollock, B. M., and E. E. Roos. 1972. Seed and

seedling vigor. In *Seed biology,* Vol. 1. T. T. Kozlowski, ed. New York: Academic Press.

34. Priestley, D. A. 1986. *Seed aging.* Ithaca, N.Y.: Cornell Univ. Press.

35. Roberts, E. H. 1972. Dormancy: A factor affecting seed survival in the soil. In *Viability of seeds,* E. H. Roberts, ed. London: Chapman & Hall, pp. 320–59.

36. Schopmeyer, C. S., ed. 1974. *Seeds of woody plants in the United States.* U.S. Dept. Agr. Handbook 450. Washington, D.C.: U.S. Govt. Printing Office.

37. Schubert, G. H., and R. S. Adams. 1971. *Reforestation practices for conifers in California.* Sacramento: Calif. State Div. of Forestry.

38. Smith, B. C. 1950. Cleaning and processing seeds. *Amer. Nurs.* 92(11):13–14, 33–35.

39. Steiner, J. J., and B. F. Letizia. 1986. A seed-cleaning sluice for fleshy-fruited vegetables from small plots. *HortScience* 21(4):1066–67.

40. Struve, D. K., J. B. Jett, D. L. Bramlett. 1987. Production and harvest influences on woody plant seed germination. *Acta Hort.* 202:9–21.

41. Toole, E. H. 1958. Storage of vegetable seeds. *USDA Leaflet 220* (rev.).

42. Van der Berg, H. H., and R. Hendricks. 1980. Cleaning flower seeds. *Seed Sci. and Tech.* 8:505–22.

43. Vaughn, C. E., B. R. Gregg, and J. C. Delouche. 1968. *Seed processing and handling.* State College, Miss.: Miss. State Univ. Seed Technology Library.

44. Wells, J. S. 1985. *Plant propagation practices* (2nd ed.). Chicago: American Nurseryman Publ. Co.

45. Wyman, D. 1953. Seeds of woody plants. *Arnoldia* 13:41–60.

46. Young, J. A., and C. G. Young. 1986. *Collecting, processing and germinating seeds of wildland plants.* Portland, Oreg.: Timber Press.

SUPPLEMENTARY READING

CANTLIFFE, D. J., ed. 1979. Symposium on seed quality: An overview of its relationship to horticulturists and physiologists. *HortScience* 15:764–89.

GORDON, A. G., and D. C. F. ROWE. 1982. *Seed manual for ornamental trees and shrubs.* Bull. 59. London: Forestry Commission.

HAWTHORN, L. R., and L. H. POLLARD. 1954. *Vegetable and flower seed production.* New York: Blakiston Co.

JUSTICE, O. L., and L. N. BASS. 1978. *Principles and practices of seed storage.* U.S. Dept. Agr. Handbook 506. Washington, D.C.: U.S. Govt. Printing Office.

KOZLOWSKI, T. T., ed. 1972. *Seed biology,* Vols. 1, 2, 3. New York: Academic Press.

LEE, D., 1987. Seed collection and handling. *Proc. Inter. Plant Prop. Soc.* 37:61–65.

ROBERTS, E. H., ed. 1972. *Viability of seeds.* London: Chapman & Hall.

Seed World, published monthly.

SCHOPMEYER, C. S., ed. 1974. *Seeds of woody plants in the United States.* U.S. Dept. Agr. Handbook 450. Washington, D.C.: U.S. Govt. Printing Office.

THOMSON, J. R. 1979. *An introduction to seed technology.* New York: John Wiley.

U.S. DEPT. OF AGRICULTURE. 1961. *Seeds: Yearbook of agriculture.* Washington, D.C.: U.S. Govt. Printing Office.

6

Principles of Propagation by Seed

A seed is a ripened ovule. At the time of separation from the parent plant it consists of an **embryo** and **stored food supply,** both of which are encased in a protective **covering.** The activation of the metabolic machinery of the embryo leading to the emergence of a new seedling plant is known as **germination.** This chapter describes the various conditions that determine the success of germination and initial growth of the seedling.

THE GERMINATION PROCESS

For germination to be initiated, three conditions must be fulfilled (*31, 67*):

First, the seed must be **viable;** that is, the embryo must be alive and capable of germination.

Second, the seed must be subjected to the appropriate environmental conditions: available **water,** proper **temperature** regimes, a supply of **oxygen,** and sometimes **light.**

Third, any **primary dormancy** condition present within the seed (*35*) must be overcome. Internal processes leading to removal of primary dormancy are collectively known as **after-ripening** and result from the interaction of the environment with the specific primary dormancy condition. After-ripening requires a period of time and sometimes specific methods of seed handling. Even in the absence of primary dormancy and/or if the seeds are subjected to adverse environmental conditions, a **secondary dormancy** can develop and further delay the period when germination takes place (*26, 68, 75*).

Stages of Germination

Germination is divided into several consecutive but overlapping stages (*19, 31*):

Stage 1—Activation

Imbibition of water. Water is absorbed by the dry seed and the moisture content increases rapidly at first, then levels off (see Fig. 6–1). Initial absorption involves the imbibition of water by colloids of the dry seed. Water softens the seed coverings and causes hydration of the protoplasm. The seed swells and the seed coats may break. Imbibition is a physical process and can take place even in dead seeds.

In triggering germination, water absorption shows three stages: (a) an initial increase to 40 to 60 percent of water (fresh weight basis) equivalent to 80 to 120 percent dry weight (water content/initial dry weight), (b) a slow (lag) period after which the radicle emerges (germination), followed by (c) a further increase to 170 to 180 percent (dry weight basis) as the seedling grows (Fig. 6–1).

Synthesis of enzymes. As the seed absorbs water, enzyme activity begins to appear within a matter of hours. Activation results in part from reactivation of stored enzymes, previously formed during development of the embryo, and, in part, from synthesis of new enzymes as germination starts (*19, 69*). Synthesis requires the presence of specifically programmed RNA molecules (see box). Some of these appear to have been formed during seed development, conserved during the ripening process and available to initiate germination. Others are apparently formed after germination starts. Energy for these processes is obtained from the high-energy phosphate bonds in adenosine triphosphate (ATP) present in mitochondria. Some ATP is conserved in the dormant seed and reactivated as the seed absorbs moisture.

Cell elongation and emergence of the radicle. The first visible evidence of germination is emergence of the radicle, which results from enlargement of cells rather than from cell division (*17, 19, 56*). Emergence of the radicle can occur within a few hours or a few days after the initiation of germination and marks the end of stage 1.

Stage 2—Digestion and Translocation

Fats, proteins, and carbohydrates stored in the endosperm, cotyledons, perisperm, or female gametophyte (conifers) are digested to simpler chemical substances, which are translocated to the growing points of the embryo axis. The existing cell systems have been activated and the protein-synthesizing system is functioning to produce new enzymes, structural materials, regulator compounds, hormones, and nucleic acids to carry on the cell functions and synthesize new materials. Water uptake and respiration now continue at a steady rate.

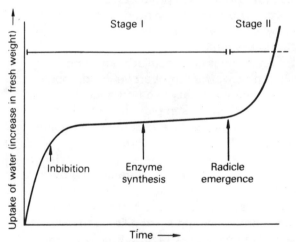

FIGURE 6–1 Water uptake by a dry seed identifies three major phases of germination initiation (stage 1). Phase 1 is characterized by imbibition of water into the seed. Phase 2 is characterized by enzyme activation and synthesis. Phase 3 is characterized by emergence of the radicle and subsequent surge in water uptake as germination proceeds. Radicle emergence is considered to represent the first visible evidence of germination. (Reprinted by permission with revision from J. D. Bewley and M. Black, 1978, Physiology and biochemistry of seeds in relation to germination, Vol. 1, *Development, germination and growth,* Springer-Verlag, Berlin.)

Stage 3—Seedling Growth

In the third stage, development of the seedling plant begins with cell division at the two ends of the embryo axis, followed by the expansion of the seedling structures. The initiation of cell division in the growing points appears to be independent of the initiation of cell elongation (*17, 56*).

The embryo consists of an **axis** bearing one or more **seed leaves,** or **cotyledons.** The growing point of the root, the **radicle,** emerges from the base of the embryo axis. The growing point of the shoot, the **plumule,** is at the upper end of the em-

BIOLOGICAL CONTROL
IN GERMINATION AND DORMANCY

Control of embryogenesis, germination, and dormancy have revolved around three different biological concepts: (a) the transfer of specific genetic information from the genes to the proteins to specific metabolic systems characteristic of particular stages of development; (b) the balance and interplay among different endogenous hormone systems, involving primarily abscisic acid, gibberellin, and cytokinin; and (c) phase transition in cell membranes at specific temperatures (*19*).

(a) Embryogenesis originates in the single-celled **zygote** (see Chaps. 1 and 3), which is described as being **totipotent,** that is, it contains all the genetic information needed to produce the embryo and the seedling plant and its subsequent development. How the information codes of the gene specified by the DNA (deoxyribonucleic acid) of the cell are transferred into embryo and seedling development determines the pattern and unique outcomes of the germination sequence for particular plants (*29*). These pathways are mediated through the activity of biologically active compounds known as RNAs (ribonucleic acids), differing from DNA in specific molecular structure. Specific stages in this sequence involve the **transcription** of specific DNA sequences to **transfer RNA** (tRNA), the processing to **messenger RNA** (mRNA), transport from the nucleus to the cytoplasm to join into structures (**ribosomes**) where it is known as **ribose RNA** (rRNA), and degradation and **translation** to direct the synthesis of specific proteins. Proteins are complex nitrogen-containing compounds that act as enzymes, serve as intermediates to other complex metabolites, and become storage products and/or structural compounds.

(b) Much research indicates that endogenous hormones are directly involved in various aspects of RNA transcription and translation and may turn on and off various key metabolic steps, most of which are not understood (*88*). For instance, the presence of cytokinin and gibberellins is associated with embryo enlargement phases of development, whereas inhibitors (particularly abscisic acid) along with dehydration have been associated with the maintenance of the embryonic phase and the prevention of premature germination. Germination initiation, breakdown of storage products, and seedling growth have been associated with gibberellins, whereas cytokinins are effective in neutralizing the inhibitors.

Sometimes an applied hormone can have a direct effect in overcoming or inducing dormancy, as described on page 148. In other cases, for example, in embryo dormancy of apple (*27*), a hormone may have only a partial effect and all three are involved.

(c) Further research emphasized biophysical effects of **cell membranes** in growth inhibition, activation, and dormancy control. Evidence is provided in the correlation of cell and membrane changes from liquid to gels at specific temperatures (*19*), shifts which are similar to dormancy responses to temperature. Inner layers of the testa and the remnants of the endosperm and/or nucellus have restricting effects that are released with alteration or removal of the membranes, shifting temperatures, or subjecting seeds to anaerobiosis (*11*).

bryo axis, above the cotyledons. The seedling stem is divided into the section below the cotyledons—the **hypocotyl**—and the section above the cotyledons—the **epicotyl.**

Once growth begins from the embryo axis, fresh weight and dry weight of the new seedling plant increase but total weight of storage tissue decreases. The respiration rate, as measured by oxygen uptake, increases steadily with advance in growth. Storage tissues of the seed eventually cease to be involved in metabolic activities except in plants of which the cotyledons emerge from the ground and become active in photosynthesis. Water absorption increases steadily as new roots explore the germination medium and the fresh weight of the seedling plant increases.

The initial growth of the seedling follows one of two patterns. In one type—**epigeous** germination—the hypocotyl elongates and raises the cotyledons above the ground. In the other type—**hypogeous** germination—the lengthening of the hypocotyl does not raise the cotyledons above the ground, and only the epicotyl emerges (Figs. 6–2 and 6–3).

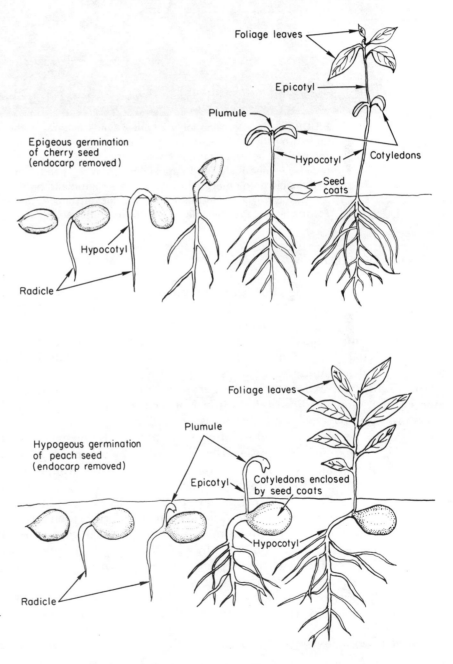

FIGURE 6–2 Seed germination in dicotyledonous plants. *Top:* Epigeous germination of cherry. The cotyledons are above ground. *Below:* Hypogeous germination of peach. The cotyledons remain below ground.

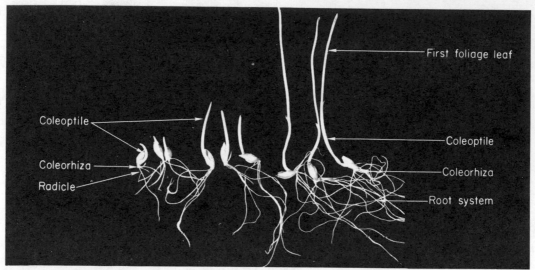

FIGURE 6–3 Germination of barley seeds illustrates the special pattern shown by the grass (Gramineae) family, a monocotyledon. See Fig. 3–12 for seed structure. The coleoptile and coleorhiza enclose the plumule and radicle, respectively, and appear first from the seed. The first foliage leaf and radicle then emerge through these structures to produce the stems and roots. (From H. T. Hartmann, W. J. Flocker, and A. M. Kofranek, 1981. *Plant science,* Prentice-Hall, Englewood Cliffs, N.J.)

QUALITY OF SEEDS

A method of judging viability is essential in successful seed propagation. A dead or dying seed is characterized by a gradual decline in vigor, and necrosis or injuries may appear in localized areas of the seed coat. But the difference between a live seed and a dead one is not always distinct (*63, 100*). **Viability** is expressed by the **germination percentage,** which indicates the number of seedlings produced by a given number of seeds. Additional characteristics of high-quality seed are prompt germination, vigorous seedling growth, and normal appearance (*2*). **Vigor** of seed and seedling is an important attribute of quality but is difficult to measure (*90, 92*). Low germination percentage, low germination rate, and low vigor are often associated. Low germination can be due to genetic properties of certain cultivars (*43*), incomplete seed development on the plant, injuries during harvest, improper processing (*41*) and storage (*104*), disease, and aging. Loss in viability is usually preceded by a period of declining vigor (*121*).

Seeds with low vigor may not be able to withstand unfavorable conditions in the seed bed, such as attacks by disease organisms. The seedlings may lack the strength to emerge if the seeds are planted too deeply or if the soil surface is crusted. Field survival of low-vigor seeds is apt to be less than a laboratory germination percentage test would indicate.

Measurement of Seed Quality

If one measures the time sequence of germination of a given lot of seeds, or the emergence of seedlings from a seed bed, one usually finds a pattern like the germination curve shown in Fig. 6–4. There is an initial delay in the start of germination, then a rapid increase in the number of seeds that germinate, followed by a decrease in their rate of appearance. When viability is less than 100 percent, the end point may not be exact.

Germination is measured on two parameters—the **germination percentage** and the **germination rate.** Vigor may be indicated by these measurements, but seedling growth rate and morphological appearance also must be considered.

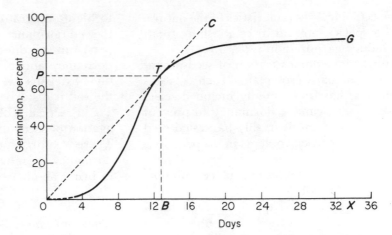

FIGURE 6–4 Typical germination curve of a sample of germinating seeds. After an initial delay, the number of seeds germinating increases, then decreases. Such a curve can be used to measure germination value. (Redrawn from Czabator (*38*).)

Sometimes abnormally growing seedlings result from low seed quality (*63*).

Statements of germination percentage should involve a time element, indicating the number of seedlings produced within a specified length of time. Germination rate can be measured by several methods. One determines the number of days required to produce a given germination percentage. Another method calculates the average number of days required for radicle or plumule emergence as follows:

Mean days =

$$\frac{N_1T_1 + N_2T_2 + \ldots + N_xT_x}{\text{total number of seeds germinating}}$$

N values are the numbers of seeds germinating within consecutive intervals of time; *T* values indicate the times between the beginning of the test and the end of the particular interval of measurement. Kotowski (*79*) has used the reciprocal of this formula multiplied by 100 to determine a **coefficient of velocity.** Gordon (*54*) has suggested the term **germination resistance** as the time (hours or days) to average germination, based on seeds that germinate.

Czabator (*38*) has suggested another measurement for seeds of woody perennials in which germination may be slow: the **germination value** (*GV*). It includes both the germination rate and percentage. To calculate *GV*, a germination curve, as shown in Figure 6–4, must be obtained by periodic counts of radicle or plumule emer-

gence. The important values on the curve are **T**—the point at which the germination rate begins to slow down—and **G**—the final germination percentage. These points divide the curve into two parts—a rapid phase and a slow phase. Peak value (*PV*) is the germination percentage at *T,* divided by the days to reach that point. Mean daily germination (*MDG*) is the final germination percentage divided by number of days to reach final germination. For example:

$$
\begin{aligned}
GV &= PV \times MDG \\
&= \frac{68}{13} \times \frac{85}{34} \\
&= 5.2 \times 2.5 \\
&= 13.0
\end{aligned}
$$

DORMANCY: REGULATION OF GERMINATION

When a seed is separated from the plant, it invariably has **primary dormancy** (see Chap. 3). This not only prevents immediate germination but also regulates the time, conditions, and place that germination will occur. In nature, different kinds of primary dormancy have evolved to aid survival of the species (*78, 118, 120, 131*) by programming the time of germination for particularly favorable times in the annual seasonal cycle.

Secondary dormancy is a further survival mechanism that can be induced under unfavorable environmental conditions and may further delay the time that germination occurs. Knowledge of the ecological characteristics of the natural habitat of a species can aid in establishing treatments to induce germination (*102, 132*).

In cultivation, domestication of seed-propagated cultivars of many crop plants, such as grains and vegetables, has undoubtedly included selection for sufficient primary dormancy to prevent immediate germination of freshly harvested seed but not enough to cause problems in propagation. Dormancy facilitates storage, transport, and handling. After-ripening changes take place with normal dry storage handling of most agricultural, vegetable, and flower seed to allow germination to proceed whenever the seeds are subjected to normal germinating conditions. Problems can occur when seed testing is attempted on freshly harvested seeds. Seeds of some species are sensitive to high-temperature and light conditions related to seed dormancy (see p. 115). Many weed seeds persist in soil due to either primary or secondary dormancy and provide "seed banks" that produce extensive weed seed germination whenever the soil is disturbed (*19*).

Practical problems occur with nursery propagation of seeds of many tree and shrub species. These require specific treatments to overcome dormancy by satisfying the requirements needed to bring about germination (see Chaps. 18 and 19).

Kinds of Primary Seed Dormancy

Propagators of cultivated plants have long recognized these germination-delaying phenomena and

---

CONCEPTS OF DORMANCY

Dormancy in a general sense has been defined as "*a temporary suspension of visible growth of any plant structure containing a meristem*" (*82*). It includes growth cessation due to both internal (physiological) and external (environmental) factors. Seed technologists, however, have restricted seed dormancy to internal conditions that *prevent germination after the seed is subjected to favorable environmental conditions* (*5*). If germination occurs immediately upon exposure to favorable conditions, the seed is said to be **quiescent.**

A seed is unique in being a combination of tissues of two plant generations, the embryo (and endosperm) being the new generation, whereas the enclosing tissues are from the parental generation. Seed dormancy is more than a cessation of growth but may include the transition from an embryonic phase of development to a juvenile seedling phase. Consequently, the point of reference in applying dormancy terminology to seeds is the triggering of germination in the entire embryo, including both root and shoot elongation.

A system of dormancy terminology has been proposed (*83*) that could be applied to any plant structure, including seeds. Table 6–1 lists the kinds of seed dormancy described in this chapter and relates them to this terminology.

Ecodormancy. Dormancy due to one or more unsuitable factors of the environment which are *nonspecific* in their effect. In seeds this term is equivalent to "quiescence" (*5, 19, 26, 67, 123*).

Paradormancy. Dormancy due to physical factors or biochemical signals originating *external* to the affected structure for the initial reaction, as in apical dominance or bud scale effects. In the seed, the control would come from any of the enclosing structures surrounding the embryo, not restricted to biochemical signals. This category could be identified by prompt germination and normal seedling growth following excision of the embryo.

Endodormancy. Dormancy regulated by physiological factors *inside* the affected structure. (Rest period in buds would be an example.) In seeds, this type is present if embryo excision fails to produce either prompt germination or normal seedling growth.

TABLE 6-1

Kinds of seed dormancy and examples, as described in this chapter, with corresponding dormancy categories

Kind of Seed Dormancy	Description	General terminology (83)		
		Eco-dormancy	Para-dormancy	Endo-dormancy
I. Nondormancy				
None	Quiescence	×		
II. Primary Dormancy				
Physical	Seed coat		×	
Mechanical	Seed coat		×	
Inhibitor	Seed coat		×	
Morphological	Undeveloped embryo		×	×
Physiological	Active membranes		×	
Thermodormancy	Active membranes	×	×	
Photodormancy	Active membranes	×	×	
Intermediate	Combination		×	
Embryo	Internal		×	×
Epicotyl	Internal		×	×
Double	Combinations		×	×
III. Secondary Dormancy				
Various		×	×	×

have learned to manipulate different kinds of seed dormancy through the adoption of appropriate pregermination and handling procedures discovered by trial and error (see Chap. 7).

Much scientific thought has gone into defining a uniform terminology for different kinds of seed dormancy. An historically early system for seeds was formulated by Crocker in 1916 (*35, 36*) who described seven kinds based primarily on treatments to overcome them. Subsequently, Nikolaeva (*97*) has defined a system based predominantly upon physiological controls of dormancy. Atwater (*6*) has shown that morphological characteristics, including both seed morphology and types of seed covering, characteristic of taxonomic plant families could be associated with dormancy categories particularly significant in seed testing.

Seed Coat Dormancy

Physical dormancy (seed coat dormancy). Dormancy is produced by seed coverings that are impervious to water. This type can preserve the dry seed for many years, even at warm temperature. Germination can be induced by any

method that can soften or scarify the covering (Fig. 6–5).

Physical dormancy is a genetic characteristic of certain plant families, including *Leguminoseae, Malvaceae, Cannaceae, Geraniaceae, Chenopodiaceae, Convolvulaceae,* and *Solanaceae.* Among cultivated crops, hardseededness is found chiefly in the herbaceous legumes, including clover and alfalfa, as well as many woody legumes (*Robinia, Acacia, Sophora,* etc.). Hardseededness is also increased by environmental (dry) conditions during seed maturation, and environmental conditions during seed storage. Drying at high temperatures during ripening will increase hardseededness. Harvesting slightly immature seeds and preventing them from drying can reduce or overcome hardseededness in some cases.

Impermeability of the seed coat is due to a layer of palisade-like **macrosclereid** cells, especially thick-walled on their outer surfaces and coated with a layer of waxy, cuticular substances (*103*) (see Fig. 6–5). Disintegration of the "caps" of such cells, or mechanical stress separating the cells, may allow water to enter and produce germination (*25, 87*). In some legume species, the

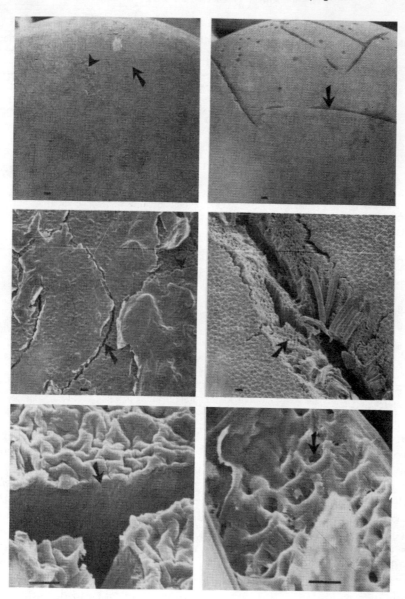

FIGURE 6–5 Scanning electron micrographs of Kentucky coffee-tree (*Gymnocladus dioicus* L.) seed coverings after different treatments. *Left top, middle, bottom:* Untreated seed (Bar length = 15×, 300×, and 3000×, respectively. *Right top, middle, bottom:* 150 minutes of acid scarification at the same range of magnification, showing destruction of macrosclereid cells, exposing lumens.

point of attachment (**hilum**) of the seed acts as a one-way valve during ripening by opening to allow water to escape in a dry atmosphere but closing in a moist atmosphere to prevent water uptake (*66*). In *Albizia lophantha* (*40*) a small opening (**strophiole**) near the hilum is sealed with a cork-like plug, which can be dislodged with vigorous shaking or impaction (*57*) or by exposure to dry heat as in a fire (*40*).

In nature, impervious seed coats are softened by action of microorganisms in the soil during warm periods of the season or by passage through the digestive tracts of birds and mammals (*36*). They may be broken through mechanical abrasion, alternate freezing and thawing, and in some species by fire. In cultivation any method to break, soften, abrade, or remove the seed coverings is immediately effective (Chap. 7).

Mechanical dormancy (hard seed coats). Some seed enclosing structures, such as shells of walnut (*36*), pits of stone fruits (*97*), and stones of olive (*34*), are too strong to allow embryo expansion during germination. Water may be ab-

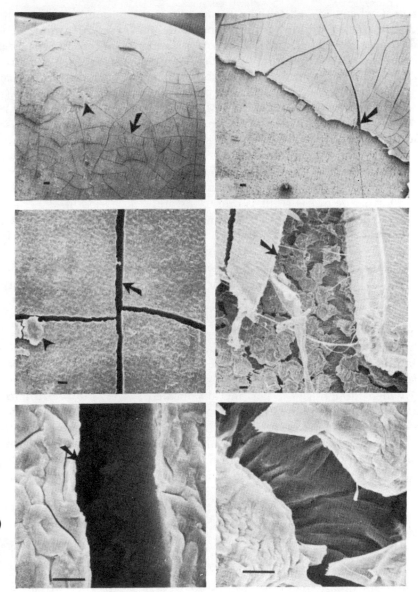

FIGURE 6-5 *(cont.) Left top, middle, bottom:* Liquid N_2 scarification shows cracked surfaces plus remnants of waxy surface. *Right top, middle, bottom:* Immersion in boiling water for 1 minute shows destruction of seed coats and separation of macrosclereid cells. Bar length (top left and right) = 100 μm; (middle left and right) = 10 μm; and (bottom left and right) = 5 μm. (Courtesy of Liu, Khatamian, and Fretz (*87*).)

sorbed but the difficulty arises in the cementing material that holds the dehiscent layers together, as shown in walnut. Softening primarily comes from soil microorganisms which are favored by nonsterile media and warm temperatures (*36*).

Chemical Dormancy (or Inhibitor Dormancy)

Chemicals that accumulate in fruit and seed covering tissues during development and remain with the seed after harvest can be shown to act as

germination inhibitors (*49*). Proving their function as germination controls does not necessarily follow, however. Nevertheless, germination can sometimes be improved by prolonged leaching with water, removing the seed coverings, or both (*45, 97*). Some examples are the following:

1. Fleshy fruits, or juices from them, can strongly inhibit seed germination. This occurs in citrus, cucurbits, stone fruits, apples, pears, grapes, and tomatoes. Likewise, dry fruits and fruit coverings, such as the hulls of guayule, *Penni-*

setum ciliare, wheat, as well as the capsules of mustard (*Brassica*), can inhibit germination. Some of the substances associated with inhibition are various phenols, coumarin, and abscisic acid.

2. Specific seed germination inhibitors play a role in the ecology of certain desert plants (*78, 128, 129*). Inhibitors are leached out of the seeds by heavy soaking rains which would in turn provide sufficient soil moisture to ensure survival of the seedlings. Since a light rain shower is insufficient to cause leaching, such inhibiting substances have been referred to as "chemical rain gauges."

3. Dormancy in iris seeds is due to a water- and ether-soluble germination inhibitor in the endosperm, which can be leached from the seeds with water or avoided by embryo excision (*4*).

4. Inhibitors are described as causing the rudimentary and undeveloped embryo, described in morphological dormancy. Their effect must be counteracted before embryo development proceeds (*6, 97*).

Inhibitors have been found in the seeds of such families as Polygonaceae, Chenopodiaceae (*Atriplex*), Portulaceae (*Portulaca*), and other species in which the embryo is peripherally located (see beet seed, Fig. 3–8). Likewise, seeds of a group of such families as Crucifereae (mustard), Linaceae (flax), Violaceae (violet), and Labiteae (*Lavendula*) have a thin seed coat with a mucilaginous inner layer that contains inhibitors (*6*).

Morphological (Rudimentary and Undeveloped) Dormancy

Dormancy occurs in some seeds in which the embryo is not fully developed at the time of seed dissemination. Enlargement of the embryo occurs after the seeds have imbibed water and before germination begins. The process of embryo enlargement is favored by a period of warm temperatures.

1. Atwater (*6*) has distinguished between two groups of embryos that are found in herbaceous flower crops. Seeds of some species have **rudimentary** embryos with little more than a proembryo embedded in a massive endosperm. These are found in various families, such as Ranunculaceae (anemone, ranunculus), Papaveraceae (poppy, *Romneya*), and Araliaceae (ginseng, fatsia). Germination-inhibiting chemicals also occur in the endosperm and become active at high temperatures. Effective aids for inducing germination include (a) exposure to temperatures of 15°C (59°F) or below, (b) exposure to alternating temperatures, and (c) treatment with chemical additives such as potassium nitrate or gibberellic acid.

A second category includes seeds with **undeveloped** embryos which are torpedo shaped and up to one-half the size of the seed cavity. Important families and species in this category include *Umbellifereae* (carrot), *Ericaceae* (rhododendron, heather), *Primulaceae* (cyclamen, primula), and *Gentianaceae* (gentian). Other conditions, such as semipermeability of the inner seed coats and internal germination inhibitors, may be involved. A temperature of about 20°C (68°F) favors germination, as does gibberellic acid treatment.

2. Certain temperate zone woody plants such as holly (*Ilex*) and snowberry (*Symphoricarpus*) have rudimentary embryos but, in addition, have other types of dormancy, such as hard seed coats and dormant embryos, which must be overcome (*97*) for germination to occur.

In seeds of some species, subsequent chilling is also required for germination after the warm embryo development period. Various temperate zone trees fall into this category, including *Fraxinus* (ash) and *Euonymous* species (*97*).

3. Various tropical species, many of which are monocots, have seeds with undeveloped embryos that require an extended period at high temperatures for germination to take place. For example, seeds of various palm species ordinarily require storage for several years to germinate, but this period can be reduced to three months by holding the seeds at temperatures of 38 to 40°C (100 to 104°F), or to 24 hours by excising the embryos and germinating them aseptically. Gibberellic acid (1000 ppm) has accelerated germination in palm seed, but a seed coat treatment is needed to assure penetration of the chemical (*96*). Other examples include *Actinidia*, whose seed requires two months' warmth, and *Annona squamosa* seed, which requires three months' warmth (*97*).

4. Orchids have both rudimentary embryos and undeveloped seeds. They are not considered dormant in the same sense as others in this cate-

gory and are prepared for germination by special aseptic methods as discussed in Chap. 17.

Physiological Dormancy

This term refers to a general type of primary dormancy that exists in many, if not most, freshly harvested seeds of herbaceous plants. This type of dormancy is often transitory and tends to disappear during dry storage (5, 19, 92, 97, 117), so that it generally is gone before germination. Consequently, it is primarily a problem with seed-testing laboratories that need immediate germination. In seed-testing laboratories such seeds respond to various short-term treatments, including short periods of chilling, alternating temperatures, and treatment with potassium nitrate and/ or gibberellic acid (see p. 148).

For most cultivated cereals, grasses, vegetables, and flower crops, physiological dormancy may last for one to six months and disappears with dry storage during normal handling procedures. For many noncultivated plants physiological dormancy may not only last longer but can develop into secondary dormancy, particularly if the moist seeds are buried in the ground.

Physiologically dormant seeds tend to have more specific environmental requirements for germination than when they become nondormant. Freshly harvested seeds of cocklebur or amaranth, for example, germinate only at high temperatures, about 30°C (86°F). In freshly harvested seeds of other species, such as some cultivars of lettuce and celery, germination is inhibited at temperatures above 25°C (77°F). Such heat sensitivity is called **thermodormancy** (see p. 127). [This type has also been called **relative dormancy** by other authors (19).]

Seeds of many species that have temperature sensitivity also have light sensitivity (**photodormancy**). Seeds of some plants, including lettuce and many flower crops, require light to germinate, whereas others require darkness (see p. 128).

Control of physiological dormancy appears to reside in the semipermeability of the physiologically active seed coverings (inner seed coat, remnants of nucellus and/or endosperm, perisperm, or endosperm) which surround the embryo (50).

These layers allow water uptake but apparently control other aspects which are not well understood.

A great deal of research has been done to determine the mechanism for the control of dormancy and germination in these seeds. Early research with cocklebur showed that the semipermeable membranes restricted gaseous exchange, limiting oxygen uptake and preventing carbon dioxide escape (50). Later experiments showed that the membranes also restricted the leaching of germination inhibitors from within the embryo which kept the embryo dormant (127). In lettuce, experiments have shown that the two-cell-layered endosperm is sufficiently resistant to prevent penetration by the dormant embryo (19). Current concepts are that germination is controlled through the interchange of (a) endogenous inhibitors and hormones, including gibberellins, cytokinin, and abscisic acid (see p. 119) and (b) specific environmental requirements, such as temperature (see p. 127) and light (see p. 129).

Intermediate Dormancy

Intermediate dormancy is a term used primarily with various conifer species whose seeds respond to chilling (see below) but do not have an absolute requirement (97, 111). Chilling greatly increases the rate, but seeds will eventually germinate. These seeds have a large storage tissue surrounding the embryo (see Fig. 3–8). The control of dormancy appears to be in the enclosing seed coverings (80, 97) since the embryo itself is capable of prompt germination if excised from the seed.

Physiologically Deep Dormancy (Embryo Dormancy)

Embryo dormancy involves controls within the embryo itself. It is characterized by a requirement for a period of one to three months' chilling while in an imbibed and aerated state. Dormant embryos are most common in seeds of trees and shrubs and some herbaceous plants of the temperate zone (36). Here seeds ripen in the fall, overwinter in the moist leaf litter on the ground, and germinate in the spring. Nursery propagators have known since early times that such seeds required **moist-chilling** (*32, 84, 97, 123*). This requirement led to the horticultural practice of **stratification,** in which seeds are placed between layers of moist sand or soil in boxes (or in the ground) and exposed to chilling temperatures, either out-of-doors or in refrigerators (see Chap. 7).

Since germination in seeds with embryo dormancy is controlled by both the seed covers and endogenous conditions within the embryo, it represents a combination of both para- and endodormancy (*82*). Endodormancy appears to be biologically equivalent to the "rest period" of buds of temperate-zone plants. Evidence for seed cover control is that removal of the seed coats and other coverings can shorten the stratification requirement and sometimes induce immediate germination (Fig. 6–5). Evidence for a dormant embryo is that the excised embryo usually will not germinate normally and the seedling produced may be abnormal. The relative response is the basis for the "excised embryo" viability test (*36, 97*) (see p. 140). Responses include enlargement and greening of the cotyledons; short thickened radicle growth, no epicotyl development, or lack of normal root systems. Typically, unchilled excised embryos develop into **physiological dwarfs** (*36, 52*) (Fig. 6–6).

Biological changes that take place progressively within the seed during moist-chilling are included in the general term of **after-ripening.**

FIGURE 6–6 Physiological dwarfing of epicotyls of almond seedlings. Left shows normal development.

These require moisture, aeration, chilling temperature, and time. Many studies have been made of the gross changes occurring during the moist-chilling treatment without a clear understanding of the fundamental control process emerging.

Moisture. The dry dormant seed absorbs moisture by imbibition to around 50 percent (*84, 97*). In some seeds, a hard bony endocarp enclosing the seed reduces water uptake, provides inhibitors, and delays the initiation of after-ripening (*45*). Sometimes mechanically removing the covers, subjecting the seeds to warm moist nonsterile conditions prior to germination, leaching, and early harvest without drying prior to stratification, reduces stratification time. The seed moisture content should remain relatively constant during stratification. Dehydration stops the after-ripening process (*59*) and the seeds may revert to secondary dormancy (*124*). When the end of the chilling period is reached, the seeds absorb water rapidly (*84*), the seed coats "crack," and the radicle eventually emerges, sometimes even at low temperatures (see Fig. 6–13A). Drying at this stage can cause injury to the seed.

Aeration. The amount of oxygen needed is related to temperature (*33*). At high temperatures the moist seed coverings of dormant, imbibed seeds restrict oxygen uptake because of (a) low oxygen solubility in water, and (b) oxygen fixation by phenolic substances in the seed coats. At chilling temperatures, however, the embryo's oxygen requirement is low and oxygen is generally adequate. If the temperature is increased, the oxygen requirement of the embryo increases, the solubility of oxygen is less, and the amount fixed by the phenols increases.

Temperature. Temperature is the single most important factor controlling after-ripening of seeds with dormant embryos. The most effective temperature regimes for moist-chilling are similar to those during the winter and early spring of the natural environment of the species. Temperatures somewhat above freezing [2 to 7°C (35 to 45°F)] are generally most effective (i.e., shows fastest rate of after-ripening) with a slower rate at higher and lower temperatures with a minimum of −5°C (23°F) (*112*). Higher temperatures can

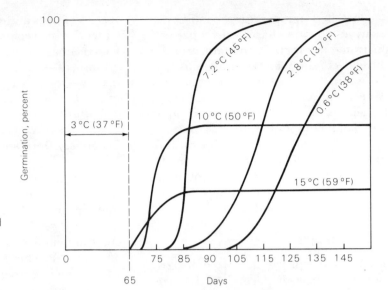

FIGURE 6–7 Effect of temperature during germination (radicle emergence) on both the rate and percentage of germination in apple seeds previously stratified at 3°C for 65 days. (Data from deHaas and Schander (*39*).)

slow the after-ripening of the most dormant seeds but can speed up the sprouting of those approaching germination (Fig. 6–7). Above a particular maximum temperature, known as the **compensation temperature,** secondary dormancy can develop (*1, 110, 111, 113, 124*). For apple, this point has been determined to be 17°C (62°F) (*1*), but it apparently varies with individual species (*115*) and different stages of after-ripening (*116*).

Toward the end of the after-ripening period, the maximum temperature for germination gradually increases and the minimum temperature gradually decreases. This period has been called *postdormancy* (*93*).

Physiological dwarfing in excised nonafter-ripened embryos has been shown to result from exposure of the apical meristem to warm germination temperatures (*99*). In peaches, temperatures of 23 to 27°C (73 to 81°F) and higher produced symptoms of physiological dwarfing, but at lower temperatures the seedlings grew normally. In almonds, exposing incompletely stratified seed to high temperatures subsequently induced physiological dwarfing in the seedling.

Pinching out the apex can circumvent dwarfing by forcing lateral growth from non-dwarfed lower nodes. Dwarfing has also been offset by exposing seedlings to long photoperiods or continuous light (*52, 81*), provided that this action is taken before the apical meristem becomes fully dormant. Repeated application of gibberellic

acid has also overcome dwarfing (*14, 15, 52*). Some experiments have shown that systematic removal of the cotyledons from the dormant embryo can induce germination and overcome physiological dwarfing, suggesting the existence of endogenous inhibitors present within the cotyledons (*19*).

Time. The time required to after-ripen seeds with dormant embryos results from the interaction of (a) the genetic characteristics of the seed population (*70, 71, 112, 131*), (b) the conditions during seed development sometimes (*123*), (c) the environment of the seedbed, and (d) the management of seed handling. There is a correlation between the seed chilling requirements and the bud chilling requirements of the plants from which the seeds were taken (*101*). In studies with almond, a high quantitative correlation was shown between the mean of the seedling populations and the mean of both the seed and pollen parents (*70*) but a low correlation between the individual seed and the buds of the new plant coming from the embryo (*71*). This difference suggests that the dormancy involves both a genetic component within the embryo and a maternal component from the seed parent. As a result a great deal of variability in individual seed germination time can occur within a given seed lot and between different seed lots of the same species collected in different years and different locations.

Under controlled temperature conditions, germination curves show a characteristic pattern (*39, 110*), illustrated in Fig. 6–7. In temperate climates under field conditions, the most effective regimes are a warm fall, cool wet winter, and a cool spring where soil temperatures gradually increase as the season advances. The concept of postdormancy is an adaptation to this pattern (*93*). If premature periods of high temperatures occur in nature, secondary dormancy can prevent germination during a dry and possibly hot summer period. In nurseries these conditions can delay germination until the second spring.

Epicotyl Dormancy

Some seeds have separate after-ripening requirements for the radicle, hypocotyl, and epicotyl (*16, 36, 97*). These species fall into two subgroups.

1. Seeds that initially germinate during a warm period of one to three months to produce root and hypocotyl growth but then require one to three months' chilling to enable the epicotyl to grow. This group includes various lily (*Lilium*) species, *Viburnum* spp., peony (*Paeonia*), black cohosh (*Cimifusa racemosa*), and *Hepatica acutiloba*.

2. Seeds that require a chilling period to after-ripen the embryo, followed by a warm period for the root to grow, then a second cold period to stimulate shoot growth. In nature, such seeds require two full growing seasons to complete germination. Examples include *Trillium* and certain other native perennials of the temperate zone (see Chap. 19).

Double Dormancy

Double dormancy combines two (or more) kinds of dormancy, such as seed coat dormancy and a dormant embryo (some tree legumes), or a rudimentary embryo and a dormant embryo (*Ilex*). To produce germination all blocking conditions must be eliminated in proper sequence. Seed coats must be modified to allow water to penetrate to the embryo; after-ripening of the embryo can then take place. Warm followed by cold stratification generally overcomes these situations.

This type of dormancy is characteristic of species of trees and shrubs in families having seeds with hard seed coats but whose plants grow in cold winter areas. In nature, various agents of the environment—those that affect physical dormancy—soften the seed coat when the seed falls to the ground, then the seeds are chilled as they overwinter.

Secondary Dormancy

In nature, primary dormancy is an adaptation to control the time and conditions for seed germination. Secondary dormancy is a further adaptation to prevent germination of an imbibed seed if other environmental conditions are not favorable (*19, 35, 68, 75*). These conditions can include unfavorably high temperatures, temperatures too low, prolonged darkness (**skotodormancy**), prolonged white light (**photodormancy**), prolonged far-red light, water stress, and anoxia. These conditions are particularly involved in the seasonal rhythms and prolonged survival of weed seeds in soil (*19*).

Induction of secondary dormancy is illustrated by experiments with freshly-harvested seeds of lettuce (*75*). If germinated at 25°C (77°F) the seeds require light, but if imbibed with water for two days in the dark, excised embryos germinate immediately, illustrating that only primary dormancy was present. If imbibition continues for as long as eight days, however, excised embryos will not germinate since they have then developed secondary dormancy. Release from secondary dormancy can be induced by chilling, sometimes by light, and in various cases, treatment with germination-stimulating hormones, particularly gibberellic acid.

Secondary dormancy can come into play in some instances in cultivated crops but prolonged dry storage may prevent its occurrence. Seeds with a dormant embryo undergoing moist-chilling may be affected if shifted to high temperatures too quickly (see p. 145). The term could apply to hardseededness that could develop in storage in seeds of some species, such as beans and other legumes.

Control of Dormancy and Germination

Much experimental evidence supports the concept that specific endogenous growth promoting and inhibiting compounds are involved directly in control of seed development, dormancy, and germination (*19, 21, 73, 84, 101*). Evidence for hormone involvement comes from correlations of hormone concentration with specific developmental stages, effects of applied hormones, and the relationship of hormones to metabolic activities.

Specific Germination Hormones

Gibberellins.   Gibberellins (GA) comprise the class of hormones most directly implicated in the control and promotion of seed germination (see Fig. 6-8). While there are many molecular variations of gibberellin, the one most widely used experimentally and commercially is gibberellic acid (GA_3), but GA_{4+7} is also active. These compounds occur at relatively high concentrations in developing seeds but usually drop to a lower level in mature dormant seeds, particularly in dicotyledonous plants. Applied gibberellins can relieve certain types of dormancy, including physiological dormancy, photodormancy, and thermodormancy.

Gibberellins appear to play a role in two different stages of germination. One occurs at the initial enzyme induction in their transcription from the chromosomes. The second is at Stage III in the activation of reserve food mobilizing systems. In barley, for instance, gibberellins appear in the embryo with imbibition, are translocated to the three- to four-cell-layered aleurone surrounding the endosperm and induce *de nova* alpha-amylase enzyme synthesis. The alpha-amylase enzyme then moves to the endosperm. Starch is converted to sugar, which is then translocated to the growing points to provide energy for seedling development.

Abscisic acid (ABA).   This naturally occurring compound is an important growth-regulating compound not only in seed germination but in plant growth in general (*125, 126*). ABA appears to play a role in preventing "precocious germination" of the developing embryo in the ovule (*51, 69*). High inhibitor levels have been considered responsible for the lack of development in rudimentary embryos (*6*) (see p. 114).

ABA tends to increase with maturation of the fruit and may prevent vivipary and induce primary dormancy. It has been isolated from the seed coats of dormant peach, walnut, apple, rose, and plum, but it decreases during stratification (Fig. 6-13).

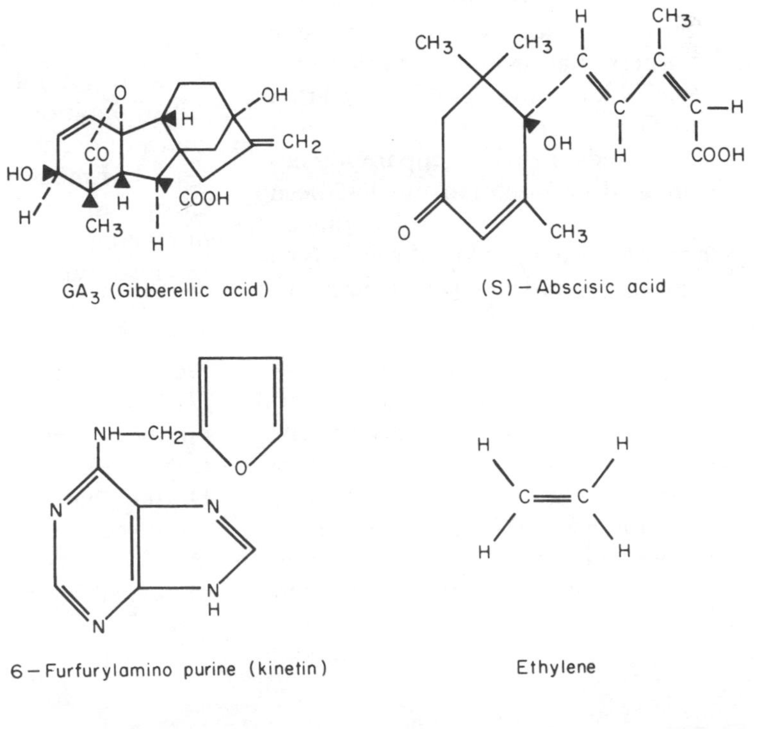

GA₃ (Gibberellic acid)

(S) — Abscisic acid

6 — Furfurylamino purine (kinetin)

Ethylene

FIGURE 6-8 Chemical structure of plant growth regulators involved in germination.

Application of ABA can inhibit germination of nondormant seed and offset the effects of applied gibberellic acid. In general, inhibition is temporary and disappears when seeds are shifted to an ABA-free solution.

Cytokinins. These naturally occurring compounds (*119*) have the basic chemical structure of an N^6-substituted adenine (see Fig. 6–8). Synthetic cytokinins available for experimental use include benzyladenine, kinetin, and others. In addition, cytokinin activity is shown by such compounds as thiourea and diphenylurea (*119*) (see also p. 148).

Cytokinin activity tends to be high in developing fruits and seeds, but decreases and becomes difficult to detect as the seeds mature. In seed germination, cytokinin is believed to offset the effect of inhibitors, notably ABA. It has been described, therefore, as playing a "permissive" role in germination in allowing gibberellic acid to function (*73*). It is believed to be active, therefore, at a different germination stage than gibberellins.

Ethylene. Ethylene gas (*72*) is an important, naturally occurring hormone involved in many aspects of plant growth. Response to ethylene treatment of dormant seeds of snowberry (*Symphoricarpos*), honeysuckle (*Lonicera*), and similar species, as well as seeds of corn and other cereals, was demonstrated many years ago. Ethylene production from germinating bean and pea seed was shown in 1935. Later work demonstrated that ethylene is a natural germination-promoting agent for certain kinds of seeds. Ethylene apparently has a limited role in seed germination but has been shown to stimulate germination in subterranean clover (*Trifolium subterranean*) (*48*), Virginia-type peanut (*Arachis hypogaea*), and witchweed (*Striga asiatica*) (*47*).

Other compounds. Certain other compounds are known to stimulate seed germination, but their role is not clear. Use of **potassium nitrate** has been an important seed treatment in seed-testing laboratories for many years without a good explanation for its action. **Thiourea** overcomes certain types of dormancy, such as the seed coat inhibiting effect of deep embryo-dormant *Prunus* seeds as well as the high-temperature inhibition of lettuce seeds (*119*). The effect of thiourea

may be due to its cytokinin activity in overcoming inhibition. Two other naturally occurring substances, *fusicoccin* and *cotylenin* (*74*), have been reported to mimic the combination of GA plus cytokinin.

Interactions of Hormones and Dormancy

Dormancy may involve interactions among specific hormones, as described in the following experimental examples.

1. **Rudimentary embryos** (see p. 114) (Fig. 3–8A) (*60, 61*). Imbibed seeds of *Trollius ledebouri* have inhibitors in the testa which prevent expansion of the embryo (*61*) (Fig. 6–9). Germination begins with rupture of the seed coverings, enlargement of the embryo, radicle growth, and digestion of proteins in the endosperm. This inhibitory effect can be removed by leaching and/or testa removal or overcome by application of gibberellin.

2. **Physiological dormancy.** Figures 6–10 and 6–11 show the interaction of exogenous hormones with light (photodormancy) (*73, 76*) and temperature (thermodormancy) (*114*) with 'Grand Rapids' lettuce seed. GA promotes germination, ABA inhibits it, and cytokinin counteracts the effect of the ABA. The relationship is an example of a "permissive" effect of cytokinin to allow GA stimulation of germination by offsetting ABA inhibition. Figure 6–12 illustrates dormancy control in cocklebur (*76*).

3. **Embryo dormancy.** The hormonal changes that occur during the two to three months of chilling of seeds with dormant embryos have received much attention.

In the peach (*Prunus persica*), two separate aspects of germination are affected: (a) initiation of radicle elongation, and (b) epicotyl elongation (*134, 135*). Figure 6–13 combines data from a number of experiments to show a three-phase pattern of seed response to chilling at 5°C (41°F). During phase 1 (0 to 30 days) no germination response occurs in intact seeds during chilling, after transfer to 25°C (77°F) at weekly intervals, or if presoaked with GA. In phase 2 (30 to 45 days) seeds being chilled do not germinate but seeds transferred to 25°C (77°F) show increasing ger-

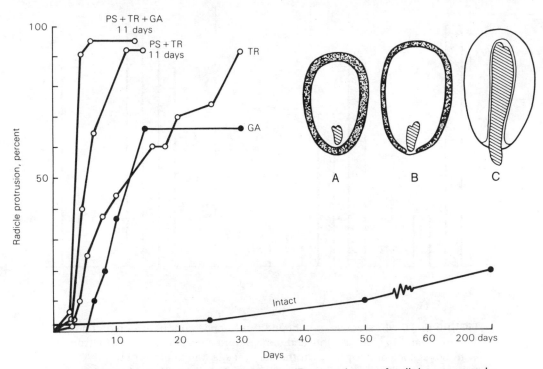

FIGURE 6-9 Germination in *Trollius* (Ranunculaceae family), an example of a seed with a rudimentary embryo. Germination in a dormant seed (a) is preceded by dissolution of the enclosing seed covering (b) and elongation of the embryo (c). These changes take place slowly (60 days) during germination on filter paper in a germinator. If the seeds are treated with GA, germination time in the germinator was shortened but not all seeds germinated. If the testae were removed (TR), germination was complete within 30 days. However, if the seeds were presoaked for 11 days and testae removed, germination was complete within 10 days. If the seeds with the last treatment were also treated with GA, germination was complete within five days. The results are interpreted that endogenous inhibitors were present in the testa and embryo to delay germination.

mination response, and GA presoaking about doubles the rate of germination. In phase 3 (45 to 75 days) the seeds begin to germinate at chilling temperatures at a rapid rate (*53*) and seedlings show normal growth.

Removal of the testa at any time results in immediate germination, but if the testa is removed prior to 10 weeks (70 days) of chilling and placed at 25°C (77°F), dwarfing results (*53*). Some research associates dwarfing with high germination temperature (*99*); others dispute this relationship (*134*).

Hormone concentrations are correlated to the responses described. Freshly harvested peach seed (*42, 86*), as well as other species, including walnut (*89*), plum (*85*), apple (*11*), and hazelnut (*133*), have a high concentration of ABA in both testae and a lesser amount in the cotyledons. ABA concentration drops to near zero during phase 1. Treatment of excised embryos with ABA prevents germination. Physical contact of the testa with the embryo axis has been sufficient to inhibit germination and/or to produce dwarfing (*60, 85*). Experimental treatment of intact peach seeds with a

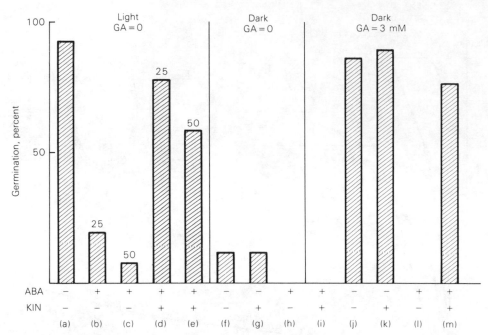

FIGURE 6-10   Interaction of light and three applied hormones on germination of 'Grand Rapids', a light- and temperature-sensitive cultivar of lettuce with physiological dormancy. In the light, seeds germinate (a) but ABA inhibits germination in proportion to concentration, i.e., 25 mM (b) and 50 mM (c). Kinetin overcomes the ABA inhibition (d, e). Germination is inhibited in the dark (f) and kinetin does not overcome the dark inhibition (g). ABA completely inhibits germination in the light (h, i). GA^3 3 mM overcomes dormancy, with or without kinetin (k). ABA overcomes the GA stimulation (l), but kinetin counteracts ABA (m), allowing GA to act. (Redrawn from Khan et al. (*77*).)

cytokinin benzylaminopurine (BAP) has overcome the inhibiting effect of the testa and allowed germination to occur (*108*).

The concentration of gibberellin-like compounds is low in phase 1 in intact seeds held at chilling temperatures (first 30 days) but shows a sharp increase in phase 2, indicating that the ability to synthesize gibberellins is either present (*53*) or there is a change from an inactive form to a free form [as shown in apple (*20*)]. Introduction of an inhibitor of gibberellin synthesis (paclobutrazol) at the beginning of phase 2 decreased gibberellin synthesis sharply but only slightly decreased the germination percentage. Epicotyl and seedling elongation was strongly inhibited, indicating a separation between the radical and epicotyl response during chilling.

These results support the concept that inhibitors (undoubtedly ABA) are present in the testa as well as the cotyledons in the dormant seed. These disappear during the early stages of dormancy (or are neutralized by cytokinins). Gibberellins are either synthesized at the chilling temperatures or are converted to an available (or unbound form), allowing radicle emergence (germination) to take place at warmer temperatures (*27, 91*). Epicotyl elongation is a more localized phenomenon which either has a higher threshold for gibberellin or involves a different control system.

Research with filbert (*Corylus avellana*) seeds has illustrated separation of growth-inhibiting and growth-promoting hormonal systems in control of germination. At the time of ripening, the intact

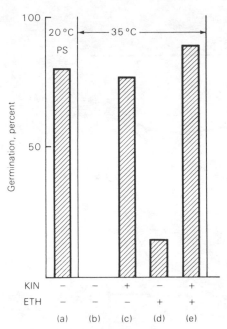

FIGURE 6–11 Hormonal effects on high-temperature dormancy 'Grand Rapids' lettuce seed. Seed pre-soaked for 24 hours at 20°C overcame thermodormancy to allow germination to proceed during subsequent exposure to 35°C (a). Nonpresoaked seed remained dormant at high temperature (b), but kinetin removed most of the dormancy (c). Ethylene had limited but significant effect without (d) or with kinetin (e). (Redrawn from Sharples (*114*).

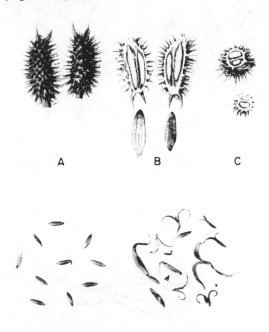

−Kinetin   + Kinetin

FIGURE 6–12 Germination of cocklebur (*Xanthium*). Two burs are shown at A, each of which contains two seeds (B and C); the smaller one is dormant but the other is not. Early experiments showed that low permeability to gases was a dormancy inducing factor. Later experiments (*127*) showed that the smaller, dormant, seed contains two water-insoluble inhibitors that prevent germination. If these are removed by leaching, or if the seed is subjected to high oxygen pressure, then germination will occur. The two seeds also differ in seed coat strength and the germinating forces required to rupture them. Treatment with kinetin or ethylene (*74*) will stimulate germination of both seeds, while abscisic acid will inhibit it. (From Khan et al. (*76*).)

seed is dormant but the embryo is quiescent. A significant amount of abscisic acid can be detected in the seed covering (*133*) as well as a detectable amount of gibberellin in the embryo (*106*). When the seed is dried following harvest, the embryo becomes dormant, and gibberellin levels decrease significantly (*105*). Stratification for several months is required for germination. The gibberellin level remains low during this chilling period but increases after the seeds are placed at warm temperatures when germination begins (Fig. 6–14, *right*).

Gibberellic acid applied to the dormant seed (*23*) can replace the chilling requirement (Fig. 6–14, *left*). ABA applied with gibberellin offsets the effect of GA and prevents germination (*107*).

ENVIRONMENTAL FACTORS AFFECTING SEED GERMINATION

Water

Water content is a major controlling factor programming the embryonic phase to the juvenile seedling phase, controlling seed longevity (Chap. 5), initiating germination, and ensuring the survival and health of the seedling. During storage it

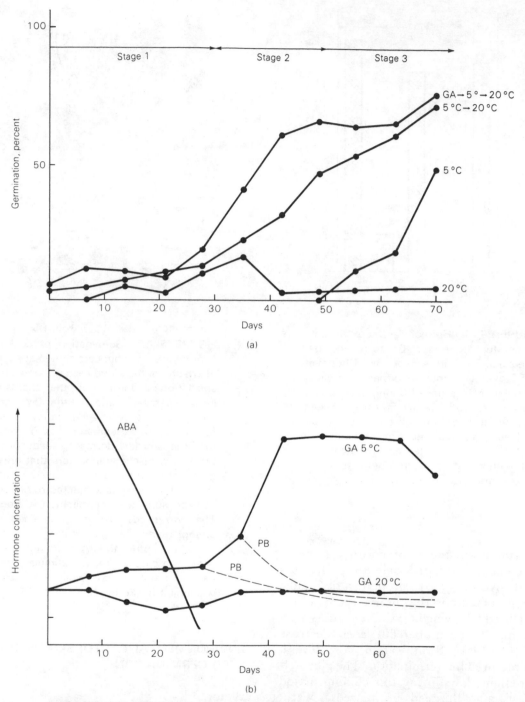

FIGURE 6–13 (A) Germination trends during stratification of peach (*53*). Germination was low in imbibed seed held continuously at warm temperatures [i.e., 20°C (77°F)], but germination began to occur rapidly after 55 days in seed held continously at cool temperatures. When stratified seed was shifted to warm termination temperatures, sprouting began to occur at about 30 days with increasing amounts with increasing stratification time.

(*cont.*)

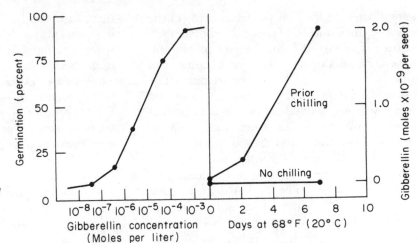

FIGURE 6–14 Interactions of gibberellin, stratification, and germination in filbert seeds. (Reproduced by permission from A. W. Galston and P. S. Davies, *Control mechanisms in plant development*, Prentice-Hall, Englewood Cliffs, N.J., 1970.)

is convenient to consider water on the fresh weight basis (water content/fresh weight). During germination it may be convenient to use the dry weight basis (water content/dry weight of seed).

Water Potential

The absorptive power of the dry seed (*19*) during stage 1 is measured by its **water potential,** which in a dry seed can be as low as -100 MPa. *Negative* potential comes from the **matric potential** (i.e., colloidal structure of the embryo, storage areas, and seed coverings) and the **osmotic potential** (solute concentration) of the living cells of the seeds. Opposed to these are the **pressure (turgor) potential** of the cell wall, which has a *positive* potential. Water moves into the seed following a gradient from high (outside the seed) to low potential (inside the seed).

> Water movement has been measured in units of barometric pressure called a bar, but later usage is for the term "megapascal (MPa)," where -1 MPa $= -10$ bar. Pure water equals 0, whereas water with solutes present has a negative potential. Water moves from high potential (least negative) to low potential (most negative).

The rate of water movement into the seed is also dependent on similar properties of the germination medium. The matric potential measures the ability of the water to move by capillarity through the pores of the soil or other germination medium to the seed. Rate of movement depends upon (a) the pore structure (texture) of the germi-

An initial GA soak increased the germination rate. (Paclobutrazol, an inhibitor of gibberellin synthesis, applied to seeds at 28 days decreased final germination percentage by 36, 25, and 27 if shifted to the warm temperature after 7, 14, and 28 days, respectively, of additional chilling. Elongation growth of the seedling was severely curtailed by paclobutrazol.) (B) Correlations to hormone concentrations in seeds. Many separate studies show that ABA is initially high in the dormant seeds, but the levels drop sharply within the first 30 days. Data from Lin and Boe (*85*), Lipe and Crane (*86*), and others is superimposed on the graph. In the peach study (*53*), GA concentration was low during the first 30 days but increased dramatically after 30 to 45 days of stratification. Paclobutrazol sharply reduced GA concentration when applied at 28 and 35 days. (Redrawn from Gianfagna and Rachmiel (*53*).)

nation medium, (b) the soil packing, and (c) the closeness and distribution of the soil–seed contact. As moisture is removed from the soil by the imbibing seed the area nearest the seed becomes dry and must be replenished by water in pores farther away. Consequently, a firm, fine-textured seedbed closely compacted to the seed is important in maintaining a uniform moisture supply.

Osmotic potential in the soil solution depends upon the presence of solutes (salts). Excess soluble salts (high salinity) in the germination medium may exert strong negative pressure (exosmosis) and counterbalance the osmotic pressure in the seeds. Salts may also produce specific toxic effects. These may inhibit germination and reduce seedling stands (7). Such salts originate in the soil and other materials used in the germination medium, the irrigation water, or excessive fertilization. Since the effects of salinity become more acute when the moisture supply is low and the concentration of salts thereby increased, it is particularly important to maintain a high moisture supply in the seedbed, where the possibility of high salinity exists. Surface evaporation from subirrigated beds can result in the accumulation of salts at the soil surface even under conditions in which salinity would not be expected. Planting seeds several inches below the top edge of a sloping seedbed can minimize this hazard (18).

Water stress can reduce the germination percentage (44, 58). Most kinds of seed germinate equally well over the range of available soil moisture, from field capacity (FC) to permanent wilting percentage (PWP) (7). Germination of some seeds, particularly those with dormancy problems (e.g., beet, lettuce, endive, or celery), is inhibited as moisture levels are decreased. Such seeds apparently contain inhibitors and require leaching. Seeds of other species (e.g., spinach), when exposed to excess water, produce extensive mucilage that restricts oxygen supply to the embryo (64); an inhibitor is also present (6). In these cases germination improves with less moisture.

Moisture stress strongly reduces the rate of seedling emergence from a seedbed. This decline in emergence rate occurs as the available moisture decreases to a level approximately halfway through the range from FC to PWP (Fig. 6–15) (7, 44, 58). Once the seed germinates and the radicle emerges, the water supply to the seedling depends on the ability of the root system to grow into the germination medium and the ability of the new roots to absorb water.

Priming and Seed Soaking

This process refers to procedures designed to initiate germination before planting in order to shorten the time of emergence, improve the uniformity of stand, and circumvent some adverse conditions in the seedbed.

Soaking. Seeds are sometimes soaked in water before planting to speed up germination. Seeds of most herbaceous species could benefit

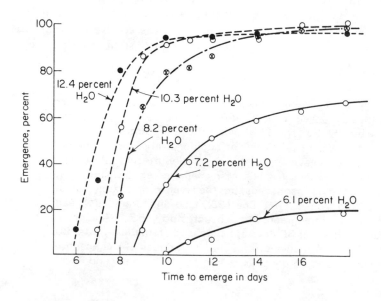

FIGURE 6–15 Effect of different amounts of available soil moisture on the germination (emergence) of 'Sweet Spanish' onion seed in Pachappa fine sandy loam. (From Ayres (7).)

from eight hours of soaking but may be injured by soaking periods of 24 hours or more. Large seeds are particularly vulnerable to prolonged soaking (*98*). Excess water may be trapped between cotyledons and suffocate the embryo (*64*). Harmful results have been attributed to the effects of microorganisms and to a reduced oxygen supply (*12, 13*). If soaking is to be prolonged, the water should be changed at least once every 24 hours. Dormant seeds of woody plants can be leached for longer periods without injury (*98*).

Fluid drilling (55, 109). In this process seeds are pregerminated under controlled conditions. They are then planted with special machines to protect the seed from drying and injury. These seeds may be mixed in special gels (see p. 149).

Osmotic priming (osmoconditioning). Osmotic priming uses a pretreatment with an external high osmotic solution to decrease the osmotic potential within the seed. This allows metabolic activities leading to germination but prevents or delays the emergence of the radicle (*22, 28, 64, 65, 74, 77*). An inert compound, polyethylene glycol (PEG 6000) is usually used, but some systems use salt solutions of NaCl or KNO_3, although these may be toxic to some kinds of seeds (*3, 24*).

Temperature

Temperature is, perhaps, the most important environmental factor that regulates the timing of germination, partly due to dormancy control and/or release and partly due to climate adaptation. Temperature control is also essential in subsequent seedling growth. Dry, unimbibed seeds can withstand extremes of temperature. For disease control, seeds can be placed in boiling water for short periods without killing them. In nature, brush fires are often effective in overcoming dormancy without damaging the seeds.

Effects on Germination

Temperature affects both germination percentage and germination rate (*79*). Germination rate is invariably low at low temperature but increases gradually as temperature rises, similar to a chemical rate-reaction curve (*78*). Above an optimum level, where the rate is most rapid, a decline occurs as the temperatures approach a lethal limit where the seed is injured.

Germination percentage, unlike the germination rate, may remain relatively constant, at least over the middle part of the temperature range, if sufficient time is allowed for germination to occur.

Three temperature points (minimum, optimum, and maximum), varying with the species, are usually designated for seed germination (*46*). **Minimum** is the lowest temperature for effective germination. **Maximum** is the highest temperature at which germination occurs. Above this the seed either is injured or goes dormant. **Optimum** temperatures for seed germination fall within the range at which the largest percentage of seedlings are produced at the highest rate. The optimum for nondormant seeds of most plants is between 25 and 30°C (77 and 86°F).

Seeds of different species, whether cultivated or native, can be categorized into temperature-requirement groups. These are related to their climatic origin:

Cool-temperature tolerant. Seeds of many kinds of plants, mostly native to temperate zones, will germinate over a wide temperature range from about 4.5°C (40°F) (or sometimes near freezing) up to the lethal limit—from 30°C (86°F) to about 40°C (104°F). The optimum germination temperature is usually about 24 to 30°C (77 to 86°F). Examples include broccoli, carrot, cabbage, alyssum, and others.

Cool-temperature requiring. Seeds of some cool-season species and cultivars adapted to a Mediterranean climate require low temperatures and fail to germinate at temperatures higher than about 25°C (77°F). Species of this group tend to be winter annuals in which germination is prevented in the hot summer, but takes place in the cool fall when winter rains commence. This category involves thermodormancy (also called *relative dormancy*) (*19*) (see p. 115), which is present in many species with physiologically dormant seeds. It tends to disappear with after-ripening at dry storage. Examples include various vegetables, such as celery, lettuce, and onion, as well as some flower seed—coleus, cyclamen, freesia, primula, delphinium, and others (6).

Warm-temperature requiring. Seeds of another broad group fail to germinate below about 10°C (50°F) (asparagus, sweet corn, and tomato) or 15°C (60°F) beans, eggplant, pepper, and cucurbits). These species originated primarily in subtropical or tropical regions. Other species, such as lima bean, cotton, soybean, and sorghum, are also susceptible to "chilling injury" when exposed to temperatures of 10 to 15°C (50 to 60°F) during initial imbibition. Planting in a cold soil can injure the embryo axis and result in abnormal seedlings (62, 100).

Alternating temperatures. Fluctuating day-night temperatures give better results than constant temperatures for both seed germination and seedling growth. Use of fluctuating temperatures is a standard practice in seed-testing laboratories, even for seeds not requiring it. The alternation should be a 10°C (18°F) difference (122). This requirement is particularly important with dormant, freshly harvested seeds (5). Seeds of a few species will not germinate at all at constant temperatures. It has been suggested that one of the reasons imbibed seeds deep in the soil do not germinate is that soil temperature fluctuations disappear with increasing soil depth (102).

Effects on Seedling Growth

The optimum temperature may shift after germination begins since seedling growth tends to have different temperature requirements than seed germination. In the nursery or laboratory the usual practice is to shift the seedlings to a somewhat lower temperature regime following germination in order to prepare the plants for transplanting and to reduce disease problems in the seed bed.

If seeds are germinated and the seedlings grown at high temperatures, it is important that other environmental conditions be favorable. Plants should have increased light, preferably long photoperiods, adequate fertilization, and

sterile conditions to eliminate disease pathogens. Increased carbon dioxide is also a useful component of this system.

Aeration

Exchange of gases between the germination medium and the embryo is essential for rapid and uniform germination. Oxygen (O_2) is essential for the respiratory processes in the germinating seed. Oxygen uptake can be measured shortly after imbibition of water begins. Rate of oxygen uptake is an indicator of germination progress and has been suggested as a measure of seed vigor (79). In general, O_2 uptake is proportional to the amount of metabolic activity taking place.

Oxygen supply is limited where there is excessive water in the soil medium. Poorly drained outdoor seed beds, particularly after heavy rains or irrigation, can have the pore spaces of the soil so filled with water that little oxygen is available to the seeds. The amount of oxygen in the germination medium is affected by its low solubility in water and its slow diffusability into the medium. Thus, gaseous exchange between the germination medium and the atmosphere, where the O_2 concentration is 20 percent, is reduced significantly by soil depth and, in particular, by a hard crust on the surface, which can limit oxygen diffusion (58).

Carbon dioxide (CO_2) is a product of respiration and under conditions of poor aeration can accumulate in the soil. At lower soil depths increased CO_2 may inhibit germination to some extent but probably plays a minor role, if any, in maintaining dormancy. In fact, high levels of CO_2 can be effective in overcoming dormancy in some seeds (78).

Seeds of different species vary in their ability to germinate at very low oxygen levels, as occurs under water (94, 95). Seeds of some water plants germinate readily under water, with germination inhibited in air. Rice seeds can germinate in a shallow layer of water. At low oxygen levels, rice seedlings, however, develop differently than those of other monocots. Shoot development is stimulated and the plumule grows to extend up through the water into the air; root growth is suppressed and poor anchorage results unless the water layer is drained away (*30*).

Light

Light has been recognized since the mid-nineteenth century as a germination-controlling factor (*37*). Recent research demonstrates that light acts in both dormancy induction and release and is a mechanism that adapts plants to specific niches in the environment often interacting with temperature. Light can involve both **quality** (wavelength) and **photoperiod** (duration).

Light is recognized to be a factor in the following situations:

1. Certain epiphytic plants, such as mistletoe (*Viscum album*) and strangling fig (*Ficus aurea*), have an absolute requirement for light and lose viability in a few weeks without it.

2. Most of the light-sensitive species fall into the category of physiological dormancy (including most grasses, various herbaceous vegetable and flower species, and various weed and native species). Light-sensitive seeds are characterized by being small in size, in which a shallow depth of planting would be an important factor favoring survival. Otherwise, if covered too deeply, the epicotyl may not penetrate the soil. Some important flower crops include alyssum, begonia, calceolaria, coleus, *Kalanchoe,* primrose, and *Saintpáulia* (*10*).

3. Many conifer seeds with intermediate dormancy have light sensitivity (see Table 7–1).

4. Germination is inhibited by light in a small number of species, such as *Phacelia, Nigelia, Allium, Amaranthus,* and *Phlox.* Some of these are desert plants where survival would be enhanced if the seeds were located at greater depths, where adequate moisture might be assured. Some flower crops are listed as dark requiring, including calendula, delphinium, pansy, annual phlox, and annual verbena (*10*).

5. Photoperiodism affects seeds of some woody plants as eastern hemlock (*Tsuga canadensis*) (*115*) and birch (*19*).

Light Quality

The basic mechanism of light sensitivity in seeds involves a photochemically reactive pigment called **phytochrome,** widely present in plants (*19, 117*). Exposure of the imbibed seed to **red light** (660 to 760 nm) causes the phytochrome to change to phytochrome$_{fr}$ (or $\mathbf{P}_{fr}$), which stimulates germination. Exposure of the seed to **far-red** light (760 to 800 nm) causes a change to the alternate form ($\mathbf{P}_r$), which inhibits germination. Both of these changes are instantaneous and can be repeated indefinitely, the last treatment being the one that is effective. In darkness a slow change to $\mathbf{P}_r$ occurs and prevents germination.

The membranes of the seed coats and/or the endosperm appear to act as the light sensors; if removed, the light control disappears. A light requirement can be offset by cool temperatures and sometimes by alternating temperatures. The control mechanism is complex and involves both (a) an initial physical reaction utilizing lower than physiological moisture levels, and (b) a physiological reaction requiring full imbibition and linkage to the hormonal regulation of the seed. Treatments with hormones can offset the light effect, as illustrated in Figs. 6–10 and 6–11.

Use of artificial lighting should take into account the potential effect of rays of particular wavelengths (*19*). White fluorescent lamps tend to be rich in the red rays and favorable to germination, whereas incandescent lamps tend to be rich in the infrared (far-red) rays and could result in dormancy.

In nature. The light quality reaching the seed can have an impact even during development. Green fruits transmit light rich in infrared rays, which can induce dormancy in the seeds as they mature (*19*). Experiments have shown that seeds of *Arapidopsis* are dormant if the plant is exposed to incandescent light before harvest and nondormant if fluorescent lights are used.

Furthermore, seeds of some plants (*Chenopodium album*) are dormant if plants are exposed to long days and nondormant if exposed to short days (*19*). In natural sunlight, red wavelengths dominate over far-red at a ratio of 2:1, so that the phytochrome tends to remain in the active $\mathbf{P}_{fr}$ form. Under a foliage canopy, far-red is dominant and the red/far-red ratio may be as low as 0.12:1.00 to 0.70:1.00, which can inhibit seed germination (*102*). Red light penetrates less deeply into the soil than far-red, so that the red/far-red ratio becomes lower with depth until eventually darkness is complete (*118*). Imbibed light-sensitive seeds buried in the soil will remain dormant until such time as the soil is cultivated or disturbed so as to expose them to light. Similarly, seedling survival is not favored if the seed germinates in close proximity to other plants, where there would be intense competition for light, nutrients, and water by the already established plant population.

In cultivation. Light sensitivity in cultivated cultivars is primarily a property of freshly harvested, physiologically dormant seeds and tends to disappear during after-ripening in dry storage as these seeds lose their primary dormancy. Consequently, light sensitivity is a problem encountered mainly in seed-testing laboratories for most cultivated crops. Light is a standard aspect of the test. Nevertheless, light exposure may be beneficial for some cultivars as lettuce as well as some flower seeds (*10*) and conifers (*111*).

Light sensitivity can be induced in secondary dormancy by exposing imbibed nonsensitive seeds to conditions inhibiting germination, such as high temperature, high osmotic pressure, or germination-inhibiting gases (*130*). Light requirements should be met in seed-priming procedures.

Light and Seedling Growth

Light of a relatively high intensity is desirable to produce sturdy, vigorous plants, particularly if transplanting is involved. Low light intensity results in etiolation and reduced photosynthesis and poor seedling survival if transplanted.

High light intensity, on the other hand, often results in high temperatures that produce heat injury to the seedling, particularly at the soil level, in a manner resembling "damping-off" fungi attacks. Shading is desirable for many kinds of plants during their early seedling growth out-of-doors to avoid heat injury. Use of supplementary artificial light is described in Chap. 2.

Disease Control during Seed Germination

The control of disease during seed germination is one of the most important tasks of the propagator. The most universally destructive pathogens are those resulting in "damping-off," which may cause serious loss of seeds, seedlings, and young plants (*9*). In addition, there are a number of fungus, virus, and bacterial diseases that are seed-borne and may infect certain plants (*8*). In such cases, specific methods of control are required during propagation. (See the discussion of sanitation in Chap. 2, p. 37.)

Damping-Off

Damping-off is a term long used to describe the death of small seedlings resulting from attacks by certain fungi, primarily *Pythium ultimum* and *Rhizoctonia solani,* although other fungi—for example, *Botrytis cinerea* and *Phytophthora* spp.—may also be involved. Mycelia from these organisms occur in soil, in infected plant tissues, or on seeds, from which they contaminate clean soil and infect clean plants. *Pythium* and *Phytophthora* produce spores that are moved about in water.

The environmental conditions prevailing during the germination period will affect the growth rate of both the attacking fungi and the seedling. For instance, the optimum temperature for the growth of *Pythium ultimum* and *Rhizoctonia solani* is between approximately 20 to 30°C (68 and 86°F), with a decrease in activity at both higher and lower temperatures. Seeds that have a high minimum temperature for germination (warm-season plants) are particularly susceptible to damping-off, because at lower or intermediate temperatures (less than 23°C or 75°F) their growth rate is low at a time when the activity of the fungi is high. At high temperatures, not only do the seeds germinate faster, but also the activity of the fungi is less. Field planting of such seeds should be delayed until the soil is warm. On the other hand, seeds of cool-season plants germinate (although slowly) at temperatures of less than 13°C (55°F), but since there is little or no activity of the fungi, they can escape the effects of damping-off. As the temperature increases, their susceptibility increases, because the activity of the fungi is relatively greater than that of the seedling.

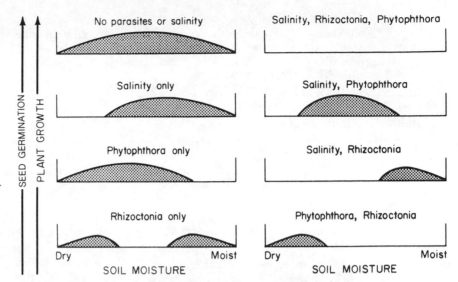

FIGURE 6-16 Interaction of pathogens and environment in affecting seed germination and seedling growth (*9*). (Redrawn from K. F. Baker, ed. 1957. The U. C. system for producing healthy container-grown plants. *Calif. Agr. Exp. Sta. Man. 23.*)

The control of damping-off involves two separate procedures: (a) the complete elimination of the pathogens during propagation, and (b) the control of plant growth and environmental conditions, which will minimize the effects of damping-off or give temporary control until the seedlings have passed their initial vulnerable stages of growth.

If damping-off begins after seedlings are growing, it may sometimes be controlled by treating that area of the medium with a fungicide. The ability to control attacks depends on their severity and on the modifying environmental conditions (see Chap. 2).

Symptoms resembling damping-off are also produced by certain unfavorable environmental conditions in the seedbed. Drying, high soil temperatures, or high concentrations of salts (see Fig. 6-16) in the upper layers of the germination medium can cause injuries to the tender stems of the seedlings near the ground level. The collapsed stem tissues have the appearance of being "burned off." These symptoms may be confused with those caused by pathogens. Damping-off fungi can grow in concentrations of soil solutes high enough to inhibit the growth of seedlings. Where salts accumulate in the germination medium, damping-off can thus be particularly serious.

REFERENCES

1. Abbott, D. L. 1955. Temperature and the dormancy of apple seeds. *Rpt. 14th Inter. Hort. Cong.* 1:746–53.

2. Abdul-Baki, A. A. 1980. Biochemical aspects of seed vigor. *HortScience* 15(6):765–71.

3. Akers, S. W., and K. E. Holley. 1986. SPS: A system for priming seeds using aerated polyethylene glycol or salt solutions. *HortScience* 21(3):529–31.

4. Arditti, J., and P. R. Pray. 1969. Dormancy factors in iris (Iridaceae) seeds. *Amer. Jour. Bot.* 56(3):254–59.

5. Assoc. Off. Seed Anal. 1975. Rules for seed testing. *Jour. Seed Tech.* 3(3):1–126.

6. Atwater, B. R. 1980. Germination, dormancy and morphology of the seeds of herbaceous ornamental plants. *Seed Sci. and Tech.* 8:523–73.

7. Ayers, A. D. 1952. Seed germination as affected by soil moisture and salinity. *Agron. Jour.* 44: 82–84.

8. Baker, K. F. 1972. Seed pathology. In *Seed biology,* Vol. 2, T. T. Kozlowski, ed. New York: Academic Press.

9. Baker, K. F., and P. A. Chandler. 1957. Development and maintenance of healthy planting stock. In *The U.C. system for producing healthy container-grown plants,* K. F. Baker, ed. Calif. Agr. Exp. Sta. Man. 23, pp. 217–36.

10. Ball, V., ed. 1985. *Ball red book: Greenhouse growing* (14th ed.). Reston, Va.: Reston Publ. Co.

11. Barthe, P., and C. Bulard. 1983. Anaerobiosis and release from dormancy in apple embryos. *Plant Physiol.* 72:1005–10.

12. Barton, L. V. 1950. Relation of different gases to the soaking injury of seeds. *Contrib. Boyce Thomp. Inst.* 16(2):55–71.

13. ———. 1952. Relation of different gases to the soaking injury of seeds, II. *Contrib. Boyce Thomp. Inst.* 17(1):7–34.

14. ———. 1956. Growth response of physiologic dwarfs of *Malus arnoldiana* Sarg. to gibberellic acid. *Contrib. Boyce Thomp. Inst.* 18:311–17.

15. Barton, L. V. and C. Chandler. 1957. Physiological and morphological effects of gibberellic acid on epicotyl dormancy of tree peony. *Contrib. Boyce Thomp. Inst.* 19:201–14.

16. Baskin, J. M., and C. C. Baskin. 1985. Epicotyl dormancy in seeds of *Cimicifuga racemosa* and *Hepatica acutiloba. Bull. Torrey Bot. Club* 112:253–57.

17. Berlyn, G. P. 1972. Seed germination and morphogenesis. In *Seed biology,* Vol. 3, T. T. Kozlowski, ed. New York: Academic Press.

18. Berstein, L., A. J. MacKenzie, and B. A. Krantz. 1955. The interaction of salinity and planting practice on the germination of irrigated row crops. *Proc. Soil Science Soc. Amer.* 19:240–43.

19. Bewley, J. D., and M. Black. 1985. *Seeds: Physiology of development and germination.* New York: Plenum Press.

20. Bianco, J., S. Lassechere, and C. Bulard. 1984. Gibberellins in dormant embryos of *Pyrus malus* L. cv Golden Delicious. *Plant Physiol.* 116:185–88.

21. Black, M. 1980/1981. The role of endogenous hormones in germination and dormancy. *Israel Jour. Bot.* 29:181–92.

22. Bodsworth, S., and J. D. Bewley. 1981. Osmotic priming of seeds of crop species with polyethylene glycol as a means of enhancing early and synchronous germination at cool temperatures. *Can. Jour. Bot.* 59:672–76.

23. Bradbeer, J.W., and N.J. Pinfield. 1967. Studies in seed dormancy. III. The effects of gibberellin on dormant seeds of *Corylus avellana* L. *New Phytol.* 66:515–23.

24. Bradford, K. J. 1986. Manipulation of seed water relations via osmotic priming to improve germination under stress conditions. *HortScience* 21(5):1105–12.

25. Brant, R. E., G. W. McKee, and R. W. Cleveland. 1971. Effect of chemical and physical treatment on hard seed of Penngift crown vetch. *Crop Sci.* 11:1–6.

26. Brown, R. Germination. 1972. In *Plant physiology,* Vol. 7C, F. C. Steward, ed. New York: Academic Press.

27. Bulard, C. 1985. Intervention by gibberellin and cytokinin in the release of apple embryos from dormancy: A reappraisal. *New Phytol.* 101:241–49.

28. Cantliffe, D. J., K. D. Shuler, and A. C. Guedes, 1981. Overcoming seed thermodormancy in a heat sensitive romaine lettuce by seed priming. *HortScience* 16(2):196–98.

29. Chaleff, R. S. 1981. *Genetics of higher plants: Applications of cell culture.* Cambridge: Cambridge Univ. Press.

30. Chapman, A. L., and M. L. Peterson. 1962. The seedling establishment of rice under water in relation to temperature and dissolved oxygen. *Crop. Sci.* 2:391–95.

31. Ching, Te May. 1972. Metabolism of germinating seeds. In *Seed biology,* Vol. 2, T. T. Kozlowski, ed. New York: Academic Press.

32. Comé, D. 1980/1981. Problems of embryonal dormancy as exemplifed by apple embryo. *Israel Jour. Bot.* 29:145–57.

33. Comé, D., and T. Tissaoui. 1973. Interrelated effects of imbibition, temperature, and oxygen on seed germination. In *Seed ecology,* W. Heydecker, ed. University Park, Pa.: Pennsylvania State Univ. Press.

34. Crisosto, C., and E. G. Sutter. 1985. Role of the endocarp in 'Manzanillo' olive seed germination. *Jour. Amer. Soc. Hort. Sci.* 110(1):50–52.

35. Crocker, W. 1916. Mechanics of dormancy in seeds. *Amer. Jour. Bot.* 3:99–120.

36. ———. 1948. *Growth of plants.* New York: Reinhold.

37. ———. 1930. Effect of the visible spectrum upon the germination of seeds and fruits. In *Biological effects of radiation.* New York: McGraw-Hill, pp. 791–828.

38. Czabator, F. 1962. Germination value: An index combining speed and completeness of pine seed germination. *For. Sci.* 8:386–96.

39. DeHaas, P. G., and H. Schander. 1952. Keimungsphysiologische Studien an Kernobst. I. Kamen and Keimung, *Z. f. Pflanz.* 31(4):457–512.

40. Dell, B. 1980. Structure and function of the strophiolar plug in seed of *Albizia lophantha*. *Amer. Jour. Bot.* 67(4):556–61.

41. Delouche, J. C. 1980. Environmental effects on seed development and seed quality. *HortScience* 15(6):13–18.

42. Diaz, D. H., and G. C. Martin. 1972. Peach seed dormancy in relation to endogenous inhibitors and applied growth substances. *Jour. Amer. Soc. Hort. Sci.* 97(5):651–54.

43. Dickson, M. H. 1980. Genetic aspects of seed quality. *HortScience* 15(6):771–74.

44. Doneen, L. D., and J. H. MacGillivray. 1943. Germination (emergence) of vegetable seed as affected by different soil conditions. *Plant Phys.* 18:524–29.

45. du Toit, H. J., G. Jacobs, and D. K. Strydom. 1979. Role of the various seed parts in peach seed dormancy and initial seedling growth. *Jour. Amer. Soc. Hort. Sci.* 104(4):490–92.

46. Edwards, T. J. 1932. Temperature relations of seed germination. *Quart. Rev. Biol.* 7:428–43.

47. Eplee, R. E. 1975. Ethylene, a witchweed seed germination stimulant. *Weed Sci* 23(5):433–36.

48. Esashi, Y., and A. C. Leopold. 1969. Dormancy regulation in subterranean clover seeds by ethylene. *Plant Phys.* 44:1470–72.

49. Evenari, M. 1949. Germination inhibitors. *Bot. Rev.* 15:153–94.

50. Evenari, M., and G. Newman. 1952. The germination of lettuce seed. II. The influence of fruit coats, seed coat and endosperm upon germination. *Bul. Res. Council, Israel* 2:75–78.

51. Finkelstein, R. R., and M. L. Crouch. 1987. Hormonal and osmotic effects on developmental potential of maturing rapeseed. *HortScience* 22(5):797–800.

52. Flemion, F., and E. Waterbury. 1945. Further studies with dwarf seedlings of non-after-ripened peach seeds. *Contrib. Boyce Thomp. Inst.* 13:415–422.

53. Gianfagna, T. J., and S. Rachmiel. 1986. Changes in gibberellin-like substances of peach seed during stratification. *Physiol. Plant.* 66:154–58.

54. Gordon, A. G. 1973. The rate of germination. In *Seed ecology*, W. Heydecker, ed. University Park, Pa.: Pennsylvania State Univ. Press, pp. 391–410.

55. Gray, D. 1981. Fluid drilling of vegetable seeds. *Hort. Rev.* 3:1–27.

56. Haber, A. H., and H. J. Luippold. 1960. Separation of mechanisms initiating cell division and cell expansion in lettuce seed germination. *Plant Phys.* 35:168–73.

57. Hamly, D. H. 1932. Softening the seeds of *Melilotus alba*. *Bot Gaz.* 93:345–75.

58. Hanks, R. S., and F. C. Thorp. 1956. Seedling emergence of wheat as related to soil moisture content, bulk density, oxygen diffusion rate and crust strength. *Proc. Soil Sci. Soc. Amer.* 20:307–10.

59. Haut, I. C. 1932. The influence of drying on after-ripening and germination of fruit tree seeds. *Proc. Amer. Soc. Hort. Sci.* 29:371–74.

60. Hepher, A., and J. A. Roberts. 1985. The control of seed germination in *Trollius ledebouri:* The breaking of dormancy. *Planta* 166:314–20.

61. Hepher, A., and J. A. Roberts. 1985. The control of seed germination in *Trollius ledebouri* model of seed dormancy. *Planta* 166:321–28.

62. Herner, R. C. 1986. Germination under cold soil conditions. *HortScience* 21(5):1118–22.

63. Heydeker, W. 1972. Vigour. In *Viability of seeds,* E. H. Roberts, ed. Syracuse, N.Y.: Syracuse Univ. Press.

64. ———. 1977. Stress and seed germination: An agronomic view. In *The physiology and biochemistry of seed dormancy and germination,* A. A. Khan, ed. Amsterdam: North-Holland Publishing Co.

65. Heydecker, W., and B. M. Gibbins, 1978. Attempts to synchronize seed germination. *Acta Hort.* 72:79–92.

66. Hyde, E. O. C. 1956. The function of some Papilionaceae in relation to the ripening of the seed and permeability of the testa. *Ann. Bot.* 18:241–56.

67. Jann, R. C., and R. D. Amen. 1977. What is germination? In *The physiology and biochemistry of seed dormancy and germination,* A. A. Khan, ed. Amsterdam: North-Holland Publishing Co., pp. 7–28.

68. Karssen, C. M. 1980/1981. Environmental

conditions and endogenous mechanisms involved in secondary dormancy of seeds. *Israel Jour. Bot.* 29:45–64.

69. Kermode, A. R., J. D. Bewley, J. Dasgupta, and S. Misra. 1986. The transition from seed development to germination: A key role for desiccation? *HortScience* 21(5):1113–18.

70. Kester, D. E. 1969. Pollen effects on chilling requirements of almond and almond hybrid seeds. *Jour. Amer. Soc. Hort. Sci.* 94:318–21.

71. Kester, D. E., P. Raddi, and R. Asay. 1977. Correlations of chilling requirements for germination, blooming and leafing within and among seedling populations of almond. *Jour. Amer. Soc. Hort. Sci.* 102(2):145–48.

72. Ketrick, D. L. 1977. Ethylene and seed germination. In *The physiology and biochemistry of seed dormancy and germination,* A. A. Khan, ed. Amsterdam: North-Holland Publishing Co., pp. 156–78.

73. Khan, A. A. 1971. Cytokinins: Permissive role in seed germination. *Science* 171:853–59.

74. ———. 1977. Preconditioning, germination and performance of seeds. In *The physiology and biochemistry of seed dormancy and germination,* A. A. Khan, ed. Amsterdam: North-Holland Publishing Co.

75. ———. 1980/1981. Hormonal regulation of primary and secondary dormancy. *Israel Jour. Bot.* 29:207–24.

76. Khan, A. A., C. E. Heit, E. C. Waters, C. C. Anojulu, and L. Anderson. 1971. Discovery of a new role for cytokinins in seed dormancy and germination. In *Search.* New York Agr. Exp. Sta. (Geneva) 1(9):1–12.

77. Khan, A. A., N. H. Peck, and C. Samiry. 1980/1981. Seed osmoconditioning: Physiological and biochemical changes. *Israel Jour. Bot.* 29:133–44.

78. Koller, D. 1972. Environmental control of seed germination. In *Seed biology,* Vol. 2, T. T. Kozlowski, ed. New York: Academic Press.

79. Kotowski, F. 1926. Temperature relations to germination of vegetable seeds. *Proc. Amer. Soc. Hort. Sci.* 23:176–84.

80. Kozlowski, T. T., and A. C. Gentile. 1959. Influence of the seed coat on germination, water absorption and oxygen uptake of eastern white pine seed. *For. Sci.* 5:389–95.

81. Lammerts, W. E. 1943. Effect of photoperiod and temperatures on growth of embryo-cultured peach seedlings. *Amer. Jour. Bot.* 30:707–11.

82. Lang, G. A. 1987. Dormancy: A new universal terminology. *HortScience* 22(5):817–20.

83. Lang, G. A., J. D. Early, G. C. Martin, and R. L. Darnell. 1987. Endo-, para-, and ecodormancy: Physiological terminology and classification for dormancy research. *HortScience* 22(3):371–77.

84. Lewak, S. 1985. Hormones in seed dormancy and germination. In *Hormonal regulation of plant growth and development,* S. S. Purohit, ed. Dordrecht: Martinus Nishoff, pp. 95–144.

85. Lin, C. F., and A. A. Boe. 1972. Effects of some endogenous and exogenous growth regulators on plum seed dormancy. *Jour. Amer. Soc. Hort. Sci.* 97:41–44.

86. Lipe, W., and J. C. Crane. 1966. Dormancy regulation in peach seeds. *Science* 153:541–42.

87. Liu, N. Y., H. Khatamian, and T. A. Fretz. 1981. Seed coat structure of three woody legume species after chemical and physical treatments to increase seed germination. *Jour. Amer. Soc. Hort. Sci.* 106(5):691–94.

88. Martin, G. C. 1987. Apical dominance. *HortScience* 22(5):824–33.

89. Martin, G. C., H. Forde, and M. Mason. 1969. Changes in endogenous growth substances in the embryo of *Juglans regia* during stratification. *Jour. Amer. Soc. Hort. Sci.* 94:13–17.

90. Mathews, S., and A. A. Powell. 1986. Environmental and physiological constraints on field performance of seeds. *HortScience* 21(5):1125–28.

91. Mathur, D. D., G. A. Couvillon, H. M. Vines, and C. H. Hendershott. 1971. Stratification effects of endogenous gibberellic acid (GA) in peach seeds. *HortScience* 6:538–39.

92. McDonald, M. B., Sr. 1980. Assessment of seed quality. *HortScience* 15:784–88.

93. Meyer, M. M., Jr. 1987. Rest, postdormancy, and woody plant seed germination. *Proc. Inter. Plant Prop. Soc.* 37:330–35.

94. Morinaga, T. 1926. Germination of seeds under water. *Amer. Jour. Bot.* 13:126–31.

95. ———. 1926. The favorable effect of reduced oxygen supply upon the germination of certain seeds. *Amer. Jour. Bot.* 13:150–65.

96. Nagao, M. A., K. Kanegawa, and W. S. Sakai. 1980. Accelerating palm seed germination with gibberellic acid, scarification, and bottom heat. *HortScience* 15(2):200–201.

97. Nikolaeva, M. G. 1977. Factors affecting the seed dormancy pattern. In *The physiology and biochemistry of seed dormancy and germination,* A. A. Khan, ed. Amsterdam: North-Holland Publishing Co., pp. 51–76.

98. Norton, C. R. 1986. Germination under flooding: Metabolic implications and alleviation of injury. *HortScience* 21(5):1123–25.

99. Pollock, B. M. 1962. Temperature control of physiological dwarfing in peach seedlings. *Plant Phys.* 37:190–97.

100. Pollock, B. M., and E. E. Roos. 1972. Seed and seedling vigor. In *Seed biology,* Vol. 3, T. T. Kozlowski, ed. New York: Academic Press.

101. Powell, L. E. 1987. Hormonal aspects of bud and seed dormancy temperature-zone woody plants. *HortScience* 22(5):845–50.

102. Roberts, E. H. 1972. Dormancy: A factor affecting seed survival in the soil. In *Viability of seeds,* E. H. Roberts, ed. Syracuse, N.Y.: Syracuse Univ. Press.

103. Rolston, M. P. 1978. Water impermeable seed dormancy. *Bot. Rev.* 44:365–96.

104. Roos, E. E. 1980. Physiological, biochemical and genetic changes in seed quality during storage. *HortScience* 15:781–84.

105. Ross, J. D., and J. W. Bradbeer. 1968. Concentrations of gibberellin in chilled hazel seeds. *Nature* 220:85–86.

106. ———. 1971. Studies in seed dormancy. V. The concentrations of endogenous gibberellins in seeds of *Corylus avellana* L. *Planta* (Berl.) 100:288–302.

107. ———. 1971. Studies in seed dormancy. VI. The effects of growth retardants on the gibberellin content and germination of chilled seeds of *Corylus avellana* L. *Planta* (Berl.) 100:303–8.

108. Rouskas, D., J. Hugard, R. Jonard, and P. Villemur. 1980. Physiologie végétale. Contribution à l'étude de la germination des graines de Pecher (*Prunus persica* Batsch.) cultivar INRA GF 305: effets de la benzyl-amino-purine (BAP) et les gibberellines GA$_3$ et GA$_{4+7}$ sur la levée de dormance embryonnaire et l'absence des anomalies foliaires observées sur les plantes issues de graines non stratifiées. *C. R. Acad. Sci. Paris Ser D* 291:861–64.

109. Salter, P. J. 1978. Techniques and prospects for 'fluid' drilling of vegetable crops. *Acta Hort.* 72:101–8.

110. Schander, H. 1955. Keimungsphysiologische Studien an Kernobst. III. Sortenvergleichende Untersuchungen über die Temperature-ansprüche stratifizierten Saatgutes von Kernobst und über die Reversibilitat der Stratifikationsvorgange, *Z. f. Pflanz.* 35:89–97.

111. Schopmeyer, C. S., ed. 1974. *Seeds of woody plants in the United States.* U.S. Dept. Agr. Handbook 450. Washington, D.C.: U.S. Govt. Printing Office.

112. Seeley, S. D., and H. Damavandy. 1985. Response of seed of seven deciduous fruits to stratification temperatures and implications for modeling. *Jour. Amer. Soc. Hort. Sci.* 110(5):726–29.

113. Semeniuk, P., and R. N. Stewart. 1962. Temperature reversal of after-ripening of rose seeds. *Proc. Amer. Soc. Hort. Sci.* 80:615–21.

114. Sharples, G. C. 1973. Stimulation of lettuce seed germination at high temperatures by ethephon and kinetin. *Jour. Amer. Soc. Hort. Sci.* 98(2):207–9.

115. Stearns, F., and J. Olson. 1958. Interactions of photoperiod and temperature affecting seed germination in *Tsuga canadensis. Amer. Jour. Bot.* 45:53–58.

116. Stewart, R. N., and P. Semeniuk. 1965. The effect of the interaction of temperature with after-ripening requirement and compensating temperature on germination of seed of five species of *Rosa. Amer. Jour. Bot.* 52:755–60.

117. Taylorson, R. B., and S. B. Hendricks. 1977. Dormancy in seeds. *Ann. Rev. Plant Phys.* 28:331–54.

118. Thomas, H. 1972. Control mechanisms in the resting seed. In *Viability of seeds,* E. H. Roberts, ed. Syracuse, N.Y.: Syracuse Univ. Press.

119. Thomas, T. H. 1977. Cytokinins, cytokinin-active compounds and seed germination. In *The physiology and biochemistry of seed dormancy and germination,* A. A. Khan, ed. Amsterdam: North-Holland Publishing Co., pp. 111–24.

120. Thompson, P. A. 1973. Geographical adaptation of seeds. In *Seed ecology,* W. Heydecker, ed. University Park, Pa.: Pennsylvania State Univ. Press.

121. Toole, E. H., V. K. Toole, and E. A. Gorman. 1948. Vegetable seed storage as affected by temperature and relative humidity. *USDA Tech. Bul. 972.*

122. USDA. 1952. *Manual for testing agricultural and vegetable seeds.* U.S. Dept. Agr. Handbook 30. Washington, D.C.: U.S. Govt. Printing Office.

123. Villiers, T. A. 1972. Seed dormancy. In *Seed bi-*

ology, Vol. 2, T. T. Kozlowski, ed. New York: Academic Press, pp. 220–82.

124. Visser, T. 1956. Some observations on respiration and secondary dormancy in apple seeds. *Proc. Koninkl. Akad. van Wetens,* Series C 59:314–24.

125. Walton, D. C. 1980. Biochemistry and physiology of abscisic acid. *Ann. Rev. Plant Phys.* 31:453–89.

126. ———. 1980/1981. Does ABA play a role in seed germination? *Israel Jour. Bot.* 29:168–80.

127. Wareing, P. F., and H. A. Foda. 1957. Growth inhibitors and dormancy in *Xanthium* seed. *Phys. Plant.* 10(2):266–80.

128. Went, F. W. 1949. Ecology of desert plants. II. The effect of rain and temperature on germination and growth. *Ecology* 30:1–13.

129. Went, F. W., and M. Westergaard. 1949. Ecology of desert plants. III. Development of plants in the Death Valley National Monument, California. *Ecology* 30:26–38.

130. Wesson, G., and P. F. Wareing. 1969. The induction of light sensitivity in weed seeds by burial. *Jour. Exp. Bot.* 20(63):414–25.

131. Westwood, M. N., and H. O. Bjornstad. 1948. Chilling requirement of dormant seeds of fourteen pear species as related to their climatic adaptation. *Proc. Amer. Soc. Hort. Sci.* 92:141–49.

132. Willemsen, R. W. 1975. Effect of stratification temperature and germination temperature on germination and the induction of secondary dormancy in common ragweed seeds. *Amer. Jour. Bot.* 62(1):1–5.

133. Williams, P. M., J. D. Ross, and J. W. Bradbeer. 1973. Studies in seed dormancy. VII. The abscisic acid content of the seeds and fruits of *Corylus avellana* L. *Planta (Berl.)* 110:303–10.

134. Zigas, R. P., and B. G. Coombe. 1977. Seedling development in peach, *Prunus persica* (L.) Batsch. I. Effects of testas and temperature. *Aust. Jour. Plant Physiol.* 4:349–58.

135. Zigas, R. P., and B. G. Coombe. 1977. Seedling development in peach, *Prunus persica* (L.) Batsch. II. Effects of plant growth regulators and their possible role. *Aust. Jour. Plant Physiol.* 4:359–69.

SUPPLEMENTARY READING

AMEN, R. D. 1968. A model of seed dormancy. *Bot. Rev.* 34:1–31.

BEWLEY, J. D., and M. BLACK. 1985. *Seeds: Physiology of development and germination.* New York: Plenum Press.

CROCKER, W., and L. V. BARTON. 1953. *Physiology of seeds.* Waltham, Mass.: Chronica Botanica.

HEYDECKER, W., ed. 1973. *Seed ecology: Proc. 19th Easter School in Agricultural Science, University of Nottingham, 1972.* University Park, Pa.: The Pennsylvania State Univ. Press.

KHAN, A. A., ed. 1977. *The physiology and biochemistry of seed dormancy and germination.* Amsterdam: North-Holland Publishing Co.

KOZLOWSKI, T. T., ed. 1972. *Seed biology,* Vols. 1, 2, 3. New York: Academic Press.

MAYER, A. M. 1980/1981. Control mechanisms in seed germination. *Israel Jour. Bot.* 29:1–4.

MAYER, A. M., and A. POLJAKOFF-MAYBER. 1975. *The germination of seeds* (2nd ed.). New York: Macmillan.

Proceedings of symposium. 1986. Seed germination under environmental stress. *HortScience* 21(5):1103–28.

Proceedings of symposium. 1987. Mechanisms of rest and dormancy. *HortScience* 22(5):815–50.

Proceedings of symposium. 1987. Synthetic seed technology for the mass cloning of crop plants: Problems and perspectives. *HortScience* 22(5):795–814.

ROBERTS, E. H., ed. 1972. *Viability of seeds.* Syracuse, N.Y.: Syracuse Univ. Press.

RUBENSTEIN, I., R. L. PHILLIPS, G. E. GREEN, and B. G. GENGENBACH. 1979. *The plant seed: Development, preservation and germination.* New York: Academic Press.

SCHOPMEYER, C. S., ed. 1974. *Seeds of woody plants in the United States.* U.S. Dept. Agr. Handbook 450. Washington, D.C.: U.S. Govt. Printing Office.

TAYLORSON, R. B., and S. B. HENDRICKS. 1977. Dormancy in seeds. *Ann. Rev. Plant Phys.* 28:331–54.

7
Techniques of Propagation by Seed

Seed propagation involves careful management of germination conditions and facilities and a knowledge of the requirements of individual kinds of seeds. Success depends upon fulfilling the following conditions:

1. *Using seed of proper genetic characteristics to produce the cultivar, species, or provenance desired.* This can be accomplished by obtaining seed from a reliable dealer, buying certified seed, or—if producing one's own—following the principles of seed selection described in Chap. 4.

2. *Using good-quality seed.* It should germinate rapidly and vigorously to withstand possible adverse conditions in the seedbed.

3. *Manipulating seed dormancy.* Accomplish by applying pregermination treatments or proper timing of planting. In the absence of specific knowledge of seed requirements, the propagator should try to duplicate the natural environmental conditions associated with germination of seed of this particular kind of plant.

4. *Supplying proper environment to the seeds and resulting seedlings.* This includes supplying sufficient water, proper temperature, adequate oxygen, and either light or darkness (depending upon the kind of seed) to the seeds and resulting seedlings until they are well established. A proper environment also includes control of diseases and insects and prevention of excess salinity.

SEED TESTING

Good-quality seed has the following characteristics: It is genetically true to species, cultivar, or provenance; capable of high germination; free from disease and insects; and free from mixture with other crop seeds, weed seeds, and inert and extraneous material. The germination capability and purity of the seed can be determined by conducting a seed test on a small representative sample drawn from the seed lot in question (*5, 45, 81*).

In the United States, state laws regulate the shipment and sale of agricultural and vegetable seeds within that state. Seeds entering interstate commerce or those sent from abroad are subject to the Federal Seed Act adopted in 1939 (*4*). Such regulations require labeling by the shipper of commercially produced seeds as to: name and cultivar; origin; germination percentage; and the percentage of pure seed, other crop seed, weed seed, and inert material. The regulations set minimum standards of quality, germination percentage, and freedom from weed seeds. Shipment and sale of tree seed is regulated by law in some states (*69*) and in most European countries.

Seed testing provides information to meet legal standards, determines seed quality, and establishes the rate of sowing for a given stand of seedlings. It is desirable to retest seeds that have been in storage for a prolonged period.

Procedures for testing agriculture and vegetable seed in reference to the Federal Seed Act are given by the U.S. Department of Agriculture (*4, 81*). The Association of Official Seed Analysts also publishes procedures for testing these seeds in addition to procedures for testing seeds of many flower, tree, and shrub species (*5*). International rules for testing seeds of many tree, shrub, agricultural, and vegetable species are published by the International Seed Testing Association (*44*). The Western Forest Tree Seed Council has also published testing procedures for tree seed (*84*).

Sampling

The first step in seed testing is to obtain a uniform sample representing the entire lot under consideration. Equally sized **primary** samples are taken from evenly distributed parts of the seed lot, such as a sample from each of several sacks in lots of less than five sacks or from every fifth sack with larger lots. The seed samples are thoroughly mixed to make a **composite** sample. A representative portion is used as a **submitted sample** for testing. This sample is further divided into smaller lots to produce a **working sample** (i.e., the sample upon which the test is actually to be run). The amount of seed required for the work-

ing sample varies with the kind of seed and is specified in the Rules for Seed Testing.

Purity Determination

Purity is the percentage by weight of the "pure seeds" present in the sample. **Pure seed** refers to the principally named kind, cultivar, or type of seed present in the seed lot. After the working sample has been weighed, it is divided visually into (a) the pure seed of the kind under consideration; (b) other crop seed; (c) weed seed; and (d) inert material, including seed-like structures, empty or broken seeds, chaff, soil, stones, and other debris (see Fig. 7–1). In some cases, it is possible to check the genuineness of the seed or trueness to cultivar or species by visual inspection. Often, however, identification cannot be made except by growing the seeds and observing the plants. At the time of making the purity test, the number of pure seeds per pound can be calculated. These data are necessary as a guide to seeding rates.

Moisture Determination

Moisture content is found by the loss of weight when a sample is dried under standardized conditions (*78*). Oven drying at 130°C (266°F) for one to four hours is used for many kinds of seeds. For oily seeds 103°C (217°F) for 17 hours is used (*78*) and for some seeds which lose oil at that temperature (e.g., fir, cedar, beech, spruce, pine, hemlock) a toluene distillation method is used. Various kinds of electronic meters can be used for quick moisture tests (*11*).

Viability Determination

Viability can be determined by several tests, the **direct germination, excised embryo,** and **tetrazolium** tests being the most important. In the direct germination test the **germination percentage** is determined by the percent of normal seedlings produced by the pure seed (the kind under consideration). To produce a good test, it is desirable to use at least 400 seeds picked at random and divided into lots of one hundred each. If any two of these lots differ by more than 10 percent, a retest should be carried out. Otherwise, the average of

FIGURE 7-1 Purity of seeds is determined by visual examination of individual seeds in a weighed sample taken from the larger lot in question. Impurities may include other crop seed, weed seed, and inert, extraneous material. (Courtesy E. L. Erickson Products, Brookings, S.D.)

the four tests becomes the germination percentage (see Fig. 7–2).

Direct Germination Tests

In a standard germination test the seeds are placed under optimum environmental conditions of light and temperature to induce germination (Fig. 7–3). The conditions required to meet legal standards are specified in the rules for seed testing, which may include type of test, environmental conditions, and length of test.

Various techniques are used for germinating seed in seed-testing laboratories. Small seeds are placed on germination trays (not galvanized steel, which contains toxic zinc salts). Plastic boxes, paraffined cardboard boxes, or covered glass petri dishes also are useful containers. Absorbent paper is cut into small pieces (*blotters*) and small seeds are placed on top or between two layers. Other media are absorbent cotton, paper towels (five thicknesses), filter paper (five layers), and for large seeds, a sand, vermiculite, perlite, or soil (16 mm; ⅝ in.) layer. Containers are placed in germinators in which temperature, moisture, and light are controlled. To discourage the growth of microorganisms, all materials and equipment should be kept scrupulously clean, sterilized when possible, and the water amount carefully regulated. No water film should form around the seeds; neither should the germination medium be so wet that a film of water appears when that medium is

FIGURE 7-2 Germination testing of seeds. *Top:* 100 seeds from the sample to be tested are placed on a moistened blotter. In this case placing the seeds evenly and quickly is made possible by an automatic vacuum counter. *Below:* After one or more weeks in a germinator the number of germinated seeds is counted. Note that this test consists of four lots of 100 seeds each. (Courtesy E. L. Erickson Products, Brookings, S.D.)

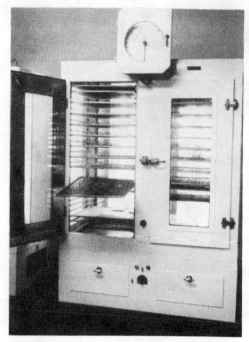

FIGURE 7-3 Commercial seed germinator with light and temperature control for testing viability. (Courtesy C. E. Heit.)

pressed with a finger. Relative humidity in the germinator should be 90 percent or more to prevent drying. Containers with sand should be kept tightly closed. Water should not be added during the test.

The **rolled towel** test is commonly used for testing cereal grains. Several layers of moist paper towelling, about 2.8 by 3.6 cm (11 by 14 in.) in size, are folded over the seeds, then rolled into cylinders and placed vertically in a germinator.

A germination test usually runs from one to four weeks but could continue for three months for some slow-germinating tree seeds with dormancy. A *first count* may be taken at one week and germinated seeds discarded with a formal count taken later. At end of the test, seeds are divided into (a) normal seedlings, (b) hard seeds, (c) dormant seeds, and (d) abnormal seedlings plus dead or decaying seeds. A normal seedling generally should have a well-developed root and shoot, although the criteria for a "normal seedling" vary with different kinds of seeds. "Abnormal seedlings" can be caused by age of seed or poor storage conditions; insect, disease, or mechanical injury;

overdoses of fungicides; frost damage; mineral deficiencies (manganese and boron in peas and beans); or toxic materials sometimes present in metal germination trays, substrata, or tap water. Any ungerminated seed should be examined to determine the possible reason. "Hard seeds" have not absorbed water. Dormant seeds are those that are firm, swollen, and free from molds, but show erratic sprouting, or none.

Under seed testing rules certain environmental requirements to overcome dormancy may be specified routinely for many kinds of agricultural, vegetable, and flower seeds (*5, 44*). Tree and shrub seeds often require special pregermination treatments before tests are run (Table 7-1).

Excised-Embryo Test

The excised-embryo test is used to test the seed viability of woody shrubs and trees whose dormant embryos require long periods of afterripening before true germination will take place (*25, 36*). In this test the embryo is excised from the seed and germinated alone (see Fig. 7-4).

The seeds are soaked for one to four days by one of the following methods until they are completely swollen: (a) in slowly running water, (b) in standing water below 15°C (59°F), or (c) in standing water at about 25°C (68°F), with at least two changes of water daily.

Storing seeds in moist peat for three days to two weeks at cool temperatures is also satisfactory in preparing seeds for excision. The excision must be done carefully to avoid injury to the embryo. Any hard, stony seed coverings, such as the endocarp of stone fruit seeds, must first be removed.

The moistened seed coats are cut with a sharp scalpel, razor blade, or knife, under clean but nonsterile conditions with sterilized instruments, preferably under a sheet of glass. The embryo is carefully removed. If a large endosperm is present, the seed coats may be slit and the seeds covered with water, and after about a half hour the embryo will float out or can easily be removed.

Procedures for germinating excised embryos are similar to those for germinating intact seeds. Petri dishes with a moist substratum, such as blotting or filter paper, are used. The embryos are placed on the filter paper so that they do not

TABLE 7–1

Types of germination requirements for tree and shrub seeds (with examples) when tested in the laboratory

Group 1: seeds that germinate within a wide temperature range and without light exposure

Beefwood (*Casuarina glauca*)
Italian cypress (*Cupressus semipervirens*)
Many species of eucalyptus
Honeylocust (*Gleditia triacanthos*)[a]

Some spruce species (*Picea abies, P. asperata, P. polita*)
Chinese and Siberian elm (*Ulmus parvifolia, U. pumila*)

Group 2: seeds that have specific temperature requirements but do not require light

20 to 30°C (68 to 86°F) diurnally alternating:
  Catalpa
  Ailanthus
  Red pine (*Pinus resinosa*) or 25°C (77°F) constant
10 to 30°C (50 to 86°F) diurnally alternating:
  Mountain mahogany (*Cercocarpus ledifolius*)
  Cliffrose (*Cowania stansburiana*)
  Antelope bitterbush (*Purshia tridentata*)

20°C (68°F) constant:
  Several pine species (*Pinus cembroides P. halepensis, P. pinea*)
  Lilac (*Syringa vulgaris*)
  Arborvitae (*Thuja orientalis*)

Group 3: seeds that germinate in 7 to 12 days within a wide temperature range if exposed to artificial light

Several spruce species (*Picea engelmannii, P. mariana, P. omerika*)
Several pine species (*Pinus banksiana, P. nigra, P. mugo var. mughus, P. rigida, P. sylvestris, P. ponderosa scopulorum*)

Group 4: seeds that germinate in 14 to 28 days if exposed to artificial light and to warm alternating temperatures of 20 to 30°C (68 to 86°F); seeds may also respond to moist-chilling

Birch (*Betula*)
Elm (*Ulmus americana*)
Larch (*Larix sibirica*)
Mulberry (*Morus alba, M. nigra*)
Liquidambar styraciflua
Some spruce series (*Picea glauca, P. orientalis, P. rubens, P. sitchensis*)

Some pine species (*Pinus densiflora, P. echinata, P. elliotti, P. taeda, P. thunbergii, P. virginiana*)
Rhododendron
Sequoia and *Sequoiadendron*
Thuja plicata, T. occidentalis

Group 5: seeds that require 3 to 4 weeks moist-chilling at 3°C (37°F) before germination at alternating 20 to 30°C (68 to 86°F) (except as noted) temperatures in light for 2 to 4 weeks

Fir species (*Abies balsamea, A. fraseri, A. grandis, A. homolepsis, A. procera*)
Cedrus species 20°C (68°F)
Some pines species (*Pinus flexilis, P. glabra, P. leucodermis, P. strobus*)

Some sources of Douglas fir (*Pseudotsuga menziesii*)
Rosa multiflora 10 to 30°C (50 to 86°F)
Eastern hemlock (*Tsuga canadensis*) 15°C (59°F)
Sumac (*Rhus aromatica*)[a]

Group 6: seeds that require 2 to 6 months moist-chilling as a minimum requirement prior to germination; some also have other dormancy problems; embryo excision or a tetrazolium test may be useful in determining germinative capacity

Sensitive to high germination temperature 20°C (68°F):
  Maple (*Acer* spp.)
  Apple (*Malus* spp.)
  Pear (*Pyrus* spp.)
  Peach, cherry, etc. (*Prunus* spp.)
  Yew (*Taxus* spp.)

Not sensitive to high germination temperature:
  Some pine species (*Pinus cembra, P. lambertiana, P. monticola, P. peuce*)

Source: C.E. Heit (*40*).
[a]Must be treated also for hard seed coats.

FIGURE 7–4 The excised embryo method of testing seed germination. *Top:* Germination of apple seeds showing the range of vigor from strong to weak to dead. Actual germination percentages for the four lots were (left to right): 100, 70, 44, and 0. *Below:* Seed vigor and germination of four peach stocks (left to right): strong seed, vigorous growth (80 percent viability); good seed, fair vigor (52 percent viability); old seed, weak (18 percent viability); and dead seed. (Courtesy C. E. Heit.)

touch. The dishes are kept in the light at a temperature of 18 to 22°C(64 to 74°F). At higher temperatures, molds may develop and interfere with the test. The time required for the test varies from three days to three weeks.

Nonviable embryos become soft, turn brown, and decay within two to ten days; viable embryos remain firm and show some indication of viability, depending upon the species. Types of response that occur include spreading of the cotyledons, development of chlorophyll, and growth of the radicle and plumule. The rapidity and degree of development gives some indication of the vigor of the seed.

Tetrazolium Test

The tetrazolium test is a biochemical method in which viability is determined by the red color appearing when the seeds are soaked in a 2,3,5-triphenyltetrazolium chloride (TTC) solution. Living tissue changes the TTC to an insoluble red compound (chemically known as formazan); in nonliving tissue the TTC remains uncolored. The test is positive in the presence of dehydrogenase enzymes. This test was developed in Germany by Lakon (*49*), who referred to it as a topographical test, since loss in embryo viability begins to appear at the extremity of the radicle, epicotyl, and cotyledon tips. The reaction takes place equally

well in dormant and nondormant seed. Results can be obtained within 24 hours, sometimes in two or three hours. TTC is soluble in water, making a colorless solution. Although the solution deteriorates with exposure to light, it will remain in good condition for several months if stored in a dark bottle. The solution should be discarded if it becomes yellow. A 0.1 to 1.0 percent concentration is commonly used. The pH should be 6 or 7. The TTC test is used primarily to obtain rapid results for both nondormant and dormant seeds.

The test distinguishes between living and dead tissues within a single seed and can indicate weakness before germination is actually impaired. Necrotic areas may be attacked by pathogenic organisms, and seeds with such dead tissues may decay during stratification or give reduced germination under unfavorable soil conditions. In the hands of a skilled technologist this test can be used for seed-quality evaluation and as a tool in seed research (*52, 63*).

On the other hand, this test may not adequately measure certain types of injury which could lead to seedling abnormality—for instance, an overdose of chemicals, seedborne diseases, frost or heat injury. Standardized procedures and skills are required for evaluating results.

Although details vary with different kinds of seeds, in general, the following procedures must be followed (*5, 44, 63*):

1. Any hard covering such as an endocarp, wing, or scale must be removed. Tips of dry seeds of some plants, such as *Cedrus,* should be clipped.

2. Seeds should first be soaked in water in the dark; moistening activates enzymes and facilitates cutting or removal of seed coverings. Seeds with fragile coverings, such as snap beans or citrus, must be softened slowly on a moist medium to avoid fracturing.

3. Most seeds require preparation for TTC absorption. Embryos with large cotyledons, such as *Prunus,* apple, and pear, often comprise the entire seed, requiring only seed coat removal. Other kinds of seed are cut longitudinally to expose the embryo (corn and large seeded grasses, larch, some conifers); or transversely one-fourth to one-third at the end away from the radicle (small-seeded grasses, juniper, *Carpinus, Cotoneaster, Crataegus, Rosa, Sorbus, Taxus*). Seed coats can be removed, leaving the large endosperm intact (some pines, *Tilia*). Some seeds (legumes, timothy) require no alteration prior to the tests.

4. Seeds are soaked in the TTC solution for 2 to 24 hours. Cut seeds require a shorter time; those with exposed embryos somewhat longer; intact seeds 24 hours or more.

5. Interpretation of results depends upon the kind of seed and its morphological structure. Completely colored embryos indicate good seed. Conifers must have both the megagametophyte and embryo stained. In grass and grain seeds only the embryo itself colors, not the endosperm. Seeds with declining viability may have uncolored spots, or be unstained at the radicle tip and the extremities of the cotyledons. Nonviability depends upon the amount and location of necrotic areas, and correct interpretation depends upon standards worked out for specific seeds (52).

6. If the test continues too long, even tissues of known dead seeds become red due to respiration activities of infecting fungi and bacteria. Likewise, the solution itself can become red because of such contamination.

X-Ray Analysis

X-ray photographs of seeds (46, 73) can be used as a rapid test for seed soundness (2). X-ray photographs do not normally measure seed viability but provide an examination of the inner structure for mechanical disturbance, absence of vital tissues, such as embryo or endosperm, insect infestation, cracked or broken seed coats, and shrinkage of interior tissues (possibly a sign of age; see Figure 7–5).

Standard x-ray equipment is used. A convenient Plexiglass seed holder with 100 seed compartments can be made by drilling 100 holes 15 mm in diameter in a 5-mm-thick Plexiglass plate. Then glue a 0.1-mm-thick Mylar sheet to the underside. Place one dry seed in each compartment and expose for ½ to 3 minutes at 15 to 20 kilovolts tube potential. The plate can be processed immediately and the status of seeds determined. Seeds with dimensions less than 2 mm are too small to show details. Since x-rays do not injure the seed, further tests for viability can be conducted on the same batch (3).

Other tests for seed viability have been developed utilizing contrast agents, such as solutions of certain salts or heavy metals (73), followed by x-ray photography. In one test, for example, seeds are soaked in water for 16 hours, then trans-

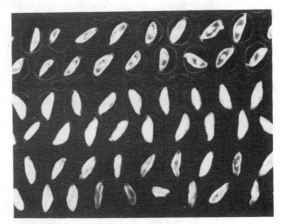

FIGURE 7–5 X-radiograph of *Abies procera* seed. Seed has been segregated to illustrate empty seed (top two rows), normally filled seed with distinct embryos (center two rows), and seed infested with chalcid fly larvae (bottom two rows). (Courtesy Jay Allison (3).)

ferred to a concentrated (20 to 30 percent) barium chloride solution for one to two hours, washed to remove all excess material, and dried. The salts will not enter living cells because of their semipermeability but can penetrate the dead cells of damaged seeds because of membrane destruction and thus produce an exposure on the film. This penetration differentiates between living and dead seeds, or portions of seeds. Fairly consistent and reliable correlations with germination percentages can be obtained with fresh conifer seeds, although consistent results may not occur with stored seed (*23*).

Other Seed Tests

Several other kinds of seed tests are used in addition to those already described (*78*).

Verification of species and cultivar, in some cases, is possible by visual observation of the external seed characteristics; more accurate verification requires observations of seedlings produced either in the laboratory or in field plots.

Seed health tests are used to obtain observations on the presence of pathogens. These specialized tests require training in plant pathology methods.

Estimating relative seed vigor is important, and standard tests are available for special situations (*17, 56*). **Germination speed** provides a general index: A seedling count at one week is compared to the final germination count (*17*). Previously, tree seed technologists have used the term **germination energy,** which is the germination percentage achieved at the maximum germination rate (*72*). The **germination value,** as described in Chap. 6, is a useful method that includes vigor measurements.

TREATMENTS TO OVERCOME SEED DORMANCY

Softening Seed Coats and Other Coverings

Scarification is any process of breaking, scratching, mechanically altering, or softening the seed coverings to make them permeable to water and gases.

Mechanical Scarification

Mechanical scarification is simple and effective with seeds of many species if suitable equipment is available (Fig. 7–6). These seeds are dry after such treatment and may be stored or planted immediately by mechanical seeders. Scarified seeds are more susceptible to injury from pathogenic organisms, however, and will not store as well as comparable nonscarified seeds.

Chipping hard seed coats by rubbing with sandpaper, cutting with a file, or cracking with a hammer or a vise are simple methods useful for small amounts of relatively large seed. For large-scale mechanical operations, special scarifiers are used. Small seeds of legumes, such as alfalfa and clover, are often treated in this manner to increase germination (*17*). Seeds may be tumbled in drums lined with sandpaper or in concrete mixers, combined with coarse sand or gravel (*72*) (Fig. 7–6). The sand or gravel should be of a different size than the seed to facilitate subsequent separation.

Scarification should not proceed to the point at which the seeds are injured. To determine the optimum time, a test lot can be germinated, the seeds may be soaked to observe swelling, or the seed coats may be examined with a hand

FIGURE 7–6 Seed scarifier for ornamental seeds. Used to scarify hard seed coats.

lens. The seed coats generally should be dull but not so deeply pitted or cracked as to expose the inner parts of the seed.

Hot Water Scarification

Drop the seeds into four to five times their volume of hot water 77 to 100°C (170 to 212°F). The heat source is immediately removed, and the seed soaked in the gradually cooling water for 12 to 24 hours. Following this the unswollen seeds can be separated from the swollen ones by suitable screens and either retreated or subjected to some other treatment. The seeds should usually be planted immediately after the hot-water treatment; some kinds of seed have been dried and stored for later planting without impairing the germination percentage, although the germination rate was reduced (see Fig. 6–5).

Acid Scarification

Dry seeds are placed in containers and covered with concentrated sulfuric acid (specific gravity 1.84) in a ratio of about one part seed to two parts acid. The amount of seed treated at any one time should be restricted to no more than about 10 kg (22 lb) (*60*) to avoid uncontrollable heating. Containers should be glass, earthenware, or wood—not metal or plastic. The mixture should be stirred cautiously at intervals during the treatment to produce uniform results and to prevent accumulation of the dark, resinous material from the seed coats that is sometimes present. Since stirring tends to raise the temperature, vigorous agitation of the mixture should be avoided to prevent injury to the seeds. The time of treatment may vary from as little as 10 minutes for some species to six hours or more for other species. Since treatment time may vary with different seed lots, making a preliminary test on a small lot is recommended prior to treating large lots (*39, 60*). With thick-coated seeds that require long periods, the progress of the acid treatment may be followed by drawing out samples at intervals and checking the thickness of the seed coat. When it becomes paper thin, the treatment should be terminated immediately.

At the end of the treatment period the acid is poured off, and the seeds are washed to remove the acid. Glass funnels are useful in removing the acid from small lots of seed. Placing seeds in a large amount of water with a small amount of baking soda (sodium bicarbonate) will neutralize any adhering acid; or the seeds can be washed for 10 minutes in running water. The acid-treated seeds can either be planted immediately when wet, or dried and stored for later planting.

Large seeds of most legume species respond to the simple sulfuric acid treatment, but variations are required for some species (*60*). Some roseaceaous seeds (*Cotoneaster, Rosa*) have hard pericarps that are best treated partially with acid and then given warm stratification. A third group, such as *Hamamelis* and *Tilia,* have very "tough" pericarps that may first need to be treated with nitric acid and then with sulfuric acid.

Warm Moist Scarification

Keeping seeds in a nonsterile, moist, warm medium (e.g., in nonpasteurized sandy soil) for several months can soften seed coats through microorganism activity. This treatment can be provided by planting hard seeds in summer or early fall while soil temperatures are warm. Usually, it is most effective for seeds having double dormancy where the warm moist scarification precedes cold stratification during the winter.

High-Temperature Scarification

Seeds of certain species of native plants with hard seed coats germinate extensively after a forest or range fire. Seed coats are modified by the high temperatures. Closed cone species of pine, for example, *Pinus radiata,* also respond to high temperatures in fires by melting the enclosing resins that seal the cone. This action allows release of the seeds which are then able to germinate (*72*).

Harvesting Immature Fruits

Extracting seeds from immature fruits improves germination in certain tree species by avoiding the development of hard seed coats. Such seeds must be planted immediately without drying.

Stratification

Stratification is a method of handling dormant seeds in which the imbibed seeds are subjected to a period of chilling to after-ripen the embryo. The term originated because nurseries placed seeds in stratified layers interspersed with a moist medium, such as soil or sand, in out-of-doors pits during winter. The term **moist-chilling** has been used as a synonym for stratification. However, with double dormancy situations, frequently a **warm-moist stratification** of several months followed by a moist-chilling stratification is required.

Refrigerated Stratification

Dry seeds should be fully imbibed with water prior to refrigerated stratification. Twelve to 24 hours of soaking at warm temperatures may be sufficient for seeds without hard seed coats or coverings. Longer periods are required, with aeration, for seeds enclosed in hard endocarp or pericarp; soaking for three days to a week or more may be necessary. Leaching seeds in running water also can be used, or alternately soaking seeds for 12 hours, then draining for 12 hours (*50*).

After soaking, seeds are usually mixed with a moisture-retaining medium for the stratification period. Almost any medium that holds moisture, provides aeration, and contains no toxic substances is suitable. These include well-washed sand, peat moss, chopped or screened [0.6 to 1.0 cm (¼ to ⅜ in.) with smaller parts discarded] sphagnum moss, vermiculite, and composted sawdust. Fresh sawdust may contain toxic substances. A good medium is a mixture of 1 coarse sand:1 peat or 1 perlite:1 peat, moistened and allowed to stand 24 hours before use. Any medium used should be moist but not so wet that water can be squeezed out.

Seeds are mixed with one to three times their volume of the medium or they may be stratified in layers, alternating with similarly sized layers of the medium. Suitable containers are boxes, cans, glass jars with perforated lids, or other containers that provide aeration, prevent drying, and protect against rodents. Polyethylene bags are excellent containers either with or without media. Stratification of seeds in a plastic bag without a surrounding medium has been called *naked chilling* (*19*). A fungicide may be added as a seed protectant.

The usual stratification temperature is 0 to 10°C (32 to 50°F). At higher temperatures seeds often heat or sprout prematurely. Lower temperatures (just above freezing) delay sprouting.

The time required for stratification depends on the kind of seed, and sometimes upon the individual lot of seed as well (Fig. 7–7). For seeds of most species, one to four months is sufficient for low-temperature stratification. During this time the seeds should be examined periodically; if they are dry, the medium should be remoistened. When sprouting begins, the seeds should be planted or moved to lower storage temperatures. The seeds to be planted are removed from the containers and separated from the medium, using care to prevent injury to the moist seeds. A good method is to use a screen that allows the medium to pass through while retaining the seeds. The seeds are usually planted without drying to avoid injury and reversion to secondary dormancy. Some success has been reported for partially drying previously stratified seeds, holding them for a

FIGURE 7–7 Walnut (*Juglans hindsii*) seeds after stratification ready for planting. Seeds on left may not have enough chilling to sprout. Seeds in center are in a favorable condition. Seeds on right are too advanced and radicles are likely to be injured in planting.

time at low temperatures, then planting them "dry" without injury or loss of after-ripened condition. Beech and mahaleb cherry seeds were successfully dried to 10 percent, then held near freezing (76). Similarly, stratified fir (*Abies*) seed has been dried to 20 to 35 percent, then stored for a year at low temperatures (24) after stratification.

Outdoor Stratification

Where refrigerated storage is not available, stratification may be done by storing outdoors, either in pits several feet deep or in raised beds enclosed in wooden frames (2, 72). Essentially the same seed preparation procedures are used but outdoor winter chilling and natural rainfall provide the required chilling temperature and moisture. However, the seeds need to be protected against freezing, drying, and rodents (75).

When seeds are stratifying in a trench, a layer of wire netting should be placed on the bottom, sides, and top of the trench to protect the seeds from rodents. A layer of clean sharp sand or gravel (10 cm; 4 in.) is placed on the bottom, then the seeds are mixed with the medium or placed in alternating layers of seed and medium. The trench is then filled and covered with the netting.

Outdoor Planting

As an alternative to container stratification, seeds requiring a cold treatment may be planted out-of-doors directly in the seedbed, cold frame, or nursery row at a time of the year when the natural environment provides the necessary conditions for after-ripening. Several different categories of seeds can be handled in this way with good germination in the spring following planting. Some wild flower seeds fall into this category (13).

Seeds must be planted early enough in the fall to allow them to become imbibed with water and to get the full benefit of the winter chilling period. The seeds generally germinate promptly in the spring when the soil begins to warm up, but while the soil temperature is still low enough to inhibit damping-off organisms and to avoid high-temperature inhibition.

Seeds with a hard endocarp, such as *Prunus* species (the stone fruits, including cherries, plums, and peaches), show increased germination

if planted early enough in the summer or fall to provide one to two months of warm temperatures prior to the onset of chilling (50). Seeds that require high temperatures followed by chilling can thus be planted in late summer to fulfill their warm-temperature requirements, followed by the subsequent winter period that satisfies the chilling requirement after preliminary testing.

For other seeds having hard coverings (such as juniper and some *Magnolia* species), germination can be facilitated if the fruit is harvested when ripe and the seeds planted immediately without drying (60, 82). Once the seeds become dry, the seed coats harden and germination may be delayed, perhaps until the second spring. Seeds that ripen early in the growing season and lose viability rapidly should be collected and planted in spring or summer as soon as they mature. Where effective treatments are not known, the propagator should attempt to reproduce the natural seeding habits of the plant and provide the germinating conditions of its natural environment. When seeds remain for a long period in an outdoor nursery or seedbed prior to germination, they must be protected from drying, adverse weather conditions, rodents, birds, diseases, and competition from weeds. Herbicides can be used for weed control.

Leaching

The purpose of leaching is to remove inhibitors by soaking seeds in running water or by placing them in frequent changes of water. The length of leaching time is 12 to 24 hours. If longer periods are involved, the water should be changed every 12 hours to provide oxygen to the submerged seeds (although dormant seeds of woody plants are unlikely to be injured). When feasible, it is desirable to leach under running water.

Combinations of Treatments

Many woody plant species have more than one type of dormancy, for example, a hard seed coat plus a dormant embryo. For such species multiple treatments are required. Treatments must be given in sequence, first, one to soften the seed coat to allow uptake of water and, second, a chilling period to overcome embryo dormancy.

Laboratory Treatments to Overcome Dormancy

Seed laboratories use various methods to initiate germination in viability testing. The same treatments might apply in propagation, although in many cases, as with seeds of cereals, vegetables, and flower crops, the seeds lose their dormancy during dry storage prior to planting.

Prechilling. For prechilling, place imbibed seeds at 5 to 10°C (37 to 42°F) for five to seven days before germination is attempted.

Predrying. In predrying, dry seeds are subjected to 37 to 40°C (99 to 104°F) for five to seven days prior to germination.

Daily alternation of temperature. When alternate temperatures are required, the usual combinations are 15 to 30°C (59 to 86°F), or 20 to 30°C (68 to 86°F), with seeds held at the lower temperatures for 16 hours and at the higher temperature for eight hours. Such temperature fluctuations are found normally out-of-doors at certain seasons of the year and often greatly promote germination.

Light exposure. Exposure to light can stimulate germination of many kinds of seeds, depending upon age, previous handling, and accompanying temperature. Light sensitivity is strongest just following harvesting and tends to disappear with dry storage. Consequently, light influence is quite important in seed testing. The Rules for Seed Testing (5) specify light exposure for testing seeds of 42 percent of the 493 species cited. These include most grasses and conifers, many flowers, and a number of vegetables, including lettuce and endive.

Light should be provided by cool-white fluorescent lamps at an intensity of at least 75 to 125 fc (806 to 1345 lux) for at least eight hours daily. Seeds must have imbibed moisture at the time of light exposure and should be placed on top of the medium.

Potassium nitrate. Many freshly harvested dormant seeds germinate better after soaking in a potassium nitrate solution. Seeds are placed in germination trays or petri dishes and the substratum moistened with 0.2 percent potassium nitrate. For Kentucky bluegrass (*Poa pratensis*) or Canada bluegrass (*P. compressa*) a 0.1 percent solution should be used. If they are rewatered, tap or distilled water is used rather than additional nitrate solution.

SEED TREATMENTS TO FACILITATE GERMINATION

Hormones and Other Chemical Stimulants

Gibberellins

Gibberellins (see Fig. 6–8) comprise a group of plant hormones that have significant activity in seed physiology. While there are many kinds of natural gibberellins in plants, **gibberellic acid** (GA_3) is the one most widely used for exogenous applications. GA_3 treatments can overcome physiological dormancy in various seeds and stimulate germination in seeds with dormant embryos.

In seed-testing laboratories, the germination medium is usually moistened with 500 mg/liter GA_3, although ranges of 200 to 1000 mg/liter are used. At concentrations above 800 mg/liter, addition of a buffer solution[1] is recommended. K-salt formulation of GA is readily water soluble. For large seeds, a 12-hour soak in 500 to 1000 mg/liter GA_3 is recommended.

Cytokinins

Cytokinins (see Fig. 6–8) are natural growth hormones that stimulate germination of some kinds of seed by overcoming germination inhibitors. A commercial preparation, **kinetin** (6-furfurylaminopurine), is available. Dissolve first in a small amount of HCl, then dilute with water. Other available synthetic cytokinins are **BAP** (6-benzylaminopurine) and **PBA** [6-benzylamino) - 9 - (2-tetrahydropyranyl) - 9H - purine]; these are more active for higher plants than is kinetin. These materials stimulate germination and

---

[1]Dissolve 1779.9 mg of $NaHPO_4 \cdot 2H_2O$ and 1379.9 mg of $NaH_2PO_4 \cdot 2H_2O$ in 1 liter of distilled water (44).

overcome high-temperature dormancy of certain kinds of seeds, such as lettuce, where seeds were soaked in 100 mg/liter kinetin solutions for three hours. For larger seed BAP at 200 mg/liter for 24 hours has been used. Large-scale treatments should be preceded by trials at varying concentrations. Cytokinins are sometimes effective in promoting germination when used in combination with gibberellic acid and with ethylene-producing compounds.

Ethylene

Ethylene is a phytohormone (plant hormone) (see Fig. 6–18). Applied to some kinds of seed, such as cocklebur, ethylene stimulates germination. Commercial ethylene-generating chemicals, such as ethephon (Ethrel), are available to supply ethylene.

Sodium Hypochlorite

Sodium hypochlorite stimulates germination of rice seed, apparently by overcoming a water-soluble inhibitor in the hull (62). Use 1 gallon of commercial concentrate to 100 gal of water.

Seed Priming Systems (SPS)

The concept of this system (also known as osmoconditioning) is to control the hydration of seeds to permit pre-germination metabolic activities to proceed but to prevent actual emergence of the radicle (12, 41, 48). After treatment the seeds are dried and later planted by usual methods. Control is exerted through soaking the seeds at 15 to 20°C (60 to 78°F) in aerated solutions of high osmotic strengths for 7 to 21 days, rinsing with distilled water, air drying at 25°C (78°F), and then storing until use. The principal compounds used have been 20 to 30 percent polyethylene glycol (PEG 6000) (41, 61) or various salt solutions, including KNO_3, K_2HPO_4, NaCl, and others (12). This procedure has the potential to improve emergence under field conditions, to shorten the time of emergence, and to overcome certain environmental constraints in the field, including cold injury and thermodormancy.

Variation exists in the SPS requirements among different species, cultivars, and seed lots in choice of priming agent, water potential, temperature, and duration of treatment. Preliminary tests must be carried out. Equipment to carry out screening tests have been described (1). A problem with the method is to devise large-scale equipment to be used in commercial procedures (10). There is also the problem of short shelf life. SPS seed can be stored for only limited periods of time before seed viability drops off greatly.

Infusion

Organic solvent infusion is a method for incorporating chemicals, such as growth regulators, fungicides, insecticides, antibiotics, and herbicidal antidotes (47) into seeds by means of organic solvents. Seeds are immersed for one to four hours in an acetone and dichloromethane solution which contains the chemical to be infused. The solvent is then removed by evaporation, and the seeds are spread out in a shallow container for drying in a vacuum desiccator for one to two hours. The incorporated chemical is then absorbed directly into the embryo upon soaking in water.

Fluid Drilling

Fluid drilling (31) is a system involving the treatment and pre-germination of seeds followed by their sowing suspended in a gel. Seeds are pregerminated under conditions of aeration, light, and optimum temperatures for the species. Among the procedures that can be used are (a) germinating seeds in trays on absorbent blotters covered with paper, or (b) placing seeds in water in glass jars or plastic columns through which air is continuously bubbled and fresh water continuously supplied. Growth regulators, fungicides, and other chemicals (29) can potentially be incorporated into the system. Chilling (10°C; 50°F) of thermodormant celery seeds for 14 days has produced short, uniform radicle emergence without injury (28). Pregerminated seeds of various vegetables have been stored for 7 to 15 days at temperatures of 1 to 5°C (34 to 41°F) in air or aerated water. Separating out germinated seeds by density separation has improved the uniformity and increased overall stand (77).

Various kinds of gels are commercially available. Among the materials used are sodium alginate, hydrolyzed starch–polyacrylonitrile, guar gum, synthetic clay, and others. Special machines are needed to deposit the seeds and gel into the seedbed.

PROTECTION OF SEEDS AGAINST PATHOGENS

Three types of seed treatments are used to control diseases: disinfestation, disinfection, and seed protection (*8, 55*).

THERMOTHERAPY

Hot-water treatment can be used as a disinfectant procedure. Dry seeds are immersed in hot water (49 to 57°C; 120 to 135°F) for 15 to 30 minutes, depending upon the species (*6, 7*). After treatment, the seeds are cooled and spread out in a thin layer to dry. To prevent injury to the seeds, temperature and timing must be regulated precisely; a seed protectant should subsequently be used, and old, weak seeds should not be treated. Hot water is effective for specific seedborne diseases of vegetables and cereals, such as *Alternaria* blight in broccoli and onion, and loose smut of wheat and barley.

Aerated steam (see Chap. 2) is an alternate method that is less expensive, easier to handle, and less likely to injure seeds (*7*). Seeds are treated in special machines in which steam and air are mixed and drawn through the seed mass to raise the temperature of the seeds rapidly (about two minutes) to the desired temperature. The treatment temperature and time vary with the organism to be controlled and kind of seed. Usually the treatment is 30 minutes, but may be as little as 10 or 15 minutes. Temperatures range from 46 to 57°C (105° to 143°F). At the end of treatment, temperatures must be lowered rapidly to 32°C (88°F) by evaporative cooling continuing until dry. Holding seeds in moisture saturated air at room temperature for one to three days prior to treatment will improve effectiveness.

Disinfestants eliminate organisms present on the surface of the seed. Materials that have this sole action are useful if the seeds or embryos are to be grown in aseptic culture or in some type of sterile medium. Materials listed in this chapter as disinfectants or protectants are also likely to be good disinfestants.

Disinfectants eliminate organisms within the seed itself. Treatments of this type include hot water, formaldehyde, and aerated steam.

Protectants are materials applied to the seeds to protect them from pathogenic fungi in the soil. These materials may also be applied as a soil drench either before or after seed planting. Numerous materials for seed treatment are available commercially under various trade names. The manufacturer's directions should be followed carefully, since improper use can cause injury or death to the seedlings of some plants. And since some protectants are hazardous to the person using them, they should be handled carefully to avoid contact with the skin or breathing of the dust. In many cases, special machines must be used for treating the seeds.

The three basic methods of application are (a) the dry or dust method, (b) the liquid method, and (c) the "slurry" method. In the latter the seed is immersed in a thick water suspension of the chemicals.

Other seed treatments include the application of insecticides, or combinations of insecticides and fungicides, plus bird and rodent repellents. *All state and federal regulations governing the use of such materials on crop plants should be followed carefully.*

SEED PROPAGATION SYSTEMS

Seed propagation is carried out with three basic systems: (a) **field seeding** in the location where the plant is to remain, (b) planting in **field nurseries** and transplanting to a permanent location, and (c) planting in **protected conditions,** as in a greenhouse, cold frame, or similar structure, and then transplanting to the permanent location. Seed propagation under protected conditions can also include direct seeding of bedding plants into small liner containers or direct seeding into cells or plugs for later transplanting.

Field Seeding of Herbaceous Cultivars

Direct field seeding is used for commercial field planting of agronomic crops (grains, legumes, forages, fiber crops, oil crops), many vegetable crops, and lawn grasses. The method may be used for home vegetable and flower gardeners and hobbyists. Compared to transplants, directly seeded plants properly handled are less expensive and can grow continuously without the check in growth produced by bare root transplanting. Frequently, direct-field-seeded vegetable and other crops are precontracted for processing, whereas the more expensive transplants are targeted as an early fresh market crop. On the other hand, there are many potential field problems that must be overcome to provide the proper environmental conditions for good uniform germination. Likewise, cold weather may decrease growth. Seeding rates are critical to provide proper plant spacing for optimum development of the crop. If final plant density is too low, yields will be reduced because the number of plants per unit area is low; if too high, the size and quality of the finished plants may be reduced by competition among plants for available space, sunlight, water, and nutrients.

To maximize direct-seeding success, the following is required:

1. *Prepare the seedbed.* A good seedbed should have a loose but fine physical texture that produces close contact between seed and soil so that moisture can be supplied continuously to the seed. Such a soil should provide good aeration but not too much, or it dries too rapidly. The surface soil should be free of clods and of a texture that will not form a crust. The subsoil should be permeable to air and water with good drainage and aeration. Adequate soil moisture should be available to carry the seeds through the germination and early seedling growth stages, but the soil should not be waterlogged or anaerobic (without oxygen). A medium loam texture, not too sandy and not too fine, is best. A good seedbed is one in which three-fourths of the soil particles (aggregates) range from 1 to 12 mm in diameter (*42*).

Seedbed preparation requires special machinery for field operations and spading and raking or rototilling equipment for small plots. Add-ing organic or soil amendments may be helpful, but these should be thoroughly incorporated and have time to decompose. Seedbed preparation may include fumigation and other soil treatments to control harmful insects, nematodes, disease organisms, and weed seeds (see Chap. 2).

2. *Choose correct planting time.* Planting time is determined by the germination temperature requirements of the seed and the need to meet production schedules. These are determined according to the individual crop and vary with the particular kind of seed. Too early planting of seeds requiring warm soil temperatures can result in slow and uneven germination, disease problems, and "chilling" injury to seedlings of some species, causing growth abnormalities. Too-high soil temperatures can result in excessive drying, injury, or death to the seedlings, or induction of thermodormancy in the case of heat-sensitive seeds such as lettuce, celery, and various flower seeds (see Chap. 20).

3. *Select high-quality seed.* Quality is based on seed testing data. A low rate of sowing requires seeds of high quality that produce not only high germination rates but also vigorous, uniform, healthy seedlings.

4. *Pretreat seeds.* It is usually desirable to plant seeds that have been treated with fungicides (see p. 130). Such treated seeds must be color coded for identification. Pregerminated seeds for fluid drilling and primed seeds (see p. 126) are available for some species. These treatments can speed up germination, increase uniformity, and offset some environmental hazards in the seedbed. **Pelleted** seeds (*43*) are seeds coated with an adhesive inert material, such as clay, which makes them uniformly round for use in precision planters.

5. *Plant at proper depth.* Depth of planting is a critical factor that determines the rate of emergence and perhaps stand density. If too shallow, the seed may be in the upper surface that dries out rapidly; if too deep, emergence of the seedling is delayed. Depth varies with the kind and size of seed and, to some extent, the condition of the seedbed and the environment at the time of planting. When exposure to light is necessary, seeds should be planted shallowly. *A rule of thumb is to plant seeds to a depth approximately three to four times their diameter.*

6. *Determine proper rate of sowing.* The rate of

sowing is critical in direct sowing in order to produce a desired plant density. It can be estimated by the following formula:

$$\begin{array}{c}\text{Weight of} \\ \text{seeds per} \\ \text{unit area}\end{array} = \frac{\text{density (plants/unit area) desired}}{\begin{array}{c}\text{number of} \\ \text{seeds/unit} \\ \text{weight} \\ \text{(seed count)}\end{array} \times \begin{array}{c}\text{germination} \\ \text{percent*}\end{array} \times \begin{array}{c}\text{purity} \\ \text{percent*}\end{array}}$$

This rate is a minimum and should be adjusted to account for expected losses in the seedbed, determined by previous experience at that site. Many seed companies will help producers in setting up spacing requirements for direct-seeding precision planting equipment.

Rates will vary with the spacing pattern. Field crops or lawn seeds may be *broadcast* (i.e., spaced randomly over the entire area) or *drilled* at given spaces. Other field crops, particularly vegetables, are *row planted,* so that the rate per linear distance in the row must be determined. Crops may be grown in rows on *raised beds,* particularly in areas of low rainfall where irrigation is practiced and excess soluble salts may accumulate to toxic levels through evaporation. Overhead sprinkling and planting seed below the crest of sloping seedbeds may eliminate or reduce this problem.

Seed tapes are available for flower and vegetable cultivars predominately for home gardens. Properly spaced seeds are attached to a plastic tape. The tapes are buried in shallow furrows and disintegrate from moisture as the seeds germinate.

7. *Provide post-planting care.* Prevent soil from drying out and developing a crust. This is primarily a function of seedbed preparation but may be aided by light sprinkling, shading, and covering with light mulch. With row planting, excess seed is planted, then the plants are **thinned** to the desired spacing. Thinning is expensive and time consuming and could be reduced or eliminated by precision planting.

Indoor Sowing for Production of Transplants

Seedling production is used extensively to produce flowers and vegetables for outdoor trans-

planting. Historically, this method has been used to extend the growing season by producing plantlets under protection for transplanting to the field as soon as the danger of spring frosts is over or placed under individual protectors to avoid freezing. This procedure also avoids some of the environmental hazards of germination and allows plants to be placed directly into a final spacing. Optimum germination conditions are provided in greenhouses, cold frames, or other structures to ensure good seedling survival and uniformity of plants.

Seedling growing has become an extensive bedding plant industry to produce small ornamental plants for instant home, park, and building landscaping, as well as vegetable plants for home gardening (9, 54). Commercial vegetable growing also relies heavily on the production of transplants, involving highly mechanized operations beginning with seed germination and ending with transplanting machines which place individual plants into the field.

Containers. Seeds are planted in a germination flat or container, and later the germinated seedlings are "pricked out" and transplanted to develop either in a transplant flat at wider spacing or in individual containers, where they remain until transplanted out-of-doors (see Chaps. 2 and 10).

Facilities. Indoor seedling production occurs in several types of structures, including greenhouses, cold frames, and hotbeds, as described in Chap. 2. Some bedding plant operations have special **growth rooms** where seed flats are placed on carts or shelves in an enclosed area and subjected to controlled environments (*66*). These facilities are used during the germination period, which usually lasts for the first two to three weeks. The flat is then moved to the greenhouse or growing area, where the small seedlings are transplanted.

Growth rooms need controlled lighting (daylength and intensity) and temperature (duration, level) and may include control of relative humidity, carbon dioxide fertilization, irrigation, and fertilization (*66*). Under proper conditions a grower may produce not only rapid uniform germination but also healthy seedlings with excellent

---

* Expressed as a decimal.

ability to transplant without any check in plant growth.

Media. Germination media for herbaceous bedding plants must retain moisture, yet provide good drainage and aeration. The pH should be neutral and soluble salts should be low. Some nutrients must be present, particularly phosphorus and calcium. Regular soil mixes, for example, equal parts soil, sand, and peat, might be used. These have been supplanted to a large extent by special nonsoil mixes made up of such components as peat moss, perlite, ground or shredded bark, and vermiculite, fortified by mineral nutrients or slow-release fertilizers (see Chap. 2). These mixes are available commercially and are used in large quantities by bedding plant producers. The mix should have excellent drainage to inhibit damping-off and reduce crusting. About 6 mm (¼ in.) of sterile sand, gently worked into the upper surface of the medium, might be used to avoid damping-off due to better aeration/moisture conditions. Sphagnum moss inhibits damping-off and is an effective germination medium. Small seeds should have a finer and more compact medium than is used for larger seeds. Seed flats and media should be pasteurized, and seeds should be treated for disease organisms (see Chap. 2).

Seed flats should be filled completely, the medium worked carefully into the corners, and the excess medium removed with a straight board drawn across the top. The medium is then tamped with a block to provide a uniformly firm seedbed to a level of about 12 mm (½ in.) below the top of the flat. The flat should be watered from above or presoaked and drained. Some media, such as vermiculite, should not be compacted.

Seeds may be broadcast over the surface or planted in rows. Advantages of row planting are reduced damping-off, better aeration, easier transplanting, and less drying out.

Planting at too high a density encourages damping-off, makes transplanting more difficult, and produces weaker, nonuniform seedlings. Suggested rates are 1000 to 1200 seeds per 29 × 54 cm (11 × 22 in.) flat for small-seeded species (e.g., petunia) and 750 to 1000 for larger seeds. Small seeds are dusted on the surface; medium seeds are covered lightly to about the diameter of

the seed. Larger seeds may be planted at a depth of two to three times their minimum diameter.

Moisture. Maintaining proper moisture conditions during the germination period is essential so that the medium neither dries out at any time nor remains so wet that damping-off becomes a problem. The seed flats may be held under polyethylene tents (see Fig. 2–12), or in small operations, covered with glass or plastic to keep the surface from drying out. An alternative method is to use a fine intermittent mist spray set for about eight seconds of water every 10 minutes during the day. Covered flats should not be exposed directly to sunlight, as excessive heat buildup injures the seedling.

Temperature. In general, a daily alternating temperature 20 to 30°C (68 to 86°F) is optimum for seeds of *warm temperature-requiring* and *cool temperature-tolerant* species. The alternations are 20°C for 16 hours and 30°C for eight hours. For seeds of *cool temperature-requiring* species, the germination temperature should be about 20°C (68°F) or lower. Light should be provided for germination of some kinds of seed.

Transplanting. The time required for germination can vary from about one to three weeks. When the first true leaves have appeared, the seedlings should be transplanted. Direct seeded flats should be shifted to a lower temperature for the seedlings of most kinds of plants to harden properly.

Transplanting flats or containers are filled with a growing medium and handled in the same manner as the seed flats. Holes are made in the medium at the correct spacing with a small **dibble.** The roots of each small seedling are inserted into a hole, and medium is pressed around them to provide good contact. Dibble boards are often used to punch holes for an entire flat at once. As soon as the flat is filled, it is thoroughly watered.

Seedling growing. Once seeds have germinated and the seedlings have been transplanted, the principal objectives of production are to prevent damping-off and to develop stocky, vigorous plants capable of further transplanting with little check in growth. The usual procedure in production is to shift the flats to lower temperatures (10°C or less) and to expose them to good light.

High temperatures and low light tend to produce spindly, elongated plants that will not survive transplanting. Such growth is termed "stretching."

Once root systems have developed sufficiently to grow into the medium, irrigation can be scheduled to keep the medium somewhat dry on the surface but moist underneath. Such irrigation helps prevent damping-off and produces sturdy seedlings.

Mechanized Seedling Production

Seed plugs. Millions of bedding plants are produced annually in greenhouses under carefully controlled environmental conditions for optimizing germination and plant growth. This has become possible mainly through the development of the **plug system** (*9, 16, 21, 35, 64, 79, 83*). A **plug** is a very small ball of artificial medium contained in a small cup or cell 1.6 (dia.) × 2.5 (ht.) cm (⅝ × 1 in.), of which several hundred are made of a single sheet of polystyrene, Styrofoam, or other suitable material. Sizes may range down to 1 × 1 cm (⅜ × ⅜ in.). These small cups of artificial germination medium (see Chap. 2) are sowed mechanically with a special seed sower (one seed per cup), then irrigated and fertilized with automatic misting and injector equipment (Fig. 7–8).

Seed plug trays are placed in greenhouses under mist or fog or in special germinators at optimum temperature and moisture conditions. It is important to have high-quality seed. Pelletting and seed priming is useful to improve germination rate, uniformity, and mechanical handling. Depending upon the seed, germination occurs within 5 to 7 days and the small plugs are transplanted into seed packs or pots, either mechanically or manually (Fig. 7–9).

Stages. Four morphological stages of seedling growth have been recognized (*64*):

Stage 1: radicle emergence
Stage 2: cotyledon spread
Stage 3: unfolding of three or four leaves
Stage 4: more than four leaves

Providing precise environmental control for each of the stages is essential in plug production.

FIGURE 7–8 *Above:* Plastic plug sheets being filled by automatic seeders—one seed to each cell which already contains a germination mix. *Below:* Plug cells containing seeds being covered by a uniform sand.

Adequate moisture must be maintained, but too much reduces the oxygen available to the seeds. Fine misting or fog is excellent. High temperature and moisture are essential for stage 1 but can be reduced in stage 2 and in later stages (see Table 7–2). Light (approximately 450 fc) and supplemental fertilization is important in stages 1 and 2. Fertilization (N, K, P) is particularly important in stage 3 but must be monitored carefully.

Transplanting. Stage 4 is a seedling almost ready to transplant. Temperatures should be lowered to 15°C (60 to 62°F) for 2 to 2½ weeks. Ammonium fertilization should be avoided at this

FIGURE 7-9 *Above:* Sheets of planted plugs placed over bottom heat (21°C; 70°F) for seed germination. *Below:* Small seedling plugs that have been transplanted into larger "6-packs" and placed over bottom heat (hot water in plastic pipes) for growing on and hardening off.

stage. It is better to use calcium or potassium nitrate. A plug dislodger is used to remove the plugs two to three hours after watering. Place into dibbled holes in packs or pots prepared with growing medium.

Other systems.   Figure 7–10 shows the elements of another completely mechanized system for field establishment of vegetable plants (*80*).

Field Seeding of Trees and Shrubs

Field seeding of forest trees is accomplished in reforestation either through natural seed dissemination or planting. Costs and labor requirements of direct seeding are lower than those for transplant-

ing seedlings, provided soil and site conditions favor the operation (*18*). The major difficulty is the very heavy losses of seeds and young plants that result from predation by insects, birds, and animals and from drying, hot weather, and disease (*71, 72*). A proper seedbed is essential and an open mineral soil with competing vegetation removed is best. The soil may be prepared by burning, disking, or furrowing. Seeds may be broadcast by hand or by special planters, or drilled with special seeders. Seeds should be coated with a bird and rodent repellent.

In certain situations, trees and shrubs can be directly seeded into a natural setting for landscape purposes (*15*). Dig a 4- to 5-in. hole, carefully pulverizing soil if compacted. On a rocky slope, place seeds in a crevice or pocket of soil. If on a slope, make a slight backslope to avoid erosion and to prevent the seed hole from being covered with loose soil from above. Place 1 gram of slow-release fertilizer in the bottom of the hole. Replace soil, leaving a slight depression for the seed. Plant 2 to 20 seeds, and cover with pulverized soil to 3.2 to 12.7 mm (⅛ to ½ in.) deep, depending upon size. Contact herbicides or mulches around the seed site are essential to remove weed competition. Mulches can be coarse organic materials (wood chips, coarse bark) or sheet materials (uncoated and polyethylene-coated mulching paper, black polyethylene film, asphalt roofing paper, or kraft building paper). With the coarse organic mulches use a tin can (size 2½), milk carton, or asphalt or kraft paper collar around the seed site. Cut a hole in the center of the sheet mulches to allow the seedlings to emerge. Small wire cages may be needed to cover the site for rodent and bird protection.

Production of Woody Plant Seedlings in Containers

Production of seedling trees and shrubs in containers involves two main activities (a) the production of landscape or fruit or nut tree plants, and (b) the mass propagation of small seedlings for reforestation. Landscape and fruit tree seedlings may be started in germination flats. Some nurseries have utilized direct seeding into plug-tray systems (*65*), avoiding the conventional propagation flats, then transplanted to either a

TABLE 7–2

Requirements for seed germination during plug propagation of three popular bedding plants

	Petunia	Pansy	Impatiens
		Stage 1	
Temperature	80°F (27°C)	77°F (25°C)	80°F (27°C)
Moisture	100%, 1.5 vpd[a]	100%, 0.5 vpd	100%, 1.0 vpd
Light	14.3 Wm^{-2}[b]	12.6 Wm^{-2}	14.3 Wm^{-2}
Fertilizer	25–75 ppm KNO 3 1 application (1–3 days)	25–50 ppm KNO 3 (1–7 days)	None
		Stage 2	
Temperature	75°F (24°C)	66°F (18°C)	75°F (24°C)
Moisture	85%, 5.5 vpd	75%, 6.0 vpd	75%, 5.5 vpd
Light	14.3 Wm^{-2}	14.3 Wm^{-2}	14.3 Wm^{-2}
Fertilizer	50 ppm 20–10–20 (3–7 days)	None	None

Source: Ref. 64.

[a]vpd = vapor pressure deficit

[b]Wm^{-2} = Watts/square meter, converted from fc assuming lights are fluorescent cool-white lamps.

transplant flat or a small pot. Later they are moved to slightly larger containers, or transplanted directly into the containers where they will remain until transplanted out-of-doors (Fig. 7–11).

With this system root pruning is essential to induce a desirable root system. Root pruning should be done at the first transplanting, soon after the roots reach the bottom of the flat (*33, 34*). Further root pruning should take place when the liner flat is transplanted, which should be done when the roots protrude through the pot about an inch. These roots are removed. Any delay in transplanting at either stage, or omission of pruning, will increase the incidence of poor root systems. Copper or plastic screen-bottomed flats (*27*) or supports stimulate formation of branch roots.

A second activity involving woody plant production is the mass propagation of small seedlings for direct planting in reforestation (*53, 78*), for planting into outdoor beds for an additional one or two years' growth (*70*), and for ornamental shrub production (*60*). Seeds are invariably planted directly into various kinds of small individualized containers—usually narrow and relatively deep. These may be plastic containers from which the seedling plug is removed and planted (see Fig. 2–15) or they may be containers made

of substances such as peat or fiber blocks, which are planted with the seedling (see Fig. 2–16).

Outdoor Seedbeds for Woody Perennials

Outdoor seedbeds where seeds are planted closely together in beds are used extensively for growing conifers and deciduous plants for forestry (*2, 72*) and for ornamentals (*14, 20, 30, 51, 60*) and to provide rootstock liners for some fruit and nut tree species (*32, 50, 68, 75*) (Fig. 7–12).

1. *Prepare the site and soil.* Nursery production requires a fertile, well-drained soil of medium to light texture. Preparation for planting may include rotation with other crops and incorporation of a green manure crop or animal manure (*74*). Preplant fumigation and weed control are essential aspects of most nursery operations.

Weed control can be facilitated by careful seedbed preparation, cultivation, and chemical sprays. Three types of chemical controls are available. **Preplant** fumigation with methyl bromide, chloropicrin, Vapam, steam, or the like is effective and also kills disease organisms and nematodes. **Pre-emergence** herbicides are applied to the soil before the weed seeds emerge. **Post-emergence** herbicides can be applied as soon as the

FIGURE 7–10   Mechanized seedling production utilizing the Speedling system. *Upper left:* Mechanized seeder. *Upper right:* Open greenhouse with automatic watering. *Lower left:* Transplant with characteristic root ball. *Lower right:* Mechanized transplanting from special flats. (Courtesy George Todd, Sr. The Speedling system, *Proc. Inter. Plant Prop. Soc.* 31:612–16, 1981.)

FIGURE 7–11   Plug propagation of forest tree seedlings. *Top:* Single seeds are planted in individual cells of plastic blocks. *Below:* Seedling after rooting. Note the root system which utilizes ''air pruning'' to stimulate root branching.

FIGURE 7–12 Growing seedlings in seedbeds. Olive (*olea europaea*) seedlings in Italy planted thickly in unshaded beds.

weed seedlings have emerged. A wide range of selective and nonselective commercial products are available. Such materials should be used with caution, however, since improper use can cause injury to the young nursery plants. Not only should directions of the manufacturer be followed, but preliminary trials should be made before large-scale use.

2. *Prepare the seedbed.* A common size of seedbed is 1.1 to 1.2 m (3½ to 4 ft) wide with the length varying according to the size of the operation. Beds may be raised to ensure good drainage, and in some cases, sideboards are added after sowing to maintain the shape of the bed and to provide support for glass frames or lath shade. Beds are separated by walkways 0.45 to 0.6 m (1½ to 2 ft) wide. North–south orientation gives more even exposure to light than east-west.

Seed may be either broadcast over the surface of the bed or drilled into closely spaced rows with seed planters. For economy, seeds should be planted as close together as feasible without overcrowding, which increases damping-off and reduces vigor and size of the seedlings (*38*), resulting in thin, spindly plants and small root systems. Seedlings with these characteristics do not transplant well (*37*).

3. *Plant the seeds.* Seeds are planted in the nursery in the fall, spring, or summer, depending on the dormancy conditions of the seed, the temperature requirements for germination, the management practices at the nursery, and the location of the nursery (in a cold-winter or a mild-winter area). Planting time varies for several general categories of seed (*40, 51, 59, 71*):

a. Certain species (e.g., apple, pear, *Prunus,* yew) require moist-chilling (stratification) and are adversely affected by high germination temperatures (Table 7–1), which produce secondary dormancy. Germination temperatures of 10 to 17°C (50 to 62°F) are optimum. Seeds of these species could be planted in the fall with germination taking place in late winter or early spring. This procedure is particularly desirable in mild-winter areas. Or seeds can be stratified during winter storage and planted as early in the spring as permitted by climate. This procedure is more likely to be used in cold-winter areas where fall-planted germinating seed might be injured by very low temperatures.

A modification of the fall planting procedure is useful for seeds that have hard seed coats plus dormant embryos. Seeds are planted in summer or early fall to allow six to eight weeks of warm stratification in the seedbed prior to the winter chilling (*50*). Another modification is collecting seeds before they are fully ripe and planting without allowing them to dry.

b. Many kinds of seeds, including most conifers (pine, fir, spruce) and many deciduous hardwood species, benefit from moist-chilling, do not germinate until soil temperatures have warmed up, and are not inhibited by high soil temperatures. Optimum germination temperatures are 20 to 30°C (68 to 86°F). Seeds of these species would not germinate in the cool fall season or in early spring. Such seeds can be handled either by fall planting or by spring planting following the required chilling treatments (stratification).

c. Seeds of some species (e.g., black locust, Colorado and Norway spruce, Chinese and Siberian elm, European larch, bristlecone pine, and Douglas fir) are able to germinate over a wide range of temperatures 15 to 32°C (60 to 90°F). Seeds of some of these species require moist-chilling or have hard coats requiring special treatment. If fall-planted in cold areas, seeds of this group may germinate prematurely when the seedlings can be injured by winter cold. Spring planting is recommended, although the actual time is not as critical as that for the other two groups.

d. Seeds of some species (maple, poplar, aspen, elm, cottonwood) ripen in spring or early summer. Such seeds should be planted immediately, as their viability declines rapidly.

4. *Determine seeding rates.* The optimum seed density depends primarily on the species but also on the nursery objectives. If a high percentage of the seedlings are to reach a desired size for field planting, low densities might be desired, but if the seedlings are to be transplanted into other beds for additional growth, higher densities (with smaller seedlings) might be more practical. Once the actual density is determined, the necessary rate of sowing can be calculated from data obtained from a germination test and from experience at that particular nursery.

The following formula is useful in calculating the rate of seed sowing (*26, 51, 72*):

$$\begin{array}{l}\text{Weight of} \\ \text{seeds to} \\ \text{sow per} \\ \text{area}\end{array} = \frac{\text{density (plants/unit area) desired}}{\begin{array}{c}\text{purity} \\ \text{percent*}\end{array} \times \begin{array}{c}\text{germi-} \\ \text{nation} \\ \text{percent*}\end{array} \times \begin{array}{c}\text{field} \\ \text{factor*}\end{array} \times \begin{array}{c}\text{number of} \\ \text{seeds/unit} \\ \text{weight} \\ \text{(seed count)}\end{array}}$$

* Expressed as a decimal.

Field factor is a correction term applied based on the expected losses which experience at that nursery indicates will occur with that species. It is a percentage expressed as a decimal.

Seeds can be planted by (a) broadcasting by hand or seeders, (b) hand spacing (larger seeds), or (c) drilling by hand with push drills, or drilling with tractor-drawn precision drills. Seeds of a particular lot should be thoroughly mixed before planting to ensure that the density in the seedbed will be uniform. Treatment with a fungicide for control of damping-off is desirable. Small conifer seeds may be pelleted for protection against disease, insects, birds, and rodents. Depth of planting varies with the kind and size of the seed. In general, a depth of three to four times the diameter of the seed is satisfactory. Seeds can be covered by soil, coarse sand, or by various mulches.

Soil firming may be done to increase contact of seed and soil. It is carried out with a tamper, hand roller, or tractor-drawn roller either before sowing or immediately afterward. Rodent and bird protection may be necessary.

5. *Provide aftercare.* During the first year in the seedbed, the seedlings should be kept growing continuously without any check in development. A continuous moisture supply, cultivation or herbicides to control weeds, and proper disease and insect control contribute to successful seedling growth. Fertilization with nitrogen is usually necessary, particularly when a mulch has been applied, since decomposition of organic material can produce nitrogen deficiency. In the case of tender plants, glass frames can be placed over the beds, although for most species a lath shade is sufficient. With some species, shade is necessary throughout the first season; with others, shade is necessary only during the first part of the season. Sprinkling with water to reduce ground temperature during the hot part of the day is sometimes useful (*22*).

Seedlings remain one to three years and are then transplanted to a permanent location. For some species the plants may be shifted to a transplant bed after one year and then grown for a period of time at wider spacing. This basic procedure is used to propagate millions of forest tree seedlings, both conifer and deciduous species.

Seedlings produced in a seedbed nursery are often designated by numbers to indicate the length of time in a seedbed and the length of time in a transplant bed. For instance, a designation of 1–2 means a seedling grown one year in a seedbed and two years in a transplant bed or field. Similarly, a designation of 2–0 means a seedling produced in two years in a seedbed and none in a transplant bed.

Nursery Row Production

Planting directly in separate nursery rows (Fig. 7–9) is one of the primary methods used to propagate rootstocks of most fruit and nut tree species (*32, 68, 75*). Cultivars are budded or grafted to the seedlings in place (see Chaps. 12 and 13). The method is also used to propagate shade trees and ornamental shrubs, either as seedlings or on rootstocks as budded selected cultivars.

As described in the preceding section, seedbeds may be used to produce liners, which are small plants dug after one year and then ''lined out'' in a nursery row, used for growing woody ornamentals to a salable size or as rootstocks of deciduous fruit trees. This procedure is used where the growing season is short, or for species such as apple, pear, and cherry which grow relatively slowly and are not large enough for budding in one year.

Deciduous fruit, nut, and shade tree propagation usually begins by planting seeds or liners in nursery rows. Where plants are to be budded or grafted in place, the width between rows is about 1.2 m (4 ft) and the seeds are planted 7.6 to 10 cm (3 to 4 in.) apart in the row (see Fig. 7–9). Seeds known to have low germinability must be planted closer together to get the desired stand of seedlings. Large seed (walnut) can be planted 10 to 15 cm (4 to 6 in.) deep, medium-sized seed (apricot, almond, peach, and pecan) about 7.6 cm (3 in.), and small seed (myrobalan plum), about 3.8 cm (1½ in.). Spacing may vary with soil type. If germination percentage is low and a poor stand results, the surviving trees may, because of the wide spacing, grow too large to be suitable for budding. Plants to be grown to a salable size as seedlings without budding could be spaced at shorter intervals and in rows closer together (Fig. 7–13).

Fall planting of fruit and nut tree seeds is commonly used in mild-winter areas such as California (*32*). Seeds are planted 2.5 to 3.6 cm (1 to 1½ in.) deep and 10 to 15 cm (4 to 6 in.) apart depending on size, and then covered with a ridge of soil 15 to 20 cm (6 to 8 in.) deep, in which the seeds remain to stratify over winter. The soil ridge is removed in the spring just before seedling emergence. Herbicide control of weeds and protection of the seeds from rodents become important considerations during these procedures.

FIGURE 7–13 Direct seeding in nursery row, illustrated by peach seedlings grown for rootstocks. These seedling plants will be budded to peach cultivars.

TRANSPLANTING SEEDLING MATERIAL TO PERMANENT LOCATIONS

The final step in seedling production is transplanting to a permanent location (*57, 58, 67*). Seedlings may be transplanted either **bare-root** (vegetable transplants or deciduous fruit, nut and shade trees), or in cells or **modular containers** (bedding plants, vegetables, forest trees), or **balled** and **burlapped** (evergreen trees) or **containerized** (ornamental shrubs and trees).

Bare-root transplanting invariably results in some root damage and transplant shock, both of which check growth. With vegetable plants these may result in premature seedstalk formation, increased susceptibility to disease, and reduced yield potential. Handling prior to transplanting should involve **hardening-off,** which involves a controlled growth cessation. This causes accumulation of carbohydrates, making the plant better able to withstand adverse environmental conditions. This effect can be achieved by temporarily withholding moisture, reducing temperature, and gradually shifting from protected to outdoor conditions over a period of a week to 10 days.

Ornamental and Vegetable Bedding Plants

During the transition to the new site, deterioration must be prevented if the plants are bare-root. Following planting, conditions must be provided for rapid root regeneration. Planting should be done as soon as possible. If not, bare-root plants can be kept (no more than 7 to 10 days) in moist, cool (10°C; 50°F) storage. Maintain high humidity but avoid direct watering to prevent disease.

Modular transplants (plugs or other containers) should be kept under conditions to prevent drying of the roots. Expose to light. Keep temperatures at 18 to 20°C (65 to 70°F).

Field beds should be moderately well pulverized although not necessarily finely prepared, well watered but not saturated (*67*). Transplanting is done in the field by hand or by machine. Afterward a good amount of irrigation should be applied to increase moisture to the roots and settle the soil but not saturate it. A starter solution containing fertilizers should be applied, but if the soil is dry, should best be diluted. Temporary shade may be used for the first few days.

Trees and Shrubs

Transplanting of bare-root evergreen forest trees follows principles similar to those described. Seedling plants should be dug in the nursery in the fall after proper physiological "hardening-off" (*71*). Seedlings are packed into moisture-retaining material (vermiculite, peat moss, sawdust, shingletow) and kept in low-temperature (30 to 35°F), humid (at least 90 percent RH) storage. Polyethylene bags without moisture material are satisfactory. Some kinds of sawdust can be toxic, particularly if fresh. Bare-root nursery stock of deciduous plants and container-grown stock are handled as described for rooted cuttings in Chap. 10.

REFERENCES

1. Akers, S. W., and K. E. Holley. 1986. SPS: A system for priming seeds using aerated polyethylene glycol or salt solutions. *HortScience* 21(3):529–31.

2. Aldous, J. R. 1972. *Nursery practice.* For. Comm. Bul. 43. London: Her Majesty's Stationery Office.

3. Allison, C. J. 1980. X-ray determination of horticultural seed quality. *Proc. Inter. Plant Prop. Soc.* 30:78–86.

4. Anonymous. 1975. Seed testing regulations under the Federal Seed Act. *U.S. Dept. Agr., Agr. Mark. Ser.,* 184 pp.

5. Assoc. Off. Seed Anal. 1978. Rules for seed testing. *Jour. Seed Tech.* 3(3):1–126.

6. Baker, K. F. 1962. Thermotherapy of planting material. *Phytopathology* 52:1244–55.

7. ———. 1972. Seed pathology. In *Seed biology,* Vol. 2, T. T. Kozlowski, ed. New York: Academic Press.

8. Baker, K. F., and P. A. Chandler. 1957. Development and maintenance of healthy planting stock. In *The U.C. system for producing healthy container-grown plants,* K. F. Baker, ed. Calif. Agr. Exp. Sta. Man. 23, pp. 217–36.

9. Ball, V., ed. 1985. *Ball red book: Greenhouse growing* (14th ed.). Reston, Va.: Reston Publ. Co.

10. Bewley, J. D., and M. Black. 1985. *Seeds: Physiology of development and germination.* New York: Plenum Press.

11. Bonner, F. T. 1974. Seed testing. In *Seeds of woody plants in the United States,* C. S. Schopmeyer, ed. U.S. Dept. Agr. Handbook 450, Washington, D.C.: U.S. Govt. Printing Office, pp. 136–52.

12. Bradford, K. J. 1986. Manipulation of seed water relations via osmotic priming to improve germination under stress conditions. *HortScience* 21(5):1105–12.

13. Brumback, W. E. 1985. Propagation of wildflowers. *Proc. Inter. Plant Prop. Soc.* 35:542–47.

14. Carville, L. 1978. Seed bed production in Rhode Island. *Proc. Inter. Plant Prop. Soc.* 28:114–17.

15. Chan, F. J., R. W. Harris, and A. T. Leiser. 1971. Direct seeding of woody plants in the landscape. *Univ. of Calif. Ext. Ser. AXT-n27.*

16. Cooley, J. 1985. Vegetable plant raising using Speedling transplants. *Proc. Inter. Plant Prop. Soc.* 35:468–71.

17. Copeland, L. O. 1976. *Principles of seed science and technology.* Minneapolis: Burgess.

18. Deer, H. J., and W. F. Mann, Jr. 1971. *Direct seedling pines in the South.* U.S. Dept. Agr. Handbook 391. Washington, D.C.: U.S. Govt. Printing Office.

19. Delong, S. K. 1985. Custom seed preparation for optimum conifer production. *Proc. Inter. Plant Prop. Soc.* 35:259–63.

20. Dirr, M. A., and C. W. Heuser, Jr. 1987. *The reference manual of woody plant propagation: From seed to tissue culture.* Athens, Ga.: Varsity Press.

21. Eastburn, D. P. 1984. The plug potential. *Florists Rev.* 174:6–29.

22. Eden, C. J. 1962. Conifer seed—from cone to seed bed. *Proc. Plant Prop. Soc.* 12:208–14.

23. ———. 1965. Use of X-ray technique for determining sound seed. *U.S. Forest Service, Tree Planters' Notes* 72:25–27.

24. Edwards, D. G. W. 1986. Special prechilling techniques for tree seeds. *Jour. Seed Tech.* 10(2):151–71.

25. Flemion, F. 1938. A rapid method for determining the viability of dormant seeds. *Contrib. Boyce Thomp. Inst.* 9:339–51.

26. Fordham, D. 1976. Production of plants from seed. *Proc. Inter. Plant Prop. Soc.* 26:139–45.

27. Frolich, E. F. 1971. The use of screen bottom flats for seedling production. *Proc. Inter. Plant Prop. Soc.* 21:79–80.

28. Furatani, S. C., B. H. Zandstra, and H. C. Price. 1985. Low temperature germination of celery seeds for fluid drilling. *Jour. Amer. Soc. Hort. Sci.* 110(2):149–53.

29. Ghate, S. R., S. C. Phatak, and K. M. Batal. 1984. Pepper yields from fluid drilling with additives and transplanting. *HortScience* 19(2):281–83.

30. Gordon, A. G., and D. C. F. Rowe. 1982. *Seed manual for ornamental trees and shrubs.* Bull. 59. London: Forestry Commission.

31. Gray, D. 1981. Fluid drilling of vegetable seeds. *Hort. Rev.* 3:1–27.

32. Hall, T. 1975. Propagation of walnuts, almonds and pistachios in California. *Proc. Inter. Plant Prop. Soc.* 25:53–57.

33. Harris, R. W., W. B. Davis, N. W. Stice, and D. Long. 1971. Root pruning improves nursery tree quality. *Jour. Amer. Soc. Hort. Sci.* 96:105–9.

34. ———. 1971. Influence of transplanting time in nursery production. *Jour. Amer. Soc. Hort. Sci.* 96:109–10.

35. Hartnett, G. P. 1985. New ideas in the use of plug systems. *Proc. Inter. Plant Prop. Soc.* 35:263–68.

36. Heit, C. E. 1955. The excised embryo method for testing germination quality of dormant seed. *Proc. Assn. Off. Seed Anal.* 45:108–17.

37. ———. 1964. The importance of quality, germinative characteristics and source for successful seed propagation and plant production. *Proc. Inter. Plant Prop. Soc.* 14:74–85.

38. ———. 1967. Propagation from seed. 5. Control of seedling density. *Amer. Nurs.* 125(8):14–15, 56–59.

39. ———. 1967. Propagation from seed. 6. Hard-seededness, a critical factor. *Amer. Nurs.* 125(10):10–12, 88–96.

40. ———. Undated. Thirty five years testing of tree and shrub seed. New York Agr. Exp. Sta. (Geneva), mimeographed leaflet.

41. Heydecker, W., and P. Coolbear. 1977. Seed treatments for an improved performance—survey and attempted prognosis. *Seed Sci. and Technol.* 5:353–425.

42. Hoyle, B. J., H. Yamada, and T. D. Hoyle. 1972. Aggresizing—to eliminate objectionable soil clods. *Calif. Agr.* 26(11):3–5.

43. Ibarbia, E. 1987. Coated seed benefits trans-

plants and direct seeding. *Amer. Veg. Grower (Western ed.)* 35(9):14–16.

44. International Seed Testing Association. 1985. International rules for seed testing. *Seed Sci. and Technol.* 13:299–513.

45. Justice, O. L. 1972. Essentials of seed testing. In *Seed biology,* Vol. 3, T. T. Kozlowski, ed. New York: Academic Press.

46. Kamra, S. K. 1964. The use of x-rays in seed testing. *Proc. Inter. Seed Testing Assn.* 29:71–79.

47. Khan, A. A. 1978. Incorporation of bioactive chemicals into seeds to alleviate environmental stress. *Acta Hort.* 83:225–34.

48. Koranski, D. S. 1988. Primed seed: A step beyond refined seed. *Grower Talks* 51(9):24–29.

49. Lakon, G. 1949. The topographical tetrazolium method for determining the germinating capacity of seeds. *Plant Phys.* 24:389–94.

50. Lawyer, E. M. 1978. Seed germination of stone fruits. *Proc. Inter. Plant Prop. Soc.* 28:106–9.

51. Macdonald, B. 1986. *Practical woody plant propagation for nursery growers,* Vol. 1. Portland, Oreg.: Timber Press.

52. MacKay, D. B. 1972. The measurement of viability. In *Viability of seeds,* E. H. Roberts, ed. Syracuse, N.Y.: Syracuse Univ. Press.

53. Maclean, N. M. 1968. Propagation of trees by tube technique. *Proc. Inter. Plant Prop. Soc.* 18:303–9.

54. Mastalerz, J. W. 1976. *Bedding plants.* University Park, Pa.: Pennsylvania Flower Growers.

55. Maude, R. B. 1978. Seed treatments for pest and disease control. *Acta Hort.* 83:205–12.

56. McDonald, M. B., Jr. 1980. Assessment of seed quality. *HortScience* 15:784–88.

57. McKee, J. M. T. 1981. Physiological aspects of transplanting vegetables and other crops. I. Factors which influence re-establishment. *Hort. Abstracts* 51(5):265–72.

58. ———. 1981. Physiological aspects of transplanting vegetables and other crops. II. Methods used to improve transplant establishment. *Hort. Abstracts* 51(6):355–68.

59. McMillan-Browse, P. D. A. 1978. Stratification—a detail of technique. *Proc. Inter. Plant Prop. Soc.* 28:191–92.

60. ———. 1979. *Hardy woody plants from seed.* London: Grower Books.

61. Mechel, B. E., and M. R. Kauffman. 1973. The osmotic potential of polyethylene glycol 6000. *Plant Phys.* 51:914–16.

62. Mikkelson, D. S., and M. N. Sinah. 1961. Germination inhibition in *Oryza sativa* and control by preplanting soaking treatments. *Crop Sci.* 1:332–35.

63. Moore, R. P. 1973. Tetrazolium staining for assessing seed quality. In *Seed ecology,* W. Heydecker, ed. London: Butterworth, pp. 347–66.

64. Ohio State University. 1987. *Tips on growing bedding plants.* Ohio Coop. Ext. Ser. and Ohio State Univ. MM 265, AGDEX 200/15. Columbus, Ohio: Ohio State Univ.

65. Pinney, T. S., Jr. 1986. Update of GROPLUG® system. *Proc. Inter. Plant Prop. Soc.* 36:577–81.

66. Poynter, M. J. 1978. Building and using a growing room for seed germination of bedding plants. *Proc. Inter. Plant Prop. Soc.* 28:109–14.

67. Price, H. C., and B. H. Zandstra. 1988. Maximize transplant performance. *Amer. Veg. Grower* 36(4):10–16.

68. Rom, R. C., and R. F. Carlson, eds. 1987. *Rootstocks for fruit crops.* New York: John Wiley.

69. Rudolf, P. O. 1965. State tree seed legislation. *U.S. Forest Service, Tree Planters' Notes* 72:1–2.

70. Schalla, S. L., and K. Doughton. 1978. Transplanting the Douglas fir plug. *Proc. Inter. Plant Prop. Soc.* 28:177–84.

71. Schubert, G. H., and R. S. Adams. 1971. *Reforestation practices for conifers in California.* Sacramento: Calif. State Div. of Forestry.

72. C. S. Shopmeyer, ed. 1974. *Seeds of woody plants in the United States.* U.S. For. Ser. Agr. Handbook 450. Washington, D.C.: U.S. Govt. Printing Office.

73. Simak, M. 1957. The x-ray contrast method for seed testing. *Medd. f. Stat. skogsforssr. Inst.* 47(4):1–22.

74. Steavenson, H. 1979. Maximizing seedling growth under midwest conditions. *Proc. Inter. Plant Prop. Soc.* 29:66–71.

75. Stuke, W. 1960. Seed and seed handling techniques in production of walnut seedlings. *Proc Plant Prop. Soc.* 10:274–77.

76. Suszka, B. 1978. Germination of tree seed stored in a partially after-ripened condition. *Acta Hort.* 83:181–88.

77. Taylor, A. G., and T. J. Kenny. 1985. Improvement of germinated seed quality by density separation. *Jour. Amer. Soc. Hort. Sci.* 110(3):347–49.

78. Thomson, J. R. 1979. *An introduction to seed technology.* New York: John Wiley.

79. Tinus, R. W., and S. E. McDonald. 1979. How to grow tree seedlings in containers in greenhouses. *USDA For. Ser. Gen. Tech. Rpt. RM-60.*

80. Todd, G., Sr. 1981. The Speedling system. *Proc. Inter. Plant Prop. Soc.* 31:612–16.

81. U.S. Department of Agriculture. 1952. *Manual for testing agriculture and vegetable seeds.* U.S. Dept. Agr. Handbook 30. Washington, D.C.: U.S. Govt. Printing Office.

82. Vanstone, D. E. 1978. Basswood (*Tilia americana* L.) seed germination. *Proc. Inter. Plant Prop. Soc.* 28:566–69.

83. Welsh, T. E. 1984. Plugs: Use and future in New Zealand. *Proc. Inter. Plant Prop. Soc.* 34:375–80.

84. Western Forest Tree Seed Council. 1966. *Sampling and service testing western conifer seeds.* Portland, Oreg.: Western Forestry and Conservation Association, 36 pp.

SUPPLEMENTARY READING

ALDOUS, J. R. 1972. *Nursery practice.* For. Comm. Bul. 43. London: Her Majesty's Stationery Office.

ASSOCIATION OF OFFICIAL SEED ANALYSTS. Proceedings of annual meetings.

BALL, V., ed. 1985. *Ball red book: Greenhouse growing* (14th ed.). Reston, Va.: Reston Publ. Co.

GRAY, D. 1981. Fluid drilling of vegetable seeds. *Hort. Rev.* 3:1–27.

INTERNATIONAL PLANT PROPAGATORS' SOCIETY. Proceedings of annual meetings.

INTERNATIONAL SEED TESTING ASSOC. *Jour. Seed Technology.* Annual.

LORENZ, O. A., and D. N. MAYNARD. 1988. *Knotts handbook for vegetable growers.* New York: John Wiley.

MASTALERZ, J. W. 1976. *Bedding plants.* University Park, Pa.: Pennsylvania Flower Growers.

MCMILLAN-BROWSE, P. D. A. 1979. *Hardy woody plants from seed.* London: Grower Books.

PROFESSIONAL PLANT GROWERS ASSOCIATION. Proceedings of annual meetings.

SCHOPMEYER, C. S., ed. 1974. *Seeds of woody plants of the United States.* U.S. For. Ser. Agr. Handbook 450. Washington, D.C.: U.S. Govt. Printing Office.

STOECKELER, J. H., and P. E. SLABAUGH. 1965. *Conifer nursery practice in the prairie states.* U.S. Dept. Agr. Handbook 279. Washington, D.C.: U.S. Govt. Printing Office.

THOMSON, J. R. 1979. *An introduction to seed technology.* New York: John Wiley.

TINUS, R. W., and S. E. MCDONALD. 1979. How to grow tree seedlings in containers in greenhouses. *USDA For. Ser. Gen. Tech. Rpt. RM-60.*

U.S. DEPT. OF AGRICULTURE. 1952. *Manual for testing agricultural and vegetable seeds.* U.S. Dept. Agr. Handbook 30. Washington, D.C.: U.S. Govt. Printing Office.

———. 1961. *Seeds: Yearbook of agriculture.* Washington, D.C.: U.S. Govt. Printing Office.

8

Source Selection and Management in Vegetative Propagation

Vegetative, or asexual, propagation is used to produce a plant identical in genotype with the source (mother) plant. New roots and/or shoots are regenerated on stems, leaves, or roots (**cuttings** and **layers**) as described in Chaps. 9, 10, and 14. Roots and shoots are united into a single plant by **grafting** or **budding** as described in Chaps. 11, 12, and 13. Some plant species grow and reproduce naturally by special vegetative structures, such as runners, bulbs, corms, rhizomes, tubers, and other propagules, as described in Chaps. 14 and 15 under **separation** and **division**.

Vegetative propagation is possible because living cells contain genetic information in their nuclei necessary to reproduce the entire plant. This property is called **totipotency** (*88*). Division of vegetative cells occurs in the apical meristems (Fig. 1–1). As the plant grows and develops, these meristematic cells become programmed to differentiate into specific cell types that make up the stems, leaves, roots, and eventually flowers and reproductive organs. The orderly progression of these events occurs in ontogenetic (life cycle) stages, described as **embryonic, juvenile**, or **mature** (adult), and results in the seedling life cycle, either sexual or apomictic, as described in Chap. 3. In annual and biennial plants, these cycles are completed in one or two yearly cycles. In perennials, the plant continues to grow and reproduce by consecutive (annual) or overlapping (biennial) vegetative to reproductive phases (*48, 49*) (Fig. 3–1).

Traditional methods of vegetative propagation have gradually expanded to include **micropropagation**, in which new plants are produced from very small structures (embryos, shoot tips, meristems) in aseptic culture as described in Chaps. 16 and 17. New plants of some species can be regenerated vegetatively from protoplasts, cells, callus, microspores, and various isolated tissues. Associated with this technology are novel methods of plant breeding, including the ability to isolate and transfer fragments of DNA representing specific genes from one vegetative (somatic) cell to another or to combine two somatic cells together outside of the normal plant reproductive processes. The latter procedures are categorized as **genetic engineering** (*39, 52*). Such technologies have become an important aspect of modern propagation and plant breeding and are included in the term **biotechnology**.

REASONS FOR USING VEGETATIVE PROPAGATION

Convenience and Ease of Propagation

Some important food crops can be propagated more easily, more conveniently, or more economically by vegetative propagation methods than by seed. One of the major historical advances in early day, primitive agriculture was the discovery that some important food crops, such as grape, olive, and fig, could be propagated by thrusting pieces of their woody stems into the ground to produce roots and subsequently, new plants (*113*). Other important crops (*84*), including potato, yam, banana, and sugarcane, could be propagated readily by separation or division of the tuber, bulbs, rhizomes, or other natural vegetative structures (Chaps. 14 and 15).

Vegetative propagation is not a natural phenomenon for most plant species and special techniques and facilities have been developed to facilitate propagation. For example, pomological crops, such as apple, pear, peach, and cherry, do not root readily as cuttings. Consequently, vegetative propagation of these crops awaited the development of grafting and budding (*113*).

The history of horticulture has been characterized in part by the development of progressively advanced technology for vegetative propagation. Controlled environment facilities (e.g., greenhouses), rooting hormones, mist propagation, etc. (*60*), have resulted in an ever-widening range of genetic materials amenable to vegetative propagation.

As a general statement, vegetative propagation is less economical (more expensive per plant) than propagation of the same plant by seed. Its primary economic benefit for most species is the high value placed on vegetatively propagated cultivars.

Selection and Maintenance of Clones

Large genetic advances can be made in a single step by selecting a single unique superior plant from a seedling population and reproducing it asexually by vegetative propagation (*58*). The resulting population of plants has the same basic genotype as the original seedling plant and is called a **clone** (*83, 87, 92, 107*). This category is one of the basic cultivars defined in the *Code of Nomenclature for Horticultural Plants* (see Chap. 1).

The genotypes of most woody and herbaceous perennials in cultivation are highly heterozygous and have long breeding cycles. Vegetative propagation is important for these crops because their genetic improvement by breeding is slow (*8, 58, 68*). Clonal propagation is a highly efficient method for immediately ''fixing'' genetic variation, in contrast to the sequence of generations required for seedling populations. Clonal selection is therefore analogous to F_1 hybrid breeding (see Chap. 4). Like hybrids, the unique characteristics of the clonal cultivar are immediately lost if it is propagated sexually by seed (see Chaps. 1 and 4).

Some desirable horticultural cultivars are seedless and must by necessity be propagated vegetatively. Examples include seedless grapes, bananas, navel oranges, and figs.

Since members of a clone are genetically uniform, they are also highly uniform phenotypically. That is, the plants will have the same uniform appearance, size, blossoming time, fruit maturity, etc. This uniformity is the basis for the highly standardized production practices that are characteristic of modern commercial horticulture.

Many clones of horticultural interest have been discovered and perpetuated by humans as named cultivars (*65, 67*). The 'Cabernet Sauvignon' and 'Sultana' (known also as 'Thompson Seedless') grapes have been perpetuated asexually as millions of plants for about 2000 years with little change. This is true also for some fig cultivars. The 'Bartlett' pear (also known as 'Williams Bon Chretien') originated as a seedling in England in 1770 and has been maintained vegetatively ever since. The 'Delicious' apple originated about

1870 in Jesse Hiatt's apple orchard in Peru, Iowa. Through vegetative propagation, not only have millions of identical 'Delicious' apple trees been grown throughout the world but new cultivars have arisen as color "mutations" or "budsports" from the original clone. The latter process is described further in this chapter.

Genetic uniformity within monoclonal plantations can be a disadvantage. Members of a clone have uniform susceptibility to insects, diseases, or other environmental hazards. Such a **monoculture** (*26, 112*) can be vulnerable to the introduction of a new pest into a locality or to the introduction of the cultivar to a new environment where conditions and pests occur that did not exist in the original growing area.

Combination of More Than One Genotype Into a Single Plant

Vegetative propagation by grafting makes it possible to unite more than one genotype and combine desirable features of both into a composite plant. Most fruit orchards are combinations of a seedling (or clonal) rootstock chosen for root characteristics, combined with a cultivar chosen for its fruiting characteristics (Chap. 11). Grafting can also result in special growth forms such as dwarfed trees or weeping plants having a hanging growth habit grafted to the top of long upright trunks. Tree roses are examples. Other examples are provided in Chap. 11.

Shortening the Time of Reproductive Maturity

Vegetative propagation of plants in the mature (adult) growth phase reduces the long nonflowering juvenile characteristics of most seedling plants. Vegetatively propagated fruit and nut cultivars invariably will flower at a younger age than comparable seedling plants (Table 8-1). Similarly, seed orchards for forest trees for controlled production of seeds (see p. 90) benefit from vegetative propagation of mature plants.

Many herbaceous perennials, such as orchids and bulb species (Chap. 15), require 5 to 10 years before flowering begins, whereas plants of the mature (adult) phase will flower annually by their vegetative (bulbing) and reproductive (flow-

TABLE 8-1

Age of flower development in some woody plants in the seedling (juvenile) phase

Species	Length of juvenile period
Rose (*Rosa* sp.)	20–30 days
Grape (*Vitis*)	1 year
Stone fruits (*Prunus* spp.)	2–8 years
Apple (*Malus* spp.)	4–8 years
Citrus (*Citrus* spp.)	5–8 years
Scotch pine (*Pinus sylvestris*)	5–10 years
Ivy (*Hedera helix*)	5–10 years
Birch (*Betula pubescens*)	5–10 years
Pear (*Pyrus* spp.)	6–10 years
Sequoia (*Sequoia sempervirens*)	5–10 years
Pine (*Pinus monticola*)	7–20 years
Larch (*Larix decidua*)	10–15 years
Ash (*Fraxinus excelsior*)	15–20 years
Maple (*Acer pseudoplatanus*)	15–20 years
Arborvitae (*Thuja plicata*)	15–25 years
Douglas fir (*Pseudotsuga menziesii*)	20 years
Bristle cone pine (*Pinus aristata*)	20 years
Redwood (*Sequoiadendron giganteum*)	20 years
Norway spruce (*Picea abies*)	20–25 years
Hemlock (*Tsuga heterophylla*)	20–30 years
Sitka spruce (*Picea sitchensis*)	20–35 years
Oak (*Quercus robur*)	25–30 years
Fir (*Abies amabilis*)	30 years
Beech (*Fagus sylvatica*)	30–40 years

Source: J. R. Clark. 1983. Age-related changes in trees. *Jour. Arbor.* 9:201–5.

ering) phases, depending upon environment and management.

Control of Growth Phases and Morphology

Plants in the juvenile phases of many species have unique characteristics that differ distinctly from comparable plants in the mature phases. Many, if not most forest tree species benefit from a long juvenile period, which confers delayed flowering, better tree growth, better lumber yields, and other useful characteristics (*7*) found in seedling plants (*9, 64*). In contrast, in other species, for example, citrus, plants in the juvenile phase are thorny, excessively vigorous, produce poor fruit quality, and are unsuitable for the production of fruit (*15, 35*). As described later in this chapter, vegetatively propagated citrus plants from the mature phase do not have these characteristics.

GENETIC VERSUS EPIGENETIC PROCESSES

Seedling plants exhibit two basic types of variation. One is the **genetic** variation from plant to plant among members of a seedling population. The control of this variation resides within the genes (the **genome**) located on **chromosomes** which occur in the nucleus as long chains of **deoxyribonucleic** acid (DNA) with repeating units (nucleotides) containing sugar, phosphate, and combinations of four chemicals (bases): cytosine, guanine, adenine, and thymine. These make up the genetic code, which is replicated during asexual cell division (**mitosis**) and rearranged during sexual reproduction (**meiosis**). In addition, the cell has other replicable units in the cytoplasm that contain DNA, located in the plastids (the **plastome**) and mitochondria (the **chondrianome**), which can affect the phenotype (*89, 105*).

A second category of variation that is particularly significant in vegetative propagation is referred to as **epigenetic**. Epigenetic variation results from regulation of gene expression but does not involve change in the genes themselves (*17, 62, 63, 96*). Variation in gene expression begins with differentiation of cells of the embryo and/or those laid down by the apical meristems. The many different kinds of cells and their arrangement produce the diverse structures of stems, leaves, roots, and reproductive organs. Control of epigenetic variation has been hypothesized as coming about through a sequence of molecular reactions primarily involving a nucleic acid similar to DNA, which is called **ribonucleic acid** (RNA): The sequence involves transcription of RNA from DNA to mRNA

(messenger RNA) and its transport into the cytoplasm, degradation of mRNA, and translation of mRNA into **protein** (*17*). Proteins become both **structural** units for the plant as well as the **regulatory enzymes** that direct metabolism (*25*).

In the intact plant, cells in apical meristems become programmed to differentiate in response to various internal (endogenous) and external (environmental and hormonal) influences. These cells continue to perform as programmed, even after the inciting influence is removed and/or the propagule has been excised and propagated. The phenomenon acts like an internal memory and has been called **determinism** (*17, 96*). For example, an apical meristem will continue to develop into a flower once it has reached a particular level of induction and will not revert to a vegetative shoot.

During the life cycle of the seedling plant, epigenetic control produces **embryonic, juvenile, transitional,** and **mature** (adult) phases of growth and development (*13, 41, 48*). These phase changes affect the age when flowering begins and can produce significant morphological and physiological variations in different parts of the same plant (Figs. 8–1, 8–2, 8–3). Plant structures with different developmental history will exhibit different phenotypes as the result of such epigenetic differences. Epigenetic variation is reversed during the formation of a zygote (either sexual or apomictic) but is weakly stable during mitosis and vegetative propagation (*49*).

SOURCES OF VARIABILITY WITHIN CLONES

Vegetative propagation begins by taking vegetative propagules (buds, cuttings, etc.) from specific plants from which new plants are produced. Propagators sometimes refer to the two consecutive generations as the *mother plant* and the *daughter plant*. Another terminology would be *source plants* and *vegetative (clonal) progeny plants*. Two special terms, used mostly in forestry, and first introduced by Stout (*92*), are **ortet**, the seedling tree

FIGURE 8–1 Juvenile versus mature variation. *Left: Hedera helix* (ivy). The mature phase of ivy shows a bushy structure with seed pods which presents a striking comparison to the vigorous, upright vine-like vegetative form of the juvenile phase. *Right:* Comparison of *Eucalyptus*. Note the narrow-leaved form with seed pods, as compared to the more rounded leaf with different leaf phylotaxy of the juvenile form on the right.

from which the clone originated, and **ramet**, the collection of plants vegetatively propagated from it. These terms have special significance in relation to juvenility variation described later.

In general, plants of the same clone are more or less phenotypically identical, but specific kinds of variation can and will appear. If the plant appearance and performance falls within an acceptable range for that cultivar, it is considered as being **true-to-type**. If it deviates significantly from this norm, it is considered **off-type** and referred to as a **variant**.

Some variation results from effects of the environment, age, or physiological condition of the plant and does not persist during vegetative propagation. Genetic variation, on the other hand, can develop within a clone and normally persists during vegetative propagation. These develop primarily due to accidents of cell division at mitosis and can produce a permanent change in the clone. In practice, such so-called **somatic variation** may not be easy to separate by phenotype alone from variation due to other causes, including epigenetic, pathological, or environmental.

A primary objective in plant propagation is to repro-

duce a specific cultivar. Somatic or other kinds of permanent change can create a serious problem for the propagator and is usually considered undesirable. On the other hand, somatic variation can create potentially valuable new cultivars. Likewise, manipulation of the epigenetic variation in the plant can improve its propagation potentialities.

Chapter 4 dealt with the principles of variation and source selection in seed propagated species and cultivars. This chapter deals with the principles and procedures involved in source selection in vegetatively propagated plants.

Within-Clone Variability Affecting Source Selection

I. Phenotypic (developmental and environmental)
    Periphysis
II. Epigenetic (phase changes)
    Topophysis—position
    Cyclophysis—ontogenetic age of apical meristem

FIGURE 8–2 Foliage variation gradients within a seedling tree can be markers of the juvenile and mature phenotype. *Top: Eucalyptus* seedling tree showing the large, broad, sessile (no petiole) leaves at the base of the tree; leaves in the upper more mature leaves at the top are narrow. Shows variation gradients both within the tree and within a single branch. *Below left: Acacia melanoxylon.* In this species the bipinnate leaves at the base are characteristic of the ''juvenile'' phase. The ''leaves,'' which are actually expanded petioles (phyllodia) at the top of the seedling, are characteristic of the adult form. *Below right:* Juvenile leaves on many conifers tend to have acicular needles (i.e., sharp pointed), whereas the leaves on the upper part tend to be scale-like.

FIGURE 8–3 Two nursery trees of 'Kara' mandarin propagated from single buds (arrows) taken from the same tree, a nucellar seedling. Thorny tree on left grew from a bud taken from the thorny, juvenile portion of the tree; the thornless tree on right grew from a bud taken from the thornless, adult part of the tree. (Courtesy H. B. Frost. From H. J. Webber and L. D. Batchelor, *The citrus industry,* vol. 1, Univ. of California Press, Berkeley.)

III. Genetic changes and chimeras
    Mutations
        Gene and chromosomal
        Mitochrondrial
        Chloroplast
    Cultivar or species-specific variants
IV. Pathogenic
    Systemic
        Virus and virus-like
        Viroids
        Mycoplasma-like organisms
        Rickettsia-like organisms
        Nematodes
        Certain bacteria
    Nonsystemic pathogens
        Bacteria
        Fungi

Phenotypic (Nonpersistent) Variation

The appearance and behavior of a plant (i.e., its phenotype) results from the interaction of the genotype and the environment. Striking differences can occur among plants of the same genotype due to different environments in which the plants are growing, or to different management procedures to which the plant is subjected. The appearance and performance of different plants or different plant parts may be due to differences in age or stage of development.

Many examples of nonpersistent phenotypic variations exist. For instance, 'Bartlett' pears grown in Washington and Oregon tend to be longer and more "pear-shaped" than those in California, apparently due to climatic differences (*102*). Continued exposure to an unfavorable environment can lead to progressive deterioration of plants of the clone, for example, lack of sufficient winter chilling in strawberries (*12*) or an unfavorable postbloom vegetative period for some bulb species (see Chap. 15). Some plants produce leaves of different appearance when growing in shade than when growing in sun. Certain water plants produce leaves different in appearance from those produced on parts submerged in water.

Sometimes the condition of the source plant, the selection of the plant part for propagation, and the conditions during and immediately after propagation can significantly affect the performance and appearance at the time without producing a permanent change within the clone. For example, vegetative production requires the presence of vegetative growing points; if only flower buds are present on a budstick, propagation cannot occur. Maintaining healthy, vigorous stock plants as sources for propagation is useful to improve propagation but is not expected to result in a permanent increase in vigor of the plant cultivar. Selecting large, vigorous nursery trees for orchard planting might improve tree growth in the initial years but would not itself permanently change the nature of the trees in the orchard. In either case, however, the increased vigor might be an indicator of other favorable differences, such as the absence of specific viruses or the presence of genetic mutants.

The term *periphysis* (*75*) has been used to describe a carryover effect of the environment on the propagule into the progeny plants but without causing a permanent effect on the genetic nature of the plant.

Phase Variation (Cyclophysis) within Plants and Clones

The Seedling Cycle (41, 48, 49)

A juvenile phase of development that follows the embryonic phase (see Fig. 3–1). It is defined as an initial period of growth when the apical meristems will not respond to internal or external stimuli to initiate flowers. The juvenile phase is characterized by three criteria:

1. Absence of potential to shift from vegetative growth to reproductive maturity and the formation of flowers
2. Specific morphological and physiological traits, including leaf shape, thorniness, vigor, and disease resistance
3. Greater ability of the juvenile phase to regenerate adventitious roots and shoots

The nonflowering juvenile phase is expressed in distinct gradients within the framework of the seedling plant (Fig. 8–4). The base of the plant is and remains juvenile; transition and mature phase develop later and higher in the plant. The result is a paradox in terminology in that the part of the plant nearest the base of the tree is oldest in terms of chronological age but is actually the "youngest" (more juvenile) in terms of biological age. On the other hand, the outer periphery of the stems and branches is the "oldest" and most "mature" in terms of biological age but is actually the youngest in chronological age (7).

The change from juvenile to mature phases may occur gradually, with little obvious phenotypic change (**homoblastic**), or it may occur abruptly, with distinct differences between them (**heteroblastic**). Control of the juvenility–maturity phenomenon is both genetic and epigenetic. The type of changes produced and the length of the juvenile phase is inherited. Undoubtedly, these traits evolved to enhance the survival of certain species by delaying the time of flowering or by increasing their ability to compete with other

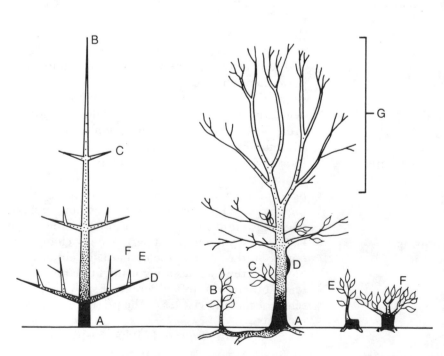

FIGURE 8–4 Juvenile–mature gradients in seedling trees represent a "cone of juvenility" near the base of the tree. *Left:* Seedling conifer tree. Juvenile root-shoot junction, A; mature apex of the tree flowers first, B. Other letters refer to shoots at different locations. Gradient in juvenile state in A > F > E > D > C > B, results from differences in "biological age" from A to each location. *Right:* Seedling deciduous tree. Juvenile root-shoot junction, A; mature flowering part of tree is in apical part, G. Juvenile structures arising from "cone of juvenility" near base of tree include B, adventitious root "sucker;" C, watersprout (epicormic); and D, sphaeroblast (adventitious). A stump sprout from pruned back tree, E, and a hedged tree, F, maintained by pruning are also juvenile. (Redrawn with permission from Bonga (7).)

BIOLOGICAL BASIS FOR JUVENILE
AND MATURE GROWTH PHASE CHANGES (*41*)

Various hypotheses have been proposed to explain the fundamental basis for juvenile and mature differences.

1. *Cellular.* One explanation is that the individual cells in the apical meristems of juvenile and mature parts of the plant are epigenetically different, stable, but change during vegetative cell division (*41*). It has been noted (*99*) that as cells divide, one daughter cell may differentiate without further division, whereas the other continues to divide. As a result, gradients could develop in the juvenile/mature potential with the number of elapsed cell divisions and the expanding architecture of the plant. Evidence for the cellular basis is that individual growing points retain the specific juvenile/mature potential when propagated. Also cell and tissue cultures may perform differently in culture.

2. *Hormonal.* Experiments have shown that a mature shoot grafted to a juvenile plant will sometimes tend toward a juvenile expression through the apparent influence of material transferred from the leaves of the juvenile plant. Also, there is indication that the roots contribute an influence that retains the juvenile phase. This can explain why, in general, shrubs, vine plants, or layered plants (see Chap. 16) tend to retain juvenility. Treatment with gibberellic acid followed by vegetative growth has produced changes from mature to a juvenile phenotype in ivy (*Hedera*) and the opposite to occur in conifers. Treatment with abscisic acid has reversed the change (*41*). Progressive invigoration (rejuvenation ?) of the shoot tip and increased rooting of a hard-to-root cherry clone have been induced by consecutive cycles of co-culture with an easy-to-root cultivar. This influence was duplicated by gibberellic acid (*80*).

3. *Cellular isolation.* A third view suggests that the apical meristems are influenced by the immediately surrounding differentiated cells. When cellular connections are severed, reversion to a juvenile condition occurs. This condition occurs in the initiation of an adventitious growing point or in the production of a microspore (*88*).

It is likely that interactions of all of these explanations may be part of the control system.

---

plants. One effect may be to protect the plant from browsing animals by thorns or unpalatable taste (*48*). One can modify epigenetic patterns through breeding and selection, such as by producing cultivars that have shorter breeding cycles or rootstocks with greater adventitious root induction potentials. Examples of phase change, or cyclophysis, are illustrated in Figs. 8–1 and 8–2 (*35*).

The Clonal Life Cycle

The relative juvenile/mature potential of the individual apical meristem (growing point or bud) in the seedling plant is determined by its position in the hierarchy of the buds of the plant (Fig. 8–1). This potential is perpetuated when the bud is removed for vegetative propagation and continues the process of phase change in the new plant. Consequently, the pattern of growth and development of a vegetatively propagated plant may differ significantly from the comparable pattern in the seedling plant. It is referred to here as the **clonal life cycle** (*48, 49*) (Fig. 8–5).

The age at which a vegetatively propagated plant reaches sexual maturity depends upon (a) the level of juvenility in the propagule, and (b) the growing conditions to which the plant is subjected. If the apical meristem is taken from a juvenile part of the plant (i.e., near its base), then the

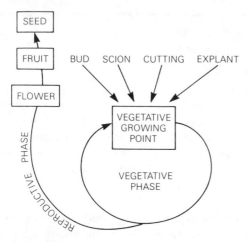

FIGURE 8-5 Model of clonal life cycle illustrating growth and development in a vegetatively propagated member of a clone. The two basic phases of development are vegetative and reproductive, not juvenile and mature. All of the plant tissue is in the adult phase.

new growth may remain juvenile for some time. Buds taken from the upper (mature) part of the plant are likely to produce progeny plants that are fully mature. Consequently, the population of plants produced in the first vegetative generation may be distinctly variable from each other and from the original seedling plant, depending upon the original location of the propagule on the source plant. Fruit and nut cultivars that have been vegetatively propagated serially for a great many years have invariably reached a stable mature phenotype which shows no evidence of the juvenile phase. On the other hand, species such as ivy (*Hedera*), grown for their ornamental vegetative characteristics, may permanently retain the juvenile phase through selection of source material.

The terms **ortet** and **ramet** described previously were developed because variation often resulted after the initial propagation in woody perennials. The term *ortet* is used for the original seedling plant, and *ramet* for the progeny plants. Often the phenological characteristics are not the same in the ramets as in the ortet, such as in change in bloom date in walnut (*61*) or loss of rooting potential (*70*).

There are many significant examples of the

mature phase expression in horticulture and forestry (*81*). Most grafted cultivars require a shorter time to flower and fruit than does the comparable seedling plant. Citrus plants are a classic example in which the standard commercial plants are the mature phase, which bears at a young age, has moderate vigor, lacks thorns, and has good-quality fruit as compared to the juvenile apomictic form (*15*). Similar but less striking differences can be shown for apple, pear, and walnut. The mature form of ivy (*Hedera*) is a tree-like woody plant with distinct leaves and phyllotaxy dramatically different from the juvenile vine type (see Fig. 8-1). The mature clones of many conifers grow horizontally (plagiotropic) and show more bushiness, branching, and early cone formation (*9*).

As the clone is serially propagated, mature characteristics tend to become stabilized and "fixed" to provide the "true-to-type" phenotype of the cultivar. Such plants grow and develop in consecutive cycles of **vegetative** and **reproductive** phases (Fig. 8-5). It is not appropriate to refer to these as juvenile and mature growth phases. In reality, the entire plant is in the mature (adult) phase.

As individual plants age, they tend to show reduced growth rates and to exhibit a change in type of branching. This "aging" is physiological in nature, however, and normally is reversed by *renewal* and *invigoration* of vegetative shoots. This process occurs during vegetative propagation of the clonal cycle and the life of the clone is considered to be more or less indefinite. In the individual plant, "aging" can be controlled by pruning, nutritional treatment, and/or crop control. **Physiological aging** should be distinguished from **ontogenetic aging,** that is, "juvenile to mature" (*31*). The latter is reversed through a *rejuvenation* process during seedling propagation (zygote formation) and adventitious bud formation (*49*) and can be induced by various treatments described in further paragraphs.

Loss or reduction of the ability to regenerate adventitious roots on stems is also closely associated with the attainment of maturity (see Chap. 9). This characteristic is particularly important in trees and accounts for the reason that the development of fruit and nut tree cultivars is so closely associated historically with budding and grafting procedures. It has now been demonstrated convincingly that the phase change from juvenile to mature (as it affects the rooting phenomenon) can be reversed (**rejuvenation**) during vegetative growth, at least in part. Conditions under which this occurs include the consecutive recycling of microplants in aseptic culture (*69, 80*) and consecutive grafting to juvenile rootstocks (*9,31*) (see Fig. 9–21). Scions with mature *squamiform* needles of cypress, arborvitae, and redwood developed shoots with juvenile *acicular* (spinelike) needles, which could be used as a marker for rejuvenation (*31*).

Control of Juvenile–Mature Changes

Progressive juvenile to mature changes. Practices that induce early maturity include the following:

1. Continuous vegetative growth to allow the apical meristems to "grow" into the mature phase (*15, 101, 111*). This should be followed by treatments to initiate flower inductions appropriate for the species.
2. Repeated vegetative propagation choosing propagules from apical parts of the seedling plant.
3. Grafting or budding to dwarfing rootstocks. This may not be successful unless the scions or buds are approaching the mature phase.

Maintenance or selection of juvenile phase. Methods for preserving the juvenile state include:

1. Collect material and propagate from juvenile tissue. This may involve selection from the base of the seedling plant from material that shows the juvenile form, such as ivy (*Hedera*), which remains juvenile when in a horizontal growth pattern (see Fig. 3–2).
2. Collect material from root sprouts (**suckers**) or from watersprouts (**epicormic** shoots) arising from latent buds near the base of the seedling, trees, or from **sphaeroblasts**, which are masses of shoots from adventitious buds that naturally occur in some species. Examples are sequoia, redwood (*9*), some quince, and some apple cultivars (*94*). Such

shoots are adventitious and usually exhibit juvenile traits (see Fig. 8–4).
3. Cut back the plant to produce "stump sprouting" from the juvenile zone of the plant (*31*). Many species have juvenile buds that are suppressed by normal vegetative growth (see Fig. 8–4).
4. Grow stock plants in hedge rows which are kept continuously invigorated by pruning (*9, 57, 64*). It is likely that the success of mound and trench layering (p. 416) is due in part to retention of a juvenile condition.
5. In vitro cloning from embryos or very young seedlings (*9*) (see Chap. 16). This method is useful in some situations in forestry.

Reversions from mature to juvenile phase (rejuvenation). Once a plant has attained the mature phase, either as a seedling plant or as a clone, it tends to remain stable. Rejuvenation to the juvenile stage occurs during sexual and apomictic reproduction (*49*). Rejuvenation during vegetative propagation can also occur and has been achieved in several ways.

1. Production of adventitious shoots as on roots (*41, 48*), in callus (*94*), or in tissue culture systems (see Chap. 16).
2. Consecutive grafting to seedling rootstocks (*9, 31*). Removal or reduction of mature leaves may be important to the process (*41*) (see Fig. 8–6) (see also p. 226).
3. Consecutive subculturing of apical meristems in micropropagation (*69, 80*). This procedure may require the optimization of the growing medium to achieve vigorous growth (*31*) (see p. 480).
4. Hormone treatment, primarily gibberellic acid, which can be reversed by treatment with abscisic acid (*41*).

Topophysis

Topophysis is defined as the effect of the position on the plant of the propagule on the type of vegetative growth subsequently shown by the vegetative progeny (*66*). Plants of certain species produced by cuttings taken from upright shoots (**orthotropic**) will produce plants in which the

FIGURE 8–6   Rejuvenation of *Eucalyptus* x *trautii* by serial grafting of scions taken from a 10-year-old tree cleft grafted onto 6- to 10-month-old rootstock. Six successive serial grafts were needed before successful rooting of scions occurred. *Left to right:* Initial cleft graft (note the narrow leaves, similar to those shown in Fig 8–2); third serial graft showing morphological changes in leaves of one of the shoots; sixth serial graft with a rejuvenated plant with juvenile leaf characteristics and from which scions could be successfully made into cuttings and rooted. Consecutive grafting to a seedling rootstock. (From Siniscalco and Pavolettoni.)

shoots grow vertically. Plant produced from cuttings taken from lateral shoots (**plagiotropic**) will grow in a horizontal direction. This phenomenon is illustrated in Fig. 8–7 with coffee, in which this phenomenon creates problems in "true-to-type" propagation. It also occurs extensively with conifers, in which horizontal or prostrate forms may be obtained by taking cuttings from horizontally growing shoots. Such response might be desirable for ornamental use but creates problems in forestry propagation. Norfolk Island pine (*Araucaria*) is a widely grown example of this phenomenon. In some plants, such as *Sequoia*, pruning has restored the orthotropic growth habit (*9*).

GENETIC VARIATION IN VEGETATIVELY PROPAGATED PLANTS

Mutations

Mutations are genetic modifications that produce permanent changes in the genotype of the plant. Mutations may affect nuclear genetic material (**chromosomes**) or cytoplasmic genes (**plastids** and **mitochondria**). Chromosomal changes may result from rearrangements of the four bases in DNA (**point mutations**) or from **deletions, duplications, translocations,** and **inversions** of parts of some chromosomes. Changes may be the result of the addition or subtraction of individual

FIGURE 8–7   *Left:* Coffee plant propagated from a vertically oriented cutting maintains an upright (orthotropic) growth habit. *Right:* Coffee plant propagated from a lateral shoot produces a horizontal (plagiotropic) growth habit. Plants in many species exhibit this growth phenomenon, requiring care in selecting cutting material.

chromosomes (**aneuploidy**) or from the multiplication of entire sets of chromosomes (**polyploidy**).

Plastid mutations usually result in loss of chlorophyll, which results either in **albino** plants or in **variegated** plants that have both albino and green sectors present (*89, 99*). Changes in ploidy may result in giant, vigorous, low-producing ''sports'' in some plants which are due to changes from the diploid ($2n$) to the tetraploid ($4n$) state (*21, 27*).

Horticulturally, mutations within clones have been observed through the sudden appearance of branches or whole plants with marked change in a specific characteristic. These are referred to as **bud-sports** or **bud-mutations** and have sometimes given rise to important new cultivars (see Fig. 8–8).

Mutations are preserved during cell division (mitosis), and the production of new cells derived from the original mutant cell effectively results in a new clone within the original plant. The mutant may become visible within the original clone or it may remain latent and become apparent only in the next vegetative generation, or after segregating in the next seedling generation.

Latent mutations will become visible in the same clone when (a) the clone of mutated cells occupies a major part of a growing point to produce a branch, or a sector within a branch; or (b) the mutation affects a trait conspicuous enough to become observable by visual inspection, for example, leaf or fruit color, size, growth habit or vigor, time of bloom or maturity, etc. Many bud-mutations are mixtures of mutated and nonmutated tissue which occupy distinct layers or sectors of the plant. These combinations, called **chimeras**, are described in more detail in the box.

Chimeras may be horticultural curiosities or economically important plants. Plants with variegated foliage, as found in *Citrus, Vitis, Pelargonium, Chrysanthemum, Hydrangea, Dahlia, Coleus, Euonymus, Bouvardia, Sanseviera,* and others, are examples of chimeras. In these plants the plastids in part of the leaf tissue lack the capacity to produce chlorophyll, whereas other leaf cells are normal. The resulting pattern shows distinct green and white (or yellow) areas in the leaf.

Red-colored apples and pears and pink-fruited grapefruit, fruit cultivars that bloom later, or fruit that ripens earlier or later than the original clone (*67*), are examples of bud mutations that are chimeras. The seedless 'Washington Navel' orange probably arose as a bud-mutation of the seedy Brazilian orange 'Laranja Selecta' (*65*).

Fruit chimeras such as the orange in Fig. 8–9 are frequently observed. Apples with both sweet and sour flesh, or peaches with fuzzy and smooth (nectarine) surfaces have been discovered (*22*). Red forms of apple (*67*) and potatoes are chimeras in which the red skin color mutation is present only in the outer layer or epidermis, the inner tissue being the original nonmutated form. Likewise, thornless blackberry cultivars exist in which the thornlessness is restricted to the outer stem layer (*20*). Chromosomal chimeras (cytochimeras) may occur in which part of the tissue of the stem is diploid (i.e., normal chromosome number for the species), and cells in other parts of the stem are polyploid (i.e., some larger number of chromosomes) (*22, 27, 76*).

FIGURE 8–8 Almond (*Prunus dulcis*) tree in which one branch has conspicuously later bloom. This represents a ''bud-sport'' where a mutation in genes controlling time of bloom has evidently occurred.

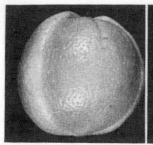

FIGURE 8-9 Mericlinal chimera in 'Washington Navel' orange. Left portion of the fruit has developed from the mutation, producing a thicker rind that is yellow instead of orange in color. This example is an undesirable mutant.

Many, if not most, genetic mutations have some adverse effects on the organism (28, 82, 85). These changes are likely to affect the reproductive system and may produce effects deleterious to performance and productivity. If unusually conspicuous, a mutated plant would be eliminated automatically. Latent mutant tissue may remain undetected. Propagation material with latent mutations may lead to long-term differences among sources of the same clone, which might not be detected except through extensive clonal testing (16, 54, 104, 108).

Rate of Mutation

The frequency of mutations affecting selection in vegetative propagation depends upon the rate of mutation and the probability of detection. Natural mutations occur randomly at very low rates and the probability of detection is low. Nevertheless, where propagation of important commercial cultivars involve literally millions of plants, the probability for some mutations to occur spontaneously somewhere in the clone is high. Since the natural rate of mutation can be increased by exposure to various mutagenic agents (x-rays, gamma rays, and certain chemicals), as utilized in mutation breeding (1, 14, 68), the possibility exists that exposure of plants to environmental mutagens can increase natural rates.

Increased mutation rates are associated with certain propagation systems, particularly those involving callus production, where tendencies occur for polyploidy (29).

Potential for Detection

The detection of new mutations may be a problem in normal growth patterns of a single plant even in mutation breeding, where rates are high because latent mutant tissue may be confined to specific sectors. The rate of detection can be increased significantly by systematically pinching back shoots to force out lateral branches, and propagating with single buds in consecutive generations (1, 14). These procedures are normally done in vegetative propagation by budding (Chap. 13), and are usually repeated through consecutive generations of propagation (see "Pedigree System," p. 190). Consequently, a nursery propagation system is a highly efficient method of screening for mutations, whether useful or inferior, even though the overall natural frequency may be low.

A further problem in source selection may occur because a deleterious mutant or other change (28, 50, 85) may not be detectable when it first develops in the original plant. The variant is thus identifiable only by its secondary distribution in the vegetative progeny. Detection may not occur until the plant reaches reproductive maturity (28, 85). High frequency of such variants in any source material can be a serious problem to the propagator and to the ultimate recipient of the products of this propagation.

Chimeras

Mutations occur initially in single cells but have only small effects until the mutated cells divide and become a major part of the growing point and eventually a significant part of the shoot (89, 99). Because of the unique arrangement of cells of the apical meristem (shoot apex), the shoot invariably becomes a combination of layers of mutated and nonmutated tissue. The term used to designate a plant that is composed of tissue of more than one genotype is a **chimera**. In most cases chimeras can be classified from their orderly patterns of cell distribution (99) as follows:

Periclinal

This most common type of chimera is composed of layers in which the mutated tissue exists as a relatively thin layer one or several cells thick, usually the epidermis, completely covering a core of nonmutated tissue. The plant has the characteristics of the outside layer and the inner tissue is not expressed. Periclinal chimeras are relatively stable if propagation involves perpetuation of the apical meristem (see Fig. 8–10). Propagation using adventitious shoots that arise endogenously (as in root cuttings) would cause a rearrangement

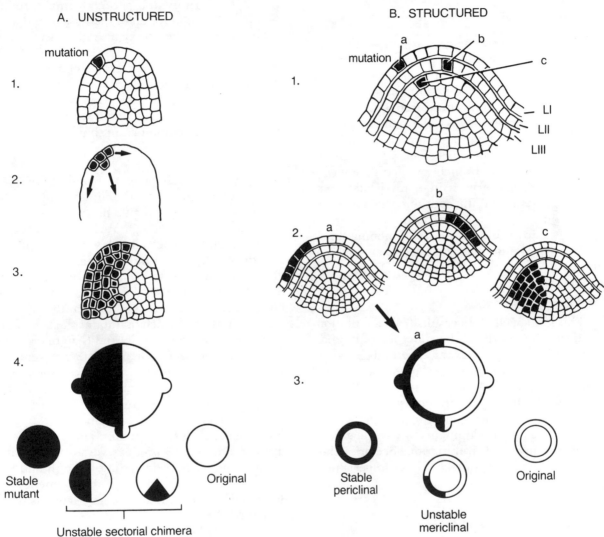

FIGURE 8–10 Chimeral structures and origins. *Left:* Unstructured apical meristems (ferns, most gymnosperms, roots and young embryos of angiosperms) give rise to sectorial chimeras because the cell initials in the shoot apex arise from a few cells which subsequently divide in both anticlinal and periclinal directions. A mutation in a single cell (1) results in a sector (2, 3) which produces a sectorial chimera. *Right:* Structured apical meristems (most angiosperms, some gymnosperms) have three (or more) distinct layers (L-I, L-II, L-III). A mutation can arise in a single cell of any one of the layers (a, b, c), which can then develop as single layers or a core of mutated tissue (2) to produce (3), a stable periclinal chimera, an unstable mericlinal chimera, and a stable reversion to the original type. Only the development of a mutation in the L-I layer is shown. See text.

of the tissues of the chimera and a **reversion** to the nonmutated genotype of the inner tissue. Likewise, the mutant genotype would not be inherited because both male and female sex cells arise from inner tissue layers.

Mericlinal

This type of arrangement is similar to periclinal except that the mutant tissue of the outside layers involves only a segment of the surface. This type of chimera is also very common and is usually the transitional type produced immediately after a mutation occurs. As the chimeral shoot continues to grow, the production of lateral buds in different segments in the periphery of the growing points gives rise to branches with different cellular arrangements of periclinal, mericlinal, and nonmutated tissue. With continued consecutive propagation, shoots with primary mutations gradually shift into subclones of stable nonmutants, stable periclinal mutants, and unstable mericlinal mutants.

Sectorial

This kind of chimera results from a mutation that affects all the cells of the shoot instead of being restricted to specific layers. Sectorial chimeras are rare and can only exist in plants where the apical meristem arises from very few cells (see ''Anatomical Origin of Chimeras,'' below) or from mutations that develop very early in the embryonic stages of development, where relatively few cells are involved in initiation. Sectorial chimeras normally revert quickly either to periclinal or mericlinal chimeras.

Anatomical Origin of Chimeras

The growing points of plant shoots and roots consist of meristematic cells that are capable of undergoing cell division (Fig. 8–10). The chimeral pattern depends on the arrangement of the cells at the growing point and the location of the mutation (*22, 99*). There are three basic types of growing points.

Type 1.   In ferns, a single apical cell divides to produce one daughter cell that continues to divide to elongate the shoot, plus a second daughter cell that remains to make up the body of the plant. A mutation in such a growing point would produce a sectorial chimera because all the new tissue would come from the single daughter cell.

Type 2. In gymnosperms, a small group of dividing cells produce all of the tissues and a mutation in one of these may also give rise to a sectorial chimera. In angiosperms, the apical meristem of the root comes from a single main cell and, if mutated, would produce a sectorial chimera in the root. An adventitious shoot arising from mutated root tissue would subsequently produce a complete mutant plant, but such events are rare.

Type 3. The shoot apex of angiosperm plants consists of a series of overlapping **histogenic layers,** each of which remains more or less distinct because the cells from separate layers divide in the same direction. The outer layer (referred to as L-I) divides by **anticlinal** (right-angled) divisions and the daughter cells eventually become the epidermis or the outer layers of the plant. A second layer (L-II) also divides predominantly by anticlinal cell divisions (which maintains the layers), but occasionally exhibits **periclinal** (lengthwise) divisions that can shift cells into other layers. The L-II layer produces several tissues, such as the cortex, a part of the vascular bundles, and (usually) the reproductive structures. Deeper cells of the apical meristem (L-III or sometimes an L-IV) divide more or less at random and layers are indistinct. These cells produce variable amounts of inner ''core'' tissue, including the pith and parts of the vascular cylinder. This arrangement is also described as a **tunica** (layers)—**corpus** (body structure) (see Fig 8–11).

Leaf **variegation** of angiosperm plants commonly results from plastid mutants that develop into chimeras of green and white tissues (or other pigments) (*90, 99, 105*). The arrangement within the leaf and pattern of variegation involves the same L-I, L-II, and L-III layers but can present interesting variable patterns of variegation because of the pattern of leaf development. Color distribution can be observed morphologically by magnification of the cells from specific layers (Fig. 8–12). Because gymnosperm needles develop from a limited number of cell initials, variegation

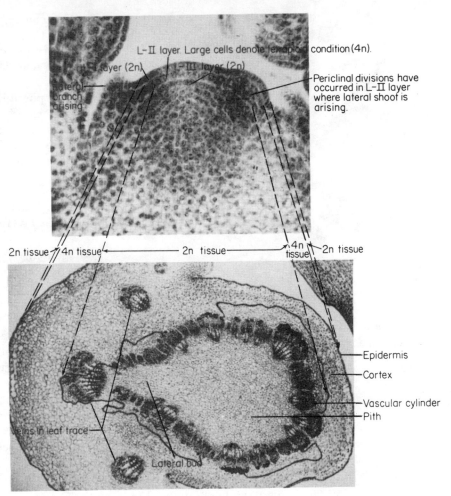

FIGURE 8-11 Chimeras may originate by *mutation in one of the histogenic layers* of the apical meristem. Photomicrographs illustrate a chimera in peach showing an enlarged, longitudinal section of the shoot tip (top) and a cross section of the stem under the tip (below). *Top:* Three histogenic cell layers in the tip are identified as L-I, L-II, and L-III (see text). In this example the L-I is diploid ($2n$), the L-II is tetraploid ($4n$), and the L-III is diploid ($2n$). Note that the L-I layer is a single row of cells, but that the L-II layer has increased in thickness on each side because of periclinal divisions at the point of leaf initiation. *Below:* The epidermis of the stem is $2n$ because it was derived from the L-I layer. The cortex and part of the vascular system are $4n$, being derived from the L-II layer. The remainder of the vascular system and the pith are $2n$, being derived from the L-III layer. (From Derman (*22*).)

in gymnosperms does not involve layers but occurs in sectors (Fig. 8–12).

Nonchimeral Variation within Clones

Some types of variation occurring within clones do not occur in a chimera arrangement. Some variants in coleus, for example, appear due to distribution of hormones in the plant. Likewise, "color breaks" in tulip and mosaic patterns can be due to viruses. Certain "pattern" genotypes occur in *Peperomia* and begonia (*105*). Various fruit and nut species have source-related genetic problems, such as *noninfectious bud failure* in almond (*49, 50*), *crinkle* in cherry (*71, 103*), and *yellows* in strawberry (*32*). Some variations are due to unique kinds of genetic units, known as **transposable elements** (*72*).

Graft Chimeras

While most chimeras occur naturally, they can also be established artificially by grafting (*6, 53*). In a young, grafted plant, if the scion is cut back severely, almost to the stock, an adventitious bud may sometimes arise from the callus at the junction of the stock and scion. Shoots resulting from such an origin are chimeras in which the cells of the two graft components remain genetically independent regardless of how intermingled they become.

Various graft chimeras have been reported in the horticultural literature. Many years ago Winkler (*109, 110*) artificially produced numerous graft chimeras of tomato (*Lycopersicon esculentum*) on black nightshade (*Solanum nigrum*) and black nightshade on tomato, both of which are

FIGURE 8–12 Variegated pink 'Eureka' lemon, a type of chimera. (From H. T. Hartmann, W. J. Flocker, and A. M. Kofranek, *Plant science*, Prentice-Hall, Englewood Cliffs, N.J., 1981.)

easily grafted and produce callus readily (see Fig. 8–13). Winkler gave such mixed shoots the name ''chimera'' after the mythological monster that was part lion and part dragon.

Natural graft chimeras have been known to exist and have been perpetuated vegetatively. The ''bizzaria'' orange, a fruit half-orange and half-citron, was discovered in 1644 in Florence, Italy. It apparently originated from an adventitious bud arising from callus at the junction of a sour orange

(*Citrus aurantium*) graft on a citron (*Citrus medica*) rootstock (*98, 99*). Similarly, medlar (*Mespilus germanica*) grafted on hawthorn (*Crataegus monogyna*) has led to several chimeras known as hawmedlars (*6, 40*).

The possibility of developing chimeras artificially by *mixed callus* cultures has been demonstrated in tobacco, adventitious shoots being produced in a chimeral arrangement from two separate species (*59*).

DISEASE, PESTS, PATHOGENS, AND VEGETATIVE PROPAGATION

Management of disease in propagation stock involves two different concepts of control (*3–5*). The term **disease-free** indicates the absence of disease symptoms or effects, even though a pathogen may be present. Various environmental conditions, cultural practices, or tolerances of the cultivar may mask visual disease symptoms. Plants may be propagated containing pathogens that might later, under favorable conditions, show visible symptoms of the disease. Some environmental growing conditions, such as cool temperatures or reduced moisture, have been used in the past to control the expression of disease.

The term **pathogen-free** indicates the absence of pathogens on or in the propagation stock and the prevention of infection of plants grown during propagation. Absolute freedom from all pathogens is not necessarily achieved and may not be a practical ideal (*93*). In the absence of visible disease symptoms, the presence of a pathogen is detected through specific indexing tests. Conse-

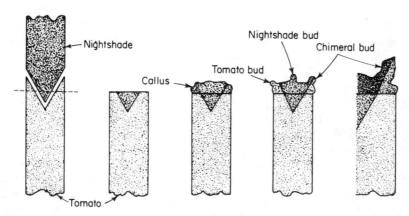

FIGURE 8–13 Stages in Winkler's method of producing a graft-chimera shoot between the nightshade and tomato. (Adapted from W. N. Jones, *Plant chimeras and graft hybrids*. Courtesy Methuen & Co., London, Publishers.)

quently, pathogens might be present but not detected by the tests. The pathogen-free concept has meaning only with respect to specific pathogens detected by the specific indexing tests implemented.

Pathogens of Importance in Propagation (*30*)

Fungi

Fungi are organisms that do not have chlorophyll but must obtain organic nutrients either from dead organic matter (**saprophytic**) or from living tissue (**parasitic**). In some cases fungi can live on both dead and living tissue (**facultative**), as compared to living on only dead or only living tissue (**obligate**).

Fungi produce long, branching, thread-like mycelia made up of hyphae, which invade the substrate to absorb nutrients and water. Some fungi also produce **sclerotia** (dense, compact masses of hyphae), which allow them to survive during periods of unfavorable environments.

Some of the many detrimental fungi associated with propagation are *Phytophthora* spp., *Pythium* spp., and *Rhizoctonia* spp. (*44*). Control measures include fumigation, pasteurization, and fungicide treatments, coupled with sanitation (see Chap. 2) and environment control. Use of pathogen-free propagation stock is the most effective control measure (*56, 79*).

Certain beneficial fungi grow in a symbiotic relationship in root cells of many higher plants in a combination known as **mycorrhiza**. Mycorrhizal symbiosis exists under most natural (uncultivated) conditions with all important horticultural, agronomic, and forestry crops. Their role is beneficial and in some cases essential (*77, 106*). Elimination of mycorrhiza by fumigation has resulted in zinc deficiency (*55*).

Bacteria

A **bacterium** is a one-celled organism with a cell wall. Bacteria occur universally and include both beneficial and pathogenic forms. They usually invade the plant through open wounds or natural openings where free moisture exists. Dormant spores allow bacteria to survive dry periods and to be carried on the surface of propagating material or tools. The major sources of the bacte-

rial pathogens in propagation are contaminated stock plants and soil-containing potting mixes. Bacteria may remain as latent infecting agents until a plant or propagule is placed in an environment favorable for their growth. Because infection is often associated with wounds, bacteria can be particularly harmful during propagation where cuts into plant tissues are necessary.

Bacterial pathogens that cause serious propagation problems include *Erwinia chrysanthemi, Erwinia carotovora, Pseudomonas* spp. (*44*), and *Xanthomonas pelargonii* (which is vascular). Crown gall (*Agrobacterium tumefasciens*) attacks many nursery crops by producing large tumerous galls that make the plants unmarketable.

Bacteria are difficult to control. Eliminating sources of contamination, sterilizing tools, and pasteurizing soils are important sanitation procedures. The use of antibiotics or other bactericides is sometimes useful for economic control.

Viruses

Viruses are submicroscopic organisms made up of nucleic acids (RNA or DNA) encased in an outer sheath of protein (*56*). The organisms live within plant cells as **obligate parasites,** multiplying and growing in collaboration with the chromosomes of the cell or in the bodies of insect vectors. Viruses are very small but can be observed in some cases by electron microscopy.

Although some viruses, such as *tobacco mosaic virus*, can be spread mechanically by contact, most move only from plant to plant via vectors, mostly aphids, leafhoppers, thrips, mites, nematodes, and sometimes pollen and infected seeds. However, the most important means of spread is by propagation material (*86*). Viruses move within the plant in the phloem and can cross a graft union to infect a scion cultivar grafted onto an infected stock, or vice versa. Any clone that has been grown continuously and propagated for a long period without intensive isolation probably has become infected with one or more viruses.

Different species and cultivars differ in their relative tolerance or susceptibility to different virus diseases (*19, 32, 71, 103*). The effect of a virus is sufficiently severe in some plants as to kill them or make them so stunted and unproductive that they are uneconomical to grow. Other viruses affect fruit and flower quality and make them unmarketable. Others affect chlorophyll and produce chlorosis, albinos, and mosaic and various variegation patterns. Some, such as strains of *Prunus* ring spot virus, produce an initial shock, after which the plant recovers with no symptoms and little obvious effect on subsequent plant appearance, production, or growth (*73*). Differences of opinion exist as to the seriousness of such latent viruses (*19, 42*). However, if tolerant themselves, infected plants are reservoirs of infections to other potential susceptible cultivars.

Rather commonly, plant viruses produce symptoms in the plant at cool (22° C or less; 70° F or less) temperatures which are not detectable when it is grown at higher temperatures (*34*). For this reason, potato propagation stock is produced in areas having cool climates, such as the northern tier of states in the United States, where diseased plants, particularly those affected by mosaic virus, can easily be detected.

Multiple infections of several virus components may be involved in many (if not most) virus diseases. For instance, three latent viruses—A, B, and C—if present individually in strawberry may cause slight or no symptoms in a given clone, but will produce a severe reaction when present in combinations of two or three (*32*). This situation indicates the danger of distributing clones carrying a single "nonpathogenic" virus, for which later infections by additional viruses can have serious consequences.

Virus control is primarily through programs for production and distribution of virus-tested material (see p. 191) and/or through control of vectors.

Mycoplasma-like Organisms (MLOs) (30)

Mycoplasmas are extremely small one-celled parasitic organisms, intermediate between bacteria and viruses in size. They live in the phloem of susceptible plants and in specific phloem-feeding leafhoppers (*Homoptera*). These organisms multiply rapidly, utilizing the plant's food for their own growth. MLOs also disrupt translocation of food materials in the phloem, causing the plant to wilt and start yellowing. Hormonal disturbances can occur in infected plants that result in "witches'

brooms" and other malformations, such as conversion of flower parts to leafy structures.

Plant diseases caused by mycoplasma include *aster yellows* in strawberries and certain vegetables and ornamentals, blueberry *stunt* disease, *pear decline*, and *X-disease* in stone fruits. Differential susceptibility to mycoplasma-like pathogens between the scion cultivar and the rootstock have caused graft failure disorders, such as occur in *pear decline*. These organisms may be moved from plant to plant by insect vectors, usually leafhoppers or psylla, and are carried in propagating material. A similar organism, known as *Spiroplasma citri* Saglio et al., causes *stubborn disease* in citrus.

Fastidious Bacteria (30) (Rickettsia-like Organisms) (71)

These are single-celled bacteria-like organisms that do not have a nucleus. They cannot be cultured in standard media and in nature occur in the xylem and phloem of susceptible plants and in the bodies of leafhopper (*Homoptera*) vectors. They are carried in seed and vegetative propagating material. *Club leaf* in clover and *greening* in citrus are caused by phloem-inhabiting fastidious bacteria. *Pierce's disease* in grape, *almond leaf scorch, phony peach, plum leaf scald*, and *leaf scorch* of various trees are caused by xylem-inhabiting fastidious bacteria. Disease symptoms include leaf scorch, wilting, small fruit, and decline. Short-term heat treatment (*37*) can control the organism in seeds and scions (see p. 150).

Viroids (23)

Viroids are extremely small agents that cause such infectious diseases as *spindle-tuber* of potato, *citrus exocortis, chrysanthemum mottle, chrysanthemum stunt*, probably avocado *sunblotch* (*97*), and *cadang-cadang* in coconut. The viroid structure appears to be a single, free RNA molecule of low molecular weight. A viroid is about one-tenth the size of the smallest virus. Some viroids, such as citrus exocortis, are transmitted mechanically on pruning shears as well as by grafting. Special procedures and sensitive indicator plants may be needed to study viroids because symptoms are often slow to develop and difficult to detect. Viroids are carried in vegetative propagation material, but some are transmitted by seeds. Some multiply most rapidly at high temperatures.

Detection methods for viroids involve extraction and purification of nucleic acids of the cell, followed by detection of unique RNA viroid species by electrophoresis (30, 56). Viroids possess a unique protein and cannot be identified by serology or by most virus-testing procedures.

Nematodes (30)

Plant parasitic nematodes are microscopic (0.5 to 3 mm; 0.02 to 0.125 in.), eel-like worms that attack roots, stems, foliage, and inflorescences. They may be present on the surface of the root (**ectoparasitic**) or may penetrate the root before feeding (**endoparasitic**). Some produce distinct galls. Nematodes affect plants by stunting growth, predisposing toward infection by other fungi and transmission of certain viruses. Nematodes are significant in propagation because their presence in roots of plants and in bulbs and other structures can be carried to progeny plants.

Propagators control nematodes by using nematode-resistant material and by sanitation practices, including fumigation or pasteurization of propagation mixes (see Chap. 2).

Insects and Mites (30)

These organisms result in plant damage caused by chewing or sucking. In propagation, sucking insects, such as leafhoppers (*Homoptera*) and mites, are vectors for the transmission of specific viruses, rickettsia-like organisms (RLO), and mycoplasma-like organisms (MLOs).

Methods to Detect Pathogens in Propagation Material

Culture Indexing

The principle of culture indexing (Fig. 8–14) is to place pieces of the plant in aseptic culture using a medium favoring the growth of the pathogen (5, 44, 74, 79). Pathogens are then detected. This method is used commercially in production systems for various ornamentals such as chrysanthemum, carnation, and pelargonium.

Virus Indexing

The primary method of detecting viruses is to transmit them by grafting or budding to a sensitive indicator plant, which then develops identifiable symptoms (32, 56, 71, 103) within a certain length of time. One procedure is to place a bud from both the test plant and the indicator plant onto the same rootstock plant. The plant is then grown under environmental conditions favoring symptom expression, preferably in the greenhouse. The virus not only moves across the union but also through an intervening section of neutral tissue (33) and causes symptoms to appear.

Leaf grafting has been used in strawberry (11) (Fig. 8–15). The necrotic reaction of highly sensitive tissue of the 'Shirofugan' cherry to the *Prunus ring spot virus* (71, 103) (Fig. 8–15) is a widely used test that requires only about a month to complete. Certain viruses can be detected in herbaceous hosts by mechanical transfer of sap.

Serology

Serology identifies unique proteins associated with particular pathogens. The ELISA test (**enzyme-linked immunosorbent assay**) has been developed for essentially any pathogen or virus with a unique protein pattern. Purified extract of the pathogen is injected into a rabbit or other animal. Specific antibodies produced in the animal's blood react against the protein of the virus. When an extract containing the same protein is tested against these antibodies, no reaction will occur; only a foreign protein will cause a reaction. This method is sensitive and highly specific (18).

Electrophoresis

Identification of certain viroids is possible by extracting and purifying the unique RNA species and observing characteristic bands on **polyacrylamide gel electrophoresis** (PAGE) (56). Similar procedures can be used to identify RNA type viruses by a procedure producing double-stranded RNA (ds-RNA) (45, 56).

Molecular Biology (56)

Monoclonal antibodies. Cell hybrids (hydrimas) have been produced by fusing mouse

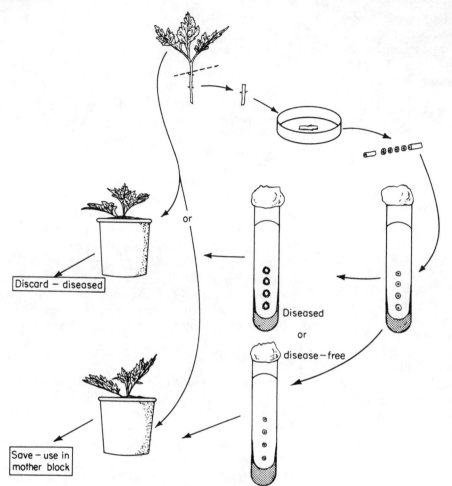

FIGURE 8-14 Testing propagating sources for disease organisms (fungi, bacteria) as illustrated for *Verticillium* wilt in chrysanthemum cuttings. (From Baker and Chandler (*5*), *California Division of Agricultural Sciences Manual 23.*)

spleen and mouse cancer cells which can produce antibodies to specific reactants, such as a virus. Monoclonal antibodies can be produced that react to very specific strains of viruses.

Hybridization analysis (*56*).   This advanced technique has the potential of detecting an unknown viroid or virus by the laboratory hybridization of a radioactive DNA probe of the known virus with the unknown virus. It is based on the recognition of similar DNA and RNA sequences of the DNA probe and the unknown.

Procedures for Eliminating Pathogens from Plant Parts

Selection of Uninfected Parts

Some parts of a plant may be infected and others not. Soilborne organisms, such as *Phytoph-*thora spp., can be avoided by taking only tip cuttings from tall-growing stock plants. Likewise, using the apical portion of vegetative shoots and discarding lower portions can often avoid tissue infected with organisms that cause vascular wilt (*Fusarium, Verticillium*, and *Phytophthora*) (*5*).

Shoot Apex Culture

The very small apical dome of a growing point is often free of virus and other pathogens even if the rest of the plant is infected. Excision and aseptic culture of this small pathogen-free segment can be the start of clean nuclear stock. There are many examples of plants where shoot apex culture has been used to eliminate viruses (*67, 78, 95*):

Flower crops: amaryllis, carnation, chrysanthemum, dahlia, freesia, geranium, gladio-

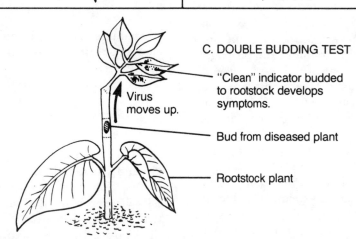

A. STRAWBERRY

Diseased scion

Nonaffected scion

Healthy indicator

Diseased

Healthy

B. SHIROFUGEN CHERRY

Shirofugen branches

Buds from test plants

30 days later

Virus positive shows necrosis and gumming.

Virus negative heals in.

C. DOUBLE BUDDING TEST

Virus moves up.

"Clean" indicator budded to rootstock develops symptoms.

Bud from diseased plant

Rootstock plant

FIGURE 8–15 Indexing for virus transmission. *Top left:* Excised leaf method for transmitting a virus from a symptomless plant to a sensitive indicator. A terminal leaf from the test plant is removed, a petiole of the indicator plant is split, and the excised test leaf is inserted. The graft union is wrapped with latex tape. Leaves on the indicator stock plant shows symptoms. *Top right:* Shirofugen cherry indexing. Buds are removed from a test plant and inserted in sequence (usually three per test) in a 'Shirofugen' flowering cherry, which is very sensitive to certain fruit tree viruses, such as the prunus ring spot complex. If the test plant is infected, the bud will die; gum and necrotic tissue will appear around the bud shield within a month. *Below:* Double budding test for fruit trees. A rootstock plant is grown in a container in the greenhouse. A bud from the test plant is placed into the base of the plant and allowed to heal. At the same time a bud from a sensitive indicator plant is budded into the rootstock plant. If the test plant is diseased, the virus will move across the graft union, move through the conducting system of the rootstock, and infect the indicator, which should develop specific symptoms.

lus, iris, lily, narcissus, nerine, orchid spp.

Fruit crops: apple, banana, citrus, gooseberry, grape, pineapple, raspberry, stone fruits, strawberry

Vegetables: brussels sprouts, cauliflower, garlic, potato, rhubarb, sweet potato

Other: cassava, hops, sugarcane, taro

For woody plant cultivars difficult to start as microcuttings, in vitro micrografting has been used. Since shoot-tip culture may not eliminate pathogens in all cases, follow-up index tests should be done for verification (*10, 46, 78, 91*).

High-Intensity Heat Treatments of Short Duration

This procedure has been used to free many kinds of plants or plant parts, bulbs, and seeds from fungi, bacteria, and nematodes (*2, 3, 5*). The plant is subjected to temperatures high enough and long enough to destroy the pathogen but not so high as to kill the plant. For example, the causal agent for *Pierce's disease* in grapevines can be eliminated by immersion of the propagating wood in hot water for three hours at 45° C (113°F) (*37*). Less drastic treatments are used where all the planting stock is to be treated, as, for instance, in commercial planting of gladiolus corms.

Treatment may vary with different plants from 43.5 to 57° C (110 to 135°F) for 1/2 to 4 hours. Methods include hot water soaking, or exposure to hot air or to aerated steam. Hardening-off vegetative material or reducing the moisture content in seeds prior to such heat treatment generally enhances plant survival. Damage to heat-treated seeds can be alleviated by holding seed in polyethylene glycol 6000 (*3*).

Low-Intensity Heat Treatments of Long Duration

This procedure will free many kinds of plants from virus diseases (*46, 56, 78, 91, 95*). Plants are grown in containers until they are well established and have ample carbohydrate reserves; then they are held in a heat chamber at 37 to 38°C (98 to 100°F) for two to four weeks or longer (Fig. 8–16). Buds are taken from the

FIGURE 8–16 Heat chamber for conducting thermotherapy of plant materials to eliminate viruses. Plants are grown at approximately 100°F for four to six weeks. At the conclusion of this period, buds are removed to be indexed for the continued presence of viruses.

treated plants and inserted into "virus-free" rootstocks, or cuttings may be taken and rooted. Alternatively, buds may be grafted onto a clean rootstock prior to treatment.

Combinations of Heat Treatment and Shoot Apex Culture

In some plants certain viruses are heat tolerant and are not eliminated by the usual heat treatments. Shoot apex culture alone may not eliminate viruses from all plants. A combination of the two methods has been successful in eliminating the viruses. Infected plants are heat-treated at 38 to 40°C (100 to 103°F) for four to six weeks. Then apical shoot tips about 0.33 mm long are aseptically removed from the heat-treated plants and grown in vitro in test tubes containing a nutrient medium.

Chemical Treatment of Propagation Stock

This procedure can sometimes eradicate externally carried pathogens. For example, white calla lily can be freed of the *Phytophthora* root fungus by soaking the rhizomes for an hour in a weak formaldehyde solution (*5*).

Growing Seedlings (Apomictic and Nonapomictic)

Most viruses are not transmitted through seeds (although there are exceptions). Seedlings may be used to produce new clones to replace older cultivars that have deteriorated from virus infections. In species where they occur, such as citrus, apomictic seedlings have been used to regenerate "old" virus-infected clones by the development of nucellar seedlings (see p. 70) (*15*). An advantage of seedling rootstocks in fruit and nut crops is that they are mostly virus-free. Most new cultivars developed as seedlings in breeding programs are virus-free.

SOURCE SELECTION IN CLONALLY PROPAGATED CULTIVARS

Most vegetatively propagated cultivars originate as individual plants of a seedling population. Superior seedling "selections" (term used by horticultural plant breeders) are developed into a clone by vegetative propagation to provide own-rooted or grafted plants for evaluation. Advanced selections are multiplied and distributed for use. Figure 8–17 shows the general sequence of steps that takes place in the initial selection of the clone as a cultivar and the subsequent selection of sources for propagation.

Whatever the basis or reason for selection of the cultivar, a **propagation source** must be provided and multiplied. Propagation then takes place in a sequential pattern in both time (vertical) and space (horizontal) and provides an historical vegetative pedigree for the cultivar.

How variation develops within the clone has a great deal to do with the success of the cultivar. The first part of this chapter describes the major causes of clonal variation. The last part of the chapter decribes specific programs designed to control this variation and to produce pathogen-free, true-to-type propagation source material.

Early horticulturists spoke of cultivars "running out" and "aging" and the need to rejuvenate through seed propagation (*51, 66*). It is now recognized that this "running-out" phenomenon is not an inherent "aging" process of the clone but results from disease infections, primarily vi-

---

PHENOTYPIC VS. GENOTYPIC SELECTION

Visual inspection is an essential part of source selection in establishing "trueness-to-name" and "trueness-to-type." **Trueness-to-name** refers to the proper labeling of the cultivar. **Trueness-to-type** is a general term that defines an acceptable standard or range of variability for the characteristics and performance for plants of the particular cultivar. **Phenotypic selection** refers to selection based only on visual inspection of the source plant(s). **Genotypic selection** is based on the visual inspection of vegetative progeny plants propagated from that source. Phenotypic selection is reliable for many situations when conducted by competent observers, but it cannot distinguish between variations which are strictly phenotypic and those which produce a permanent change in the progeny plants. Furthermore, phenotypic selection cannot identify latent mutants or determine genetic or epigenetic performance.

Phenotypic or latent genetic variation can be separated by testing vegetative propagules in new environments and comparing performance to a known plant of the same clone. This procedure constitutes a **vegetative progeny test**. The concept is the same as that used for phenotypic and genotypic selection in seed propagation (see p. 87) but tests for somatic rather than segregational variation that results from the sexual cycle.

---

ruses and other systemic pathogens, although accumulated unrecognized somatic variations might also be present (*100*). To a certain extent the cultivars that survive are those that tolerate adverse environmental conditions. Nevertheless, intensive propagation of most clonally propagated cultivars of horticultural significance demonstrates the need to control the spread of pathogens in infected sources during propagation.

Historically, it appears that the great advances made in horticultural clonal propagation have been accompanied by the co-evolution of clones, vectors, viruses, and virus-like pathogens.

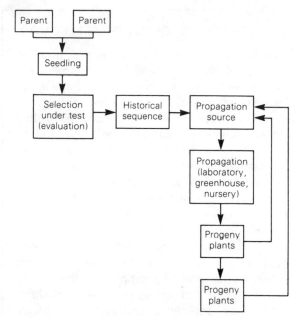

FIGURE 8-17 Flowchart showing the pedigree system of source analysis.

In nature, seed reproduction tends to screen out and/or select against such pathogens. Plant breeders and propagators have unwittingly aided the accumulation of systemic pathogens by collecting and bringing together many different kinds of plants into a single location, as in nurseries, breeding collections, and repositories and by widely disseminating clones from one environment to another (*86*). The process of systemic infection is significantly increased by combining more than one genotype by budding and grafting (*38, 86*). The problem has been further enhanced by the concentration of the genetic material of important horticultural industries into a few widely planted clones to produce monocultures of limited numbers of cultivars.

Single Plant (Clonal) Selection

A single plant may be used as a source for propagation if only limited numbers of propagules are needed. This may occur when propagating minor commercial cultivars, establishing collections, transporting through quarantine barriers, or beginning a nuclear stock program. Because such a

plant becomes the sole representative of that clone in future propagation, it constitutes a new **source clone.** If the plant originates as a recognizable mutation, the vegetative progeny can become recognized as a new clone and could be introduced as a new cultivar (see p. 177). Equally significant, the plant may be an inferior unrecognized variant or be infected with a serious virus or other pathogen.

It is important to select as a source plant one with a known pedigree having some history of superior performance (**source pedigree analysis**). It should be subjected to careful **visual inspection** (**phenotypic selection**) at appropriate stages of development and indexed for significant viruses or other pathogens of that crop. Even after propagation, visual inspections of the progeny plants (**genotypic selection**) should continue at least through the first vegetative generation.

Pedigree Selection

Many commercial cultivars are propagated by mass selection from commercial material of that clone. In this procedure, a number of separate source plants may be used with limited numbers of propagules from each plant. These plants may be part of the commercial cycle which is retained to provide a source (cuttings, bulbs, tubers, divisions, buds, scions) for the next generation of propagation. In other cases, source material may come from special plants maintained as a **source block** in a particular part of the nursery. Commercial **source orchards** or **vineyards** with good production records may be used to provide buds, scions, or cuttings of fruit, nut, and vine crops.

Source blocks or orchards eventually become too old or otherwise unsuitable to continue to provide propagules and new source blocks or orchards are established or selected. Invariably, these new plants are vegetative progeny of the original source. This can then be referred to as a **progeny source block** (or **progeny source orchard**). The significance of the relationship is that any somatic variant, no matter what frequency, is likely to be passed on to the progeny generation at about the same rate as it occurred in the original source. Maintaining accurate records of sources and propagation sequences may enable a propagator to analyze the entire source pedigree to its origin in case of the unexpected occurrence of a new variant.

Selection in the pedigree source system is based primarily on visual inspection of the source plants (phenotypic selection) and avoidance or roguing of "off-type" plants. Sometimes, records of commercial performance are available, but these would involve the entire planting and not individual source plants. The success or failure of pedigree selection depends upon (a) the frequency of off-type variants or latent infection, (b) the ability to detect them by visual inspection or other tests, and (c) the extent to which variability can be tolerated in the progeny plants.

For many important horticultural crops (potato, sweet potato, sugarcane, banana, strawberry, grapes, fruit trees, etc.) that have been intensively propagated as clones, this system has resulted historically in their eventual contamination by viruses and other pathogens, which in some instances have threatened the viability of the entire industry. For this reason, **nuclear stock programs** have been developed (*36, 43*).

Nuclear Stock Selection

Nuclear stock programs include the following steps:

1. Initial selection of a nucleus of individual source plant(s).
2. Maintenance of nuclear stock in special blocks with safeguards against reinfection and genetic change.
3. A system of commercial propagation and distribution whereby source material is multiplied and disseminated without reinfection and/or genetic change. Source identity must be maintained so that the unexpected occurrence of a new variant in the progeny plants can be traced to its origin and corrected.

Figure 8–18 shows a generalized scheme for a nuclear stock program.

Step I—Initial Selection of Source-Clone

The procedures vary with the unique requirements of individual crops. One procedure is to screen large numbers of plants of the cultivar to find individuals that are free of pathogens, utilizing methods such as culture indexing and/or virus indexing (described on p. 187). Candidate plants are tested for genetic potential.

A second approach is, first, to select an individual plant for its genetic potentiality, as de-

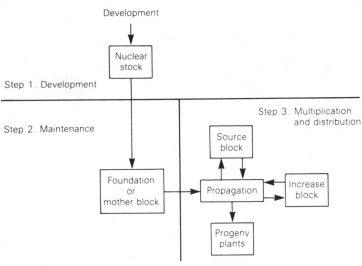

FIGURE 8–18 Flowchart showing the relationships in a nuclear stock system for selection, maintenance, and distribution of true-to-type, virus-tested propagation material.

Development

Nuclear stock

Step 1. Development

Step 2. Maintenance

Step 3. Multiplication and distribution

Foundation or mother block

Propagation

Source block

Increase block

Progeny plants

Includes:

Visual inspection of source (phenotypic selection)

Indexing (as required at each step)

Visual inspection of progeny (genotypic selection)

scribed for individual plant selection (p. 190), and second, to test for latent viruses or other pathogens. In both cases, the nuclear stock would proceed to step II. If infected, plants would be treated to rid the plant of the pathogen (described on p. 188) before advancing to step II.

Step II—Maintenance of Foundation Stock

Nuclear stock which has been selected as "true-to-name," "true-to-type," and "virus-tested" is maintained in limited numbers under protected conditions to prevent reinfection. The planting is usually referred to as a **foundation** (or **mother**) **block** and individual plants may be **registered** by plant location. Preservation of the clone is based on proper identification, careful labeling, isolation, sanitation, periodic visual inspections, and indexing.

Isolation is necessary to separate the plants from infective agents. For vector-less diseases, a minimum requirement is to separate the propagation source block from the propagation area and with sufficient spacing and prominent labels to decrease the chance of mechanical mixture and incorrect plant identification. For insect- or nematode-vectored diseases, isolation can probably best be achieved by growing plants in containers in an insect-proof screenhouse or in a greenhouse with restricted entry. In the open, insect- or wind-pollinated plants should be separated by at least ½ mile from potential sources of pollen-vectored viruses. As an alternative, flowers should be removed before blooming to prevent cross-pollination.

Step III—Distribution and Multiplication during Commercial Propagation

A limited amount of foundation propagating material is provided from a foundation (or mother) block. It is then multiplied to provide a nursery source block sufficiently large to provide propagules for commercial propagation. Figure 8–18 includes the use of a specific **source block** (stock block, mother block, or scion orchard) which is developed, maintained, and used by the propagator solely for the production of cuttings, scions, buds, or other propagules for propagation.

The scheme includes provision for the limited use of a commercial planting as an **increase block** to be used to supplement the mother block in periods of high demand. Such a block involves only a single vegetative generation away from either the foundation or the mother block.

In some plants (e.g., potato and strawberry) a number of increase generations will be required to build up a volume of plants for commercial use. During this period, roguing of any off-type plants is important and regular indexing is recommended.

Plants originating from propagules obtained from the foundation block, the mother block, or an increase block would be eligible for commercial distribution. However, to reestablish new mother blocks, all material would be obtained only from the foundation blocks.

Quarantines and Movement of Vegetatively Propagated Material (30)

Propagation of vegetatively propagated material may include shipment from country to country or from state to state in the United States. This may involve commercial shipments, or movement of germplasm. Because of the potential risk of transporting dangerous pests on the material or systemically infected clonal material, such as viruses, specific regulations and sometimes quarantines are in place to control such movement (30). Soil on the plant is not usually allowed because of possible nematodes, and other pests. Plants are inspected before entry and are sometimes treated. Restrictions may be put on the use and destination. Post-entry inspections are required, and specific indexing for viruses may also be required.

It is important that propagators be familiar with any national, state, and local regulations that affect distribution of their products.

THE PLANT PATENT LAW

An amendment to the U.S. Patent Law (24) was enacted in 1930 which enabled the originators of new plant forms in the United States to obtain patents on them. This amendment added impetus to the development and introduction of new and improved plants by establishing the possibility of monetary rewards to private plant breeders and alert horticulturists for valuable plant introductions. It has stimulated growers of fruit and ornamentals to be on the lookout for improved plant forms and has handsomely paid numerous fortunate or observant persons for their finds.

A patent can be obtained on "any distinct and new variety of plant, including cultivated sports, mutants, hybrids, and newly found seedlings, other than a tuber-propagated plant or a plant found in an uncultivated state." For a new plant to be patentable, it must be asexually reproduced and propagated commercially, as by cuttings, layering, budding, or grafting. The plant patent law also includes microorganisms. Protection is also given to some seed-propagated cultivars (see Chap. 4).

Characteristics which would cause a plant to be "distinct and new" and thus patentable include such things as growth habit; immunity from disease; resistance to cold, drought, heat, wind, or soil conditions; the color of the flower, leaf, fruit, or stem; flavor; productivity, including everbearing qualities in the case of fruits; storage qualities; form; and ease of reproduction.

The applicant for a plant patent must be the person who invented or discovered and subsequently reproduced the new cultivar of plant for which the patent is sought. If a person who was not the inventor applied for a patent, the patent, if it were obtained, would be void; in addition, such a person would be subject to criminal penalties for committing perjury. A plant found growing wild in nature is not considered patentable.

A plant patent issued to an individual is a grant consisting of the right to exclude others from propagation of the plant or selling or using the plant so reproduced. Essentially, it is a grant by the U.S. government, acting through the Patent Office, to the inventor (or his heirs or assigns) of certain exclusive rights to his invention for a term of 17 years throughout the United States and its territories and possessions. A U.S. plant patent affords no protection in other countries. The mere fact that a patent has been issued on a new plant does not imply any endorsement by the government of high quality or merit. The only implication of a patent on a plant is that it is "distinct and new."

The U.S. Plant Variety Protection Act, which became effective in 1970, extends plant patent protection to certain sexually propagated cultivars which can be maintained as "lines," such as those of cotton, alfalfa, soybeans, marigolds, bluegrass, and others (see Chap. 4).

REFERENCES

1. Abbott, A. J., and P. K. Atkin, eds. 1987. *Improving vegetatively propagated crops*. New York: Academic Press.

2. Baker, K. F. 1962. Thermotherapy of planting material. *Phytopathology* 52:1244–55.

3. Baker, K. F. 1984. Development of nursery techniques. *Proc. Inter. Plant Prop. Soc.* 34:152–64.

4. ———. 1984. The obligation of the plant propagator. *Proc. Inter. Plant Prop. Soc.* 34:195–209.

5. Baker, K. F., and P. A. Chandler. 1957. Development and maintenance of healthy planting stock. In *The U.C. system for producing healthy container-grown plants*, K. F. Baker, ed. Calif. Agr. Exp. Sta. Man. 23, pp. 217–36.

6. Baur, E. 1909. Pfropfbastarde periclinal Chi-

mäeren and Hyperchimäeren. *Ber Deuts. Bot. Ges.* 27:603–5.

7. Bonga, J. M. 1982. Vegetative propagation in relation to juvenility, maturity, and rejuvenation. In *Tissue culture of forest trees*, J. M. Bonga and D. J. Durzan, eds. Amsterdam: Elsevier.

8. ———. 1987. Clonal propagation of mature trees: Problems and possible solutions. In *Tissue culture of forest trees*. J. M. Bonga and D. J. Durzan, eds. Amsterdam: Elsevier, pp. 249–71.

9. Boulay, M. 1987. Conifer micropropagation: Applied research and commercial aspects. In *Tissue culture of forest trees*, J. M. Bonga and D. J. Durzan, eds. Amsterdam: Elsevier, pp. 185–206.

10. Boux, P., and P. Druart. 1980. Micropropaga-

tion, an industrial propagation method of quality plants true-to-type at a reasonable price. In *Plant cell cultures: Results and perspectives*, F. Sala et al., eds. Amsterdam: Elsevier/North Holland Biomedical Press, pp. 265–69.

11. Bringhurst, R. S., and V. Voth. 1956. Strawberry virus transmission by grafting excised leaves. *Plant Disease Rpt.* 40(7):596–600.

12. Bringhurst, R. S., V. Voth, and D. van Hook. 1960. Relationship of root starch content and chilling history to performance of California strawberries. *Proc. Amer. Soc. Hort. Sci.* 75:373–81.

13. Brink R. A. 1962. Phase change in higher plants and somatic cell heredity. *Quart. Rev. Bio.* 37(1):1–22.

14. Broerties, C., and A. M. Van Harten. 1987. Application of mutation breeding methods. In *Improving vegetatively propagated crops*, A. J. Abbott and R. K. Atkin, eds. New York: Academic Press, pp. 335–48.

15. Cameron, J. W., R. K. Soost, and H. B. Frost. 1957. The horticultural significance of nucellar embryony in citrus. In *Citrus virus diseases*, J. M. Wallace, ed. Berkeley: Univ. Calif. Div. Agr. Sci., pp. 191–96.

16. Campbell, A. I., ed. 1986. Workshop on clonal selection in tree fruit. *Acta Hort.* 180:1–131.

17. Chaleff, R. S. 1981. *Genetics of higher plants: Applications of cell culture.* Cambridge: Cambridge Univ. Press.

18. Clark, M. F., and A. N. Adams. 1977. Characteristics of the microplate method of enzyme-linked immunosorbent assay for the detection of plant viruses. *Jour. Gen. Virol.* 34:475–83.

19. Converse, R. H. 1985. Latent viruses: Harmful or harmless? *HortScience* 20(5):845–48.

20. Darrow, G. M. 1931. A productive thornless sport of the Evergreen blackberry. *Jour. Hered.* 22:404–6.

21. Darrow, G. M., R. A. Gibson, W. E. Toenjes, and H. Dermen. 1948. The nature of giant apple sports. *Jour. Hered.* 39:45–51.

22. Dermen, H. 1960. Nature of plant sports. *Amer. Hort. Mag.* 39:123–73.

23. Diener, T. O. 1979. *Viroids and viroid diseases.* New York: John Wiley.

24. Donahue, L. J. 1980. Plant patents and legalities. *Proc. Inter. Plant Prop. Soc.* 30:414–21.

25. Durzan, D. J., and F. C. Steward. 1983. Nitrogen metabolism. In *Plant Physiology.* Vol. VIII, F. C. Steward, ed. New York: Academic Press, pp. 55–265.

26. Duvick, D. N. 1978. Risks of monoculture via clonal propagation. In *Propagation of higher plants through tissue culture; a bridge between research and application*, K. W. Hughes et al., eds. Springfield, Va.: U.S. Dept. of Energy Tech. Infor. Center, pp. 73–84.

27. Einset, J., and C. Pratt. 1954. "Giant" sports of grapes. *Proc. Amer. Soc. Hort. Sci.* 63:251–56.

28. ———. 1959. Spontaneous and induced apple sports with misshapen fruit. *Proc. Amer. Soc. Hort. Sci.* 73:1–8.

29. Evans, D. A., and J. E. Bravo. 1986. Phenotypic and genotypic stability of tissue cultured plants. In *Tissue culture as a plant production system for horticultural crops*, R. H. Zimmerman, R. J. Griesbach, F. A. Hammerschlag, and R. H. Lawson, eds. Amsterdam: Martinus Nijhoff Publishers, pp. 73–94.

30. Foster, J. A. 1988. Regulatory actions to exclude pests during the international exchange of plant germplasm. *HortScience* 23(1):60–66.

31. Franclet, A., M. Boulay, F. Bekkaoui, Y. Fouret, B. Verschoore-Martouzet, and N. Walker. 1987. Rejuvenation. In *Tissue culture of forest trees*, J. M. Bonga and D. J. Durzan, eds. Amsterdam: Elsevier, pp. 232–48.

32. Frazier, N. W., ed. 1970. *Virus diseases of small fruits and grapevines.* Berkeley: Univ. Calif. Div. Agr. Sci.

33. Fridlund, P. R. 1980. Maintenance and distribution of virus-free fruit trees. In *Proc. conf. on nursery production of fruit plants through tissue culture*, R. H. Zimmerman, ed. U.S. Dept. Agr. Sci. and Education Administration, ARR-NE-11, pp. 11–22.

34. ———. 1970. Temperature effects on virus disease symptoms in some *Prunus, Malus*, and *Pyrus* cultivars. *Wash. Agr. Exp. Sta. Bul. 726.*

35. Frost, H. B. 1938. Nucellar embryony and juvenile characters in clonal varieties of citrus. *Jour. Hered.* 29:423–32.

36. Goff, L. M., 1986. Certification of horticultural crops—a state perspective. In *Tissue culture as a plant production system for horticultural crops*, R. H. Zimmerman, R. J. Griesbach, F. A. Hammerschlag, and R. H. Lawson, eds. Dordrecht: Martinus Nijhoff Publishers, pp. 139–46.

37. Goheen, A. C., G. Nyland, and S. K. Lowe. 1973. Association of a rickettsia-like organism with Pierce's disease of grapevines and alfalfa

dwarf and heat therapy of the disease in grape-vines. *Phytopathology* 63:341–45.

38. Goheen, A. C. 1981. Grape pathogens and prospects for controlling grape diseases. *Proc. Grape and Wine Cent. Symp.* Davis, Calif.: University of California.

39. Goodman, R. M., H. Hauptli, A. Crossway, and V. C. Knauf. 1987. Gene transfer in crop improvement. *Science* 236:48–54.

40. Haberlandt, G. 1930. Das Wesen der *Crataegomespili* Sitzber. *Preuss. Akad. Wiss.* 20:374–94.

41. Hackett, W. P. 1985. Juvenility, maturation and rejuvenation in woody plants. In *Horticultural Reviews,* Vol. 7, J. Janick, ed. Westport, Conn.: AVI Publ. Co., pp. 109–55.

42. Hamilton, R. I. 1985. Using plant viruses for disease control. *HortScience* 20(5):848–52.

43. Hansen, A. J. 1985. An end to the dilemma—virus-free all the way. *HortScience* 20(5):852–59.

44. Joiner, N., ed. 1981. *Foliage plant production.* Englewood Cliffs, N.J.: Prentice-Hall.

45. Jordan, R. L. 1986. Diagnosis of plant viruses using double-stranded RNA. In *Tissue culture as a plant production system for horticultural crops,* R. H. Zimmerman, R. J. Griesbach, F. A. Hammerschlag, and R. H. Lawson, eds. Dordrecht: Martinus Nijhoff Publishers, pp. 125–34.

46. Kartha, K. K. 1986. Production and indexing of disease-free plants. In *Plant tissue culture and its agricultural applications,* L. A. Withers and P. G. Alderson. London: Butterworth, pp. 219–38.

47. Kester, D. E. 1976. Noninfectious bud failure in almond. In *Virus diseases and noninfectious disorders of stone fruits in North America.* U.S. Dept. Agr. Handbook 437. Washington, D.C.: U.S. Govt. Printing Office, pp. 278–82.

48. ———. 1976. The relationship of juvenility to plant propagation. *Proc. Inter. Plant Prop. Soc.* 26:71–84.

49. ———. 1983. The clone in horticulture. *HortScience* 18(6):831–37.

50. Kester, D. E., and R. N. Asay. 1978. Variability in noninfectious bud-failure in 'Nonpareil' almond. I. Location and environment. *Jour. Amer. Soc. Hort. Sci.* 103:377–82.

51. Knight, T. A. 1795. Observations on the grafting of trees. *Phil. Trans. Roy. Soc., London* 85:290.

52. Kosuge, T., C. P. Meredith, and A. Hollaender, eds. 1983. *Genetic engineering of plants: An agricultural perspective.* New York: Plenum Press.

53. Krenke, N. P. 1933. *Wundkompensation Transplantation und Chimären bei Pflanzen* (translated from Russian). Berlin: Springer.

54. Lacey, C. N. D., and A. I. Campbell. 1987. Selection, stability and propagation of mutant apples. In *Improving vegetatively propagated crops,* A. J. Abbott and R. K. Atkin, eds. New York: Academic Press, pp. 349–62.

55. Lambert, D. H., R. F. Stouffer, and H. Cole, Jr. 1979. Stunting of peach seedlings following soil fumigation. *Jour. Amer. Soc. Hort. Sci.* 104:433–35.

56. Lawson, R. H. 1986. Pathogen detection and elimination. In *Tissue culture as a plant production system for horticultural crops,* R. H. Zimmerman, R. J. Griesbach, F. A. Hammerschlag, and R. A. Lawson, eds. Dordrecht: Martinus Nijhoff Publishers, pp. 97–118.

57. Libby, W. J., and J. V. Hood. 1976. Juvenility in hedged radiata pine. *Acta Hort.* 56:91–98.

58. Libby, W. J., and R. M. Rauter. 1984. Advantages of clonal forestry. *For. Chron.,* pp. 145–49.

59. Marcotrigiano, M. 1985. Synthesizing plant chimeras as a source of new phenotypes. *Proc. Inter. Plant Prop. Soc.* 35:582–86.

60. Marsden, M. E. 1955. The history of vegetative propagation. *Rpt. 14th Inter. Hort. Cong.,* Vol. 2, pp. 1157–64.

61. McGranahan, G., and H. I. Forde. 1985. Relationship between clone age and selection trait expression in mature walnuts. *Jour. Amer. Soc. Hort. Sci.* 110(5):692–95.

62. Meins, F., Jr., and A. N. Binns. 1979. Cell determination in plant development. *BioScience* 29:221–25.

63. ———. 1978. Epigenetic clonal variation in the requirement of plant cells for cytokinin. In *The clonal basis for development,* S. Subtelny et al., eds. New York: Academic Press, pp. 185–201.

64. Menzies, M. J. 1985. Vegetative propagation of radiata pine. *Proc. Inter. Plant. Prop. Soc.* 35:383–89.

65. Miller, E. V. 1954. The natural origins of some popular varieties of fruit. *Econ. Bot.* 8:337–48.

66. Molisch, H. 1938. *The longevity of plants* (1928). Lancaster, Pa.: E. Fulling (English translation).

67. Moore, J. M., and J. Janick, eds. 1975. *Advances in fruit breeding.* West Lafayette, Ind.: Purdue Univ. Press.

68. Moore, J. M. and J. Janick, eds. 1983. *Methods of fruit breeding.* West Lafayette, Ind.: Purdue Univ. Press.

69. Mullins, M. G., Y. Nair, and P. Sampet. 1979. Rejuvenation *in vitro*: Induction of juvenile characters in an adult clone of *Vitis vinifera. Ann. Bot.* 44:623–27.

70. Nelson, S. H. 1977. Loss of productivity in clonal apple rootstocks. *Proc. Inter. Plant. Prop. Soc.* 27:350–55.

71. Nemeth, M. 1986. *Virus, mycoplasma and rickettsia diseases of fruit trees.* Dordrecht: Martinus Nijhoff Publishers.

72. Nevers, P., N. S. Shepherd, and H. Saedler. 1986. Plant transposable elements. In *Advances in botanical research.* J. A. Callow, ed. Orlando, Fla.: Academic Press, pp. 103–203.

73. Nyland, G., R. M. Gilmer, and J. D. Moore. 1976. "Prunus" ring spot group. In *Virus diseases and noninfectious disorders of stone fruits in North America.* U.S. Dept. Agr. Handbook 437. Washington, D.C.: U.S. Govt. Printing Office, pp. 104–32.

74. Oglevee-O'Donovan, W. 1986. Production of culture virus-indexed geraniums. In *Tissue culture as a plant production system for horticultural crops*, R. H. Zimmerman, R. J. Griesbach, F. A. Hammerschlag, and R. H. Lawson, eds. Dordrecht: Martinus Nijhoff Publishers, pp. 119–24.

75. Oleson, P. O. 1978. On cyclophysis and topophysis. *Silvae Genetica* 27:173–78.

76. Olmo, H. P. 1952. Breeding tetraploid grapes. *Proc. Amer. Soc. Hort. Sci.* 59:285–90.

77. Powell, C. L., and D. J. Bagyaraj, eds. 1984. *Mycorrhizae.* Boca Raton, Fla.: CRC Press.

78. Quak, F. 1977. Meristem culture and virus-free plants. In *Applied and fundamental aspects of plant cell, tissue, and organ culture*, J. Reinert and Y. P. S. Bajaj, eds. Berlin, Springer-Verlag, pp. 598–615.

79. Raju, B. C., and J. C. Trolinger. 1986. Pathogen indexing in large-scale propagation of florist crops. In *Tissue culture as a plant production system for horticultural crops.* R. H. Zimmerman, R. J. Griesbach, F. A. Hammerschlag, and R. H. Lawson, eds. Dordrecht: Martinus Nijhoff Publishers, pp. 135–38.

80. Ranjit, M., and D. E. Kester. 1988. Micropropagation of cherry rootstocks: II. Invigoration and enhanced rooting of '46-1 Mazzard' by co-culture with "Colt." *Jour. Amer. Soc. Hort. Sci.* 113(1):150–54.

81. Schaffelitzky de Muckadell, M. 1959. Investigations on aging of apical meristems in woody

plants and its importance in silviculture. *Forstl. Forsogov. Danm.* 25:310–455.

82. Shamel, A. D., C. S. Pomeroy, and R. E. Caryl. 1929. Bud selection in the Washington Navel orange; progeny tests of limb variations. *USDA Tech. Bul. 123.*

83. Shull, C. H. 1912. "Phenotype" and "clone." *Science n. s.* 35:182–83.

84. Simmonds, N. W., ed. 1979. *Evolution of cultivated crops.* London: Longman.

85. Soost, R. K., J. W. Cameron, W. P. Bitters, and R. G. Platt. 1961. Citrus bud variation. *Calif. Citrograph* 46:176, 188–93.

86. Stace-Smith, R. 1985. Role of plant breeders in dissemination of virus diseases. *HortScience* 20(5):834–37.

87. Stern, W. T. 1943. The use of the term "clone." *Jour. Roy. Hort. Soc.* 74:41–47.

88. Steward, F. C., and A. D. Krikorian. 1978. Problems and potentialities of cultured plant cells in retrospect and prospect. In *Plant cell and tissue culture: Principles and applications*, W. R. Sharp et al., eds. Columbus, Ohio: Ohio State Univ. Press, pp. 221–62.

89. Stewart, R. N. 1978. Ontogeny of the primary body in chimeral forms of higher plants. In *The clonal basis of development*, S. Subtelny et al., eds. New York: Academic Press, pp. 131–60.

90. ———. 1965. The origin and transmission of a series of plastogene mutants in *Dianthus* and *Euphorbia. Genetics* 52:925–47.

91. Stone, O. M. 1978. The production and propagation of disease free plants. In *Propagation of higher plants through tissue culture: A bridge between research and application*, K. W. Hughes et al., eds. Springfield, Va.: U.S. Dept. of Energy Tech. Inform. Center, pp. 25–34.

92. Stout, A. B. 1940. The nomenclature of cultivated plants. *Amer. Jour. Bot.* 27:339–47.

93. Stout, G. L. 1962. Maintenance of "pathogen-free" planting stock. *Phytopathology* 52:1255–58.

94. Stoutemyer, V. T. 1937. Regeneration in various types of apple wood. *Iowa Agr. Exp. Sta. Res. Bul.* 220:308–52.

95. Styer, D. J., and C. K. Chin. 1983. Meristem and shoot-tip culture for propagation, pathogen elimination, and germplasm preservation. In *Horticultural reviews*, Vol. 5, J. Janick, ed. Westport, Conn.: AVI Publ. Co., pp. 221–27.

96. Sussex, I. M. 1983. Determination of plant organs and cells. In *Genetic engineering of plants: An*

agricultural perspective, T. Kosuge, C. P. Meredith, and A. Hollaender, eds. New York: Plenum Press, pp. 443–51.

97. Symons, R. H. 1980. Rapid indexing of sunblotch viroid in avocados and of exocortis viroid in citrus. *Proc. Inter. Plant Prop. Soc.* 30:578–83.

98. Tanaka, T. 1927. Bizzarria—a clear case of periclinal chimera. *Jour. Gen.* 18:77–85.

99. Tilney-Bassett, R. A. E. 1986. *Plant chimeras.* London: Edward Arnold.

100. Tincker, M. A. H. 1945. Propagation, degeneration, and vigor of growth. *Jour. Roy. Hort. Soc.* 70:333–37.

101. Tolley, I. S. 1975. A technique for the accelerated production of commercially acceptable citrus clones from seed. *Proc Inter. Plant Prop. Soc.* 25:294–97.

102. Tufts, W. P., and C. J. Hansen. 1931. Variation in shape of Bartlett pears. *Proc. Amer. Soc. Hort. Sci.* 28:627–33.

103. U.S. Dept. Agr. 1976. *Virus and other disorders with virus-like symptoms of stone fruits in North America.* U.S. Dept. Agr. Handbook 10. Washington, D.C.: U.S. Govt. Printing Office.

104. van Oosten, H. J., and H. H. van der Borg, eds. 1977. Symposium on clonal variation in apple and pear. *Acta Hort.* 75:1–185.

105. Vaughn, K. 1983. Chimeras: Problems in propagation. *HortScience* 18(6):845–48.

106. Verkade, S. D. 1986. Mycorrhizal inoculation during plant propagation. *Proc. Inter. Plant Prop. Soc.* 36:613–18.

107. Webber, H. J. 1903. New horticultural and agricultural terms. *Science* N.S., 18:501–2.

108. Whiting, J. R., and W. J. Hardie. 1981. Yield and compositional differences between selections of grapevine cv. Cabernet Sauvignon. *Amer. Jour. Enol. and Vit.* 32:212–18.

109. Winkler, H. 1907. Uber Pfropfbastarde und pflanzliche Chimären. Ber. Deuts Bot. Ges. 25:568–76.

110. ————. 1910. Uber die Nachkommenschaft der *Solanum* Pfropfbastarde und die Chromosomenzahlen ihrer Keimzellen. *Zeits. f. Bot.* 2:1–38.

111. Zimmerman, R. H. 1971. Flowering in crabapple seedlings: Methods of shortening the juvenile phase. *Jour. Amer. Soc. Hort. Sci.* 96(4):404–11.

112. Zobel, B. J., G. van Wyk, and P. Stahl. 1987. *Growing exotic forests.* New York: John Wiley.

113. Zohary, D., and P. Spiegel-Roy. 1975. Beginnings of fruit growing in the old world. *Science* 187(4174):319–27.

SUPPLEMENTARY READING

Abbott, A. J., and P. K. Atkin, eds. 1987. *Improving vegetatively propagated crops.* New York: Academic Press.

Baker, K. F., ed. 1957. *The U.C. system for producing healthy container-grown plants.* Calif. Agr. Exp. Sta. Man. 23.

Broertjes, C., and A. M. Van Harten. 1987. Application of mutation breeding methods. In *Improving vegetatively propagated crops*, A. J. Abbott and R. K. Atkin, eds. New York: Academic Press, pp. 335–48.

Campbell, A. I., ed. 1986. Workshop on clonal selection in tree fruit. *Acta Hort.* 180:1–131.

Converse, R. H., ed. 1987. *Virus diseases of small fruits.* U.S. Dept. Agr. Handbook 631. Washington, D.C.: U.S. Govt. Printing Office.

Doorenbos, J. 1965. Juvenile and adult phases in woody plants. In *Handbuch der Pflanzenphysiologie*, Vol. 15 (Part I). Berlin: Springer-Verlag, pp. 1222–35.

Frazier, N. W., ed. 1970. *Virus diseases of small fruits and grapevines.* Berkeley: Univ. Calif. Div. Agr. Sci.

Kirk, J. T. O., and R. A. E. Tilney-Bassett. 1978. *The plastids.* Amsterdam: Elsevier.

Klekowski, E. J., Jr. 1988. *Mutation, developmental selection and plant evolution*, New York: Columbia Press.

Kneen, O. H. 1948. Plant patents enrich our world. *National Geographic* 93:357–78.

Koenig, R., ed. 1980. Fifth international symposium on virus diseases of ornamental plants. *Acta Hort.* 110:334.

Marston, M. E. 1955. The history of vegetative propagation. *Rept. 14th Inter. Hort. Cong.*, Vol. 2, pp. 1157–64.

Nemeth, M. 1986. *Virus, mycoplasma and rickettsia diseases of fruit trees.* Dordrecht: Martinus Nijhoff Publishers.

Nielson-Jones, W. 1969. *Plant chimeras* (2nd ed.). London: Methuen & Co.

Nyland, G., and A. C. Goheen. 1969. Heat therapy

of virus disease of perennial plants. *Ann. Rev. Phytopathol.* 7:331–54.

POSNETTE, A. F., ed. 1963. *Virus diseases of apples and pear.* Tech. Comm. 30. Bucks, England: Commonwealth Agricultural Bureaux, Farnham Royal.

Proceedings of symposium. 1983. The clone in horticulture. *HortScience* 18(6):829–48.

Proceedings of symposium. 1988. Genetic considerations in the collection and maintenance of germplasm. *HortScience* 23(1):78–97.

Proceedings of symposium. 1988. Temperate fruit crops: History and management of genetic resources. *HortScience* 23(1):49–75.

Proceedings of symposium. 1985. Virus diseases: A dilemma for plant breeders. *HortScience* 20(5):831–59.

SIMMONDS, N. W., ed. 1979. *Evolution of cultivated crops.* Longman: London.

ZIMMERMAN, R. H., W. P. HACKETT, and R. P. PHARIS. 1985. Hormonal aspects of phase change and precocious flowering. In *Hormonal regulation of development,* R. P. Pharis and D. M. Reid, eds. Berlin: Springer-Verlag, p. 79.

9

Anatomical and Physiological Basis of Propagation by Cuttings

In propagation by **stem** and **leaf-bud cuttings** (**single-eye cuttings**), it is only necessary that a new adventitious root system[1] be formed, since a potential shoot system (a bud) is already present. **Root cuttings** must initiate both a new shoot system—from an adventitious bud[2]—as well as new adventitious roots or extending the existing root piece. Likewise, **leaf cuttings** must initiate both a new root and a new shoot system.

This capacity for regenerating the entire plant structure, a property of essentially all the plant's living cells, is demonstrated in the various cell and tissue systems described in Chaps. 16 and 17. This capacity depends on two fundamental characteristics of plant cells. One is **totipotency**, which means that each living plant cell contains the genetic information necessary for reconstituting all the plant parts and functions. The second is **dedifferentiation,** the capability of previously developed, differentiated cells to return to a meristematic condition and develop a new growing point. Since these two characteristics are more pronounced in some cells and plant parts than in others, the propagator must do some manipulation to provide the proper conditions for plant regeneration.

ADVENTITIOUS ROOT FORMATION

Adventitious roots form naturally on various kinds of plants. For example, "brace" roots develop on corn, pandanus, and other monocots, arising from the intercalary regions at the base of internodes. Banyan trees (Fig. 9-1) produce huge aerial roots that grow to and into the ground.

[1]**Adventitious roots** arise from any plant part other than by the normal development and ontogony of the seedling root and its branches; adventitious roots can also be regenerated from root-pruned seedlings.

[2]**Adventitious buds** (and shoots) are those arising from any plant part other than terminal, lateral, or latent buds on stems.

FIGURE 9–1 The ultimate in adventitious root production is shown on this banyan tree (*Ficus benghalensis*) at Hilo, Hawaii. Such roots arise from branches, extend downward, and grow into the soil.

Plants that are regenerated from rhizomes, bulbs, and other such structures also develop adventitious roots.

Adventitious roots are of two types: **preformed roots** and **wound roots** (Figs. 9–4, 9–5, 9–7, and 9–8). Preformed roots develop naturally on stems while they are still attached to the parent plant and may emerge prior to severing the stem piece. Wound roots develop only after the cutting is made, in response to the wounding effect in preparing the cutting. In effect, they are considered to be formed **de novo** (anew) (*50*). When a cutting is made, living cells at the cut surfaces are injured and exposed, and the **wound healing response** begins (*42*). In the subsequent healing and root regeneration process, three steps occur:

1. As the outer injured cells die, a necrotic plate forms, seals the wound with a corky material (suberin), and plugs the xylem with gum. This plate helps to protect the cut surfaces from desiccation and pathogens.

2. Living cells behind this plate begin to divide after a few days and a layer of parenchyma cells (callus) form a wound periderm.

3. Certain cells in the vicinity of the vascular cambium and phloem begin to divide and initiate adventitious roots.

The developmental anatomical changes that occur in *de novo* adventitious root formation of wound roots can be divided into four stages:

1. **Dedifferentiation** of specific differentiated cells
2. **Formation of root initials** from certain cells near vascular bundles, or vascular tissue, which have become meristematic by dedifferentiation
3. Subsequent development of these root initials into organized **root primordia**
4. Growth and **emergence** of the root primordia outward through other stem tissue plus the formation of vascular (conducting) connections between the root primordia and the vascular tissues of the cutting itself

Less commonly, five histologically distinct stages of de novo rooting were observed in *Vigna* hypocotyl cuttings (*260*). Likewise, there are three stages identified in the development of *Pinus radiata* seedling root primordia: (a) pre-initiative phase with no histological changes, (b) an initiative phase forming a meristematic locus, and (c) a post-initiative phase with meristemoids differentiating into root primordia (*266, 267*).

The precise location inside the stem where adventitious roots originate has intrigued plant anatomists for centuries. Probably the first study of this phenomenon was made by a French dendrologist, Duhamel du Monceau, in 1758 (*63*); a great many subsequent studies have covered a wide range of plant species (*91, 206, 303*).

In **herbaceous plants** adventitious roots usually originate just outside and between the vascular bundles (*239*), but the tissues involved at the site of origin can vary widely depending upon the kind of plant (Fig. 9–2). For example, in the tomato, pumpkin (*234*), and mung bean (*20*), adventitious roots arise in the phloem parenchyma; in *Crassula* they arise in the epidermis (*214*); and in coleus they originate from the pericycle (*35*).

Adventitious roots in castor bean (*Ricinus communis*) cuttings arise between the vascular bundles, as shown in Fig. 9–3. Root initials in carnation cuttings arise in a layer of parenchymatous cells inside a fiber sheath; the developing root tips, upon reaching this band of impenetrable fi-

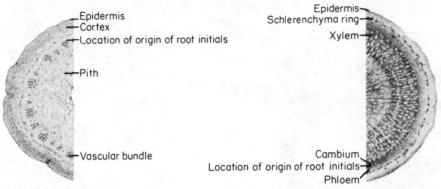

FIGURE 9-2 Stem cross sections showing the usual location of origin of adventitious roots. *Left:* Young, herbaceous, dicotyledonous plant. *Right:* Young, woody plant.

ber cells, do not push through it, but turn downward, emerging from the base of the cutting (*274*).

In **woody perennial plants**, where one or more layers of secondary xylem and phloem are present, adventitious roots in stem cuttings usually originate from living parenchyma cells, primarily in the young, secondary phloem (Fig. 9–4), but sometimes from such other tissues as vas-

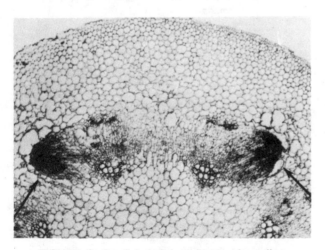

FIGURE 9-3 Adventitious root primordia (arrows) arising laterally and adjacent to vascular bundles in the castor bean (*Ricinus communis*), a herbaceous plant. Vascular connections will develop between the adventitious roots and the plant's vascular bundles. Epidermis is at top, pith at bottom. (From Priestley and Swingle (*239*).)

cular rays, cambium, phloem, lenticels, or pith (*90, 206*) (Table 9–1).

Generally, the origin and development of adventitious roots takes place **endogenously** next to and just outside the central core of vascular tissue. Many easy-to-root woody plant species develop adventitious roots from phloem ray parenchyma cells. Figure 9–5 depicts first anticlinal division of a phloem ray cell during dedifferentiation and a young root primordium elongating through the cortex. Upon emergence from the stem (Fig. 9–6), the adventitious roots have developed a root cap as well as a complete vascular connection with the originating stem.

The time for root initials to develop after cuttings are placed in the propagating bed varies widely. In one study (*274*) they were first observed microscopically after three days in chrysanthemum, five days in carnation (*Dianthus caryophyllus*), and seven days in rose (*Rosa*). Visible roots emerged from the cuttings after 10 days for the chrysanthemum, but three weeks were required for the carnation and rose.

Phloem ray parenchyma cells in juvenile (easy-to-root) cuttings of *Ficus pumila* undergo anticlinal cell division (dedifferentiation) and root primordia formation more quickly than mature (difficult-to-root) plants under optimal auxin treatments (Table 9–2). Once primordia are formed there is a comparable time period (seven to eight days) between root primordia elongation (emergence) and maximum rooting in both the easy and difficult-to-root plants (*50*). This also

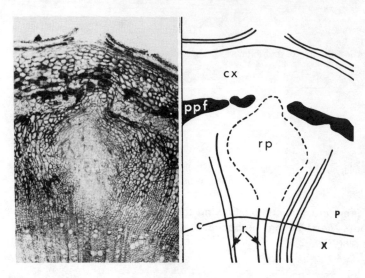

FIGURE 9-4 Tissues involved in adventitious root formation in 'Brompton' plum hardwood cuttings. cx, cortex; ppf, primary phloem fibers; rp, root primordium; p, phloem; r, rays; c, cambium; x, xylem. (Courtesy A. Beryl Beakbane, East Malling Research Station (*12*).)

TABLE 9-1

Origin of wound-induced **de novo** adventitious roots in stems of woody plants

Origin	Genera
Cambial and ray	
Cambial and phloem portions of ray tissues	*Acanthopanax, Chamaecyparis, Cryptomeria, Cunninghamia, Cupressus, Metasequoia*
Medullary rays	*Vitis*
Cambium	*Acanthus, Lonicera*
Fascicular cambium	*Clematis*
Phloem ray parenchyma	*Ficus, Hedera*
Secondary phloem in association with a ray	*Malus* (Malling stocks), *Camellia,* 'Brompton' *plum*
Phloem area close to the cambium	*Pistacia*
Cambium and inner phloem ray also in leaf gap	*Griselinia*
Bud and leaf gaps	
Outside the cambium in small groups	*Rosa, Cotoneaster, Pinus, Cephalotaxus, Larix, Sciadopitys, Malus*
Pericycle	*Acanthus*
Callus, internal	
Irregularly arranged parenchymatous tissues	*Abies, Juniperus, Picea, Sequoia*
Callus, external	
Callus tissues (external)	*Abies, Cedrus, Cryptomeria, Ginkgo, Larix, Pinus, Podocarpus, Sequoia, Sciadopitys, Taxodium, Pinus*
Bark and basal callus	*Citrus*
Within callus at base of cutting	*Pseudotsuga*
Other	
Hyperhydric outgrowth of the lenticels	*Tamarix*
Margin of differentiating resin duct or parenchyma within the inner cortex	*Pinus*

Source: M. B. Jackson (ed.). 1986. New root formation in plants and cuttings. Martinus Nijhoff Publishers. Dordrecht, The Netherlands.

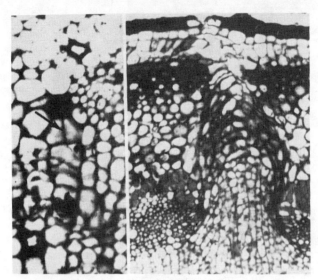

FIGURE 9-5 Developmental stage of rooting in mature cutting of *Ficus pumila*. First anticlinal division of phloem ray cell occurs during dedifferentiation (*left*) (see arrow), and a young root primordium elongates through the cortex (*right*) (*50*).

was reported with *Agathis australis* where primordia formation was variable in cuttings from different aged stock plants, but once root primordia formed, root emergence consistently occurred within a three-to-four-week period (*206, 326, 327*).

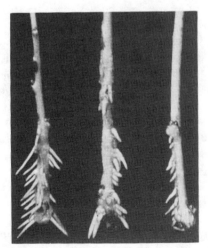

FIGURE 9-6 Emergence of adventitious roots in plum stem cuttings. Observe the tendency of the roots to form in longitudinal rows, which appear directly below buds.

TABLE 9-2

Time of adventitious root formation in juvenile and mature leaf-bud cuttings of *Ficus pumila* treated with IBA

	Juvenile	*Mature*
Anticlinal cell divisions of ray parenchyma	Day 4	Day 6
Primordia	Day 6	Day 10
First rooting[a]	Day 7	Day 20
Maximum rooting[b]	Day 14	Day 28

Source: Ref. 50.

[a]Based on 25 percent or more cuttings with roots protruding from stem.

[b]Based on 100 percent rooting and maximum root number.

Preformed, or **latent root initials** (*206*) generally lie dormant until the stems are made into cuttings and placed under environmental conditions favorable for further development and emergence of the primordia as adventitious roots. In *Populus* × *robusta* they form in stems in midsummer and then emerge from cuttings made the following spring (*269*). In some species primordia develop into aerial roots on the intact plant and become quite prominent (Fig. 9-7). Such preformed root initials occur in a number of easily rooted genera, such as willow (*Salix*), hydrangea (*Hydrangea*), poplar (*Populus*), jasmine (*Jasminum*), currant (*Ribes*), citron (*Citrus medica*), and others (*91*). The position of origin of these preformed root initials is similar to *de novo* adventitious root formation (Table 9-3) (*192, 206*). In some of the clonal apple and cherry rootstocks and in old trees of some apple and quince cultivars, these pre-

FIGURE 9-7 Preformed aerial roots at a node of *Ficus pumila*.

TABLE 9–3

Origin of preformed root initials (primordia, burrknots, and/or rootgerms) in stems of woody plants

Origin	Genera
Wide rays	*Populus*
Medullary rays, three double series associated with buds	*Ribes*
Nodal and connected with wide radial bands of parenchyma	*Salix*
Internodal medullary rays	*Salix*
Cambial ring in branch and leaf gap; 1° and 2° medullary rays	*Malus*
Bud gap	*Cotoneaster*
Median and lateral leaf trace gaps at node	*Lonicera*
Parenchymatous cells in divided bud gap	*Cotoneaster*
Medullary ray	*Citrus*
Phloem ray parenchyma	*Hydrangea*
Cambial region of an abnormally broad ray	*Acer, Chamaecyparis, Fagus, Fraxinus, Juniperus, Populus, Salix, Taxus, Thuja, Ulmus*

Source: M. B. Jackson (ed.). 1986. New root formation in plants and cuttings. Martinus Nijhoff   Publishers. Dordrecht, The Netherlands.

FIGURE 9–8 Burr knots developing from preformed root initials at the base of shoots of the 'Colt' cherry rootstock. (Photographed at East Malling Research Station, England.)

formed latent roots cause swellings, called **burr knots** (*251*), as shown in Fig. 9–8. Species with preformed root initials generally root rapidly and easily, but cuttings of many species without such root initials root just as easily.

In willow, latent root primordia can remain dormant, embedded in the inner bark, for years if the stems remain intact on the tree (*36*). Their location can be observed by peeling off the bark and noting the protuberances on the wood, with corresponding indentations on the inside of the bark that was removed.

CALLUS

Callus usually develops at the basal end of the cutting when it is placed under environmental conditions favorable for rooting. Callus is an irregular mass of parenchyma cells in various stages of lignification. Callus growth proliferates from cells at the base of the cutting in the region of the vascular cambium, although cells of the cortex and pith may also contribute to its formation (Table 9–1).

Frequently, roots appear through the callus, leading to the belief that callus formation is essential for rooting. The formation of callus and the formation of roots are independent of each other, even though they both involve cell division; that they often occur simultaneously is due to their dependence upon similar internal and environmental conditions.

In some species, however, callus formation apparently is a precursor of adventitious root formation. Origin of adventitious roots from callus tissue has been associated with difficult-to-root species (Table 9–1) (*50, 147*). For example, in *Pinus radiata* (*34*), *Sedum* (*333*), and *Hedera helix* (*92*) (adult phase), adventitious roots originate in the callus tissue that has formed at the base of the cutting—and from "tracheary nests" in callus of *Ficus pumila* (Fig. 9–9).

STEM STRUCTURE AND ROOTING

The development of a continuous **sclerenchyma ring** (Fig. 9–2) between the phloem and cortex, exterior to the point of origin of adventitious roots, which is often associated with maturation,

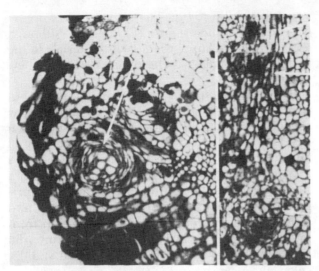

FIGURE 9-9 *Left:* Root primordia (see arrow) in callus originating in the vicinity of differentiating tracheary elements which have been described as "callus xylem" or "tracheary nests." *Right:* Callus primordia distinguished by "tracheary nests" (see arrows) connected to the main vascular system of *Ficus pumila* (50).

possibly constitutes an anatomical barrier to rooting. In studies of olive stem cuttings (41) and difficult-to-root mature *Hedera helix* (93) and *Ficus pumila* (50), such a ring was associated with recalcitrant cuttings, while easily-rooted types were characterized by discontinuity or fewer cell layers of this scelerenchyma ring (12).

While a sheath of lignified tissue in stems may in some cases act as a mechanical barrier to root emergence, there are so many exceptions that this certainly cannot be a primary cause of rooting difficulty. Moreover, auxin treatments and rooting under mist (205) cause considerable cell expansion and proliferation in the cortex, phloem, and cambium, resulting in breaks in continuous sclerenchyma rings, yet in difficult-to-root cultivars of several fruit species there is still no formation of root initials.

Easily rooted carnation cultivars have a band of sclerenchyma present in the stems, yet the developing root primordia emerge from the cuttings by growing downward and out through the base (274). In other plants in which an impenetrable ring of sclerenchyma could block root emer-

gence this same possibility is open. Rooting is more likely related to the actual formation of root initials rather than to the mechanical restriction of a sclerenchyma ring barring root emergence (50, 254, 325).

LEAF CUTTINGS

Many plant species, including both monocots and dicots, can be propagated by leaf cuttings (109). Although the origin of new shoots and new roots in leaf cuttings is quite varied, they generally develop from primary or secondary meristems, the latter type being the most common.

> **Primary (preformed) meristems** are groups of cells directly descended from embryonic cells that have never ceased to be involved in meristematic activity.
>
> **Secondary (wound) meristems** are groups of cells that have differentiated and functioned in some previously differentiated tissue system and then dedifferentiate into new meristematic zones (de novo), resulting in the regeneration of new plant organs.

Leaf Cuttings with Primary Meristems

In detached leaves of *Bryophyllum*, small plants arise from the notches around the leaf margin (see Fig. 10-17). These small plants originate from so-called foliar "embryos," formed in the early stages of leaf development from small groups of cells at the edges of the leaf. As the leaf expands, a foliar embryo develops until it consists of two rudimentary leaves with a stem tip between them, two root primordia, and a "foot" that extends toward a vein (138, 332). As the leaf matures, cell division in the foliar embryo ceases, and it remains dormant. If the leaf is detached and placed in close contact with a moist rooting medium, the young plants rapidly break through the leaf epidermis and become visible in a few days. Roots extend downward, and after several weeks many new independent plants form while the original leaf dies. The new plants thus develop from latent primary meristems—from cells that have not assumed mature characteristics. Such production of new plants from leaf cuttings by the renewed activity of primary meristems is found, too, in other species, as the piggyback plant (*Tolmiea*) and walking fern (*Camptosorus*).

Leaf Cuttings with Secondary Meristems

In leaf cuttings of such plants as *Begonia rex, Sedum,* African violet (*Saintpaulia*), snake plant (*Sansevieria*), *Crassula,* and lily, new plants may develop from secondary meristems arising from mature cells at the base of the leaf blade or petiole as a result of wounding.

In *Lilium longiflorum* and *L. candidum,* the bud primordium originates in parenchyma cells in the upper side of the bulb scale, whereas the root primordium arises from parenchyma cells just below the bud primordium. Although the original scale serves as source of food for the developing plant, the vascular system of the young bulblet is independent of that of the parent scale, which eventually shrivels and disappears (*313*).

In African violet new roots and shoots arise de novo by the formation of meristematic cells from previously differentiated cells in the leaves. The roots are produced endogenously from thin-walled cells lying between the vascular bundles. The new shoots arise from cells of the subepidermis and the cortex immediately below the epidermis and are generally **exogenous** in origin. Roots first emerge, form branch roots, and continue to grow for several weeks before adventitious buds and their subsequent development into adventitious shoots occurs. Root initiation and development is independent of adventitious bud and

shoot formation (*310*). The same process occurs with many begonia species (Fig. 9–10). Although the original leaf supplies nutrient materials to the young plant, it does not become a part of the new plant (*225*).

In several species, for example, sweet potato, *Peperomia,* and *Sedum,* new roots and new shoots on leaf cuttings arise in callus tissue which develops over the cut surface through activity of secondary meristems. The petiole of leaf cuttings of *Sedum* forms a considerable pad of callus within a few days after the cuttings are made. Root primordia are organized within the callus tissue, and shortly thereafter four or five roots develop from the parent leaf. Following this, stem primordia arise on a lateral surface of the callus pad and develop into new shoots (*333*).

Adventitious roots form on leaves much more readily than do adventitious buds. In some plants, such as the India rubber fig (*Ficus elastica*) and the jade plant (*Crassula argentea*), the cutting must include a portion of the old stem containing an axillary bud because although adventitious roots may develop at the base of the leaf, an adventitious shoot is not likely to form. In fact, rooted leaves of some species will survive for years without producing an adventitious shoot. Treatments with cytokinins such as benzyladenine, PBA, or kinetin may, however, initiate buds and shoots (*24, 51*).

ROOT CUTTINGS

Development of adventitious shoots, and in many cases adventitious roots, must take place if new

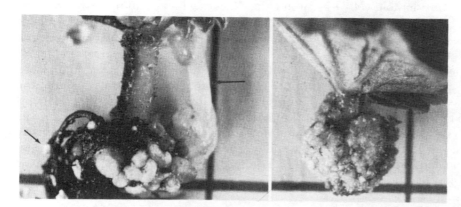

FIGURE 9–10 *Left:* Adventitious shoot (upper arrow) and roots (lower arrow) from a leaf cutting of Rieger begonia. At high cytokinin concentration only buds and bud-like tissue are visible (*right*) (roots formed but were removed before the photograph was taken) (*51*).

plants are to be regenerated from root pieces (root cuttings) (*248*). In some species, adventitious buds form readily on roots of intact plants, producing suckers. When roots of such species are dug, removed, and cut into pieces, buds are even more likely to form. In young roots, such buds may arise endogenously in the pericycle near the vascular cambium (*255, 330*). The developing buds first appear as groups of thin-walled cells having a prominent nucleus and a dense cytoplasm (*72, 308*). In old roots buds may arise exogenously in a callus-like growth from the phellogen; or they may appear in a callus-like proliferation from ray tissue. Bud primordia may also develop from wound callus tissue that proliferates from the cut ends or injured surfaces of the roots (*239*), or they may just arise at random from cortex parenchyma (*249*).

Regeneration of new root meristems on root cuttings is often more difficult than the production of adventitious buds (*26*). New roots may not always be adventitious and can develop from latent root initials contained in old branch roots and present on the root piece (*258*). Generally, such branch roots arise from differentiated cells of the pericycle or endodermis, or both, adjacent to the central vascular cylinder (*18*). Adventitious root initials have been observed to arise in the region of the vascular cambium in roots.

ROOT CUTTING PROPAGATION OF CHIMERAL PLANTS

One of the chief advantages claimed for asexual propagation is the exact reproduction of all characteristics of the parent plant. With root cuttings, however, this generalization does not always hold true. In periclinal chimeras, in which the cells of the outer layers are of a different genetic makeup from those of the inner tissues, the production of a new plant by root cuttings results in a plant that is different from the parent. This is well illustrated in the thornless boysenberry, and the 'Thornless Evergreen' trailing blackberry, in which stem or leaf-bud cuttings produce plants that retain the (mutated) thornless condition, but root cuttings develop into (normal, nonmutated) thorny plants.

Regeneration of new plants from root cuttings takes place in different ways, depending upon the species. In the most common type, the root cutting first produces an adventitious shoot, and later produces roots, often from the base of the new shoot rather than from the original root piece itself. Sometimes these adventitious shoots can be removed and rooted as stem cuttings when treated with a rooting hormone (*249*). In other plants, a well-developed root system has formed by the time the first shoots appear. Root cuttings of some species form a strong adventitious shoot, but no new roots develop, and the cutting eventually dies. In certain species, root cuttings produce a strong new root system, but no adventitious shoot arises, so the cutting finally dies (*156*).

Root cuttings taken from very young seedling trees are much more successful than those taken from older trees. Failures in the latter case are apparently due to the inability of the root pieces to regenerate a new root system. This inability is probably related to the phenomenon of juvenility, which is also involved in root formation in stem cuttings.

Not all underground portions of plants are roots—some, like rhizomes, are stem tissue. New plants developing from shoots arising from such underground stem tissue could not, therefore, be said to have been propagated by root cuttings. A list of plants commonly propagated by root cuttings is given in Chap. 10.

POLARITY

The polarity inherent in shoots and roots is shown dramatically in rooting cuttings (Figs. 9–11 and 9–12). Stem cuttings form shoots at the **distal** end (nearest the shoot tip), and roots at the **proximal** end (nearest the crown or junction of the shoot and root system). Root cuttings of many species form roots at the distal end and shoots at the proximal end. Changing the position of the cuttings with respect to gravity does not alter this tendency (*23*) (Fig. 9–12).

In Vöchting's early studies (*312*) on the polarity of regeneration in plants, he pointed out that stem tissue is strongly polarized. The theory was then advanced that this property could be attributed to the individual cellular components,

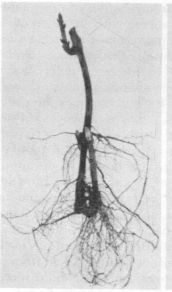

FIGURE 9-11 Results of planting a cutting of red currant (*Ribes sativum*) upside down (reversed polarity). *Left:* Several months after starting. *Right:* One year later. The shoot from the center bud with new roots at its base has become the main plant and is growing with correct polarity. The shoot from the top bud, while still alive, has failed to develop normally.

since no matter how small the piece, regeneration was consistently polar. Polarity is observed in leaf cuttings even though roots and shoots arise at the same position, usually the base of the cutting (See Fig. 9-10).

In mature pecan trees, root formation from root cuttings is greatest the closer the root piece is to the crown (stem/root junction) of the tree. This positional response is probably related to the more proximal root pieces having closer proximity to the "cone of juvenility" than more distal root cuttings.

When tissue segments are cut, the physiological unity is disturbed. This must cause a redistribution of some substance, probably auxin, thus accounting for the different responses observed at previously adjacent surfaces. The correlation of polarity of root initiation with auxin movement has been noted in several instances (*208, 250, 263, 291, 316*). It is also known that the polarity in auxin transport varies in intensity among different tissues. The polar movement of auxins is an active transport process and apparently is a secretion activity, which occurs in phloem parenchyma cells (*193*).

FIGURE 9-12 Polarity of root regeneration in grape hardwood cuttings. Cuttings at left were placed for rooting in an inverted position, but roots still developed from the morphologically basal (proximal) end. Cuttings on right were placed for rooting in the normal, upright orientation with roots forming at the basal end.

PHYSIOLOGICAL BASIS OF ADVENTITIOUS ROOT AND SHOOT INITIATION

Plant Growth Substances

Certain concentrations of naturally occurring compounds having hormonal properties are more favorable than others for adventitious root initiation. Much study has been given to determining these relationships. In distinguishing between

plant hormones[3] and **plant growth regulators**[4] it can be said that all hormones regulate growth but not all growth regulators are hormones. Various classes of growth regulators, such as auxins, cytokinins, gibberellins, and ethylene, as well as inhibitors, such as abscisic acid and phenolics, influence root initiation. Of these, auxins have the greatest effect on root formation in cuttings. In addition to these groups, other naturally occurring materials that are not well defined, such as various inhibitors and promoters, may have a less direct part in adventitious root initiation (*187*).

Auxins

In the mid-1930s and later, studies of the physiology of auxin action showed that auxin was involved in such varied plant activities as stem growth, adventitious root formation (*184, 188, 294, 296, 324*), lateral bud inhibition, abscission of leaves and fruits, and activation of cambial cells.

Indole-3-acetic acid (IAA) was identified as a naturally occurring compound having considerable auxin activity (*182, 292*). Synthetic indoleacetic acid (IAA) (see Fig. 9–23) was subsequently tested for its activity in promoting roots on stem segments, and in 1935 investigators (*294*) demonstrated the practical use of this material in stimulating root formation on cuttings. About the same time it was shown (*291, 338*) that two synthetic materials, **indolebutyric acid** (IBA) and **napththaleneacetic acid** (NAA) (see Fig. 9–23), were even more effective than the naturally occurring IAA for rooting. It has been confirmed many times that auxin is required for initiation of adventitious roots on stems, and indeed, it has been shown that divisions of the first root initial cells are dependent upon either applied or endogenous auxin (*58, 89, 111, 209, 285, 295, 298*).

There are considerable advantages of integrating biochemical, physiological, and developmental anatomy approaches in rooting studies. Adventitious root formation is a developmental process involving sequences of histological events, with each stage having different requirements for growth substances (auxins, cytokinins, gibberellic acid, etc.). Thus establishing the time intervals of adventitious root initiation and development has made possible the correlation of sequential physiological and histological events in rooting (*25, 49, 76, 266, 267*). As an example, physiological studies (*73, 74, 217*) with pea cuttings have shown the role of auxins in the intricate developmental process of root initiation. Root formation and development in pea cuttings was found to occur in two basic stages:

1. **An initiation stage** in which root meristems are formed. This stage could be further divided into:
   a. **An auxin-active stage,** lasting about four days, during which auxin must be supplied continuously for roots to form, coming either from a terminal bud or from applied auxin (if the cutting has been decapitated) (*74, 217*). This stage is followed by
   b. **An auxin-inactive stage.** Withholding auxin during this stage (which lasts about four days) does not adversely affect root formation.
2. **A root elongation and growth stage**, during which the root tip grows outward through the cortex, finally emerging from the epidermis of the stem (see Fig. 9–6). A vascular system then develops in the new root primordia and becomes connected to adjacent vascular bundles. At this stage there is no response to applied auxin.

Cytokinins

Various natural and synthetic materials such as **zeatin, kinetin, 2iP, PBA**, and **6-benzyl adenine** have cytokinin activity. Generally a high auxin:low cytokinin favors adventitious root formation and a low auxin:high cytokinin favors adventitious bud formation (*137*) (Fig. 9–13). Cuttings of species with high native cytokinin lev-

---

[3] *Plant hormones (phytohormones)* are organic compounds, other than nutrients, produced by plants which, in low concentrations, regulate plant physiological processes. They ususally move within the plant from a site of production to a site of action.

[4] *Plant growth regulators* are either synthetic compounds or plant hormones that modify plant physiological processes. They regulate growth by mimicking hormones, by influencing hormone synthesis, destruction, or translocation, or (possibly) by modifying hormonal action sites.

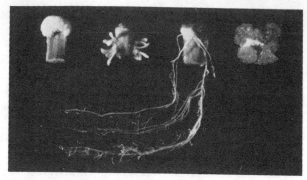

FIGURE 9–13 Effects of adenine sulfate (a cytokinin) and indoleacetic acid (auxin) on growth and organ formation in tobacco stem segments. *Far left:* Control. *Central left:* Adenine sulfate, 40 mg per liter. Bud formation with decrease in root formation. *Center right:* Indoleacetic acid, 0.02 mg per liter. Root formation with prevention of bud formation. *Far right:* Adenine sulfate, 40 mg per liter plus indoleacetic acid, 0.02 mg per liter. Growth stimulation but without organ formation. (Courtesy Folke Skoog.)

els have been more difficult to root than those with low cytokinin levels (*227*). Applied synthetic cytokinins normally inhibit root initiation in stem cuttings (*157, 235, 256*). However, cytokinins at very low concentrations, when applied to decapitated pea cuttings at an early developmental stage (*75*), or to begonia leaf cuttings (*137*), promoted root initiation, while higher concentrations inhibited initiation (see Fig. 9–14). Application to pea cuttings at a later stage in root initiation did not show such inhibition. The influence of cyto-

kinins in root initiation may thus depend upon the particular stage of initiation and the concentration (*25, 49, 62, 267*).

It has been suggested that the few cases of rooting success using "aboveground" exogenous applications indicate that cytokinins may play a protective role rather than a direct role on rooting (*307*). Cytokinins may also be indirectly involved in rooting through effects on rejuvenation and carbohydrate loading (i.e., accumulation of carbohydrates at cutting base) (*307*).

Leaf cuttings provide good test material for studying auxin-cytokinin relationships, since such cuttings must initiate both roots and shoots. Cytokinin at relatively high concentrations (13 mg/l) promoted bud formation and inhibited root formation of *Begonia* leaf cuttings under aseptic conditions, while auxins, at high concentrations, stimulated the opposite effect. At low concentrations (about 2mg/l), IAA promoted bud formation, enhancing the cytokinin influence. Also, at low concentrations (about 0.8 mg/l), kinetin stimulated the effect of IAA on root promotion. Too high a cytokinin concentration applied to leaf cuttings maximizes adventitious bud formation, but reduces the quality of new shoots (Figs. 9–10 and 9–14); from a horticultural standpoint, adventitious shoot quality is an important criterion in regenerating new plants from leaf cuttings (*51*).

It seems that the considerable seasonal changes in the regenerative ability of *Begonia* leaf cuttings are due to a complex interaction of temperature, photoperiod and light intensity, controlling the levels of endogenous auxins and other growth regulators (*140*).

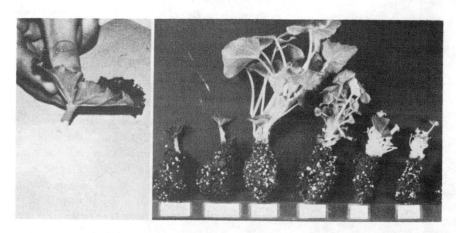

FIGURE 9–14 *Left:* Application of the cytokinin PBA mixed with talc to leaf cutting petiole base of Rieger begonias. *Right:* Concentrations of talc–applied cytokinin ranged from 0 to 1 percent with 0.01 percent (100 mg/l PBA) producing the optimum shoot number. At higher concentrations more adventitious buds are produced, but they fail to develop a commercially acceptable shoot system (*51*).

Buds are initiated in the leaf notches of *Bryophyllum* at an early stage in leaf formation. Shoots arising from the leaves originate from these buds. Cytokinin applications have a stimulatory effect on bud development while auxin applications inhibit bud development but stimulate root formation (*46, 138*)

Gibberellins (GA)

The gibberellins (Figure 6–8) are a group of closely related, naturally occurring compounds first isolated in Japan in 1939 and known principally for their effects in promoting stem elongation. At relatively high concentrations (i.e., 10^{-3} M) they have consistently inhibited adventitious root formation (*261*). There is evidence that this inhibition is a direct local effect that prevents the early cell divisions involved in transformation of differentiated stem tissues to a meristematic condition (*31*). Gibberellins have a function in regulating nucleic acid and protein synthesis and may be suppressing root initiation by interfering with these processes, particularly transcription (*121, 181*). At lower concentrations (10^{-11} to 10^{-7} M), however, gibberellin has promoted root initiation in pea cuttings, especially when the stock plants were grown at low irradiances (*119*).

In *Begonia* leaf cuttings, gibberellic acid was noted (*141*) to inhibit both adventitious bud and root formation, probably by blocking the organized cell divisions that initiate formation of bud and root primordia. Inhibition of root formation by gibberellin is dependent on the developmental stage of rooting. With herbaceous materials, inhibition is usually greatest when GA is applied three to four days after cutting excision (*121*). However, woody plant species such as *Salix* (*111*) and *Ficus* (*49*) were not adversely affected by GA during root initiation, but after primordia were initiated GA caused reduction in cell number in older established primordia, which was deleterious to root formation.

Abscisic Acid (ABA), Growth Retardants

Reports on the effect of abscisic acid on adventitious root formation are contradictory (*11, 38, 139, 242*), apparently depending upon the concentration and upon the environmental and nutritional status of the stock plants from which the cuttings are taken. Shoot growth retardants have been used to enhance rooting based on the rationale that they (a) antagonize GA synthesis or activity (GA is normally inhibitory to rooting), or (b) reduce shoot growth, resulting in less competition and consequently more assimilates are available for rooting at cutting bases (*55*). Synthetic antigibberellins and inhibitors of GA biosynthesis include chlormequat chloride (CCC), paclobutrazol (PP333, Bonzi), XE-1019 (a triazole growth retardant related to PP333), morphactins, ancymidol (Arest), gonadotropins, daminozide (SADH, Alar), etc. (*171, 194, 243*). Triiodobenzoic acid (TIBA) retards shoot elongation of conifers and also inhibits rooting by reducing basipetal auxin transport (*111*). Growth retardants that antagonize GA and/or reduce shoot elongation generally promote rooting (*55*). However, the mode of action of how these compounds enhance rooting is not well understood.

GROWTH REGULATORS THAT INFLUENCE ROOT INITIATION

Auxins are compounds that can induce elongation in decapitated *Avena* (oat) coleoptile tips and generally have a structural requirement of an aromatic ring, acidic side chain, and a spatial relationship (charge separation) between the positive charge ring structure and negative carboxyl side chain.

Cytokinins are plant growth regulators that promote cytokinesis (cell division) in the presence of optimal auxin levels.

Gibberellins are compounds with an ent-giberellane ring structure that can stimulate cell division and cell elongation or both in intact plants.

Abscisic Acid is a sesquiterpene with specific steric (chemical structure) requirements that can inhibit growth. It also helps regulate dormancy, stomatal control, tuberization, and other plant functions.

Ethylene is a gas involved in the regulation of ripening, abscission, dormancy, and other growth processes.

Ethylene (C₂H₄)

Ethylene can enhance, reduce, or have no effect on adventitious root formation (*262*). In 1933 Zimmerman and Hitchcock (*335*) showed that applied ethylene at about 10 ppm causes root formation on stem and leaf tissue as well as the development of pre-existing latent roots on stems. These and other workers (*338*) also showed about the same time that auxin applications can regulate ethylene production, and suggested that auxin-induced ethylene may account for the ability of auxin to cause root initiation. Centrifuging *Salix* cuttings in water, or just soaking them in hot or cold water, stimulates ethylene production in the tissues as well as root development, suggesting a possible casual relationship between ethylene production and subsequent root development (*171, 173*). Centrifugation seems to increase the water content of the cutting, which may block ethylene diffusion out of the cutting, the increased ethylene thus stimulating root formation (*174*).

Ethylene promotion of rooting occurs more frequently in intact plants than cuttings, herbaceous rather than woody plants, and plants having preformed root initials. A large body of evidence suggests that endogenous ethylene is not directly involved in auxin-induced rooting of cuttings (*223*).

The Effects of Buds and Leaves

Duhamel du Monceau (*63*) explained in 1758 the formation of adventitious roots on stems on the basis of the downward movement of sap. In extending this concept, Sachs, the German plant physiologist, postulated (*253*) in 1882 the existence of a specific root-forming substance manufactured in the leaves, which moves downward to the base of the stem, where it promotes root formation. It was shown (*192*) by van der Lek in 1925 that strongly sprouting buds promote the development of roots just below the buds in cuttings of such plants as willow, poplar, currant, and grape. It was assumed that hormone-like substances were formed in the developing buds and transported through the phloem to the base of the cutting, where they stimulated root formation.

The existence of a specific root-forming factor was first determined by Went (*324*) in 1929 when he found that if leaf extracts from the *Acalypha* plant were applied back to *Acalypha* or to *Carica* tissue, they would induce root formation. Bouillenne and Went (*27*) found substances in cotyledons, leaves, and buds that stimulated rooting of cuttings; they called this material "rhizocaline."

Bud Effects on Rooting

In Went's pea test (*324*) for root-forming activity of various substances, it is significant that the presence of at least one bud on the pea cutting was essential for root production. A budless cutting would not form roots even when treated with an auxin-rich preparation. This finding indicated again that a factor other than auxin, presumably one produced by the bud, was needed for root formation. In 1938, Went postulated that specific factors other than auxin were manufactured in the leaves and were necessary for root formation. Later studies (*73, 216*) with pea cuttings confirm this observation. For roots to form, the presence of an actively growing shoot tip (or a lateral bud) is necessary during the first three or four days after the cuttings are made (*81*). But after the fourth day the shoot terminals and buds could be removed without interfering with subsequent root formation.

Removal of the buds from cuttings in certain plants will stop root formation almost completely, especially in species without preformed root initials (*192*). In some plants, if a ring of bark is removed down to the wood just below a bud, root formation is reduced, indicating that some influence travels through the phloem from the bud to the base of the cutting, where it is active in promoting root initiation. It has been shown (*80*) that if hardwood cuttings are taken in midwinter when the buds are in the rest period,[5] they have no stimulating effect on rooting, but if the cuttings are made in early fall or in the spring, when the buds are active and not in the "rest" influence, they show a strong root-promoting effect.

---

[5]The "rest period" is a physiological condition of the buds of many woody perennial species beginning shortly after the buds are formed. While in this condition, they will not expand into flowers or leafy shoots even under suitable growing conditions. After exposure to sufficient cold, however, the "rest" influence is broken, and the buds will develop normally with the advent of favorable growing temperatures.

With cuttings of apple and plum rootstocks, the capacity of the shoots to regenerate roots increased during the winter, reaching a high point just before bud-break in the spring; this is believed to be associated with a decreasing level of bud dormancy following winter chilling (*152*).

Studies (*247*) with Douglas fir cuttings showed a pronounced relationship between bud activity and rooting of cuttings. Rooting was least in September and October when bud dormancy was highest. Rooting was greatest in December and January (if auxin was applied), and in February and March (when no auxin was added), after winter cold had removed bud dormancy.

Leaf Effects on Rooting

It has long been known that the presence of leaves on cuttings exerts a strong stimulating influence on root initiation (see Figs. 9–15 and 9–16).

The stimulatory effect of leaves on rooting in stem cuttings is nicely shown by studies (*244*) with the avocado. Cuttings of difficult-to-root cultivars under mist soon shed their leaves and die, whereas leaves on the cuttings of cultivars that rooted are retained as long as nine months. In this study, after five weeks in the rooting bed, there was five times more starch in the base of the easily

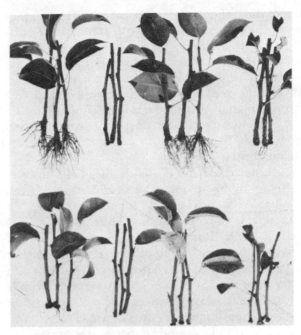

FIGURE 9–16 Effect of leaves, buds, and applied auxin on adventitious root formation in leafy 'Old Home' pear cuttings. *Top:* Cuttings treated with auxin (indolebutyric acid at 4000 ppm for five seconds). *Bottom:* Untreated cuttings. Left to right: with leaves; leaves removed; buds removed; one-fourth natural leaf area. (Courtesy W. Chantarotwong.)

rooted cuttings than there was at the beginning of the tests.

Carbohydrates translocated from the leaves undoubtedly contribute to root formation. However, the strong root-promoting effects of leaves and buds are probably due to other, more direct factors (*30*). Leaves and buds are known to be powerful auxin producers, and the effects are observed directly below them, showing that polar apex-to-base transport is involved.

Cuttings of certain clones are easily rooted but cuttings of other, closely related, clones root with considerable difficulty. It would seem that grafting a leafy portion of the easily rooted clone onto a basal stem portion of the difficult-to-root clone and then preparing the combination as a cutting, would cause it to root readily. The rooting factors provided by the leaves or buds of the easily rooted clone could perhaps stimulate root-

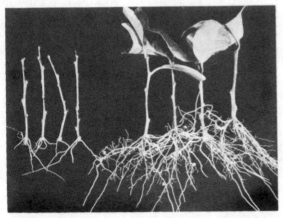

FIGURE 9–15 Effect of leaves on cuttings of 'Lisbon' lemon. Both groups were rooted under intermittent mist and were treated with indolebutyric acid at 4000 ppm by the concentrated-solution-dip method.

ing of the difficult-to-root basal part. Such experiments have been tried but with conflicting results. Van Overbeek and others (*305, 306*) rooted the difficult-to-root 'Purity' (white) hibiscus by previously grafting onto it sections of the easily rooted 'Brilliant' (red) hibiscus, then applying auxins. Auxin-treated 'Purity' cuttings, ungrafted, did not root, the leaves soon dropping. Auxin-treated ungrafted 'Brilliant' cuttings rooted readily and retained their leaves.

Later experiments with the same hibiscus cultivars by Ryan and others (*252*) gave different results. With cuttings held for six weeks, they obtained 100 percent rooting with 'Brilliant', 90 percent with 'Purity', and 84 percent with cuttings made from 'Brilliant' on 'Purity' grafts. It is possible that in van Overbeek's studies, difficulty in rooting 'Purity' cuttings was due primarily to the early loss of leaves under the conditions of their experiments. Leaves of the 'Purity' cultivar may stimulate rooting just as well as those of the 'Brilliant' if they can be retained long enough.

Bouillenne and Bouillenne-Walrand (*27*) proposed that "**rhizocaline**" be considered as a complex of three components: (a) a specific factor, translocated from the leaves, and characterized chemically as an orthodihydroxy phenol; (b) a nonspecific factor (auxin), which is translocated and is found in biologically low concentrations; and (c) a specific enzyme located in cells of certain tissues (pericycle, phloem, cambium), which is probably of the polyphenol-oxidase type.

They further proposed that the ortho-dihydroxy phenol reacts with auxin wherever the required enzyme is present, giving rise to the complex "rhizocaline," which may be considered one step in a chain of reactions leading to root initiation. Libbert (*195*) also developed evidence showing that auxin, forming a complex with a mobile factor "X," would result in root initiation, but he rejected the concept that a nonmobile enzyme was involved. He asserted that auxin itself acts to cause dedifferentiation of cells, determining the site of root formation.

Rooting Cofactors (Auxin Synergists)

Hess (*143, 144*) isolated various **rooting cofactors** from cuttings, using chromatography together with mung bean (*Phaseolus aureus*) bioassay techniques. He worked with cuttings of the easily rooted juvenile form and the difficult-to-root mature form of English ivy (*Hedera helix*). He also used easy and difficult-to-root cultivars of chrysanthemum and of the red-flowered and white-flowered forms of *Hibiscus rosa-sinensis*. These cofactors are naturally occurring substances that appear to act synergistically with indoleacetic acid in promoting rooting. The easily rooted forms of plants he has worked with have a larger content of such co-factors than the difficult ones.

One of these cofactors (No. 4) represents a group of active substances, tentatively characterized as oxygenated terpenoids. Another (No. 3) was identified as isochlorogenic acid (*144*). Further work showed that there were three lipid-like root-promoting compounds in juvenile ivy tissue that contain functional alcohol and nitrile groups. All three were colorless and unstable, breaking down to orange-yellow compounds and losing their root-promoting activity (*146*).

In testing the biological activity of compounds structurally related to cofactor 3, Hess (*143*) found that the phenolic compound catechol reacts synergistically with indoleacetic acid in root production in the mung bean bioassay. Since, as he points out, catechol is readily oxidized to a quinone, and since the mung bean itself is a good source of phenolase, it may be that oxidation of an *ortho-dihydroxy phenol* is one of the first steps leading to root initiation, as suggested earlier by Bouillenne and Bouillenne-Walrand (*24*). In addition, various other compounds have been found to react synergistically with auxin in promoting rooting (*96*) (see Fig. 9–17).

One of the postulated rooting cofactors could possibly be abscisic acid which can promote root initiation (*39*), perhaps by antagonizing gibberellic acid which, at certain concentrations, inhibits root formation.

Rooting cofactors were found in hardwood cuttings of 'Crab C' and 'Malling 26' apple rootstocks by the mung bean bioassay (*4, 38*). Increased cofactor activity was found when cuttings of 'Malling 26' were subjected to elevated temperatures 18.5°C (65°F) during the winter storage period, a practice known to increase rooting. Hardwood stem tissue of the easily rooted 'Malling 106' was found to show strong root promoting factors in the mung bean bioassay, whereas in the difficult-to-root 'Malling 2' such factors were present in lower amounts and rooting inhibitors appeared.

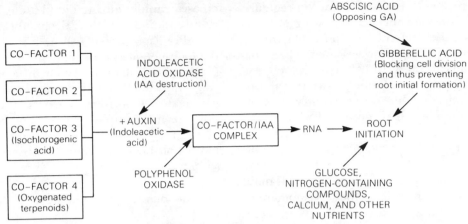

FIGURE 9–17 Hypothetical relationships of various components leading to adventitious root initiation. In addition, specific root-inhibiting factors may be present, which interfere with root development.

Rooting cofactors were found in five maple (*Acer*) species (*181*). One extract (partially characterized as a phenolic compound and/or weak acid) was more stimulatory than IAA on mung bean cuttings and stimulated root initiation in softwood cuttings of *A. saccharinum* and *A. griseum*. This is in contrast to rooting cofactors of the easy-to-root juvenile *Hedera helix*, which had no effect on rooting of the mature, difficult-to-root phase.

Fadl and Hartmann (*79, 80*) isolated an endogenous root-promoting factor from basal sections of hardwood cuttings of an easily rooted pear cultivar ('Old Home'). This highly active root-promoting material appeared only in cuttings having buds and treated with indolebutyric acid. It was found in largest amounts about 10 days after the cuttings had been made, treated with IBA, and placed in the rooting medium. Extracts from basal segments of similar cuttings of a difficult-to-root cultivar ('Bartlett'), treated with IBA, did not show this rooting factor. It did not appear either in 'Old Home' cuttings which had the buds removed or at a time of year when the buds were in a deep "rest" condition. Tests using UV spectrum analysis and infrared spectroscopy indicated that this rooting factor is a complex structure of high molecular weight and possibly is a condensation product between the applied auxin and a phenolic substance produced by the buds.

The action of these phenolic compounds in root promotion could be, at least partly, in protecting the root-inducing, naturally occurring auxin—indoleacetic acid—from destruction by the enzyme, indoleacetic acid oxidase (*61, 78*). This possibility was suggested also in rooting (*106*) juvenile *Hedera helix*, in which the phenol, catechol, showed remarkable synergism with IAA in promoting rooting. However, when naphthaleneacetic acid (NAA), which is not affected by IAA oxidase, was used as the auxin, NAA alone was as effective as NAA plus catechol. This finding implies that catechol was protecting IAA from destruction by the enzyme.

Using in vitro techniques, phloroglucinol (1,3,5-trihydroxybenzene), a phenolic material, was shown (*162*) to act synergistically with indolebutyric acid in stimulating adventitious root initiation in 'Malling 9' and other, but not all, apple rootstocks (*337*). In these studies, shoot cultures developing in the presence of phloroglucinol subsequently rooted better than shoots grown in its absence. However, the interaction of IBA and phloroglucinol in promoting rooting in vitro, depends in the apple at least, upon the growth phase of the cutting material (juvenile or adult), as well as the concentrations of each compound (*321*). Phloroglucinol has also been shown to stimulate rooting in *Rubus* (*160, 161*) and in *Prunus* (*170*) species.

Jarvis (*163*) attempted to integrate the biochemical and developmental anatomy of adventitious root formation by examining the four developmental stages of rooting (Fig. 9–18). His premise was (a) that the high concentrations of auxin needed to initiate rooting are inhibitory to later organization of the primordium and its subsequent growth, hence the importance of regulating endogenous auxin concentration with the IAA oxidase/peroxidase enzyme complex playing a central role (i.e., IAA oxidase metabolizes or breaks down auxin); and (b) IAA oxidase activity is controlled by phenolics (*o*-diphenols are inhibitory to IAA oxidase), while borate complexes with *o*-diphenols would result in greater IAA oxidase activity and hence reduction of IAA to levels that are optimal for the later organizational stages of rooting.

How auxins conjugate with amino acids may also play an important role in the developmental sequencing of rooting. Indoleacetylaspartic acid was identified as the primary amino acid conjugate of IAA (in IAA-treated mung beans) and in-

creased rapidly during the first day of root induction/initiation and then declined (*226*).

Predisposition of cells to initiate root primordia (competence) is dependent on active enzymes (*14*) and/or substrate which difficult-to-root cuttings may lack for synthesis of auxin–phenolic conjugates. Controversy exists since auxin–phenol–enzyme complexes have not been found in vivo, and promoter–inhibitor systems of rooting have not been universally observed in plants (*5, 9*).

Endogenous Rooting Inhibitors

Cuttings of certain difficult-to-root plants may fail to root because of naturally occurring rooting in-

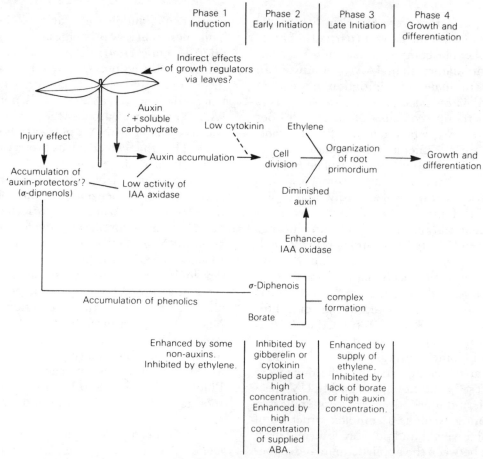

FIGURE 9–18 Hypothesized scheme of Jarvis (*163*) which proposes the role of phenolics, IAA oxidase/peroxidase, borate, and phytohormones in the four developmental stages of adventitious root production. (From M. B. Jackson (ed.). 1986. New root formation in plants and cuttings. Martinus Nijhoff Publishers. Dordrecht, The Netherlands.)

hibitors. This was found many years ago (*273*) to be the case with grapes, in which chromatographic studies suggested the presence of two inhibitors associated with rooting response. Leaching the cuttings with water enhanced the quantity and quality of roots. An inhibitor released into the water during leaching had a detrimental effect on rooting cuttings of the easily rooted *Vitis vinifera*. Shy-rooting cuttings of *V. berlandieri* seemed to possess a high inhibitor content.

'Bartlett' pear hardwood cuttings are difficult to root under treatments which give good rooting of 'Old Home' pear hardwood cuttings (*64*). Extracts taken from cuttings of both cultivars 20 days after being treated with IBA and placed in a rooting medium showed distinctly different amounts of inhibitors and promoters.

Other evidence implicating rooting inhibitors is given in studies (*45, 232*) in Australia with the difficult-rooting adult tissues of *Eucalyptus grandis*, which contained compounds that blocked adventitious root formation. In these studies three such inhibitors were found and considerable information was developed concerning their structure and physical properties. They were determined to be naturally occurring derivatives of the 2,3-dioxa-bicyclo [4,4,0] decane system. These inhibitors were not present in the easily rooted juvenile tissue of *E. grandis*. Mature difficult-to-root chestnut cuttings lack rooting cofactors, but have two inhibiting compounds which have been identified as derivatives of ellagic acid (*311*). These inhibitors were absent in easy-to-root juvenile and etiolated cuttings.

From studies (*15, 17*) with cuttings taken from dahlia cultivars it was determined that in plants whose cuttings were difficult to root, inhibitors formed in the roots and moved upward, accumulating in the shoots, subsequently interfering with root formation. In cultivars whose cuttings rooted easily, inhibitor levels were low.

Classes of Plants in Respect to Ease of Rooting

Plants can be divided into three classes in regard to adventitious root initiation:

1. Those in which the tissues provide all the various native substances, including auxin, essential for root initiation. When cuttings are made and placed under proper environmental conditions, rapid root formation occurs.

2. Those in which the naturally occurring cofactors are present in ample amounts but auxin is limiting. With the application of auxin, rooting is greatly increased.

3. Those that lack the activity of one or more of the internal cofactors, although natural auxin may or may not be present in abundance. External application of auxin gives little or no response, owing to lack of the effects of one or more of the naturally occurring materials essential for root formation.

Concerning the last group, Haissig (*112*) postulates that lack of root initiation in response to applied auxin (or even to native auxin) may be due to one or more of the following:

1. Lack of necessary enzymes to synthesize the root-inducing auxin–phenol conjugates
2. Lack of enzyme activators
3. Presence of enzyme inhibitors
4. Lack of substrate phenolics
5. Physical separation of enzyme reactants due to cellular compartmentalization

The fundamental basis of the formation of adventitious roots remains one of the least understood of plant functions. Despite years of active research, the primary chemical stimulus to dedifferentiation and root initial formation (the critical steps of adventitious root formation) and the subsequent organization of meristemoids (meristematic centers of cells actively dividing into root primordia) remain unknown.

Biochemical Changes in the Development of Adventitious Roots

Once adventitious roots have been initiated in cuttings, considerable metabolic activity takes place as new root tissues are developed and the roots grow through and out of the surrounding stem tissue to become external functioning roots. Protein synthesis and RNA production were both shown indirectly to be involved in adventitious root development in etiolated stem segments of

Salix tetrasperma (*159*). The fact that auxin action requires the presence of nutritional factors (glucose) is due to the requirement of a carbon source for the biosynthesis of nucleic acids and proteins. To date, it is not clear to what extent RNA metabolism is altered within that small pool of cells actually involved in root initiation (*164*). More definitive studies need to include microautoradiographic and histochemical approaches.

Some significant studies (*219, 220*) have been made of the biochemical changes taking place during the *development* of preformed root initials in hydrangea into emerging roots. These studies followed, in particular, the changing patterns of DNA and enzyme levels as the roots developed.

Root initials were found to originate in the phloem ray parenchyma, with roots emerging 10 to 12 days after the cuttings were made. The total protein content of the root initials increased over 100 percent in the first four days after the cuttings were made but there was no pronounced increase in the DNA content of the cells until the sixth day. Apparently, then, considerable protein synthesis preceded large-scale replication of nuclear DNA (*180*).

Pronounced increases in enzyme activity took place during adventitious root formation in hydrangea (*219*); the enzymes peroxidase, cytochrome oxidase, succinic dehydrogenase, and starch-hydrolyzing enzymes were observed in the phloem and xylem ray cells of the vascular bundles. Then, during root development, enzyme activity shifted from the vascular tissues to the periphery of the bundles. Such increases in enzyme activity occurred two to three days after the cuttings were made.

Starch was found to disappear from the endodermis, phloem and xylem rays, and pith in the region of the developing root primordia, apparently being utilized as a carbohydrate food source. Starch evidently plays an important nutritional role in adventitious root development.

In the development of adventitious roots on IBA-treated plum cuttings, it was determined (*29*), using radioactive CO_2 applied to leaves, that as soon as callus and roots started forming, pronounced sugar increases—and starch losses—occurred at the base of the cuttings, the ^{14}C appearing in sucrose, glucose, fructose, and sorbitol.

The developing callus and roots apparently act as a "sink" for the movement of soluble carbohydrates from the top of the cutting.

ADVANCES IN ASEXUAL PROPAGATION

Developments in somatic seed technology (see p. 477), artificially inducing roots by nonpathogenic *Agrobacterium*, and the potential transformation of cells using a disarmed **plasmid**[6] from a root-producing bacterium or from an auxin-inducing fragment of the **T-DNA**[7] may play important roles in the vegetative propagation of plants.

In the future, it may be possible to propagate vegetatively plants that do not breed true from seed by using somatic seeds (*264*). In effect, this is artificial seed propagation. A somatic seed is produced from an asexual (vegetative) embryo grown from callus which is encased in an artificial, protective coat. Literally thousands of encapsulated somatic embryos can be produced clonally from a single plant. Commercially important conifer species (*Pinus, Pseudotsuga,* and *Picea*) have been regenerated from cell suspension cultures by somatic polyembryogenesis, and these asexual embryos (embryoids) encapsulated with artificial seed coats (*64, 65*) (see Chap. 16). The potential is to take difficult-to-root species and avoid conventional vegetative propagation problems by utilizing clonally produced artificial seeds.

Likewise, there is progress being made by using root-inducing bacteria to enhance root regeneration of bare root almond stock (*283, 284*) and in vitro rooting of difficult-to-root apple (*230*).

In studies of tobacco plants transformed with root-inducing [Ri T-DNA] of *Agrobacterium rhizogenes*, spontaneous rooting of the transformed tobacco explants was due to genes which increased auxin sensitivity of the tissue to form roots; root-

---

[6]"Plasmids" are small molecules of extrachromosomal DNA that carry only a few genes and occur in the cytoplasm of a bacterium.

[7]"T-DNA" is the portion of the root-inducing (Ri) plasmid (i.e., from *Agrobacterium rhizogenes*) which is inserted into the plant genome (i.e., of a difficult-to-root species) and stabilized; hence this normally difficult-to-root species is potentially "transformed" to an easy-to-root clone.

ing of transformed plants was not due to genes which regulated auxin production, or to a substantially altered balance of endogenous hormones (*272*). In other studies with nonrooting mutants of tobacco, sensitivity to auxin was due to general alteration of the cellular response to auxin and was not correlated with an increased rate of conjugation of auxins by these tissues or by disruption of auxin transport (*33*). Thus, there are possible implications that the lack of cell competency in difficult-to-root species may be due to lack of cell sensitivity to auxin rather than to suboptimal level of endogenous auxin. However, there is no evidence to date that difficult-to-root species have been genetically ''transformed'' to easy rooters.

FACTORS AFFECTING REGENERATION OF PLANTS FROM CUTTINGS

Great differences exist among plants of different species and cultivars in the rooting ability of cuttings taken from them. Empirical trials with each cultivar are necessary to determine these differences. This has already been done with most plants of economic importance. Stem cuttings of some cultivars root so readily that the simplest facilities and care give high rooting percentages. On the other hand, cuttings of many cultivars or species have yet to be rooted. Cuttings of some ''difficult'' cultivars can be rooted only if various influencing factors are taken into consideration and maintained at the optimum condition. Environmental factors to be discussed in this section are of great importance to this group, and the attention given to them makes the difference between success or failure in obtaining satisfactory rooting. These factors are:

Selection of Cutting Material from Stock Plants

1. Environmental conditions and physiological status of the stock plant: (a) water stress, (b) temperature, (c) light (intensity, photoperiod, quality), (d) stock plant etiolation, (e) CO_2 enrichment, (f) carbohydrates, (g) mineral nutrition, (h) girdling

2. Rejuvenation and conditioning of stock plants prior to cutting (preseverance)
3. Type of wood selected
4. Seasonal timing

Treatment of Cuttings

1. Storage of cuttings
2. Growth regulators
3. Mineral nutrition of cuttings
4. Leaching of nutrients
5. Fungicides
6. Wounding

Environmental Conditions During Rooting

1. Water relations
2. Temperature
3. Light (irradiance, photoperiod, quality)
4. Accelerated growth techniques (AGT)
5. Photosynthesis of cuttings
6. Rooting medium

The remainder of this chapter discusses these factors in detail.

Selection of Cutting Material

Environmental and Physiological Status of the Stock Plant

Water stress. The physiological condition of stock plants is a function of genotype (species, cultivars) and environmental conditions (water, temperature, light, CO_2, and nutrition). To avoid water stress, plant propagators often emphasize the desirability of taking cuttings early in the morning when the plant material is in a turgid condition. There is experimental evidence to support this view. Studies with both cacao (*77*) and pea (*241*) cuttings showed reduced rooting when the cuttings were taken from stock plants having a water deficit. Most likely endogenous abscisic acid and ethylene levels of cuttings are affected by drought stress conditions (*215*).

Temperature. Information on temperature interactions with stock plant water relations, irradiance, and CO_2 is limited. Research has shown that there is a complex interaction of temperature and stock plant photoperiod on the level of endogenous auxins and other hormones (*140*). However, the air temperature of stock plants (12 to 27°C; 54 to 81°F) appears to play only a minor role in ease of rooting of cuttings (*215*).

Light. **Light duration** (photoperiod or daylength), **illuminance** (intensity or photon flux), and **spectral quality** (wavelength) influence stock plant condition and subsequent rooting of cuttings (*66*). Light is a contributing factor in the seasonal variation of rooting ability of cuttings (*238*). The effect of stock plant irradiance levels is controversial and rooting can be inhibited, promoted, or not affected (*120, 215*). Possible explanations for light inhibition of rooting include inadequate IAA synthesis or increased breakdown, photoinactivation of factors promoting root formation, inhibition of rooting cofactor synthesis, increased peroxidase activity, and formation of histological barriers (*66*).

There is some evidence that the **photoperiod** under which the stock plants are grown exerts an influence on the rooting of cuttings taken from them (*40, 101, 169, 215*). This could be a photosynthetic or morphogenic effect. If the photoperiod influences photosynthesis, it may be related to carbohydrate accumulation, with best rooting obtained under photoperiods promoting carbohydrate increase, although in some cases (*123, 275*) stock plants held under short photoperiods have produced the best rooting cuttings. In this latter case a photomorphogenic effect is suggested.

The influence of stock-plant photoperiod on rooting of cuttings appears related to the photoperiodic response of flowering, dormancy, and/or senescence (*215*). Long-day conditions (sufficient hours of light to satisfy the critical photoperiod) have been used with some short-day flowering cultivars of chrysanthemum; where flowering is antagonistic to rooting, the long-day conditions promote vegetative growth and enhance rooting (*84*). Likewise, with some woody perennials where the onset of dormancy shuts down vegetative growth and/or reduces rooting, propagators can manipulate stock plants by extending the photoperiod with low irradiance artificial lighting.

The conflicting reports on the influence of **light quality** on stock plants and subsequent rooting of cuttings is attributed to the effect of red and far-red light on rooting (i.e., via the phytochrome system) and distribution of light to the shoot system (*215*). In one test (*282*) when stock plants were exposed for six weeks to light sources of different quality before taking the cuttings, those from plants exposed to blue light rooted most readily.

The preformed root primordia found in the stems of Lombardy poplar (*Populus nigra* var. *italica*) develop and emerge if the cuttings are placed in darkness, but they fail to emerge if the cuttings are exposed to light each day. Red light (about 680 nm) is more inhibitory than blue, green, or far-red light (*257*).

Stock plant etiolation. By definition **etiolation**[8] is the total exclusion of light; however, plant propagators also use this term when forcing new stock plant shoot growth under conditions of heavy shade. Propagules are then taken from the new growth and often root more readily. **Banding** is a localized light exclusion pretreatment which excludes light from that portion of a stem that will be used as the cutting base (*10, 212*). Banding can be applied to etiolated shoots or applied to light-grown shoots which are still in the softwood stage. In the latter case a band of Velcro or black adhesive tape is said to **blanch** the underlying tissues since the stock-plant shoot accomplished its initial growth in light prior to banding. **Shading** refers to any stock-plant growth under reduced light conditions (*169*). (See Chap. 10 for techniques of banding, blanching, etiolation, and shading.)

Successful propagation practices for the exclusion of light during stock-plant growth are banding/blanching, etiolation, etiolation + banding, and shading (Table 9–4). The vegetative propagation practices of air layering and mounding (stooling) also entail exclusion of light from that portion of the stock plant to be rooted, prior to removing propagules (see Chap. 14).

Interestingly, etiolation can increase rooting of selected difficult-to-root species, whereas ap-

---

[8] Etiolation is the development of plants or plant parts in the absence of light. This results in such characteristics as small unexpanded leaves, elongated shoots, and lack of chlorophyll, resulting in a yellowish or whitish color.

TABLE 9–4

Enhanced rooting of cuttings from stock-plant etiolation, shading, and banding treatments

Treatment	Species	
Banding/blanching	Acer platanoides	Platanus occidentalis
	Tilia cordata	Rhododendron cvs.
	Pinus elliottii	Rubus idaeus
Etiolation	Artocarpus heterophyllus	Syringa vulgaris cvs.
	Bryophyllum tubiflorum	Malus sylvestris
	Camphora officinarum	Mangifera indica
	Clematis spp.	Persea americana
	Corylus maxima	Prunus domestica
	Cotinus coggygria	Rubus idaeus 'Meeker'
	Polygonum baldschuanicum	Tilia tomentosa
Etiolation plus banding	Acer spp.	Pinus strobus
	Betula papyrifera	Malus × domestica
	Carpinus betulus	Persea americana
	Castanea mollissima	Pistachia vera
	Corylus americana	Syringa vulgaris
	Pinus spp.	Carpinus betulus
	Quercus spp.	Tilia spp.
	Hibiscus rosa-sinensis cvs.	
Shading	Crassula argentea	Rhododendron spp.
	Schefflera arboricola	Rosa spp.
	Hibiscus rosa-sinensis	Euonymus japonicus
	Picea sitchensis	

Source: Maynard and Bassuk (212, 213).

plied auxin alone is not effective (87). Etiolation greatly enhances a stem's sensitivity to auxin. Translocatable factors produced distal to (above) an etiolated segment also enhance the etiolation effect (10). Etiolation has been associated with change in phenolic substances that may act as rooting cofactors or inhibitors of IAA oxidase (142). Finally, there are anatomical differences in stem tissue that may increase root primordia initiation potential due to greater undifferentiated parenchyma cells and lack of mechanical barriers. One hypothesis is that the reduced production of lignin [for structural support cells (sclereids, fibers)] may alter availability of phenolic metabolites, which instead of forming lignin are channeled to enhance root initiation (48).

Carbon dioxide enrichment. With many species carbon dioxide enrichment of the stock plant environment has increased the number of cuttings that can be harvested from a given stock plant, but there is considerable variation of rooting response among species. Principal reasons for increased cutting yields are increased photosynthesis, higher relative growth rate, and greater lateral branching of stock plants (215). Any benefits of CO_2 enrichment have been limited to greenhouse-grown stock plants during conditions when greenhouse vents are closed and ambient CO_2 becomes a limiting factor to photosynthesis (i.e., October–March in central Europe). Without adequate light (supplementary greenhouse lighting during low-light-intensity months), CO_2 enrichment is of minimal benefit (218). (See Chap. 2.)

Carbohydrates. The relationship between carbohydrates and adventitious root formation remains controversial. Since Krause and Kraybill (185) hypothesized the role of carbohydrate-to-nitrogen (C/N) ratio on plant growth and development, rooting ability of cuttings has been discussed in relation to carbohydrate content. The carbohydrate pools of free reducing sugars (soluble carbohydrates) and storage carbohydrates (starches or insoluble carbohydrates) are important to rooting as building blocks of complex macromolecules, structural elements, and energy sources (*103, 104, 117, 286*).

Although stock plant carbohydrate content and rooting may sometimes be positively correlated, evidence does not indicate that carbohydrates have a regulatory role in rooting. A positive correlation between carbohydrate content and rooting may reveal that the supply of current photosynthate is insufficient for supporting optimal rooting (*309*). High C/N ratios in cutting tissue promote rooting but do not accurately predict the degree of rooting response (*286*). Cuttings use stored carbohydrates in root regeneration, but only in small amounts. Differences in C/N ratios are due mainly to nitrogen rather than carbohydrate content. Nitrogen has been negatively correlated to rooting (*118*), which suggests that the correlation between high C/N ratios and rooting may be due to low N levels. Rooting may be benefited by maintenance of nitrogen metabolism of stock plants (i.e., nitrogen fertility programs) such that cutting shoot development is not stimulated (by high N levels), since adventitious rooting would be at a competitive disadvantage with rapidly developing shoots for carbohydrates, mineral nutrients, and hormones (*117*).

Mineral nutrition. Cuttings of *Hypercium, Rosa,* and *Rhododendron* rooted best when stock plants were suboptimally fertilized resulting in less than maximal shoot growth. Very low nitrogen leads to reduced vigor, whereas high nitrogen caused excess vigor; either extreme is unfavorable for good rooting.

To maintain moderate nitrogen and adequate carbohydrate levels for optimal rooting of cuttings, producers can manipulate stock plants as follows:

1. Reducing nitrogen fertilizers to the stock plants, thus reducing shoot growth and allowing for carbohydrate accumulation. Any type of root restriction of the stock plants, such as occurs when they are grown in containers or close together in hedge rows, tends to prevent excessive vegetative growth and permits the accumulation of carbohydrates.
2. Using cutting material, select portions of the plant that are in the desirable nutritive stage.

For example, take lateral shoots in which rapid growth has decreased and carbohydrates have accumulated, rather than succulent terminal shoots. [But for plants showing a plagiotropic growth pattern (see p. 176), use of lateral shoots should be avoided.]

3. Selecting regions of the shoot that are known to have a high carbohydrate content. In a chemical analysis (*301*) of rose shoots of the type used for cuttings, the nitrogen content increased uniformly from the base of the shoot to the tip. Theoretically basal portions of such a shoot would therefore have the low-nitrogen/high-carbohydrate balance favorable for good rooting; however, empirical trials indicate that apical and medial hardwood cuttings of rose root as well as or better than basal cuttings (*118*).
4. For maintenance of adequate carbohydrate levels, light intensity of stock plants can be controlled by increasing light irradiance through high pressure sodium-vapor lights in greenhouse plant production. Light reduction is a common practice by using shade cloth or a lathhouse to enhance rooting; possible benefits of light reduction may be to inhibit flowering and vigorous vegetative growth, which may release more carbohydrate pools for rooting.

Studies (*233*) with grapes showed that when stock plants were grown under phosphorus, potassium, magnesium, or calcium deficiency, root formation in cuttings was poorer than in cuttings from full-nutrient plants; but with reduced nitrogen in the stock plants, root formation by the cuttings was increased. However, extreme nitrogen deficiency of the stock plants lowered rather than increased rooting. This was substantiated by tests (*134*) with geranium in which stock plants were grown at three levels of nitrogen, phosphorus, and potassium. The nitrogen nutrition of the stock plants had a greater effect on rooting response of the cuttings than did phosphorus or potassium, with the low and medium levels of nitrogen resulting in higher percentages of rooted cuttings than the high level.

For root initiation to take place, however, nitrogen is necessary for nucleic acid and protein synthesis. Below a cutoff level of nitrogen availability, root initiation is impaired; in such cases added nitrogen promotes rooting.

It is difficult to interpret the effect of nutrition on rooting, since few studies have approached mineral effects on the various stages of *de novo* adventitious root formation (*19*).

Grape cuttings taken from vines fertilized with zinc rooted in higher percentages and were of better quality than cuttings taken from untreated vines (*19*). This effect could be due to an increase in native auxin production resulting from the increased level of tryptophan (an auxin precursor) found in the treated plants. (Zinc is required for tryptophan production.) In fact, applications of synthetic tryptophan have increased rooting of vine cuttings. Beneficial effects of zinc applications to stock plants have been noted also in South Africa in the propagation of 'Marianna' plum cuttings.

High levels of manganese were found (*244*) in leaves of cuttings taken from difficult-to-root avocado cultivars, whereas cuttings from easy-to-root cultivars had a much lower manganese level (Fig. 9–19). Of eleven elements for which analyses were made in this study, only manganese showed any correlation with rooting. Manganese is known (*297*) to be an activator of the enzyme, indoleacetic acid oxidase, which destroys auxin and could lead to a reduced amount of natural auxin at the base of the cuttings, thus causing poor rooting.

Girdling. Girdling, or otherwise constricting the stem, blocks the downward translocation of carbohydrates, hormones, and other possible root-promoting factors and can result in an increase in root initiation. Using this technique on shoots prior to their removal for use as cuttings can improve rooting. This practice has been remarkably successful in some instances. For example, rooting of citrus and hibiscus cuttings was stimulated by girdling or binding the base of the shoots with wire several weeks before taking the cuttings (*165, 277*).

In cuttings from mature trees of the water oak (*Quercus nigra*), a threefold improvement in rooting was obtained when cuttings were taken from shoots that had been girdled six weeks previously, especially if a powder consisting of 1 percent each of IBA, PPZ (1-phenyl-3-methyl-5-pyrazolone), 20 percent sucrose, and 5 percent captan in talc, was rubbed into the girdling cuts (*124*). Enhanced rooting of cuttings taken from girdled ortets has also been obtained with sweet gum, slash pine, and sycamore (see Chap. 10, Figs. 10–21 and 10–22). Girdling (*276*) also caused substantial increase in a rooting cofactor above the girdle in an easily rooted hibiscus clone. Girdling just below a previously etiolated stem section was particularly effective in promoting rooting in apple cuttings (*56*).

Rejuvenation and Pre-severance Conditioning of Stock Plants

In difficult-to-root woody plant species, ease of adventitious root formation declines with age of parent stock, resulting in a propagation problem since desirable characteristics are frequently not expressed until after a plant has reached maturity. The transition from the *juvenile* to the *mature* phase has been referred to as *phase change, ontogenetic aging,* or *meristem aging.* There are progressive changes in such morphological and developmental characteristics as leaf shape, branching pattern, shoot growth, vigor, and the ability to form adventitious buds and roots (*102, 107, 108, 222*) (see Chap. 8). Experiments with apple, pear, eucalyptus, live oak, Douglas fir, and many other species have shown that the ability of cuttings to

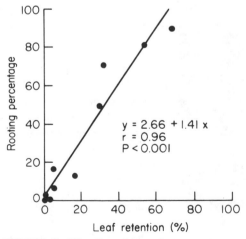

$$y = 2.66 + 1.41\,x$$
$$r = 0.96$$
$$P < 0.001$$

FIGURE 9–19 Correlation between manganese content of leaves of different avocado clones and rooting percentage. (From data of Reuveni and Raviv (*244*).)

form adventitious roots decreased with increasing age of the plants from seed, in other words, when the stock plant changed from the juvenile to the adult phase (*130*).

In a study (*293*) of rooting cuttings of certain coniferous and deciduous species known to root only with extreme difficulty, it was concluded that the most important single factor affecting root initiation was the age of the tree from which the cuttings were taken. However, it is the **biological** age, not **chronological** age, of the propagule that is most important in rooting success (see Chap. 8).

In some species such as apple, English ivy, and olive, differences in certain morphological characteristics, such as leaf size and shape, make it easy to distinguish between the adult and the lower, juvenile portions of the plant. Cuttings should be taken from the latter type for best rooting (*237*). In some kinds of deciduous trees, such as oak and beech, leaf retention late into the fall occurs on the basal parts of the tree and indicates the part (**cone of juvenility**) still in the juvenile stage. Cuttings should be taken from this type of wood (*97*).

There are horticultural and forestry practices that can rejuvenate physiologically mature stock plants, cuttings, explants (propagules used for tissue culture), etc., and frequently improve rooting success (*48, 154, 179*). The hedging or shearing treatments given *Pinus radiata* trees (Fig. 9–20) were quite effective in maintaining the rooting potential of cuttings taken from them as the trees aged, compared to nonhedged trees (*196*). Such benefits in maintaining rooting potential by hedging are probably explained by the prevention of the normal phase change from the juvenile to the adult form. Stock plants for hardwood cuttings of selected fruit tree rootstocks are maintained as hedges rather than allowed to grow to a tree form (*88*). The combined practices of hedging and mounding (stooling) have been successfully used with difficult-to-root pecan rootstock in Peru (see Fig. 14–11).

Juvenility in relation to rooting may possibly by explained by the increasing production of rooting inhibitors as the plant grows older. Stem cuttings taken from young seedlings of a number of eucalyptus species root easily, but as the stock plants become older rooting decreases dramatically. Studies (*232*) in Australia showed that there was a direct and quantitative association between such decreased rooting and the production of a rooting inhibitor in the tissues at the base of the cuttings. In easily rooted young seedling stems this inhibitor was absent, as it was absent in adult stem tissue of the easily rooted *Eucalyptus deglupta*.

Reduced rooting potential as plants age may possibly be a result of lowering phenolic levels. Phenols are postulated as acting as auxin cofactors

FIGURE 9–20 Stock plants for *Pinus radiata* cuttings maintained in a hedge form by shearing (compare with unsheared trees in background). Such "hedged" trees can yield high numbers of cuttings which maintain the high rooting percentages, root quality, and growth potential normally associated with the younger stock trees (*196*). (Photo courtesy W. J. Libby.)

or synergists in root initiation. In certain plants, lower phenolic levels were noted in mature forms than in the juvenile forms (*93*).

In rooting cuttings of difficult species it would be useful to be able to induce rejuvenation to the easily rooted juvenile stage from plants in the adult form. This has been done in several instances by the following methods:

Juvenile forms of apple can be obtained from mature trees by causing adventitious shoots to develop from root pieces, which are then made into softwood stem cuttings and rooted (*278*).

By removing terminal and lateral buds and spraying stock plants of *Pinus sylvestris* with a mixture of cytokinin, tri-iodobenzoic acid, and Alar, many interfascicular shoots can be forced out. With proper subsequent treatment, high percentages of these shoots can be rooted (*328*).

In some plants juvenile wood can be obtained from mature plants by forcing juvenile growth from **sphaeroblasts** (wartlike protuberances containing meristematic and conductive tissues sometimes found on trunks or branches). These are induced to develop by disbudding and heavily cutting back stock plants (*251*). These juvenile shoots then may be easily rooted under the usual conditions (*322*). Using the mound-layering (stooling) method on these rooted sphaeroblast cuttings produces rooted shoots that continue to possess the juvenile characteristics.

Grafting adult forms of ivy onto juvenile forms has induced a change of the adult to the juvenile stage, provided that the plants are held at fairly high temperatures (*62, 280*); such transmission of the juvenile rooting ability from seedlings to adult forms by grafting has also been accomplished in rubber trees (*Hevea brasiliensis*) (*224*), and later with serial graftage of mature difficult-to-root scions onto seedling rootstock of *Eucalyptus* × *trabutii* (Fig. 9–21).

Ringbarking, girdling, severe pruning, etiolation, banding, and blanching (*48, 155*) are mechanical means of physiologically preconditioning stock plants to stimulate "competent cells" (*113, 300*) in propagules that can dedifferentiate and develop into meristematic regions of adventitious root primordia.

Chemical manipulation with gibberellin sprays on English ivy plants in the growth phase have caused substantial stimulation of growth and reversion of some of the branches to the juvenile stage (*245*), as well as improved rooting of cuttings taken from the sprayed plants (*280*).

It is important to remember that the juvenile condition is found only in such tissues as those originating from young seedlings, those on the lower juvenile portion of mature adult seedling trees, those arising from adventitious (not latent) buds on stems, or those caused to revert to juvenility either by gibberellin treatments or by grafting to juvenile wood.

Rejuvenation of tissue in vitro has tremendous potential to increase rooting of tissue culture–derived cuttings (*105, 179, 236*). Likewise, recent advances of somatic embryogenesis and artificial seed technology may be used to restore the juvenile condition to mature plants (*64, 65*).

Type of Wood Selected

There are many choices of the type of material to use, ranging (in woody perennials) from the very succulent softwood terminal shoots of current growth to large dormant hardwood cuttings. Here, as with most of the other factors affecting rooting of cuttings, it is impossible to state any one type of cutting material that would be best for all plants. What may be ideal for one plant would be a failure for another. What has been found to hold true for certain species or cultivars, however, often may be extended to other related species or cultivars.

Differences between individual seedling plants. In rooting cuttings taken from individual plants of a species which ordinarily is propagated by seed, experience has shown that wide differences exist among individuals in the ease with which cuttings taken from them form roots. Just as seedlings differ in many respects, it is not surprising that this difference in root-forming ability should also exist. Differences in the rooting ability of various clones are, of course, recognized. Likewise, such differences should be anticipated when woody plants usually propagated by seed—such as most forest tree species—are propagated by cuttings. In rooting cuttings taken from old seedling trees of Norway spruce (*Picea abies*), white pine (*Pinus strobus*), and red maple (*Acer rubrum*)

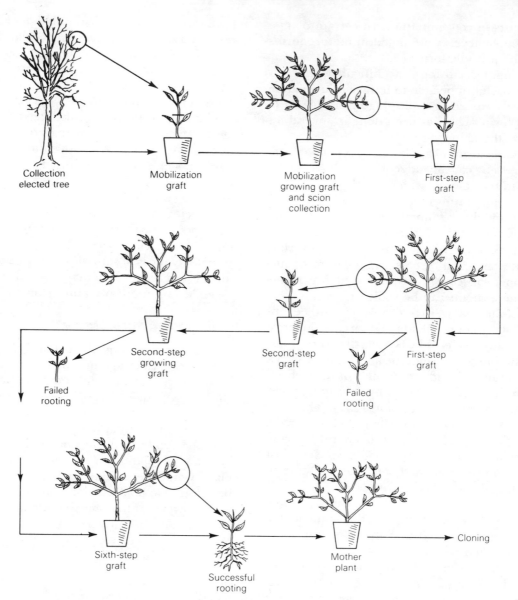

FIGURE 9–21   Scheme for rejuvenation techniques used in serial graftage of 10-year-old *Eucalyptus × trabutii* onto juvenile seedling understock. Six serial grafts were needed before mature grafted scions could be used as cuttings and rooted (*259*).

marked differences occurred in the rooting capacity of shoots taken from the individual trees (*59, 270*).

Differences between lateral and terminal shoots.   In rooting different types of softwood plum cuttings taken in the spring, there was a marked superiority in rooting of lateral shoots.

The terminals had only 10 percent rooting; laterals in active growth, 19 percent; and laterals that had ceased active growth, 35 percent.

Similarly, lateral branches of white pine and Norway spruce gave consistently higher percentages of rooted cuttings than did terminal shoots (*59, 82*). In rhododendrons, too, thin cuttings made from lateral shoots consistently give higher rooting percentages than those taken from vigorous, strong terminal shoots. In certain species, however, plants propagated as rooted cuttings taken from lateral branches may have an undesirable growth habit (see Fig. 8–7).

Differences between various parts of the shoot. In some woody plants, hardwood cuttings are made by sectioning shoots several feet long and obtaining four to eight cuttings from a single shoot. Marked differences are known to exist in the chemical composition of such shoots from base to tip. Variations in root production on cuttings taken from different portions of the shoot are often observed, with the highest rooting, in many cases, found in cuttings taken from the basal portions of the shoot. Cuttings prepared from shoots of three cultivars of the highbush blueberry (*Vaccinium corymbosum*) were significantly more successful if taken from the basal portions of the shoot rather than from terminal portions (*229*). Exceptions to this are found in rose (*118*) and other species. The number of preformed root initials in woody stems (in some plants at least) distinctly decreases from the base to the tip of the shoot (*111*). Consequently, the rooting capacity of basal portions of such shoots would be considerably higher than that of the apical parts.

In studies with a different type of wood, however, cherries (*Prunus cerasus, P. avium, P. mahaleb*) rooted under mist by softwood cuttings prepared from succulent new growth gave the following percentages of rooted cuttings: 'Stockton Morello', basal—30 percent, tip—77 percent; 'Bing', basal—0 percent, tip—100 percent; and 'Montmorency', basal—10 percent, tip—90 percent (*128*).

In woody stems a year or more in age, where carbohydrates have accumulated at the base of the shoots and, perhaps, where some root initials have formed in the basal portions, possibly under the influence of root-promoting substances from buds and leaves, the best cutting material is the basal portions of such shoots. An entirely different physiological situation exists in deciduous plants in which succulent shoots are used for softwood cuttings. Here carbohydrate storage and preformed root initials are not present. The better rooting of shoot tips may be explained by the possibility of higher concentrations of endogenous root-promoting substances arising in the terminal bud. There is also less differentiation in the terminal cuttings, with more cells capable of becoming meristematic.

This factor is of little importance, however, in cuttings of easily rooted species in which entirely satisfactory rooting is obtained regardless of the position of the cutting on the shoot.

Flowering or vegetative wood. In most plants, cuttings could be made from shoots that are in either a flowering or a vegetative condition. Again, with easily rooted species it makes little difference which is used, but in difficult-to-root species this can be an important factor. For example, in a blueberry species (*Vaccinium atrococcum*), hardwood cuttings from shoots bearing flower buds did not root nearly so well as those bearing only leaf buds (*228*). When vegetative wood was used, 39 percent of the cuttings rooted, but when cuttings contained one or more flower buds, not one rooted. Dahlia cuttings bearing flower buds are more difficult to root than cuttings having only vegetative buds (*17*).

In rhododendron, early removal of potential flower buds increased rooting, presumably through the elimination of a flower-promoting stimulus antagonistic to rooting; later flower bud removal still enhanced rooting, possibly by eliminating competition for metabolites necessary for rooting (*166*). Flowering is a complex phenomenon and can serve as a competing sink to the detriment of rooting. With many ornamental species (i.e., *Abelia, Ligustrum, Ilex*, etc.) it is commercially desirable to remove flower buds from cuttings for more rapid root development, earlier vegetative growth, and more efficient liner production (*175*).

"Heel" versus "nonheel" cuttings. In preparing cuttings, it is often recommended that a "heel" (a small slice of older wood) be retained at the base of the cutting in order to obtain maximum rooting. For hardwood cuttings of some plants, this may be true. In quince (*Cydonia oblonga*), considerably better rooting was obtained with the heel type of cutting, probably owing in this case to the presence of preformed root initials in the older wood. Narrow-leaved evergreen cut-

tings often, but not always, root more readily if a heel of old wood is retained at the base of the cuttings. However, heel cuttings are not as satisfactory as nonheel cuttings with most ornamental woody plants.

Seasonal Timing

Seasonal timing, or the period of the year in which cuttings are taken, can play an important role in rooting (*8, 132*). With many species there is an optimal period of the year for rooting (*2*). Propagators strive to "maintain the plants' momentum" by rooting during these optimal periods to maximize the rooting process and speed up the production of **liners**.[9] It is possible, of course, to make cuttings at any time during the year.

In propagating **deciduous** species, hardwood cuttings could be taken during the dormant season (from leaf fall and until spring bud development). Leafy softwood or semihardwood (greenwood) cuttings could be prepared during the growing season, using succulent and partially matured wood, respectively. The narrow- and broad-leaved evergreen species have one or more flushes of growth during the year, and cuttings can be obtained year-round in relation to these flushes of growth.

Certain species, such as privet, can be rooted readily if cuttings are taken almost any time during the year; on the other hand, excellent rooting of leafy olive cuttings under mist can be obtained during late spring and summer, whereas rooting drops almost to zero with similar cuttings taken in midwinter. Seasonal changes influenced rooting of both juvenile and mature (difficult-to-root) *Ficus pumila* cuttings; however, treating juvenile (easy-to-root) cuttings with IBA overcame the seasonal fluctuation in rooting (Fig. 9–22). Shoot RNA was found to be an index of bud activity and subsequent seasonal rooting differences. Highest shoot RNA levels and increased vascular cambial

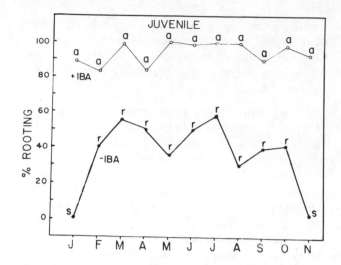

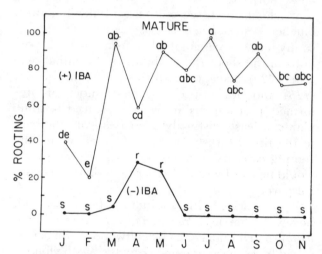

FIGURE 9–22 Seasonal fluctuation in percent rooting of juvenile (easy-to-root) and mature (difficult-to-root) *Ficus pumila* cuttings. Mature control (−IBA) plants root only from March to May, while juvenile control roots poorly from November to January. When treated with IBA (+IBA), the juvenile cuttings overcome the seasonal fluctuation in rooting, whereas mature cuttings do not (*47*).

[9]A liner is a small plant produced from a rooted cutting, tissue culture, or a seedling which is "lined-out" in a nursery. Rooted cuttings are usually transplanted into small liner pots and allowed to become established during liner production, before being retransplanted to larger containers or directly planted in the field.

activity occurred during peak rooting periods in both the easy and difficult-to-root forms.

Softwood cuttings of many deciduous woody species taken during spring or summer usually root more readily than hardwood cuttings procured in the winter. For plants difficult to root, it is thus often necessary to resort to the use of softwood cuttings. In tests (*128*) with cherries, hardwood cuttings taken in winter did not root, whereas softwood cuttings made in the spring gave satisfactory rooting with most cultivars. This situation is also well illustrated in the lilacs (*Syringa* spp.). About the only way cuttings of these plants can be rooted is to make softwood cuttings during a short period in the spring when the shoots are several inches long and in active growth. The Chinese fringe tree (*Chionanthus retusus*) is notoriously difficult to root, but by taking cuttings during a short period in midspring, high rooting percentages can be obtained (*279*).

The effect of timing is also strikingly shown by difficult-to-root deciduous azalea cuttings. These root readily if the cuttings are taken from succulent growth in early spring; by late spring, however, the rooting percentages decline rapidly (*183*). For any given species, empirical tests are required to determine the optimum time to take cuttings, which is more related to the physiological condition of the plant than to any given calendar date.

Often the effects of timing are merely a reflection of the response of the cuttings to environmental conditions at the different times of the year. When hardwood cuttings of deciduous species are taken and planted in the nursery in early spring after the rest period of the buds has been broken by winter chilling, the results are quite often a complete failure, since the buds quickly open with the onset of warm days. The newly developing leaves will start transpiring and remove the moisture from the cuttings before they have the opportunity to form roots; thus they soon die. Newly expanding buds and shoots are also competing sinks for metabolites and phytohormones, to the detriment of rooting. This has been shown with *Rosa multiflora* under an intermittent mist system where water stress was not a factor (*118*). If cuttings can be taken and planted in the fall while the buds are still in the rest period, roots may form and be well established by the time the buds open in the spring. A preplanting moist, warm (15 to 21°C; 60 to 70°F) storage period is often helpful in promoting adventitious root initiation (*12, 129, 131, 133, 152*).

Broad-leaved evergreens usually root most readily if the cuttings are taken after a flush of growth has been completed and the wood is par-

tially matured. This occurs, depending upon the species, from spring to late fall. In rooting cuttings of **narrow-leaved évergreens**, best results may be expected if the cuttings are taken during the period from late fall to late winter. With junipers, rooting was lowest during the season of active vegetative growth and highest during the dormant period. Furthermore, the low temperatures occurring at the time when such coniferous evergreens root best apparently is not a requirement, since juniper stock plants held in a warm greenhouse from early fall to midwinter produced cuttings that rooted better than those taken from out-of-door stock plants (*191*).

The season of the year in which **root cuttings** are taken is very important with some species; in others it makes no difference. Regeneration of the red raspberry (*Rubus idaeus*) from root cuttings taken from autumn to spring was almost 100 percent successful. Cuttings taken during the summer months failed to survive. On the other hand, root cuttings of horseradish were completely successful regardless of the time of year they were taken (*156*).

In many containerized ornamental nurseries, cuttings from difficult-to-root species are taken early in the propagation season, whereas cuttings of easy-to-root species are taken later in the summer. This seasonal scheduling of propagation also more efficiently utilizes propagation facilities and personnel.

Treatment of Cuttings

Handling of Cutting Material

Only cuttings of good quality should be collected for propagation. As the wise instructions to employees in a commercial propagation department go—"A cutting that is barely good enough is *never* good enough, so don't put it in the bunch!" Quality control of cuttings begins with stock plant quality control, and since many containerized ornamental nurseries no longer use stock plants, it is essential to maintain quality control of all production container plants from which propagules are taken. It is important that propagules be collected from stock plants free of viruses, bacteria, fungi, and other pathogenic organisms. Propagation is the foundation on which production horticulture hinges. Marginal quality propagules delay product turnover and create cultural and quality problems throughout the production cycle (6).

As noted earlier, many propagators prefer to collect propagules from stock plants early in the day when cuttings are still turgid. If the cuttings cannot be stuck immediately, they are misted down to reduce transpiration and held overnight in refrigeration facilities at 4 to 8°C (40 to 48°F) and generally stuck the next day.

Cuttings of some temperate-zone woody species have been stored at low temperatures for extended periods without any deleterious effects on subsequent root formation and leaf retention. Storage of *Rhododendron catawbiense* cuttings in moist burlap bags at either 21 or 2°C (70 or 36°F) for 21 days did not consistently reduce the percentage of rooting or rootball size (54). Although carbohydrate concentrations in the bases of cuttings changed with time and storage temperature, neither these changes nor changes in xylem water potential were large enough to influence subsequent rooting.

Cuttings of many tropical foliage plants are imported from Central America to be rooted and finished off in the United States or Europe. It takes 4 to 14 days to deliver cuttings from Central America to U.S. nurseries. Duration in transit may affect cutting quality due to excess respiration, light exclusion, moisture loss, pathogen invasion, and/or ethylene buildup. Croton (*Codiaeum variegatum*) cuttings had excellent quality when stored 5 to 10 days at 15 to 30°C (59 to 86°F) or 15 days at 15 to 20°C (59 to 68°F) (314).

Storage life of unrooted cuttings of geranium (*Pelargonium* × *hortorum* Bailey) was improved by high-humidity storage in polyethylene bags at 4°C (39°F) and low-intensity illumination. Prestorage applications of antitranspirants was detrimental, but soaking cutting bases in 2 to 5 percent sucrose for 24 hours prior to storage improved rooting (231). The ethylene inhibitor, silver nitrate, was more effective in maintaining storage life than silver thiosulfate, which reduced rooting (231). Abscisic acid will reduce transpiration in geranium cuttings, which may be of practical value in the shipment and storage of geranium cuttings (3).

In general, successful storage of unrooted cuttings depends on storage conditions, state of the cuttings, and species. It is important that dry matter losses and pathogens be minimized. Within the storage unit it is best to maintain nearly 100 percent humidity, and temperature should be as low as the hardiness of the given species can tolerate (13). Reduced oxygen and ethylene levels and high CO_2 [controlled atmospheric storage (CA)] help to retain rooting capacity (13). Storage duration can vary from days to several months depending on cutting food reserves, frost hardiness, and degree of lignification (woodiness of the material).

Growth Regulators

To induce adventitious roots before the use of synthetic root-promoting growth regulators (auxins) in rooting stem cuttings, many chemicals were tried with limited success (177). Zimmerman (335) showed in 1933, before auxins were discovered, that certain unsaturated gases, such as ethylene, carbon monoxide, and acetylene, stimulate initiation of adventitious roots as well as development of latent, preexisting root initials. Cuttings of many herbaceous plants respond to these gases with increased rooting.

The discovery that natural auxins, such as **indoleacetic acid** (IAA) and synthetic ones such as **indolebutyric acid** (IBA) and **naphthaleneacetic acid** (NAA), stimulated the production of adventitious roots in stem and leaf cuttings, was a major milestone in propagation history. The response, however, is not universal; cuttings of some difficult-to-root species still root poorly after treatment with auxin. Hence auxin is not always the limiting chemical component in rooting. It is possible that there is a lack of endogenous rooting cofactors, adequate enzyme system components (regulators), substrate phenolics, etc.

A practice of some Europeans and Middle Eastern gardeners in early days was to embed grain seeds into the split ends of cuttings to promote rooting. This seemingly odd procedure had a sound physiological basis, for it is now known that germinating seeds are good producers of auxin which aids in root formation in the cuttings.

Some of the phenoxy compounds having auxin activity promote root formation even at very low concentrations (148, 150). The weedkiller, **2,4-dichlorophenoxyacetic acid (2,4-D)**, is quite potent at low concentrations in inducing rooting of certain species, but it has the disadvantage of inhibiting shoot development and is very toxic to plants at higher concentrations.

Mixtures of root-promoting substances are sometimes more effective than either component alone. For example, equal parts of indolebutyric acid (IBA) and naphthaleneacetic acid (NAA), when used on a number of widely diverse species, were found to induce a higher percentage of cuttings to root and more roots per cutting than either material alone (*149*). Species are also known to react differently when treated with equal amounts of the two auxins. NAA is more effective than IBA in stimulating rooting of Douglas fir (*240*).

Adding a small percentage of certain phenoxy compounds to either IBA or NAA has, in some instances, caused excellent rooting (*150*)

and produced root systems qualitatively better than those obtained with phenoxy compounds alone. Likewise, other tests (*71*) have shown that a combination of IBA, NAA, and 2,4-D gave much better rooting than the use of any of these auxins alone.

The acid form of auxin is relatively insoluble in water, but can be dissolved in a few drops of alcohol or ammonium hydroxide before adding to water. Salts of some of the growth regulators may be more desirable than the acid in some instances, because of their comparable activity and greater solubility in water (336). The aryl esters of both IAA and IBA, and aryl amide of IBA (Fig. 9–23), have been reported to be more effective than the acids in promoting root initiation (*114, 115, 287*). Again, this is species dependent. It may be that these new formulations are less toxic to plant material than the acid form.

For general use in rooting stem cuttings of

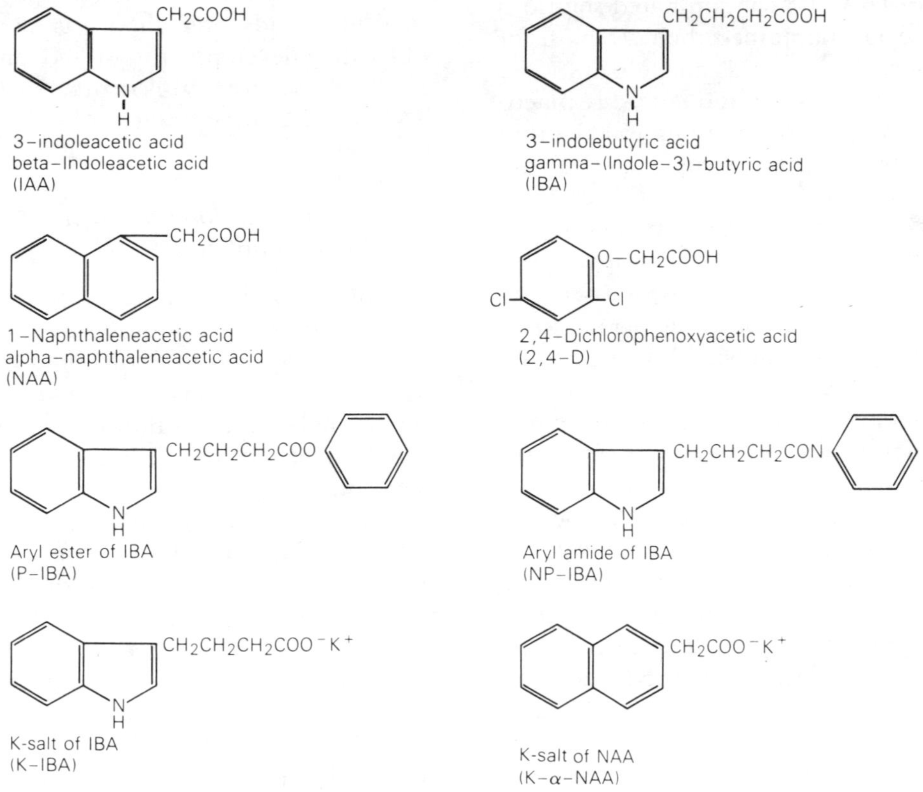

FIGURE 9-23 Structural formulas of auxins, aryl esters, aryl amides, and K-salt formulations of auxins active in promoting adventitious root initiation in cuttings.

the majority of plant species, IBA and/or NAA are recommended (43). To determine the best material and optimum concentration for rooting any particular species under a given set of conditions, empirical trials are necessary and should be repeated over several occasions, since experiments show that even apparently clear results are not always reproducible.

Application of synthetic auxins to stem cuttings at high concentrations can inhibit bud development, sometimes to the point at which no shoot growth will take place even though root formation has been adequate. Also, applications of rooting substances to root cuttings may inhibit the development of shoots from such root pieces.

There is often a question of how long the various root-promoting preparations will keep without losing their activity. Bacterial destruction of IAA occurs readily in unsterilized solutions. In sterile solutions, this material remains active for several months. A widely distributed species of *Acetobacter* destroys IAA, but the same organism has no effect on IBA. Uncontaminated solutions of NAA and 2,4-D maintained their strength for as long as a year.

IAA is sensitive to light and is readily inactivated. Concentrated IBA solutions (5000 ppm; 24.6 mM) in 50 percent isopropyl alcohol are quite stable and can be stored up to six months at room temperature in clear glass bottles under light conditions of 6 μmoles m^{-2}s^{-1} without loss in activity (246). Both NAA and 2,4-D seem to be entirely light-stable. Indoleacetic acid oxidase in plant tissue will break down IAA but has no effect on IBA or NAA.

The natural auxin flow in stem tissue is in a basipetal direction (apex to base). In early work, synthetic auxin applications were made to cuttings at the upper end to conform to the natural downward flow. As a practical matter, however, it was soon found that basal applications gave better results. Sufficient movement apparently resulted to carry the applied auxin into the parts of the cutting where it stimulated root production. In tests using radioactive IAA for rooting leafy plum cuttings, IAA was absorbed and distributed throughout leafy cuttings in 24 hours, whether application was at the apex or base (288). However, with basal application, most of the radioactivity

remained in the basal portion of the cuttings. Leafless cuttings absorbed the same amount of IAA as leafy cuttings, indicating that transpiration "pull" was not the chief cause of absorption and translocation.

In respiration studies of tissues at the basal ends of IBA-treated and of control cuttings it was found that by the time roots had formed on treated cuttings, their respiration rate was four times as great as in untreated cuttings. In addition, IBA-treated cuttings had a considerably higher level of amino acids at their bases 48 hours after treatment than untreated cuttings. This pattern continued, with nitrogenous substances accumulating in the basal part of treated cuttings, apparently mobilized in the upper part and translocated as asparagine (289).

For application methods of root-promoting growth regulators and listing of commercial auxin sources, see Chap. 10.

To induce adventitious shoots.   In propagation by leaf or root cuttings the production of adventitious buds and shoots is necessary, along with the development of adventitious roots. Treatments with **cytokinins,** such as **kinetin, BA,** or **PBA,** may be helpful (51, 137, 329) (see Fig. 9–10).

Mineral Nutrition of Cuttings During Rooting

Since rooting is a developmental process, it has been difficult to quantify the effect of nutrition on root primordia initiation versus root primordia elongation. It is important that stock plants be maintained under optimum nutrition prior to the collection of cuttings. Mobilization studies have been conducted to examine the movement of mineral nutrients into the base of cuttings during root initiation (339). The redistribution of nitrogen in stem cuttings during rooting was accelerated by auxin treatment of plum (289). However, N was not mobilized, nor was any redistribution of P, K, Ca, and Mg detected during root initiation in stem cuttings of Japanese holly (21, 22).

There are conflicting reports on mobilization since P, but not N, K, or Ca, was mobilized during root initiation in chrysanthemum cuttings (95). Although considered immobile, redistribution of Ca was reported during rooting of Japanese holly. Apparently, Ca was redistributed to support tissue development in the upper cutting sections and not for root growth and development.

The importance of N in root initiation is supported by nutrition studies on rooting of cuttings and the importance of N in nucleic acid and protein synthesis (*19*). The influence of N on root initiation and development also relates to such factors as carbohydrate availability, C/N ratio, and hormonal interactions.

Boron stimulates root production in cuttings of some species by promotion of root growth rather than to an effect on root initiation. Experiments (*135*) in rooting cuttings prepared from bean hypocotyls showed that when cuttings were placed in nutrient solutions completely lacking in boron, visible roots failed to appear, although in complete nutrient solutions or in solutions lacking in certain other trace elements, adequate rooting took place. The use of boron—in combination with IBA—increased the rooting percentage, the number and length of roots, and the speed of rooting of English holly cuttings taken in the fall. Apparently a synergistic reaction with IBA occurred, since boron alone had no effect (*319*). While the mechanism of the action of boron in stimulating rooting is not precisely known, there is evidence (*320*) to support the belief that boron is acting in oxidative processes, possibly increasing mobilization of oxygen-rich citric and isocitric acids into the rooting tissues.

In developmental studies of rooting, boron was needed for sustained cell division of phloem ray parenchyma cells during dedifferentiation (*1*). This research on mung bean suggested that boron deprivation inhibited cell division rather than DNA synthesis. Boron may also affect rooting by regulating endogenous auxin levels through enhancement of IAA oxidase activity (*163*).

Leaching of Nutrients

The development of intermittent mist revolutionized propagation, but mist can severely leach cuttings of nutrients. This is a particular problem with cuttings of difficult-to-root species which take a longer time period to root under mist. Mineral nutrients such as N, P, K, Ca, and Mg are leached from cuttings while under mist (*22, 95*). Leaching depends on the growth stage of the cutting material; leafy hardwood cuttings are reported to be more susceptible than softwood or herbaceous cuttings (*95*). Apparently, young growing tissues more quickly sequester nutrients by using them in the synthesis of cell walls and other cell components which are difficult to leach, as opposed to greater leaching in hardwood cuttings, which is attributed to a greater portion of nutrients in exchangeable forms (*95*). Nitrogen and Mn are easily leached; Ca, Mg, S, and K are moderately leached; and Fe, Zn, P, and Cl are leached with difficulty (*302*). Both leaching and mineral nutrient mobilization contribute to foliar deficiencies of cuttings (*19*).

As a whole, mist application of nutrients has not been a viable technique to maintain cutting nutrition. Nutrient mist application can inhibit rooting (*176*) and stimulate algae growth, which causes sanitation and media aeration problems (*331*).

A commercial technique is to apply moderate levels of controlled slow-release macro and micro elements to the propagation media either preincorporated into the media prior to sticking cuttings or by topdressing during propagation. These supplementary nutrients do not promote root initiation (*168*), but rather improve root development after root primordia initiation has occurred. Hence, propagation turnover occurs more quickly and plant growth is maintained by producing rooted liners that are more nutritionally fit. Optimum levels of fertilization for rooting need to be determined on a species basis.

Fungicides

Adventitious root initiation and survival of the rooted cuttings are two different phases. Often cuttings root but do not survive for very long. During the rooting and immediate postrooting period, cuttings are subject to attacks by various microorganisms. Treatments with fungicides should give some protection and result in both better survival and improved root quality. Such benefits have been demonstrated in a number of instances, but there is a question whether improvement is due to protection from fungus attacks, or stimulation of root initiation by the fungicide, or both. Studies (*122*) indicate that captan[10] may act by protecting the newly formed roots from fungal attacks and giving increased cutting survival (*83, 122, 151, 290*). Other reports (*304, 323*) also indicate considerable improvement in cutting survival and in the quality of the rooted cuttings resulting from the use of captan. Captan may be used as a powder dip following an IBA treatment, or the IBA in talc may be mixed with the captan powder. Captan is especially suitable for treating cuttings, since it does not decompose easily and has a long residual action. A talc powder combination of captan, IBA, sucrose, growth retardants, and rooting cofactors has been used successfully to root difficult-to-root woody and herbaceous species (*125*).

In a study testing 10 commercial fungicides on cuttings, Terraclor 75W and captan reduced rooting of *Peperomia caperata* (*221*), whereas the broad-spectrum chemical treatments of Subdue 5E + Benlate, and Truban 5G + Benlate 50W + Agrimycin 21W (streptomycin) resulted in greater root development. Terraclor 75W and Banrot enhanced rooting of *Buxus microphylla* var, *japonica* (*221*). Enhanced rooting associated with crop protection chemicals is more likely due to disease control than actual stimulation of root initiation (with exceptions), while reduced rooting is due to chemical phytotoxicity (*153, 268*).

The enhancement or reduction of rooting is also dependent on chemical application method (drench, soak, incorporation into propagation media), plant species, stage of plant growth, and environmental conditions. Fungicides and bactericides are important components of disease control in stock plant maintenance and treatment of cuttings prior to sticking and during propagation. It is important that the propagator conduct small-scale empirical experiments to determine the broad-spectrum disease control chemicals that best fit their particular propagation system.

Wounding

Basal wounding is beneficial in rooting cuttings of certain species, such as rhododendrons

and junipers, especially cuttings with older wood at the base. Following wounding, callus production and root development frequently are heavier along the margins of the wound. Evidently, wounded tissues are stimulated into cell division and production of root primordia (*207*) (Figs. 9-24 and 9-25). This is due, perhaps, to a natural accumulation of auxins and carbohydrates in the wounded area and to an increase in the respiration rate in the creation of a new "sink area." In addition, injured tissues from wounding would be stimulated to produce ethylene, which can promote adventitious root formation (*186, 335*).

Wounded cuttings absorb more water from the medium than unwounded, and wounding may also permit greater absorption of applied growth regulators by the tissues at the base of the cuttings. In stem tissue of some species there is a sclerenchymatic ring of tough fiber cells in the cortex external to the point of origin of adventitious roots. There is some evidence (*12, 41*) that newly formed roots may have difficulty penetrating this band of cells. A shallow wound would cut through these cells and perhaps permit outward penetration of the developing roots more readily.

FIGURE 9-24 Split base treatment to enhance root initiation in dormant apple rootstock cuttings (*207*).

---

[10]*N*-Trichloromethylmercapto-4-cyclohexene 1,2-dicarboximide.

FIGURE 9–25 *Left:* Split-base treatment. One half of the split stem base has been removed to show the nodular callus in the split. *Middle:* Split base—transverse section near the apex of the wound. The split (Sp) here is narrow and, consequently, the new cambium (NC) re-forms across the split instead of forming salients as in, *right*, split base. Note roots emerging in ranks from the wound. (From MacKenzie, Howard, and Harrison-Murray (*207*).)

Environmental Conditions during Rooting

Water Relations

Although the presence of leaves on cuttings is a strong stimulus to adventitious root and bud initiation, loss of water from leaves may reduce the water content of the cuttings to such a low level that they do not survive. The goals of propagation systems are (a) to maintain an atmosphere with low evaporative demand, so minimizing transpirational water losses from cuttings and thereby avoiding substantial tissue water deficits (cuttings without roots lack effective organs to replace transpired water lost, and cells must maintain adequate **turgor**[11] for initiation and development of roots); (b) to maintain acceptable temperatures for the regeneration metabolism needed at the cutting base, while avoiding heat stress of leaves; and (c) to maintain light levels suitable for photosynthesis and carbohydrate production for use once root initiation has occurred (*202*).

While **leaf water potential** (ψ_{leaf}) is an important parameter for measuring water status of cuttings and frequently relates to rooting, turgor (ψ_p) is more physiologically important for growth processes (*155*). Turgor has been correlated to rooting (*98, 100*). Poor rooting has been associated with water stress and the closing of stomata that occur with many plant species at the low leaf water potential of (-1.0 MPa or less) (*197*). The tissue water potential of cuttings is very dynamic and can fluctuate daily and throughout the propagation period; also, the water potential of tissue at the cutting base where roots are initiated may substantially differ from ψ_{leaf}. Hence simpler gravimetric measurements (measured by fresh weight changes in cuttings) may be a better system for determining water status of cuttings than ψ_{leaf} (*203*).

[11]**Water potential** (ψ_{water}) refers to the difference between the activity of water molecules in pure distilled water and the activity of water molecules in any other system in the plant. Pure water under these conditions has a water potential of zero. Since the activity of water in a cell is usually less than that of pure water, the water potential in a cell is usually a negative number. The magnitude of water potential is expressed in megapascals [1 megapascal (MPa) = 10 bars = 9.87 atmospheres]. Propagators can determine water potential by using a pressure chamber (pressure bomb) manufactured by PMS (Corvallis, OR) or Soil Moisture Corp. (Santa Barbara, CA). A psychrometer with a microvolt meter (LiCor, Lincoln, NE) can also be used.

Estimation of **turgor** (ψ_p) requires measurement of water potential (ψ_{water}) minus the **osmotic potential** (ψ_π), which is based on the formula $\psi_{\text{water}} = \psi_\pi - \psi_p$. Osmotic potential can also be determined by either a pressure chamber or a psychrometer.

The water status of cuttings is a balance between transpirational losses and uptake of water. Water absorption through the leaves is not the major contributor to water balance in most species. Rather, the cutting base and any foliage immersed in the propagation media are main entry points for water (*203*). Water uptake of cuttings is directly proportional to volumetric water content of the propagation media, with wetter media improving water uptake (see Fig. 9–26) (*99*). However, keep in mind that excess water reduces media aeration and can cause disease problems.

Water uptake in cuttings declines after they are inserted into propagation media. This reduced hydraulic conductivity of cuttings is apparently caused by blockage of xylem vessels and/or collapse of tracheids, which is similar to problems observed with cut flowers (158). Another advantage in wounding cuttings is to increase the contact area between the cutting base and propagation medium, thus improving water uptake of cuttings (*99, 203*).

Degree of stomatal opening can be a useful indicator to determine if a given propagation system is maintaining adequate turgor of cutting leaves. A simpler and more useful system is to measure evaporation rates directly with evaporimeters (Figs. 9–27 and 10–37) (*199, 203*). When water deficits cause stomata to close, carbohydrate gain in cuttings through photosynthesis is reduced by limiting diffusion of CO_2 into the chloroplasts and by the increased heat load on the leaf, since the major cooling effects due to stomatal

transpiration are lost. Carbon gain due to photosynthesis is probably more important after root initiation has occurred to promote rapid development of roots. It has been reported that translocation of photosynthate from leaves of intact plants continues under moderate or severe stress (*28*).

The driving force that determines the rate at which cuttings lose water is the difference in pressure between water vapor in the leaves (V_{leaf}) and that in the surrounding air (V_{air}). Commercial propagation systems aim to minimize this difference either by decreasing V_{leaf} through reducing leaf temperature (i.e., with intermittent mist) and/or by increasing the V_{air} by preventing the escape of water vapor (i.e., with an enclosed poly tent). Enclosed systems use humidification (*86*) since only the V_{air} is increased, while intermittent mist operates primarily on V_{leaf} but also provides a modest increase in V_{air}. Methods used to control water loss of leaves (*202*) are:

1. **Enclosures**—outdoor propagation under low tunnels or cold frames, or nonmisted enclosures in a glasshouse or polyhouse (shading, tent and contact polyethylene systems, wet tents).
2. **Intermittent mist**—open and enclosed mist systems
3. **Fogging systems**

Major advantages of *enclosed systems* are their simplicity and low cost. The main disadvantage is

FIGURE 9–26 Water uptake by cuttings is directly proportional to the volumetric water content of the rooting medium. Here, softwood cuttings of *Escallonia × exoniensis* are inserted in a peat–pumice mix containing 15, 20, 40 and 60 percent by volume of water and in water (*left to right*). The degree of wilting relates to the water content. (Courtesy K. Loach.)

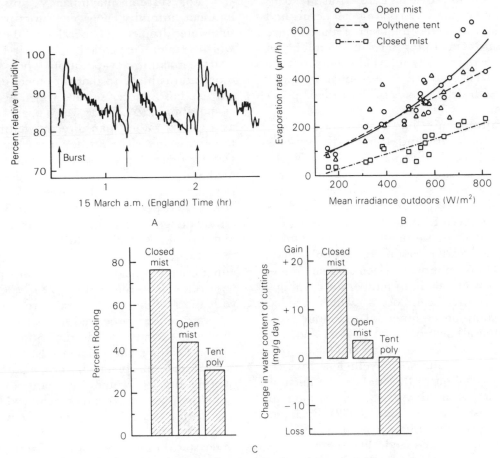

FIGURE 9-27   *Top left:* Cyclic changes in relative humidity under open-bench mist. *Top right:* Evaporation rates measured in three propagation systems: open mist, closed mist, polyethylene closed tent. *Below:* Rooting and water loss of six woody species in the three propagation systems (*201, 202*).

that they trap heat if light intensity is high. The trapped heat increases the saturation deficit of the air, and leaf temperature rises to increase the leaf-to-air vapor pressure gradient and consequently leaves lose water. Shading must be used with these systems. Polyethylene films have a low permeability to water vapor loss, but allow gas exchange. They are used to cover outdoor propagation structures as well as for closed mist systems in greenhouses. Modified polyethylene films are now available with additives of vinyl acetate, aluminum, or magnesium silicates which increase their opacity to long-wave radiation (i.e., reduce heat buildup). Polyethylene covered structures have replaced much of the traditional glass-covered cold frames (see Figs. 2-3 and 10-32).

Nonmisted enclosures in a glasshouse or polyhouse can be used for difficult-to-root species and have the advantage of avoiding nutrient-leaching problems of intermittent mist, yet affording greater environmental control than outdoor propagation. The shading system entails applying shading compounds to greenhouse roofs and/or utilizing automatically operating light-regulated shading curtains (see Fig. 2-7). Shading systems are integrated with temperature control by ventilated fogging or pad-and-fan cooling and heaters.

Contact systems entail laying polyethylene sheets directly on cuttings that are watered-in (Fig. 10-32). Leaves tend to be cooler under contact polyethylene because they are in direct contact with the polyethylene and are moistened by condensate from the cover. There is the dual benefit that some evaporative cooling can occur, and that water loss from the foliage is reduced since the condensate contributes to relative humidity of the air, rather than solely internal water from the leaf tissue. Hence, there is less internal water stress than with drier leaves of the **indoor polytent system**.

Intermittent mist has been used in propagation since the 1940s and 1950s. Mist systems minimize V_{leaf}, which lowers the leaf-to-air vapor pressure gradient and slows down leaf transpiration. Mist also lowers ambient air temperature and the cooler air consequently lowers leaf temperature by **advection**, in addition to cooling occurring through evaporation of the applied film of water (*202*). Advective cooling occurs only minimally in enclosed nonmist systems. Since intermittent mist lowers medium temperature, suboptimal temperatures can occur which reduce rooting. A common commercial technique is to use bottom (basal) heat both with indoor and outdoor mist systems (see Fig. 10–33 and Chap. 10). **Enclosed mist** utilizes polyethylene-covered structures in glasshouses that reduce the fluctuation in ambient humidity that are common to open-bench mist (Fig. 9–27 and 10–32). Enclosed mist also ensures more uniform wetting of foliage since air currents are reduced. There are advantages in using enclosed mist with difficult-to-root species compared to **open mist** or the polytent system without mist (Figs. 9–27 and 9–28).

Fog systems maximize V_{air} by raising the ambient humidity. Fog generators produce very fine water droplets of generally $< 20 \ \mu m$ diameter which remain suspended in the air for long periods to maximize evaporation. Their surface/volume ratio is high, so that the finely divided mist particle has a larger surface which also increases evaporation (*202*). Since water passes into the air as a vapor, rather than directly wetting leaf surfaces as occurs with mist, foliar leaching and overwetting of media are avoided. The propagation media are not cooled to the same degree as with mist, thus avoiding suboptimal rooting temperatures, and perhaps less basal heat is needed. A disadvantage of the fog system is generally the high initial cost and maintenance.

There are advantages to using fog over either open or enclosed mist systems, particularly with difficult-to-root plants (*125, 126, 127*) and with the acclimation and ex vitro rooting needs of tissue culture–produced liners (see Chap. 17). Propagators must decide the most cost-effective system for their particular needs.

Disease problems in propagation are frequently due to water and temperature stress. Avoiding plant stress can be achieved by harvesting cuttings early in the morning when propagules are turgid and in maintaining a desirable water status of cuttings during propagation. Disease is most severe where wilting has occurred, making damaged tissue of cuttings more susceptible to fungal attack. Development of disease orga-

FIGURE 9–28 Cuttings of *Forsythia × intermedia* 'Lynwood' rooted under open-bench mist (*left*) and in a misted, polyethylene enclosure (*right*). Note that cuttings rooted in the warmer, more humid enclosure break bud and grow faster than those under open mist. However, for some species and circumstances, too much top growth can divert carbohydrates away from developing root initials and slow root growth. (Courtesy K. Loach.)

nisms under mist might be expected, but such has not been the case; in fact, the reverse has been true. For example, in greenhouse roses, no mildew was found to develop on leaves under mist, but mildew did develop on those not under mist. This is explained by the failure of mildew spores to germinate in water (*189*). Water sprays inhibit the development of powdery mildew (*Sphaerotheca pannosa*), and it may be that other disease organisms are held in check in the same manner (*334*). It is of interest that disease is less of a problem in an enclosed mist system (England) with higher heat buildup than a conventional open mist system (*201*). During propagation preventive disease control should be practiced with regularly scheduled applications of broad-spectrum fungicides and bactericides, whereas pesticides for insect and mite problems, etc. are applied as warranted. (For details in setting up mist propagating systems, see Chap. 10.)

Temperature

Temperature of the propagation medium can be suboptimal for rooting due to the cooling effect of mist or seasonally related ambient air temperature. It is often more satisfactory and cost-effective to manipulate temperature by heating at the propagation bench level rather than heating the entire propagation house. See Chap. 10 for heating equipment and sensors.

The consensus regarding the optimum medium temperature for propagation is 18 to 25°C (65 to 77°F) for temperate climate species and 7°C (13°F) higher for warm climate species (*66, 178*). Daytime air temperatures of about 21 to 27°C (70 to 80°F) with night temperatures about 15°C (60°F) are satisfactory for rooting cuttings of most temperate species, although some root better at lower temperatures. High air temperatures tend to promote bud development in advance of root development and to increase water loss from the leaves. It is important that adequate moisture status be maintained by the propagation system so that cuttings gain the potential benefit of the higher basal temperature.

The temperatures that propagators use for rooting should be tested for their particular system and the species to be propagated. Root initiation in cuttings is temperature driven but subsequent root growth is strongly dependent on available carbohydrates. This is particularly evident in leafless hardwood cuttings in which excessive root initiation and growth can so deplete stored reserves that there are insufficient available carbohydrates for satisfactory bud growth. The same principle holds true for leafy cuttings (semihardwood, softwood, herbaceous), where shoot growth can divert carbohydrates away from developing root initials and thereby slow root growth (Fig. 9–28).

There is evidence that different stages of root formation have different temperature requirements. Optimum temperature for root **initiation** in *Forsythia* and *Chrysanthemum* was 30°C (86°F); whereas root **development** (elongation of primordia and protruding of roots from the stem cutting) was optimum at lower temperatures of 22 to 25°C (72 to 77°F). Respiration was probably reduced at the lower temperature, allowing optimum photosynthate accumulation for root development.

Light

As discussed earlier in the stock plant manipulation section, light is a contributing factor in adventitious root and bud formation of cuttings (*57, 68, 69*).

Intensity (irradiance). Evidence that cuttings of some plants root best under relatively low irradiance[12] was shown many years ago (*281*) in studies with fluorescent lamps and confirmed in studies with blueberry, weigela, forsythia, viburnum, and hibiscus cuttings (*168, 198*). However, cuttings of certain herbaceous plants, such as chrysanthemum, geranium, and poinsettia, rooted better in tests when the irradiance increased to 116 W/m^2 during trials in winter

---

[12]*Irradiance* is the relative amount of light as measured by radiant energy per unit area. For plant propagation, *photosynthetic active radiation* (PAR) is measured in the 400 to 700-nanometer (nm) waveband as watts per square meter (W/m^2) with a pyranometeric sensor or as μmol m^{-2} s^{-1} (1 μE m^{-2} s^{-1} = 6.023 × 10^{23} photons) with a quantum sensor. Many propagators still measure irradiance with a photometric sensor which determines footcandles or lux (1 footcandle = 10.8 lux). Light meters that measure irradiance with radiometric, quantum, and photometric sensors can be purchased from camera supply houses or instrument companies (i.e., Licor, Inc. of Lincoln, NE). For determining *light quality* or *wavelength*, the spectral distribution is measured with a portable spectroradiometer, which is an expensive instrument.

months. Very high irradiance (174 W/m²) damaged leaves on the cuttings, delayed rooting, and reduced root growth.

With selected temperate species under an English propagation system, acceptable light ranges were 1.5 to 3.0 MJ/m² or 20 to 100 W/m² per day (99). Propagators need to determine irradiance levels to fit their particular production systems. There is a growing trend for **modeling** propagation environments to determine optimal light, temperature, water, CO_2, and nutritional regimes. Computers can be programmed to monitor the propagation environment and adjust environmental conditions as needed through automated environmental control systems (Figs. 2–7 and 10–37.)

In working with cuttings of some plants that are difficult to root, possible enhancement of rooting may be obtained by growing the stock plants and rooting the cuttings at reduced irradiance levels, either by placing shade cloth over outdoor-grown stock plants, or by placing container-grown stock plants in a shade house. The practice of stock-plant **etiolation, banding, blanching,** and **shading** are further discussed in Chap. 10.

Daylength (photoperiod).   In some species, the photoperiod under which the cuttings are rooted may affect root initiation, long-days or continuous illumination generally being more effective than short-days (37, 281), although in other species photoperiod has no influence (271, 275).

This situation can become quite complex, however, since photoperiod can affect shoot development as well as root initiation. For example, in propagation by leaf cuttings there must be development of both adventitious roots and shoots. Using *Begonia* leaf cuttings (136), where the light intensity was adjusted so that the total light energy was about the same under both long days and short days, it was found that short days and relatively low temperatures promoted adventitious bud formation on the leaf pieces, whereas short days suppressed adventitious root formation. Roots formed best under long days with relatively high temperatures.

In rooting cuttings of 'Andorra' juniper, pronounced variations in rooting occurred during the year, but the same variations took place whether the cuttings were maintained under long days, short days, or the natural daylength (190). A number of tests have been made of the effect of photoperiod on root formation in cuttings, but the results are conflicting; hence it is difficult to make any generalization (7, 67, 70, 190, 317).

In some plants photoperiod will control growth after the cuttings have been rooted. Certain plants cease active shoot growth in response to natural decreases in daylength. This is the case with spring cuttings of deciduous azaleas and dwarf rhododendrons which had rooted and were potted in late summer or early fall. Considerably improved growth of such plants was obtained during the winter in the greenhouse if they were placed under continuous supplementary light, in comparison with similar plants subjected only to the normal short winter days. The latter plants, without added daylength, remained in a dormant state until the following spring (44, 94, 318).

High carbohydrate reserves are important for rooted cuttings (liners), since spring growth in a deciduous plant depends on reserves accumulated during the previous growing season. A night interruption lighting period to extend the natural photoperiod to maintain high carbohydrate reserves did initially enhance growth of rooted liners, but was not economically justified since growth of natural daylength liners were comparable after two years of field culture (265).

Quality.   Radiation in the orange-red end of the spectrum seems to favor rooting of cuttings more than that in the blue region (281), but there are conflicting reports as discussed in the section on stock plants.

Accelerated Growth Techniques

The forestry industry has developed accelerated growth systems to speed up the production of liners from vegetative propagules and from seed propagation. Woody perennial plants undergo cyclic growth, and many tree species experience dormancy. Liners are grown in protective culture facilities where photoperiod is extended and water, temperature, carbon dioxide, nutrition, mycorrhizal fungi, and growing media are optimized for each woody species and for each different phase of growth (Fig. 9–29).

This concept is also being used in propaga-

Selected vegetative or seed propagules
 (genetically superior)
↓
Efficient, container growing methods
↓
Protective culture (glasshouses, etc.)
↓
Programmed growth *control*
 throughout the year:
↓

> Light
> Temperature
> Mineral nutrients
> Water
> Carbon dioxide
> Growth regulators
> Mycorrhiza
> Growing media
> Competition
> Pests

↓
Production of *large* rooted cuttings or
 seedlings in months *rather* than *years*
↓
Acclimation to natural conditions
 and planting
↓
Genetic selection, hybridization, propagation
of superior strains or cultivars, and release to
nurseries

FIGURE 9-29 Components of accelerated growth techniques used in speeding up vegetative and seed propagation in the production of marketable liners.

tion of horticultural crops where supplementary lighting with high-pressure sodium vapor lamps and injection of CO_2 gas into mist water are used to enhance rooting of *Ilex aquifolium* (*85*). The promotive effects on cuttings have been attributed to enhanced photosynthesis.

Photosynthesis of Cuttings

Photosynthesis by cuttings is not an absolute requirement for root formation. This has been observed in leafy cuttings forming roots when placed in the dark (*53*) and with leafless, nonphotosyn-thetic hardwood cuttings that root. Increasing light intensity has not always promoted rooting, and net photosynthesis of unrooted cuttings is saturated at relatively low PAR (irradiance measured as photosynthetically active radiation) levels (*55*); hence high PAR will probably not increase photosynthesis and could potentially lead to desiccation of cuttings.

It has long been thought that carbohydrate content of cuttings is important to rooting, and carbohydrates do accumulate in the base of cuttings during rooting (*116*). The amount of carbohydrates accumulated at cutting bases has been correlated with photosynthetic activity (*53*), but carbohydrates can also accumulate in the upper portion of leafy cuttings until after roots have formed (*30*).

If a generalization can be made, photosynthesis in cuttings is probably more important *after* root initiation has occurred and helps aid root development and the more rapid growth of a rooted liner. Many questions still need to be answered on the role of photosynthesis in adventitious root formation (*52, 211*).

Rooting Medium

The rooting medium has four functions:

1. To hold the cutting in place during the rooting period
2. To provide moisture for the cutting
3. To permit penetration to and exchange of air at the base of the cutting
4. To create a dark or opaque environment by reducing light penetration to the cutting base

An ideal propagation medium provides sufficient porosity to allow good aeration and has a high water-holding capacity, yet is well drained and free from pathogens (*204*). Pathogens can occur in peat (*Pythium, Penicillium*, etc.) and other organic components of media. A good commercial practice it to pasteurize (by aerated steam) or chemically disinfect the medium with gas (i.e., methyl bromide). Some peat mixes are sufficiently free of pathogens and need not be pretreated. After cuttings are stuck, broad-spectrum, preventive chemical treatments are applied on a scheduled program during rooting to control damping-off organisms (*Pythium*, *Phytophthora*, *Rhizoctonia*, *Pestalotiopsis*, *Glomerella*, *Botrytis*, *Peronospora*, etc.) (*221*). (See Chap. 2.)

Rarely can a rooting response be attributed to differences in aeration due to the physical properties of the various media (*299*). Apparently, gaseous diffusion proceeds relatively freely:

1. Through the bulk of conventional propagation media
2. Because of diffusion of oxygen through the aerial portion of the cutting to its base, which supplies most of the aerobic requirements for rooting (*200*)

However, water films both within and around the base of the cutting can obstruct the free passage of oxygen to developing root initials. Thus it is very difficult to pinpoint the relationship of rooting with physical characteristics such as volumetric air and water contents of the medium (*200*). During cool winter months and/or with a closed mist system, the medium should be sufficiently loose to allow adequate aeration even when water utilization is lower.

Chemical and physical standards of a propagation media used at a successful commercial nursery are listed in Table 9–5. Rooting media used at most nurseries consist of an **organic component:** peat, softwood and hardwood barks, or sphagum moss. Sawdust, leaf mulch, and rice hulls have been used, but they oxidize readily and compact easily, which decreases pore space and aeration and have a high C/N ratio, which can result in nutritional problems after root initiation (*32*). A **coarse mineral component** is used to increase the proportion of large, air-filled pores and includes sand (avoid fine particle sands), grit, pumice, scoria, expanded shale, perlite, vermiculite, polystyrene, clay granules, or rockwool.

TABLE 9–5

Suggested rooting medium chemical and physical standard

Property	Comments
Chemical	
pH	4.5–6.5; 5.5–6.5 preferred
Buffer capacity	As high as possible
Soluble salts	400–1000 ppm (1 soil: 2 water by volume)
Cation exchange capacity	25 to 100 meg/liter
Physical	
Bulk density	0.3–0.80 g/cm³ (dry) or 0.60–1.15 g/cm³ (wet)
Air-filled porosity	15–40% by volume, ideally 20–25% range
Water-holding capacity	20–60% volume after drainage
Particle stability	Materials should resist decomposing quickly; decomposition can alter other media components

Source: Maronek, Studebaker, and Oberly (*210*).

This mineral component is added to improve aeration.

There is no ideal mix. An appropriate propagation medium depends on the species, cutting type, season, propagation system (i.e., with fog a waterlogged medium is less of a problem than with mist); the cost and availability of the medium components are other considerations.

A cutting from a nutritionally sound stock plant will generally have enough mineral reserves to initiate roots. Once roots are initiated, there are considerable advantages of applying low levels of liquid fertilizer to the medium or top dressing with Osmocote, Ficote, Nutricote, or some other suitable slow-release fertilizer. As discussed earlier, slow-release macro and microelements can also be incorporated into the medium (*315*).

REFERENCES

1. Ali, A. H. N., and B. C. Jarvis. 1988. Effects of auxin and boron on nucleic acid metabolism and cell division during adventitious root regeneration. *New Phytol.* 108:383–91.

2. Anand, V. K., and G. T. Heberlein. 1975. Seasonal changes in the effects of auxin on rooting in stem cuttings of *Ficus infectoria. Phys. Plant.* 34:330–34.

3. Arteca, R. N., D. S. Tsai, and C. Schlagnhaufer. 1985. Abscisic acid effects on photosyn-

thesis and transpiration in geranium cuttings. *HortScience* 20:370–72.

4. Ashiru, G. A., and R. F. Carlson. 1968. Some endogenous rooting factors associated with rooting of East Malling II and Malling-Merton 106 apple clones. *Proc. Amer. Soc. Hort. Sci.* 92:106–12.

5. Aung, L. H. 1972. The nature of root-promoting substances in *Lycopersicon esculentum* seedlings. *Phys. Plant.* 26:306–9.

6. Baker, K. F. 1984. The obligation of the plant propagator. *Proc. Inter. Plant Prop. Soc.* 34:195–203.

7. Baker, R. L., and C. B. Link. 1963. The influence of photoperiod on the rooting of cuttings of some woody ornamental plants. *Proc. Amer. Soc. Hort. Sci.* 82:596–601.

8. Bassuk, N. L., and B. H. Howard. 1981. Seasonal rooting changes in apple hardwood cuttings and their implications to nurserymen. *Proc. Inter. Plant Prop. Soc.* 30:289–93.

9. Bassuk, N. L., L. D. Hunter, and B. H. Howard. 1981. The apparent involvement of polyphenol oxidase and phloridzin in the production of apple rooting co-factors. *Jour. Hort. Sci.* 56:313–22.

10. Bassuk, N., and B. Maynard. 1987. Stock plant etiolation. *HortScience* 22:749–50.

11. Basu, R. N., B. N. Roy, and T. K. Bose. 1970. Interaction of abscisic acid and auxins in rooting of cuttings. *Plant and Cell Phys.* 11:681–84.

12. Beakbane, A. B. 1969. Relationships between structure and adventitious rooting. *Proc. Inter. Plant Prop. Soc.* 19:192–201.

13. Behrens, V. 1989. Storage of unrooted cuttings. In *Adventitious root formation in cuttings,* T. D. Davis, B. E. Haissig, and N. Sankhla, eds. Portland, Oreg.: Dioscorides Press.

14. Bhattacharya, N. C. 1989. Enzyme activities during adventitious rooting. In *Adventitious rooting formation in cuttings,* T. D. Davis, B. E. Haissig, and N. Sankhla, eds. Portland, Oreg.: Dioscorides Press.

15. Biran, I., and A. H. Halevy. 1973. Endogenous levels of growth regulators and their relationship to the rooting of dahlia cuttings. *Phys. Plant* 28:436–42.

16. ———. 1973. Stock plant shading and rooting of dahlia cuttings. *Sci. Hort.* 1:125–31.

17. ———. 1973. The relationship between rooting of dahlia cuttings and the presence and type of bud. *Phys. Plant.* 28:244–47.

18. Blakely, L. M., S. J. Rodaway, L. B. Hollen, and S. G. Croker. 1972. Control and kinetics of branch root formation in cultured root segments of *Haplopappus ravenii. Plant Phys.* 50:35–49.

19. Blazich, F. A. 1989. Mineral nutrition and adventitious rooting. In *Adventitious root formation in cuttings,* T. D. Davis, B. E. Haissig, and N. Sankhla, eds. Portland, Oreg.: Dioscorides Press.

20. Blazich, F. A., and C. W. Heuser. 1979. A histological study of adventitious root initiation in mung bean cuttings. *Jour. Amer. Soc. Hort. Sci.* 104(1):63–67.

21. Blazich, F. A., and R. D. Wright. 1979. Nonmobilization of nutrients during rooting of *Ilex crenata* cv. Convexa stem cuttings. *HortScience* 14:242.

22. Blazich, F. A., R. D. Wright, and H. E. Schaffer. 1983. Mineral nutrient status of 'Convexa' holly cuttings during intermittent mist propagation as influenced by exogenous auxin application. *Jour. Amer. Soc. Hort. Sci.* 108:425–29.

23. Bloch, R. 1943. Polarity in plants. *Bot. Rev.* 9:261–310.

24. Boe, A. A., R. B. Steward, and T. J. Banko. 1972. Effects of growth regulators on root and shoot development of *Sedum* leaf cuttings. *HortScience* 74(4):404–5.

25. Bollmark, M., and L. Eliasson. 1986. Effects of exogenous cytokinins on root formation in pea cuttings. *Physiol. Plant.* 68:662–66.

26. Bonnett, H. T., Jr., and J. G. Torrey. 1965. Chemical control of organ formation in root segments of *Convolvulus* cultured in vitro. *Plant Phys.* 40:1228–36.

27. Bouillenne, R., and M. Bouillenne-Walrand. 1955. Auxines et bouturage. *Rpt. 14th Inter. Hort. Cong.* 1:231–38.

28. Boyer, J. S. 1976. Photosynthesis at low water potentials. *Phil. Trans. Royal Soc. (London), Ser. B,* 273:501–12.

29. Breen, P. J., and T. Muraoka. 1973. Effect of indolebutyric acid on distribution of ^{14}C photosynthate in softwood cuttings of Marianna 2624 plum. *Jour. Amer. Soc. Hort. Sci.* 98(5):436–39.

30. ———. 1974. Effect of leaves and carbohydrate content and movement of ^{14}C-assimilate in plum cuttings. *Jour. Amer. Soc. Hort. Sci.* 99(4):326–32.

31. Brian, P. W., H. G. Hemming, and D. Lowe. 1960. Inhibition of rooting of cuttings by gibberellic acid. *Ann. Bot.* 24:407–9.

32. Bruckel, D. W., and E. P. Johnson. 1969. Ef-

fects of pH on rootability of *Thuja occidentalis*. *The Plant Propagator* 15(4):10–12.

33. Caboche, M., J. F. Muller, F. Chanut, G. Aranda, and S. Cirakoglu. 1987. Comparison of the growth promoting activities and toxicities of various auxin analogs on cells derived from wild type and a nonrooting mutant tobacco. *Plant Phys.* 83:795–800.

34. Cameron, R. J., and G. V. Thomson. 1969. The vegetative propagation of *Pinus radiata:* Root initiation in cuttings. *Bot. Gaz.* 130(4): 242–51.

35. Carlson, M. C. 1929. Origin of adventitious roots in *Coleus* cuttings. *Bot. Gaz.* 87:119–26.

36. ———. 1950. Nodal adventitious roots in willow stems of different ages. *Amer. Jour. Bot.* 37:555–61.

37. Carpenter, W. J., G. R. Beck, and G. A. Anderson. 1973. High intensity supplementary lighting during rooting of herbaceous cuttings. *HortScience* 8(4):338–40.

38. Challenger, S., H. J. Lacey, and B. H. Howard. 1965. The demonstration of root promoting substances in apple and plum rootstocks. *Ann. Rpt. E. Malling Res. Sta. for 1964,* pp. 124–28.

39. Chin, T. Y., M. M. Meyer, Jr., and L. Beevers. 1969. Abscisic acid stimulated rooting of stem cuttings. *Planta* 88:192–96.

40. Christiansen, M. V., E. N. Eriksen, and A. S. Andersen. 1980. Interaction of stock plant irradiance and auxin in the propagation of apple rootstocks by cuttings. *Sci. Hort.* 12:11–17.

41. Ciampi, C., and R. Gellini. 1958. Anatomical study on the relationship between structure and rooting capacity in olive cuttings. *Nuovo Giorn. Bot. Ital.* 65:417–24.

42. Cline, M. N., and D. Neely. 1983. The histology and histochemistry of the wound healing process in geranium cuttings. *Jour. Amer. Soc. Hort. Sci.* 108:452–96.

43. Cooper, W. C. 1944. The concentrated-solution-dip method of treating cuttings with growth substances. *Proc. Amer. Soc. Hort. Sci.* 44:533–41.

44. Crossley, J. H. 1965. Light and temperature trials with seedlings and cuttings of *Rhododendron molle. Proc. Inter. Plant Prop. Soc.* 15:327–34.

45. Crow, W. D., W. Nicholls, and M. Sterns. 1971. Root inhibitors in *Eucalyptus grandis:* Naturally occurring derivatives of the 2,3-dioxabicyclo (4,4,0) decane system. *Tetrahedron Letters 18.* London: Pergamon Press, pp. 1353–56.

46. Danckwardt-Lillieström, C. 1957. Kinetin-induced shoot formation from isolated roots of *Isatis tinctoria. Phys. Plant* 10:794–97.

47. Davies, F. T., Jr. 1984. Shoot RNA, cambial activity and indolebutyric acid effectively in seasonal rooting of juvenile and mature *Ficus pumila* cuttings. *Physiol. Plant.* 62:571–75.

48. Davies, F. T., Jr., and H. T. Hartmann. 1988. The physiological basis of adventitious root formation. *Acta Hort.* 227:113–20.

49. Davies, F. T., Jr., and J. N. Joiner. 1980. Growth regulator effects on adventitious root formation in leaf bud cuttings of juvenile and mature *Ficus pumila. Jour. Amer. Soc. Hort. Sci.* 100:643–46.

50. Davies, F. T., Jr., J. E. Lazarte, and J. N. Joiner. 1982. Initiation and development of roots in juvenile and mature leaf bud cuttings of *Ficus pumila* L. *Amer. Jour. Bot.* 69:804–11.

51. Davies, F. T., Jr., and B. C. Moser. 1980. Stimulation of bud and shoot development of Rieger begonia leaf cuttings with cytokinins. *Jour. Amer. Soc. Hort. Sci.* 105(1):27–30.

52. Davis, T. D. 1989. Photosynthesis during adventitious rooting. In *Adventitious root formation in cuttings,* T. D. Davis, B. E. Haissig, and N. Sankhla, eds. Portland, Oreg.: Dioscorides Press.

53. Davis, T. D., and J. R. Potter. 1981. Current photosynthate as a limiting factor in adventitious root formation in leafy pea cuttings. *Jour. Amer. Soc. Hort. Sci.* 106:278–82.

54. ———. 1985. Carbohydrates, water potential and subsequent rooting of stored rhododendron cuttings. *HortScience* 20:292–93.

55. Davis, T. D., and N. Sankhla. 1989. Effect of shoot growth retardants and inhibitors on adventitious rooting. In *Adventitious root formation in cuttings,* T. D. Davis, B. E. Haissig, and N. Sankhla, eds. Portland, Oreg.: Dioscorides Press.

56. Delargy, J. A., and C. E. Wright. 1978. Root formation in cuttings of apple (cv. Bramley's Seedling) in relation to ringbarking and to etiolation. *New Phytol.* 81:117–27.

57. ———. 1979. Root formation in cuttings of apple in relation to auxin application and to etiolation. *New Phytol.* 82:341–47.

58. Delisle, A. L. 1942. Histological and anatomical changes induced by indoleacetic acid in rooting cuttings of *Pinus strobus* L. *Va. Jour. Sci.* 3:118–24.

59. Deuber, C. G. 1940. Vegetative propagation of conifers. *Trans. Conn. Acad. Arts and Sci.* 34:1–83.

60. Doak, B. W. 1940. The effect of various nitrogenous compounds on the rooting of rhododendron cuttings treated with naphthaleneacetic acid. *New Zealand Jour. Sci. and Tech.* 21:336A–43A.

61. Donoho, C. W., A. E. Mitchell, and H. N. Sell. 1962. Enzymatic destruction of C^{14} labelled indoleacetic acid and naphthaleneacetic acid by developing apple and peach seeds. *Proc. Amer. Soc. Hort. Sci.* 80:43–49.

62. Doorenbos, J. 1954. Rejuvenation of *Hedera helix* in graft combinations. *Proc. Kon. Ned. Akad. Wet.* Series C 57:99–102.

63. Duhamel du Monceau, H. L. 1758. *La physique des arbres,* Vols. 1 and 2. Paris: Guerin and Delatour.

64. Durzan, D. J. 1988. Rooting in woody perennials: problems and opportunities with somatic embryos and artificial seeds. *Acta Hort.* 227:121–25.

65. Durzan, D. J., S. C. Chafer, and S. M. Lopushanski. 1973. Effects of environmental changes on sugar, tannins and organized growth in cell suspension cultures of white spruce. *Planta* 113:241–49.

66. Dykeman, B. 1976. Temperature relationship in root initiation and development of cuttings. *Proc. Inter. Plant Prop. Soc.* 26:201–7.

67. Economou, A. S., and P. E. Read. 1987. Light treatments to improve efficiency of in vitro propagation systems. *HortScience* 22:751–54.

68. Eliasson, L. 1971. Growth regulators in *Populus tremula.* II. Effect of light on inhibitor content in root suckers. *Phys. Plant.* 24:205–8.

69. ———. 1980. Interaction of light and auxin in regulation of rooting in pea stem cuttings. *Phys. Plant.* 48:78–82.

70. Eliasson, L. and L. Brunes. 1980. Light effects on root formation in aspen and willow cuttings. *Phys. Plant.* 48:261–65.

71. Ellyard, R. K. 1981. Effect of auxin combinations on the rooting of *Persoonia chameapitys* and *P. pinifolia* cuttings. *Proc. Inter. Plant Prop. Soc.* 31:251–55.

72. Emery, A. E. H. 1955. The formation of buds on roots of *Chamaenerion angustifolium* (L.) *Scop. Phytomorphology* 5:139–45.

73. Eriksen, E. N. 1973. Root formation in pea cuttings. I. Effects of decapitation and disbudding

at different development stages. *Phys. Plant.* 28:503–6.

74. ———. 1974. Root formation in pea cuttings. II. The influence of indole-3-acetic acid at different development stages. *Phys. Plant.* 30:158–62.

75. ———. 1974. Root formation in pea cuttings. III. The influence of cytokinin at different development stages. *Phys. Plant.* 30(2):163–67.

76. Eriksen, E. N., and S. Mohammed. 1974. Root formation in pea cuttings. II. Influence of indole-3-acetic acid at different developmental stages. *Phys. Plant.* 32:158–62.

77. Evans, H. 1952. Physiological aspects of the propagation of cacao from cuttings. *Proc. 13th Inter. Hort. Cong.* 2:1179–90.

78. Fadl, M. S., A. S. El-Deen, and M. A. El-Mahady. 1979. Physiological and chemical factors controlling adventitious root initiation in carob (*Ceratonia siliqua*) stem cuttings. *Egyptian Jour. Hort.* 6(1):55–68.

79. Fadl, M. S., and H. T. Hartmann. 1967. Isolation, purification, and characterization of an endogenous root-promoting factor obtained from the basal sections of pear hardwood cuttings. *Plant Phys.* 42:541–49.

80. ———. 1967. Relationship between seasonal changes in endogenous promoters and inhibitors in pear buds and cutting bases and the rooting of pear hardwood cuttings. *Proc. Amer. Soc. Hort. Sci.* 91:96–112.

81. Fann, Y. S., F. T. Davies, Jr., and D. R. Paterson. 1983. The influence of rootstock lateral buds on bench chip-budded 'Mirandy' field roses. *Scient. Hort.* 20:101–6.

82. Farrar, J. H., and N. H. Grace. 1942. Vegetative propagation of conifers. XI. Effects of type of cutting on the rooting of Norway spruce cuttings. *Can. Jour. Res., Sect. C.* 20:116–21.

83. Fiorino, P., J. N. Cummins, and J. Gilpatrick. 1969. Increased production of rooted *Prunus besseyi* Bailey softwood cuttings with pre-planting soak in benomyl. *Proc. Inter. Plant Prop. Soc.* 19:320–36.

84. Fischer, P., and J. Hansen. 1977. Rooting of chrysanthemum cuttings: Influence of irradiance during stock plant growth and of decapitation and disbudding of cuttings. *Scient. Hort.* 7:171–78.

85. French, C. J., and W. C. Lin. 1984. Seasonal variations in the effect of CO_2 mist and supplementary lighting from high pressure sodium

lamps on rooting of English holly cuttings. *Hort-Science* 19:519–21.

86. Gaffney, J. J. 1978. Humidity: Basic principles and measurement techniques. *HortScience* 13(5): 551–55.

87. Gardner, F. E. 1937. Etiolation as a method of rooting apple variety stem cuttings. *Proc. Amer. Soc. Hort. Sci.* 34:323–29.

88. Garner, R. J., and E. S. J. Hatcher. 1962. Regeneration in relation to vegetative growth and flowering. *Proc. 16th Inter. Hort. Cong.*, pp. 105–11.

89. Gasper, T., and M. Hofinger. 1989. Auxin metabolism during rooting. In *Adventitous root formation in cuttings*, T. D. Davis, B. E. Haissig, and N. Sankhla, eds. Portland, Oreg.: Dioscorides Press.

90. Ginzburg, C. 1967. Organization of the adventitious root apex in *Tamarix aphylla*. *Amer. Jour. Bot.* 54:4–8.

91. Girouard, R. M. 1967. Anatomy of adventitious root formation in stem cuttings. *Proc. Inter. Plant Prop. Soc.* 17:289–302.

92. ———. 1967. Initiation and development of aventitious roots in stem cuttings of *Hedera helix*. *Can. Jour. Bot.* 45:1883–86.

93. ———. 1969. Physiological and biochemical studies of adventitious root formation. Extractible rooting co-factors from *Hedera helix*. *Can. Jour. Bot.* 47(5):687–99.

94. Goddard, W. 1963. Forestalling dormancy and inducing continuous growth of *Azalea molle* with supplementary light for winter propagation. *Proc. Inter. Plant Prop. Soc.* 13:276–78.

95. Good, G. L., and H. B. Tukey, Jr. 1964. Leaching of nutrients from cuttings under mist. *Proc. Inter. Plant Prop. Soc.* 14:138–42.

96. Gorter, C. J. 1969. Auxin-synergists in the rooting of cuttings. *Phys. Plant* 22:497–502.

97. Grace, N. H. 1939. Rooting of cuttings taken from the upper and lower regions of a Norway spruce tree. *Can. Jour. Res.* 17(C):172–80.

98. Grange, R. I., and K. Loach. 1983. Environmental factors affecting water loss from leafy cuttings in different propagation systems. *Jour. Hort. Sci.* 58:1–7.

99. ———. 1983. The water economy of unrooted leafy cuttings. *Jour. Hort. Sci.* 58:9–17.

100. ———. 1984. Comparative rooting of eighty-one species of leafy cuttings in open and poly-ethylene-enclosed mist sytems. *Jour. Hort. Sci.* 59:15–22.

101. ———. 1985. The effect of light on the rooting of leafy cuttings. *Scient. Hort.* 27:105–11.

102. Greenwood, M. S. 1987. Rejuvenation in forest trees. *Plant Growth Regul.* 6:1–12.

103. Greenwood, M. S., and G. P. Berlyn. 1973. Sucrose: Indoleacetic acid interactions on root regeneration by *Pinus lambertiana* embryo cuttings. *Amer. Jour. Bot.* 60:42–47.

104. Greenwood, M. S., and G. P. Berlyn. 1973. Sucrose: Indole-3-acetic acid interactions on root regeneration by *Pinus lamberitana* embryo cuttings. *Amer. Jour. Bot.* 60(1):42–47.

105. Gupta, P. K., and D. J. Durzan. 1987. Micropropagation and phase specificity in mature, elite Douglas fir. *Jour. Amer. Soc. Hort. Sci.* 112:969–71.

106. Hackett, W. P. 1970. The influence of auxin, catechol, and methanolic tissue extracts on root initiation in aseptically cultured shoot apices of the juvenile and adult forms of *Hedera helix*. *Jour. Amer. Soc. Hort. Sci.* 95(4):398–402.

107. ———. 1985. Juvenility, maturation and rejuvenation in woody plants. *Hort. Rev.* 7:109–15.

108. ———. 1989. Donor plant maturation and adventitious root formation. In *Adventitious root formation in cuttings,* T. D. Davis, B. E. Haissig, and N. Sankhla, eds. Portland, Oreg.: Dioscorides Press.

109. Hagemann, A. 1932. Untersuchungen an Blattstecklingen. *Gartenbauwiss* 6:69–202.

110. Haissig, B. E. 1965. Organ formation *in vitro* as applicable to forest tree propagation. *Bot. Rev.* 31:607–26.

111. ———. 1972. Meristematic activity during adventitious root primordium development. Influences of endogenous auxin and applied gibberellic acid. *Plant Phys.* 49:886–92.

112. ———. 1973. Influence of hormones and auxin synergists on adventitious root initiation. In *Proc. I.U.F.R.O. Working Party on Reprod. Processes,* Rotorua, New Zealand.

113. ———. 1974. Influence of auxins and auxin synergists on adventitious root primordium initiation and development. *N. Zealand Jour. For. Sci.* 4:299–310.

114. ———. 1979. Influence of aryl esters of indole-3-acetic and indole-3-butyric acids on adventitious root primordium initiation and development. *Phys. Plant.* 47:29–33.

115. ———. 1983. N-phenyl indolyl-3-butyramide and phenyl indole-3-thiolobutyrate enhance ad-

ventitious root primordium development. *Phys. Plant.* 57:435–40.

116. ———. 1984. Carbohydrate accumulation and partitioning in *Pinus banksiana* seedlings and seedling cuttings. *Phys. Plant.* 61:13–19.

117. ———. 1986. Metabolic processes in adventitious rooting of cuttings. In *New root formation in plants and cuttings,* M. B. Jackson, ed. Dordrecht: Martinus Nijhoff Publishers.

118. Hambrick, C. E., F. T. Davies, Jr., and H. B. Pemberton. 1985. Effect of cutting position and carbohydrate/nitrogen ratio on seasonal rooting of *Rosa multiflora*. *HortScience* 20:570.

119. Hansen, J. 1976. Adventitious root formation induced by gibberellic acid and regulated by irradiance to the stock plants. *Phys. Plant.* 36:77–81.

120. ———. 1987. Stock plant lighting and adventitious root formation. *HortScience* 22:746–49.

121. ———. 1989. Influence of gibberellins on adventitious root formation. In *Adventitious root formation in cuttings,* T. D. Davis, B. E. Haissig, and N. Sankhla, eds. Portland, Oreg.: Dioscorides Press.

122. Hansen, C. J., and H. T. Hartmann. 1968. The use of indolebutyric acid and captan in the propagation of clonal peach and peach-almond hybrid rootstocks by hardwood cuttings. *Proc. Amer. Soc. Hort. Sci.* 92:135–40.

123. Hansen, J., L. H. Strömquist, and A. Ericsson. 1978. Influence of the irradiance on carbohydrate content and rooting of cuttings of pine seedlings (*Pinus sylvestris* L.). *Plant Phys.* 61:975–79.

124. Hare, R. C. 1977. Rooting of cuttings from mature water oak (*Quercus nigra*). *Southern Jour. Appl. For.* 1(2):24–25.

125 ———. 1981. Improved rooting powder for chrysanthemums. *HortScience* 16:90–91.

126. Harrison-Murray, R. S., B. H. Howard, and R. Thompson. 1988. Potential for improved propagation of cuttings through the use of fog. *Acta Hort.* 227:205–10.

127. Harrison-Murray, R. S., and R. Thompson. 1988. In pursuit of a minimum stress environment for rooting leafy cuttings: Comparison of mist and fog. *Acta Hort.* 227:211–16.

128. Hartmann, H. T., and R. M. Brooks. 1958. Propagation of Stockton Morello cherry rootstock by softwood cuttings under mist sprays. *Proc. Amer. Soc. Hort. Sci.* 71:127–34.

129. Hartmann, H. T., and C. J. Hansen. 1958. Ef-

fect of season of collecting, indolebutyric acid, and pre-planting storage treatments on rooting of Marianna plum, peach, and quince hardwood cuttings. *Proc. Amer. Soc. Hort. Sci.* 71:57–66.

130. ———. 1958. Rooting pear and plum rootstocks. *Calif. Agr.* 12(10):4, 14, 15.

131. Hartmann, H. T., W. H. Griggs, and C. J. Hansen. 1963. Propagation of own-rooted Old Home and Barlett pears to produce trees resistant to pear decline. *Proc. Amer. Soc. Hort. Sci.* 82:92–102.

132. Hartmann, H. T., and F. Loreti. 1965. Seasonal variation in the rooting of olive cuttings. *Proc. Amer. Soc. Hort. Sci.* 87:194–98.

133. Hatcher, E. S. J., and R. J. Garner. 1957. Aspects of rootstock propagation. IV. The winter storage of hardwood cuttings. *Ann Rpt. E. Malling Res. Sta. for 1956*, pp. 101–8.

134. Haun, J. R., and P. W. Cornell. 1951. Rooting response of geranium (*Pelargonium hortorum*, Bailey, var. Ricard) cuttings as influenced by nitrogen, phosphorus, and potassium nutrition of the stock plant. *Proc. Amer. Soc. Hort. Sci.* 58:317–23.

135. Hemberg, T. 1951. Rooting experiments with hypocotyls of *Phaseolus vulgaris* L. *Phys. Plant.* 11:1–9.

136. Heide, O. M. 1965. Photoperiodic effects on the regeneration ability of *Begonia* leaf cuttings. *Phys. Plant.* 18:185–90.

137. ———. 1965. Interaction of temperature, auxin, and kinins in the regeneration ability of *Begonia* leaf cuttings. *Phys. Plant.* 18:891–920.

138. ———. 1965. Effects of 6-benzylamino-purine and 1-naphthaleneacetic acid on the epiphyllous bud formation in *Bryophyllum*. *Planta* 67:281–96.

139. ———. 1968. Stimulation of adventitious bud formation in *Begonia* leaves by abscisic acid. *Nature* 219(5157):960–61.

140. ———. 1968. Auxin level and regeneration of *Begonia* leaves. *Planta* 81:153–59.

141. ———. 1969. Non-reversibility of gibberellin-induced inhibition of regeneration in *Begonia* leaves. *Phys. Plant.* 22:671–79.

142. Herman, D. E., and C. E. Hess. 1963. The effect of etiolation upon the rooting of cuttings. *Proc. Inter. Plant Prop. Soc.* 13:42–62.

143. Hess, C. E. 1962. Characterization of the rooting co-factors extracted from *Hedera helix* L. and *Hibiscus rosa-sinensis* L. *Proc. 16th Inter. Hort. Cong.*, pp. 382–88.

144. ———. 1968. Internal and external factors regulating root initiation. In *Root growth: Proc. 15th Easter School in Agricultural Science, University of Nottingham.* London: Butterworth.

145. Heuser, C. W. 1976. Juvenility and rooting cofactors. *Acta Hort.* 56:251-61.

146. Heuser, C. W., and C. E. Hess. 1972. Isolation of three lipid root-initiating substances from juvenile *Hedera helix* shoot tissue. *Jour. Amer. Soc. Hort. Sci.* 97(5):571-74.

147. Hiller, Charlotte. 1951. A study of the origin and development of callus and root primordia of *Taxus cuspidata* with reference to the effects of growth regulator. Master's thesis, Cornell Univ., Ithaca, N.Y.

148. Hitchcock, A. E., and P. W. Zimmerman. 1937. A sensitive test for root formation. *Amer. Jour. Bot.* 24:735-36.

149. ———. 1940. Effects obtained with mixtures of root-inducing and other substances. *Contrib. Boyce Thomp. Inst.* 11:143-60.

150. ———. 1942. Root inducing activity of phenoxy compounds in relation to their structure. *Contrib. Boyce. Thomp. Inst.* 12:497-507.

151. Hoitink, H. A. J., and A. F. Schmitthenner. 1970. Disease control in rhododendron cuttings with benomyl or thiabendazole mixtures. *Plant Dis. Rpt.* 54:427-30.

152. Howard, B. H. 1965. Increase during winter in capacity for root regeneration in detached shoots of fruit tree rootstocks. *Nature* 208:912-13.

153. ———. 1972. Depressing effects of virus infection on adventitious root production in apple hardwood cuttings. *Jour. Hort. Sci.* 47:255-58.

154. Howard, B. H., R. S. Harrison-Murray, J. Vesek, and O. P. Jones. 1988. Techniques to enhance rooting potential before cutting collection. *Acta Hort.* 227:176-86.

155. Hsiao, T., E. Aceudo, E. Feres, and D. W. Henderson. 1976. Water stress, growth and osmotic adjustment. *Phil Trans Royal Soc. (London) Ser. B* 273:479-500.

156. Hudson, J. P. 1955. The regeneration of plants from roots. *Proc. 14th Inter. Hort. Cong.* 2:1165-72.

157. Humphries, E. C. 1960. Inhibition of root development of petioles and hypocotyls of dwarf bean (*Phaseolus vulgaris*) by kinetin. *Phys. Plant.* 13:659-63.

158. Ikeda, T., and T. Suzaki. 1985. Influence of hydraulic conductance of xylem on water status in cuttings. *Can. Jour. For. Res.* 16:98-102.

159. Jain, M. K., and K. K. Nanda. 1972. Effect of temperature and some antimetabolites on the interaction effects of auxin and nutrition in rooting etiolated stem segments of *Salix tetrasperma. Phys. Plant.* 27:169-72.

160. James, D. J. 1979. The role of auxins and phloroglucinol in adventitious root formation in *Rubus* and *Fragaria* grown *in vitro. Jour. Hort. Sci.* 54:273-77.

161. James, D. J., V. H. Knight, and I. J. Thurbon. 1980. Micropropagation of red raspberry and the influence of phloroglucinol. *Scient. Hort.* 12:313-19.

162. James, D. J., and I. J. Thurbon. 1981. Shoot and root initiation *in vitro* in the apple rootstock M9 and the promotive effects of phloroglucinol. *Jour. Hort. Sci.* 56:15-20.

163. Jarvis, B. C. 1986. Endogenous control of adventitious rooting in non-woody species. In *New root formation in plants and cuttings,* M. B. Jackson, ed. Dordrecht: Martinus Nijhoff Publishers.

164. Jarvis, B. C., S. Yasmin, and M. T. Coleman. 1985. RNA and protein metabolism during adventitious root formation in stem cuttings of *Phaseolus aureus. Physiol. Plant.* 64:53-59.

165. Jauhari, O. S., and S. F. Rahman. 1959. Further investigations on rooting in cuttings of sweet lime (*Citrus limetoides*) Tanaka. *Sci. and Cult.* 24:432-34.

166. Johnson, C. R. 1970. The nature of flower bud influence on root regeneration in the *Rhododendron* shoot. Ph.D. Dissertation. Oreg. State Univ., Corvallis, Oreg.

167. Johnson, C. R., and D. F. Hamilton. 1977. Rooting of *Hibiscus rosa-sinensis* L. cuttings as influenced by light intensity and ethephon. *HortScience* 12(1):39-40.

168. ———. 1977. Effects of media and controlled-release fertilizers on rooting and leaf nutrient composition of *Juniperus conferta* and *Ligustrum japonicum* cuttings. *Jour. Amer. Soc. Hort. Sci.* 102:320-22.

169. Johnson, C. R., and A. N. Roberts. 1971. The effect of shading rhododendron stock plants on flowering and rooting. *Jour. Amer. Soc. Hort. Sci.* 96:166-68.

170. Jones, O. P., and M. E. Hopgood. 1979. The successful propagation *in vitro* of two rootstocks of *Prunus:* the plum rootstock Pixy (*P. insititia*) and the cherry rootstock F 12/1 (*P. avium*). *Jour. Hort. Sci.* 54:63-66.

171. Kawase, M. 1964. Centrifugation, rhizocaline, and rooting in *Salix alba. Phys. Plant.* 17:855-65.

172. ———. 1965. Etiolation and rooting in cuttings. *Phys. Plant.* 18:1066–76.

173. ———. 1971. Causes of centrifugal root promotion. *Phys. Plant.* 25:64–70.

174. ———. 1976. Centrifugation and rooting of cuttings. *Revista dell' Ortoflorofrutticoltura Italiana* N. 2.

175. Keever, G. J., G. S. Cobb, and D. R. Mills. 1987. Propagation of four woody ornamentals from vegetative and reproductive stem cuttings. *Ornamentals Res. Rep. 5*, Alabama Agr. Exp. Sta., Auburn University, Montgomery.

176. Keever, G. I., and H. B. Tukey, Jr. 1979. Effect of nutrient mist on the propagation of azaleas. *HortScience* 14(6):755–56.

177. Kefford, N. P. 1973. Effect of a hormone antagonist on the rooting of shoot cuttings. *Plant. Phys.* 51:214–16.

178. Kester, D. E. 1970. Temperature and plant propagation. *Proc. Inter. Plant Prop. Soc.* 20:153–63.

179. Kester, D. E. 1983. The clone in horticulture. *HortScience* 18:831–37.

180. Key, J. L. 1969. Hormones and nucleic acid metabolism. *Ann. Rev. Plant. Phys.* 20:449–74.

181. Kling, G. J., M. M. Meyer, Jr., and D. Seigler. 1988. Rooting co-factors in five *Acer* species. *Jour. Amer. Soc. Hort. Sci.* 113:252–57.

182. Kögl, F., A. J. Haagen-Smit, and H. Erxleben, 1934. Uber ein neues Auxin ("Heteroauxin") aus Harn, XI. *Mitteilung. Z. physiol. Chem* 228:90–103.

183. Kraus, E. J. 1953. Rooting azalea cuttings. *Nat. Hort. Mag.* 32:163–64.

184. Kraus, E. J., N. A. Brown, and K. C. Hamner. 1936. Histological reactions of bean plants to indoleacetic acid. *Bot. Gaz.* 98:370–420.

185. Kraus, E. J., and H. R. Kraybill. 1918. Vegetation and reproduction with special reference to the tomato. *Oreg. Agr. Exp. Sta. Bul. 149.*

186. Krishnamoorthy, H. N. 1970. Promotion of rooting in mung bean hypocotyl cuttings with Ethrel, an ethylene releasing compound. *Plant & Cell Phys.* 11:979–82.

187. Krul, W. R. 1968. Increased root initiation in Pinto bean hypocotyls with 2,4-dinitrophenol. *Plant Phys.* 43(3):439–41.

188. Laibach, F., and O. Fischnich. 1935. Künstliche Wurzelneubildung mittels Wuchsstoffpaste, *Ber. Deuts. Bot. Ges.* 53:528–39.

189. Langhans, R. W. 1955. Mist for growing plants. *Farm Res.* (Cornell Univ.) 21(3).

190. Lanphear, F. O., and R. P. Meahl. 1961. The effect of various photoperiods on rooting and subsequent growth of selected woody ornamental plants. *Proc. Amer. Soc. Hort. Sci.* 77:620–34.

191. ———. 1966. Influence of the stock plant environment on the rooting of *Juniperus horizontalis* 'Plumosa'. *Proc. Amer. Soc. Hort. Sci.* 89:666–71.

192. Lek, H. A. A., van der. 1925. Root development in woody cuttings. *Meded. Landbouwhoogesch. Wageningen* 38(1).

193. Leopold, A. C. 1964. The polarity of auxin transport in *Meristems and Differentiation*, Brookhaven Symposia in Biology, Rpt. 16. Upton, N.Y.: Brookhaven Natl. Lab., pp. 218–34.

194. Lesham, Y., and B. Lunenfield. 1968. Gonadotropin in promotion of adventitious root production on cuttings of *Begonia semperflorens* and *Vitis vinifera. Plant Phys.* 43:313–17.

195. Libbert, E. 1956. Untersuchungen über die Physiologie der Adventivewurzelbildung, I. Die Wirkungsweise einiger Komponenten des 'Rhizokalinkomplexes.' *Flora* 144:121–50.

196. Libby, W. J., A. G. Brown, and J. M. Fielding. 1972. Effect of hedging radiata pine on production, rooting, and early growth of cuttings. *New Zealand Jour. For. Sci.* 2(2):263–83.

197. Loach, K. 1977. Leaf water potential and the rooting of cuttings under mist and polythene. *Phys. Plant.* 40:191–97.

198. ———. 1979. Mist propagation: Past, present, future. *Proc. Inter. Plant Prop. Soc.* 29:216–29.

199. ———. 1983. Propagation systems in New Zealand: A means of comparing their effectiveness. *Proc. Inter. Plant Prop. Soc.* 33:291–94.

200. ———. 1985. Rooting of cuttings in relation to the propagation medium. *Proc. Inter. Plant Prop. Soc.* 35:472–85.

201. ———. 1987. Mist and fruitfulness. *Horticulture Week,* April 10, 1987, pp. 28–29.

202. ———. 1989. Controlling environmental conditions to improve adventitious rooting. In *Adventitious root formation in cuttings,* T. D. Davis, B. E. Haissig, and N. Sankhla, eds. Portland, Oreg.: Dioscorides Press.

203. ———. 1989. Water relations and adventitious rooting. In *Adventitious root formation in cuttings,* T. D. Davis, B. E. Haissig, and N. Sankhla, eds. Portland, Oreg.: Dioscorides Press.

204. Long, J. C. 1933. The influence of rooting media on the character of roots produced by cuttings. *Proc. Amer. Soc. Hort. Sci.* 29:352–55.

205. Long, W. G., D. V. Sweet, and H. B. Tukey. 1956. The loss of nutrients from plant foliage

by leaching as indicated by radioisotopes. *Science* 123:1039–40.

206. Lovell, P. H., and J. White. 1986. Anatomical changes during adventitious root formation. In *New root formation in plants and cuttings,* M. B. Jackson, ed. Dordrecht: Martinus Nijhoff Publishers.

207. MacKenzie, K. A. D., B. H. Howard, and R. S. Harrison-Murray. 1986. The anatomical relationship between cambial regeneration and root initiation in wounded winter cuttings of the apple rootstock M.26. *Annal. Bot.* 58:649–61.

208. Maini, J. S. 1968. The relationship between the origin of adventitious buds and the orientation of *Populus tremuloides* root cuttings. *Bul. Ecol. Soc. Amer.* 49:81–82.

209. Maldiney, R., F. Pelese, G. Pilate, L. Sossountzov, and E. Miginiac. 1986. Endogenous levels of abscisic acid, indole-3-acetic acid, zeatin and zeatin-riboside during the course of adventitious root formation in cuttings of *Craigella and Craigella* lateral suppressor tomatoes. *Physiol. Plant.* 68:426–30.

210. Maronek, D. M., D. Studebaker, and B. Oberly. 1985. Improving media aeration in liner and container production. *Proc. Inter. Plant Prop. Soc.* 35:591–97.

211. Massini, P., and G. Voorn. 1967. The effect of ferrodoxin and ferrous ion on the chlorophyll sensitized photoreduction of dinitrophenol. *Photochem. Photobiol.* 6:851–56.

212. Maynard, B. K., and N. L. Bassuk. 1987. Stock plant etiolation and blanching of woody plants prior to cutting propagation. *Jour. Amer. Soc. Hort. Sci.* 112:273–76.

213. ———. 1989. Etiolation and banding effects on adventitious root formation. In *Adventitious root formation in cuttings,* T. D. Davis, B. E. Haissig, and N. Sankhla, eds. Portland, Oreg.: Dioscorides Press.

214. McVeigh, I. 1938. Regeneration in *Crassula multicava. Amer. Jour. Bot.* 25:7–11.

215. Moe, R., and A. S. Andersen. 1989. Stock plant environment and subsequent adventitious rooting. In *Adventitious root formation in cuttings,* T. D. Davis, B. E. Haissig, and N. Sankhla, eds. Portland, Oreg.: Dioscorides Press.

216. Mohammed, S. 1975. Further investigations on the effects of decapitation and disbudding at different development stages of rooting in pea cuttings. *Jour. Hort. Sci.* 50:271–73.

217. Mohammed, S., and E. N. Eriksen. 1974. Root formation in pea cuttings. IV. Further studies on the influence of indole-3-acetic acid at different development stages. *Phys. Plant.* 32:94–96.

218. Molitor, H. D., and W. U. von Hentig. 1987. Effect of carbon dioxide enrichment during stock plant cultivation. *HortScience* 22:741–46.

219. Molnar, J. M., and L. J. LaCroix. 1972. Studies of the rooting of cuttings of *Hydrangea macrophylla:* Enzyme changes. *Can. Jour. Bot.* 50(2):315–22.

220. ———. 1972. Studies of the rooting of cuttings of *Hydrangea macrophylla:* DNA and protein changes. *Can. Jour. Bot.* 50(3):387–92.

221. Morgan, D. L., and P. F. Colbaugh. 1983. Influence of chemical sanitation treatments on propagation of *Buxus microphylla* and *Peperomia caperata. Proc. Inter. Plant Prop. Soc.* 33:600–7.

222. Morgan, D. L., E. L. McWilliams, and W. C. Parr. 1980. Maintaining juvenility in live oak. *HortScience* 15(4):493–94.

223. Mudge, K. W. 1989. Effect of ethylene on rooting. In *Adventitious root formation in cuttings,* T. D. Davis, B. E. Haissig, and N. Sankhla, eds. Portland, Oreg.: Dioscorides Press.

224. Muzik, T. J., and H. J. Cruzado. 1958. Transmission of juvenile rooting ability from seedlings to adults of *Hevea brasiliensis. Nature* 181:1288.

225. Naylor, E. E., and B. Johnson. 1937. A histological study of vegetative reproduction in *Saintpaulia ionantha. Amer. Jour. Bot.* 24:673–78.

226. Norcini, J. G., and C. W. Heuser. 1988. Changes in the level of [^{14}C] indole-3 acetic acid and [^{14}C] indoleacetylaspartic acid during root formation in mung bean cuttings. *Plant. Phys.* 86:1236–39.

227. Okoro, O. O., and J. Grace. 1978. The physiology of rooting *Populus* cuttings. II. Cytokinin activity in leafless hardwood cuttings. *Phys. Plant.* 44:167–70.

228. O'Rourke, F. L. 1940. The influence of blossom buds on rooting of hardwood cuttings of blueberry. *Proc. Amer. Soc. Hort. Sci.* 40:332–34.

229. ———. 1944. Wood type and original position on shoot with reference to rooting in hardwood cuttings of blueberry. *Proc. Amer. Soc. Hort. Sci.* 45:195–97.

230. Patena, L., E. G. Sutter, and A. M. Dandekar. 1988. Root induction by *Agrobacterium rhizogenes* in a difficult-to-root woody species. *Acta Hort.* 227:324–29.

231. Paton, F., and W. W. Schwabe. 1987. Storage of cuttings of *Pelargonium* × *hortorum* Bailey. *Jour. Hort. Sci.* 62:79–87.

232.   Paton, D. M., R. R. Willing, W. Nichols, and L. D. Pryor. 1970. Rooting of stem cuttings of eucalyptus: A rooting inhibitor in adult tissue. *Austral. Jour. Bot.* 18:175–83.

233.   Pearse, H. L. 1946. Rooting of vine and plum cuttings as affected by nutrition of the parent plant and treatment with phytohormones. *Sci. Bul. 249, Dept of Agr. Union of S. Afr.*

234.   Petri, P. S., S. Mazzi, and P. Strigoli. 1960. Considerazióne sulla formazióne delle radici aventìzie con particolare riguardo a: *Cucurbita pepo, Nerium oleander, Menyanthes trifoliatae, Solanum lycopersicum. Nuovo Giorn. Bot. Ital.* 67:131–75.

235.   Pierik, R. L. M., and H. H. M. Steegmans. 1975. Analysis of adventitious root formation in isolated stem explants of *Rhododendron. Scient. Hort.* 3:1–20.

236.   Pliego-Alfaro, F., and T. Murashige. 1987. Possible rejuvenation of adult avocado by graftage onto juvenile rootstocks in vitro. *HortScience* 22:1321–24.

237.   Porlingis, I. C., and I. Therios. 1976. Rooting response of juvenile and adult leafy olive cuttings to various factors. *Jour. Hort. Sci.* 51:31–39.

238.   Poulsen, A., and A. S. Anderson. 1980. Propagation of *Hedera helix:* Influence of irradiance to stock plants, length of internode and topophysis of cutting. *Phys. Plant.* 49:359–65.

239.   Priestley, J. H., and C. F. Swingle. 1929. Vegetative propagation from the standpoint of plant anatomy. *USDA Tech. Bul. 151.*

240.   Proebsting, W. M. 1984. Rooting of Douglas-fir stem cuttings: Relative activity of IBA and NAA. *HortScience* 19:854–56.

241.   Rajagopal, V., and A. S. Andersen. 1980. Water stress and root formation in pea cuttings. *Phys. Plant.* 48:144–49.

242.   Rasmussen, S., and A. S. Andersen. 1980. Water stress and root formation in pea cuttings. II. Effect of abscisic acid treatment of cuttings from stock plants grown under two levels of irradiance. *Phys. Plant.* 48:150–54.

243.   Read, P. E., and V. C. Hoysler. 1969. Stimulation and retardation of adventitious root formation by application of B-Nine and Cycocel. *Jour. Amer. Soc. Hort. Sci.* 94:314–16.

244.   Reuveni, O., and M. Raviv. 1981. Importance of leaf retention to rooting avocado cuttings. *Jour. Amer. Soc. Hort. Sci.* 106(2):127–30.

245.   Robbins, W. J. 1960. Further observations on juvenile and adult *Hedera. Amer. Jour. Bot.* 47:485–91.

246.   Robbins, J. A., M. J. Campidonica, and D. W. Burger. 1988. Chemical and biological stability of indole-3-butyric acid (IBA) after long-term storage at selected temperatures and light regimes. *Jour. Environ. Hort.* 6(2):33–38.

247.   Roberts, A. N., and L. H. Fuchigami. 1973. Seasonal changes in auxin effect on rooting of Douglas-fir stem cuttings as related to bud activity. *Phys. Plant.* 28:215–21.

248.   Robinson, J. C. 1975. The regeneration of plants from root cuttings with special reference to the apple. *Hort. Abst.* 45(6):305–15.

249.   Robinson, J. C., and W. W. Schwabe. 1977. Studies on the regeneration of apple cultivars from root cuttings. I. Propagation aspects. *Jour. Hort. Sci.* 52:205–20.

250.   ———. 1977. Studies on the regeneration of apple cultivars from root cuttings. II. Carbohydrate and auxin relations. *Jour. Hort. Sci.* 52:221–33.

251.   Rom, R. C., and S. A. Brown. 1979. Factors affecting burrknot formation on clonal *Malus* rootstocks. *HortScience* 14(3):231–32.

252.   Ryan, G. F., E. F. Frolich, and T. P. Kinsella. 1958. Some factors influencing rooting of grafted cuttings. *Proc. Amer. Soc. Hort. Sci.* 72:454–61.

253.   Sachs, J. 1880 and 1882. Stoff und Form der Pflanzenorgane. I and II. *Arb. bot. Inst. Würzburg* 2:452–88; 4:689–718.

254.   Sachs, R. M., F. Loreti, and J. DeBie. 1964. Plant rooting studies indicate sclerenchyma tissue is not a restricting factor. *Calif. Agr.* 18(9):4–5.

255.   Schier, G. A. 1973. Origin and development of aspen root suckers. *Can. Jour. For. Res.* 3:39–44.

256.   Schraudolf, H., and J. Reinert. 1959. Interaction of plant growth regulators in regeneration processes. *Nature* 184:465–66.

257.   Shapiro, S. 1958. The role of light in the growth of root primordia in the stem of Lombardy poplar. In *The physiology of forest trees,* K. V. Thimann, ed. New York: Ronald Press.

258.   Siegler, E. A., and J. J. Bowman. 1939. Anatomical studies of root and shoot primordia in 1-year apple roots. *Jour. Agr. Res.* 58:795–803.

259.   Siniscalco, C., and L. Pavolettoni. 1988. Rejuvenation of *Eucalyptus* × *trabutii* by successive grafting. *Acta Hort.* 227:98–100.

260.   Sircar, P. K., and S. K. Chatterjee. 1973. Physiological and biochemical control of meri-

stemation and adventitious root formation in *Vigna* hypocotyl cuttings. *The Plant Propagator* 19(1):1.

261. ———. 1974. Physiological and biochemical changes associated with adventitious root formation in *Vigna* hypocotyl cuttings. II. Gibberellin effects. *The Plant Propagator* 20(2):15–22.

262. ———. 1980. Effect of foliar applications of kinetin and Ethrel on adventitious root formation at the base of *Vigna* hypocotyl cuttings. *The Plant Propagator* 26(4):3–5.

263. Skoog, F., and C. Tsui. 1948. Chemical control of growth and bud formation in tobacco stem and callus. *Amer. Jour. Bot.* 35:782–87.

264. Sluis, C. J. 1984. Plant genetic engineering and biotechnologies update. *Proc. Inter. Plant Prop. Soc.* 34:189–95.

265. Smally, T. J., and M. A. Dirr. 1988. Effect of night interruption photoperiod treatment on subsequent growth of *Acer rubrum* cuttings. *HortScience* 23:172–74.

266. Smith, D. R., and T. A. Thorpe. 1975. Root initiation in cuttings of *Pinus radiata* seedlings. I. Developmental sequence. *Jour. Exp. Bot.* 26:184–92.

267. ———. 1975. Root initiation in cuttings of *Pinus radiata* seedlings. II. Growth regulator interactions. *Jour. Exp. Bot.* 26:193–202.

268. Smith, P. M. 1982. Diseases during propagation of woody ornamentals. *Proc. 21st Int. Hort. Cong.* 2:884–93.

269. Smith, N. G., and P. F. Wareing. 1972. The distribution of latent root primordia in stems of *Populus* × *robusta* and factors affecting emergence of preformed roots from cuttings. *Forestry* 45:197–210.

270. Snow, A. G., Jr. 1939. Clonal variation in rooting response of red maple cuttings. *USDA Northeastern For. Exp. Sta. Tech. Note 29.*

271. Snyder, W. E. 1955. Effect of photoperiod on cuttings of *Taxus cuspidata* while in the propagation bench and during the first growing season. *Proc. Amer. Soc. Hort. Sci.* 66:397–402.

272. Spano, L., D. Mariotti, M. Cardarelli, C. Branra, and P. Costantino. 1988. Morphogenesis and auxin sensitivity of transgenic tobacco with different complements of Ri T-DNA. *Plant. Phys.* 87:479–83.

273. Spiegel, P. 1954. Auxins and inhibitors in canes of *Vitis. Bul. Res. Coun., Israel* 4:176–83.

274. Stangler, B. B. 1949. An anatomical study of the origin and development of adventitious roots

in stem cuttings of *Chrysanthemum morifolium* Bailey, *Dianthus caryophyllus* L., and *Rosa dilecta* Rehd. Ph.D. dissertation, Cornell Univ., Ithaca, N.Y.

275. Steponkus, P. L., and L. Hogan. 1967. Some effects of photoperiod on the rooting of *Abelia grandiflora* Rehd. 'Prostrata' cuttings. *Proc. Amer. Soc. Hort. Sci.* 91:706–15.

276. Stoltz, L. P., and C. E. Hess. 1966. The effect of girdling upon root initiation: Auxin and rooting co-factors. *Proc. Amer. Soc. Hort. Sci.* 89:744–51.

277. ———. 1966. The effect of girdling upon root initiation. Carbohydrates and amino acids. *Proc. Amer. Soc. Hort. Sci.* 89:734–43.

278. Stoutemyer, V. T. 1937. Regeneration in various types of apple wood. *Iowa Agr. Exp. Sta. Res. Bul.* 220:309–52.

279. ———. 1942. The propagation of *Chionanthus retusus* by cuttings. *Nat. Hort. Mag.* 21(4):175–78.

280. Stoutemyer, V. T., O. K. Britt, and J. R. Goodin. 1961. The influence of chemical treatments, understocks, and environment on growth phase changes and propagation of *Hedera canariensis. Proc. Amer. Soc. Hort. Sci.* 77:552–57.

281. Stoutemyer, V. T., and A. W. Close. 1946. Rooting cuttings and germinating seeds under fluorescent and cold cathode light. *Proc. Amer. Soc. Hort. Sci.* 48:309–25.

282. ———. 1947. Changes of rooting response in cuttings following exposure of the stock plants to light of different qualities. *Proc. Amer. Soc. Hort. Sci.* 49:392–94.

283. Strobel, G. A., and A. Nachmias. 1985. *Agrobacterium rhizogenes* promotes the initial growth of bare rootstock almond. *Jour. Gen. Microbiol.* 131:1245–49.

284. ———. 1989. *Agrobacterium rhizogenes:* A root inducing bacterium. In *Adventitious root formation in cuttings,* T. D. Davis, B. E. Haissig, and N. Sankhla, eds. Portland, Oreg.: Dioscorides Press.

285. Strömquist, L., and J. Hansen. 1980. Effects of auxin and irradiance on the rooting of cuttings of *Pinus sylvestris. Phys. Plant.* 49:346–50.

286. Struve, D. K. 1981. The relationship between carbohydrates, nitrogen and rooting of stem cuttings. *The Plant Propagator* 27:6–7.

287. Struve, D. K., and M. A. Arnold. 1986. Aryl esters of IBA increase rooted cutting quality of red maple 'Red Sunset' softwood cuttings. *HortScience* 21:1392–93.

288. Strydom, D. K., and H. T. Hartmann. 1960. Absorption, distribution, and destruction of indoleacetic acid in plum stem cuttings. *Plant. Phys.* 35:435–42.

289. ———. 1960. Effect of indolebutyric acid on respiration and nitrogen metabolism in Marianna 2624 plum softwood stem cuttings. *Proc. Amer. Soc. Hort. Sci.* 76:124–33.

290. Thielges, B. A., and H. A. J. Hoitink. 1972. Fungicides and rooting of eastern white pine cuttings. *For. Sci.* 18(1):54–55.

291. Thimann, K. V. 1935. On an analysis of activity of two growth-promoting substances on plant tissues. *Proc. Kon. Ned. Akad. Wet.* 38:896–912.

292. ———. 1935. On the plant growth hormone produced by *Rhizopus suinus. Jour. Bio. Chem.* 109:279–91.

293. Thimann, K. V., and A. L. Delisle. 1939. The vegetative propagation of difficult plants. *Jour. Arnold Arb.* 20:116–36.

294. Thimann, K. V., and J. B. Koepfli. 1935. Identity of the growth-promoting and root-forming substances of plants. *Nature* 135:101–2.

295. Thimann, K. V., and E. F. Poutasse. 1941. Factors affecting root formation of *Phaseolus vulgaris. Plant. Phys.* 16:585–98.

296. Thimann, K. V., and F. W. Went. 1934. On the chemical nature of the root-forming hormone. *Proc. Kon. Ned. Akad. Wet.* 37:456–59.

297. Thomaszewski, M., and K. V. Thimann. 1966. Interactions of phenolic acids, metallic ions, and chelating agents on auxin-induced growth. *Plant. Phys.* 41:1443–54.

298. Tillburg. E. 1974. Levels of indole-3-acetic acid and acid inhibitors in green and etiolated bean seedlings (*Phaseolus vulgaris*). *Phys. Plant.* 31:106–11.

299. Tilt, K. M., and T. E. Bilderback. 1987. Physical properties of propagation media and their effects on rooting of three woody ornamentals. *HortScience* 22:245–47.

300. Tripepi, R. R., C. W. Heuser, and J. C. Shannon. 1983. Incorporation of tritiated thymidine and uridine into adventitious root initial cells of *Vigna radiata. Jour. Amer. Soc. Hort. Sci.* 108:469–74.

301. Tukey, H. B., and E. L. Green. 1934. Gradient composition of rose shoots from tip to base. *Plant. Phys.* 9:157–63.

302. Tukey, H. B., Jr., H. B. Tukey, and S. H. Wittwer. 1958. Loss of nutrients by foliar leaching as determined by radioisotopes. *Proc. Amer. Soc. Hort. Sci.* 71:496–506.

303. Vander der Meer, F. A. 1965. Nerfvergelingsmozaick bij kruisbessen. *Fruitteelt* 55:245–46.

304. Van Doesburg, J. 1962. Use of fungicides with vegetative propagation. *Proc. 16th Inter. Hort. Cong.* 4:365–72.

305. Van Overbeek, J., and L. E. Gregory. 1945. A physiological separation of two factors necessary for the formation of roots on cuttings. *Amer. Jour. Bot.* 32:336–41.

306. Van Overbeek, J., S. A. Gordon, and L. E. Gregory. 1946. An analysis of the function of the leaf in the process of root formation in cuttings. *Amer. Jour. Bot.* 33:100–107.

307. Van Staden, J., and A. R. Harty. 1989. Cytokinins and adventitious root formation. In *Adventitious root formation in cuttings,* T. D. Davis, B. E. Haissig, and N. Sankhla, eds. Portland, Oreg.: Dioscorides Press.

308. Vasilevskaya, V. K. 1957. The anatomy of bud formation on the roots of some woody plants. *Russian Vest. Leningr. Univ., Ser. Bio. Bul.* 1:3–21.

309. Veierskov, B. 1989. Relations between carbohydrates and adventitious root formation. In *Adventitious root formation in cuttings,* T. D. Davis, B. E. Haissig, and N. Sankhla, eds. Portland, Oreg.: Dioscorides Press.

310. Venverloo, G. J. 1976. The formation of adventitious organs. III. A comparison of root and shoot formation on *Nautilocalyx* explants. *Z. Pflanzenphysiol.* 80:310–22.

311. Vieitez, J., D. G. I. Kingston, A. Ballester, and E. Vieitez. 1987. Identification of two compounds correlated with lack of rooting capacity of chestnut cuttings. *Tree Phys.* 3:247–55.

312. Vöchting, H. 1878. *Uber Organbildung im Pflanzenreich.* Bonn: Verlag Max Cohen, pp. 1–258.

313. Walker, R. I. 1940. Regeneration in the scale leaf of *Lilium candidum* and *L. longiflorum. Amer. Jour. Bot.* 27:114–17.

314. Wang, Y. T. 1987. Effect of temperature, duration and light during simulated shipping on quality and rooting of croton cuttings. *HortScience* 22:1301–2.

315. Ward, J. D., and C. E. Whitcomb. 1979. Nutrition of Japanese holly during propagation and production. *Jour. Amer. Soc. Hort. Sci.* 104:523–26.

316. Warmke, H. E., and G. L. Warmke. 1950. The role of auxin in the differentiation of root and

shoot primordia from root cuttings of *Taraxacum* and *Cichorium. Amer. Jour. Bot.* 37:272–80.

317. Waxman, S., and J. P. Nitsch. 1956. Influence of light on plant growth. *Amer. Nurs.* 104(10): 11–12.

318. Weiser, C. J. 1963. Rooting and night-lighting trials with deciduous azaleas and dwarf rhododendrons. *Amer. Hort. Mag.* 42:95–100.

319. Weiser, C. J., and L. T. Blaney. 1960. The effects of boron on the rooting of English holly cuttings. *Proc. Amer. Soc. Hort. Sci.* 75:704–10.

320. ———. 1967. The nature of boron stimulation to root initiation and development in beans. *Proc. Amer. Soc. Hort. Sci.* 90:191–99.

321. Welander, M., and I. Huntrieser. 1981. The rooting ability of shoots raised *in vitro* from the apple rootstock A2 in juvenile and in adult growth phase. *Phys. Plant.* 53(3):301–6.

322. Wellensiek, S. J. 1952. Rejuvenation of woody plants by formation of sphaeroblasts. *Proc. Kon. Ned. Akad. Wet.* 55:567–73.

323. Wells, J. S. 1963. The use of captan in rooting rhododendrons. *Proc. Inter. Plant Prop. Soc.* 13: 132–35.

324. Went, F. W. 1934. A test method for rhizocaline, the root-forming substance. *Proc. Kon. Ned. Akad. Wet.* 37:445–55.

325. White, J., and P. H. Lovell. The anatomy of root initiation in cuttings of *Griselinia littoralis* and *Griselinia lucida. Ann Bot.* 54:7–20.

326. ———. 1984. Anatomical changes which occur in cuttings of *Agathis australis* (D. Don) Lindl. 1. Wounding responses. *Ann. Bot.* 54:621–32.

327. ———. 1984. Anatomical changes which occur in cuttings of *Agathis australis* (D. Don) Lindl. 2. The initiation of root primordia and early root development. *Ann Bot.* 54:633–45.

328. Whitehill, S. J., and W. W. Schwabe. 1975.

Vegetative propagation of *Pinus sylvestris. Phys. Plant.* 35:66–71.

329. Wildern, J. A., and R. A. Criley, 1975. Cytokinins increase shoot production from leaf cuttings of begonia. *The Plant Propagator* 21(1):7–9.

330. Wilkinson, R. E. 1966. Adventitious shoots on saltcedar roots. *Bot. Gaz.* 127:103–4.

331. Wott, J. A., and H. B. Tukey, Jr. 1967. Influence of nutrient mist on the propagation of cuttings. *Proc. Amer. Soc. Hort. Sci.* 90:454–61.

332. Yarborough, J. A. 1932. Anatomical and developmental studies of the foliar embryos of *Bryophyllum calycinum. Amer. Jour. Bot.* 19:443–53.

333. ———. 1936. Regeneration in the foliage leaf of *Sedum. Amer. Jour. Bot.* 23:303–7.

334. Yarwood, C. E. 1939. Control of powdery mildews with a water spray. *Phytopath.* 29:288–90.

335. Zimmerman, P. W. 1933. Initiation and stimulation of adventitious roots caused by unsaturated hydrocarbon gases. *Contrib. Boyce Thomp. Inst.* 5:351–69.

336. ———. 1937. Comparative effectiveness of acids, esters, and salts as growth substances and methods of evaluating them. *Contrib. Boyce Thomp. Inst.* 8:337–50.

337. Zimmerman, R. H., and O. C. Broome. 1981. Phloroglucinol and in vitro rooting of apple cultivar cuttings. *Jour. Amer. Soc. Hort. Sci.* 106(5): 648–52.

338. Zimmerman, P. W., and F. Wilcoxon. 1935. Several chemical growth substances which cause initiation of roots and other responses in plants. *Contrib. Boyce Thomp. Inst.* 7:209–29.

339. Zucconi, F., and A. Pera. 1978. The influence of nutrients and pH effects on rooting as shown by mung bean cuttings. *Acta Hort.* 79:57–62.

SUPPLEMENTARY READING

BROUWER, R., O. GASPARIKOVA, J. KOLEK, and B. C. LOUGHMAN, eds. 1981. *Structure and function of plant roots.* Dordrecht: Martinus Nijhoff Publishers.

DAVIS, T. D., B. E. HAISSIG, and N. SANKHLA, eds. 1989. *Adventitious root formation in cuttings.* Portland, Oreg.: Dioscorides Press.

INTERNATIONAL PLANT PROPAGATORS' SOCIETY. Proceedings of annual meetings.

JACKSON, M. B., ed. 1986. *New root formation in plants and cuttings.* Dordrecht: Martinus Nijhoff Publishers.

JACKSON, M. B., and A. D. STEAD. 1983. *Growth regulators in root development.* Monograph 10. Wantage, England: British Plant Growth Regulator Group.

LORETI, F., ed. 1988. Special issue on the International Symposium on Vegetative Propagation of

Woody Species. Pisa, Italy. (Sept. 3–5, 1987). *Acta Hort.* 227:1–516.

Special issue on vegetative propagation of conifers and hardwoods. 1974. *N. Zealand Jour. For. Sci.* 4(2).

SUTTON, R. F., and R. W. TINUS. 1983. *Root and root system terminology.* Forest Science Monograph. Washington, D.C.: Soc. Amer. Foresters.

TORREY, J. G., and D. T. CLARKSON, eds. 1975. *The development and function of roots.* London: Academic Press.

10

Techniques of Propagation by Cuttings

In cutting propagation, a portion of stem, root, or leaf is cut from the parent or stock plant and induced to form roots and shoots by chemical, mechanical, and/or environmental manipulation. In most cases the new independent plant produced is a clone, which is identical with the parent plant.

THE IMPORTANCE AND ADVANTAGES OF PROPAGATION BY CUTTINGS

Cuttings are the most important means of propagating ornamental shrubs—deciduous species as well as the broad- and narrow-leaved types of evergreens. Cuttings are also used widely in commercial greenhouse propagation of many florists' crops, foliage crops, and in propagating certain fruit species.

For species that can be propagated easily by cuttings, this method has numerous advantages.

Many new plants can be started in a limited space from a few stock plants. It is inexpensive, rapid, and simple, and does not require the special techniques necessary in grafting, budding, or micropropagation. There is no problem of incompatibility with rootstocks or of poor graft unions. Greater uniformity is obtained by absence of the variation which sometimes appears as a result of the variable seedling rootstocks of grafted plants. The parent plant is usually reproduced exactly, with no genetic change.

It is not always desirable, however, to produce plants on their own roots by cuttings even if it is possible to do so. It is often advantageous or necessary to use a rootstock resistant to some adverse soil condition or soil-borne organism, or to utilize available dwarfing or invigorating rootstocks. Cutting propagation, like any other asexual technique, may potentially increase disease and insect susceptibility, since the clonal plants lack the genetic diversity of most seedling-produced plants.

TYPES OF CUTTINGS

Cuttings are made from the vegetative portions of the plant, such as **stems, modified stems** (rhizomes, tubers, corms, and bulbs), **leaves,** or **roots**. Cuttings can be classified according to the part of the plant from which they are obtained:

> Stem cuttings
> > Hardwood
> > > Deciduous
> > > Narrow-leaved evergreen
> > Semihardwood
> > Softwood
> > Herbaceous
> Leaf cuttings
> Leaf-bud cuttings (single-eye or single-node cuttings)
> Root cuttings

Many plants can be propagated by several different types of cuttings with satisfactory results. The preferred type depends on individual circumstances; the least expensive and easiest method is usually selected.

For easy-to-root woody perennial plants, hardwood stem cuttings in an outdoor nursery are frequently used because of the simplicity and low cost. For more tender herbaceous species, or for those more difficult to propagate, it is necessary to resort to the more expensive and more elaborate facilities required for rooting the leafy types of cuttings. In today's containerized nurseries, a larger portion of easy and difficult-to-root species are propagated with mist, fog, or modified polyethylene sheet systems. Root cuttings of some species are also satisfactory, but cutting material may be difficult to obtain in large quantities.

In selecting cutting material it is important to use stock plants that are free from diseases, moderately vigorous, and of known identity. Propagators should avoid stock plants that have been injured by frost or drought, defoliated by insects, stunted by excessive flowering or fruiting, or by lack of soil moisture or proper nutrition, and plants that have made rank, overly vigorous growth. A marginal cutting slows down the whole production process, creates cultural problems, and produces an inferior quality plant.

A desirable practice for the propagator is the establishment of stock blocks as a source of propagating material, where uniform, true-to-type, pathogen-free mother plants can be maintained and held under the proper nutritive condition for the best rooting of cuttings taken from them. However, in most container nurseries no stock plants are maintained. Rather, propagules are collected from the container plants in production, hence the need for good cultural controls of *all* plants in production.

Stem Cuttings

Stem cuttings can be divided into four groups, according to the nature of the wood used: **hardwood, semihardwood, softwood,** and **herbaceous.** In propagation by stem cuttings, segments of shoots containing lateral or terminal buds are obtained with the expectation that under the proper conditions adventitious roots will develop and thus produce independent plants.

The type of wood, the stage of growth used in making the cuttings, the time of year when the cuttings are taken, and several other factors can be very important in securing satisfactory rooting of some plants. Information concerning these factors is given in Chap. 9, although much of this knowledge can be obtained by actual experience over the years in propagating plants.

It is imperative that propagators keep thorough records of procedures, seasonal condition of plant materials, and empirical trials utilized to achieve optimum success for their particular propagation system. The type of cutting utilized and cultural manipulation during propagation is dependent on the plant species, environmental conditions, propagation system utilized, propagation facilities, available personnel, and ultimately what is cost-effective for the producer.

Hardwood Cuttings
(Deciduous Species)

Hardwood cuttings are those made of matured, dormant hardwood after leaves have abscised and before new shoots emerge in the spring. The use of hardwood cuttings is one of the least expensive and easiest methods of vegetative propagation. Hardwood cuttings are easy to prepare, are not readily perishable, may be shipped safely over long distances if necessary, and require little or no special equipment during rooting. Hardwood cuttings are easily transplanted after rooting, and some producers report that liners produced from hardwood cuttings are larger than those from softwood cuttings (22).

Because of the low cost of hardwood cutting propagation, this method makes feasible high-density meadow orchards, consisting of precocious dwarfed fruit trees, planted many thousands to an hectare, in which there is considerable interest. Some peach cultivars, for example, can be propagated easily on a large scale from rooted hardwood cuttings (19, 26).

Hardwood cuttings are prepared during the dormant season—late fall, winter, or early spring—usually from wood of the previous season's growth, although with a few species, such as the fig, olive, and certain plum cultivars, two-year-old or older wood can be used. Hardwood cuttings are most often used in propagation of deciduous woody plants, although some broadleaved evergreens, such as the olive, can be propagated by leafless hardwood cuttings. Many deciduous ornamental shrubs are started readily by this type of cutting. Some common ones are privet, forsythia, wisteria, honeysuckle, *Salix, Populus, Cornus, Pontentilla, Sambucus*, crape myrtle, and spiraea. Rose rootstocks, such as *Rosa multiflora*, are propagated in great quantities by hardwood cuttings. A few fruit species are propagated commercially by this method—for example, fig, quince, olive, mulberry, grape, currant, gooseberry, pomegranate, and some plums. Certain trees, such as the willow and poplar, are propagated by hardwood cuttings.

The propagating material for hardwood cuttings should be taken from healthy, moderately vigorous stock plants growing in full sunlight. The wood selected should not be from extremely rank growth with abnormally long internodes, or from small, weakly growing interior shoots. Generally, hardwood cutting material is ready when you can remove the leaves without tearing the bark. Wood of moderate size and vigor is the most desirable. The cuttings should have an ample supply of stored foods to nourish the developing roots and shoots until the new plant becomes self-sustaining. Tip portions of a shoot, which are usually low in stored foods, are discarded. Central and

basal parts generally make the best cuttings, but there are exceptions (25).

Hardwood cuttings vary considerably in length—from 10 to 76 cm (4 to 30 in.). Long cuttings, when they are to be used as rootstocks for fruit trees, permit the insertion of the cultivar bud into the original cutting following rooting, rather than into a smaller new shoot arising from the original cutting.

At least two nodes are included in the cutting; the basal cut is usually just below a node and the top cut 1.3 to 2.5 cm (½ to 1 in.) above a node. However, in preparing stem cuttings of plants with short internodes, little attention is ordinarily given to the position of the basal cut, especially when quantities of cuttings are prepared and cut to length, many at a time, by a bandsaw.

When hardwood cuttings of *Rosa multiflora* are to be field-planted in raised soil beds, it is common procedure to "de-eye" or remove all lower basal axillary buds prior to sticking cuttings to prevent suckering from the base of the cutting.

Hardwood cuttings will desiccate, so it is important that they not dry out during handling and storage. After canes are cut with a bandsaw, some producers will take the bundled cuttings and dip the tops (apex) in wax. The wax helps reduce desiccation and indicates the orientation of the cuttings for late fall or spring field propagation.

The diameter of the cuttings may range from 0.6 to 2.5 or even 5 cm (¼ in. to 1 or 2 in.), depending upon the species. Three different types of

FIGURE 10-1 Types of hardwood cuttings. *Left:* Straight—the type ordinarily used. *Center:* Heel cuttings. A small piece of older wood is retained at the base. *Right:* Mallet cuttings. An entire section of the branch of older wood is retained.

FIGURE 10–2 *Left:* Hardwood cuttings of quince (*Cydonia oblonga*) prepared and ready for planting in the outdoor nursery in early spring. *Right:* Rooted cuttings after one summer's growth. (From *Univ. of Calif. Div. Agr. Sci. Leaflet 21103.*)

cuttings are shown in Fig. 10–1: the **mallet,** the **heel,** and the **straight** cutting. The mallet includes a short section of stem of the older wood, whereas the heel cutting includes only a small piece of the older wood. The straight cutting, not including any of the older wood, is the most common and gives satisfactory results in most instances. Hardwood cuttings of quince, freshly prepared and after a summer's growth in the nursery, are shown in Fig. 10–2.

Where it is difficult to distinguish between the top and base of the cuttings, it is advisable to make one of the cuts at a slant rather than at right angles. In large-scale operations, bundles of cutting material are cut to the desired lengths by band saws or other types of mechanical cutters rather than individually by hand (Fig. 10–3). For large-scale commercial operations planting the cuttings is mechanized, using equipment as illustrated in Fig. 10–4; but for planting a limited number of cuttings the method shown in Fig. 10–5 is satisfactory.

Several methods are commonly used for preparing and handling hardwood cuttings before planting:

Winter callusing. During the dormant season, make the cuttings of uniform length, tie them with heavy rubber bands into convenient-sized bundles, placing the tops all one way, and store them under cool, moist conditions until spring. The bundles of cuttings may be buried out-of-doors in sandy soil, sand, or sawdust in a well-drained location. They may be buried horizontally or in a vertical position, but upside down with the basal end of the cuttings several inches below the surface of the soil. The basal ends are somewhat warmer and better aerated than the terminal ends. This procedure tends to promote root initiation at the base, while retarding bud development at the top. At planting time in the spring, the bundles of cuttings are dug up and the cuttings planted right side up. In regions with mild winters, the bundles of cuttings are often stored during this callusing period in large boxes of moist sand, sawdust, peat moss, or shavings, either in an unheated building or out-of-doors.

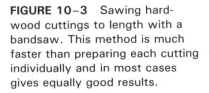

FIGURE 10–3 Sawing hardwood cuttings to length with a bandsaw. This method is much faster than preparing each cutting individually and in most cases gives equally good results.

FIGURE 10-4 Machine for large-scale planting of hardwood cuttings. Developed at the Tree Nursery Division, P. F. R. A., Indian Head, Saskatchewan, Canada, primarily for propagation of willow and poplar for shelter belt use. This four-unit machine will plant 10,000 to 12,000 cuttings per hour. (Courtesy Canada Department of Regional Economic Expansion.)

This probably would not be enough protection for the cuttings, however, in regions where severe, subzero winter temperatures are experienced. A cool, but above-freezing storage room would be satisfactory for such climates. If refrigerated rooms are available, the cuttings can be safely stored during the callusing period at temperatures of about 4.5°C (40°F) until they are ready to plant.

Direct spring planting.   It is often sufficient with easily rooted species to gather the cutting material during the dormant season, wrap it in heavy paper or slightly damp peat moss in a polyethylene bag, and store at 0 to 4.5°C (32 to 40°F) until spring. The cutting material should not be allowed to dry out or to become excessively wet

during storage. At planting time, the cuttings are made into proper lengths and planted in the nursery.

Stored cutting material should be examined frequently. If signs of bud development appear, lower storage temperatures should be used or the cuttings should be made and planted without delay. If the buds are forcing when the cuttings are planted, leaves will form and the cuttings will die, because of water loss from the leaves and depletion of stored food reserves prior to rooting.

Direct fall planting.   In regions with mild winters or reliable snow cover, cuttings can be made in the autumn and planted immediately in the nursery. Callusing and perhaps rooting may take place before the dormant season starts, or the formation of roots and shoots may occur simultaneously the following spring. Hardwood cuttings of peach and peach × almond hybrids have been successfully rooted in the nursery by this method provided that they were treated prior to planting with indolebutyric acid (IBA) and captan (*26*). Selected ornamental shrubs can be fall-propagated as far north as southern Canada without auxin pretreatment (*22*). Field rose rootstock for landscape and cut rose production are fall-planted in Spain, California, and Texas. Interestingly, the latitude where roses are field grown in these three regions is nearly identical.

Once hardwood cuttings have rooted in the field they will be dug after a growing season as rooted liners using an apparatus such as a modified potato digger or with a U-blade attached to a tractor (Fig. 10-43). With greater spacing, or in the case of landscape roses, rooted cuttings are left in the field for an additional season and sold as finished plants.

Warm-temperature callusing.   Take the cuttings in the fall while the buds are in or entering the "rest period," treat them with a root-promoting chemical, then store under moist conditions at relatively warm temperatures—18 to 21°C (65 to 70°F)—for three to five weeks to stimulate root initiation. After this, plant the cuttings in the nursery (in mild climates) or hold in cold storage (2 to 4.5°C; 35 to 40°F) until spring. Experimentally, good rooting of hardwood pear cuttings occurred when the cuttings were allowed to callus (and initiate roots) while the buds were

FIGURE 10–5 Steps in making and planting hardwood cuttings. *Top left:* Preparing the cuttings from dormant and leafless one-year-old shoots. A common length is 15 to 20 cm (6 to 8 in.) and the basal cut is generally made just below a node. *Top right:* Treating the cuttings with a root-promoting substance. On the left a bundle of cuttings is dipped in a commercial talc preparation. On the right the basal ends of the cuttings are soaked for 24 hours in a dilute solution of the chemical. With easily rooted plants such treatments are unnecessary. *Middle left:* The cuttings may be planted immediately, but with some plants it is helpful to callus the cuttings for several weeks in a box of moist shavings or peat moss before planting. *Middle right:* Planting the cuttings in the nursery row. A *dibble* (heavy, pointed, flat-bladed knife) is a useful tool for inserting the cutting and at the same time firming the soil around the previously planted cutting. *Bottom left:* The cuttings should be planted 7.6 or 10 cm (3 or 4 in.) apart and deep enough so that just one bud shows above the ground. A loose, sandy loam is best for starting hardwood cuttings. *Bottom right:* Several weeks after planting, the cuttings start to grow. They must be watered frequently if rains do not occur, and weeds must be controlled.

under the "rest" influence and, consequently, did not start growth and compete for food reserves in the cuttings (*1*).

In dealing with hardwood cuttings (which are leafless, cannot photosynthesize, and are thus surviving on stored carbohydrate reserves), it is important to avoid temperatures that cause excessive callusing and the loss of stored reserves; otherwise, rooting and field survival are poor (Fig. 10–6).

Bottom heat callusing. This method has been successful for difficult-to-root subjects such as some apple, pear, and plum rootstocks. Cuttings are collected in either the fall or late winter, the basal ends treated with IBA at 2500 to 5000 ppm, then placed upright for about four weeks in damp packing material over bottom heat at 18 to 21°C (65 to 70°F) but with the top portion of the cuttings left exposed to the cool outdoor temperatures. It is best to do this in a covered open shed for protection against excessive moisture from rains. The East Malling Research Station in England has developed (*35, 38, 39*) commercial pro-

FIGURE 10–6 *Rosa multiflora* hardwood cuttings that were simultaneously chip-budded and callused for rooting. Too high a temperature 27°C (80°F) caused profuse callusing at the base, but poor rooting and field survival.

cedures, as shown in Fig. 10–7, for propagating difficult subjects by this method. Cuttings must be transplanted before buds commence growth; this is usually done as roots begin to emerge. It is important to prevent decay in the cuttings by avoiding excessive application of water to the rooting compost. As long as the correct stimulation has been given, it is not essential to await root emergence before transplanting.

This procedure is probably best suited for regions having relatively mild winters (*30, 31, 38*). When soil or weather conditions are not suitable for planting after roots become visible, it has been satisfactory to leave the cuttings undisturbed in the rooting bed, shut off the bottom heat, then plant them in the nursery when conditions become suitable (*9*).

Plastic bag storage. Hardwood cuttings are taken during the dormant season, the bases dipped into a root-promoting material—for example, IBA at 2000 ppm, for a few seconds—then sealed in polyethylene bags which are placed in the dark at a temperature of about 10°C (50°F). Studies with this technique using peach hardwood cuttings showed 85 to 100 percent rooting after about 50 days (*64*). While high rooting can be obtained by this method, it may be difficult to obtain survival of the cuttings following transplanting.

Hardwood Cuttings (Narrow-Leaved Evergreen Species)

Narrow-leaved evergreen cuttings must be rooted under moisture conditions that will prevent excessive drying as they usually are slow to root, sometimes taking several months to a year. Some species root much more readily than others. In general *Chamaecyparis, Thuja,* and the low-growing *Juniperus* species root easily and the yews (*Taxus* spp.) fairly well, whereas the upright junipers, the spruces (*Picea* spp.), hemlocks (*Tsuga* spp.), firs (*Abies* spp.), and pines (*Pinus* spp.) are more difficult. In addition, there is considerable variability among the different species in these genera in regard to the ease of rooting of cuttings. Cuttings taken from young seedling stock plants root much more readily than those taken from older trees because of the juvenility factor. Treatments with IBA at relatively high concentrations

FIGURE 10–7 Steps in propagation by hardwood cuttings using the bottom-heat technique for difficult-to-root materials. *Upper left:* Removing 'M 26' apple cuttings from hard-pruned, vigorous hedges by cutting one-year shoots at their base. *Upper right:* 60 cm (2 ft) cuttings inserted (after IBA treatment) to a 25 cm (10 in.) depth in insulated bins filled with rooting compost [one-half coarse peat and one-half grit — ½ cm (3/16 in.) gravel and washed sand] maintained at 21°C (70°F) by bottom heat. Bins are situated in a cool building to retard bud development. *Lower left:* Root development after six weeks (shown here for plum cuttings). *Lower right:* Apple rootstock hardwood cuttings ('M 26', 'M 106', 'MM 111') after one season's growth in nursery. (Courtesy East Malling Research Station, England.)

are usually beneficial in increasing the speed of rooting, the percentage of cuttings rooted, and obtaining heavier root systems.

Narrow-leaved evergreen cuttings ordinarily are best taken between late fall and late winter (see Fig. 10–8). Rapid handling of the cuttings after the material is taken from the stock plants is important. The cuttings are usually best rooted in a greenhouse or polyhouses with relatively high light intensity and under conditions of high hu-

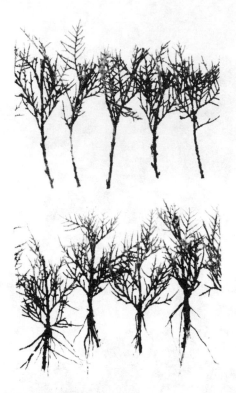

FIGURE 10–8 Narrow-leaved evergreen cuttings. *Above:* Juniper cuttings ready for sticking in rooting medium. *Below:* Cuttings after rooting. (From H. T. Hartmann, W. J. Flocker, and A. M. Kofranek, *Plant science*, Prentice-Hall, Englewood Cliffs, N.J., 1981.)

midity or very light misting, but without heavy wetting of the leaves. A bottom heat temperature of 24 to 26.5°C (75 to 80°F) has given good results. Dipping the cuttings into a fungicide helps prevent disease attacks. Sand alone is a satisfactory rooting medium, as is a 1:1 mixture of perlite and peat moss. Some individual cuttings take longer to root than others. The slower-rooting ones can be inserted again in the rooting medium, and often will root eventually.

The type of wood to use in making the cuttings varies considerably with the particular species being rooted. As shown in Fig. 10–8, the cuttings are made 10 to 20 cm (4 to 8 in.) long with all the leaves removed from the lower half. Mature terminal shoots of the previous season's growth are usually used. In some instances, as with *Juniperus chinensis* 'Pfitzeriana', older and heavier wood also can be used, thus resulting in a larger plant when it is rooted. On the other hand, some propagators use small tip cuttings, 5.0 to 6.6 cm (2 to 3 in.) long, placed very close together in a flat for rooting. In some species, as *Juniperus excelsa,* older growth taken from the sides and lower portion of the stock plant roots better than the more succulent tips. Cuttings of *Taxus* root best if they are taken with a piece of old wood at the base of the cutting; such cuttings seem less subject to fungus attacks (*80*). In certain of the narrow-leaved evergreen species, some type of basal wounding is often beneficial in inducing rooting.

Semihardwood Cuttings

Semihardwood cuttings are those made from woody, broad-leaved evergreen species, but leafy summer and early fall cuttings taken from partially matured wood of deciduous plants could also be considered as semihardwood. Cuttings of broad-leaved evergreen species are generally taken during the summer from new shoots just after a flush of growth has taken place and the wood is partially matured. Many ornamental shrubs, such as camellia, pittosporum, rhododendron, euonymus, the evergreen azaleas, and holly, are commonly propagated by semihardwood cuttings. A few fruit species, such as citrus and olive, can also be propagated in this manner.

Summer semihardwood cuttings and spring softwood cuttings (with a mallet or heel of one-year-old wood left at the base) are used for producing high-density plantings of own-rooted peach trees (*44*). IBA is applied at 1000 ppm to semihardwood (reddish-brown color) and 500 to 700 ppm to softwood cuttings (green bark). Wounding is done by bark incision or bark removal. Some advantages of using semihardwood cuttings for own-rooted peaches are elimination of graft incompatibility and bud failure, uniform growth and fruiting characteristics, straight trunk, more fibrous root system closer to the crown, improved angles of scaffold limbs, and elimination of suckering from the rootstock (*44*).

The cuttings are made 7.5 to 15 cm (3 to 6 in.) long with leaves retained at the upper end, as shown in Fig. 10–9. If the leaves are very large, they should be reduced in size to lower the water loss and to allow closer spacing in the cutting bed. The shoot terminals are often used in making cut-

FIGURE 10–9 Semihardwood cuttings as shown by euonymus. *Left:* Cuttings as prepared for rooting from partially matured wood in midsummer. *Right:* Cuttings after rooting.

tings, but the basal parts of the stem will usually root also. The basal cut is usually just below a node. The cutting wood should be obtained in the cool, early morning hours when the stems are turgid, and kept wrapped in clean moist burlap or put in large polyethylene bags. Cuttings should be kept out of the sun until they can be stuck and propagation is initiated.

It is necessary that leafy cuttings be rooted under conditions that will keep water loss from the leaves at a minimum; commercially, they are ordinarily rooted under intermittent mist sprays, fog, or under polyethylene sheets laid over the cuttings (in cool temperate climates or in the southern United States during the fall). Bottom heat and growth-regulator treatment are also beneficial. Rooting media, such as 1:1 mixture of perlite and peat moss, perlite and vermiculite, or some of the commercial mixes such as Terra-Lite Metro-Mix Growing Medium give satisfactory results.

Softwood Cuttings

Cuttings prepared from the soft, succulent, new spring growth of deciduous or evergreen species may properly be classed as **softwood cuttings.** Many ornamental woody shrubs can be started by softwood cuttings. Typical examples are the hybrid French lilacs, forsythia, magnolia, weigela, and spiraea. Other examples are shown in Fig. 10–10.

FIGURE 10–10 Softwood cuttings of several ornamental species. *Top:* Cuttings made in late spring from young shoots. *Below:* Cuttings after rooting. *Left to right: Myrtus, Pyracantha, Oleander,* and *Hebe.*

Some deciduous ornamental trees, such as the maples, can also be started in this manner. Although fruit tree species are not commonly propagated by softwood cuttings, those of apple, peach, pear, plum, apricot, and cherry will root, especially under mist.

Softwood cuttings generally root easier and quicker than the other types but require more attention and equipment. This type of cutting is always made with leaves attached. They must, consequently, be handled carefully to prevent drying, and be rooted under conditions which will avoid excessive water loss from the leaves. Temperature should be maintained during rooting at 23 to 27°C (75 to 80°F) at the base and 21°C (70°F) at the leaves for most species. Softwood cuttings produce roots in two to four or five weeks in most cases. In general, they respond well to treatments with root-promoting substances.

It is important in making softwood cuttings to obtain the proper type of cutting material from the stock plant. Such material will vary greatly, however, with the species being propagated. Extremely fast-growing, soft, tender shoots are not desirable, as they are likely to deteriorate before rooting. At the other extreme, older woody stems are slow to root or may just drop their leaves and not root. The best cutting material has some degree of flexibility but is mature enough to break when bent sharply. Weak, thin, interior shoots should be avoided as well as vigorous, abnormally thick, or heavy ones. Average growth from portions of the plant in full light is the most desirable to use. Some of the best cutting material is the lateral or side branches of the stock plant. Heading back the main shoots will usually force out numerous lateral shoots from which cuttings can be made. Softwood cuttings are 7.5 to 12.5 cm (3 to 5 in.) long with two or more nodes. The basal cut is usually made just below a node. The leaves on the lower portion of the cutting are removed, with those on the upper part retained. Large leaves should be reduced in size to lower the transpiration rate and to occupy less space in the propagating bed. All flowers or flower buds should be removed. In some nurseries where quantities of cuttings are prepared, bundles of cutting material are rapidly cut into uniform lengths by paper cutters.

The cutting material is best gathered in the early part of the day and should be kept moist, cool, and turgid at all times by wrapping in damp, clean burlap or placing in large unsealed polyethylene bags (kept out of the sun). Laying the cutting material or prepared cuttings in the sun for even a few minutes will cause serious damage. Soaking the cutting material or cuttings in water for prolonged periods to keep them fresh is undesirable. Refrigerated storage (4 to 8°C; 40 to 47°F) for one to two days is another option.

Some producers of ornamental shurbs will fall-propagate hardwood cuttings in polyhouses with heated benches under mist, and then harvest softwood cuttings from the developing hardwood shoots as the various species flush in the spring (2). Before the end of summer, both the original crop of hardwood cuttings and the bonus crop of softwood cuttings can be lined out.

Herbaceous Cuttings

Herbaceous cuttings are made from such succulent, herbaceous plants as geraniums, chrysanthemums, coleus, carnations, and many foliage crops. They are 7.5 to 12.5 cm (3 to 5 in.) long with leaves retained at the upper end, as shown in Fig. 10–11, or without leaves (Fig. 10–12). Most florists' crops are propagated by herbaceous cuttings which root easily. They are rooted under the same conditions as softwood cuttings, requiring high humidity. Bottom heat is also helpful. Under proper conditions, rooting is rapid and in high percentages. Although root-promoting substances are usually not required, they are often used to gain uniformity in rooting and development of heavier root systems. Herbaceous cuttings of some plants that exude a sticky sap, such as geranium, pineapple, or cactus, do better if the basal ends are allowed to dry for a few hours under low light conditions before they are inserted in the rooting medium. This practice tends to prevent the entrance of decay organisms.

Types of cuttings (apical versus basal) and nodal position of herbaceous cuttings can influence shoot growth and finished plant quality of rooted liners. Basal cuttings of *Hedera helix* and *Schefflera arboricola* develop longer shoots and more roots than apical cuttings. With golden pothos (*Epipremnum aureum*), a 3-cm (1-in.) or longer internode section below the node and a fraction of

FIGURE 10–11 Typical herbaceous cuttings. *Left to right:* Chrysanthemum, begonia, and geranium. It is often necessary with large-leaved plants, such as the begonia, to trim back some of the leaves to prevent wilting and to conserve space in the propagating bench.

the old aerial root should be retained on cuttings for most rapid axillary shoot development (*78*).

Leaf Cuttings

In **leaf cuttings**, the leaf blade, or leaf blade and petiole, is utilized in starting new plants. Adventitious roots and an adventitious shoot form at the base of the leaf and develop into the new plant; the original leaf does not become a part of the new plant.

One type of propagation by leaf cuttings is illustrated by *Sansevieria*. The long tapering leaves are cut into sections 7.5 to 10 cm (3 to 4 in.) long as shown in Fig. 10–13. These leaf pieces are inserted three-fourths of their length into the rooting medium, and after a period of time a new

FIGURE 10–12 Type of stem cutting consisting only of a leafless stem piece; used here in propagating the monocotyledonous plant, *Dieffenbachia maculata*. Latent buds develop into shoots along with the formation of adventitious roots.

plant forms at the base of the leaf piece, the original cutting disintegrating. The variegated form of *Sansevieria, S. trifasciata laurenti,* is an example of a periclinal chimera that will not reproduce true to type from leaf cuttings; to retain its characteristics, it must be propagated by division of the original plant.

In starting plants with fleshy leaves, such as *Begonia rex,* by leaf cuttings, the large veins are cut on the undersurface of the mature leaf, which is then laid flat on the surface of the propagating medium. The leaf is pinned or held down in some manner, with the natural upper surface of the leaf exposed. After a period of time under humid conditions, new plants form at the point where each vein was cut. The old leaf blade gradually disintegrates.

Another method, sometimes used with fibrous-rooted begonias, is to cut large, well-matured leaves into triangular sections, each containing a piece of a large vein. The thin outer edge of the leaf is discarded. These leaf pieces are then inserted upright in sand with the pointed end down. The new plant develops from the large vein at the base of the leaf piece.

New plants arise from leaves in a variety of ways. An interesting example is illustrated in Fig. 10–14, where the new plant develops at the junction of the leaf blade and petiole, even while the leaf is still growing on the mother plant.

The African violet (*Saintpaulia*) is typical of leaf cuttings that can be made of an entire leaf (leaf blade plus petiole), the leaf blade only, or just a portion of the leaf blade. The new plant forms at the base of the petiole or midrib of the leaf blade (see Figs. 9–10, 10–15, and 10–16).

An unusual type of leaf cutting is illustrated

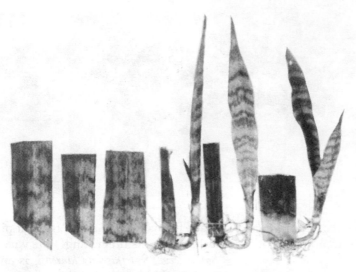

FIGURE 10–13 Leaf cuttings of *Sansevieria. Left:* The thick, leathery leaves are cut into pieces 7.5 to 10 cm (3 to 4 in.) long. To avoid trying to root upside down, the basal end can be marked by cutting on a slant as shown with two of the cuttings. *Right:* Development of the plant. The original cutting does not become a part of the new plant.

in Fig. 10–17, where many new plants or **offsets** arise at the margins of the leaf. The leaf itself eventually deteriorates. Leaf cuttings should be rooted under the same conditions of high humidity as those used for softwood or herbaceous cuttings. Most leaf cuttings root readily, but the limitation to regeneration is adventitious bud and shoot development. Hence cytokinins are frequently used to induce buds to form (*13*).

Leaf-Bud Cuttings

A **leaf-bud cutting** (**single-eye** or **single-node cutting**) consists of a leaf blade, petiole, and a short piece of the stem with the attached axillary bud (Fig. 10–18). Such cuttings are of particular value in plants where adventitious roots but not adventitious shoots are initiated from detached leaves; the axillary bud at the base of the petiole provides for the new shoot. A number of plant species such as the black raspberry (*Rubus occidentalis*), blackberry, boysenberry, lemon, camellia, maple, and rhododendron are readily started by leaf-bud cuttings, as well as many tropical shrubs and most herbaceous greenhouse plants usually started by stem cuttings.

FIGURE 10–14 Leaf cuttings of piggyback plant (*Tolmiea menziesi*). The large parent leaf (underneath) is set in a moist rooting medium in a humid location for rooting. The new plants (arrows) arise at the junction of the leaf blade and petiole.

FIGURE 10–15 Leaf cuttings of African violet (*Saintpaulia*). *Left:* Each cutting consists of a leaf blade and petiole. *Right:* Leaf cuttings after rooting. One or more new plants will form at the base of the petiole. The original leaf can be cut off and used again for rooting.

FIGURE 10-16 Inserting African violet leaf cuttings in flats of rooting medium.

FIGURE 10-18 Leaf-bud cutting used in the propagation of rhododendrons. *Left:* Cutting when made. *Center:* Root development after several weeks. *Right:* Appearance of root ball and new shoot after five months. (Courtesy H. T. Skinner.)

Leaf-bud cuttings are particularly useful when propagating material is scarce, because they will produce at least twice as many new plants from the same amount of stock material as can be started by stem cuttings. Each node can be used as a cutting. For plants with opposite leaves, two leaf-bud cuttings can be obtained from each node. Leaf-bud cuttings are best made from material having well-developed buds and healthy, actively growing leaves.

Treatment of the cut surfaces with one of the root-promoting substances should stimulate root production. The cuttings are inserted in the rooting medium with the bud 1.3 to 2.5 cm (½ to 1

FIGURE 10-17 Leaf cuttings of *Kalanchoe pinnata* (*Bryophyllum pinnata*), air plant. *Left:* New plants developing from foliar "embryos" in the notches at the margin of the leaf. *Right:* Leaves ready to lay flat on the rooting medium. They should be partially covered or pegged down to hold the leaf margin in close contact with the rooting medium.

in.) below the surface. High humidity is essential, and bottom heat is desirable for rapid rooting.

Root Cuttings

Best results with **root cuttings** are likely to be attained if the root pieces are taken from young stock plants in late winter or early spring when the roots are well supplied with stored foods but before new growth starts. Taking the cuttings during the spring when the parent plant is rapidly making new shoot growth should be avoided. Root cuttings of the Oriental poppy (*Papaver orientale*) should be taken in midsummer, the dormant period for this species. Securing cutting material in quantities for root cuttings can be quite laborious unless it can be obtained by trimming roots from nursery plants as they are dug.

It is important with root cuttings to maintain the correct polarity when planting. To avoid planting them upside down, the proximal end (nearest the crown of the plant) may be made with a straight cut and the distal end (away from the crown) with a slanting cut. The proximal end of the root piece should always be up. In planting, insert the cutting vertically so that the top is at about soil level. With many species, however, it is satisfactory to plant the cuttings horizontally 2.5 to 5 cm (1 to 2 in.) deep (Fig. 10-19), avoiding the possibility of planting them upside down.

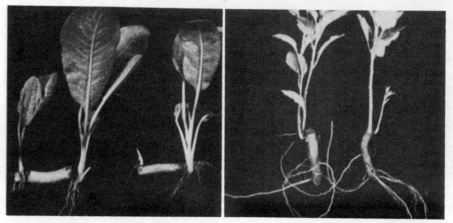

FIGURE 10–19 Propagation by root cuttings. *Left:* Horseradish (*Armoracia rusticana*) root pieces planted horizontally. *Right:* Apple (*Malus pumila*) root pieces set vertically. Adventitious buds arising from the root piece form the new shoot system.

In using root cuttings to propagate chimeras with variegated foliage, such as some *Aralias* and *Pelargoniums,* the new plants lose their variegated form. Propagation by root cuttings is very simple, but the root size of the plant being propagated may determine the best procedure.

Plants with Small, Delicate Roots

Root cuttings of plants with small, delicate roots should be started in propagation flats or cell packs in the greenhouse, hotbed, or heated polyhouse. The roots are cut into short lengths, 2.5 to 5 cm (1 to 2 in.) long, and scattered horizontally over the surface of the medium. They are covered with a layer of medium of 1 to 2 cm (½ in.) After watering, a polyethylene cover or a pane of glass should be placed over the flat to prevent drying until the plants are started. The flats are set in a shaded place. After the plants become well formed, they can be transplanted to other flats or lined-out in nursery rows for further growth.

Plants with Somewhat Fleshy Roots

Cuttings of plants with fleshy roots are best started in a flat in the greenhouse or hotbed. The root pieces should be 5 to 7.5 cm (2 to 3 in.) long and planted vertically, observing correct polarity. New adventitious shoots should form rapidly, and

as soon as the plants become well established with good root development, they can be transplanted. These root pieces can also be stuck directly in containers and held in dormant storage in cool greenhouses during the winter season, then undergo a period of active spring growth followed by midsummer planting in the field.

Plants with Large Roots, Propagated Out-of-Doors

Large root cuttings are made 5 to 15 cm (2 to 6 in.) long. They are tied in bundles, care being used to keep the same ends together to avoid planting upside down later. The cuttings are packed in boxes of damp sand, bark, sawdust, or peat moss for about three weeks and held at about 4.5°C (40°F). After this they should be planted 5 to 7.5 cm (2 to 3 in.) apart in a well-prepared nursery soil with the tops of the cuttings level with, or just below, the top of the soil.

Some deciduous shrubs grown from root pieces are converted to softwood summer cuttings by taking elongating shoots from the root pieces and rooting these softwood cuttings under mist. This conversion technique is reported to produce heavier, faster-growing plants and reduce production time by one year with *Aronia, Clethra, Comptonia, Euonymus, Spiraea,* and *Viburnum* species (*55*).

TABLE 10–1

Some species that can be propagated by root cuttings

Acanthopanax pentaphyllus (coralberry)	*Myrica pennsylvanica* (bayberry)
Actinidia deliciosa (kiwifruit)	*Papaver orientale* (oriental poppy)
Aesculus parviflora (bottle-brush buckeye)	*Phlox* spp. (phlox)
Ailanthus altissima (tree-of-heaven)	*Plumbago* spp. (leadwort)
Albizia julibrissin (silk tree)	*Populus alba* (white poplar)
Anemone japonica (Japanese anemone)	*Populus tremula* (European aspen)
Aralia spinosa (Devil's walking stick)	*Populus tremuloides* (quaking aspen)
Artocarpus altilis (breadfruit)	*Prunus glandulosa* (dwarf flowering almond)
Broussonetia papyrifera (paper mulberry)	*Pyrus calleryana* (oriental pear)
Campsis radicans (trumpet vine)	*Rhus copallina* (shining sumac)
Celastrus scandens (American bittersweet)	*Rhus glabra* (smooth sumac)
Chaenomeles japonica (Japanese flowering quince)	*Rhus typhina* (staghorn sumac)
Chaenomeles speciosa (flowering quince)	*Robinia hispida* (rose acacia)
Clerodendrum trichotomum (glory-bower)	*Robinia pseudoacacia* (black locust)
Comptonia peregrina (sweet fern)	*Rosa blanda* (rose)
Daphne genkwa (daphne)	*Rosa nitida* (rose)
Dicentra spp. (bleeding heart)	*Rosa virginiana* (rose)
Eschscholzia californica (California poppy)	*Rubus* spp. (blackberry, raspberry)
Ficus carica (fig)	*Sassafras albidum* (sassafras)
Forsythia intermedia (forsythia)	*Sophora japonica* (Japanese pagoda tree)
Hypericum calycinum (St. Johnswort)	*Stokesia laevis* (Stokes aster)
Koelreuteria paniculata (goldenrain tree)	*Symphoricarpos hancockii* (coralberry)
Liriope spp. (liriope)	*Syringa vulgaris* (lilac)
Malus spp. (apple, flowering crabapple)	*Ulmus carpinifolia* (smooth-leaved elm)

STOCK PLANTS: SOURCES OF CUTTING MATERIAL

In cutting propagation, the source of the cutting material is very important (5). The stock plants, from which the cutting material is obtained, should be:

1. True-to-name and type
2. Free of disease and insect pests
3. In the proper physiological state so that cuttings taken from them are likely to root

Several sources are possible for obtaining cutting material:

1. From plants growing in the landscape in parks, around houses or buildings, or in the wild. For nurseries propagating plants for sale, this can be a dangerous practice. While the species may be known, identification as to cultivar may be pure guesswork. In addition, such plants may be infected with virus, fungal, or bacterial diseases, which may appear later in the rooted cuttings or in the nursery plants.

2. Prunings from young nursery plants as they are trimmed and shaped. Many nurseries use prunings as the primary source of their cutting material. Sometimes, however, the trimming is not done at the proper time to root the cuttings—and the unrooted cuttings must be stored.

3. Stock plants can be especially maintained as a source of cutting material. Although such plants may occupy valuable land space, this is probably the ideal source of cutting material. The history and identity of each stock plant can be determined accurately. The health status can be controlled, and the plants can be maintained in the proper degree of nutrition and vigor. By culturally maintaining uniformity of growth in stock plants, the propagator ensures that evenly graded batches of cutting material are available during a given period. Consequently, the success rate and uniformity of rooting is that much greater.

4. It is becoming more common to use tissue culture-produced liners as sources of stock plants in the development of new cultivars and disease-indexed plants. Conventional macropropagation techniques can then be used after stock plant establishment (Fig. 10–20).

STOCK PLANT MANIPULATION

Pruning and Girdling

Annual pruning is an important aspect of stock plant management in relation to (a) maintenance of juvenility to improve rooting, (b) plant shaping for easier and faster collection of propagules (Fig. 9–20), (c) increased cutting production, (d) tim-

ing of flushes, and (e) reducing reproductive shoots (*65*).

The types of pruning (*65*) are:

Modified stooling. Plants are severely cut back to their base but not mounded with soil as with traditional stooling; this eliminates reproductive shoots and is beneficial for *Hydrangea, Senecio.*

Hard pruning. Stock plants are cut back to half their size annually; this avoids the rank growth that can occur from modified stooling and eliminates reproductive growth; useful for *Forsythia, Weigela,* heather.

Moderate pruning. Plants are cut back one-third to one-half each year and there is less dieback than with the foregoing two methods; used with *Viburnum,* deciduous azalea.

FIGURE 10–20 *Upper left:* Custom propagation greenhouse specializing in producing unrooted and rooted cuttings. Disease-indexed stock plants produced by tissue culture are maintained in isolation. Plants are replaced each one to two years. *Upper right:* Rooted poinsettia cuttings. *Lower left:* Unrooted geranium cuttings. *Lower right:* Cuttings are boxed and shipped to greenhouse producers to be finished off.

Light pruning. Light pruning implies tipping back or just normal removal of cuttings from the stock.

Hedging. The severity of pruning to maintain the hedge form is generally determined by the ease with which cuttings can be collected from the stock plant (e.g., *Berberis* and *Pyracantha* are heavily pruned, while *Eleagnus* and *Cornus* receive light trims).

Renewal pruning. A three-year schedule is used for species that do not produce suitable cutting material in the season following pruning, but propagule numbers decline if left unpruned. Third-year wood is cut back hard in the spring, stimulating a new growth flush which is unsuitable for cuttings. Cutting this new growth back by half in the second year produces a flush of growth suitable for propagules, prior to its removal in the third season.

Double pruning. Spring pruning produces a flush of cuttings for summer softwood cuttings or semihardwood fall cuttings. A second trimming in June (England) delays the softwood cutting collection period to fall; cutting production is increased, but growth is weaker than during the normal summer flush.

With *Dracaena* stock plant production, **incisions** are made above axillary buds by cutting one-third to one-half through the cane. This breaks apical dominance and induces additional buds to develop in the plant without sacrificing the apical heads. This technique further promotes greater branching during field propagation (*11*).

Girdling shoots of stock plants prior to taking cuttings has been used successfully to root slash pine, sweetgum, sycamore, and 19- to 57-year-old water oak (*27–29*). The treatment consisted of girdling shoots by removing 2.5 cm (1 in.) of bark, then applying a talc composed of IBA and PPZ [(1-phenyl-3-methyl-5-pyrazolone) K&K Laboratories, Plainview, NY] and wrapping the shoot with polyethylene film and aluminum. Once primordia become visible as small bumps in the callus, the cutting is removed from the stock plant and rooted under mist (Figs. 10–21 and 10–22).

Etiolation, Shading, Blanching, and Banding

A modification of the traditional technique of etiolation and blanching, using Velcro adhesive fabric strips as the blanching material is shown in Fig. 10–23 (*50*). This etiolation technique for softwood cutting propagation has improved rooting

---

TECHNIQUES OF SHADING, ETIOLATION, BANDING, AND BLANCHING (*50*)

1. Field-grown or containerized dormant stock plants are ready for treatment when all chilling or rest requirements have been fulfilled and the buds begin to swell—usually early spring in the field or midwinter in a greenhouse.

2. During the **shading** process, entire plants or several branches of a plant are then covered with an opaque material, usually black cloth or plastic; taking care to allow space for the new growth to extend. A wire or wooden frame may be used to support the covering.

3. Cuts should be made in the covering material or corners left slightly open near the top of the structure to allow for ventilation. A small heat buildup under the structure is desirable, but enough ventilation should be supplied so that plants are not scorched. It is not necessary or desirable to exclude 100 percent of the light. Between 95 and 98 percent light exclusion is preferable.

4. Initial growth is allowed to progress in the dark (**etiolation**) until new shoots are between 5 and 7 cm (2 to 3 in.) long, after which time the shade is gradually removed over the period of one week so as not to scorch the very tender shoots.

5. On the first day of shade removal, **banding** is initiated with the placement of self-adhesive black bands at the base of each new shoot (the future cutting base). These bands keep the base of the shoot in an etiolated condition while the tops of the shoots are allowed to turn green in the light.

6. The bands are approximately 2.5 cm wide × 2.5 cm (1 in.) long but can vary in length. The material used goes under the trade name of Velcro, which is no longer patented. It is made up of two pieces, one a "wooly" side and the other a material with "hooks" or "nubs" that adhere to the wool when they are pressed together. The Velcro band "sandwiches" the shoot so that the hooks gently pierce the surface of the shoot when they are pressed together. In addition, a rooting hormone in a talc base is added to both parts of the Velcro before banding the shoot, thereby delivering a rooting stimulus to the shoot prior to its being made into a cutting. The hooks of the Velcro aid in wounding the stem so as to get better penetration of the stem with hormone. Typically, 8000 ppm IBA in talc is used. Both pieces of Velcro are dipped in the talc preparation: the excess powder is tapped off and the Velcro band is then firmly pressed onto the shoot base.

7. Velcro bands are generally left on the shoots for four weeks, although as short as two weeks and as long as 12 weeks of banding has also been successful.

8. After four weeks, the cuttings are removed from the stock plant below (proximal to) the banded area. The bands are removed and the cutting bases are again treated with hormone before sticking them in the rooting medium. This second IBA treatment is generally a quick dip of 4000 ppm in 50 percent aqueous ethanol.

9. The area underneath the band is typically yellowish white and swollen, and occasionally root primordia can be seen already emerging.

10. Cuttings are stripped of lower leaves and placed in a rooting medium consisting of peat and perlite (1:1 by volume). Bottom heat (25°C; 77°F) is supplied during the winter months. Mist is applied intermittently beginning at seven seconds of mist every two minutes, becoming less frequent with time and amount of cloud cover. Fifty percent Saran shading is also applied over the rooting bench to keep down air temperatures around the cuttings. Greenhouse ambient temperature is maintained at 20°C (68°F), and daylength is regulated to 16 hours of daylight using 60-W incandescent bulbs spaced 1 m apart and hung 1 m above the bench.

11. The cuttings are left in the rooting bench from two to five weeks for deciduous plants and up to 12 weeks for pines.

12. After rooting, the cuttings are potted up, hardened-off from un-

der the mist, fertilized, and kept under long days (16 hours) to encourage growth. Depending on the time of year, they may be placed outside to continue growing during the late spring and summer (for winter-propagated plants) or they may be hardened off and placed in a protective overwintering structure (for summer-propagated plants).

Blanching

1. All the previous steps are followed except that stock plants are not initially covered, so new growth grows out in the light.

2. When the soft, green shoots are 5 to 7 cm long, banding with Velcro as described above proceeds in exactly the same way as for etiolation and banding.

FIGURE 10–21 *Left:* Girdled shoot of slash pine (*Pinus caribaea*) with 2.5 cm (1 in.) of bark removed, treated with an IBA, PPZ (1-phenyl-3-methyl-5-pyrazolone) talc slurry, then wrapped with plastic film and aluminum foil. *Upper right:* Root primordia visible as small bumps on the callus. *Lower right:* The cutting is removed from the stock plant and rooted under mist. Only girdled cuttings will root when removed from the stock plant. (Courtesy R. C. Hare.)

FIGURE 10-22 A comparison of girdled versus control cuttings of Formosan sweetgum (Liquidambar formosana), *left;* mature water oak (*Quercus nigra*), *center;* and sycamore (*Platanus*), *right.* Girdling and treatment with IBA and PPZ talc was done to shoots on stock plants prior to removing cuttings for rooting under mist. (Courtesy R. C. Hare.)

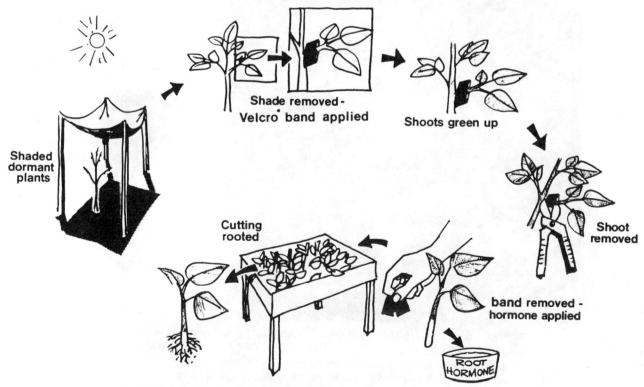

Shaded dormant plants

Shade removed - Velcro band applied

Shoots green up

Shoot removed

band removed - hormone applied

ROOT HORMONE

Cutting rooted

FIGURE 10-23 Scheme for etiolated softwood cutting propagation using Velcro fabric strips as the blanching material. (Courtesy B. K. Maynard and N. L. Bassuk.)

success of a wide range of difficult-to-root woody species (*50*). See Chap. 9 for discussion of the physiological implications of these techniques and listing of species tested.

ROOTING MEDIA

There is no universal or ideal rooting mix for cuttings. Propagation medium to use depends on the plant species (easy, moderate, or difficult-to-root), cutting type (softwood, semihardwood, hardwood), season, type of propagation system (contact poly, enclosed or open intermittent mist, fog), and cost and availability of the medium components. The key to successful propagation medium is good water management. See Chap. 9 for physical and chemical components of media and Chap. 2 for general discussion.

Propagation medium is composed of an **organic** component: peat, sphagnum moss, or softwood and hardwood barks. The **coarse mineral** component is used to increase the proportion of large, air-filled pores and drainage and includes perlite, vermiculite, expanded shale, coarse sand or grit, pumice, scoria, polystyrene, and rockwool (*54*). Rarely is mineral soil used as a propagation medium component, except in field propagation of hardwood cuttings. Most propagators use a combination of organic and mineral components (i.e., peat–perlite, peat–expanded shale, peat–vermiculite–perlite, peat–bark–sand, peat–rockwool, etc.) (*63*).

Sometimes the mineral component is used alone (i.e, sand, rockwool, Oasis cubes, perlite) or in combination (i.e., vermiculite–perlite, sand–polystyrene). Sufficient *coarse* mineral component should be added to improve aeration. It is important that sound sanitation procedures be followed, which include pasteurization of media or gas sterilization treatment with methyl bromide. Some mineral components, such as vermiculite, perlite, and rockwool, are relatively sterile due to their high-temperature exposure during manufacturing. There are a myriad of premixed commercial propagation media available, which can help reduce production steps, save time and labor, and assure better media standards [e.g., Promix BX (sphagnum peat, perlite, vermiculite), W.R. Grace Co.].

Trends in U.S. nurseries are for partial replacement of expensive peat with fine bark. In Europe, Australia, and Israel, rockwool sheets and cubes are being used for direct sticking of cuttings (Fig. 10–24). Rockwool has the advantage of being sterile, can be efficiently handled (no containers or trays to be filled), and since cuttings are directly stuck into cubes or sheets, there is minimal disturbance of roots, which avoids transplant shock, once rooted cuttings are shifted up to liner production stages. Gloves should be worn when handling the material, water management is critical, and there are some algae problems reported during propagation (*66*).

Propagation Unit Systems

Besides rockwool rooting cubes and blocks, and the standard plastic propagation flat, there are a variety of *propagation unit systems* to work with (*49*). These include Jiffy-7's (compressed peat blocks), synthetic foam media Rootcubes, Japanese paper pot containers with an expandable honeycomb form, bottomless polystyrene or plastic trays with wedge- or cylindrical cells for air pruning of roots (Speedling, Gro-Plug, etc.), and various modifications of plastic flats. Wooden rooting trays have generally been replaced by plastic and other synthetic materials. See Fig. 10–24 and Chap. 2.

Direct sticking of cuttings into small liner plastic containers for rooting, as opposed to sticking in conventional flats or rooting trays, is an important technique for utilizing personnel and materials more efficiently (*51*). Over 50 percent of cutting propagation costs are due to labor. By direct sticking, the production step of transplanting rooted cuttings and potential transplant shock due to a disturbed root system is avoided. The plant materials must be easy to root (more than 85 percent) to justify the additional propagation space required, but the labor savings and versatility of this system are substantial (Fig. 10–25).

WOUNDING

Root production on stem cuttings can be promoted by wounding the base of the cutting. This has proved useful in a number of species, such as juniper, arborvitae, rhododendron, maple, mag-

FIGURE 10-24 Various rooting media used by propagators. *Upper left:* Rockwool. *Upper right:* Polystyrene root cubes. *Center left:* Compressed peat pellet. *Center right:* Conventional cutting flat with peat-perlite mix. *Lower left:* Individualized small liner container for direct sticking. *Lower right:* Polystyrene root sheets or blocks.

nolia, and holly (*81*). Wounds may be produced in cuttings of narrow-leaved evergreen species, such as arborvitae, by stripping off the lower side branches of the cuttings.

Stripping the basal leaves off a cutting has also been used as a method of wounding (*48*). The benefits of stripping basal leaves is species–dependent; *Berberis* and *Juniperus* cuttings benefit, whereas *Spiraea*, forsythia, and weigela do not. Stripping basal leaves of cuttings reduces the propagation bench space required for some spe-

cies, allows the propagator more flexibility to work with different size propagules, may serve to improve the contact area between the cutting and media (see Chap. 9), and can potentially improve absorption of rooting compounds.

A common method of wounding—making a vertical cut with the tip of a sharp knife down each side of the cutting for an inch or two, penetrating through the bark and into the wood—may be enough. A more drastic wound is made with a razor blade device. This consists of four single-

FIGURE 10–25 Ornamental nursery propagation systems. More cuttings can be rooted per unit area in a conventional plastic rooting flat (*upper left*), but additional labor is needed in transplanting rooted cuttings into small liner pots, prior to producing a finished container or field-grown crop.

Direct sticking (direct rooting) allows cuttings to be rooted directly into small liner pots (*upper and lower right*), or into large 3.8 liter (1-gal) containers (*lower left*), which saves labor and avoids transplant shock to the root system.

edged blades soldered together along their backs. Four wound cuts are then made simultaneously with this equipment.

Larger cuttings, such as magnolias and rhododendrons, may be more effectively wounded by removing a thin slice of bark for about 2.5 cm (1 in.) from the base on two sides of the cutting, exposing the cambium but not cutting deeply into the wood (*81*). For the greatest benefit, the cuttings should be treated after wounding with one of the root-promoting compounds, either a talc or a concentrated solution dip preparation, working the material into the wounds. The device shown in Fig. 10–26 can be used to make rapid and uniform wounded cuttings.

One type of wounding to avoid is inadvert-

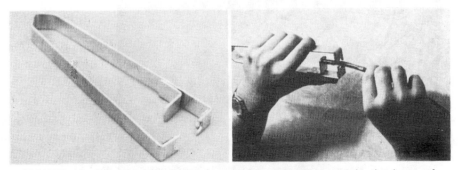

FIGURE 10–26 Tool designed for making wounding cuts in the base of cuttings to stimulate rooting. Four sharp prongs make the actual cuts as the cutting is pulled through the opening, as shown in the photo on the right.

ent crushing and damaging of basal cutting tissue with dull shears. It is from this basal tissue that both the flow of water from the propagation media must occur for cutting survival, as well as root initiation for regenerating new plants. Always use sharp, periodically disinfected pruning shears, such as Felco No 7.

TREATING CUTTINGS WITH GROWTH REGULATORS

The purpose of treating cuttings with auxin-type growth regulators (**hormones**) is to increase the percentage of cuttings that form roots, to hasten root initiation, to increase the number and quality of roots produced per cutting, and to increase uniformity of rooting. Figures 10–27 and 10–28 show examples of the benefits of such materials. Plants whose cuttings root easily may not justify the additional expense and effort of using these materials. Best use of rooting hormones is with plants whose cuttings will root but only with difficulty. The use of these substances, however, does not permit other good practices in cutting propagation, such as the maintenance of proper water relations, temperature, and light conditions, to be ignored. The value of these chemicals in propagation is well established. Although treatment of cuttings with root-promoting substances is useful in propagating plants, and can increase production efficiency time from propagule to rooted liner, the ultimate size and vigor of such treated plants are no greater than is obtained with untreated plants.

Materials

The synthetic root-promoting chemicals that have been found most reliable in stimulating adventitious root production in cuttings are **indolebutyric acid** (IBA) and **naphthaleneacetic acid** (NAA), although others can be used. IBA is probably the best material for general use, because it is generally nontoxic to plants over a wide concentration range and is effective in promoting rooting of a large number of plant species. IBA may be toxic to softwood cuttings of certain species, which leads to poor cutting regrowth and overwintering losses. These chemicals are available in commercial preparations, dispersed in talc or in liquid formulations that can be diluted with water to the proper strength (see Table 10–2).

The **K+ salt formulations** enable IBA and NAA to be dissolved in water. Otherwise, the **acid formulations** of these auxins need to be dissolved initially in alcohol (isopropyl, ethanol, methanol), acetone, DMSO (dimethyl sulfoxide),

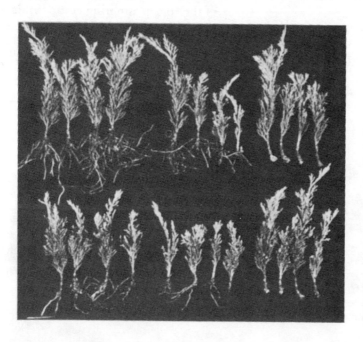

FIGURE 10–27 Effect of wounding and auxin treatment on the rooting of cuttings of *Juniperus sabina* 'Tamariscifolia' under intermittent mist in the greenhouse. *Top:* Wounded. *Below:* Not wounded. *Left:* Treated with indolebutyric acid at 4000 ppm by the concentrated dip method. *Center:* Treated with indolebutyric acid in talc, at 8000 ppm. *Right:* Not treated. Cuttings started in early spring and dug six weeks later.

FIGURE 10-28 Effect of indolebutyric acid at four concentrations on rooting of *Escallonia* leafy cuttings under mist. (From H. T. Hartmann, W. J. Flocker, and A. M. Kofranek, *Plant science,* Prentice-Hall, Englewood Cliffs, N.J., 1981.)

TABLE 10-2

Partial list of commercial rooting compounds, sources, formulations and ingredients

Trade name	Source	Formulation	Ingredients
Chryzopon	ACF Chemiefarma Maarssen, The Netherlands	Powder (talc)	0.1–8% IBA (series)
C-mone	Coor Farm Supply Services, Inc. Smithfield, NC	Liquid (isopropyl alcohol)	1% and 2% IBA
Dip 'N Grow	Alpkem Corp. Clackamas, OR	Liquid (alcohol)	1% IBA + 0.5% NAA + boron
Hormex	Brooker Chemical Corp. North Hollywood, CA	Powder (talc)	0.1–4.5% IBA (series)
Hormodin	MSD-Agvet (Merck & Co.) Rahway, NJ	Powder (talc)	0.1, 0.3, 0.8% IBA (series)
Hormo-Root	Hortus Products Co. Newfoundland, NJ	Powder (talc)	15% thiram (fungicide) plus 0.1–4.0% IBA (series)
Rhizopon	ACF Chemiefarma Maarssen, The Netherlands	Powder and water-soluble tablet forms	0.5–1.0% IAA or 0.1–0.2% NAA or 0.5–8.0% IBA (series)
Rootone	Rhone-Poulenc Ag. Co. Research Triangle Park, NC	Powder (talc)	0.1% IBA + 0.03% NAA + NAM + NAD + thiram
Roots	Wilson Laboratories, Inc. Dundas, Ontario, Canada	Liquid	0.4% IBA + ethazol (fungicide)
Synergol	Silvaperl Products Ltd. Harrogate, North Yorkshire, Great Britain		0.5% K-IBA + 0.5% K-NAA + fungicides and other additives
Woods Rooting Compound	Earth Science Products Corp. Wilsonville, OR	Liquid [ethanol + DMF (dimethylformamide)]	1% IBA + 0.5% NAA

DMF (dimethylformamide), or another solvent or carrier before water can be added. Caution should be taken when using DMSO and DMF as carriers since no EPA registration has been obtained, nor have toxicological studies been performed. With some woody plant species, the **aryl esters** of IAA and IBA and the **aryl amid** of IBA are more effective than the acid formulation in promoting root initiation (*71*). Besides commercially available powder and liquid formulations (Table 10–2), the pure chemicals are also available from chemical supply companies, so it is possible for propagators to prepare their own solutions and talc formulations.

Methods of Application

Commercial Powder Preparations

Complete directions come with the commercial materials, together with a list of plants that are likely to respond to the particular preparation. Woody, difficult-to-root species should be treated with higher concentration preparations, whereas tender, succulent, and easily rooted species should be treated with lower-strength materials. Fresh cuts should be made at the base of the cuttings shortly before they are dipped into the powder. The operation is faster if a bundle of cuttings is dipped at once rather than each cutting individually, being sure that the inner cuttings in the bundle receive as much powder as those on the outside. The powder adhering to the cuttings after they are lightly tapped is sufficient. It may be beneficial to prewet cutting bases with water so that powder adheres better. Enhanced rooting has occurred by predipping cuttings in 50 percent aqueous solutions of acetone, ethanol, or methanol prior to applying IBA talc (*37*). Some propagators have used combined treatments of auxin quick-dips (solution concentrate) followed by auxin mixed in talc applied to cutting bases.

It is advisable in using powder and liquid preparations to place a small portion of the stock material into a temporary container, sufficient for the work at hand, and discard any remaining portion after use, rather than dipping the cuttings into the entire prepared chemical stock, which may lead to its early deterioration because of contamination with moisture, fungi, or bacteria.

The cuttings should be inserted into the rooting medium immediately after treatment. To avoid brushing off the powder during insertion, a thick knife may be used to make a trench in the rooting medium before the cuttings are inserted (see Fig. 10–40).

Talc preparations have the advantage of being readily available and easy to use. Uniform results may be difficult to obtain, owing to the variability in the amount of the material adhering to the cuttings, influenced by such factors as the amount of moisture at the base of the cutting and the texture of the stem (coarse or smooth).

Dilute Solution Soaking Method

In an older procedure, the basal part—2.5 cm (1 in.)—of the cuttings is soaked in a dilute solution of the material for about 24 hours just before they are inserted into the rooting medium. The concentrations used vary from about 20 ppm for easily rooted species to about 200 ppm for the more difficult species. During the soaking period, the cuttings should be held at about 20°C (68°F), but not placed in the sun.

This is a slow, cumbersome technique and is not commercially popular. Equipment is needed for soaking cuttings, and with the long time duration, there is variability of results, with environmental changes occurring during the soaking period (*43*).

Concentrated Solution Dip Method (Quick-Dip)

In the dip method, a concentrated solution varying from 500 to 10,000 ppm or higher (0.05 to 1.0 percent) of the root-promoting chemical in 50 percent alcohol is prepared, and the basal 1/2 to 1 cm (1/5 to 2/5 in.) of the cuttings are dipped in it for a short time (usually three to five seconds, sometimes longer); then the cuttings are inserted into the rooting medium. Cuttings are most efficiently dipped as a bundle, not one by one (Fig. 10–29). With hardwood cuttings of some species, dipping just the basal cut surface gives better results than dipping 2.5 cm (1 in.) or more of the base (*39*).

Many propagators prefer the concentrated solution dip to talc application because of application ease (the base of cuttings can be dipped in *after* cuttings are stuck in the propagation media. Hence there is a time delay before auxins are absorbed through the cutting base. Only in a few limited cases with certain *Elaeagnus, Rhododendron,* and *Ilex* species does talc outperform the quick-dip method (*7*).

The same quick-dip solution can be used for many thousands of cuttings, but it must be tightly sealed when not in use because the evaporation of the alcohol will change its concentration. It is best to use only a portion of the material at a time, just sufficient for immediate needs, discarding it after use at the end of the day, rather than pouring it back into the stock solution. Stock solutions that contain a high percentage of alcohol will retain their activity almost indefinitely if kept clean.

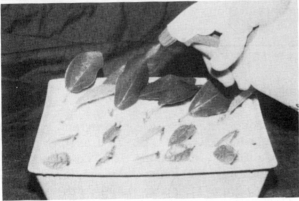

FIGURE 10-29 Application of auxin by (*top*) talc, (*center*) liquid concentrate quick-dip, or (*bottom*) spraying foliage or cutting bases.

PREPARING IBA OR NAA TALC FORMULATION

Follow the same concentration procedures as with liquid preparation (substituting 1 g of talc for 1 ml of final liquid). To make 200 g of a 5000 ppm auxin talc, dissolve 1 g of auxin in 50 to 100 ml of alcohol or acetone and then add 200 g of talc (baby powder or talc from the local pharmacy will suffice). Thoroughly stir the slurry of talc–auxin–alcohol and allow to air dry in a well-ventilated area. Rather than making serial dilutions from the highest stock solution as is done with concentrated solution dips, it is more accurate and easier with talc formulations to make up each concentration individually.

Growth regulators used in excessive concentrations for the species may inhibit bud development, causing yellowing and dropping of leaves, blackening of the stem, and eventual death of the cuttings. An effective, nontoxic concentration has been used if the basal portion of the stem shows some swelling, callusing, and profuse root production just above the base of the cutting. A concentration just below the toxic point is considered the most favorable for root promotion.

Alternative Auxin Application Methods

Rather than quick-dipping, some propagators will spray concentrate over the bases of cut-

bundles, and inadvertently removing talc from the cutting base while sticking cuttings in the medium is not a problem) (*8, 15, 37*). Greater rooting and more consistent rooting response have been reported with quick-dips than with talc, due to more uniform coverage and reduced environmental influence on chemical uptake (*16, 43*). In talc formulation, auxins (which have low solubility in the acid forms) must first go into solution

---

PREPARING AN IBA OR NAA QUICK-DIP IN 50 PERCENT ALCOHOL

Concentration: 500 to 10,000 ppm
Duration of basal dip: 3 to 5 seconds

Final concentration (ppm)	Auxin (per liter of solution)	
	mg	g
500	500	0.5
1,000	1,000	1.0
5,000	5,000	5.0
10,000	10,000	10.0

- To make a 10,000 ppm (10,000 mg/liter or 1 percent **stock solution** of auxin, dissolve 10 g of auxin in 15 to 20 ml of alcohol (ethyl, isopropyl, or methyl), then top to 1000 ml (1 liter) with 50 percent alcohol.

- To make 1 liter (1000 ml) of a 1000 ppm auxin solution from the **1 percent stock solution**, add 100 ml of the 1 percent stock solution and 900 ml of 50 percent alcohol.

- With K-IBA or K-NAA follow the same procedures, except that water is used as the solvent and no alcohol is needed.

- Avoid precipitation problems with auxin solutions by using distilled or deionized water, not tap water.

- Label solutions and color code different solution concentrations with food dyes, which can be purchased at supermarkets.

---

tings or on the foliage (Fig. 10–29). Immersing the entire cuttings into the concentrated solution dip containing a wetting agent (surfactant) has been more effective in promoting rooting in some cases than just dipping the base alone (*73*). There is an initial retardation of shoot growth, but this does not seem to be a disadvantage. Pretreatment of stock plants with foliar application of auxins prior to removing cuttings has also been used to promote rooting (*60*).

A modified banding technique using black tape (PVC) coated with IBA crystals which is wrapped around stock plant shoots prior to removing cuttings has enhanced rooting of peaches and walnut (*60*). IBA is dissolved in acetone and allowed to crystalize on glass plates holding the solution. Once the acetone is evaporated, IBA crystals adhere to the black PVC tape, which is pressed on top of the plate. The tape is then wrapped around stock plant shoots and, subsequently, roots initiate prior to shoot removal (*4*).

PREVENTIVE DISEASE CONTROL

As part of a preventive disease program, cuttings should be harvested from disease-free stock plants under nonstress conditions. It is best to collect turgid cuttings early in the day to assure optimum water conditions. It is important that the pruning shears used to collect cuttings be disinfected periodically (Fig. 10–30). Physon, isopropyl alcohol, and monochloramine are better disinfectants than sodium hypochlorite (Clorox), which is quickly inactivated when it comes in contact with organic matter (stem material, media components). Monochloramine was found to be equal in efficacy to alcohol, least corrosive, least costly, and had excellent stability under high organic contamination (*67*). See the referenced article on how a propagator can make monochloramine from local materials (*67*).

Once cuttings are collected, they should be treated with broad-spectrum fungicidal dips prior

FIGURE 10–30 Preventive disease control measures. *Upper left:* Collecting cuttings in buckets containing cups for periodically disinfecting knives and shears. *Upper center:* Bundling cuttings with rubber bands for processing. *Lower left:* Reducing stress on cuttings before sticking by maintaining them under low light, cool temperatures, and high relative humidity (note the mist line for syringing). *Lower center and right:* Soaking cuttings in a broad spectrum fungicide and bactericide prior to treating with rooting hormones and sticking.

to sticking, and/or chemical drenches during propagation. Cuttings can be dipped in a solution of Benlate (benomyl) and captan (control of damping-off organisms) and agricultural streptomycin (bacterial control) (Fig. 10–30). The cutting bases are allowed to dry (keeping the leaves wet) and then treated with auxin. Alliette and Benlate, which have systemic action and when combined give broad-spectrum control, can be rotated, along with other suitable fungicides. Always follow directions and conduct small tests to check for phytotoxicity. See the discussion of fungicides in Chaps. 2 and 9.

With hardwood cutting propagation of roses, crown gall is controlled by dipping cutting bases in an *Agrobacterium* isolate that is antagonistic to the virulent form. Mycorrhizal fungi are another biological agent that enhances root development (not root initiation) of cuttings when incorporated into propagation media. This occurs either by hormonal control or by antagonizing microflora in the media which are detrimental to rooting (77). Unfortunately, most fungicides used in propagation depress the beneficial endomycorrhizae.

Several commercial formulations of auxin include fungicides (Table 10–2), and captan and Benlate and can be mixed with commercially available or grower-made formulations of auxin (26, 29).

ENVIRONMENTAL CONDITIONS FOR ROOTING LEAFY CUTTINGS

For successful rooting of leafy cuttings, some essential environmental requirements are:

- Rooting media temperature 18 to 25°C (65 to 77°F) for temperate species and 7°C (12°F) higher for most tropical species
- Atmosphere conductive to low water loss and maintenance of turgor in leaves
- Ample, but not excessive light—1.5 to 3.0 MJ/m² per day, or 20 to 100 W/m² per day with selected temperate woody species (exceptions are with species propagated under full sun in outdoor mist beds)
- Clean, moist, well-aerated, well-drained rooting medium

A wide range of equipment is satisfactory for providing these conditions—ranging from a simple glass jar or polyethylene cover placed over a few cuttings stuck in sand, or easy-to-root cuttings placed in polyethylene rolls filled with media (Fig. 10–31) to very elaborate greenhouse propagation facilities having raised benches, automatic mist and fog systems, with computerized environmental control of relative humidity, temperature, photoperiod, light intensity, and CO_2 enrichment (Fig. 2–7) (53).

Enclosure Systems (Closed-Case Propagation)

Rooting of cuttings can be done with simple **enclosure systems (closed-case propagation)** outdoors with **low polyethylene tunnels** (sun tun-

FIGURE 10–31 Polyethylene plastic sheeting can be used for starting cuttings off easily rooted species. The basal ends of the cuttings are inserted in damp sphagnum or peat moss and rolled in the polyethylene as shown here. The roll of cuttings should then be set upright in a cool, humid location for rooting.

nels), or **cold or hot frames** covered with glass or polyethylene. Enclosed systems are also used inside a greenhouse with **contact poly** systems, where thin 1- to 3-mil poly sheets are laid in direct contact with watered-in cuttings on a raised bench or on propagation flats placed on the floor (Fig. 10–32). **Indoor polytents**, which are nonmisted polyethylene tents supported by wire or wooden frames, is another low-cost way to propagate (Fig. 10–32). Nonmisted enclosures in a greenhouse can be used to propagate difficult-to-root species and have the advantage of avoiding nutrient leaching problems of mist propagation, yet afford greater environmental controls than does outdoor propagation. See the discussion in Chap. 9.

With enclosed systems, a reduction in the water loss from leaves is accomplished by the resulting increase in relative humidity, but enclosure also tends to trap heat. Leaf tissue is not readily cooled since there is minimal air movement and evaporative cooling. To help reduce the heat load, light intensity reaching the enclosed poly system is regulated by shading, and greenhouse temperature is controlled by ventilated fogging or fan and pad cooling. Another variation of the contact poly system is to use **rooting beds on the ground out-of-doors** in full sun and lay

FIGURE 10-32 *Upper left and right:* Rooting cuttings in enclosed polyethylene tents under mist. The shade cloth shown in upper left corner can be readily pulled if light intensity becomes too high. *Below left:* Nonmisted poly tent. *Below right:* Nonmisted contact polyethylene sheet system. Note condensation on the underside of the poly sheet.

sheets of 0.63-cm (¼ in.) **Microfoam** directly on the cuttings, covering them with white 4-mil co-polymer film sealed to the ground by gravel or pieces of pipe (24). With propagation in temperate climates, the ideal cycle may be to root cuttings under contact polyethylene film in the fall and winter, and utilize intermittent mist during the spring and summer.

Intermittent Mist

Intermittent mist systems are widely used and have given propagators great flexibility in rooting softwood, semihardwood, hardwood, and herbaceous cuttings. Such sprays provide a film of water over the cuttings and media. Intermittent mist controls water loss from cuttings by reducing both leaf and surrounding air temperature via evaporative cooling, and raising relative humidity. To counteract the lower media temperatures caused by mist, bottom heat is frequently used in outdoor and indoor rooting structures (Fig. 10-33).

Open mist systems are used in both outdoor propagation in cold frames, polyethylene tunnels, and lath and shade houses, and under full sun (Figs. 10-25 and 10-33). The open mist system is also used in glass and poly-covered greenhouses

and set up on the floor area, or on, or above, the propagation bench (Fig. 10-34). A very short duration (usually 3 to 15 seconds) should be used for misting. Unless mist actually wets the leaves, rooting is likely to be unsatisfactory.

Enclosed mist systems are covered polyethylene structures inside greenhouses to reduce the fluctuation in ambient humidity and ensure more uniform coverage of mist, since air currents that disturb mist patterns are avoided (Fig. 10-32). This system has been very effective in propagating difficult-to-root species, softwood cuttings of large-leaved species (i.e., *Corylus maxima*), and broad-leaved evergreens; it is not effective for conifers. The enclosed mist system has fewer disease problems than the open mist system, since there is less mist required, less media saturation, and fewer leaching problems (42).

Mist Nozzles

The choice of mist nozzles is based on (a) cost, (b) maintenance, (c) convenience in operation, (d) availability from suppliers, (d) size of mist droplet [ideally, 50 to 100 μm (0.002 to 0.004 in.), enabling water to be suspended as a cloud for a few seconds; coarse drizzle of 80 to 150 μm should be

FIGURE 10–33 *Upper left and right:* Ground bed heating in a glass propagation house with hot water forced through solar heating panels. *Lower left:* Outdoor propagation facilities relying on bottom heat by circulating hot water through PVC pipe embedded in scoria. (Note the partially exposed PVC pipe) *Lower right:* Outdoor hot-water-heated concrete bed.

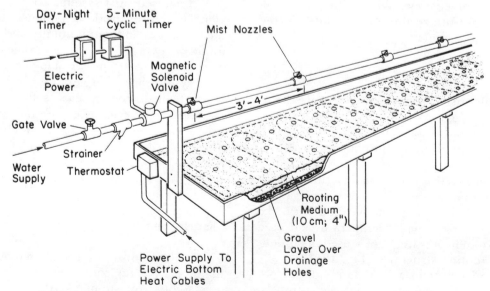

FIGURE 10–34 Basic component parts of an open intermittent mist propagating installation with electric bottom-heat cable. One timer turns the mist system on in the morning and off at night. The second is a short interval timer to provide the intermittent mist cycles.

avoided], (f) amount of water used (fine orifice mist nozzles use less water but clog-up more readily), and (g) mist pattern (sufficient coverage but avoiding overwetting media).

The two main types of nozzles are: (a) the **oil burner type**—or **pressure jet** or **whirl nozzle** and (b) the **deflection** or **anvil nozzle**.

In the oil burner type, water is forced under pressure through small grooves set on angles to each other, which produces a mist when water exits the orifice. There are improved designs operating under lower water pressures, which are nondripping and selfcleaning. Many of these nozzles have a low water output of 9 to 20 liters (2 to 5 gallons) per hour. Pressure jet nozzles have curved internal grooves, and when pressurized water is forced through the grooves, the impact at the orifice of the nozzle breaks up the water flow into mist. The Spray Systems Parasol is one such nozzle commonly used in U.S. nurseries which comes with a larger, more maintenance-free orifice (Fig. 10–35).

The deflection nozzle develops a mist when pressurized water passes through the orifice and strikes a flat surface or anvil. The larger aperture used in this type reduces clogging but uses more water. Again, there are many variations among the two principal types in orifice size and water

FIGURE 10–35 Deflection-type mist nozzles made of hard plastic (*top left*) or metal (*top center*) and pressure jet or whirl-type nozzles (*top right, bottom right*). A pneumatic pump tank (*bottom left*) is one method of maintaining sufficient water pressure so that mist (not drizzle) is produced instantaneously on demand.

efficiency. There are also some excellent hard-plastic nozzles that are less expensive and more durable than metal ones. (*72*).

Mist Controls

Intermittent mist during the daylight hours, which supplies water at intervals frequent enough to keep a film of water on the leaves but no more, gives better results than continuous mist. Since it would be impractical to turn the mist on and off by hand at short intervals throughout the day, automatic control devices are necessary. Several types are available, all operating to control a solenoid (magnetic) valve in the water line to the nozzles.

In a mist installation, especially the outdoor type, the cuttings will be damaged if the leaves are allowed to become dry for very long. Even 10 minutes without water on a hot, sunny day can be disastrous. In setting up the control system to provide an intermittent mist, every precaution should be taken to guard against accidental failure of the mist applications. This includes the use of a "normally open" solenoid valve—that is, one constructed so that if the electric power becomes disconnected, the valve is open and water passes through it. Application of electricity closes the valve and shuts off the water. If an accidental power failure occurs or any failure in the electrical control mechanism takes place, the mist remains on continuously, and no damage to the cuttings results.

Timers

Electrically operated timer mechanisms are available that operate the mist as desired. A successful type uses two timers acting together in series—one turns the entire system on in the morning and off at night; the second, an interval timer, operates the system during the daylight hours to produce an intermittent mist—at any desired combination of timing intervals, such as six seconds **ON** and two minutes **OFF**. Time clocks for regulating the application of water are preferred by many propagators because of their dependability (*32*). Some electronic timers are very versatile and can operate many banks of mist nozzles in sequence (Fig. 10–36). Timers have the disadvan-

tage of not responding to daily fluctuation in light intensity, cloud cover, relative humidity, and temperature. Although mechanically and electronically reliable, the propagator must make daily adjustments with this equipment.

> The danger of electrical shock should always be kept in mind when installing and using any electrical control unit in a mist bed where considerable water is present. The complete electrical installation should be done by a competent electrician.

Electronic Leaf

In another type of control mechanism, the so-called "electronic leaf" or "artificial leaf," two carbon electrodes are set in a small circular ebonite block and placed under the mist along with the cuttings. The alternate wetting and drying of the terminals makes and breaks the electric circuit, which, in turn, controls the solenoid valve. There are several variations in this type of control. In one, a piece of filter paper is used as the sensing material connected between two electrodes (*6, 76*). The electronic leaf would theoretically maintain a film of water on the leaves of the cuttings at all times, automatically compensating for hourly and daily changes in the leaf evaporation. The principal defect of the electronic leaf is the gradual buildup of a mineral deposit between the terminals, which will conduct electricity. The leaf must therefore be cleaned periodically. The Aquamonitor overcomes salt deposit problems since it has an air gap between the electrodes, which can be adjusted.

Screen Balance

Another type of control is based upon the weight of water. A small stainless-steel screen is attached to a lever actuating a mercury switch. When the mist is on, water collects on the screen until its weight trips the mercury switch, shutting off the solenoid. When the water evaporates from the screen, it raises, closing the switch connection, which opens the solenoid, again turning on the mist. This type of control is best adapted to regions where considerable fluctuation in weather

FIGURE 10-36 Mist can be controlled by (*top left*) producer-made systems comprised of a 24-hour time clock wired in series to 30- to 60-minute clocks which control banks of mist beds. More sophisticated electronic mist control systems are available (*top right*). A light sensor (*bottom left*) wired to a computer records radiant energy. The programmed computer then determines when shade cloth will be drawn mechanically over cutting flats being propagated with a contact poly system. A screen balance system (*middle right*) more closely approximates changing environmental conditions and leaf evaporation than do time clock systems. A simple evaporimeter (*bottom right*) made of a pipette filled with water attached to filter paper and PVC tape can be used for measuring propagation environmental conditions and in determining the adequacy of the mist being applied. (Bottom right courtesy of K. Loach.)

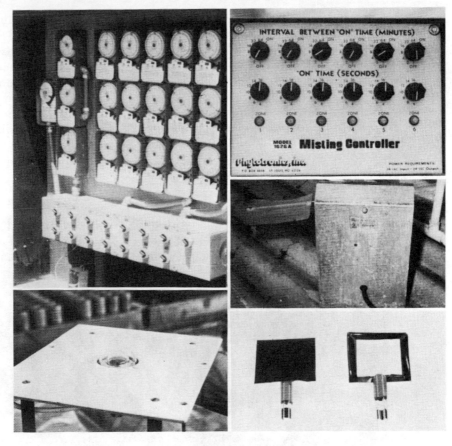

patterns may occur throughout the day, from warm and sunny to overcast, cool, and rainy; the unit compensates for changes in leaf evaporation. A Mist-A-Matic is a common screen balance unit (Fig. 10–36). These units have greater maintenance requirements than time clocks, and are prone to salt deposits, algae growth, and wind currents which distort balance accuracy.

Photoelectric Cell

Controls based upon the relationship between light intensity and transpiration rate are available. These contain a photoelectric cell which conducts current in proportion to the light intensity. It activates a magnetic counter, or charges a condenser, so that after a certain period of time the solenoid valve is opened and the mist applied. The higher the light intensity, the more frequently mist is applied. At dawn and dusk very little is used, and at night none. During cloudy days less mist is used than during bright, sunny days. Such a control system would not be well suited for outdoor mist beds, where transpiration is affected by wind movement as well as by light intensity (*57, 79*). There may be advantages in using a time clock to override the system at night since photoelectric cells are prone to overwater during cloudy weather and at night. The Mac-Penny Solarmist utilizes such a photoelectric cell system.

Computerized Controllers

There are computerized propagation controllers that can be programmed to monitor air, media, leaf cutting temperature, light intensity (radiant energy), and vapor pressure differences between air and leaf; the environmental information can then be coupled to frequency and duration of mist or fog needed. These initial computerized systems have been experimental; however, in Holland [with more than 2000 hectares (5000 acres) of glasshouses with computer-controlled climates—temperature, artificial light, shade, CO_2 enrichment] these systems are becoming a reality. Likewise, larger U.S. and English propagators are using computer-controlled systems to record environmental conditions of temperature, light, and mist in order to **model** optimum conditions for rooting cuttings (Figs. 2–7 and 10–37).

Fog Systems

Fog generators or atomizers produce very fine water droplets, less than 20 μm (about one-tenth the size of a human hair). Fog remains airborne sufficiently long for evaporation and subsequent in-

creased relative humidities of 93 to 100 percent to occur. With fog, water is suspended in the air as a vapor, whereas mist droplets (generally 50 to 100 μm) lose their suspension and fall onto the surface of leaves and media, and condense as a liquid; this liquid from mist cools the leaf surface but leaches nutrients from the leaf and can easily overwet the media. There are many advantages with fog systems; however, they are more costly than mist systems.

Fogging Equipment

Centripetal foggers for ventilated high humidity consist of a self-contained unit incorporating a large fan that forces a stream of air through

FIGURE 10–37 *Upper left and right:* ''Low-tech'' system of commercially rooting dracena cane cuttings in a polyethylene bag filled with sphagnum. *Lower left:* Some commercial nurseries utilize high-tech systems of plant modeling to determine optimal environmental conditions for rooting cuttings. *Lower right:* A data logger records environmental conditions for later evaluation.

water ejected from a spinning nozzle. The water is atomized into 30-μm droplets, which are then forced into a cool airstream through the propagation house by a fan attached to the rear of the unit. Like other fog systems, it directly humidifies the air around the cutting instead of wetting the leaf surface. Ventilated high-humidity propagation is successful only when circulation of air among the cuttings is adequate for removing heat generated from sunlight (52). Greenhouses must be shaded and good fan ventilation is essential. Best results are obtained with an oscillating humidifier that produces a large volume [10 to 50

gal (38 to 190 liters) per hour] of fog with 20- to 30-μm droplets. One such unit is the Agritech (Broadway, N.C.) (Fig. 10–38). Another variation is the Humidifan (Centre Hall, PA.), which has a single motor and is operated without a nozzle, which eliminates potential blockage problems.

With **high-pressure fogging,** water is forced under high pressure (500 to 1000 psi) through mist nozzles with very fine orifices. The water then hits an impact pin attached to the nozzle, which atomizes droplets to less than 20 μm in size, and subsequently forms a dense fog. Individ-

FIGURE 10–38 High-pressure fogging systems (*upper left and right*), centripetal foggers for ventilated high humidity (*middle right*), and pneumatic or ultrasonic humidifier nozzle systems (*bottom*) have many advantages over intermittent mist, albeit more expensive. (Bottom photo courtesy of K. Loach.)

ual nozzles typically put out 1 to 2 gal (5 to 8 liters) per hr and are spaced 2 m (6 ft) apart. One such system is the Mee (San Gabriel, Calif.) (Fig. 10–38).

Pneumatic or ultrasonic humidifier nozzles use compressed air and water. Water is disrupted by passage through a field of high-frequency sound waves generated by compressed air in a resonator located in front of the nozzle. In essence, water is being accelerated and atomized to fog. The nozzle orifice is much larger and less prone to blockage than are high-pressure fogging nozzles. Outputs range from 20 to 55 liters (5 to 14 gal) per hour. One such unit is the Sonicore ultrasonic humidifier (Ultrasonics Ltd., West Yorkshire, England) (Fig. 10–38).

Fogging Controllers

The key to successful fogging hinges on sufficient but not excessive fog, and good ventilation systems to avoid heat buildup from stagnant warm air. Relative humidity sensing is needed for fog systems. Unfortunately, accurate control of high humidity presents problems. Time clocks are not satisfactory for controlling rate of fogging. Most fog controllers operate to maintain a fixed relative humidity, which is the simplest and least expensive option, albeit less than perfect (*42*).

PREPARING THE ROOTING FRAME AND INSERTING THE CUTTINGS

The rooting frames or benches should preferably be raised or, if on the ground, equipped with

drainage tile, to assure perfect drainage of excess water. It has become popular to propagate in flats and liner containers placed on the ground of greenhouses, quonset-style houses, or outdoors on washed gravel or concrete base mist beds that have been sloped for good water drainage (Figs. 2–3 and 10–33). This avoids the high cost of propagation bench construction and utilizes space more efficiently.

The frames or flats should be deep enough so that about 10 cm (4 in.) of rooting medium can be used, and a cutting of average length—7.5 to 13 cm (3 to 5 in.)—can be inserted up to half its total length, with the end of the cutting still 2.5 cm (1 in.) or more above the bottom of the frame. The rooting medium should be watered thoroughly before the cuttings are inserted, which should be as soon as possible after they are prepared. It is very important that the cuttings be protected from drying at all stages during their preparation and insertion (Fig. 10–39).

After a section of the rooting bench or flat or small liner container has been filled with cuttings, it should be well watered to settle the rooting medium around the cuttings (see Fig. 10–40).

PREVENTING OPERATION PROBLEMS

Difficulties may arise in operating mist systems. Low water pressure can be a problem. Many propagators like to operate with a minimum mist-line water pressure of 34.5 newtons/cm^2 (50 psi) to assure that mist (not drizzle) is produced and that uniform coverage is maintained within the

FIGURE 10–39 Commercial preparation and insertion of cuttings in a large wholesale nursery. Cuttings are prepared on the right and stuck into flats for rooting on the left. Spray bottles (on left) contain a rooting hormone (indolebutyric acid) solution for spraying on the base of bundles of cuttings before inserting them in the rooting medium.

FIGURE 10-40 Steps in placing semihardwood cuttings in mist propagating beds for rooting; shown here with English holly. *Upper left:* Cutting row in rooting medium with heavy knife against board. *Upper right:* Selecting prepared cuttings for inserting. *Lower left:* Dipping cuttings for five seconds in liquid rooting ''hormone'' (auxin) preparation. *Lower right:* Sticking cuttings in rooting medium. (Courtesy Klass Ellerbrook, West Oregon Nursery, Portland, Oregon.)

specification and the spacing of the nozzle used (*46*). A small, electrically powered rotary booster pump that builds up pressure into a pressurized holding tank assures that mist can be produced instantly on demand (Fig. 10-35).

If there is much sand or debris in the water, filters should be installed in the supply line and strainers, which are checked and cleaned periodically, should always precede the solenoid valve in a mist line (Fig. 10-34). Fog systems utilize very elaborate filtration systems (particularly high-pressure fog systems with ultrafine nozzle orifices); chlorination is also included to remove algae.

Algal growth often develops a green coating on and around mist-propagation installations after an extended period of operation. This coating is not directly harmful to the cuttings but is unsightly and—because it is very slippery—can be a hazard. This slimy material is principally blue-green (*Oscillatoria, Phormidium,* and *Arthrospira*) and green (*Stichococcus* and *Chlamydomonas*) algae (*12, 18*). Allowing the water to remain off for a period each night so that the mist area can

completely dry out will generally hold the algae in check. Algicides such as Algimine, Algofen (dichlorophen), Algae-Go 36-20, or Cyprex (dodine acetate) can also be used and are effective on mosses and liverworts as well. Algae will reduce aeration of propagation media and create cultural problems.

All propagation media should have sufficient aeration and allow for good drainage. A looser mix will be needed for a closed mist system or when propagating during the cool winter months compared to the summer with its higher evaporative demands.

The **quality of water** used in mist can influence the rooting obtained (see also Chap. 2). A complete water analysis should be done (by the local municipal water department if using municipality water; otherwise, university soil labs or commercial nutrient and water analyses laboratories) to determine pH, total soluble salts, SAR (sodium absorption ratio), total carbonates, electrical conductivity, etc. High pH is commonly controlled by acidification (acid injection of sulfuric or phosphoric acid) into mist water. Water high in salts, such as sodium, or potassium carbonate, bicarbonates, or hydroxides, can be detrimental, especially when coupled with low levels of calcium (62). Adding gypsum ($CaSO_4$) to the rooting media is one way to partially offset sodium problems. Another option could be switching to a propagation system that requires less water on the foliage (i.e., a contact poly system or closed mist versus a traditional open mist with greater water demands).

It makes good sense to work with chlorinated or chloraminated mist water (67), which reduces algae problems and keeps in check some damping-off organisms. This is particularly appropriate since many U.S. nursery producers are being forced to capture and reuse runoff water.

CUTTING NUTRITION

It is essential that cuttings be taken from nutritionally healthy stock plants (56). Most cuttings have sufficient tissue nutrients to allow root initiation to occur. Intermittent mist will very rapidly deplete cutting leaves of nutrients. However, until a cutting initiates roots, its ability to pick up nutrients from propagation media is limited.

As a whole, mist application of nutrients has not been a feasible technique for maintaining cutting nutrition (69, 83). There can be problems with inhibition of rooting, and the algae formed create sanitation and media aeration problems.

A commercial technique that works with many plant species is to supply a low level of slow-release fertilizer either topdressed on the media or preincorporated (i.e., Osmocote, Ficote, etc.). Top dressing Osmocote 18-6-12 at 6.8 to 13.8 g/ m^2 (2.6 to 5.3 oz/ft²) enhanced both root and shoot development of *Ligustrum japonicum*; too high a rate will delay the rooting of cuttings. These supplementary nutrients do not promote root initiation, but rather improve root development once root primordia and elongation have occurred. Likewise, dilute liquid fertilizer can be applied to the propagation medium *after* roots have been initiated. Propagation turnover occurs more quickly and plant growth is maintained by producing rooted liners that are nutritionally more fit. However, with some species, such as *Thuja occidentalis*, there is no benefit of fertilizer application on root development (10).

CARE OF CUTTINGS DURING ROOTING

Hardwood stem cuttings or root cuttings started out-of-doors in the nursery require only the usual care given to other crop plants, such as adequate soil moisture, freedom from weed competition, and insect and disease control. Best results are obtained if the nursery is established in full sun, where shading and root competition from large trees or shrubbery does not occur.

Leafy softwood, herbaceous, or semihardwood stem cuttings and leaf-bud or leaf cuttings being rooted under high humidity require close attention throughout the rooting period. The temperature should be controlled carefully. The cuttings must not be allowed to show wilting for any length of time. Glass-covered frames, particularly the tighter polyethylene-covered structures, exposed to a few hours of strong sunlight will build up excessively high and injurious temperatures. Such structures should always be shaded by cloth screens, whitewash on the glass or poly, or some other method of reducing the light intensity.

If bottom heat is provided, thermometers should be inserted in the rooting medium to the level of the base of the cutting and checked at frequent intervals, especially at first. Excessively high temperatures in the rooting medium, even for a short time, are likely to result in death of the cuttings.

It is important to maintain humidity as high as possible in rooting leafy cuttings to keep water loss from the leaves to a minimum. Without automatic mist or fog equipment, syringing the leaves with a spray nozzle at frequent intervals is necessary, especially during hot weather. Although more time consuming, syringing with multiple light sprinklings of water each day are better than heavy soakings at less frequent intervals. A nozzle should be used that breaks the water into a fine spray. A drop of the humidity to a low level with a consequent pronounced wilting of the cuttings, if prolonged for any length of time, may so injure the cuttings that rooting will not occur, even though high-humidity conditions are subsequently resumed. However, most nursery operations use intermittent-mist, contact polyethylene sheets, or fog to overcome such problems.

Adequate drainage must be provided so that excess water can escape and not cause the rooting medium to become soggy and waterlogged. When peat or sphagnum moss is used as a component of the rooting medium, it is especially important to see that it does not become excessively wet (*59*).

It is also necessary to maintain sanitary conditions in the propagating frame. Leaves that drop should be removed promptly, as should any obviously dead cuttings. Weeds should be removed to prevent their seeding and competing with cuttings. Some preemergent herbicides, such as Ronstar and Rout, can be used with unrooted cuttings of selected species of *Rhododendron, Euonymus, Ilex,* and *Cotoneaster* (*41*). Propagation weed control begins with (a) using a clean, pasteurized, or gas-treated medium, (b) keeping the perimeters free of weeds, (c) choosing the correct herbicide to control the particular weed species, (d) following label directions, and (e) *always* conducting small trials to determine possible phytotoxicity, reduced rooting, and reduced growth of the cutting species prior to large-scale application of *any* herbicide.

Pathogens find ideal conditions in a humid, closed propagating frame with low light intensity and, if not controlled, can destroy thousands of cuttings in a short time. Disease control is done on a preventive and scheduled weekly basis, rotating fungicides for broad-spectrum control. Disease problems under mist or fog propagation conditions have not been serious. Frequent washing of leaves by aerated water can remove spores before they are able to germinate. The greater light intensity possible and good ventilation and movement of air also decrease the incidence of disease. Cuttings that are propagated under minimal stress conditions (avoidance of high heat, desiccation, and maintaining optimum water status without severe leaching) are much less susceptible to disease problems.

Pests such as mites, aphids, and mealy bugs are controlled by miticides and insecticides or by immediately roguing (discarding) infested cuttings.

Hardening-Off and Post-Propagation Care

Hardening-off is the process of gradually removing rooted cuttings from the high-relative-humidity environment of a mist, fog, or contact polysheet system. This weaning process forces the rooted cutting to become more self-sufficient in absorbing nutrients and water through the root system, in photosynthetically producing carbohydrates, and in hardening-off leaves and stems to better tolerate the stresses of lower relative humidity coupled with higher temperature and light intensity.

Cuttings *deteriorate* when they are left under mist too long after they have rooted. This reduces root quality, causes premature leaf drop, and slows down the plant's momentum, which can delay the production period and produce a poorer-quality plant. Cutting deterioration is one reason why flats or liner containers of easy-to-root and difficult-to-root species are not mixed together in the same propagation system or house—one species needs to have the mist reduced, be hardened-off and removed, while it is premature for the slower-rooting species. Hardening-off encourages more efficient secondary roots to develop from rooted cuttings. The *key to plant survival* is to reduce mist once roots start to develop, allow secondary root growth, but avoid excessive root development, which can be particularly detrimental to shoot development of leafless hardwood cuttings (*82*).

There are several ways of successfully taking the rooted cuttings from the mist conditions:

1. The cuttings may be left in place in the mist bed but with the duration of the misting periods gradually decreased, either by lessening the *on* periods and increasing the *off* periods or by leaving the misting intervals the same but gradually decreasing the time for which the mist is in operation each day.

2. In some out-of-door operations, the rooted cuttings are left in place and allowed to send their roots on through the rooting medium to the soil beneath. The propagating frame is moved to a different location to root another set of cuttings.

3. Another method is to root the cuttings in flats and move the flats after rooting to another mist frame, where they are "hardened-off" and then potted into containers as rooted liners. Cuttings may be left in the rooting medium until the dormant season, when they can be dug more safely, to be either lined out in the nursery row for further growth or potted and brought into the greenhouse. If the rooted cuttings are left in

the rooting medium for a considerable time, it is advisable to water them at intervals with a nutrient solution.

4. Some propagators **direct stick** cuttings in small containers set up in flats. Then, after rooting, the plants may be easily moved for transplanting without disturbing the roots. An alternative method is to root the cutting in a solid, block-type rooting medium, which, after rooting, permits transplanting without disturbing the roots. Such products are made from wood products, compressed peat, and synthetic materials (Fig. 10–24).

5. Another method is to pot the cuttings immediately after rooting and hold them for a time in a cool, humid, shaded location (e.g., a fog chamber, closed frame, or greenhouse).

6. With contact poly, slits can be made in the polyethylene with a knife; these are gradually increased in size and number over time to lower relative humidity.

Overwintering and/or storage problems are associated with rooted cuttings of certain deciduous woody plants, such as *Acer, Cornus, Hamamelis, Magnolia, Prunus, Rhododendron,* and *Viburnum*. Newly rooted cuttings go dormant in the fall, but die either during the winter or after budbreak in the spring (*23*). With some species it is essential that after rooting a flush of growth be put on late in the summer and the rooted liner hardened-off prior to winter. This assures winter survival. Late summer flushes of growth can be accomplished by altering daylength conditions. Fertilizer manipulation needs to be evaluated carefully (*23*). There must be sufficient time after growth flush(es) for rooted liners to harden-off before the onset of winter. In the following spring the rooted liners are then transplanted to containers or lined out in the field.

COLD STORAGE OF ROOTED AND UNROOTED LEAFY CUTTINGS

Sometimes it may be convenient to collect cuttings when nursery plants are being sheared and shaped, and store them for later propagating.

Most nurseries have refrigerated storage facilities (4 to 8°C; 40 to 47°F) for holding cuttings one to two days or longer before processing for propagation. Cuttings of *Rhododendron catawbiense* can be stored for 21 days in moist burlap bags at 2 to 21°C (36 to 70°F) with no reduction in rooting (*14*). Softwood cuttings of Kurume-type azaleas were taken in spring and held for 10 weeks in polyethylene bags at -0.5 to 4.5°C (31 to 40°F) with no adverse effect on rooting (*61*).

Many nurseries will overwinter rooted cuttings either in flats or in small liner containers protected by minimum-heat-maintained structures (i.e., quonset, polyhoop houses, greenhouses, etc.). It is possible to store rooted cuttings of certain species safely for up to five months at 1 to 4°C (34 to 39°F) in polyethylene bags (*68*). Cuttings of 31 woody ornamentals stored for six months had better survival at 0°C than at 4.5°C (32°F versus 40°F), although with some species there was no difference (*21*).

Unrooted chrysanthemum and carnation cuttings can be stored in sealed plastic bags for several weeks at -0.5°C (31°F) for subsequent rooting. In tests on the effects of storage on subsequent performance of plants, cuttings rooted after storage gave better results than those stored after rooting. Prestorage of chrysanthemum for 12 days at 10°C (50°F) enhanced rooting of cuttings compared to nonstored cuttings (Fig. 10–41). Roots were initiated while cuttings were in storage (*74*). This could enable producers to store cuttings for a more convenient time to stick, and also reduce the time needed for propagating under a mist or fog system.

Storage and subsequent rooting of carnations was better if auxin was applied *after* storage temperatures of less than 13°C (55°F), whereas auxin was more effective if applied *prior* to cuttings maintained at storage temperatures above 13°C (*75*). Again, the implications are for using storage for convenience of propagating as well as for reducing the propagation period.

Carnation cuttings, both rooted and unrooted, store well at -0.5°C (31 to 33°F) for at least five months if placed in polyethylene-lined boxes with a small amount of moist sphagnum or peat moss. The poly film should not be sealed airtight (*34*).

Some species survive cold temperature bet-

FIGURE 10-41 Effect of length of time at a prestorage temperature of 10°C (50°F) on rooting of *Chrysanthemum morifolium* 'Pink Boston'. Cuttings were stored 0 days (*left*) and 12 days (*right*). Both were propagated at the same time and evaluated after 7 days. (Courtesy P. A. Van de Pol (*74*).)

ter than others (*33*). The storage procedures for all types of nursery plants are a consideration for commercial nurseries and are an important aspect of nursery management (*45, 47*). The storage unit for unrooted cuttings should be maintained at close to 100 percent relative humidity, pathogens must be controlled (Benlate, Alliette, and other systemic fungicides should be considered), and temperature should be as low as the hardiness of the species will tolerate without impairing rooting.

HANDLING FIELD-PROPAGATED HARDWOOD CUTTINGS

Bare-Root Nursery Stock

Rooted hardwood cuttings[1] in the nursery row are usually dug during the dormant season after the

[1]The procedures described in this section also apply to nursery plants propagated as seedlings, tissue culture-produced lines, or as budded or grafted trees.

leaves have dropped. With fast-growing species, the cuttings may be sufficiently large to dig after one season's growth. Slower-growing species may require two or even three years to become large enough to transplant.

The digging should take place on cool, cloudy days when there is no wind. If possible, digging should not be done when the soil is wet, especially if it has a high clay content. Most of the soil should drop readily from the roots after the plants are removed. After the plants are dug, they should be quickly heeled-in in a convenient location, placed in cold storage, or replanted immediately in their permanent location. **Heeling-in** is to place dug, **bare-rooted** deciduous nursery plants close together in trenches with the roots well covered. This is a temporary provision for holding the young plants until they can be set out in their permanent location (see Fig. 10-42).

Commercial nurseries often store quantities of deciduous plants for several months through the winter in cool, dark rooms with the roots protected by damp bark, wood shavings, shingle tow, or some similar material. Nursery stock to be kept for extended periods should be held under environmentally controlled conditions with high relative humidity and temperatures of 0 to 2°C (32 to 35°F). If bare-root liners leaf out during storage, they should be stripped of their leaves

FIGURE 10-42 Temporary storage of nursery stock by "heeling-in" in raised beds filled with damp wood shavings sufficiently deep to cover the roots.

FIGURE 10–43 Digging machines specially constructed for undercutting nursery stock in commercial nurseries. Photo on left shows large U-shaped blade which can be lowered to travel 0.3 to 0.6 m (1 to 2 ft) below the soil surface to cut roots. Vibrating "lifter" behind the cutting blade facilitates the movement of the blade through the soil.

prior to transplanting in the field (*3*). Water stress is the chief factor limiting transplanting success. The use of antitranspirants or dipping roots in hydrogels is not nearly as effective as defoliating plants (*3*).

If only a few plants are to be removed from the nursery row, they can be dug with shovels, but in large-scale nursery operations some type of mechanical digger, such as is shown in Fig. 10–43, is generally used. This digger "undercuts" the plants. A sharp U-shaped blade travels 30 to 60 cm (1 to 2 ft) below the soil surface under the nursery row, cutting through the roots. Sometimes a horizontal, vibrating, "lifting" blade is attached to, and travels behind, the cutting blade. This slightly lifts the plants out of the soil, making them easy to pull by hand.

Balled and Burlapped Stock

Unless very small, plants of broad- or narrow-leaved evergreen species usually are not handled successfully bare-root, as is done with dormant and leafless deciduous plants. The presence of leaves on evergreen plants requires continuous contact of the roots with soil. Therefore, large,

salable plants of broad- or narrow-leaved evergreens, and occasionally deciduous plants, are either grown in containers or dug and sold "balled and burlapped" (B&B). By the latter method, the plants are removed from the soil by carefully digging a trench around each individual plant or using a digging machine such as a Vermeer spade. The soil mass around the roots is sometimes tapered at top and bottom, resulting in a ball of soil in which the roots are embedded. It is important that the soil be at the proper moisture level—not too wet and not too dry—otherwise, it will fall apart. The ball is tipped gently onto a large square of burlap, which is then pulled tightly around the ball, pinned with nails, and wrapped with heavy twine. When done properly, the burlap holds the soil to the roots adequately, and the plant can be moved safely for considerable distances and replanted successfully (Fig. 10–44). Some field-grown B&B plants are **shifted** into larger, rigid-plastic containers to ease handling and marketing, and to increase the sale value.

An alternative to ball and burlapping is to produce plants in **grow bags** which are filled with mineral soil and placed in predug holes in the field. The synthetic woven material of the bags

FIGURE 10-44 Steps in ball and burlapping nursery stock, the procedure being used here in digging tree roses. Plant is carefully removed with a ball of soil adhering to the roots (*top left*). Soil must be at proper moisture level so it does not fall apart. Burlap is wrapped around soil ball and tied with heavy cord (*top right*). Plants ready to be moved and transplanted (*below*). This method is used for all types of evergreen plants that have not been grown in containers.

limits most root penetration, and directs root growth to occur within the bag (over 90 percent of the root system of conventional bare-root and balled and burlapped plants are lost during digging). Since the bag is placed in the ground, there is greater insulation of the root system to high and low temperatures (versus above ground containerized crops), and the bag can be pulled out of the field, potentially reducing labor costs of traditional techniques. This system does not work with all species, but has merits.

Production of plants in containers has replaced much of the traditional nursery field production methods, primarily because of handling ease, improved plant marketability, greater cultural control, and faster product turnover. However, many plant species are still field-grown because of their growth habits, lower production cost, and in climates experiencing low temperatures which limit survival of containerized crops without overwintering structural protection (see Fig. 2-21).

REFERENCES

1. Ali, N., and M. N. Westwood. 1966. Rooting of pear cuttings as related to carbohydrates, nitrogen, and rest period. *Proc. Amer. Soc. Hort. Sci.* 88:145–50.

2. Alward, T. M., 1984. Softwood cuttings taken from developing hardwood cuttings. *Proc. Inter. Plant Prop. Soc.* 34:535–36.

3. Askew, J. C., C. H. Gilliam, H. G. Ponder, and G. J. Keever. 1985. Transplanting leafed-out bare root dogwood liners. *HortScience* 20:219–21.

4. Avanzato, D., and P. Cappellini. 1988. Anatomical investigations on IBA treated walnut shoots. *Acta Hort.* 227:155–59.

5. Baldwin, I., and J. Stanley. 1981. How to manage stock plants. *Amer. Nurs.* 153(8):16, 74–80.

6. Bean, G., E. S. Trickett, and D. A. Wells. 1957. Automatic mist control equipment for the rooting of cuttings. *Jour. Agr. Eng. Res.* 2:44–48.

7. Berry, J. B. 1984. Rooting hormone formulations: A chance for advancement. *Proc. Inter. Plant Prop. Soc.* 34:486–91.

8. Bonaminio, V. P. 1983. Comparison of quick-dips with talc for rooting cuttings. *Proc. Inter. Plant Prop. Soc.* 33:565–68.

9. Carlson, R. F. 1966. Factors influencing root formation in hardwood cuttings of fruit trees. *Mich. Quart. Bul.* 48:449–54.

10. Chong, C. 1982. Rooting response of cuttings of two cotoneaster species to surface-applied Osmocote slow-release fertilizer. *The Plant Propagator* 28(3):10–12.

11. Cialone, J. 1984. Developments in dracaena production. *Proc. Inter. Plant Prop. Soc.* 34:491–94.

12. Coorts, G. D., and C. C. Sorenson, 1968. Organisms found growing under nutrient mist propagation. *HortScience* 3(3):189–90.

13. Davies, F. T., Jr., and B. C. Moser. 1980. Stimulation of bud and shoot development of Rieger begonia leaf cuttings with cytokinins. *Jour. Amer. Soc. Hort. Sci.* 105(1):27–30.

14. Davis, T. D., and T. R. Potter. 1985. Carbohydrates, water potential and subsequent rooting of stored rhododendron cuttings. *HortScience:* 20:292–93.

15. Dirr, M. A. 1982. What makes a good rooting compound? *Amer. Nurs.* 155(8):33–40.

16. ———. 1983. Comparative effects of selected rooting compounds on the rooting of *Photinia × fraseri. Proc. Inter. Plant Prop. Soc.* 33:536–40.

17. Donovan, D. M. 1976. A list of plants regenerating from root cuttings. *The Plant Propagator* 22(1):7–8.

18. Durrell, L. W., and R. Baker. 1959. Algae causing clogging of cooling systems. *Colo. Flower Growers Assn. Bul. 111,* Apr.–May.

19. Erez, A., and Z. Yablowitz. 1981. Rooting of peach hardwood cuttings for the meadow orchard. *Scient. Hort.* 15:137–44.

20. Flemer, W., III. 1961. Propagating woody plants by root cuttings. *Proc. Inter. Plant Prop. Soc.* 11:42–47.

21. Flint, H. L., and J. J. McGuire. 1962. Response of rooted cuttings of several woody ornamental species to overwinter storage. *Proc. Amer. Soc. Hort. Sci.* 80:625–29.

22. Fourrier, B. 1984. Hardwood cutting propagation at McKay nursery. *Proc. Inter. Plant Prop. Soc.* 34:540–43.

23. Goodman, M. A., and D. P. Stimart. 1987. Factors regulating overwinter survival of newly propagated stem tip cuttings of *Acer palmatum* Thunb. 'Bloodgood' and *Cornus florida* L. var. rubra. *HortScience* 22:1296–98.

24. Gouin, F. R. 1981. Vegetative propagation under thermoblankets. *Proc. Inter. Plant Prop. Soc.* 30:301–5.

25. Hambrick, C. E., F. T. Davies, Jr., and H. B. Pemberton. 1985. Effect of cutting position and carbohydrate/nitrogen ratio on seasonal rooting of *Rosa multiflora. HortScience* 20:570.

26. Hansen, C. J., and H. T. Hartmann. 1968. The use of indolebutyric acid and captan in the propagation of clonal peach and peach-almond hybrid rootstocks by hardwood cuttings. *Proc. Amer. Soc. Hort. Sci.* 92:135–40.

27. Hare, R. C. 1976. Rooting of American and Formosan sweetgum cuttings taken from girdled and nongirdled cuttings. *Tree Planters Notes* 27(4):6–7.

28. ———. 1976. Girdling and applying chemicals promote rapid rooting of sycamore cuttings. *U.S. For. Serv. Res. Note SO*-202, p. 3.

29. ———. 1977. Rooting of cuttings from mature water oak. *S. Jour. Appl. Forest.* 1(2):24–25.

30. Hartmann, H. T., W. H. Griggs, and C. J. Hansen. 1963. Propagation of ownrooted Old Home and Bartlett pears to produce trees resistant to pear decline. *Proc. Amer. Soc. Hort. Sci.* 82:92–102.

31. Hartmann, H. T., C. J. Hansen, and F. Loreti. 1965. Propagation of apple rootstocks by hardwood cuttings. *Calif. Agr.* 19(6):4–5.

32. Hess, C. E., and W. E. Snyder. 1953. A simple and inexpensive time clock for regulating mist in plant propagation procedures. *Proc. Plant Prop. Soc.* 3:56–61.

33. Hocking, D., and R. D. Nyland. 1971. Cold storage of coniferous seedlings. *AFRI Res. Rpt. 6,* Syracuse Univ. Col. Forestry, Syracuse, N.Y.

34. Holley, W. D., and R. Baker. 1963. *Carnation production.* London: Grower Books.

35. Howard, B. H. 1968. The influence of 4 (indolyl-3) butyric acid and basal temperature on the rooting of apple rootstock hardwood cuttings. *Jour. Hort. Sci.* 43:23–31.

36. ———. 1971. Propagation techniques. *Scient. Hort.* 23:116–26.

37. ———. 1985. Factors affecting the response of leafless winter cuttings of apple and plums to IBA applied in powder formulation. *Jour. Hort. Sci.* 60:161–68.

38. Howard, B. H., and R. J. Garner. 1965. High temperature storage of hardwood cuttings as an aid to improved establishment in the nursery. *Ann. Rpt. E. Malling Res. Sta. for 1964,* pp. 83–87.

39. Howard, B. H., and N. Nahlawi. 1970. Dipping depth as a factor in the treatment of hardwood cuttings with indolybutyric acid. *Ann. Rpt. E. Malling Res. Sta. for 1969,* pp. 91–94.

40. Inose, K. 1971. Pumice as a rooting medium. *Proc. Inter. Plant Prop. Soc.* 21:82–83.

41. Johnson, J. R., and J. A. Meade. 1986. Preemergent herbicide effect on the rooting of cuttings. *Proc. Inter. Plant Prop. Soc.* 36:567–70.

42. Loach, K. 1989. Controlling environmental conditions to improve adventitious rooting. In *Adventitious root formation in cuttings,* T.D. Davis, B. E. Haissig, and N. Sankhla, eds. Portland, Oreg.: Dioscorides Press.

43. ———. 1988. Hormone applications and adventitous root formation in cuttings—a critical review. *Acta Hort.* 227:126–33.

44. Lohnes, J. P. 1986. Propagation of peaches and nectarines by softwood and semi-hardwood cuttings. *The Plant Propagator* 32:7–10.

45. Lutz, J. M., and R. E. Hardenberg. 1968. *The commercial storage of fruits, vegetables, and florist and nursery stocks.* USDA-ARS Agr. Handbook 66. Washington, D.C.: U.S. Govt. Printing Office.

46. Macdonald, B. 1986. *Practical woody plant propagation for nursery growers,* Vol. 1. Portland, Oreg.: Timber Press.

47. Mahlstede, J. P., and W. E. Fletcher. 1960. *Storage of nursery stock.* Washington, D.C.: Amer. Assn. Nurs., pp. 1–62.

48. Maronek, D. M., D. Studebaker, T. McCloud, V. Black, and R. St. Jean. 1983. Stripping vs. nonstripping on rooting of woody ornamental cuttings-grower results. *Proc. Inter. Plant Prop. Soc.* 33:388–97.

49. Maunder, C. 1983. A comparison of propagation unit systems. *Proc. Inter. Plant Prop. Soc.* 33:233–38.

50. Maynard, B. K., and N. L. Bassuk. 1987. Stockplant etiolation and blanching of woody plants prior to cutting propagation. *Jour. Amer. Soc. Hort. Sci.* 112:273–76.

51. Merker, R. 1985. System of direct stick propagation. *Proc. Inter. Plant Prop. Soc.* 35:182–83.

52. Milbocker, D. C. 1987. The use of humidifan in propagation. *Proc. Inter. Plant Prop. Soc.* 37:513–18.

53. Molnar, J. M., and W. A. Cumming. 1968. Effect of carbon dioxide on propagation of softwood, conifer, and herbaceous cuttings. *Can. Jour. Plant Sci.* 48:595–99.

54. Noland, D. A., and D. J. Williams. 1980. The use of pumice and pumice–peat mixtures as propagation media. *The Plant Propagator* 26(4):6–7.

55. Orndorff, C. 1987. Root pieces as a means of propagation. *Proc. Inter. Plant Prop. Soc.* 37:432–35.

56. Petersen, F. 1961. Current methods in the selection and production of nursery stock. *Proc. Inter. Plant Prop. Soc.* 11:235–40.

57. Petersen, H. 1962. A photoelectric timing control for mist application. *Proc. 15th Inter. Hort. Cong.* 3:273–79.

58. Pike, A. V. 1972. Propagation by roots. *Horticulture* 50(5):56, 57–61.

59. Pokorny, F. A. 1965. An evaluation of various equipment and media used for mist propagation and their relative costs. *Ga. Agr. Exp. Sta. Bul. n.s. 139.*

60. Preece, J. E. 1987. Treatment of stock plant with plant growth regulators to improve propagation success. *HortScience* 22:754–59.

61. Pryor, R. L., and R. N. Stewart. 1963. Storage of unrooted azalea cuttings. *Proc. Amer. Soc. Hort. Sci.* 82:483–84.

62. Raabe, R. D., and J. Vlamis. 1966. Rooting fail-

ure of chrysanthemum cuttings resulting from excess sodium or potassium. *Phytopath.* 56:713–17.

63. Sabalka, D. 1986. Propagation media for flats and for direct sticking: What works? *Proc. Inter. Plant Prop. Soc.* 36:409–13.

64. Scaramuzzi, F. 1965. Nuòva tècnica per stimolare la radicazióne delle talle légnose di ramo (A new technique for stimulating rooting in hardwood cuttings). *Riv. Ortoflorofrutticoltura Ital.* 49:101–4.

65. Scott, M. A. 1987. Management of hardy nursery stock plants to achieve high yields of quality cuttings. *HortScience* 22:738–41.

66. Scott, R. 1984. Propagation of *Camellia japonica* using horticultural rockwool. *Proc. Inter. Plant Prop. Soc.* 34:48–50.

67. Skimina, C. A. 1984. Use of monochloramine as a disinfectant for pruning shears. *Proc. Inter. Plant Prop. Soc.* 34:214–20.

68. Snyder, W. E., and C. E. Hess. 1956. Low temperature storage of rooted cuttings of nursery crops. *Proc. Amer. Soc. Hort. Sci.* 67:545–48.

69. Sorenson, D. C., and G. D. Coorts. 1968. The effect of nutrient mist on propagation of selected woody ornamentals. *Proc. Amer. Soc. Hort. Sci.* 92:696–703.

70. Stoutemyer, V. T. 1968. Root cuttings. *The Plant Propagator* 14(4):4–5.

71. Struve, D. K., and M. A. Arnold. 1986. Aryl esters of IBA increase rooted cutting quality of red maple 'Red Sunset' softwood cuttings. *HortScience* 21:1392–93.

72. Sumner, P. E. 1987. Uniformity analysis of various types of mist propagation nozzles. *Proc. Inter. Plant Prop. Soc.* 37: 522–28.

73. Van Braght, J., H. van Gelder, and R. L. M. Pierik. 1976. Rooting of shoot cuttings of ornamental shrubs after immersion in auxin-containing solutions. *Scient. Hort.* 4:91–94.

74. Van de Pol, P. A. 1988. Partial replacement of the rooting procedure of *Chrysanthemum morifolium* cuttings by pre-rooting storage in the dark. *Acta Hort.* 226: in press.

75. Van de Pol, P. A., and J. V. M. Vogelezang. 1983. Accelerated rooting of carnation 'Red Baron' by temperature pretreatment. *Scient. Hort.* 20:287–94.

76. Vanstone, F. H. 1959. Equipment for mist propagation developed at the N.I.A.E. *Ann. Appl. Bio.* 47:627–31.

77. Verkade, S. D. 1986. Mycorrhizal inocculation during plant propagation. *Proc. Inter. Plant Prop. Soc.* 36:613–18.

78. Wang, Y. T., and C. A. Boogher. 1988. Effect of nodal position, cutting length, and root retention on the propagation of golden pothos. *HortScience* 23:347–49.

79. Waxman, S., and J. H. Whitaker. 1960. A light-operated interval switch for the operation of a mist system. *Prog. Rpt. 40. Storrs (Conn.) Agr. Exp. Sta.*

80. Wells, J. S. 1952. Pointer on propagation. Propagation of *Taxus. Amer. Nurs.* 96(11):13, 37–38, 43.

81. ———. 1962. Wounding cuttings as a commercial practice. *Proc. Plant Prop. Soc.* 12:47–55.

82. Whalley, D. N., and K. Loach. 1981. Rooting of two genera of woody ornamentals from dormant, leafless (hardwood) cuttings and their subsequent establishment in containers. *Jour. Hort. Sci.* 56:131–38.

83. Wott, J. A., and H. B. Tukey, Jr. 1967. Influence of nutrient mist on the propagation of cuttings. *Proc. Amer. Soc. Hort. Sci.* 90:454–61.

SUPPLEMENTARY READING

American Nurseryman Magazine. Chicago: American Nurseryman Publ. Co.

DIRR, M. A., and C. W. HEUSER, JR. 1987. *The reference manual of woody plant propagation.* Athens, Ga.: Univ. Press.

INTERNATIONAL PLANT PROPAGATORS' SOCIETY. Proceedings of annual meetings.

JOINER, J. N. 1981. *Foliage plant production.* Englewood Cliffs, N.J.: Prentice-Hall.

LAMB, J. G. D., J. C. KELLY, and P. BOWBRICK. 1985. *Nursery stock manual.* London: Grower Books.

MACDONALD, B. 1986. *Practical woody plant propagation for nursery growers,* Vol. 1. Portland, Oreg.: Timber Press.

MCMILLAN-BROWSE, P. 1979. *Plant propagation.* New York: Simon & Schuster.

WELCH, H. J. 1970. *Mist propagation and automatic watering.* London: Faber & Faber.

11

Theoretical Aspects of Grafting and Budding

The origins of grafting can be traced back to ancient times. There is evidence that the art of grafting was known to the Chinese at least as early as 1000 B.C. Aristotle (384–322 B.C.) discussed grafting in his writings with considerable understanding. During the days of the Roman Empire grafting was very popular, and methods were precisely described in the writings of that era. Paul the Apostle, in his Epistle to the Romans, discussed grafting between the "good" and the "wild" olive trees (Romans 11:17–24).

The Renaissance period (1350–1600 A.D.) saw a renewed interest in grafting practices. Large numbers of new plants from foreign countries were imported into European gardens and maintained by grafting. By the sixteenth century the cleft and whip grafts were in widespread use in England and it was realized that the cambium layers must be matched, although the nature of this tissue was not then understood and appreciated. Propagators were handicapped by a lack of a good grafting wax; mixtures of wet clay and

dung were used to cover the graft unions. In the seventeenth century many orchards were planted in England, the trees all being propagated by budding and grafting.

Early in the eighteenth century Stephen Hales, in his studies on the "circulation of sap" in plants, approach-grafted three trees and found that the center tree stayed alive even when severed from its roots. Duhamel, about the same time, studied wound healing and the uniting of woody grafts. The graft union at that time was considered to act as a type of filter changing the composition of the sap flowing through it. Thouin (*176*) in 1821 described 119 methods of grafting and discussed changes in growth habit resulting from grafting. Vöchting (*186*) in the late nineteenth century continued Duhamel's earlier work on the anatomy of the graft union. Development of some of the early grafting techniques have been reviewed by Wells (*194*).

Liberty Hyde Bailey in *The Nursery Book* (*6*), published in 1891, described and illustrated the methods of grafting and budding commonly used in the United States and Europe at that time. The methods used today differ very little from those described by Bailey.

TERMINOLOGY

Grafting is the art of connecting two pieces of living plant tissue together in such a manner that they will unite and subsequently grow and develop as one plant. As any technique that will accomplish this could be considered a method of grafting, it is not surprising that innumerable procedures for grafting are described in the literature on this subject. Through the years several distinct methods have become established that enable the propagator to cope with almost any grafting problem at hand. These are described in Chap. 12 with the realization that there are many variations of each and that there are other, somewhat different, forms that could give the same results. Figure 11–1 illustrates a grafted plant and the parts involved in the graft.

Budding is similar to grafting except that the scion (see below) is reduced in size to usually contain only one bud. The various budding methods are described in Chap. 13.

The **scion** is the short piece of detached shoot containing several dormant buds, which, when united with the stock, comprises the upper portion of the graft and from which will grow the stem or branches, or both, of the grafted plant. It should be of the desired cultivar and free from disease.

The **stock** (*rootstock, understock*) is the lower portion of the graft, which develops into the root system of the grafted plant. It may be a seedling, a rooted cutting, or a layered plant. If the grafting is done high in a tree, as in topworking, the stock may consist of the roots, trunk, and scaffold branches.

The **interstock** (*intermediate stock, interstem*) is a piece of stem inserted by means of two graft unions between the *scion* and the *rootstock*. An interstock is used for several reasons, such as to avoid an incompatibility between the stock and scion, to make use of a winter-hardy trunk, or to take advantage of its growth-controlling properties.

Vascular cambium is a thin tissue of the plant located between the bark (phloem) and the wood (xylem). Its cells are meristematic; that is, they are capable of dividing and forming new cells. For a successful graft union, it is essential that the cambium of the scion be placed in close contact with the cambium of the stock.

Callus is a term applied to the mass of

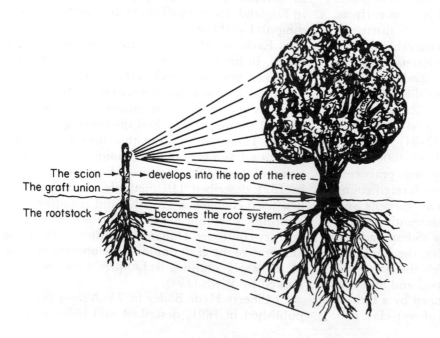

The scion → develops into the top of the tree
The graft union →
The rootstock → becomes the root system

FIGURE 11–1 In grafted plants the entire shoot system consists of growth arising from one (or more) buds on the scion. The root system consists of an extension of the original rootstock. The graft union remains at the junction of the two parts throughout the life of the plant.

parenchyma cells that develops from and around wounded plant tissues. It occurs at the junction of a graft union, arising from the living cells of both scion and stock. The production and interlocking of these parenchyma (or callus) cells constitute one of the important steps in the healing process of a successful graft.

REASONS FOR GRAFTING AND BUDDING

Grafting and budding serve many different purposes:

1. Perpetuating clones that cannot be readily maintained by cuttings, layers, division, or other asexual methods
2. Obtaining the benefits of certain rootstocks
3. Changing cultivars of established plants
4. Hastening the reproductive maturity of seedling selections in hybridization programs
5. Obtaining special forms of plant growth
6. Repairing damaged parts of trees
7. Studying virus diseases

Each of these reasons is discussed in detail below.

Perpetuating Clones That Cannot Be Readily Maintained by Cuttings, Layers, Division, or Other Asexual Methods

Cultivars of some groups of plants, including most fruit and nut species and many other woody plants, such as eucalyptus and spruce, are not propagated commercially by cuttings, because they cannot be rooted at all or in satisfactory percentages by the methods now available. Additional individual plants often can be started by layering or division, but for propagation in large quantities, it is necessary to resort to budding or grafting scions of the desired cultivar on rootstock plants with which they are compatible.

Since these plants are highly heterozygous, seed propagation will not maintain the clone. In established cultivars, buds and scions are invariably taken from the *adult* growth phase to avoid effects of juvenility delays in fruiting.

ROOTSTOCK CATEGORIES

Rootstocks can be divided into two groups: **seedling** and **clonal.**

Seedling rootstocks, those developed from germinated seeds, have certain advantages. Production of seedlings is relatively simple and economical and is well adapted to mass propagation methods (see Chap. 7). Most seedling plants do not retain viruses occurring in the parent plant (although some viruses are seed-transmitted). In some instances (e.g., plum rootstocks) the root system developed by seedlings tends to grow deeper and to be more firmly anchored than rootstocks grown from cuttings.

However, seedling rootstocks have the disadvantage of genetic variation, which may lead to variability in the growth and performance of the scion of the grafted plant. Such variation is most likely to occur if the seed is obtained from unknown, unselected sources. The seed source plants may be unusually heterozygous or may have been cross-pollinated with related species. Within the same species some seed sources produce much better rootstock plants than others. With careful testing, individual plants in a species could be selected to be a new mother-tree seed source, or the start of a clonal rootstock.

Variability among seedling rootstocks can be reduced by careful selection of the parental seed source as to identity, and by its protection from cross-pollination. Variation can be reduced if all nursery trees of the same age are dug from the nursery row at one time, and small or obviously off-type seedlings or budded trees discarded. In most nurseries the young trees are graded by size, all those of the same grade being sold together. The practice of retaining slow-growing seedlings or budded trees for an additional year's growth is undesirable as it tends to perpetuate variability. In the United States many fruit orchards if grown on uniform seedling rootstocks show no more variability resulting from the rootstock used than from unavoidable environmental differences in the orchard, principally soil variability.

Historically, **clonal rootstocks** received much attention in European countries, especially England, in regard to their development and use. These rootstocks can be propagated vegetatively either by stool layering, rooted cuttings, or by aseptic tissue culture methods. Each individual rootstock plant is the same genetically as all the other plants in the clone and can be expected to have identical growth characteristics in a given environment.

Micropropagation of clonal rootstocks makes possible the production of great numbers of such plants, upon which the scion cultivar can be grafted or budded. Apomictic seedlings (see p. 70) can also be a source of clonal rootstocks.

Clonal rootstocks are desirable not only to produce uniformity but—equally important—to preserve their special characteristics and the specific influences they have on scion cultivars, such as disease resistance, growth, or flowering habit.

To maintain rootstock influence, deep planting of the nursery tree—which may lead to "**scion rooting**"—must be avoided, as illustrated in Fig. 11–2. The deeper the graft union below the soil surface, the higher the incidence of scion rooting is likely to be (*26*).

The combinations of different clonal rootstocks with different scion cultivars allow much refinement in the performance of grafted trees. Each particular scion–stock combination requires thorough testing, however, before its performance is established and can be predicted.

In propagating and using clonal rootstocks it is very important in nursery propagation that only pathogen-free material be obtained, as any diseases present in such stocks are maintained and spread, along with the rootstock material. On such "clean" rootstock material only disease-free scion material should be grafted or budded.

Obtaining the Benefits of Certain Rootstocks

Many plant cultivars selected for their desirable fruit or ornamental qualities do not have comparably suitable root systems but require grafting onto other roots to give satisfactory plants. For many kinds of plants, rootstocks are available that tolerate unfavorable conditions, such as heavy, wet soils, or resist soil-borne insect or disease organisms (*97*) better than the plant's own roots.[1] Also, for some species size-controlling rootstocks are available that can cause the composite grafted plant to have exceptional vigor or to become dwarfed. Some rootstocks, particularly in citrus species, give better size and quality of the fruit of the scion cultivar than do others (*149*).

Special rootstocks for glasshouse vegetable crops are used in Europe to avoid root diseases such as *fusarium* and *verticillium* wilt (*87*). In the Netherlands, forcing types of greenhouse cucumbers are grafted onto *Cucurbita ficifolia*, and commercial tomato cultivars are grafted onto vigorous F_1 hybrid, disease-resistant rootstocks (*168*).

Double Working

More than two kinds of plants can be combined together in a vertical arrangement. In addition to the rootstock and scion, one may insert a third kind between them by grafting. Such a section is termed an **intermediate stem section,** an **intermediate stock,** or **interstock.**

There are several reasons for using double-working in propagation.

1. The interstock makes it possible to avoid certain kinds of incompatibility.

---

[1]Detailed discussions of the rootstocks available for the various fruit and ornamental species are given in Chaps. 18 and 19.

FIGURE 11-2 Scion roots of an 'Old Home' pear grafted on quince. (A) Original quince roots. (B) Scion roots arising from the 'Old Home' pear above the graft union. These have assumed the major support of the tree. The dwarfing influence due to the quince roots has disappeared.

2. The interstock may possess a particular characteristic (such as disease resistance or cold-hardiness) not possessed by either the rootstock or the scion, which makes the interstock valuable for use as the main framework of the tree.

3. A certain scion cultivar may be required for disease resistance in cases where the interstock characteristics are the chief consideration, such as in the control of leaf blight on *Hevea* rubber trees (*97*).

4. The interstock may have a definite influence on the growth of the tree. For example, when a stem piece of the dwarfing 'Malling 9' apple stock is inserted between a vigorous rootstock and a vigorous scion cultivar, it reduces growth of the composite tree and stimulates flowering in comparison with a similar tree propagated without the interstem piece (*145*) (see Fig. 11-24).

Nurseries supplying trees on seedling or clonal rootstocks, or with a clonal interstock, should identify such stocks on the label just as they do for the scion cultivar.

Changing Cultivars of Established Plants (Topworking)

A fruit tree, or an entire orchard, may be of an undesirable cultivar. It could be unproductive, or an old cultivar whose fruits are no longer in demand; it could be one with poor growth habits, or possibly one that is susceptible to prevalent diseases or insects. As long as a compatible type is used, the framework of the tree may be regrafted to a more desirable cultivar, if one exists (*130*) (see Fig. 12–28).

In an orchard of a single cultivar of a species requiring cross-pollination, provision for adequate cross-pollination can be obtained by topworking scattered trees throughout the orchard to a proper pollinizing cultivar. Or if a single tree of a species such as the sweet cherry is unfruitful because of lack of cross-pollination, a branch of the tree may be grafted to the proper pollinizing cultivar.

A single pistillate (female) plant of a dioecious (pistillate and staminate flowers borne on separate individual plants) species, such as the hollies (*Ilex*), may be unfruitful because of the lack of a nearby staminate (male) plant to provide proper pollination. This problem can be corrected by grafting a scion taken from a staminate plant onto one branch of the pistillate plant.

Of interest to the home gardener is the fact that several cultivars of almost any fruit species can be grown on a single tree of that species by topworking each primary scaffold branch to a different cultivar. In a few cases, different species can be worked on the same tree. For example, on a single citrus tree it would be possible to have oranges, lemons, grapefruit, mandarins, and limes; or on a peach root system could be grafted plum, almond, apricot, and nectarine. Some different cultivars (or species), however, grow at different degrees of vigor, so careful pruning is required to cut back the most vigorous cultivar on the tree to prevent it from becoming dominant over the others.

Hastening Reproductive Maturity of Seedling Selections

In various fruit-breeding projects, the young seedling selections, if left to grow on their own roots, may take five to ten or more years to grow out of the juvenile phase and come into bearing. It has sometimes been possible to hasten the onset of maturity by grafting scions from the seedling shoots onto large, established trees (*183*) or onto certain dwarfing rootstocks (*184*). Such grafting does not eliminate seedling juvenility, rather it takes advantage of an existing large root system of the rootstock plant to advance the change toward maturity.

It is possible to graft several seedling selections on one mature tree, but this could be inadvisable owing to the danger of virus contamination coming from the old tree, or perhaps from one or more of the seedling scions, and spreading to the others.

Sometimes desirable new seedling plants produced in breeding programs never attain the ability to grow well on their own roots, but when grafted on a vigorous, compatible rootstock, develop into plants of the desired form and stature (*38*). Crosses of *Syringa laciniata* × *S. vulgaris,* for example, produce Chinese lilac hybrids, but many of the seedlings lack sufficient vigor to live more than two or three years. However, if they are budded on the tree lilac, *Syringa amurensis japonica,* they survive and grow vigorously (*156*).

Obtaining Special Forms of Plant Growth

By grafting certain combinations together it is possible to produce unusual types of plant growth, such as "tree" roses or "weeping" cherries or birches. For this purpose, an upright growing trunk is grafted at a height of 1.2 to 1.8 m (4 to 6 ft) to a cultivar or species that has a drooping or hanging growth habit. Often an intermediate upright growing stock must be used to obtain the height needed before grafting on the "weeping" top scion cultivar (see Fig. 11–3).

Cactus is easily grafted to produce unusual plant forms, as shown in Fig. 11–4.

FIGURE 11–3 How "weeping" plant forms may be obtained by grafting. An upright growing cultivar is grafted at the top by a side graft with another cultivar having a hanging growth pattern.

Repairing Damaged Parts of Trees

Occasionally, the roots, trunk, or large limbs of trees are severely damaged by winter injury, cultivation implements, diseases, or rodents. By the use of bridge-grafting, or inarching, such damage can be repaired and the tree saved. This is discussed in detail in Chap. 12.

Studying Virus Diseases

Virus diseases can be transmitted from plant to plant by grafting. This characteristic makes possible testing for the presence of the virus in plants that may carry the pathogens but show few or no symptoms. By grafting scions or buds from a plant suspected of carrying the virus onto an indicator plant known to be highly susceptible, which

FIGURE 11-4 Grafted ornamental cactus. An easily rooted cultivar is used as the rootstock and an unusual attractive type is used as the scion. These grafts are made in large quantities in Japan and shipped to wholesale nurseries in other countries for rooting, potting, and growing until ready for sale in retail outlets.

shows prominent symptoms, detection is easily accomplished (*110*). This procedure is known as *indexing* (see Chap. 8).

In order to detect the presence of a latent virus in a symptomless carrier, it is not necessary to use only combinations which make a permanent, compatible graft union. For example, the 'Shirofugen' flowering cherry (*Prunus serrulata*) is used to detect viruses in peach, plum, almond, and apricot. Cherry does not make a compatible union with these species, but a temporary, incompatible union is a sufficient bridge for virus transfer (*94*).

NATURAL GRAFTING

One occasionally sees branches that have become grafted together naturally following a long period of being pressed together without disturbance.

English ivy (*Hedera helix*) forms such grafts, and detailed studies have been made of translocation in natural grafts of this species (*113*). Not so obvious but of much greater significance and occurrence, particularly in stands of forest species, such as pine, hemlock, oak, and Douglas fir, is the natural grafting of roots (*58, 109, 172*).

Such grafts are most common between roots of the same tree or between roots of trees of the same species. Grafts between roots of trees of different species are rare. In the forest, living stumps sometimes occur, kept alive because their roots have become grafted to those of nearby intact, living trees allowing the exchange of nutrients, water, and metabolites (*108*). The anatomy of natural grafting of aerial roots has been studied; the initial contact is established by the formation and fusion of epidermal hairs (*143*). Such natural grafting permits transmission of fungi, viruses, and mycoplasms from infected trees to their neighbors (49). This can be important in closely set orchard and nursery plantings of fruit trees, where numerous root grafts could occur and result in the slow spread of pathogens throughout the planting. Natural root grafting is a potential source of error in virus indexing procedures where virus-free and virus-infected trees are grown in close proximity (61). In addition, fungal pathogens such as those causing oak wilt and Dutch elm disease can be spread by such natural root connections.

FORMATION OF THE GRAFT UNION

A number of detailed studies have been made of the healing of graft unions, mostly with woody plants (35, 50, 107, 147, 158, 164, 170, 174), but some have been done with herbaceous plants (103, 117, 119). Briefly, the usual sequence of events in the healing of a graft union is as follows (Fig. 11–5):

1. **Lining up of vascular cambiums.** Freshly cut scion tissue capable of meristematic activity is brought into secure, intimate contact with similar freshly cut stock tissue in such a manner that the cambial regions of both are in close proximity. Temperature and humidity conditions must be such as to promote growth activity in the newly exposed, and surrounding, cells.

2. **Wound healing response.** The formation of necrotic material from the cell contents and cell walls of cut scion and stock cells.

3. **Callus bridge formation.** The outer exposed layers of undamaged cells in the cambial region of both scion and stock produce parenchyma cells that soon intermingle and interlock, filling up the spaces between scion and stock; this is called **callus tissue** (Figs. 11–6 and 11–7).

4. **Cambium formation.** Certain cells of this newly formed callus in line with the cambium layer of the intact scion and stock differentiate into new cambium cells.

5. **Vascular tissue formation.** These new cam-

bium cells produce new vascular tissue, xylem toward the inside and phloem toward the outside, thus establishing a secondary vascular connection between the scion and stock, a requisite of a successful graft union.

The healing of a graft union can be considered as the healing of a wound (14). Such injury to tissue as would occur if the cut end of a branch were split longitudinally would heal quickly if the split pieces were bound tightly together. New parenchyma cells would be produced by abundant proliferation from cells of the cambium region of both pieces, forming callus tissue. Some of the newly produced parenchyma cells differentiate into cambium cells, which subsequently produce xylem and phloem.

If, between the two split pieces, one interposed a third, detached, piece cut so that a large number of its cells in the cambial region could be placed in intimate contact with cells of the cambial region of the two split pieces, proliferation of parenchyma cells from all cambial areas would soon result in complete healing, with the foreign, detached piece joined completely between the two original split pieces. A graft union is essentially a healed wound, with an additional, foreign, piece of tissue incorporated into the healed wound.

This added piece of tissue, the scion, will not resume its growth successfully, however, unless vascular connection has been established so that it may obtain water and mineral nutrients. In addition, the scion must have a terminal meristematic region—a bud—to resume shoot growth and, eventually, to supply photosynthates to the root system.

In the healing of a graft union, the parts of the graft that are originally prepared and placed in close contact do not themselves move about or grow together. The union is accomplished entirely by cells that develop *after* the actual grafting operation has been made.

In addition, it should be stressed that in a graft union *there is no intermingling of cell contents* (i.e., no fusion of cell contents or protoplasts). Cells produced by the stock and by the scion each maintain their own distinct identity.

Considering in more detail the steps involved in the healing of a graft union, we may say

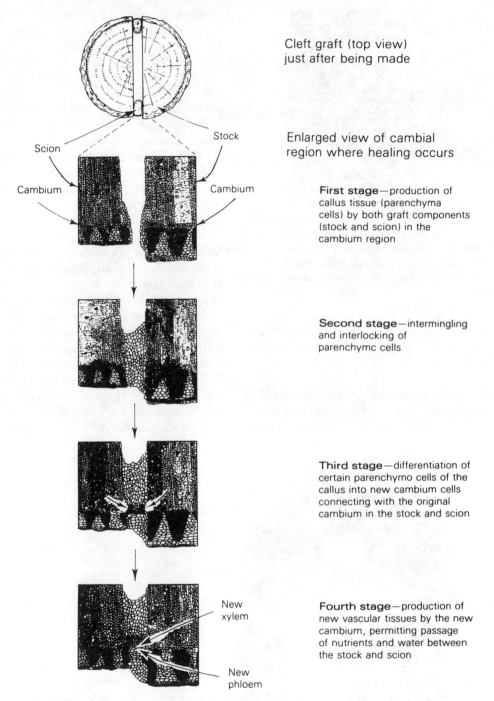

Cleft graft (top view)
just after being made

Scion   Stock

Cambium   Cambium

Enlarged view of cambial
region where healing occurs

First stage—production of
callus tissue (parenchyma
cells) by both graft components
(stock and scion) in the
cambium region

Second stage—intermingling
and interlocking of
parenchymc cells

Third stage—differentiation of
certain parenchymo cells of the
callus into new cambium cells
connecting with the original
cambium in the stock and scion

New
xylem

New
phloem

Fourth stage—production of
new vascular tissues by the new
cambium, permitting passage
of nutrients and water between
the stock and scion

FIGURE 11–5 Diagrammatic developmental sequence during the healing
of a graft union as illustrated by the cleft graft.

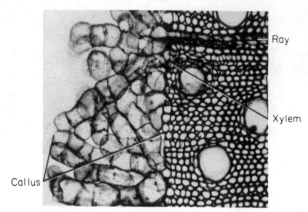

FIGURE 11–6 Callus production from incompletely differentiated xylem, exposed by excision of a strip of bark. × 120. (Reprinted with permission from K. Esau, *Plant Anatomy,* John Wiley & Sons, Inc., New York, 1953.)

that the first one listed below is a preliminary step, but nevertheless, it is essential and one over which the propagator has control.

1. *Establishment of intimate contact of a considerable amount of the cambial region of both stock and scion under favorable environmental conditions.* Temperature

conditions that will cause cell activity are necessary. Usually, temperatures from 12.8 to 32°C (55 to 90°F), depending upon the species, are conducive to rapid growth. Outdoor grafting operations should thus take place at a time of year when such favorable temperatures can be expected and when the plant tissues, especially the cambium, are in a naturally active state. These conditions generally occur during the spring months. Temperature levels under greenhouse and bench grafting situations can, of course, be readily controlled, thereby permitting greater reliability of results and permit grafting over a longer period of time.

The new callus tissue arising from the cambial region is composed of thin-walled, turgid cells, which can easily become desiccated and die. It is important for the production of these parenchyma cells that the air moisture around the graft union be kept at a high level. This explains the necessity of thoroughly waxing or sealing the graft union or placing root grafts in a moist medium to maintain a high degree of tissue hydration.

It is important, too, that the region of the graft union be kept as free as possible from pathogenic organisms. The thin-walled parenchyma cells, under relatively high humidity and temper-

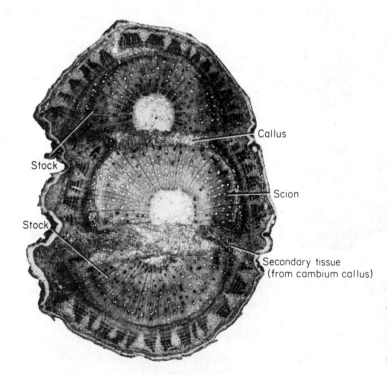

FIGURE 11–7 Cross section of a *Hibiscus* wedge graft showing the importance of callus development in the healing of a graft union. Cambial activity in the callus has resulted in the production of secondary tissues which have joined the vascular tissues of stock and scion. × 10. (Reprinted with permission from K. Esau, *Plant Anatomy,* John Wiley & Sons, Inc., New York, 1953.)

ature conditions, provide a favorable medium for growth of fungi and bacteria, which are exceedingly detrimental to the successful healing of the union. Prompt covering of the graft union helps prevent such infection.

It is essential that the two original graft components be held together firmly by some means, such as wrapping, tying, or nailing, or better yet, by wedging (as in the cleft or notch grafts) so that the parts will not move about and dislodge the interlocking parenchyma cells after proliferation has begun.

The statement is often made that for successful grafting the cambium layers of stock and scion must be "matched." Although this is desirable, it is unlikely that complete matching of the two cambium layers is, or ever can be, attained. In fact, it is only necessary that the cambial regions be close enough together so that the parenchyma cells from both stock and scion produced in this region can become interlocked. A slight crossing of the stock and scion cambium layers insures callus interlocking. It is in the region of the cambium that the essential callus production is the highest. Two badly matched cambial layers may delay union or, if extremely mismatched, prevent the graft union from taking place. In studies (127) of grafting monocotyledonous plants, it was found that a cambium layer is not necessarily required for a successful graft union but that any meristematic tissue would generate callus tissue to lead to the formation of a union between stock and scion.

2. *Formation of necrotic material from the cell contents and cell walls of cut scion and stock cells (a wound healing response).* In the actual cutting of cells of both scion and stock during the preparation of the graft the cells are killed at each surface at least one cell layer deep. Much of the necrotic material disappears or it may remain in pockets between subsequently formed living parenchyma cells.

3. *Production and interlocking of parenchyma cells (callus tissue) by both stock and scion.* Underneath the dead cells the living cells show increased cytoplasmic activity with, in some plants at least, a pronounced accumulation of **dictyosomes**[2] along the

[2]Dictyosomes in plant cells are a series of flattened plates or double lamellae—one of the component parts of the Golgi apparatus.

graft interfaces (see Fig. 11–8) (*116–119*). These dictyosomes appear to secrete materials into the cell wall space between the graft components via vesicle migration to the plasmalemma, resulting in a rapid adhesion between parenchymatous cells at the graft interface.

From these living cells new parenchyma cells (**callus**) proliferate in one to seven days from both stock and scion, coming from the parenchyma of the phloem rays and the immature parts of the xylem. The actual cambial layer itself seems to take little part in this first development of the callus (*91, 153, 162*).

In grafting scions on established stocks, the stock produces most of the callus. These parenchyma cells comprising the spongy callus tissue penetrate the thin necrotic layer within two or three days and soon fill the space between the two components of the graft (scion and stock), becoming intimately interlocked and providing some mechanical support, as well as allowing for lim-

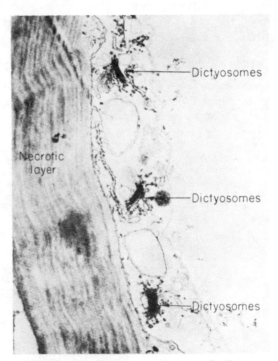

FIGURE 11–8 Accumulation of dictyosomes along the cell walls adjacent to the necrotic layer at six hours after grafting in the compatible autograft in *Sedum telephoides.* × 17,500. (Courtesy R. Moore and D. B. Walker (*119*).)

ited passage of water and nutrients between the stock and the scion. For a time, between the callus arising from the stock and that arising from the scion, there is a more or less continuous brown line consisting of dead and crushed cells remaining between the two tissues of the graft. This line of cells is gradually resorbed, however, and disappears. At the final stages of healing, the cells of the outer layer of callus become suberized (*174*). Previously existing vascular elements (tracheids, vessels) are sealed off with a deposit of gum (*16*).

4. *Production of new cambium through the callus bridge.* At the edges of the newly formed callus mass, parenchyma cells touching the cambial cells of the stock and scion differentiate into new cambium cells within two to three weeks after grafting. This cambial formation in the callus mass proceeds farther and farther inward from the original stock and scion cambium, and on through the **callus bridge,** until a continuous cambial connection forms between stock and scion. Figure 11–9 shows the development of a cambial bridge through the callus in a mango "chip bud" graft union.

5. *Formation of new xylem and phloem from the new vascular cambium in the callus bridge.* The newly formed cambial layer in the callus bridge begins typical cambial activity, laying down new xylem

toward the inside and phloem toward the outside, along with the original vascular cambium of the stock and scion, continuing this throughout the life of the plant.

In the formation of new vascular tissues following cambial continuity, it appears that the type of cells formed by the cambium is influenced by the cells of the stock adjacent to the cambium. For example, xylem ray cells are formed where the cambium is in contact with xylem rays of the stock, and xylem elements where they are in contact with xylem elements (*140*). Leaves developing on the scion—but not on the stock—have been observed to exert a strong stimulus to induce the differentiation of vascular tissue across the graft interface (*171*).

The new xylem tissue originates from the activities of the scion tissues rather than from that of the stock. This origin is demonstrated in "ring grafting," in which a ring of bark from a young tree is removed and replaced by a ring of bark from another tree (*158*). Researchers using bark rings of the 'Scugog' apple, which has purple xylem tissue, found that subsequent xylem growth of the tree following grafting was entirely purple in color just under the 'Scugog' ring of bark, whereas in the remainder of the tree the xylem remained white (*199*).

Production of new xylem and phloem thus permits vascular connection between the scion

FIGURE 11–9 Cambial bridge developing through callus in mango "chip bud" graft union–after 12 days. Sc, scion; St, stock; Ph, phloem; Xy, xylem; C, cambium; Ca, callus; LC, laticiferous canal. (Courtesy J. Soule (*170*).)

and the stock. It is essential that this stage be completed before much new leaf development arises from buds on the scion. Otherwise, the enlarging leaf surfaces on the scion shoots will have little or no water to offset that lost by transpiration, and the scion will quickly become desiccated and die. It is possible, however, even though vascular connections fail to occur, enough translocation can take place through the parenchyma cells of the callus to permit survival of the scion. In grafts of *Vanilla* orchid, a monocot, scions survived and grew for two years with only parenchyma union (*127*).

A somewhat different developmental sequence has been found to take place in tobacco (*37*) and cotton (*86*). Here, xylem tracheary elements or phloem sieve tubes, or both, form directly by differentiation of callus into these vascular elements. A cambium layer subsequently forms between the two vascular elements. Apparently parenchyma cells, which make up the callus, can easily differentiate into tracheid-like elements.

Buds, as would occur on the scions of grafts, are effective in inducing differentiation of vascular elements in the tissues on which they are grafted. This bud influence has been shown by inserting a bud into a piece of *Cichorium* root, consisting only of old vascular parenchyma, and observing, as shown in Fig. 11–10, that under the influence of auxin produced by the bud, the old parenchyma cells differentiate into groups of conducting elements (*24, 56, 164*).

Induction of vascular tissues in callus, as would be formed in the healing of a graft union, is under the control of materials originating from growing points of shoots. Furthermore, the stimulus from such shoot apices to produce xylem formation can be replaced by appropriate concentrations of such hormones as auxins, cytokinins, and gibberellins, as well as by sugars (*178*). It has been demonstrated (*196*) that sugar concentrations of 2.5 to 3.5 percent, plus IAA or NAA at 0.5 mg/liter, will cause the induction of xylem and phloem in the callus, with a cambium in between.

THE HEALING PROCESS IN T- AND CHIP BUDDING

In T-budding, the bud piece usually consists of the periderm, cortex, phloem, cambium, and often some xylem tissue. Attached externally to this is a lateral bud subtended, perhaps, by a leaf petiole. In budding, this piece of tissue is laid against the exposed xylem and cambium of the stock, as shown diagrammatically in Fig. 11–11.

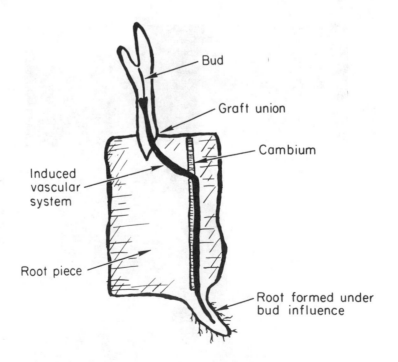

FIGURE 11–10 Stylized depiction of the induction of vascular tissue by grafting a bud on to a piece of *Chicorium* root tissue. (Redrawn from Gautheret (*56*).)

CROSS SECTION THROUGH INSERTED T-BUD

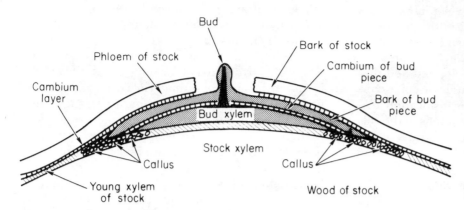

FIGURE 11-11 Tissues involved in healing of an inserted T-bud as prepared with the wood (xylem) attached to the bud piece. Healing occurs when callus cells developing from the young xylem of the stock intermingle with callus cells forming from exposed cambium and young xylem of the T-bud piece. As the bark is lifted on the stock for insertion of the bud piece it detaches by separation of the youngest xylem cells.

Detailed studies of the healing process in T-budding have been made for the rose (20), citrus (106), and apple (124). Figure 11-12 shows a longitudinal section through a healed T-bud.

When the bud piece is removed from the budstick, and when the flaps of bark on either side of the "T" incision on the stock are raised, the cambium and newly formed xylem and phloem in these tissues are usually destroyed, owing to their very tender, succulent nature.

In the apple, when the bark of the stock is lifted in the budding operation, the separation occurs in the young, undifferentiated xylem. The entire cambial zone remains attached to the inside of the bark flaps. Very shortly after the bud shield is inserted a necrotic plate of material develops from the cut cells. Next, after about two days, callus parenchyma cells start developing from the rootstock xylem rays and break through the necrotic plate. Some callus parenchyma from the bud scion ruptures through the necrotic area in a similar manner. As additional callus is produced it surrounds the bud shield and holds it in place. The callus originates almost entirely from the rootstock tissue, mainly from the exposed surface of the xylem cylinder. Very little callus is pro-

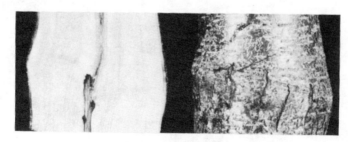

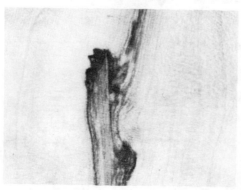

FIGURE 11-12 Healing at the bud union in a three-year-old T-budded almond tree. *Top left:* Longisection through bud union. *Top right:* External appearance of union showing slight swelling. *Below:* Detail at union showing location of original inserted bud and the "buried" remains of the budded rootstock. Note clean line at union and good vascular connection.

duced from the sides of the bud shield. Callus proliferation continues rapidly for two to three weeks until all internal air pockets are filled. Following this, a continuous cambium is established between the bud and the rootstock. The callus then begins to lignify, and isolated tracheary elements appear. Lignification of the callus is completed about 12 weeks after budding (*124, 188*).

In the rose, about three days after budding, the terminal cells of the broken xylem rays and adjacent cambial derivatives on the exposed surface of the stock begin to enlarge and divide, leading to the production of callus strands. In the same manner, callus strands develop from terminal cells of broken phloem rays and adjacent young secondary phloem cells on the cut surface of the inner side of the bud piece. Within 14 days the space between the stock and the bud piece is completely filled with callus, which has developed mainly from the proliferating immature secondary xylem of the stock and the immature secondary phloem of the bud piece. During the second week, short areas of cambium cells appear in this newly developed callus tissue. By the tenth day, a completed band of cambium tissue extends over the face of the stock and is joined to the uninjured cambium on either side of the bud piece.

After cambial continuity is completed, vascular tissue connection soon becomes established between the bud and stock. In T-budding, then, the primary union is between the surface of the phloem on the inner face of the shield and the meristematic xylem surface of the stock. A secondary type of union may occur, however, at the edges of the shield piece, as it would in chip budding (*16*).

The various stages and time intervals involved in the healing of the union in T-budding of citrus have been determined as follows (*106*):

Stage of Development	*Approximate Time After Budding*
1. First cell division	24 hours
2. First callus bridge	5 days
3. Differentiation of cambium	
a. In the callus of the bark flaps	10 days
b. In the callus of the shield	15 days
4. First occurrence of xylem tracheids	
a. In the callus of the bark flaps	15 days
b. In the callus of the shield	20 days
5. Lignification of the callus completed	
a. In the bark flaps	25 to 30 days
b. Under the shield	30 to 45 days

Anatomical studies (*167*) have been made comparing the healing process in T and chip budding. Early union formation between 'Lord Lambourne' apple scion and 'Malling 26' rootstock showed a more rapid and complete union of xylem and cambial tissues of scion and rootstock after chip budding compared to T-budding, probably due to much closer matching of the scion tissue to the rootstock stem. Other studies were made comparing T- and chip-bud healing with 'Crimson King' maple on *Acer platanoides* rootstock, 'Cox's Orange Pippin' apple on 'Malling 26' rootstock, 'Conference' pear on 'Quince A' roots, and 'Rubra' linden on *Tilia platyphyllos* roots, all with similar results.

FACTORS INFLUENCING THE HEALING OF THE GRAFT UNION

As anyone experienced in grafting or budding knows, the results are often inconsistent, an excellent percentage of ''takes'' occurring in some operations, whereas in others the results are disappointing. A number of factors can influence the healing of graft unions.

Incompatibility

One of the symptoms of incompatibility in grafts between distantly related plants is a complete lack, or a very low percentage, of successful unions (see page 326). Grafts between some plants known to be incompatible, however, will initially make a satisfactory union, even though the combination eventually fails (see Fig. 11–17.)

Kind of Plant

Some plants are much more difficult to graft than others even when no incompatibility is involved. Difficult ones, for example, are the hickories, oaks, and beeches. Nevertheless, such plants, once successfully grafted, grow very well with a perfect graft union. In topgrafting apples and pears, even the simplest techniques usually give a good percentage of successful unions, but in topgrafting certain of the stone fruits, such as peaches and apricots, much more care and attention to details are necessary. Strangely enough, topgrafting peaches to some other compatible species, such as plums or almonds, is more successful than reworking them back to peaches. Many times one method of grafting will give better results than another, or budding may be more successful than grafting, or vice versa. For example, in topworking the native black walnut (*Juglans hindsii*) to the Persian walnut (*Juglans regia*) in California, the bark graft method is more successful than the cleft graft.

Some easily grafted plants, such as apple, form a "wound gum" plugging exposed xylem elements after the grafting operation, thus preventing excessive desiccation and death of tissues. Other plants, such as the walnut, in which graft unions heal with difficulty, form such "wound gum" very slowly, and in these desiccation and death of tissues in the area of the graft union may be extensive (*16*).

Some species, such as the Muscadine grape (*Vitis rotundifolia*), mango (*Mangifera indica*), and *Camellia reticulata,* are so difficult to propagate by the usual grafting or budding methods that "approach grafting," in which both partners of the graft are maintained for a time on their own roots, is often used. This variation among plant species and cultivars in their grafting ability is probably related to their ability to produce callus parenchyma, which is essential for a successful graft union.

Temperature, Moisture, and Oxygen Conditions during and following Grafting

Certain environmental requirements must be met for callus tissue to develop.

Temperature

Temperature has a pronounced effect on the production of callus tissue (Fig. 11–13). In apple grafts little, if any, callus is formed below 0°C (32°F) or above about 40°C (104°F). Even around 4°C (40°F), callus development is slow and meager, and at 32°C (90°F) and higher, callus production is retarded, with cell injury becoming more apparent as the temperature increases, until death of the cells occurs at 40°C. Between 4 and 32°C, however, the rate of callus formation increases directly with the temperature (*166*). In such operations as bench grafting, callusing may be allowed to proceed slowly for several months by storing the grafts at relatively low temperatures, 7 to 10°C (45 to 50°F), or if rapid callusing is desired, they may be kept at higher temperatures for a shorter time.

Following bench grafting of grapes, a temperature of 24 to 27°C (75 to 80°F) is about optimum; 29°C (85°F) or higher results in profuse formation of a soft type of callus tissue that is easily injured during the planting operations. Below

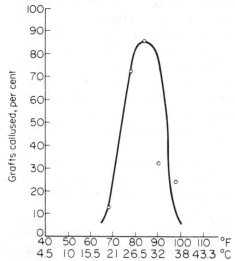

FIGURE 11–13 Influence of temperature on the callusing of walnut (*Juglans*) grafts. Callus formation is essential for the healing of the graft union. Maintaining an optimum temperature following grafting is very important for successful healing of walnut grafts. (Adapted from data of Sitton (*166*).)

20°C callus formation is slow, and below 15°C (60°F) it almost ceases.

Outdoor grafting operations performed late in the spring when excessively high temperatures may occur, often result in failure. Tests (*67*) of walnut top-grafting in California during very hot weather in May showed that whitewashing the area of the completed graft union definitely promoted healing of the union. Whitewash reflects the radiant energy of the sun, thus resulting in lower bark temperatures. In addition, in these tests, scions placed on the north and east sides of the stub survived much better than those on the south and west, probably because of the lower temperatures resulting from their shaded position.

Moisture

Since the parenchyma cells comprising the important callus tissue are thin-walled and tender, with no provision for resisting desiccation, it is obvious that if they are exposed to drying air for very long, they will be killed. This was found (*165*) to be the case in studies of the effect of humidity on healing of apple grafts. Air moisture levels below the saturation point inhibited callus formation, the rate of desiccation of the cells increasing as the humidity dropped. In fact, the presence of a film of water against the callusing surface was much more conducive to abundant callus formation than just maintaining the air at 100 percent relative humidity.

Highly turgid cells are more likely to give proliferation of callus than those in a wilt condition. In vitro studies (*46*) of stem pieces of ash (*Fraxinus excelsior*) have shown that callus production on the cut surfaces was markedly reduced as the water potential (turgidity) decreased.

Unless the adjoining cut tissues of a completed graft union are kept by some means at a very high humidity level, the chances of successful healing are rather poor. With most plants, thorough waxing of the graft union, which retains the natural moisture of the tissues, is all that is necessary. Often root grafts are not waxed but stored in a moist (not wet) packing material during the callusing period. Damp peat moss or wood shavings are good media for callusing, providing adequate moisture and aeration.

Oxygen

It has been shown (*165*) that oxygen is necessary at the graft union for the production of callus tissue. This would be expected since rapid cell division and growth are accompanied by relatively high respiration, which requires oxygen. For some plants, a lower percentage of oxygen than is found naturally in air is sufficient, but for others healing of the union is better if it is left unwaxed but placed in a well-moistened medium. This may indicate that the latter plants have a high oxygen requirement for callus formation. Waxing restricts air movement and oxygen may become a limiting factor, thus inhibiting callus formation. This situation seems to be true for the grape in which, usually, the union is not covered with wax or other air-excluding materials during the callusing period.

There is some evidence (*25*) that light will inhibit callus development. In vitro callus cultures of black cherry (*Prunus serotina*) became much larger in darkness than they did in light.

Growth Activity of the Stock Plant

Some propagation methods, such as T-budding and bark grafting, depend upon the bark "slipping," which means that the cambium cells are actively dividing, producing young thin-walled cells on each side of the cambium. These newly formed cells separate easily from one another, so the bark "slips."

Initiation of cambial activity in the spring results from the onset of bud activity since, shortly after the buds start growth, cambial activity can be detected beneath each developing bud, with a wave of cambial activity progressing down the stems and trunk. This stimulus is due to production of auxin and gibberellins originating in the expanding buds (*189*).

In budding seedlings in the nursery in late summer it is important that they have an ample supply of soil moisture just before and during the budding operation. If they should lack water during this period, active growth is checked, cell division in the cambium stops, and it becomes difficult to lift the bark flaps to insert the bud.

There is evidence to show that callus prolif-

eration—essential for a successful graft union—
occurs most readily at the time of year just before
and during "bud break" in the spring, due to
auxin gradients, diminishing through the summer
and into fall. Increasing callus proliferation takes
place again in late winter, but this is not dependent upon breaking of bud dormancy (*173*).

At certain periods of high growth activity in
the spring, plants exhibiting strong root pressure
(such as the walnut, maple, and grape) show excessive sap flow or "bleeding" when cuts are
made preparatory to grafting. Grafts made with
such moisture exudation around the union will
not heal properly. Such "bleeding" at the graft
union can be overcome by making slanting knife
cuts below the graft around the tree. They should
be made through the bark and into the xylem to
permit such exudation to take place below the
graft union. When potted rootstock plants to be
grafted (for example, in species of *Fagus, Betula,*
or *Acer*) show excessive root pressure, they should
be put in a cool place with reduced watering until
the "bleeding" stops.

On the other hand, potted rootstock plants,
such as junipers or rhododendrons, when first
brought into a warm greenhouse in winter for
grafting, are dormant, and grafting should be delayed until the rootstock plants have been held for
several weeks at 15 to 18°C (60 to 65°F) and new
roots start to form; then the rootstock plant is
physiologically active enough for the union to
heal.

When the rootstock plant is physiologically
overactive (excessive root pressure and "bleeding"), or underactive (no root growth being
made), some form of side-graft, in which the rootstock top is not at first removed, should be used.
In situations in which the rootstock is neither
overactive nor underactive, one of the many
forms of topgrafting, in which the top of the stock
is completely removed at the time the graft is
made, is likely to be successful (*47*).

Propagation Techniques

Sometimes the techniques used in grafting are so
poor that only a small portion of the cambial regions of the stock and scion are brought together.
Although healing occurs in this region and growth
of the scion may start, after a sizable leaf area de-

velops and high temperatures and high transpiration rates occur, sufficient movement of water
through the limited conducting area cannot take
place, and the scion subsequently dies. Other er-
rors in grafting technique, such as poor or delayed
waxing, uneven cuts, or use of desiccated scions
can, of course, result in grafting failure.

Poor grafting techniques, although they may
delay adequate healing for some time—weeks or
months—do not in themselves cause any permanent incompatibility. Once the union is adequately healed, growth can proceed normally.

Virus Contamination, Insects, and Diseases

Using virus-infected propagating materials in
nurseries can reduce bud "take" as well as the
vigor of the resulting plant (*137*). In stone fruit
propagation the use of budwood free of ring spot
virus has consistently given improved percentages
of "takes" over infected wood.

Topgrafting olives in California is seriously
hindered in some years by attacks of the American
plum borer (*Euzophera semifuneralis*), which feeds
on the soft callus tissue around the graft union,
resulting in the death of the scion. In England,
nurseries are often plagued with the red bud borer
(*Thomasiniana oculiperda*), which feeds on the callus
beneath the bud-shield in newly inserted T-buds
causing them to die (*55*).

Sometimes bacteria or fungi gain entrance at
the wounds made in preparing the graft or bud
unions. For example, it was found that a rash of
failures in grafts of *Cornus florida* 'Rubra' on *C.
florida* stock was due to the presence of the fungus
Chalaropsis thielavioides (*31*). Chemical control of
such infections materially aids in promoting healing of the unions (*45*).

In South and Central America, rubber
(*Hevea*) trees are propagated by a modification of
the patch bud. A major cause of budding failures
in such warm, humid areas has been infection of
the cut surfaces by a fungus, *Diplodia theobromae,*
but control of this infection can be obtained by
fungicidal treatments (*96*). In topgrafting mangos
in Florida, control of the fungus diseases anthracnose and scab is essential for success. This is done
by spraying the rootstock trees and the source of
scionwood regularly with copper fungicides before
grafting is attempted (*130*).

Plant Growth Regulators in the Healing of Graft Unions

Trials with the use of growth substances, particularly auxin, applied to tree wounds or to graft unions have given variable results in promoting subsequent healing (*105, 125, 135, 164*). In tissue culture studies, however, a definite relationship has been found to exist between callus production (which is essential for graft healing) and the levels of certain applied growth substances, particularly kinetin and auxin (*46, 52, 126, 134*). There is some evidence, too, that abscisic acid stimulates callus production, especially when applied to tissues in combination with auxin (*1*), or with kinetin (*15*).

Some practical benefit from the use of combinations of auxin and kinetin or abscisic acid plus auxin in stimulating callus formation and subsequent healing of graft unions may be possible.

POLARITY IN GRAFTING

Proper polarity is essential if the graft union is to be permanently successful. In all commercial grafting operations correct polarity is strictly observed. As a general rule, as shown in Fig. 11–14, in grafting two pieces of stem tissue together, the morphologically proximal end of the scion should be inserted into the morphologically distal end of the stock. But in grafting a piece of stem tissue on a piece of root, as is done in normal root grafting,

the proximal end of the scion should be inserted into the proximal end of the root piece.

> The **proximal** end of either the shoot or the root is that nearest the stem-root junction of the plant. The **distal** end of either the shoot or the root is that furthest from the stem-root junction of the plant and nearest the tip of the shoot or root.

Should a scion be inserted with reversed polarity—"upside down"—in bridge-grafting, for example, it is possible for the two graft unions to be successful and the scion to stay alive for a time. But, as seen in Fig. 11–15, the reversed scion does not increase from its original size, whereas the scion with correct polarity enlarges normally.

In **nurse-root grafting,** the rootstock may purposefully be grafted to the scion with reversed polarity. Union will occur, and the root will supply water and mineral nutrients to the scion but the scion is unable to supply necessary organic materials to the rootstock and the stock eventually dies. In nurse-root grafting, the graft union is purposely set well below the ground level, and the scion itself produces adventitious roots, which ultimately become the entire root system of the plant.

In T-budding or patch budding, the rule for observance of correct polarity is not as exacting. Buds can be inserted with reversed polarity and still make permanently successful unions. As

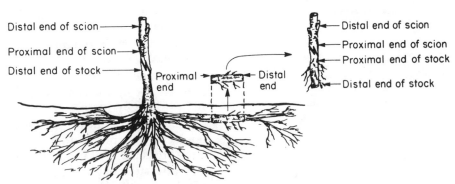

FIGURE 11–14   Polarity in grafting. In top grafting, the proximal end of the scion is attached to the distal end of the stock. In root grafting, however, the proximal end of the scion is joined to the proximal end of the stock.

FIGURE 11–15 Bridge graft on a pear tree five months after grafting. Center scion was inserted with reversed polarity. Although the scion is alive it has not increased from its original size. The two scions on either side have grown rapidly.

shown in Fig. 11–16, **inverted T-buds** start growing downward, then the shoots curve and grow upward (*155*). In the inverted bud piece, the cambium is capable of continued functioning and growth. In the xylem, phloem, and fibers formed from cambial activity, there is a twisting configuration which apparently allows for normal translocation and water conduction (*32*).

FIGURE 11–16 Two-year-old 'Stayman Winesap' apple budded on 'McIntosh' seedling by inverted T-buds. Note development of wide-angle crotches. (Courtesy Arnold Arboretum, Jamaica Plain, Mass.)

LIMITS OF GRAFTING

Since one of the requirements for a successful graft union is the close matching of the callus-producing tissues near the cambium layers, grafting is generally confined to the dicotyledons in the angiosperms, and to the gymnosperms, the cone-bearing plants. Both have a vascular cambium layer existing as a continuous tissue between the xylem and the phloem. In the monocotyledonous plants of the angiosperms, which do not have a vascular cambium, grafting is more difficult, with a low percentage of "takes." There are cases of successful graft unions between the stem parts of monocots. By making use of the meristematic properties found in the intercalary tissues (located at the base of internodes), successful grafts have been obtained with various grass species as well as the large tropical monocotyledonous vanilla orchid (*127, 128*).

Before a grafting operation is started, it should be determined that the plants to be combined are capable of uniting and producing a permanently successful union. There is no definite rule that can predict exactly the ultimate outcome of a particular graft combination except that *the more closely the plants are related botanically, the better the chances are for the graft union to be successful.* However, there are numerous exceptions to this rule.

Grafting within a Clone

A scion can be grafted back on the plant from which it came, and a scion from a plant of a given clone can be grafted to any other plant of the same clone. For example, a scion taken from an 'Elberta' peach tree could be grafted successfully to any other 'Elberta' peach tree in the world.

Grafting between Clones within a Species

In the tree fruit and nut crops different clones within a species can almost always be grafted together without difficulty and produce satisfactory trees. However, in some conifer species, notably Douglas fir (*Pseudotsuga menziesii*), incompatibility problems have arisen in grafting together individuals of the same species, such as selected *P. menziesii* clones onto *P. menziesii* rootstock seedlings (*35*).

Grafting between Species within a Genus

For plants in different species but in the same genus, grafting is successful in some cases but unsuccessful in others. Grafting between most species in the genus *Citrus,* for example, is successful and widely used commercially. The almond (*Prunus amygdalus*), the apricot (*Prunus armeniaca*), the European plum (*Prunus domestica*), and the Japanese plum (*Prunus salicina*)—all different species— are grafted commercially on the peach (*Prunus persica*), a still different species, as a rootstock. But on the other hand, almond and apricot, both in the same genus, cannot be intergrafted successfully. The 'Beauty' cultivar of Japanese plum (*Prunus salicina*) makes a good union when grafted on the almond, but another cultivar of *P. salicina*, 'Santa Rosa', cannot be successfully grafted on the almond. In another example of this inconsistency, some almond cultivars (which are clones) are highly successful when grafted on the 'Marianna 2624' plum as a rootstock, while other cultivars are not (*93*). Thus, compatibility between species in the same genus depends upon the particular genotype combination of stock and scion.

Reciprocal interspecies grafts are not always successful. For instance, 'Marianna' plum (*Prunus cerasifera* × *P. munsoniana*) on peach (*Prunus persica*) roots makes an excellent graft combination, but the reverse—grafts of the peach on 'Marianna' plum roots—either soon die or fail to develop normally (*2, 102*). In another example, many Japanese plum (*Prunus salicina*) cultivars can be successfully grafted on the European plum (*P. domestica*), but the reverse, the European plum grafted on the Japanese plum, is unsuccessful (*75*).

Grafting between Genera within a Family

When the plants to be grafted together are in the same family but in different genera, the chances of a successful union become more remote. Cases can be found in which such grafts are successful and used commercially, but in most instances such combinations are failures.

Trifoliate orange (*Poncirus trifoliata*) is used commercially as a dwarfing stock for the orange (*Citrus sinensis*), in a different genus. The quince (*Cydonia oblonga*) has long been used as a dwarfing rootstock for certain pear (*Pyrus communis*) cultivars. The reverse combination, quince on pear, though, is unsuccessful. Intergeneric grafts in the nightshade family, Solanaceae, are quite common. Tomato (*Lycopersicon esculentum*) can be grafted successfully on Jimson weed (*Datura stramonium*), tobacco (*Nicotiana tabacum*), potato (*Solanum tuberosum*), and black nightshade (*Solanum nigrum*). The evergreen loquat (*Eriobotrya japonica*) can be grafted on quince roots (*Cydonia oblonga*), a deciduous tree, giving a dwarfed loquat plant.

Grafting between Families

Successful grafting between plants of different botanical families is usually considered to be impossible but there are reported instances (*200*) in which it has been accomplished. These are with short-lived, herbaceous plants, though, for which the time involved is relatively short. Grafts, with vascular connections between the scion and stock, were successfully made (*132*) using white sweet clover, *Melilotus alba* (Leguminosae), as the scion and sunflower, *Helianthus annuus* (Compositae), as the stock. Cleft grafting was used, with the scion inserted into the pith parenchyma of the stock. The scions continued growth with normal vigor for more than five months. As far as is known, however, there are no instances in which woody perennial plants belonging to different families have been successfully and permanently grafted together.

GRAFT INCOMPATIBILITY

The ability of two different plants, grafted together, to produce a successful union and to develop satisfactorily into one composite plant is termed **compatibility** (*150*). The opposite, of course, would be **incompatibility.** The distinction between a compatible and an incompatible graft union is not clearcut. On one hand, stock and scion of closely related plants unite readily and grow as one plant. On the other hand, stock and scion of unrelated plants grafted together are likely to fail completely in uniting. Many graft combinations lie between these two extremes in that they unite initially, with apparent success (*21*) (Figs. 11–17 and 11–18), but gradually develop distress symptoms with time, due either to failure at the union or to the development of abnormal growth patterns (*16, 52*). A strong, well-healed union and one showing incompatibility symptoms are illustrated in Fig. 11–19. Nelson (*129*) has developed an extensive survey of incompatibility in horticultural plants which should be consulted before attempting graft combinations between species whose graft reactions are unknown to the grafter.

FIGURE 11–18   Apple grafts (right portion of tree) five months after being grafted on a pear tree. This is an incompatible combination; the apple grafts eventually died although they initially made a strong growth.

Symptoms of Incompatibility

Graft union malformations resulting from incompatibility can usually be correlated with certain external symptoms. The following symptoms

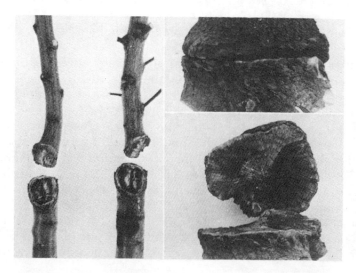

FIGURE 11–17   Breakage at the graft union resulting from incompatibility. *Left:* One-year-old nursery trees of apricot on almond seedling rootstock. *Right:* 15-year-old 'Texas' almond tree on seedling apricot rootstock which broke off cleanly at the graft union—a case of "delayed incompatibility" symptoms.

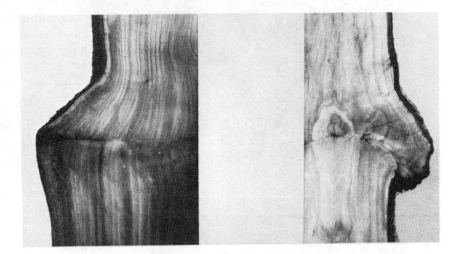

FIGURE 11-19 Radial sections through two fruit-tree graft unions. *Left:* A strong, well-knit compatible graft union. *Right:* A weak, poorly connected union, showing incompatibility symptoms.

have been associated with incompatible graft combinations:

1. Failure to form a successful graft or bud union in a high percentage of cases.
2. Yellowing foliage in the latter part of the growing season, followed by early defoliation. Decline in vegetative growth, appearance of shoot dieback, and general ill health of the tree.
3. Premature death of the trees, which may live only a year or two in the nursery.
4. Marked differences in growth rate or vigor of scion and stock.
5. Differences between scion and stock in the time at which vegetative growth for the season begins or ends.
6. Overgrowths at, above, or below the graft union.
7. Graft components breaking apart cleanly at the graft union.

An isolated case of one or more of the above symptoms (except for the last) does not necessarily mean the combination is incompatible; some of these symptoms can result from unfavorable environmental conditions, such as lack of water or some essential nutrient, attacks by insects or diseases, or poor grafting or budding techniques (*3, 51, 77, 98*).

Incompatibility is clearly indicated by *the breaking off of trees at the point of union, particularly when they have been growing for some years and the break is clean and smooth, rather than rough or jagged.* This break may occur in a year or two after the union is made—for instance, in the apricot on almond roots (see Fig. 11–17).

In certain other cases (e.g., some apricot cultivars grafted on myrobalan plum roots), such breakage may not take place until the trees are full-grown and bearing crops (*43, 48, 141*). The occurrence of this situation in a single instance provides justification for stating that the particular combination is incompatible.

Scion overgrowth at the graft union is sometimes associated with incompatibility but it also occurs in compatible unions. This symptom is not a reliable indication of incompatibility (*2, 16*) (Fig. 11–20).

Types of Incompatibility

There are three types of graft incompatibility. In her studies of incompatibility, Mosse (*123*) placed the then-known cases into two categories, which she called **localized** and **translocated.** A third category can be termed **virus-induced.**

Localized incompatibility includes combinations in which the incompatibility reactions apparently depend upon actual contact between stock and scion. Separation of the components by insertion of a **mutually compatible** interstock overcomes the incompatibility symptoms. In the incompatible combination the union structure is often mechanically weak, with continuity of cam-

FIGURE 11-20 It is possible for the scion to overgrow the stock and yet develop into a large, strong tree. Such overgrowth is shown here for the Caucasian wingnut (*Pterocarya fraxinifolia*) grafted on *Pterocarya stenoptera*.

TYPES OF GRAFT INCOMPATIBILITY WITH EXAMPLES

Localized

'Bartlett' pear on quince roots.

Translocated

'Hale's Early' peach on 'Myrobalan B' plum roots

'Nonpareil' almond on 'Marianna 2624' plum roots

Peach cultivars on 'Marianna 2624' plum roots

Virus-induced

Citrus quick decline

Pear decline

Walnut black line

---

gree of anatomical disturbance at the graft union. Root starvation eventually results, owing to translocation difficulties across the defective graft union.

The best example in this category is that of 'Bartlett' ('Williams') pear grafted directly on quince rootstock. With the use of a "compatibility bridge" of 'Old Home' or ('Beurré Hardy') pear as an interstock, the three-part combination is completely compatible, and satisfactory tree growth takes place (*122, 140, 141, 191*).

Translocated incompatibility includes certain cases in which the incompatible condition is *not* overcome by the insertion of a mutually compatible interstock because, apparently, some labile influence can move across it. This type involves phloem degeneration, and can be recognized by the development of a brown line or necrotic area in the bark. Consequently, restriction of movement of carbohydrates occurs at the graft union—accumulation above and reduction below. Reciprocal combinations may be compatible. In the various combinations in this category the range of bark tissue breakdown can extend from virtually no union at all, to a mechanically weak union with distorted tissues, to a strong union with tissues normally connected.

An example of a combination in this category is that of 'Hale's Early' peach grafted onto 'Myrobalan B' plum rootstock (*76*). This forms a weak union in which distorted tissues occur. Abnormal quantities of starch accumulate at the base of the peach scion. If the mutually compatible 'Brompton' plum is used as an interstock between the 'Hale's Early' peach and the 'Myrobalan B' rootstock, the incompatibility symptoms still persist, with an accumulation of starch in the 'Brompton' interstock. However, in other studies (*78*), in which peach/myrobalan plum grafts were made, but using small seedlings at the cotyledonary stage, no signs of incompatibility appeared even after the grafted trees were 13 years old. In the same study all trees of this combination grafted at the usual stage in the nursery showed incompatibility symptoms one year after grafting. Possibly some factor responsible for the incompatibility symptoms is not present in the juvenile seedling tissues.

In another example of translocated incompatibility, the combination of 'Nonpareil' almond

bium and vascular tissues broken, although there are cases in which the union is strong and the tissues joined normally. Often external symptoms develop slowly, appearing in proportion to the de-on 'Marianna 2624' plum shows complete phloem breakdown, although the xylem tissue connections are quite satisfactory. However, another cultivar, the 'Texas' almond, on 'Marianna 2624' plum produces a good, compatible combination. Inserting a 6-in. piece of 'Texas' almond as an interstock between the 'Nonpareil' almond and the 'Marianna' plum stock fails to overcome the incompatibility between these two components. Bark disintegration occurs at the normally compatible 'Texas' almond/'Marianna' plum graft union, owing, presumably, to some factor translocated in the phloem from the 'Nonpareil' scion above, through the 'Texas' interstock, to the 'Marianna' plum rootstock (*93*).

Peach grafted on 'Marianna' plum generally forms an incompatible union. Although peach buds unite readily with this plum stock and grow satisfactorily the first season, later an enlargement appears above the graft union, followed by wilting of the peach leaves and death of the tree. Anatomical studies (*102*) showed this to be a case of incompatibility in which good xylem connections developed at the graft union but the phloem tissues failed to unite. Death of the roots resulted, with subsequent wilting and death of the peach top. If leafy shoot growth was retained on the 'Marianna' plum stock to nourish the roots, however, the trees could be kept alive indefinitely, a situation found in other incompatibility cases also (*193*).

Virus-induced incompatibility cases apparently are widespread and more are continually being found.

Graft union failures resembling incompatibility symptoms can be due to pathogens. In certain cases abnormalities first attributed to stock-scion incompatibility were later found to be due to latent virus (or mycoplasma-like pathogens) introduced by grafting from a resistant partner to a susceptible partner (*28, 44, 108*). Figure 11–21 shows such an occurrence in apple.

An example of virus involvement is found in citrus. At one time incompatibility was blamed for the difficulty encountered in budding sweet orange (*Citrus sinensis*) onto sour orange (*C. aurantium*) roots in South Africa (1910) and in Java (1928), even though this combination was a commercial success in other parts of the world. The incompatibility was believed to be due to the pro-

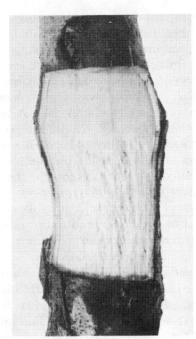

FIGURE 11–21 Latent viruses in the scion portion of a graft combination may cause symptoms to appear in a susceptible rootstock following grafting. Here "stempitting" virus symptoms have developed in the sensitive 'Virginia Crab" apple rootstock. The wood of the scion cultivar–above the graft union–is unaffected. (Courtesy H. F. Winter (*197*).)

duction of some substance by the scion that was toxic to the stock (*179*). In the light of subsequent studies stimulated by the development of orange "tristeza" or "quick decline" in Brazil and California, it is clear that the toxic substance from the sweet orange scions was a virus, tolerated by the sweet orange, but lethal to the sour orange roots (*12, 192*).

The "pear decline" disease, developing first in Italy and later in western North America, killed hundreds of thousands of pear trees in California alone (*131, 161*). Early studies (*7*) related the trouble to the rootstock used and indicated it to be an "induced incompatibility," with the causal factor unknown. 'Bartlett' trees on oriental pear roots, such as *Pyrus pyrifolia,* were highly susceptible to the problem, while trees on *P. communis* roots were not. Subsequent research showed that pear decline was not due to an inherent stock-scion incompatibility but was associated with what was then thought to be a virus (*89*), transmitted by an insect vector, *Psylla pyricola*. The 'Bartlett' scion cultivar and the *P. communis* roots were observed to be resistant to decline, but the rootstock, *P. pyrifolia,* was highly susceptible, showing a phloem degeneration just below the graft union. Later studies (*82, 83*) showed the infective agent to be a mycoplasma-like organism rather than a virus. For the decline condition to appear, the concurrent presence of the infective bodies, pear psylla, and a susceptible rootstock are necessary. The mycoplasma-like organisms were found to be sensitive to tetracycline antibiotics; thus, remission of the symptoms is possible by tree trunk injections (*133*).

Delayed incompatibility was formerly believed to be the cause of the so-called "black-line" of walnuts; black-line occurs in certain Persian walnut orchards in Oregon, California, and France where cultivars of *Juglans regia* are grafted onto seedling rootstocks of *J. hindsii* (northern California black walnut) or onto Paradox roots (*J. hindsii × J. regia*) (*111, 160*). Affected trees grow normally, bearing good crops, until they reach maturity—15 to 20 or more years of age. When the trouble appears, a thin layer of cambium and phloem cells dies at the graft union. The dead tissue gradually extends around the tree at the graft union until the tree is girdled; the vertical width of the dead area may reach 30 cm (12 in.). Such girdling kills the tree above the union, but the rootstock usually develops sprouts and remains alive. It is now known that the problem is not due to incompatibility but to the cherry leaf roll virus, which is transmitted by pollen. When the virus in the tree reaches the graft union, the very sensitive *J. hindsii* cells undergo a toxic reaction at the union, causing the girdling to form (*114*).

Causes and Mechanisms of Incompatibility

Although incompatibility is clearly related to genetic differences between stock and scion, the mechanisms by which particular cases are expressed are not clear. With the large numbers of genetically different plant materials that can be combined by grafting, a wide range of different physiological, biochemical, and anatomical systems are brought together, with many possible interactions, both favorable and unfavorable. Several proposals have been advanced in attempts to explain incompatibility, but the evidence supporting most of them is generally inadequate and often conflicting.

One possible mechanism is *physiological and biochemical differences between stock and scion*. This hypothesis is supported by studies of Gur et al. with incompatible combinations of certain pear cultivars on quince rootstock (*62–64*). The experimental evidence supports the following conclusions:

1. When certain pear cultivars are grafted onto quince roots, a cyanogenic glucoside, prunasin—normally found in quince, but not in pear tissues—is translocated from the quince into the phloem of the pear. The pear tissues break down the prunasin in the region of the graft union, with hydrocyanic acid as one of the decomposition products. This enzymatic breakdown is hastened by high temperatures. In addition, different pear cultivars vary in their ability to decompose the glucoside.

2. The presence of the hydrocyanic acid leads to a lack of cambial activity at the graft union, with pronounced anatomical disturbances in the phloem and xylem at the union resulting. The phloem tissues are gradually destroyed at and above the graft union. Conduction of water and materials is seriously reduced in both xylem and phloem.

3. A reduction of the levels of the sugar reaching the quince roots leads to further decomposition of prunasin, liberating hydrocyanic acid and killing large areas of the quince phloem.

4. A water-soluble and readily diffusable inhibitor of the action of the pear enzyme (which breaks down the glucoside) occurs in the various pear cultivars, although they differ in their content of this inhibitor. This may explain why certain pear cultivars are compatible and others incompatible with quince rootstock.

Heuser (*80, 81*), working with various *Prunus* species, has found that the cyanogenic glycosides, amygdalin and prunasin, and their toxic breakdown products, benzaldehyde and cyanide, can inhibit the growth of callus from *Prunus* understocks which have shown graft incompatibility problems. This leads to the possibility that the production of benzaldehyde and cyanide may be causal agents in some graft failures among *Prunus* species since callus production is needed for the healing of graft unions.

Electron microscope studies (*18, 19*) have been made of compatible and incompatible pear–quince grafts, with detailed observations of cell-wall structure at the graft unions. Examination of the cell walls of compatible graft combinations showed the concentration of lignin at the line of union to be as high as that in cell walls not a part of the union. In incompatible combinations, however, the adjoining cell walls of the two components contained no lignin, with the parts either not connected or interlocked only by cellulose fibers. From these studies it was concluded that the **lignification** processes of cell walls are involved in the formation of strong unions in pear–quince grafts. Reactions that inhibited the formation of lignin and the establishment of a mutual middle lamella between the two components resulted in weak unions.

In a study (*119*) at the cellular level of an obviously incompatible combination, *Sedum telephoides* (CRASSULACEAE) on *Solanum pennellii* (SOLANACEAE), the initial stages of graft union healing were similar to those occurring in a compatible combination (*Sedum* on *Sedum*). In the incompatible *Sedum* on *Solanum* graft, however, after 48 hours *Sedum* cells at the graft interface deposited an insulating layer of suberin along the cell wall, which later underwent a lethal cellular senescence and collapse, to form a necrotic layer of increasing thickness (*118, 119*) (see Fig. 11–22). Associated with this cellular senescence in *Sedum* cells a dramatic increase in a hydrolytic enzyme, acid phosphatase, has been observed (*120*). Rather than callus interlocking, cambial formation, and vascular connection, the thick necrotic layer prevented cellular connection, which led to scion desiccation and eventual death. Interestingly, the *Solanum* stock did not show the rejection response that the *Sedum* scion did.

In tests comparing the tensile strength of compatible *Sedum* on *Sedum* grafts with incompatible *Sedum* on *Solanum* grafts, that of the former increased steadily for 11 days after grafting when it leveled off, whereas the tensile strength of the incompatible grafts decreased, being correlated with the continued cellular necrosis of *Sedum* cells bordering the graft interface (*121*).

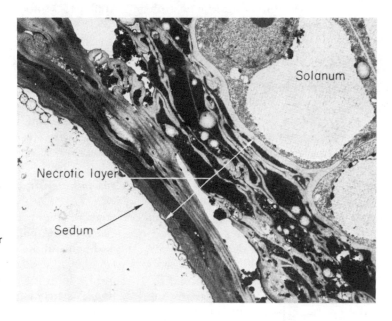

FIGURE 11–22 The graft interface of an incompatible graft between *Sedum telephoides* and *Solanum pennellii* at eight days after grafting. Lethal cellular senescence in *Sedum* has resulted in the formation of a necrotic layer of collapsed cells that separates the two graft partners. × 5000. (Courtesy R. Moore and D. B. Walker (*115*).)

Predicting Incompatible Combinations

To be able to predict in advance of grafting whether or not the components of the proposed scion–stock combination are compatible would be of tremendous value. Progress has been made by Santamour et al. in developing an electrophoresis test using cambial peroxidase banding patterns of scion and stock. If they match, the combination may be said to be compatible; if they do not, incompatibility may be predicted. A simple electrophoretic diagnostic system being developed would allow nursery people to perform such procedures themselves (*150–152*).

Correcting Incompatible Combinations

If a graft combination known to be incompatible has been made and is discovered before the tree dies or breaks off at the union, it is possible, in graft combinations of the localized type, to correct this condition by bridge grafting with a mutually compatible stock, if such exists. If a tree has been mistakenly propagated on a rootstock known to show symptoms of incompatibility with the scion cultivar, and with the probability that the tree will eventually break off at the graft union, it is also possible to correct this condition by inarching with seedlings of a compatible rootstock. If breakage does not occur until the inarches are strong enough to support the tree, it may thus be possible to save it, the inarched seedlings becoming the main root system.

SCION–STOCK (SHOOT–ROOT) RELATIONSHIPS

Combining two (or more, in the case of interstocks) different plants (genotypes) into one plant by grafting—one part producing the top and the other part the root—can produce growth patterns that are different from those that would have occurred if each component part had been grown separately. Some of these effects have major horticultural value, while others are detrimental and should be avoided. These altered characteristics may result from (a) incompatibility reactions; (b) the fact that one of the graft partners possesses one or more specific characteristics not found in the

other, for example, resistance to certain diseases, insects, or nematodes—or tolerance of certain adverse weather or soil conditions; or (c) specific interactions between the stock and the scion that alter size, growth, productivity, fruit quality, or other horticultural attributes.

In practice it may be difficult to separate which of the three kinds of influencing factors is dominant in any given graft combination growing in a particular environment.

Effects of the Rootstock on the Scion Cultivar

Size and Growth Habit

Size control, and sometimes an accompanying change in tree shape, is one of the most significant rootstock effects. Apparently, the rootstock alters the vigor of a given scion cultivar. By proper rootstock selection in apples, the complete range of tree size—from very dwarfed to very large—has been obtained with a given scion cultivar grafted to different rootstocks.

That specific rootstocks can be used to influence the size of trees has been known since ancient times. Theophrastus—and later the Roman horticulturists—made use of dwarfing apple rootstocks that could be easily propagated. The name "Paradise," which refers to a Persian park or garden—"pairidaeza"—was applied to dwarfing apple stocks about the end of the fifteenth century (*23*).

A wide assortment of size-controlling rootstocks has now been developed for certain of the major tree fruit crops (*17*). Most notable is the series of vegetatively propagated apple rootstocks collected and developed at the East Malling Research Station in England, beginning in 1912. These were classified into four groups according, primarily, to the degree of vigor imparted to the scion cultivar: *dwarf, semidwarf, vigorous,* and *very vigorous,* as compared to apple seedling rootstocks (*70, 115, 138*). Similarly, the size-controlling effects of the rootstock on sweet cherry (*Prunus avium*) scion cultivars has been known since the early part of the eighteenth century (*79*). Mazzard (*P. avium*) seedling rootstocks produce large, vigorous, long-lived trees, whereas *P. mahaleb* seedlings, as a rootstock, tend to produce smaller trees of shorter life (*42*). However, individual seedlings of these species, increased and maintained as clones, can produce distinct rootstock effects different from that of the whole species. Rootstock effects on tree size and vigor are recognized also in citrus, pear, and other species (10, 53). A discussion of specific rootstocks for the various fruit and nut crops is given in Chap. 18.

Prediction of rootstock effects cannot be made with certainty without considering the entire system in which it is used, including the particular cultivar used as the scion top, which can modify the rootstock influence. Each combination of scion and rootstock must be tested before any conclusion regarding the behavior of the composite tree can be reached (69). The environment in which the particular combination is to be grown also must be taken into account. If grafted plants are grown under optimum conditions the differences in performance of those on strong-growing rootstocks versus those on weaker stocks may be minimized. Good soil and cultural conditions are required for successful tree performance when very dwarfing rootstocks are used, as well as structural support (i.e., staking or trellising), due to a restricted root system.

Alterations in the normal shape of the tree are often associated with the dwarfing effect caused by certain rootstocks. A low and spreading, rather than upright, form is illustrated by grafts of the 'McIntosh' apple on the apomictic, semidwarfing rootstock *Malus sikkimensis* (155). Such effects may possibly be due to changes in levels of certain hormones (auxins, gibberellins) in the tree.

Symptomless viruses that occur in plants may exert a dwarfing influence. If no harmful symptoms appear, useful dwarfing could be produced in this manner (60). Removing viruses from some dwarfing clonal stocks by heat treatment may decrease their dwarfing influence.

It would be very useful in developing new clonal rootstocks from seedlings to be able to predict whether such stocks would be dwarfing or invigorating. Studies (8, 104) in England indicate that apple rootstocks known to produce dwarf trees have a high proportion of bark to wood in the lateral roots, whereas stocks causing increased vigor in the scion cultivar have a lower proportion of bark to wood. Also, much of the functional wood tissue of roots of dwarfing apple stocks was composed of living cells, whereas in nondwarfing, vigorous rootstocks, the wood consisted of a relatively large amount of lignified tissue without living cell contents (i.e., larger vessel/tracheid system for water transport). However, similar studies (177) in Italy with grafted apple and pear trees did not show such a relationship.

Fruiting

Fruiting precocity, fruit bud formation, fruit set, and yield of a tree can be influenced by the rootstock used. In general, fruiting precocity is associated with dwarfing rootstocks and slowness to start fruiting with vigorous rootstocks. Long-term yield studies, conducted in England, involving several apple rootstocks showed that the results varied according to tree age and tree spacing (138, 139). Trees on 'Malling 9' roots planted 3.6 by 3.6 m (12 by 12 ft) apart showed highest accumulated yield per tree up to ten years of age because of their early bearing. By the tenth year this yield was surpassed by trees on the moderately vigorous 'Malling 4' roots. By the fifteenth year this was surpassed by trees on the vigorous 'Malling 1' roots, but by the twentieth year yield of all trees was superseded by trees grown on the very vigorous 'Malling 16' roots.

Vigorous, strongly growing rootstocks in some cases result in a larger and more vigorous plant that produces greater crops over a long period of years. On the other hand, trees on dwarfing stocks may be more fruitful, and if closely planted, produce higher yields per acre, especially in the early years of bearing.

The presence of the graft union itself may stimulate earlier and perhaps heavier bearing. For instance, in studies with citrus (85) five rootstocks—sour orange, sweet orange, trifoliate orange, grapefruit, and rough lemon—all started fruiting two seasons earlier when budded to themselves than when unbudded, although in each case the trees were about the same size.

If there is an imperfect graft union, such as the partial incompatibility that occurs in some combinations, a reduction in translocation at the graft union can have a girdling effect and thus lead to increased fruitfulness.

Rootstock influence can vary greatly with different kinds of plants. In growing Oriental persimmon (*Diospyros kaki*) cultivars the rootstock seems to have a direct effect on flower production and fruit set. In tests (*159*) using the 'Hachiya' persimmon, trees on *D. lotus* roots produced more flowers but matured fewer fruits than similar trees on *D. kaki* roots, while trees of the same cultivar on *D. virginiana* roots produced so few flowers that crops were very poor.

In grapes, where yield is dependent upon the vigor of the current season's growth, the rootstock used can be a strong influencing factor (*68*). Large yield increases of certain American types (*Vitis labrusca*) of grapes were obtained (*185*) when they were grafted on vigorous rootstocks, in comparison with own-rooted plants. Over a six-year period, yield of 'Concord' vines was increased from 30 to 150 percent, depending upon the rootstock used. On the other hand, with European type (*V. vinifera*) scions, use of 'Dog Ridge' (*V. champini*), an extremely vigorous rootstock cultivar, on fertile soils can lead to such strong growing vines that they become unproductive.

Size, Quality, and Maturity of Fruit

There is considerable variation among plant species in regard to the effect of the rootstock on fruit characteristics on the scion cultivar. In a grafted tree no transmission of characteristics of the fruit which the stock would produce is encountered in fruit of the scion cultivar. For example, quince, commonly used as a pear rootstock, has fruits with a pronounced tart and astringent flavor, yet this flavor does not appear in the pear fruits. The peach is often used as a rootstock for apricot, yet there is no indication that the apricot fruits have taken on any characteristics of peach fruits.

Although there is no intermingling of fruit characteristics between the stock and the scion, certain rootstocks can affect fruit quality of the scion cultivar. A good example of this is the "black-end" defect of pears. 'Bartlett', 'Anjou', and some other pear cultivars on several different rootstocks often produce fruits that are abnormal at the calyx end. The injury consists of blackened flesh, which in severe cases cracks open. Sometimes the calyx end of the fruit is hard and protruding. Such fruit is worthless commercially. It has been shown (*40, 74*) that this trouble develops when the trees are propagated on certain root-

stocks, such as *Pyrus pyrifolia,* but only rarely when the French pear, *P. communis,* is used. This trouble affects only the fruit; no symptoms of adverse tree growth appear.

The development of black-end fruits disappears if trees on *P. pyrifolia* rootstock are inarched with *P. communis* seedlings, and—after the inarches are able to support the tree—the original *P. pyrifolia* roots are cut away. Black-end fruits continue to appear unless the original connection of the 'Bartlett' top with the *P. pyrifolia* roots is broken. Whether the *P. pyrifolia* roots are producing substances toxic to the fruit, or whether there is some other interaction, has not been established.

In citrus, striking effects of the rootstock appear in fruit characteristics of the scion cultivar (*11*). If sour orange (*Citrus aurantium*) is used as the rootstock, fruits of sweet orange, tangerine, and grapefruit are smooth, thin-skinned, and juicy, with excellent quality, and they store well without deterioration. Sweet orange (*C. sinensis*) rootstocks also result in thin-skinned, juicy, high-quality fruits. Citrus fruits on grapefruit (*C. paradisi*) stocks are usually excellent in size, grade, and quality if heavy fertilization is provided. But when rough lemon (*C. limon*) is used as the rootstock, the fruits are often thick-skinned, somewhat large and coarse, inferior in quality, and low in both sugar and acid. Such effects of rootstock on fruit quality may be due to the vigor of the rootstock.

Fruit size of both 'Washington Navel' and 'Valencia' orange is strongly influenced by the rootstock. The largest navel orange fruits are produced on sour orange stocks and the smallest on the Palestine sweet lime. The largest 'Valencia' oranges are associated with the dwarfing trifoliate orange stock, whereas sweet orange rootstocks produce the smallest fruits.

While many such tests have shown that the various characteristics of citrus fruit are affected by the rootstock used, the underlying physiological mechanisms remain unknown.

Tomato (*Lycopersicon esculentum*) grafted on Jimson weed (*Datura stramonium*) roots had been used at one time in the southern part of the United States, owing to the resistance of these roots to nematodes. As it was known that Jimson weed contains poisonous alkaloids, concern was felt about whether such alkaloids might not be translocated from the rootstock to the tomato fruits. Tests showed that this was the case. The alkaloid content of tomato fruits from grafted plants ranged from 3.9 to 28.6 mg per kilogram of fresh fruit, whereas ungrafted control plants had zero alkaloid content (*100*).

There are other similar examples of translocation of compounds between stock and scion in intergeneric grafts in Solanaceae. In a series of reciprocal grafts between tomato and tobacco, nicotine was found in tomato scions when they were grown on tobacco rootstocks. But when tobacco scions were grown on tomato rootstocks, tobacco alkaloid production was greatly reduced, most of the alkaloid synthesis being at the graft union and in the tobacco stem immediately above the union. Such localization was not found in tomato scions on tobacco roots (*41, 169*).

Miscellaneous Effects of the Rootstock on the Scion Cultivar— Winter-Hardiness, Disease Resistance, and Time of Fruit Maturity

In citrus the rootstock used can affect the cold-hardiness of the scion cultivar. During killing freezes in the winter of 1950–1951 in the Texas Rio Grande Valley, young grapefruit trees on 'Rangpur' lime roots survived much better than those on rough lemon or sour orange, whereas trees on 'Cleopatra' mandarin were the most severely damaged (*33*). Survival following a severe winter freeze in Florida in 1962 showed that a wide range in tree hardiness in oranges and grapefruit could be attributed to the rootstock used (*53*). The rootstock can affect the rate of maturity of the scion wood as it hardens-off in the fall.

Different rootstocks respond differently to soil conditions, thus resulting in an altered effect on the behavior of the scion cultivar (Fig. 11–23). Almond and myrobalan plum roots tolerate excess boron in the soil better than 'Marianna' plum roots. Thus, in this case, there was fairly good growth of 'French' prune trees when they were grown on almond and myrobalan plum roots under conditions in which the trees were severely injured when grafted on 'Marianna' plum or apricot roots (*66*).

The four rootstocks (plum, peach, apricot, and almond) commonly used for stone fruits differ

FIGURE 11–23 The tolerance of fruit trees to toxic amounts of certain elements may be influenced by the rootstock used. These 'French' prune trees on four different rootstocks were irrigated with water containing different levels of boron. Rootstocks were: (*top left*) almond seedlings; (*top right*) myrobalan plum seedlings; (*below left*) apricot seedlings; (*below right*) 'Marianna' plum cuttings. Irrigation water containing the following five concentrations of boron were used for each rootstock (left to right): 0.5 ppm (tap water); 2 ppm; 3 ppm; 5 ppm; 10 ppm. (Courtesy C. J. Hansen.)

markedly in their response to adverse soil conditions (*43*) and salinity, and consequently can affect growth of the scion cultivar. For example, trees with myrobalan plum as the rootstock are the most tolerant of excessive soil moisture, followed by peach or apricot roots, almond being the most susceptible to injury from such conditions.

In citrus, considerable variability exists in the tolerance of the various rootstocks to adverse soil conditions. The choice of rootstock upon which to work the scion cultivar is very important. In Texas, for example, the severity of lime-induced chlorosis symptoms in trees grown on calcareous soils is greatly influenced by the rootstock used. Tests of grapefruit on 36 different rootstocks showed that with four of the stocks no chlorosis appeared, but severe chlorosis developed with 13 of the rootstocks (*34*).

It is known that some rootstocks are more tolerant than others to adverse soil pests, such as nematodes (*Meloidogyne* spp.) or oak root fungus (*Armillaria mellea*). The growth of the scion cultivar is subsequently influenced by the rootstock through the latter's relative ability to withstand such adverse conditions.

The above examples illustrate cases in which the behavior of the scion cultivar is affected by the rootstock used, which in turn can be traced to reactions of the rootstock to certain adverse soil situations.

Effects of the Scion Cultivar on the Rootstock

Although there is a tendency to attribute all cases of dwarfing or invigoration of a grafted plant to the rootstock, the effect of the scion on the behavior of the composite plant may be as important as that of the rootstock. Unquestionably, however, the scion, the interstock, the rootstock, and the graft union itself all interact to influence each other and determine the over-all behavior of the plant. In certain combinations, however, a particular member of the combination could have a marked influence no matter what part of the plant it becomes. For instance, a dwarfing stock will exert a dwarfing influence on the entire plant whether used as rootstock, intermediate stock, or scion.

Effect on the Vigor of the Rootstock

Vigor is the major effect of the scion on the stock, just as it was in the case of rootstock effect on scion cultivar. If a strongly growing scion cultivar is grafted on a weak rootstock, the growth of the rootstock will be stimulated so as to become larger than it would have been if left ungrafted. Conversely, if a weakly growing scion cultivar is grafted on a vigorous rootstock, the growth of the rootstock will be lessened from what it might have been if left ungrafted. In citrus, for example, when the scion cultivar is less vigorous than the

rootstock cultivar, it is the scion cultivar rather than the rootstock that determines the rate of growth and ultimate size of the tree (*84*).

That the scion influences the growth of the rootstock was recognized at least as early as the middle of the nineteenth century (*57*). It has long been known, particularly with apples, that the size, nature, and form of the root system that develops from the seedling rootstocks of grafted trees can be affected by the cultivar of the scion (*163, 182*). Different scion cultivars may cause a characteristic root growth pattern to develop in the rootstock. For example, if apple seedlings are budded with the 'Red Astrachan' apple, a very fibrous root system with few taproots develops. If other similar seedlings are budded with 'Oldenburg' or 'Fameuse', the subsequent root system is not fibrous but has a two- or three-pronged deep taproot system (*73*). In fact, nurseries propagating apples often can identify many of the scion cultivars by the appearance of the root system of the grafted nursery trees.

Effect on Cold-Hardiness of the Rootstock

In some species at least, the cold-hardiness of a certain rootstock can be influenced by the particular scion cultivar grafted on it. This effect is not due necessarily to a winter-hardy scion imparting hardiness to the rootstock. Rather, it is probably related to the degree of maturity attained by the rootstock, certain scion cultivars tending to prolong growth of the roots long into the fall so that insufficient maturity of the root tissues is reached by the time killing low winter temperatures occur. The rootstock, if left ungrafted, or grafted to a scion cultivar that stops growth in early fall, may mature its tissues sufficiently early so as to develop adequate winter-hardiness.

Effects of an Intermediate Stock on Scion and Rootstock

The ability of certain dwarfing clones, inserted as an interstock between a vigorous top and vigorous root, to produce dwarfed and early-bearing fruit trees has been known for centuries and is used commercially to propagate dwarfed trees. It is reported (*136*) that one of the earliest records (*95*) of such procedures was given in 1681 in England, advocating the use of the Paradise apple as an interstock to induce precocity in apple trees grown on crabapple rootstocks. Figure 11–24 shows, for example, the degree of dwarfing induced in apples by various interstocks.

Tests (*175*) made to determine whether intermediate stocks of various apple cultivars inserted between the apple rootstock 'Malling 2' and the scion cultivar 'Jonathan' or 'Delicious' had any effect on the behavior of the scion showed that in every case a depression of growth occurred in comparison with cases in which the scion cultivar itself was used as the interstock.

An intermediate dwarfing stem piece seems to have a built-in mechanism that causes reduced growth in the rootstock as well as in the scion top (*136*). Repeated comparisons of the influence exerted on the scion cultivar by the rootstock and an interstock show that although both have an influence, the rootstock's effect is greater (*180, 187*).

Dwarfing of apple trees by the use of a dwarfing interstock, such as 'Malling 9', has been widely used commercially for many years. This method has the advantage of allowing the use of well-anchored, vigorous seedlings as the rootstock rather than a brittle, poorly anchored dwarfing clone. However, excessive suckering from the roots may occur due to the dwarfing interstock, even in rootstock types that normally do not sucker freely.

This interstock effect could, in some cases,

FIGURE 11–24   Effect of interstock on size of six-year-old Cox's Orange Pippin apple grafted on the strong-growing 'MM 104' roots. *Top left:* Cox/ 'M.9'/'MM.104'. *Top right:* Cox/'M.27'/'MM.104'. *Lower left:* Cox/ 'MM.104'/'MM.104'. *Lower right:* Cox/'M.20'/'MM.104'. (Courtesy M. S. Parry, W. S. Rogers, and the Editors, *Journal of Horticultural Science* (*136*).)

be due to the introduction of an additional graft union with the possibility of translocation restrictions. Imperfect graft unions are indicated as a cause of the dwarfing exerted on orange trees by a lemon interstock. In contrast to the dwarfing situation in apples with a 'Malling 9' interstock, the lemon itself is strong-growing.

On the other hand, there is evidence that the observed effects of the interstock are due directly to an influence of the interstock piece rather than to abnormalities at the graft union (*157*). The dwarfing effect of 'Malling 9' seems to be due to something more than restrictions at the graft union, since 'Malling 9' is an early-bearing, dwarf tree itself.

Studies (*145*) have shown quantitatively that the initial response obtained from a 'Malling 9' interstock in composite 'Starking Delicious'/ 'M. 9'/'M. 16' trees was early and heavy flowering. This was followed—as a result of the heavy cropping—by a reduction in tree size. The degree of the response obtained was proportional to the length of the 'Malling 9' interstock. This finding supports earlier evidence that the interstock effect is due to a direct influence of the stock, since increasing the length of the interstock intensifies its effect (*39, 59*). However, studies with 'Old Home' pear as an interstock between 'Bartlett' and quince showed no effects resulting from different interstock lengths (*195*).

Possible Mechanisms for the Effects of Stock on Scion and Scion on Stock

The nature of the rootstock–scion relationship is very complex and probably differs among genetically different combinations. The fundamental mechanisms by which stock and scion influence each other are not well understood. Some of the explanations offered for the observed effects are speculative, often conflicting, and sometimes not well substantiated. Several theories have been advanced as possible explanations for the interaction between stock and scion.

One suggested mechanism is that the rootstock influences are the result of *translocation effects* rather than the *absorbing ability* of the root system. That the stem portion of the tree has, to some extent at least, a definite influence is shown by experiments (*13*) in which commonly used rootstock materials were used as intermediate stocks between a vigorous root system and the scion cultivar. The expected effects were still present, although to a lesser degree, even though the materials were used as interstocks rather than as the entire root-absorbing system. This same influence on the tree was noted if the intermediate stock tissue was reduced to just a ring of bark (*146*).

By this reasoning, the uniformity of trees produced on vegetatively propagated rootstocks is due to the uniformity of the *stems* of such stocks, whereas the variable stems of seedlings are responsible for the variation encountered in the growth of the scion cultivar. This is especially so if they are grafted or budded high on the stem so that there is more opportunity for stem influence to be exerted. But if the grafting is done on the roots of seedlings (those with relatively passive influence)—as is done in root grafting—this variability does not appear, owing to the absence of influence contributed by the stem section. Therefore a fairly uniform group of trees could be produced, even on seedling rootstocks. This situation has been noted with cherries. Orchards consisting of nursery trees that were low-worked on 'Mazzard' seedlings were quite uniform, whereas orchards from nursery trees budded high on such seedlings were quite variable.

However, some workers in England concluded from their experiments (*71, 187*) that the *root system* itself, rather than the *stem* of the rootstock, plays the major role in rootstock effects on the scion. Beakbane and Rogers (*9*), after experiments conducted with apple trees for 19 years, concluded that the characteristic rootstock influence was due to the roots themselves. This effect did not depend upon the presence of a piece of rootstock stem, although it did tend to increase the rootstock influence.

Ramirez and Tabuenca (*142*) may have resolved such contradictions in stem and root influence by showing that if the scion or interstock is to influence root morphology, it must be a type that has dominating characteristics.

Chandler (*29*) and Gardner, Bradford, and Hooker (*54*), discussing the subject in general terms, asserted that the effects of stock on scion and scion on stock can be explained by *physiological factors,* chiefly the influences due to *changes in vigor.*

Chandler pointed out that when the scion is the more vigorous part of the combination, the carbohydrate supply to the roots should be greater. And since certain roots supply and are supplied by certain branches, it would be expected that the branching habit of the top would influence the branching habit of the roots, thus explaining the different root types obtained in using different scion cultivars. If trees of different cultivars were pruned to exactly the same number and distribution of branches, then the difference in rootstock growth associated with the different scion cultivars might not occur.

Rootstock effects are not invariably related to vigor. Such things as flowering, fruit setting, fruit size, and fruit color or quality may be affected by the rootstock used even on trees showing an equal amount of vigor. For example, the marked effect of the rootstock on the development of black-end in pears is certainly not due to an alteration in vigor.

Tukey and Brase (*182*) concluded from their studies that no one part of a grafted tree could be considered to have complete control, but that all—rootstock, interstock, and scion—influenced the growth of the whole, although generally the rootstock had the dominant role.

Although no completely satisfactory explanation exists of how the three genetically different components of a grafted plant—rootstock, interstock, and scion—interact to influence the growth, flowering, and fruiting responses of the composite plant, three approaches can be considered: (a) *nutritional uptake and utilization,* (b) *translocation of nutrients and water,* and (c) *alterations in endogenous growth factors.*

Nutrition

It could possibly be that dwarfing rootstocks cause small trees by starvation effects. This has not been the case, however. Dwarf trees often contain higher concentrations of organic and mineral nutrients than vigorous ones (*30, 144*). The fruitful condition existing in young apple trees worked on the dwarfing 'Malling 9' roots was found to be associated with accumulation of starch in the shoots early in the season (*30*). Such an early starch storage would be expected to be favorable for the initiation of flower bud primordia.

Nonfruitful trees on the vigorous 'Malling 12' roots failed to show such a starch accumulation. The increased supply of water and nutrients from the vigorous roots would stimulate production of new growth rather than retard growth and would not allow for the accumulation of carbohydrates, as would be the case with weaker, dwarfing rootstocks.

Mineral absorption by the various rootstocks, which is made available for use by the scion cultivar, can explain certain rootstock influences. For example, the very vigorous 'Shalil' peach root system grown at low nutrient levels was able to pick up and furnish the scion cultivar with a greater supply of nutrients and water than the less vigorous 'Lovell' rootstock. Under these conditions, the scion cultivar showed better growth on 'Shalil' than on 'Lovell' roots. But when the salt level was higher and included toxic chloride ions, the greater accumulation of such salts, subsequently translocated to the scion cultivar, caused injury and growth depression. In this case, the difference in vigor of two rootstocks caused opposite effects under two different soil conditions—a low-salt and a high-salt condition (*72*).

Studies (*4, 5*) in England with the dwarfing 'Malling 9' and 'Malling 27' have provided evidence that trees on these stocks are smaller because they are restricted to fewer growing points, which continue growth for a shorter period. The even slower growth rate of the root system so limits the trees that they are unable to make use of photosynthates completely in continued growth.

Translocation

The fact that interstocks of such dwarfing apple clones as 'Malling 9' will cause a certain amount of dwarfing could indicate that translocation is involved, owing either to partial blockage at the graft unions or to a reduction in movement of water or nutrient materials (or both) through the interstock piece itself.

One explanation advanced (*190*) for the dwarfing effect of certain rootstocks is supported by studies concerning the efficiency of water conductivity of the graft union. There are indications that the graft union does introduce an additional resistance to the flow of water. This resistance was greater in unions of which the 'Malling 9' stock was one of the components. Certain of the growth characteristics of trees on 'Malling 9' roots, such as small leaves, short internodes, and the early cessation of seasonal shoot growth, are those generally associated with a slight water deficit in the tree. As mentioned before, however, although a restricted graft union may contribute to the dwarfing influence of certain rootstocks, the primary influence probably lies more in the nature of the growth characteristics of such stocks.

In a study of the translocation of radioactive phosphorus (^{32}P) and calcium (^{45}Ca) from the roots to the tops of one-year 'McIntosh' apple trees grown in solution culture, it was shown that over three times as much of both elements was found in the scion top when the vigorous 'Malling 16' root was used in comparison with the dwarfing 'Malling 9' (22). This may indicate a superior ability of the vigorous stock to absorb and translocate mineral nutrients to the scion in comparison with the dwarfing stock. Or it may only mean that the 'Malling 9' roots, with their high percentage of living tissue, formed a greater "sink" for these materials, retaining them in the roots.

Endogenous Growth Factors

The idea was proposed by Sax (154, 157) that some of the growth alterations noted when certain interstocks are used, or when a ring of bark in the trunk of young trees is inverted, may be due to interference with the normal translocation of natural growth substances and nutrients from the leaves to the roots. It is known that certain factors do exist (e.g., vitamin B$_1$) that are necessary for root growth, and anything that would stop or reduce the flow of such substances would limit growth of the rootstock and subsequently dwarf the entire tree.

Dwarfing rootstocks may exhibit their characteristic effects because of their own low production of endogenous growth promoters (auxins and gibberellins) or their inability to conduct or utilize such substances produced by the scion. Young trees on dwarfing stocks can make vigorous growth for a year or two in the nursery (181) when growth-regulating materials may still be present in sufficient amounts, but several years later

dwarfing develops, possibly because growth promoters decrease.

There is evidence (65) showing that the amount of indoleacetic acid (a growth-promoting auxin) that is destroyed by enzymes in root and shoot bark tissue of various apple rootstocks is correlated inversely with the scion vigor induced by the rootstock. In a study (112) of leaf extracts from various size-controlling apple rootstocks, it was found that those giving the greatest dwarfing contained materials that stimulated oxidative breakdown of indoleacetic acid.

In a comparison of endogenous growth-regulating factors in the bark of the dwarfing 'Malling 9' apple rootstock with those in the bark of the invigorating 'Malling 16', it was noted (101) that 'Malling 9' contained lower amounts of growth-promoting materials—but more growth inhibitors—than did 'Malling 16'.

Tissue extracts from four size-controlling apple rootstocks 'Malling 16' (invigorating), 'Malling 1' (semi-invigorating), 'Malling 7' (semidwarfing), and 'Malling 9' (dwarfing), showed increasing levels of an endogenous inhibitor—in the same order—which apparently was abscisic acid (198).

There is the possibility, too, that different levels of endogenous gibberellin, as a growth promoter, may account, in part, for the size-controlling characteristics of various rootstocks. It is known that roots do produce gibberellin and that the amounts of gibberellin in the transpiration stream are sufficient to have a decided growth-controlling influence (90). Extracts of the dwarfing 'Malling 9' apple rootstocks were found (88) to contain lower amounts of gibberellic acid-like substances than extracts of the more invigorating 'Malling 1' and 'Malling 25'. Such low gibberellin levels in the more dwarfing stock could be due either to lower production or to more rapid destruction. The differential dwarfing due to different rootstocks thus could possibly be accounted for, in part, by different levels of gibberellin translocated from the roots to the shoot system (27). Injections of gibberellic acid into grafted apple trees have given increasing stimulation to the top growth as rootstock vigor decreased, while injections of the growth inhibitor, abscisic acid, caused shoot growth to cease (148).

Grafting studies (*99*), where complete rings of the dwarfing 'Malling 26' apple bark were grafted into 'Gravenstein' stems grafted on the invigorating 'M111' roots resulted in dwarfed trees, gave rise to the following postulation of the dwarfing mechanism in apples: Auxin produced by the shoot tip moves downward through the phloem. The amount arriving in the root influences root metabolism, including the amount and kind of cytokinins synthesized and translocated upward from the roots to the shoot system via the xylem. These cytokinins influence the amount of shoot growth.

In these studies the grafted bark ring of the dwarfing 'Malling 26' apple caused reduction in the downward movement of auxins, subsequently reducing cytokinin production in the roots; this lack of cytokinins available for upward movement reduced top growth, eventually giving a dwarfed tree.

The various dwarfing and invigorating types of rootstocks apparently either contain different amounts of naturally occurring growth-promoting and growth-inhibiting materials or can affect the amounts passing through their tissues. They could thus not only control their own growth, but similarly affect growth of their graft partners—which seems to be a good explanation for the mechanisms involved in size-controlling rootstocks.

However, it is likely that there are interactions of all three factors: nutrition, translocation, and growth substances on stock/scion effects or, in some species, one or two of the factors may dominate.

REFERENCES

1. Altman, A., and R. Goren. 1971. Promotion of callus formation by abscisic acid in citrus bud cultures. *Plant Phys.* 47:844–46.

2. Amos, J., T. N. Hoblyn, R. J. Garner, and A. Witt. 1936. Studies in incompatibility of stock and scion. I. Information accumulated during twenty years of testing fruit tree rootstocks with various scion varieties at East Malling. *Ann. Rpt. E. Malling Res. Sta. for 1935,* pp. 81–99.

3. Argles, G. K. 1937. A review of the literature on stock-scion incompatibility in fruit trees, with particular reference to pome and stone fruits. *Imp. Bur. of Fruit Prod. Tech. Comm. 9.*

4. Avery, D. J. 1969. Comparisons of fruiting and deblossomed maiden apple trees and of non-fruiting trees on a dwarfing and an invigorating rootstock. *New Phytol.* 68:323–36.

5. ———. 1970. Effects of fruiting on the growth of apple trees on four rootstock varieties. *New Phytol.* 69:19–30.

6. Bailey, L. H. 1891. *The nursery book.* New York: Rural Publishing Company.

7. Batjer, L. P., and H. Schneider. 1960. Relation of pear decline to rootstocks and sieve tube necrosis. *Proc. Amer. Soc. Hort. Sci.* 76:85–97.

8. Beakbane, A. B. 1953. Anatomical structure in relation to rootstock behavior. *Rpt. 13th Inter. Hort. Cong.,* Vol. 1, pp. 152–58.

9. Beakbane, A. B., and W. S. Rogers. 1956. The relative importance of stem and root in determining rootstock influence in apples. *Jour. Hort. Sci.* 31:99–110.

10. Bitters, W. P. 1950. Citrus rootstocks for dwarfing. *Calif. Agr.* 4(2):5–14.

11. ———. 1961. Physical characters and chemical composition as affected by scions and rootstocks. Chapter 3 in *The orange: Its biochemistry and physiology,* W. B. Sinclair, ed. Berkeley: Univ. Calif. Div. Agr. Sci.

12. Bitters, W. P., and E. R. Parker. 1953. Quick decline of citrus as influenced by top-root relationships. *Calif. Agr. Exp. Sta. Bul. 733.*

13. Blair, D. S. 1938. Rootstock and scion relationship in apple trees. *Sci. Agr.* 19:85–94.

14. Bloch, R. 1952. Wound healing in higher plants. *Bot. Rev.* 18:655–79.

15. Blumenfield, A., and S. Gazit. 1969. Interaction of kinetin and abscisic acid in the growth of soybean callus. *Plant Phys.* 45:535–36.

16. Bradford, F. C., and B. G. Sitton. 1929. Defective graft unions in the apple and pear. *Mich. Agr. Exp. Sta. Tech. Bul. 99.*

17. Brase, K. D., and R. D. Way. 1959. Rootstocks and methods used for dwarfing fruit trees. *N. Y. State Agr. Exp. Sta. Bul. 783.*

18. Buchloh, G. 1960. The lignification in stock-scion junctions and its relation to compatibility. In *Phenolics in plants in health and disease,* J. B. Pridham, ed. Long Island City, N.Y.: Pergamon Press.

19. ———. 1962. Verwachsung und Verwachsungsstörungen als Ausdruck des Affinitätsgrades bei Propfungen von Birnenvarietäten auf *Cydonia oblonga. Beit Biol. Pfl.* 37:183–240.

20. Buck, G. J. 1953. The histological development of the bud graft union in roses. *Proc. Amer. Soc. Hort. Sci.* 62:497–502.

21. Buck, G. J., and B. J. Heppel. 1970. A budgraft incompatibility in *Rosa. Jour. Amer. Soc. Hort. Sci.* 95(4):442–46.

22. Bukovac, M. J., S. H. Wittwer, and H. B. Tukey. 1958. Effect of stock-scion interrelationships on the transport of ^{32}P and ^{45}Ca in the apple. *Jour. Hort. Sci.* 33:145–52.

23. Bunyard, E. A. 1920. The history of the Paradise stocks. *Jour. Pom.* 2:166–76.

24. Camus, G. 1949. Recherches sur le rôle des bourgeons dans les phénomènes de morphogénèse. *Revue Cytol. et Biol. Vég.* 11:1–199.

25. Caponetti, J. D., G. C. Hall, and R. E. Farmer, Jr. 1971. In vitro growth of black cherry callus: Effects of medium, environment, and clone. *Bot. Gaz.* 132(4):313–18.

26. Carlson, R. F. 1967. The incidence of scion-rooting of apple cultivars planted at different soil depths. *Hort. Res.* 7(2):113–15.

27. Carr, D. J., D. M. Reid, and K. G. M. Skene. 1964. The supply of gibberellins from the root to the shoot. *Planta* 63:382–92.

28. Cation, D., and R. F. Carlson. 1962. Determination of virus entities in an apple scion/rootstock test orchard. *Quart. Bul. Mich. Agr. Exp. Sta., Rpt.* I, 43(2):435–43, 1960. *Rept. II,* 45(17): 159–66.

29. Chandler, W. H. 1925. *Fruit growing.* Boston: Houghton Mifflin.

30. Colby, H. L. 1935. Stock-scion chemistry and the fruiting relationships in apple trees. *Plant Phys.* 19:483–98.

31. Collins, R. P., and S. Waxman. 1958. Dogwood graft failures. *Amer. Nurs.* 108(8):12.

32. Colquhoun, T. T. 1929. Polarity in *Casuarina paludosa. Trans. and Proc. Roy. Soc. South Australia* 53:353–58.

33. Cooper, W. C. 1952. Influence of rootstock on injury and recovery of young citrus trees exposed to the freezes of 1950–51 in the Rio Grande Valley. *Proc. 6th Ann. Rio Grande Valley Hort. Inst.,* pp. 16–24.

34. Cooper, W. C., and E. O. Olson. 1951. Influence of rootstock on chlorosis of young Red Blush grapefruit trees. *Proc. Amer. Soc. Hort. Sci.* 57:125–32.

35. Copes, D. A. 1969. Graft union formation in Douglas-fir. *Amer. Jour. Bot.* 56(3):285–89.

36. ———. 1970. Initiation and development of graft incompatibility symptoms in Douglas-fir. *Silvae Genet.* 19:101–7.

37. Crafts, A. S. 1934. Phloem anatomy in two species of *Nicotiana,* with notes on the interspecific graft union. *Bot. Gaz.* 95:592–608.

38. Crane, M. B., and E. Marks. 1952. Pear-apple hybrids. *Nature* 170:1017.

39. Dana, M. N., H. L. Lantz, and W. E. Loomis. 1962. Effects of interstock grafts on growth of Golden Delicious apple trees. *Proc. Amer. Soc. Hort. Sci.* 81:1–11.

40. Davis, L. D., and W. P. Tufts. 1936. Black end of pears III. *Proc. Amer. Soc. Hort. Sci.* 33:304–15.

41. Dawson, R. F. 1942. Accumulation of nicotine in reciprocal grafts of tomato and tobacco. *Amer. Jour. Bot.* 29:66–71.

42. Day, L. H. 1951. Cherry rootstocks in California. *Calif. Agr. Exp. Sta. Bul. 725.*

43. ———. 1953. Rootstocks for stone fruits. *Calif. Agr. Exp. Sta. Bul. 736.*

44. Dimalla, G. G., and J. A. Milbrath. 1965. The prevalence of latent viruses in Oregon apple trees. *Plant Dis. Rpt.* 49(1):15–17.

45. Doesburg, J. van. 1962. Use of fungicides with vegetative propagation. *Rpt. 16th Inter. Hort. Cong.,* pp. 365–72.

46. Doley, D., and L. Leyton. 1970. Effects of growth regulating substances and water potential on the development of wound callus in *Fraxinus. New Phytol.* 69:87–102.

47. Dorsman, C. 1966. Grafting of woody plants in the glasshouse. *Proc. 17th Inter. Hort. Cong.* 1: 366.

48. Eames, A. J., and L. G. Cox. 1945. A remarkable tree-fall and an unusual type of graft union failure. *Amer. Jour. Bot.* 32:331–35.

49. Epstein, A. H. 1978. Root graft transmission in tree pathogens. *Ann. Rev. Phytopathol.* 16:181–92.

50. Evans, G. E., and H. P. Rasmussen. 1972. Anatomical changes in developing graft unions of *Juniperus. Jour. Amer. Soc. Hort. Sci.* 97(2):228–32.

51. Evans, W. D., and R. J. Hilton. 1957. Methods of evaluating stock/scion compatibility in apple trees. *Can. Jour. Plant Sci.* 37:327–36.

52. Fujii, T., and N. Nito. 1972. Studies on the compatibility of grafting of fruit trees. I. Callus fusion between rootstock and scion. *Jour. Jap. Soc. Hort. Sci.* 41(1):1–10.

53. Gardner, F. E., and G. H. Horanic. 1963. Cold tolerance and vigor of young citrus trees on various rootstocks. *Proc. Fla. State Hort. Soc.* 76:105–10.

54. Gardner, V. R., F. C. Bradford, and H. D. Hooker, Jr. 1939. *Fundamentals of fruit production* (2nd ed.). New York: McGraw-Hill.

55. Garner, R. J., and D. H. Hammond. 1939. Studies in nursery technique. Shield budding. Treatment of inserted buds with petroleum jelly. *Ann. Rpt. E. Malling Res. Sta. for 1938,* pp. 115–17.

56. Gautheret, R. J. 1947. La culture des tissus végétaux. *Proc. 6th Inter. Cong. Exp. Cytol.* pp. 437–49.

57. Goodale, S. L. 1846. Influence of the scion upon the stock. *Horticulture* 1:290.

58. Graham, B. F., Jr., and F. H. Bornmann. 1966. Natural root grafts. *Bot. Rev.* 32(3):255–92.

59. Grubb, N. H. 1939. The influence of intermediate stem pieces in double-worked apple and pear trees. *Scient. Hort.* 7:17–23.

60. Guengerich, H. W., and D. F. Milliken. 1966. Bud transmission of dwarfing in sweet cherry. *Plant Dis. Rpt.* 50:367–68.

61. ———. 1965. Root grafting, a potential source of error in apple indexing. *Plant Dis. Rpt.* 49:39–41.

62. Gur, A. 1957. The compatibility of the pear with quince rootstock. *Spec. Bul. 10,* Agr. Res. Sta., Rehovot (Israel), pp. 1–99.

63. Gur, A., and E. Lifshitz. 1968. The role of the cyanogenic glycoside of the quince in the incompatibility between pear cultivars and quince rootstocks. *Hort. Res.* 8:113–34.

64. Gur, A., and R. M. Samish. 1965. The relation between growth curves, carbohydrate distribution, and compatibility of pear trees grafted on quince rootstocks. *Hort. Res.* 5:81–100.

65. Gur, A., and R. M. Samish. 1968. The role of

auxins and auxin destruction in the vigor effect induced by various apple rootstocks. *Beitr. Biol. Pflanz.* 45:91–111.

66. Hansen, C. J. 1948. Influence of the rootstock on injury from excess boron in French (Agen) prune and President plum. *Proc. Amer. Soc. Hort. Sci.* 51:239–44.

67. Hansen, C. J., and H. T. Hartmann. 1951. Influence of various treatments given to walnut grafts on the percentage of scions growing. *Proc. Amer. Soc. Hort. Sci.* 57:193–97.

68. Harmon, F. N. 1949. Comparative value of thirteen rootstocks for ten vinifera grape varieties in the Napa Valley in California. *Proc. Amer. Soc. Hort. Sci.* 54:157–62.

69. Hartmann, H. T. 1958. Rootstock effects in the olive. *Proc. Amer. Soc. Hort. Sci.* 72:242–51.

70. Hatton, R. G. 1927. The influence of different rootstocks upon the vigor and productivity of the variety budded or grafted thereon. *Jour. Pom. and Hort. Sci.* 6:1–28.

71. ———. 1931. The influence of vegetatively raised rootstocks upon the apple, with special reference to the parts played by the stem and root portions in affecting the scion. *Jour. Pom. and Hort. Sci.* 9:265–77.

72. Hayward, H. E., and E. M. Long. 1942. Vegetative responses of the Elberta peach on Lovell and Shalil rootstocks to high chloride and sulfate solutions. *Proc. Amer. Soc. Hort. Sci.* 41:149–55.

73. Hedrick, U. P. 1915. Stocks for fruits. *Rpt. N.Y. State Fruit Growers Assn.,* pp. 84–94.

74. Heppner, M. 1927. Pear black-end and its relation to different rootstocks. *Proc. Amer. Soc. Hort. Sci.* 24:139.

75. Heppner, M., and R. D. McCallum. 1927. Grafting affinities with special reference to plums. *Calif. Agr. Exp. Sta. Bul. 438.*

76. Herrero, J. 1955. Incompatibilidad entre patrón e injerto. II, Effecto de un intermediario en la incompatibilidad entre melocotonero y mirobalán. *An. Aula Dei* 4:167–72.

77. ———. 1951. Studies of compatible and incompatible graft combinations with special reference to hardy fruit trees. *Jour. Hort. Sci.* 26:186–237.

78. Herrero, J., and M. V. Tabuenca. 1969. Incompatibilidad entre patrón e injerto. X. Comportamiento de la combinación melocotonero/mirobálan injertado en estado cotiledónor. *An. Estac, exp. Aula Dei* 19:937–45.

79. Hesse, H. 1710. *Teutscher Gärtner.* Leipsig.

80. Heuser, C. W. 1984. Graft incompatibility in

woody plants. *Proc. Inter. Plant Prop. Soc.* 33:407–12.

81. Heuser, C. W. 1987. Graft incompatibility: Effect of cyanogenic glycoside on almond and plum callus growth. *Proc. Inter. Plant Prop. Soc.* 36:91–97.

82. Hibino, H., G. H. Kaloostian, and H. Schneider. 1971. Mycoplasma-like bodies in the pear psylla vector of pear decline. *Virology* 43:34–40.

83. Hibino, H., and H. Schneider. 1970. Mycoplasma-like bodies in sieve tubes of pear trees affected with pear decline. *Phytopath.* 60:499–501.

84. Hodgson, R. W. 1943. Some instances of scion domination in citrus. *Proc. Amer. Soc. Hort. Sci.* 43:131–38.

85. Hodgson, R. W., and S. H. Cameron. 1935. On bud union effect in citrus. *Calif. Citrog.* 20(12):370.

86. Homes, J. 1965. Histogenesis in plant grafts. In *Proc. Inter. Conf. Plant Tissue Cult.,* P. R. White and A. R. Groves, eds. Amer. Inst. Biol. Sci., pp. 553.

87. Honma, S. 1977. Grafting eggplants. *Scient. Hort.* 7:207–11.

88. Ibrahim, I. M., and M. N. Dana. 1971. Gibberellin-like activity in apple rootstocks. *HortScience* 6(6):541–42.

89. Jensen, D. D., W. H. Griggs, C. Q. Gonzales, and H. Schneider. 1964. Pear decline virus transmission by pear psylla. *Phytopath.* 54:1346–51.

90. Jones, O. P., and H. J. Lacey. 1968. Gibberellin-like substances in the transpiration stream of apple and pear trees. *Jour. Exp. Bot.* 19:526–31.

91. Juliano, J. B. 1941. Callus development in graft union. *Philippine Jour. Sci.* 75:245–51.

92. Keane, F. W. L., and J. May. 1963. Natural root grafting in cherry and spread of cherry twisted-leaf virus. *Can. Plant Dis. Sur.* 43(2):54–60.

93. Kester, D. E., C. J. Hansen, and C. Panetsos. 1965. Effect of scion and interstock variety on incompatibility of almond on Marianna 2624 rootstock. *Proc. Amer. Soc. Hort. Sci.* 86:169–77.

94. Kunkel, L. O. 1938. Contact periods in graft transmission of peach viruses. *Phytopath.* 28:491–97.

95. Langford, G. T. 1681. *Plain and full instructions to raise all sorts of fruit trees that prosper in England.* London: J. M. at the Rose and Crown.

96. Langford, M. H., J. B. Carpenter, W. E. Manis, A. M. Gorenz, and E. P. Imle. 1954. *Hevea* diseases of the Western Hemisphere. *Plant Dis. Rpt. Suppl. 225.*

97. Langford, M. H., and C. H. T. Townsend, Jr. 1954. Control of South American leaf blight of *Hevea* rubber trees. *Plant Dis. Rpt. Suppl. 225.*

98. Lapins, K. 1959. Some symptoms of stock-scion incompatibility of apricot varieties on peach seedling rootstock. *Can. Jour. Plant Sci.* 39:194–203.

99. Lockard, R. G., and G. W. Schneider. 1981. Stock and scion growth relationships and the dwarfing mechanism in apple. *Hort. Rev.* 3:315–75.

100. Lowman, M. S., and J. W. Kelley. 1946. The presence of mydriatic alkaloids in tomato fruit from scions grown on *Datura stramonium* rootstock. *Proc. Amer. Soc. Hort. Sci.* 48:249–59.

101. Martin, G. C., and E. A. Stahly. 1967. Endogenous growth regulating factors in bark of EM IX and XVI apple trees. *Proc. Amer. Soc. Hort. Sci.* 91:31–38.

102. McClintock, J. A. 1948. A study of uncongeniality between peaches as scions and the Marianna plum as a stock. *Jour. Agr. Res.* 77:253–60.

103. McCully, M. E. 1983. Structural aspects of graft development. In *Vegetative compatibility responses in plants,* R. Moore, ed. Waco, Tex.: Baylor Univ. Press.

104. McKenzie, D. W. 1961. Rootstock-scion interaction in apples with special reference to root anatomy. *Jour. Hort. Sci.* 36:40–47.

105. McQuilkin, W. E. 1950. Effects of some growth regulators and dressings on the healing of tree wounds. *Jour. For.* 48(9):423–28.

106. Mendel, K. 1936. The anatomy and histology of the bud-union in citrus. *Palest. Jour. Bot. (R),* 1(2):13–46.

107. Mergen, F. 1954. Anatomical study of slash pine graft unions. *Quart. Jour. Fla. Acad. Sci.* 17:237–45.

108. Milbraith, J. A., and S. M. Zeller. 1945. Latent viruses in stone fruits. *Science* 101:114–15.

109. Miller, L., and F. W. Woods. 1965. Root grafting in loblolly pine. *Bot Gaz.* 126:252–55.

110. Miller, P. W. 1952. Technique for indexing strawberries for viruses by grafting to *Fragaria vesca. Plant Dis. Rpt. 36.*

111. ———. 1965. The etiology of blackline in grafted Persian walnuts. *Plant Dis. Rpt.* 49:954.

112. Miller, S. R. 1965. Growth inhibition produced

by leaf extracts from size controlling apple rootstocks. *Can. Jour. Plant Sci.* 45(6):519–24.

113. Millner, M. E. 1932. Natural grafting in *Hedera helix. New Phytol.* 31:2–25.

114. Mircetich, S. M. J., J. Refsguard, and M. E. Matheron. 1980. Blackline of English walnut trees traced to graft-transmitted virus. *Calif. Agr.* 34(11–12):8–10.

115. Montgomery, H. B. S. 1963. *Fruit tree raising.* Bul. 135. London: Minist. Agr., Fish., and Foods.

116. Moore, R. 1981. Graft compatibility/incompatibility in higher plants. *What's new in plant phys.* 12(4):13–16.

117. ———. 1982. Graft formation in *Kalanchoe blossfeldiana. Jour. Exp. Bot.* 33:533–40.

118. Moore, R., and D. B. Walker. 1981. Graft compatibility-incompatibility in plants. *BioScience* 31(5):389–91.

119. ———. 1981. Studies of vegetative compatibility-incompatibility in higher plants. I. A structural study of a compatible autograft in *Sedum telephoides* (Crassulaceae). II. A structural study of an incompatible heterograft between *Sedum telephoides* (Crassulaceae) and *Solanum penellii* (Solanaceae). *Amer. Jour. Bot.* 68(6):820–42.

120. ———. 1981. Studies of vegetative compatibility-incompatibility in higher plants. III. The involvement of acid phosphatase in the lethal cellular sensescence associated with an incompatible heterograft. *Protoplasma* 109:317–34.

121. ———. 1983. Studies of vegetative compatibility-incompatibility in higher plants. IV. The development of tensile strength in a compatible and an incompatible graft. *Amer. Jour. Bot.* 70(2):226–31.

122. Mosse, B. 1958. Further observations on growth and union structure of double-grafted pear on quince. *Jour. Hort. Sci.* 33:186–93.

123. ———. 1962. *Graft-incompatibility in fruit trees.* Tech. Comm. 28. East Malling, England: Comm. Bur. Hort. and Plant. Crops.

124. Mosse, B., and M. V. Labern. 1960. The structure and development of vascular nodules in apple bud unions. *Ann. Bot.* 24:500–507.

125. Muller-Stoll, W. R. 1938. Versuche über die Verwenbarkeit der B-indolylessigsaure als Verwachsungsförderunges Mittel in der Rebenveredlung. *Angew. Bot.* 20:218–38.

126. Murashige, T., and F. Skoog. 1962. A revised medium for rapid growth and bioassays with tobacco tissue cultures. *Phys. Plant.* 15:473–97.

127. Muzik, T. J. 1958. Role of parenchyma cells in graft union in *Vanilla* orchid. *Science* 127:82.

128. Muzik, T., and C. D. LaRue. 1954. Further studies on the grafting of monocotyledonous plants. *Amer. Jour. Bot.* 41:448–55.

129. Nelson, S. H. 1968. Incompatibility survey among horticultural plants. *Proc. Inter. Plant Prop. Soc.* 18:343–407.

130. Nelson, R., S. Goldweber, and F. J. Fuchs. 1955. Top-working for mangos. *Fla. Grower and Rancher,* Jan., p. 45.

131. Nichols, C. W., H. Schneider, H. J. O'Reilly, T. A. Shalla, and W. H. Griggs. 1960. Pear decline in California. *Bul. Calif. State Dept. Agr.* 49:186–92.

132. Nickell, L. G. 1946. Heteroplastic grafts. *Science* 108:389.

133. Nyland, G., and W. J. Moller. 1973. Control of pear decline with a tetracycline. *Plant Dis. Rpt.* 57:634–37.

134. Overbeek, J. van. 1966. Plant hormones and regulators. *Science* 152:721–31.

135. Parkinson, M., and M. M. Yeoman. 1982. Graft formation in cultured explanted internodes. *New Phytol.* 91:711–19.

136. Parry, M. S., and W. S. Rogers. 1968. Dwarfing interstocks: Their effect on the field performance and anchorage of apple trees. *Jour. Hort. Sci.* 43:133–46.

137. Posnette, A. F. 1966. Virus diseases of fruit plants. *Proc. 17th Inter. Hort. Cong.* 3:89–93.

138. Preston, A. P. 1958. Apple rootstock studies: Thirty-five years' results with Cox's Orange Pippin on clonal rootstocks. *Jour. Hort. Sci.* 33:194–201.

139. ———. 1958. Apple rootstock studies: Thirty-five years' results with Lane's Prince Albert on clonal rootstocks. *Jour. Hort. Sci.* 33:29–38.

140. Proebsting, E. L. 1928. Further observations on structural defects of the graft union. *Bot. Gaz.* 86:82–92.

141. ———. 1926. Structural weaknesses in interspecific grafts of *Pyrus. Bot. Gaz.* 82:336–38.

142. Ramirez, D., and M. C. Tabuenca. 1964. The reciprocal effects of M. IX and M. XVI apples (English summary). *Ann. Estac. Exp. Aula Dei* 7:164–74.

143. Rao, A. N. 1966. Developmental anatomy of natural root grafts in *Ficus globosa. Austral. Jour. Bot.* 14:269–76.

144. Rao, Y. V., and W. E. Berry. 1940. The carbo-

hydrate relations of a single scion grafted on Malling rootstocks IX and XIII. A contribution to the physiology of dwarfing. *Jour. Pom.* 18: 193–225.

145. Roberts, A. N., and L. T. Blaney. 1967. Qualitative, quantitative, and positional aspects of interstock influence on growth and flowering of the apple. *Proc. Amer. Soc. Hort. Sci.* 91:39–50.

146. Roberts, R. H. 1949. Theoretical aspects of graftage. *Bot. Rev.* 15:423–63.

147. Robitaille, R. H., and R. F. Carlson. 1970. Graft union behavior of certain species of *Malus* and *Prunus*. *Jour. Amer. Soc. Hort. Sci.* 95(2):131–34.

148. Robitaille, H., and R. F. Carlson. 1971. Response of dwarfed apple trees to stem injections of gibberellic and abscisic acids. *HortScience* 6(6):539–40.

149. Samish, R. M. 1962. Physiological approaches to rootstock selection. *Advances in horticultural sciences and their applications,* Vol. 2. Long Island City, N.Y.: Pergamon Press.

150. Santamour, F. S., Jr. 1988. Graft compatibility in woody plants: An expanded perspective. *Jour. Environ. Hort.* 6(1):27–32.

151. ———. 1988. Graft incompatibility related to cambial peroxidase isozymes in Chinese chestnut. *Jour. Environ. Hort.* 6(2):33–39.

152. Santamour, F. S., Jr., A. J. McArdle, and R. A. Jaynes. 1986. Cambial isoperoxidase patterns in *Castanea. Jour. Environ. Hort.* 4(1):14–16.

153. Sass, J. E. 1932. Formation of callus knots on apple grafts as related to the histology of the graft union. *Bot. Gaz.* 94:364–80.

154. Sax, K. 1954. The control of tree growth by phloem blocks. *Jour. Arn. Arb.* 35:251–58.

155. ———. 1950. Dwarf trees. *Arnoldia* 10:73–79.

156. ———. 1950. The effect of the rootstock on the growth of seedling trees and shrubs. *Proc. Amer. Soc. Hort. Sci.* 56:166–68.

157. ———. 1953. Interstock effects in dwarfing fruit trees. *Proc. Amer. Soc. Hort. Sci.* 62:201–4.

158. Sax, K., and A. Q. Dickson. 1956. Phloem polarity in bark regeneration. *Jour. Arn. Arb.* 37:173–79.

159. Schroeder, C. A. 1947. Rootstock influence on fruit set in the Hachiya persimmon. *Proc. Amer. Soc. Hort. Sci.* 50:149–50.

160. Serr, E. F., and H. I. Forde. 1959. Blackline, a delayed failure at the union of *Juglans regia* trees propagated on other *Juglans* species. *Proc. Amer. Soc. Hort. Sci.* 74:220–31.

161. Shalla, T. A., and L. Chiarappa. 1961. Pear decline in Italy. *Bul. Calif. State Dept. Agr.* 50: 213–17.

162. Sharples, A., and H. Gunnery. 1933. Callus formation in *Hibiscus rosasinensis* L. and *Hevea brasiliensis* Mull. Arg. *Ann. Bot.* 47:827–39.

163. Shaw, J. K. 1915. The root systems of nursery apple trees. *Proc. Amer. Soc. Hort. Sci.* 12:68–72.

164. Shimomura, T., and K. Fuzihara. 1977. Physiological study of graft union formation in cactus. II. Role of auxin on vascular connection between stock and scion. *Jour. Japan. Soc. Hort. Sci.* 45:397–406.

165. Shippy, W. B. 1930. Influence of environment on the callusing of apple cuttings and grafts. *Amer. Jour. Bot.* 17:290–327.

166. Sitton, B. G. 1931. Vegetative propagation of the black walnut. *Mich. Agr. Exp. Sta. Tech. Bul. 119.*

167. Skene, D. S., H. R. Shepard, and B. H. Howard. 1983. Characteristic anatomy of union formation in T- and chip-budded fruit and ornamental trees. *Jour. Hort. Sci.* 58(3):295–99.

168. Smith, J. W. M., and P. Proctor. 1965. Use of disease resistant rootstocks for tomato crops. *Exp. Hort.* 12:6–20.

169. Solt, M. L., and R. V. Dawson. 1958. Production, translocation, and accumulation of alkaloids in tobacco scions grafted on tomato rootstocks. *Plant Phys.* 33:375–81.

170. Soule, J. 1971. Anatomy of the bud union in mango (*Mangifera indica* L.). *Jour. Amer. Soc. Hort. Sci.* 96(3):380–83.

171. Stoddard, F. L., and M. E. McCully. 1980. Effects of excision of stock and scion organs on the formation of the graft union in coleus: A histological study. *Bot. Gaz.* 141:401–2.

172. Stone, E. L., J. E. Stone, and R. C. McKittrick. 1973. Root grafting in pine trees. *Food and Life Sci. Quart.* 6(2):19–21.

173. Sussex, I. M., and Mary E. Clutter. 1959. Seasonal growth periodicity of tissue explants from woody perennial plants *in vitro. Science* 129:836–37.

174. Thiel, K. 1954. Untersuchungen zur Frage des Unverträglichkeit bei Birnenedelsorten auf Quitte A (*Cydonia* E. M. A). *Gartenbauwiss* 1(19): 127–59.

175. Thomas, L. A. 1954. Stock and scion investigations. X. Influence of an intermediate stempiece upon the scion in apple trees. *Jour. Hort. Sci.* 29:150–52.

176. Thouin, A. 1821. *Monographie des greffes, ou description technique* (in Royal Hort. Soc. Library, London).

177. Tomaselli, R., and E. Refatti. 1960. The non-existence of constant relationship between root anatomy and vigor in grafted apple and pear trees. *Atti Ist. bot., Univ. Pavia Ser. 5* 18:130–40.

178. Torrey, J. G., D. E. Fosket, and P. K. Hepler. 1971. Xylem formation: A paradigm of cytodifferentiation in higher plants. *Amer. Sci.* 59:338–52.

179. Toxopeus, H. J. 1936. Stock-scion incompatibility in citrus and its cause. *Jour. Pom. and Hort. Sci.* 14:360–64.

180. Tukey, H. B. 1943. The dwarfing effect of an intermediate stem-piece of Malling IX apple. *Proc. Amer. Soc. Hort. Sci.* 42:357–64.

181. ———. 1941. Similarity in the nursery of several Malling apple stock and scion combinations which differ widely in the orchard. *Proc. Amer. Soc. Hort. Sci.* 39:245–46.

182. Tukey, H. B., and K. D. Brase. 1933. Influence of the scion and of an intermediate stem-piece upon the character and development of roots of young apple trees. *N.Y. (Geneva) Agr. Exp. Sta. Tech. Bul. 218.*

183. Tydeman, H. M. 1937. Experiments on hastening the fruiting of seedling apples. *Ann. Rpt. E. Malling Res. Sta. for 1936,* pp. 92–99.

184. Tydeman, H. M., and F. H. Alston. 1965. The influence of dwarfing rootstocks in shortening the juvenile phase of apple seedlings. *Ann. Rpt. E. Malling Res. Sta. for 1964,* pp. 97–98.

185. Vaile, J. E. 1938. The influence of rootstocks on the yield and vigor of American grapes. *Proc. Amer. Soc. Hort. Sci.* 35:471–74.

186. Vöchting, H. 1892. Veber transplantation am pflanzenköper.

187. Vyvyan, M. C. 1938. The relative influence of rootstock and of an intermediate piece of stock stem in some double-grafted apple trees. *Jour. Pom. and Hort. Sci.* 16:251–73.

188. Wagner, D. F. 1969. Ultrastructure of the bud graft union in *Malus.* Ph.D. dissertation, Iowa State Univ., Ames, Iowa.

189. Wareing, P. F., C. E. A. Hanney, and J. Digby. 1964. The role of endogenous hormones in cambial activity and xylem differentiation. In *The formation of wood in forest trees,* M. H. Zimmerman, ed. New York: Academic Press.

190. Warne, L. G. G., and Joan Raby. 1939. The water conductivity of the graft union in apple trees, with special reference to Malling rootstock No. IX. *Jour. Pom. and Hort. Sci.* 16:389–99.

191. Waugh, F. A. 1904. The graft union. *Mass. Agr. Exp. Sta. Tech. Bul. 2.*

192. Webber, H. J. 1943. The "tristeza" disease of sour orange rootstock. *Proc. Amer. Soc. Hort. Sci.,* 43: 160–68.

193. Wellensiek, S. J. 1949. The prevention of graft-incompatibility by own foliage on the stock. *Meded. Landbouwhoogesch. Wageningen* 49:255–72.

194. Wells, R. B. 1986. An historical review of grafting techniques. *Proc. Inter. Plant Prop. Soc.* 35: 96–101.

195. Westwood, M. N., and H. O. Bjornstad. 1972. Length of Old Home interstem makes little growth difference. *Oreg. Orn. Nurs. Dig.* 16(1): 3–4.

196. Wetmore, R. H., and J. P. Rier. 1963. Experimental induction of vascular tissues in callus of angiosperms. *Amer. Jour. Bot.* 50:418–30.

197. Winter, H. F. 1963. Prevalence of latent viruses in Ohio apple trees. *Ohio Farm and Home Res.* 48:58–59, 63.

198. Yadava, U. L., and D. F. Dayton. 1972. The relation of endogenous abscisic acid to the dwarfing capability of East Malling apple rootstocks. *Jour. Amer. Soc. Hort. Sci.* 97(6):701–5.

199. Yeager, A. F. 1944. Xylem formation from ring grafts. *Proc. Amer. Soc. Hort. Sci.* 44:221–22.

200. Zebrak, A. R. 1937. Intergeneric and interfamily grafting of herbaceous plants. *Timirjazey Seljskohoz Akad.* 2:115–33. *Herb. Abst.* 9:675. 1939.

SUPPLEMENTARY READING

ARGLES, G. K. 1937. A review of the literature on stock–scion incompatibility in fruit trees with particular reference to pome and stone fruits. *Imp. Bur. Fruit Prod., Tech. Comm. 9.*

BEAKBANE, A. B. 1956. Possible mechanisms of rootstock effect. *Ann. of App. Bio.* 44:517–21.

BRASE, K. D., and R. D. WAY. 1959. Rootstocks and methods used for dwarfing fruit trees. *N.Y. Agr. Exp. Sta. Bul. 783.*

DANIEL, L. 1927, 1929. *Etudes sur la greffe,* Vols. 1 and 2. Rennes, Paris: Imprimerie Oberthur.

ESAU, K. 1977. *Anatomy of seed plants* (2nd ed.). New York: John Wiley.

FEUCHT, W. 1987. Graft incompatibility of tree crops: An overview of the present scientific status. *Acta Hort.* 227:33–41.

GARDNER, V. R., F. C. BRADFORD, and H. D. HOOKER. 1939. Propagation. Section 6 in *Fundamentals of fruit production* (2nd ed.). New York: McGraw-Hill.

HATTON, R. G. 1930. The relationship between scion and rootstock with special reference to the tree fruits. *Jour. Roy. Hort. Soc.* 55:169–211.

LOCKARD, R. G., and G. W. SCHNEIDER. 1981. Stock and scion growth relationships and the dwarfing mechanism in apple. *Hort. Rev.* 3:315–75.

MOORE, R., ed. 1983. *Vegetative compatibility responses in plants.* Waco, Tex.: Baylor Univ. Press.

MOORE, R., and D. B. WALKER. 1981. Graft compatibility–incompatibility in plants. *BioScience* 31(5): 389–91.

MOSSE, B. 1962. Graft incompatibility in fruit trees. *Commonwealth Agr. Bur. Tech. Comm. 28.*

NELSON, S. H. 1968. Incompatibility survey among horticultural plants. *Proc. Inter. Plant Prop. Soc.* 18:343–93.

ROBERTS R. H. 1949. Theoretical aspects of graftage. *Bot. Rev.* 15:423–63.

ROGERS, W. S., and A. B. BEAKBANE. 1957. Stock and scion relations. *Ann. Rev. Plant Phys.* 8:217–36.

SIMONS, R. K. 1987. Compatibility and stock–scion interactions as related to dwarfing. In *Rootstocks for fruit crops,* R. C. Rom and R. F. Carlson, eds. New York: John Wiley.

TUBBS, F. R. 1973. Research fields in the interaction of rootstocks and scions in woody perennials. I and II. *Hort. Abst.* 43:247–53, 325–35.

12

Techniques of Grafting

Over the years, ever since the early horticulturists learned that certain plants, hard to root by cuttings, could be propagated by grafting or budding, a myriad of grafting and budding techniques have been developed. In the *Grafters' Handbook,* Garner has enumerated and described many of these, although many more have come and gone. (See Suppl. Readings.)

In this chapter we describe the most important grafting methods. Among them a person with an ability to do somewhat delicate work with a sharp knife can find a technique and be able to perform a successful grafting operation, meeting most any grafting need. Much of the success in grafting depends, however, not only on constructing a technically correct graft, but in properly caring for the graft after it is made, doing the grafting operation at the proper time of year, and subsequently caring for the growing scions following grafting.

For any successful grafting operation, pro-
ducing a plant as shown in Fig. 12–1, there are five important requirements:

1. *The stock and scion must be compatible.* They must be capable of uniting. Usually, but not always, plants closely related, such as two apple cultivars, can be grafted together. Distantly related plants, such as an oak tree and an apple tree, cannot be used to make a successful graft combination. (See Chap. 11 for a discussion of this factor.)

2. *The cambial region of the scion must be placed in intimate contact with that of the stock.* The cut surfaces should be held together tightly by wrapping, nailing, wedging, or some similar method. Rapid healing of the graft union is necessary so that the scion may be supplied with water and nutrients from the stock by the time the buds start to open.

3. *The grafting operation must be done at a time when the stock and scion are in the proper physiological*

FIGURE 12–1 Cultivars of the Persian (English) walnut (*Juglans regia*) grafted on *J. hindsii* rootstocks. The characteristics of these two species remain distinctly different after grafting, exactly to the junction of the graft union.

stage. Usually, this means that the scion buds are dormant while, at the same time, the cut tissues at the graft union are capable of producing the callus tissue necessary for healing of the graft. For deciduous plants, dormant scionwood is collected during the winter and kept inactive by storing at low temperatures. The rootstock plant may be dormant or in active growth, depending upon the grafting method used.

4. *Immediately after the grafting operation is completed, all cut surfaces must be protected from desiccation.* This is done by covering the graft union with tape or grafting wax, or by placing the grafts in moist material or in a covered grafting frame.

5. *Proper care must be given the grafts for a period of time after grafting.* Shoots coming from the stock below the graft will often choke out the desired growth from the scion. Or, in some cases, shoots from the scion will grow so vigorously that they break off unless staked and tied or cut back.

METHODS OF GRAFTING

Whip Grafting

Whip grafting, shown in Figs. 12–2 and 12–3, is particularly useful for grafting relatively small material, 6 to 13 mm (¼ to ½ in. in diameter). It is

highly successful if done properly because there is considerable cambial contact. It heals quickly and makes a strong union. Preferably, the scion and stock should be of equal diameter. The scion should contain two or three buds with the graft made in the smooth internode area below the lower bud.

The cuts made at the top of the stock should be exactly the same as those made at the bottom of the scion. First, a smooth, sloping cut is made, 2.5 to 6 cm (1 to 2½ in.) long. The longer cuts are made when working with large material. This first cut should be made preferably with one single stroke of the knife, so as to leave a smooth, flat surface. To do this, the knife must be razor sharp. Wavy, uneven cuts made with a dull knife will not result in a satisfactory union.

On each of these cut surfaces, a reverse cut is made. It is started downward at a point about one-third of the distance from the tip and should be about one-half the length of the first cut. To obtain a smooth-fitting graft, this second cut should not just split the grain of the wood but should follow along under the first cut, tending to parallel it.

The stock and scion are then inserted into each other, with the tongues interlocking. It is extremely important that the cambium layers match along at least one side, preferably along both sides. The lower tip of the scion should not overhang the stock, as there is a likelihood of the formation of large callus knots. In some species, such callus overgrowths are often mistaken for crown

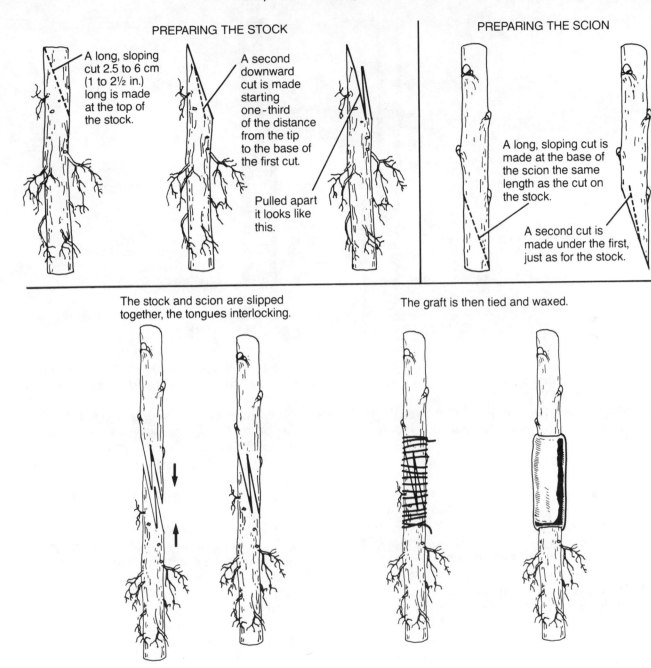

PREPARING THE STOCK

A long, sloping cut 2.5 to 6 cm (1 to 2½ in.) long is made at the top of the stock.

A second downward cut is made starting one-third of the distance from the tip to the base of the first cut.

Pulled apart it looks like this.

PREPARING THE SCION

A long, sloping cut is made at the base of the scion the same length as the cut on the stock.

A second cut is made under the first, just as for the stock.

The stock and scion are slipped together, the tongues interlocking.

The graft is then tied and waxed.

FIGURE 12-2 Whip graft. This method is widely used in grafting small plant material and is especially valuable in making root grafts as illustrated here.

gall knots, caused by bacteria. The use of scions larger than the stock should be avoided for the same reason. If the scion is smaller than the stock, it should be set at one side of the stock so that the cambium layers will be certain to match along that side. If the scion is much smaller than the stock, the first cut on the stock consists only of a slice taken off one corner.

After the scion and stock are fitted together, they should be held securely in some manner until

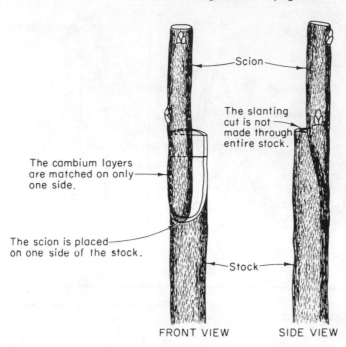

The cambium layers are matched on only one side.

The scion is placed on one side of the stock.

Scion

The slanting cut is not made through entire stock.

Stock

FRONT VIEW   SIDE VIEW

FIGURE 12-3 Method of making a whip graft when the scion is considerably smaller than the stock.

the pieces have united. There are a number of possible ways of doing this.

1. If the unions are very well made with a tight, snug fit, it is possible that no additional wrapping or tying is needed, but it is safer to provide some type of wrapping. If not wrapped, the grafts must be protected from drying by burying in moist sand, peat moss, or sawdust until the union has healed. Or they may be planted directly in the nursery with the union below soil level. If the whip graft is used in topworking, the exposed union must be protected in some manner.

2. With a secure fit it may be sufficient to omit tying and merely cover the union with hot grafting wax, which will secure the pieces to some extent and give good protection against drying. This method is not recommended for inexperienced grafters.

3. A common method is to wrap the union with budding rubbers or older materials such as raffia or waxed string (No. 18 knitting cotton). After wrapping, the entire union can be covered with grafting wax. Waxing may be omitted if the grafts are to be protected from drying by burying in moist sand or peat moss, or if the grafts are planted imme-

diately with the union below the soil surface. Grafts wrapped with budding rubbers and covered with soil should be inspected later, since the rubber decomposes very slowly below ground and may cause a constriction at the graft union.

4. A practice widely used is to wrap the grafts with some type of adhesive tape. A special nursery tape is available. The tape is drawn tightly around the graft union with the edges slightly overlapping. This holds the parts together very well and prevents drying, thus eliminating the need for waxing. If just one thickness of tape is used, it will decompose sufficiently fast (if the union is below ground) that no constriction of growth will develop. If used above ground, the tape should be cut in three to four weeks, after the graft has healed. The use of tight wrapping material such as this is especially recommended when difficulty is encountered with the formation of excessive callus.

5. Plastic tapes are available for wrapping grafts. They are used just as adhesive tape, although they are not adhesive. The final turn of the tape is secured by slipping it under the previous turn. This tape has some elasticity. Also, it deteriorates more slowly

below ground than above. These tapes should later be removed to prevent girdling.

Splice Grafting

This method is the same as the whip graft except that the second, or "tongue," cut is not made in either the stock or scion. A simple slanting cut of the same length and angle is made in both the stock and the scion. These are placed together and wrapped or tied as described for the whip graft. The splice graft is simple and easy to make. It is particularly useful in grafting plants that have a very pithy stem or that have wood that is not flexible enough to permit a tight fit when a tongue is made as in the whip graft.

Side Grafting

There are numerous variations of the side graft. As the name suggests, the scion is inserted into the side of the stock, which is generally larger in diameter than the scion. This method has proven useful for large-scale propagation of nursery trees (31).

Stub Graft

The stub graft is useful in grafting branches of trees that are too large for the whip graft yet not large enough for other methods such as the cleft or bark graft. For this type of side graft, the best stocks are branches about 2.5 cm (1 in.) in diameter. An oblique cut is made into the stock branch with a chisel or heavy knife at an angle of 20 to 30 degrees. The cut should be about 2.5 cm (1 in.) deep, and at such an angle and depth that when the branch is pulled back the cut will open slightly but will close when the pull is released.

The scion should contain two or three buds and be about 7.5 cm (3 in.) long and relatively thin. At the basal end of the scion, a wedge about 2.5 cm (1 in.) long is made. The cuts on both sides of the scion should be very smooth, each made by one single cut with a sharp knife. The scion must be inserted into the stock at an angle as shown in Fig. 12–4 so as to obtain maximum contact of the cambium layers. The grafter inserts the scion into the cut while the upper part of the stock is pulled backward, using care to obtain the best cambium

contact. Then the stock is released. The pressure of the stock should grip the scion tightly, making tying unnecessary, but if desired, the scion can be further secured by driving two small flat-headed wire nails [20 gauge, 1.5 cm (⅝ in.) long] into the stock through the scion. Wrapping the stock and scion at the point of union with nursery tape also may be helpful. After the graft is completed the stock may be cut off just above the union. This must be done very carefully or the scion may become dislodged. The entire graft union must be thoroughly covered with grafting wax, sealing all openings. The tip of the scion also should be covered with wax (42).

Side-Tongue Graft

The side-tongue graft, shown in Fig. 12–5, is useful for small plants, especially some of the broad- and narrow-leaved evergreen species. The stock plant should have a smooth section in the stem just above the crown of the plant. The diameter of the scion should be slightly smaller than that of the stock. The cuts at the base of the scion are made just as for the whip graft. Along a smooth portion of the stem of the stock, a thin piece of bark and wood, the same length as the cut surface of the scion, is completely removed. Then a reverse cut is made downward in the cut on the stock, starting one-third of the distance from the top of the cut. This second cut in the stock should be the same length as the reverse cut in the scion. The scion is then inserted into the cut in the stock, the two tongues interlocking, and the cambium layer(s) matching. The graft is wrapped tightly, using one of the methods described for the whip graft.

The top of the stock is left intact for several weeks until the graft union has healed. Then it may be cut back above the scion gradually or all at once. This forces the buds on the scion into active growth.

Side-Veneer Graft (Spliced Side Graft)

The side-veneer variation of side grafting, shown in Fig. 12–6, is widely used, especially for grafting small potted plants, such as seedling evergreens, to named cultivars. A shallow downward and inward cut from 25 to 38 mm (1 to 1½ in.) long is made in a smooth area just above the

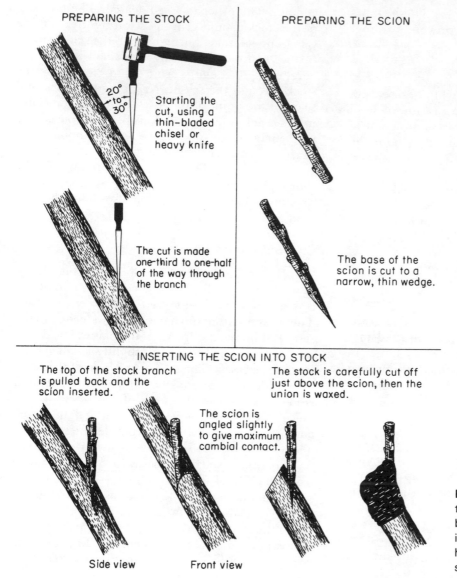

PREPARING THE STOCK

20° to 30°

Starting the cut, using a thin-bladed chisel or heavy knife

The cut is made one-third to one-half of the way through the branch

PREPARING THE SCION

The base of the scion is cut to a narrow, thin wedge.

INSERTING THE SCION INTO STOCK

The top of the stock branch is pulled back and the scion inserted.

The scion is angled slightly to give maximum cambial contact.

The stock is carefully cut off just above the scion, then the union is waxed.

Side view   Front view

FIGURE 12-4 Steps in preparing the side, or stub, graft. A thin-bladed chisel as illustrated here is ideal for making the cut, but a heavy butcher knife could be used satisfactorily.

crown of the stock plant. At the base of this cut, a second short inward and downward cut is made, intersecting the first cut, so as to remove the piece of wood and bark. The scion is prepared with a long cut along one side and a very short one at the base of the scion on the opposite side. These scion cuts should be the same length and width as those made in the stock so that the cambium layers can be matched as closely as possible.

After inserting the scion, the graft is tightly wrapped with waxed or paraffined string, with budding rubbers, or with nursery adhesive tape.

The graft may or may not be covered with wax, depending upon the species. A common practice in side grafting small potted plants of some of the woody ornamental species is to plunge the grafted plants into a damp medium, such as peat moss, so that it just covers the graft union. The newly grafted plants may be placed for healing in a mist propagating house or set in grafting cases. The latter are closed boxes with a transparent cover, which permits retention of a high humidity around the grafted plant until the union has healed. The grafting cases are kept closed for a

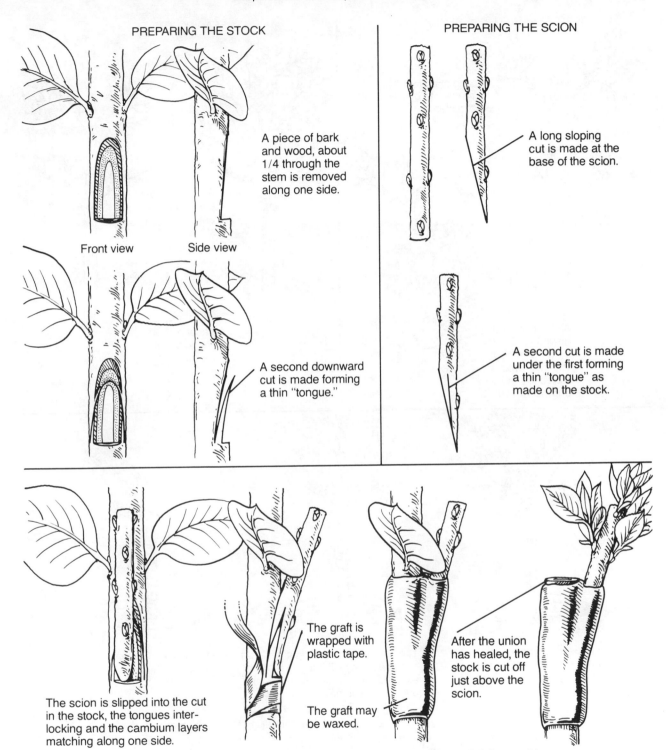

PREPARING THE STOCK

A piece of bark and wood, about 1/4 through the stem is removed along one side.

Front view   Side view

A second downward cut is made forming a thin "tongue."

PREPARING THE SCION

A long sloping cut is made at the base of the scion.

A second cut is made under the first forming a thin "tongue" as made on the stock.

The scion is slipped into the cut in the stock, the tongues interlocking and the cambium layers matching along one side.

The graft is wrapped with plastic tape.

The graft may be waxed.

After the union has healed, the stock is cut off just above the scion.

FIGURE 12–5 Side-tongue graft. This method is very useful for grafting broad-leaved evergreen plants. Final tying may be done with budding rubbers, plastic tape, or waxed string, just before waxing.

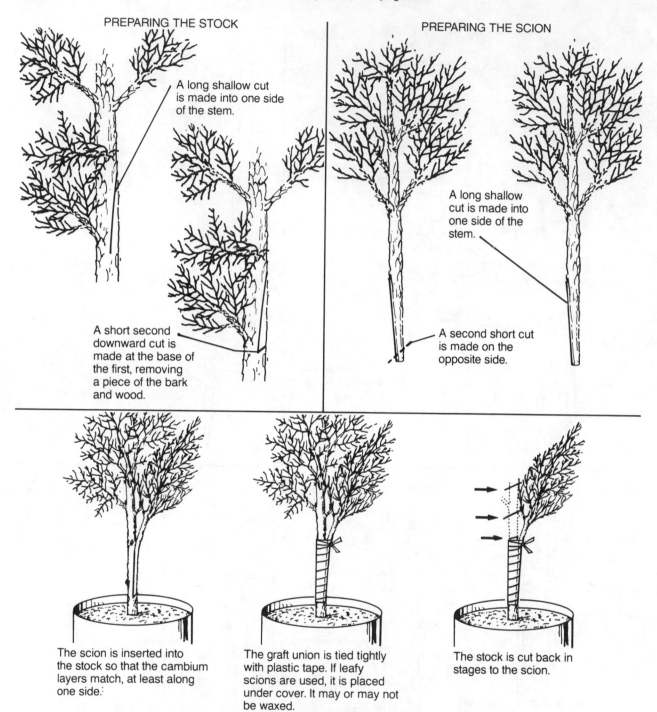

PREPARING THE STOCK

A long shallow cut is made into one side of the stem.

A short second downward cut is made at the base of the first, removing a piece of the bark and wood.

PREPARING THE SCION

A long shallow cut is made into one side of the stem.

A second short cut is made on the opposite side.

The scion is inserted into the stock so that the cambium layers match, at least along one side.

The graft union is tied tightly with plastic tape. If leafy scions are used, it is placed under cover. It may or may not be waxed.

The stock is cut back in stages to the scion.

FIGURE 12–6 Steps in making the side-veneer graft. This method is widely used in propagating narrow-leaved evergreen species that are difficult to start by cuttings.

week or so after the grafts are put in, then gradually opened over a period of several weeks; finally, the cover is taken off completely.

After the union has healed, the stock can be cut back above the scion either in gradual steps or all at once. The side-veneer graft is widely used for propagating both conifer and deciduous trees and shrubs, many thousands being propagated each year by this method (*46*).

Cleft Grafting

Cleft grafting is one of the oldest and most widely used methods of grafting, being especially adapted to topworking trees, either in the trunk of a small tree or in the scaffold branches of a larger tree (Figs. 12–7 and 12–8). Cleft grafting is useful also for smaller plants, as in crown grafting established grapevines or camellias. In topwork-

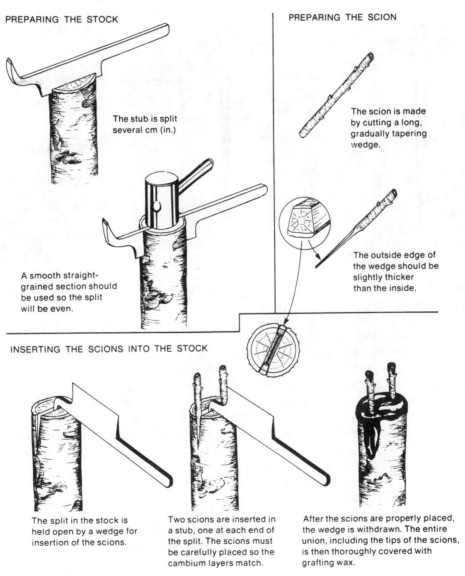

PREPARING THE STOCK

The stub is split several cm (in.)

A smooth straight-grained section should be used so the split will be even.

PREPARING THE SCION

The scion is made by cutting a long, gradually tapering wedge.

The outside edge of the wedge should be slightly thicker than the inside.

INSERTING THE SCIONS INTO THE STOCK

The split in the stock is held open by a wedge for insertion of the scions.

Two scions are inserted in a stub, one at each end of the split. The scions must be carefully placed so the cambium layers match.

After the scions are properly placed, the wedge is withdrawn. The entire union, including the tips of the scions, is then thoroughly covered with grafting wax.

FIGURE 12–7 Steps in making the cleft graft. This method is very widely used and is quite successful if the scions are inserted so that the cambium layers of stock and scion match properly.

CORRECT   INCORRECT

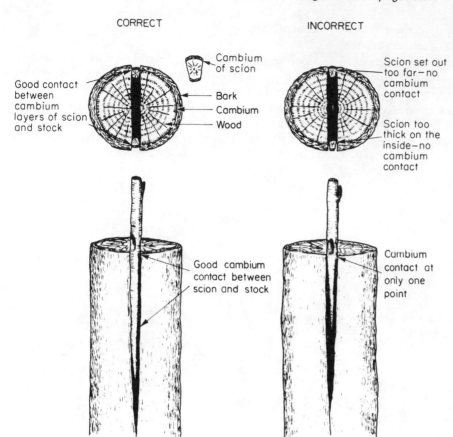

Good contact
between
cambium
layers of scion
and stock

Cambium
of scion

Bark
Cambium
Wood

Scion set out
too far—no
cambium
contact

Scion too
thick on the
inside—no
cambium
contact

Good cambium
contact between
scion and stock

Cambium
contact at
only one
point

FIGURE 12–8 In making the cleft graft, the proper placement of the scions is very important. The correct way of doing this is shown on the left. Scions inserted as shown on the right probably would not grow.

ing trees, this method should be limited to stock branches about 2.5 to 10 cm (1 to 4 in.) in diameter and to species with fairly straight-grained wood that will split evenly. Although cleft grafting can be done any time during the dormant season, the chances for successful healing of the graft union are best if the work is done in early spring just when the buds of the stock are beginning to swell but before active growth has started. If cleft grafting is done after the tree is in active growth, it is likely that the bark of the stock will separate from the wood, causing difficulties in obtaining a good union. When this separation occurs, the loosened bark must be firmly nailed back in place. The scions should be made from dormant, one-year-old wood. Unless the grafting is done early in the season (when the dormant scions can be collected and used immediately), the scionwood should be collected in advance and held under refrigeration until it is used. In sawing off the

branch for this and other topworking methods, the cut should be made at right angles to the main axis of the branch.

In making the cleft graft, a heavy knife, such as a butcher knife, or one of several special cleft grafting tools, is used to make a vertical split for a distance of 5 to 8 cm (2 to 3 in.) down the center of the stub to be grafted. This split is made by pounding the knife in with a hammer or mallet. It is very important to saw the branch off in such a position that the end of the stub which is left is smooth, straight-grained, and free of knots for at least 15 cm (6 in.). Otherwise, the split may not be straight, or the wood may split one way and the bark another. The split should be in a tangential rather than radial direction in relation to the center of the tree. This permits better placement of the scions for their subsequent growth. Sometimes the cleft is made by a longitudinal saw cut rather than by splitting. After a good, straight split is made, a screwdriver, chisel, or the wedge part of the cleft-grafting tool is driven into the top of the split to hold it open.

Two scions are inserted, one at each side of the stock where the cambium layer is located. The scions should be 8 to 10 cm (3 to 4 in.) long, about 10 to 13 cm ($\frac{3}{8}$ to $\frac{1}{2}$ in.) in thickness, and should have two or three buds. The basal end of each scion should be cut into a gently sloping wedge about 5 cm (2 in.) long. It is not necessary that the end of the wedge come to a point. The side of the wedge which is to go to the outer side of the stock should be slightly wider than the inside edge. Thus, when the scion is inserted and the tool is removed, the full pressure of the split stock will come to bear on the scions at the position where the cambium of the stock touches the cambium layer on the outer edge of the scion. Since the bark of the stock is almost always thicker than the bark of the scion, it is usually necessary for the outer surface of the scion to set slightly in from the outer surface of the stock in order to match the cambium layers.

> In all types of grafting the scion must be inserted right side up. That is, the points of the buds on the scion should be pointing upward and away from the stock. Unless this rule is observed, the graft will not be successful.

The long, sloping wedge cuts at the base of the scion should be smooth, made by a single cut on each side with a very sharp knife. Both sides of the scion wedge should press firmly against the stock for their entire length. A common mistake in cutting scions for this type of graft is to make the cut on the scion too short and the slope too abrupt, so that the only point of contact is at the top. Slightly shaving the sides of the split in the stock will often permit a smoother contact.

After the scions are properly made and inserted, the tool is withdrawn, using care not to disturb the scions. They should be held so tightly by the pressure of the stock that they cannot be pulled loose by hand. No further tying or nailing is needed unless very small stock branches have been used, in which case the top of the stock can

be wrapped tightly with string or adhesive tape to hold the scions in place more securely.

Thorough waxing of the completed graft is essential. The top surface of the stub should be entirely covered, permitting the wax to work into the split in the stock. The sides of the grafted stub should be well covered with wax as far down the stub as the length of the split. The tops of the scions should be waxed but not necessarily the bark or buds of the scion. Two or three days later all the grafts should be inspected and rewaxed where openings appear. Lack of thorough and complete waxing in this type of graft is almost certain to result in failure.

Wedge Grafting

Wedge grafting is illustrated in Fig. 12–9. If properly done the wedge graft gives excellent results. Like the cleft graft it can be made in late winter (in mild climates) or early spring before the bark begins to slip (separates easily from the wood).

The diameter of the stock to be grafted is the same as for the cleft graft—5 to 10 cm (2 to 4 in.), and the scions also are the same size—10 to 13 cm (4 to 5 in.) long and 10 to 13 mm ($\frac{3}{8}$ to $\frac{1}{2}$ in.) in thickness.

A sharp, heavy, short-bladed knife is used for making a V-wedge in the side of the stub, about 5 cm (2 in.) long. Two cuts are made coming together at the bottom and as far apart at the top as the width of the scion. These cuts extend about 2 cm ($\frac{3}{4}$ in.) deep into the side of the stub. After these cuts are made, a screwdriver is pounded downward behind the wedge chip from the top of the stub to knock out the chip, leaving a V opening for insertion of the scion. The base of the scion is trimmed to a wedge shape exactly the same size and shape as the opening. With the two cambium layers matching, the scion is tapped downward firmly into place and slanting outward slightly at the top so that the cambium layers cross. If the cut is long enough and gently tapering, the scion should be so tightly held in place that it would be difficult to dislodge.

In a 5-cm (2-in)-wide stub, two scions should be inserted 180° apart; in a 10-cm (4-in.) stub, three scions should be used, 120° apart. After all scions are firmly tapped into place, all cut sur-

PREPARING THE STOCK

A heavy sharp knife is pounded into the side of the stub to make two cuts to form a V.

A screwdriver is used to flip out the V-shaped chip, leaving a space for the insertion of the scion.

PREPARING THE SCION

The scion should be about 10 to 13 cm (4 to 5 in.) long, 10 to 12 mm (3/8 to 1/2 in.) thick, and with 2 or 3 healthy vegetative buds. The basal ends should be cut to a V-shaped wedge, matching the opening in the stock.

Front view   Side view

INSERTING THE SCIONS INTO THE STOCK

The scion is gently tapped into the V-shaped opening in the stock, matching the cambium layers at a slight angle so that the cambium of stock and scion cross.

Scion should be inserted at an angle so that the cambium layers of stock and scion are closely matched, barely crossing each other.

After scions are in place all cut surfaces are thoroughly covered with grafting wax.

FIGURE 12–9 Wedge grafting.

faces, including the tips of the scion, should be waxed thoroughly.

Bark Grafting

Bark grafting is rapid, simple, readily performed by amateurs, and if properly done, gives a high percentage of "takes." It requires no special equipment and can be performed on branches ranging from 2.5 cm (1 in.) up to 30 cm (1 ft) or more in diameter. The latter size is not recommended as it is difficult to heal over such large

stubs before decay-producing organisms attack. The bark graft, since it depends on the bark separating readily from the wood, can only be done after active growth of the stock has started in the spring. As dormant scions must be used, it is necessary to gather the scionwood for deciduous species during the dormant season and hold it under refrigeration until the grafting operation is done. For evergreen species, freshly collected scionwood can be used. In the bark graft, scions are not as securely attached to the stock as in some of the other methods and are more susceptible to wind breakage during the first year even though healing has been satisfactory. Therefore, the new shoots arising from the scions probably should be staked during the first year, or cut back to about half their length, especially in windy areas. After a few years' growth, the bark graft union is as strong as the unions formed by other methods. Two modifications of the bark graft are described next.

Bark Graft (Method 1)

Several scions are inserted into each stub. For each scion, a vertical knife cut 2.5 to 5 cm (1 to 2 in.) long is made at the top end of the stub through the bark to the wood. The bark is then lifted slightly along both sides of this cut, in preparation for the insertion of the scion. The scion should be of dormant wood, 10 to 13 cm (4 to 5 in.) long, containing two or three buds, and be 6 to 13 mm ($\frac{1}{4}$ to $\frac{1}{2}$ in.) in thickness. One cut about 5 cm (2 in.) long is made along one side at the base of the scion. With large scions, this cut extends about one-third of the way into the scion, leaving a "shoulder" at the top. The purpose of this shoulder is to reduce the thickness of the scion to minimize the separation of bark and wood after insertion in the stock. The scion should not be cut too thin, however, or it will be mechanically weak and break off at the point of attachment to the stock. If small scions are used, no shoulder is necessary. On the side of the scion opposite the first long cut, a second, shorter cut is made, as shown in Fig. 12–10, thereby bringing the basal end of the scion to a wedge shape. The scion is then inserted between the bark and the wood of the stock, centered directly under the vertical cut through the bark. The longer cut on the scion is placed against the wood, and the shoulder on the scion is brought down until it rests on top of the stub. The scion is then ready to be fastened in place. Nail the scion into the wood, using two nails per scion. Flat-headed nails 15 to 25 mm ($\frac{5}{8}$ to 1 in.) long, of 19 or 20 gauge wire, depending on the size of the scions, are satisfactory. The bark on both sides of the scion also should be nailed down securely or it will tend to peel back from the wood.

Another method commonly used with soft-barked trees, such as the avocado, is to insert all the scions in the stub and then hold them in place by wrapping string, adhesive tape, or waxed cloth around the stub. This is more effective than nailing in preventing the scions from blowing out but probably does not give as tight a fit. Both nailing and wrapping are advisable for maximum strength. If a wrapping material is used, it must be checked later to prevent constriction if this is occurring. After the stub has been grafted and the scions fastened by nailing or tying, all cut surfaces, including the end of the scions, should be covered thoroughly with grafting wax.

Bark Graft (Method 2)

In this method, as shown in Fig. 12–11, *two* knife cuts about 5 cm (2 in.) long are made through the bark of the stock down to the wood, rather than just one. The distance between these two cuts should be exactly the same as the width of the scion. The piece of bark between the cuts should be lifted and the terminal two-thirds cut off. The scion is prepared with a smooth slanting cut along one side at the basal end completely through the scion. This cut should be about 5 cm (2 in.) long but *without* the shoulder, in contrast to method 1. On the opposite side of the scion, a cut about 13 mm ($\frac{1}{2}$ in.) long is made, forming a wedge at the base of the scion. The scion should fit snugly into the opening in the bark with the longer cut inward and with the wedge at the base slipped under the flap of remaining bark. Rapid healing can be expected because both sides of the scion are touching undisturbed bark and cambium cells, which is not the case in the other type of bark graft.

The scion should be nailed into place with two nails, the lower nail going through the flap of bark covering the short cut on the back of the scion. If the bark along the sides of the scion should accidentally become disturbed, it must be nailed back into place. Method 2 is well adapted for use with thick-barked trees, such as walnuts and pecans, on which it is not feasible to insert the scion under the bark.

Approach Grafting

The distinguishing feature of approach grafting is that two independent, self-sustaining plants are

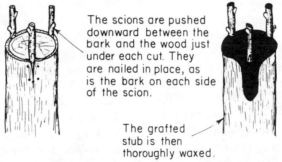

PREPARING THE STOCK

A vertical cut 2.5 to 5 cm (l to 2 in.) long is made through the bark to the wood.

The bark on both sides of the cut is slightly separated from the wood.

PREPARING THE SCION

The scion is cut as shown below, a long cut with a shoulder on one side, and a shorter cut on the opposite side.

Side view   Back view   Front view

(This side is placed next to the wood of the stock.)

INSERTING THE SCIONS INTO THE STOCK

The scions are pushed downward between the bark and the wood just under each cut. They are nailed in place, as is the bark on each side of the scion.

The grafted stub is then thoroughly waxed.

FIGURE 12–10 Steps in preparing the bark graft (method 1). In grafting some thick-barked plants the vertical cut in the bark is unnecessary, the scion being inserted between the bark and wood of the stock.

grafted together. After a union has occurred, the top of the stock plant is removed above the graft and the base of the scion plant is removed below the graft. Sometimes it is necessary to sever these parts gradually rather than all at once. Approach grafting provides a means of establishing a graft union between certain plants in which successful graft unions are difficult to obtain. Approach grafting is usually performed with one or both of the plants to be grafted growing in a container. Rootstock plants in containers may be placed ad-

joining an established plant which is to furnish the scion part of the new, grafted plant (see Fig. 12–12).

This type of grafting should be done at times of the year when growth is active and rapid healing of the graft union will take place. As in other methods of grafting, the cut surfaces should be securely fastened together, then covered with grafting wax to prevent drying of the tissues. Three useful methods of making approach grafts are described below and illustrated in Fig. 12–13.

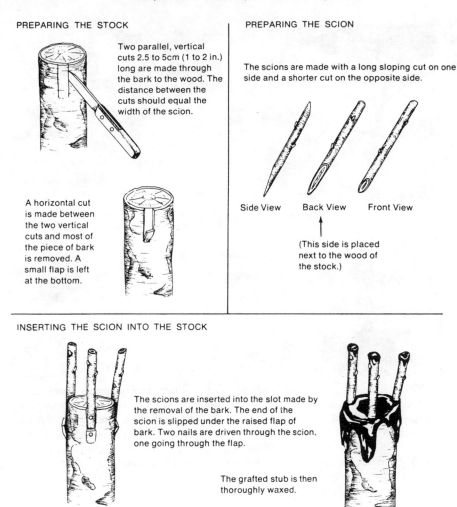

PREPARING THE STOCK

Two parallel, vertical cuts 2.5 to 5cm (1 to 2 in.) long are made through the bark to the wood. The distance between the cuts should equal the width of the scion.

A horizontal cut is made between the two vertical cuts and most of the piece of bark is removed. A small flap is left at the bottom.

PREPARING THE SCION

The scions are made with a long sloping cut on one side and a shorter cut on the opposite side.

Side View   Back View   Front View

(This side is placed next to the wood of the stock.)

INSERTING THE SCION INTO THE STOCK

The scions are inserted into the slot made by the removal of the bark. The end of the scion is slipped under the raised flap of bark. Two nails are driven through the scion, one going through the flap.

The grafted stub is then thoroughly waxed.

FIGURE 12–11 Bark graft, method 2.

Spliced Approach Graft

In the spliced approach graft the two stems should be approximately the same size. At the point where the union is to occur, a slice of bark and wood 2.5 to 5 cm (1 to 2 in.) long is cut from both stems. This cut should be the same size on each so that identical cambium patterns will be made. The cuts must be perfectly smooth and as nearly flat as possible so that when they are pressed together there will be close contact of the cambium layers. The two cut surfaces are then bound tightly together with string, raffia, or nursery tape. The whole union should then be covered with grafting wax. After the parts are well united, which may require considerable time in some cases, the stock above the union and the scion below the union are cut, and the graft is then completed. It may be necessary to reduce the leaf area of the scion if it is more than the root system of the stock can sustain. Figure 12–12 shows the use of this method with camellias.

Tongued Approach Graft

The tongued approach graft is the same as the spliced approach graft except that after the first cut is made in each stem to be joined, a second cut—downward on the stock and upward on the scion—is made, thus providing a thin tongue

FIGURE 12–12　Approach grafting used in obtaining a desirable scion culti-var on a seedling camellia. *Top:* Seedling plant in container is set close to a large plant of the desired cultivar. The graft union is made and tightly wrapped with adhesive tape. *Below:* After the graft union has healed, which may take several months, the stock plant is cut off above the graft union, and the scion is severed from the parent plant just below the graft union. Approach grafting is sometimes necessary for plants very difficult to graft by other methods.

on each piece. By interlocking these tongues a very tight, closely fitting graft union can be obtained.

Inlay Approach Graft

The inlay approach graft may be used if the bark of the stock plant is considerably thicker than that of the scion plant. A narrow slot, 7.5 to 10 cm (3 to 4 in.) long, is made in the bark of the stock plant by making two parallel knife cuts and removing the strip of bark between. This can be done only when the stock plant is actively growing and the bark "slipping." The slot should be exactly as wide as the scion to be inserted. The stem of the scion plant, at the point of union, should be given a long, shallow cut along one side, of the same length as the slot in the stock plant and deep enough to go through the bark and slightly into

the wood. This cut surface of the scion branch should be laid into the slot cut in the stock plant and held there by nailing with two or more small, flat-headed wire nails. The entire union must then be thoroughly covered with grafting wax. After the union has healed, the stock can be cut off above the graft and the scion below the graft.

Inarching

Inarching is similar to approach grafting in that both stock and scion plants are on their own roots at the time of grafting; it differs in that the top of the new rootstock plant usually does not extend above the point of the graft union as it does in approach grafting. Inarching is generally considered to be a form of "repair grafting," used to repair roots damaged by cultivation implements, rodents, or disease. It can be used to very good

Spliced approach graft

Tongued approach graft

Inlay approach graft

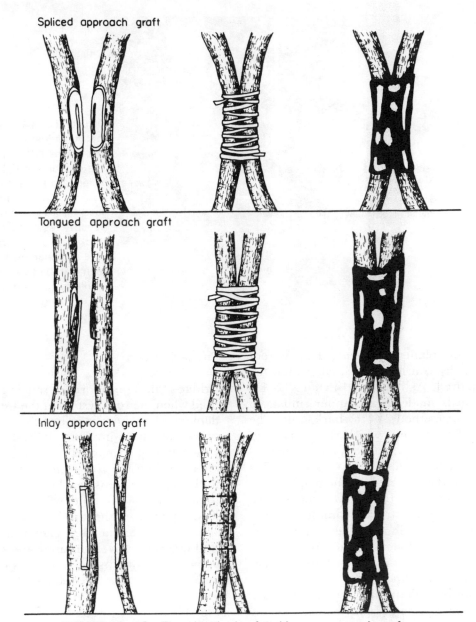

FIGURE 12-13   Three methods of making an approach graft.

advantage in saving a valuable tree or improving its root system (Fig. 12–14).

Seedlings (or rooted cuttings) planted beside the older, damaged tree, or suckers arising near its base, are grafted into the trunk of the tree to provide a new root system to supplant the damaged roots. The seedlings to be inarched into the tree should be spaced about 13 to 15 cm (5 to 6 in.) apart around the circumference of the tree if the damage is extensive. A damaged tree will usu-

ally stay alive for some time unless the injury is very severe. A satisfactory procedure for inarching is to plant seedlings of a compatible species around the tree during the dormant season. Then the grafting operation can be done as active growth commences in early spring. Inarching old, weakly growing trees with strong, vigorous seedling rootstocks has on some occasions (22) proved beneficial in promoting renewed active growth of the old trees.

FIGURE 12-14 *Left:* Inarches that have just been inserted. The one on the left has been waxed. The one on the right has been nailed into place and is ready for waxing. *Right:* Inarching can be used for invigorating established trees by replacing a weak rootstock with a more vigorous one. Here a Persian walnut tree has been inarched with Paradox hybrid seedlings (*Juglans hindsii* × *J. regia*) seedlings.

The seedling plants to provide the new root system are usually considerably smaller than the tree to be repaired. As illustrated in Fig. 12–15, the graft union is made in a manner similar to that described for method 2 of the bark graft. The upper end of the seedling, which should be 6 to 12 mm (¼ to ½ in.) thick, is given a long shallow cut along the side for 10 to 15 cm (4 to 6 in.) This cut should be on the side next to the trunk of the tree and made deep enough to remove some of the wood, thus exposing two strips of cambium tissue. At the end of the seedling another shorter cut, about 13 mm (½ in.) long, is made on the side opposite the long cut; this makes a sharp, wedge-shaped end on the seedling stem.

A long slot is made in the trunk of the older tree by removing a piece of bark the exact width of the seedling and just as long as the cut surface made on the seedling. A small flap of bark is left at the upper end of the slot, under which the wedge end of the seedling is inserted. Then the seedling is nailed into the slot with four or five small, flat-headed wire nails. The nail at the top of the slot should go through the flap of bark and through the end of the seedling. If any of the bark of the tree along the sides of the seedling should accidentally be pulled loose, it is necessary to nail it back in place. After nailing, the entire area of the graft union should be thoroughly waxed.

Bridge Grafting

Bridge grafting is a form of repair grafting, and is used when the root system of the tree has not been damaged but there is injury to the trunk. Sometimes cultivation implements, rodents, disease, or winter injury damage a considerable trunk area, often girdling the tree completely. If the damage to the bark is extensive, the tree is almost certain to die, because the roots will be deprived of their food supply from the top of the tree. Trees of some species, such as the elm, cherry, and pecan, can heal over extensively injured areas by the development of callus tissue. But trees of most species with severely damaged bark should be bridge grafted if they are to be saved, as illustrated in Fig. 12–16.

The bridge grafting operation is best performed in early spring just as active growth of the tree is beginning and the bark is slipping easily. The scions to be used should be taken when dormant from one-year-old growth, 6 to 12 mm (¼ to ½ in.) in diameter, of the same or compatible species, and held under refrigeration until the grafting work is to be done. In an emergency, one may successfully perform bridge grafting late in the spring, using scionwood whose buds have already started to grow. Remove the developing buds or new shoots.

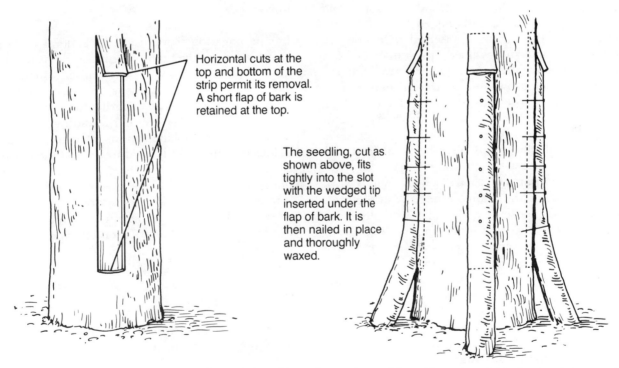

PREPARING THE TREE TO BE INARCHED

Vertical cuts about 15 cm (6 in.) long are made through the bark near the base of the tree to be inarched.

PREPARING THE SEEDLING FOR INARCHING

The upper portion of the seedling is cut about 1/3 through on the side adjacent to the tree to be inarched. The cut is about 15 cm (6 in.) long.

On the opposite side a short cut is made in the tip of the seedling bringing it to a wedge.

View of the cut surface to be placed against the wood of the tree.

Horizontal cuts at the top and bottom of the strip permit its removal. A short flap of bark is retained at the top.

The seedling, cut as shown above, fits tightly into the slot with the wedged tip inserted under the flap of bark. It is then nailed in place and thoroughly waxed.

FIGURE 12-15 Steps in inarching a large plant, with smaller ones planted around its base.

FIGURE 12–16 Injured trunk of a cherry tree successfully bridge grafted by a modification of the bark graft.

The first step in bridge grafting is to trim the wounded area back to healthy, undamaged tissue by removing dead or torn bark. Then every 5 to 7.5 cm (2 to 3 in.) around the injured section a scion is inserted, attached at both the upper and lower ends into live, undamaged bark. It is important that the scions be inserted right side up. If they are put in reversed, they may make a union and stay alive for a year or two, but the scions will not grow and enlarge in diameter as they would if inserted correctly. Figure 12–17 shows the details of making a satisfactory type of bridge graft.

After all the scions have been inserted, the cut surfaces must be thoroughly covered with grafting wax, particular care being taken to work the wax around the scions, especially at the graft unions. The exposed wood of the injured section may also be covered with grafting wax to prevent the entrance of decay organisms and excessive drying of the wood, which is important, being the path for upward movement of water and nutrients in the tree.

TOOLS AND ACCESSORIES FOR GRAFTING

Special equipment needed for any particular method of grafting has been illustrated along with the description of the method. There are some pieces of equipment, however, that are used in all types of grafting.

Knives

For propagation work, the two general types of knives used are the budding knife and the grafting knife (Fig. 12–18). Where a limited amount of either budding or grafting is done, the budding knife can be used satisfactorily for both operations. The knives have either a folding or a fixed blade. The fixed-blade type is stronger, and if a holder of some kind is used to protect the cutting edge, it is probably the most desirable. A well-built, sturdy knife of high-carbon steel is essential if much grafting work is to be done. To do good work, the knife must be kept razor sharp. Grafting knives are available for either right- or left-handed people.

Grafting Waxes

Grafting wax has two chief purposes. (a) It seals over the graft union, thereby preventing the loss of moisture and death of the tender, exposed cells of the cut surfaces of the scion and stock. These cells are essential for callus production and healing of the graft union. (b) It prevents the entrance of various decay-producing organisms that may lead to wood rotting.

PREPARING THE STOCK

All dead and damaged bark around the wound is trimmed back to live healthy tissue.

Cuts are made in the bark at top and bottom of the wound, just as for Method 2 of the bark graft. The slots in the bark should be the same width as the scions to be inserted.

PREPARING THE SCIONS

One long, slanting cut is made at each end of the scion, with both cuts on the same side.

A second, short, slanting cut is made on the back side of the scion, bringing the ends to a sharp wedge. Buds can be trimmed off the scions if desired.

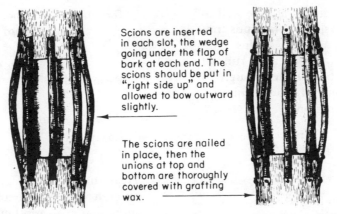

Scions are inserted in each slot, the wedge going under the flap of bark at each end. The scions should be put in "right side up" and allowed to bow outward slightly.

The scions are nailed in place, then the unions at top and bottom are thoroughly covered with grafting wax.

FIGURE 12–17 A satisfactory method of making a bridge graft, using a modification of method 2 of the bark graft.

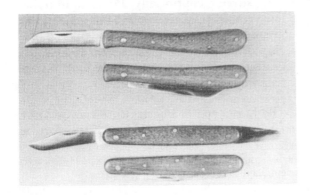

FIGURE 12–18 Types of folding knives used in plant propagation. *Top:* Grafting knife. *Below:* Budding knife. The blunt part on the right of this knife is used in T-budding to open the bark flap for insertion of the shield bud piece.

HOW TO SHARPEN A KNIFE PROPERLY

To sharpen the grafting knife, the initial grinding may be done with a medium-grit stone, but a hard, fine-grit stone should be used for the final sharpening. The stone should be wet with water or oil during sharpening. Do not use a carborundum stone, because it is too abrasive and will grind off too much metal. Most prefer to use knives beveled only on one side, the back side being flat, whereas others prefer a knife beveled on both sides. In sharpening the knife, hold it so that only the edge of the blade touches the stone in order that a stiff edge for cutting can be obtained. Use the whole width of the stone so that its surface will remain flat. A correctly sharpened knife of high-quality steel should retain a good edge for several days' work, with only occasional stropping on a piece of leather.

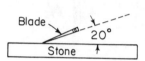

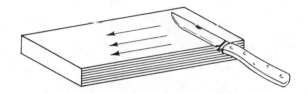

An ideal grafting wax should adhere well to the plant surfaces, not be washed off by rains, not be so brittle as to crack and chip during cold weather, or so soft that it will melt and run off during hot days, but still be pliable enough to allow for swelling of the scion and growth enlargement of the stock without cracking.

Hot Waxes

The ingredients used in hot waxes are resin, beeswax, either raw linseed oil or tallow, lampblack or powdered charcoal, and fish glue. The proportions of each ingredient may vary somewhat without affecting the results appreciably. The purpose of the lampblack is to give some color to the otherwise colorless wax so that it will show more readily when the grafts have been well covered. Lampblack also imparts a more workable consistency to the wax, eliminating some of the stickiness and stringiness. After application, the dark-colored wax tends to absorb heat from the sun and thus remain soft and pliable. During hot weather, however, the wax may become too soft. Due to the difficulties in obtaining the materials and preparing the wax, hot waxes are not used as extensively as are the cold waxes.

Directions for Making Hot Grafting Wax (Hard Type)

Ingredients:

Resin[1]	2.25 kg (5 lb)
Beeswax	337 g (¾ lb)
Raw linseed oil	235 ml (½ pt)
Lampblack	28 g (1 oz)
Fish glue[2]	42 g (1½ oz)

Heat the glue in a double boiler with just enough water to dissolve it. Melt the other ingredients in another container and allow the mixture to cool but still remain fluid. Add the glue slowly to the partly cooled mixture, stirring continually. Pour out into shallow greased pans or wooden boxes lined with greased paper. Allow to harden. To use, chip or break into small lumps. Reheat in a grafting wax melter and apply with a small paintbrush.

The hard type of hot grafting wax solidifies upon cooling and must be reheated just before be-

[1]Available from Pacific Coast Chemical Co., 2424 4th Street, Berkeley, Calif. 94710 U.S.A.

[2]Available from Nicholson & Co., Linden, N.J. 07036 U.S.A.

ing applied to the graft union. It is important that this wax be at the right temperature when it is used. If the wax is boiling, it will injure the plant tissues. At the other extreme, if the wax is too cool, it will not flow easily into all the crevices in the bark, thus leaving openings for the entrance of air. *The wax should be hot enough to flow easily, yet not be boiling.*

For heating the wax, any small burner is satisfactory. A brush is used to apply the wax, but provision should be made to suspend it from the side of the container for if the brush rests at the bottom, the heat will burn the bristles. Homemade grafting wax melters can easily be constructed (*39*).

Cold Waxes

A commercially prepared type of grafting wax consisting of an emulsion of asphalt and water is available. It has proved quite satisfactory and is widely used (*41*). This material is about 50 percent water, which evaporates after application, leaving a coat of asphalt over the graft. For the remaining wax to be thick enough to protect the graft adequately, the original application should be fairly heavy. Since this material is water soluble until it dries, rains occurring within 24 hours after application are likely to wash it off. It would then, of course, be necessary to rewax the grafts immediately. If grafting is done during rainy weather, it is advisable to use the hot type of wax, which is not affected by rains.

The emulsions of this type of cold wax are broken down by freezing, so it is very important that the containers be stored in a warm place during cold weather. *Do not confuse these materials with roofing compounds, which have a similar appearance but are entirely unsuitable for use as grafting wax.*

For small-scale operations, grafting waxes are available in aerosol applicator cans. Several repeated applications of the wax are generally necessary to give sufficient coverage for adequate protection.

Tying and Wrapping Materials

Some of the grafting methods, particularly the whip graft, require that the graft union be held together by tying until the parts unite. Tying can be done in several ways—the simplest would be an older technique of merely tying with ordinary string and covering with grafting wax. For large-scale operations, waxed string is convenient because it will adhere to itself and to the plant parts without tying. It should be strong enough to hold the grafted parts together yet weak enough to be broken by hand.

A special nursery adhesive tape is manufactured that is similar to surgical adhesive tape but lighter in weight and not sterilized. It is more convenient to use than waxed cloth tape. Adhesive tape is useful for tying and sealing whip grafts. When using any kind of tape or string for wrapping grafts, it is important not to use too many layers or the material may eventually girdle the plant unless it is cut. When this type of wrapping is covered with soil, it usually rots and breaks before damage can occur, unless it has been wrapped in too many layers. It is best to observe such wrappings carefully and cut them after the graft has healed to prevent constriction.

Plastic polyethylene or polyvinyl chloride (PVC) grafting and budding tapes also are available. Since they are not adhesive, they must be secured by using a slipknot at the last turn. These plastic tapes are slightly elastic and will allow for some diameter growth of the grafts, but must eventually be removed. Masking tape is a satisfactory wrapping material also.

Parafilm tape has been used with successful results to wrap graft unions rapidly (*5*) and for chip-budding roses. This material is a waterproof, flexible, stretchable, thermoplastic film with a paper backing. The film is removed from the paper and two layers are placed over the graft union and pressed into place with the fingers.

Grafting Machines

Several machines or devices have been developed to prepare graft and bud unions and a few have been widely used, especially in propagating grapevines (*2, 3*).

One of the most successful machines is the hand-operated device shown in Fig. 12–19, originally developed in France and called the L. B. Grafting Tool.[3] This device makes a type of wedge graft, cutting out a long **V**-notch in the

---

[3]Available in the United States from Heitz Wine Cellars, 500 Taplin Road, St. Helena, Calif. 94574.

FIGURE 12–20 Machine in operation for cutting notches in ends of grape scion and stocks for making the type of graft illustrated in Fig. 12–21. The blades in front are making cuts in the stocks; those in the back are making cuts for the scions.

FIGURE 12–19 A type of wedge grafting made by a French grafting device. *Top left:* Appearance of graft as made by the cutting blades (before wrapping with grafting tape). *Top right:* Apple graft union after one year's growth in the nursery. *Below:* Grafting device in use. Grafts can be made with this device much faster than by the whip graft method.

Another machine[5] widely used in California is shown in Fig. 12–20. It is powered by an electric motor and has two sets of saw blades on a single shaft for cutting notches, one in the base of the scion and one at the top end of the rootstock, as shown in Fig. 12–21. Three persons work together, one cutting out the scions, one cutting the rootstocks, and one assembling the two. After the stock and scion have been pushed together, the graft fit is so tight that no wrapping or stapling is necessary.

A third device, widely used for grapes, is the Pfropf-Star grafting machine[6] manufactured in Germany. It cuts through both stock and scion, one laid on top the other, making an omega-shaped cut and leaving the two parts interlocked. The device is available with either foot-operated or electromagnetic-operated mechanism. It is the fastest of the three machines but the graft union is not held together as tightly as the other two and it is not as suitable for tree fruits as the other two devices.

rootstock and a corresponding long, tapered cut at the base of the scion. Although the cuts fit together very well, the operation is slow because the graft union must be either tied with a budding rubber or stapled together. This machine has been used successfully in propagating both grapes and fruit trees. A similar, but portable, device useful for making top grafts is manufactured in New Zealand.[4]

---

[4]Available from Raggett Industries, 10 Cheesman Road, P.O. Box 1286, Gisborne, New Zealand.

[5]Available from Spink Refrigeration Co., 677 St. Helena Highway South, St. Helena, Calif. 94574 U.S.A.

[6]Manufactured by E. & H. Wahler, 7056 Weinstadt-Schnait, Buchhaldenstrabe 21, Stuttgart, Germany. Available in the United States from Tree & Vine Management Co., P.O. Box 311, Reedley, Calif. 93654 U.S.A.

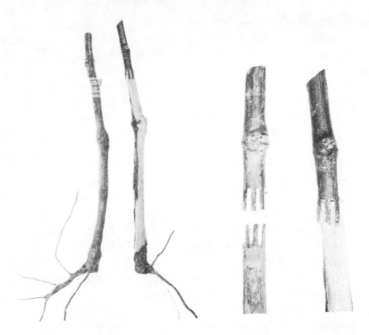

FIGURE 12–21 Grapes propagated by machine grafting. Small, one-budded scions are grafted on rooted cuttings during the dormant season. The graft union is wrapped with a budding rubber, then allowed to callus before planting.

SELECTION AND HANDLING OF SCIONWOOD

Kind of Wood

Since field or bench grafting of deciduous species takes place in late winter or early spring, the use of scionwood that grew the previous summer is necessary.

In selecting such scion material the following points should be observed:

1. For most species, the wood should be one year old or less (current season's growth). Avoid including older growth, although with certain species, such as the fig or olive, two-year-old wood is satisfactory, or even preferable, if it is of the proper size.

2. Healthy, well-developed vegetative buds should be present. Avoid wood with flower buds. Usually, vegetative buds are narrow and pointed, whereas flower buds are round and plump (see Fig. 13–1).

3. The best type of scion material is vigorous (but not overly succulent), well-matured, hardened shoots from the upper part of the tree, which have made 60 to 90 cm (2 to 3 ft) of growth the previous summer. Such growth develops on relatively young, well-grown, vigorous plants; high production of scion material can be promoted by pruning the plant back heavily the previous winter. Water sprouts from older trees sometimes make satisfactory scionwood, but suckers arising from the base of grafted trees should not be used since they may consist of rootstock material. A satisfactory size is from 0.6 to 1.2 cm (¼ to ½ in.) in diameter.

4. The best scions are obtained from the center portion or from the basal two-thirds of the shoots. The terminal sections, which are likely to be too succulent, pithy, and low in stored carbohydrates, should be discarded. Well matured wood with short internodes should be selected.

Source of Material

Scionwood should be taken from source plants of the correct cultivar known to be pathogen-tested and genetically true-to-type (see Chap. 8). Virus diseased, undesirable sports, chimeras, and virus-like genetic disorders must be avoided. Source plants may be of two basic types:

1. Plants in an orchard, vineyard, or landscape, growing under conditions where the

flowering, fruiting, and growth habits can be observed. For fruit-bearing species, it is best to take propagation material from bearing plants where production history is known. Visual inspection, however, may not reveal the true condition of the proposed source plant and, to be sure, appropriate indexing and progeny tests would be required (see Chap. 8).

2. In commercial nurseries special scion blocks, where plants are grown particularly for propagation may be maintained. Such plants are handled differently than they would be for producing a crop. For example, fruit trees may be pruned back each year to produce a large annual supply of long, vigorous shoots well-suited for scionwood. Such special blocks would usually be handled to conform to registration and certification programs and would be subject to isolation, indexing, and inspection requirements. In addition, it is important to maintain source identity of scion material through the entire propagation sequence, so that over a period of time proper sources of the various cultivars can be identified and maintained.

Collection and Handling

For deciduous plants to be grafted in early spring, the scionwood can be collected almost any time during the winter season when the plants are fully dormant (6). In climates with severe winters, the wood should not be gathered when it is frozen. Any wood that shows freezing injury should not be used. Where considerable winter injury is likely, it is best to collect the scionwood and put it in cold storage after leaf fall but before the onset of winter.

Storage

Scionwood collected prior to grafting must be properly stored. It should be kept slightly moist and at a low enough temperature to prevent development of the buds. A common method is to wrap the wood, in bundles of 25 to 100 sticks, in heavy, waterproof paper or in polyethylene sheets or bags. A small amount of some clean, slightly moist material, such as sawdust, wood shavings,

or peat moss, should be sprinkled through the bundle. Sand should not be used, because it will adhere to the scionwood and dull the edge of the knife during the grafting operation. If the packing material is wet, various fungi may develop and damage the buds, even at low storage temperatures. The packing should be barely moist; scionwood is more apt to be damaged by being too wet than by being too dry. If it is to be stored for a prolonged period, the bundle should be examined every few weeks to see that the wood is not becoming either dried out or too wet. When the buds show signs of swelling, the wood should either be used for grafting immediately or moved to a lower storage temperature.

Polyethylene plastic sheeting is a good material for wrapping scionwood. It allows the passage of oxygen and carbon dioxide, which are exchanged during the respiration process of the stored wood, but retards the passage of water vapor. Therefore, if this type of wrapping is sealed, no moist packing material is needed; the natural moisture in the wood is sufficient since little will be lost. Polyethylene bags are useful for storing small quantities of scionwood. All bundles should be labeled accurately.

The temperature at which the wood is stored is important. If it is to be kept only two or three weeks before grafting, the temperature of the home refrigerator—about 5°C (40°F)—is satisfactory. If stored for a period of one to three months, scionwood should be held at about 0°C (32°F) (5) to keep the buds dormant. However, buds of some species, such as the almond and sweet cherry, will start growth after about three months even at this temperature. Do not store scionwood in a home freezer because the very low temperatures, about −18°C (0°F), may injure the buds.

If cold storage facilities are not available, the scionwood can be kept for a period of time in cold winter regions by burying the bundles in the ground, below frost level, on the north side of a building or tall hedge. Drainage should be provided so that excess water does not remain around the scions and cause the buds to deteriorate.

Storage of scions should not be attempted if succulent, herbaceous plants are being grafted; such scions should be obtained at the time of grafting and used immediately. Certain broad-leaved evergreen species, such as camellias, olives, and citrus, can be grafted in the spring before much active growth starts without previous collection and storage of the scionwood. It is taken directly from the tree as needed, using the basal part of the shoots containing dormant, axillary buds. The leaves are removed at the time of collection.

Attempting to use scionwood in which the buds are starting active growth is almost certain to result in failure. In such cases, the buds quickly leaf out before the graft union has healed; consequently the leaves withdraw water from the scions by transpiration, and cause the scions to die. In addition, the strong competing sink of a developing shoot can interfere with graft union formation.

In topworking pecans (4) it was found that good results could be obtained by using precut scions; that is, scions were cut in advance by skilled persons at a convenient time, then held in cold storage in polyethylene bags for periods up to nine days before inserting in the graft unions. Grafting success was reduced only slightly by the use of precut scions.

GRAFTING CLASSIFIED BY PLACEMENT

Grafting may be classified according to the part of the plant on which the scion is placed—a root, the crown (the junction of the stem and root at the ground level), or various places in the top of the plant.

Root Grafting

In root grafting the rootstock seedling, rooted cutting, or layered plant is dug up, and the roots are used as the stock for the graft. The entire root system may be used (**whole-root graft**—Figs. 12–22 and 12–23), or the roots may be cut up into small pieces and each piece used as a stock (**piece-root graft**). Both methods give satisfactory results. As the roots used are relatively small [0.6 to 1.3 cm (¼ to ½ in.) in diameter], the whip graft is generally used. Root grafting is usually performed indoors during the late winter or early spring. The scionwood collected previously is held in storage, while the rootstock plants are also dug in the late fall and stored under cool (1.5 to 4.5°C; 35 to

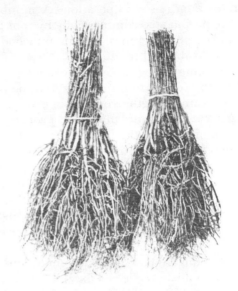

FIGURE 12–22 Bundles of one-year-old apple seedlings of a size suitable for whole root bench grafting.

40°F) and moist conditions until the grafting is done.

The term **bench grafting** is sometimes given to this process, because it is often performed with dormant scions and stocks at benches by skilled

FIGURE 12–23 Whole root apple grafts made by the whip or tongue method and wrapped with adhesive nursery tape.

grafters as a large-scale operation. A number of plants are propagated commercially by root grafting—apples, pears, grapes, and many woody ornamental shrubs and trees (*16,17*).

In making root grafts, the root pieces should be 7.5 to 15.0 cm (3 to 6 in.) long and the scions about the same length, containing two to four buds. After the grafts are made and properly tied, they are bundled together in groups of 50 to 100 and stored for callusing in damp sand, peat moss, or other packing material. They may be placed in a cool cellar or under refrigeration at approximately 7°C (45°F) for about two months. The callusing period for apples can be shortened to around 30 days if the grafts are stored at a temperature of about 21°C (70°F) and at a high humidity. To use this higher callusing temperature the material should be collected in the fall and the grafts made before any cold weather has overcome the rest period of the scion buds. After the unions are well healed, the grafts must be stored at cool temperatures—2 to 4°C (35 to 40°F)—to overcome the "rest period" of the buds and to hold them dormant until planting (*24*). The grafts are lined-out in early spring in the nursery row directly from the low temperature storage conditions. For general callusing purposes, temperatures from 7 to 21°C (45 to 70°F) are the most satisfactory. By the proper regulation of temperature, callusing processes may be accelerated by increased temperature or retarded by decreased temperature so that, within reasonable limits, a desired degree of callus formation may be had within a given length of time.

With some plants the graft union should be kept warm, 24 to 27°C (75 to 82°F), but the roots and the buds on the scion should be kept cool, about 7°C (45°F) to prevent premature growth before the graft union has callused and healed together. An ingenious system (Fig. 12-24) has been developed for whip grafting filberts, which are notoriously difficult to graft, that overcomes this problem. The graft union itself is kept warm by a plastic pipe to which electric heating cables are taped or which contains hot water. The graft union is laid against the pipe, but the scion and roots protrude into areas having lower temperatures. This "hot callusing" system, when used outdoors in late winter or early spring, has increased the grafting "takes" in filberts from 10

FIGURE 12-24   "Hot grafting" system for root grafting difficult plants. The graft union is placed in a slot in a large plastic pipe. Inside the large pipe is a smaller pipe through which thermostatically-controlled hot water flows or to which electric heating cables are taped. Insulating material laid over this pipe retains the heat. The protected roots and scions protrude into areas having lower temperatures, which retards their development. (Courtesy H. B. Lagerstedt and USDA (*29, 30*).)

percent to more than 90 percent. It has also been used successfully in root grafting apples, pears, peaches, and plums (*30*). Some aeration of the callusing grafts is required, so that airtight containers should not be used (*38*).

As soon as the ground can be prepared in the spring, the grafts are lined out in the nursery row 10 to 15 cm (4 to 6 in.) apart. They should be planted before growth of the buds or roots begins. If growth starts before the grafts can be planted, they should be moved to lower temperatures (−1 to 2°C; 30 to 35°F). The grafts are usually planted deep enough so that the graft union is just below the ground level, but if the roots are to arise only from the rootstock, the graft should be planted with the union well above the soil level. It is very important to prevent **scion rooting** where certain definite influences, such as dwarfing or disease resistance, are expected from the rootstock.

After one summer's growth, the grafts should be large enough to transplant to their permanent location. If not, the scion may be cut back to one or two buds, or headed-back somewhat to force out scaffold branches, and then allowed to grow a second year. With the older root system a strong, vigorous top is obtained the second year.

Nurse-Root Grafting

Under certain conditions, it is desirable to have a stem cutting of a difficult-to-root species on its own roots. One way this can be done is by making a root graft, using the plant to be grown on its own roots as the scion and a root of a compatible species as the stock. The scion may be made longer than usual and the graft planted deeply with the major portion of the scion below ground. **Scion rooting** can be promoted by rubbing a rooting stimulant, such as indolebutyric acid, into several vertical cuts made through the bark at the base of the scion, just above the graft union. This is done just before planting, and the grafts are set deeply so that most of the scion is covered with soil (*27, 28*). After one or two seasons of growth many of the scions have roots. The temporary nurse rootstock is then cut off and the top reduced in proportion to the root system. The rooted scion is replanted to grow on its own roots.

Scion rooting is often better if the nurse root is buried deeply enough so that a new shoot of the current season's wood, developing from a bud on the scion, will be rooted, rather than trying to root the older, lignified tissue of the original scion. This is essentially a form of mound layering. As the new scion shoot grows, soil is gradually mounded up around it to a height of 13 to 15 cm (5 to 6 in.), although the terminal leaves are at no time covered (*32*).

Several methods of handling eliminate the necessity of digging up the graft and cutting off the rootstock. (a) The rootstock piece will eventually die if it is grafted onto the scion in an inverted position (Fig. 12–25) (*32*). The union heals and the inverted stock piece sustains the scion until scion roots are formed, but the stock fails to receive food from the scion and eventually dies, thus leaving the scion on its own roots. (b) In a second method an incompatible rootstock is used. Hence, if the graft is planted deeply, scion roots will gradually become more important in sustaining the plant, and the incompatible rootstock will finally cease to function. An example of this is apple scion on pear rootstock. (c) In another method the base of the scion, just above the graft union, is bound with some type of wrapping material to girdle and cut off the rootstock (Fig. 12–26). Excellent results have been obtained with ordinary budding rubber strips (0.016 gauge) (*7*). Budding rubbers disintegrate within a month when exposed to sun and air; when buried in the soil, they will last as long as two years, allowing sufficient

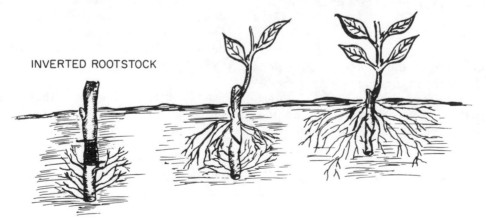

INVERTED ROOTSTOCK

FIGURE 12–25 The "nurse-root" graft is a temporary graft used to induce the scion to develop its own roots. The nurse root sustains the plant until the scion roots form, then it dies. An easy method is shown here for preparing a nurse-root graft by inverting the rootstock piece.

FIGURE 12–26 Nurse-root graft. A 'Malling 9' apple scion was root-grafted to an apple seedling nurse root. Just above the graft union the scion was wrapped with a budding rubber strip (see arrow). After two years in the nursery, vigorous scion roots were produced. The budding rubber has effectively constricted development of the seedling nurse root, which can now be broken off and discarded. (Courtesy D. S. Brown.)

time for the scion to become rooted. Yet, owing to the slow deterioration of the rubber below ground, the rootstock is finally girdled and cut off.

Crown Grafting

A graft union made at the root-stem transition region—the "crown" of the plant—on an established rootstock is termed a **crown graft.** Methods commonly used in crown grafting include the whip, side, cleft, wedge, and bark graft, the choice depending upon the species and size of the rootstock.

Crown grafting of deciduous plants is done from late winter to late spring. In each species grafting should take place shortly before new growth starts. The scions should be prepared from well-matured, dormant wood of the previous season's growth.

If the graft is above the soil level, the union

must be well wrapped to hold it solid and prevent desiccation. However, when the operation is performed just below, at, or just above the soil level, it is possible to cover the graft union, or even the entire scion, with soil and thus eliminate the necessity for waxing. In all cases the union should be tied securely with tape to hold the grafted parts together until healing takes place.

Double-Working (Budding or Grafting)

A double-worked plant has three parts, all different genetically: the rootstock, the interstock, and the scion, or fruiting, top (*18*) (see Fig. 12–27). Such a plant has two unions, one between the rootstock and interstock, and one between the interstock and the scion. The interstock may be less than 25 mm (1 in.) in length or extensive enough to include the trunk and secondary scaffold branches of a tree.

Double-working is used for various purposes. See Chap. 11.

Examples of double-working are (a) the propagation of 'Bartlett' pears on quince as a dwarfing rootstock by using a mutually compatible interstock, such as 'Old Home' or 'Hardy' pear (Fig. 12–27), and (b) the propagation of dwarfed apple trees consisting of the scion cultivar grafted onto a 'M 9' or 'M 27' interstem piece that is grafted onto a more vigorous rootstock, such as 'MM 106', 'MM 111', or apple seedlings (*10*).

Several methods are used for developing double-worked nursery trees. The grafting in these techniques can be done by machine grafting devices as illustrated in Figs. 12–19 and 12–20, or by use of the whip graft.

1. Rootstock "liners," either seedlings, clonal rooted cuttings, or rooted layers, are set out in the nursery row in early spring. These are then fall-budded with the interstock buds, growth from which, a year later, is fall-budded with the scion cultivar buds. Three years are required to produce a nursery tree by this method.

2. The interstock piece is bench-grafted onto the rooted rootstock—either a seedling or a clonal stock—in late winter. After callusing, the grafts are lined-out in the nursery row in

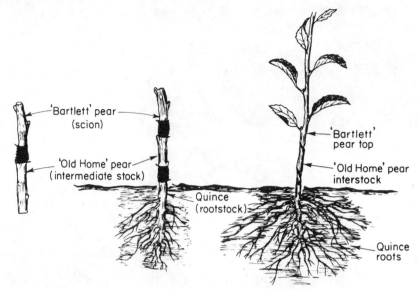

FIGURE 12-27 There are three distinct parts and two graft unions in a double-worked plant, illustrated here by a 'Bartlett' pear grafted on an 'Old Home' pear which in turn is grafted on quince as a rootstock. The 'Old Home' pear, in this case, is the intermediate stock, or interstock.

'Bartlett' pear (scion)

'Old Home' pear (intermediate stock)

Quince (rootstock)

'Bartlett' pear top

'Old Home' pear interstock

Quince roots

the spring. These are then fall-budded to the scion cultivar. By this method the nursery tree is propagated in two years.

3. A variation of method 2 is to prepare, by bench grafting, two graft unions—the scion grafted to the interstock, and the interstock grafted to the rooted rootstock. After callusing, the completed graft, with two unions, is lined-out in the nursery row. Depending upon growth rate, a nursery tree can be obtained in one or two years.

4. The scion piece is bench-grafted onto the interstem piece in late winter or early spring; then, after callusing, this component is grafted—as shown in Fig. 12–27—onto rootstocks that have been grown in place in the nursery row. A nursery tree is obtained in one or two years by this method.

5. Double-shield budding, in which the double-working is done in one operation by budding. A nursery tree is produced in one year, or if growth is slow, two years after budding (see p. 406).

6. The interstock shoots still on the plant can be T-budded in late summer with the scion buds inserted about 15 cm (6 in.) apart. These heal into place and during late winter the budded interstock shoots are cut apart with a scion bud at the terminal end of each piece. The budded (shoot) pieces are then whip grafted onto the rooted seedling stocks. After they have callused, the completed graft, consisting of rootstock and interstem piece with scion bud in place, is ready for planting in the nursery row (*15*).

Topworking (Topgrafting or Topbudding)

Topworking is used primarily to change the cultivar of an established plant—tree, shrub, or vine—either by grafting (Figs. 12–28 and 12–29) using one of the methods described earlier in this chapter, or by budding (see Chap. 13). This procedure may be preferred to removal and replacement of the entire plant, since return to flowering and fruiting is faster with topworking than with a new nursery plant, particularly if the topworked plant is young, healthy, and well cared for. Plants that are old, diseased, or of a short-lived species are not satisfactory candidates for topworking.

Topworking can be used in the propagation of young ornamental nursery trees (*14*). Scions of the desired ornamental cultivars are grafted onto the appropriate compatible rootstock at a height of 3 to 6 ft by either the cleft, whip, or bark grafts in late winter or early spring. Considerable care is required to shape the trees during the summer but by fall, with good growing conditions, they can be dug and sold.

Sometimes virus diseases may be introduced

FIGURE 12–28 Extensive topworking operation. Young apple trees have been top-grafted to a more profitable cultivar. Each tree has a small "nurse" branch.

into the new topworked plant, from either the stock or the scion. Because such diseases interfere with growth, flowering, and fruiting, one should know the virus status of both components before considering topworking.

Preparation for Topworking

Topgrafting is usually done in the spring, shortly before new growth starts. The exact time depends upon the method to be used. The cleft, side, whip, and wedge grafts can be done before the bark is slipping, but the bark graft must be done when the bark is slipping, preferably just as the buds of the stock tree are starting to grow. Topbudding can be done with the T-, patch-, or chip-bud methods.

It is usually advisable to obtain an ample amount of good-quality scionwood prior to grafting and store it under the proper conditions, although for broad-leaved evergreens, such as avocado or citrus, scionwood can be collected at the time of the grafting operation.

In preparing for topworking, one must decide, for each individual stock tree, how many scaffold branches should be used (usually three to five). If many grafts are made high in the tree in the small, secondary scaffold branches (**frameworking**), the tree will return to bearing earlier than it will if fewer and larger limbs are grafted lower in the tree. Frameworking, however, is expensive and necessitates pruning out all new shoots from the stock below the grafts as they develop.

The branches to be grafted should be well distributed around the tree and up and down the main trunk, avoiding branches with weak, narrow crotches. All others can be removed unless one or more nurse branches are used. Figure 12–30 shows a worker preparing the branches for grafting.

Topworking is an extremely severe pruning operation for the tree and results in a considerable imbalance between the root system and the top. However, deciduous trees soon recover and the new growth from the scions and from latent buds on the stock restores a balance without damage, providing the tree is healthy and vigorous and the grafting is done when the tree is dormant or shortly after growth starts in the spring. However,

FIGURE 12–29 Proper method of topworking trees. *Left:* Scions, which have been inserted into fairly small branches, are starting to grow. Tree has been white-washed to prevent sunburn injury. *Right:* Same tree about six weeks later. Stakes have been nailed to branches; the shoots developing from scions are tied to these stakes to prevent their breaking off in winds.

FIGURE 12–30 Sawing off a branch in preparation for topgrafting so that no bark-tearing occurs. *Top left:* The first cut is made in a smooth area starting on the under side of the branch and continuing about one-third of the distance through. *Top right:* The second cut is made starting from the upper part of the branch and cutting downward. It should be back 2.5 or 5 cm (1 or 2 in.) on the branch from the first cut. *Below left:* After the branch breaks off and falls, the second cut is continued. *Below right:* Final smooth cut, ready for grafting.

under some situations, not always well understood, the tree will be adversely affected by the heavy cutting back, and will show intense leaf burning and may even fail to survive. This situation is most likely to occur when the grafting is delayed until after new spring growth is well underway, where all foliage and new shoots are stripped off the stock plants, or where high temperatures prevail shortly after grafting.

Also when all branches of the tree are cut back and grafted, considerable energy goes into the scion growth, producing vigorous, succulent shoots that may break off if not properly pruned or supported. In addition, such rank, succulent growth is very susceptible to winter freezing damage in cold climates.

To avoid the problems cited above it is advisable to retain some of the foliage on the stock plant as well as small lower branches and even large limbs to be held temporarily as **nurse branches.** Such nurse limbs should be pruned back fairly severely, otherwise the grafted scions may not make adequate growth.

It is essential that nurse limbs be left when

topworking broad-leaved evergreen trees, such as citrus or olive. If they are retained on the south and west parts of the tree they will shade the grafted portions and reduce the chances of sunburn injury to the exposed branches.

A practice often recommended, especially for older trees, is to topwork them during a two-year period, grafting perhaps two main branches on the northeast side of the tree the first year and retaining two branches on the southwest as nurse branches. The second year these branches are topworked.

Topworking is most successful when done on relatively young trees where the branches to be grafted are no larger than 7.5 to 10 cm (3 to 4 in.) in diameter and are relatively close to the ground. When attempting to topwork large, old trees, it is often necessary to go high up in the trees to find branches with a diameter as small as 10 cm. If the grafting is done on such branches, the new top is inconveniently high for the various orchard operations, such as thinning and harvesting. The other alternative is cutting off the branches or main trunk close to the ground and inserting the scions into wood 30 cm (1 ft) or more in diameter. Although many scions can be inserted around the tree between the bark and the wood by the bark-graft method and may grow well for several years, it is quite likely, as shown in Fig. 12–31, that wood rot will develop in the center of the stub before the growth of the scions can heal it over. Also, some scions are not mechanically held in place securely and may be blown out by strong winds after they reach considerable size.

It is important that the branch to be topworked be cut off in such a location that the region just below the cut is smooth and free from knots or small branches, so that there will be a satisfactory place for inserting the scions. The branches are best cut off about 23 to 30 cm (9 to 12 in.) from the main trunk to keep the tree headed low. The branch should not be cut off more than a few hours before the grafting is to be done.

In Preparation for Topworking (Topgrafting)

1. Do the work in the spring when the trees are dormant or shortly after growth starts.
2. Select for grafting three to five well-placed scaffold branches no larger than about 10 cm

FIGURE 12–31 An improper method of topworking trees. This walnut tree was cut off close to the ground, and a number of scions were inserted around its circumference by the bark-graft method. Two of the scions grew successfully, but the remainder failed. Most of the original tree is dead; healing of such a large cut is almost impossible. Better methods of topgrafting trees are shown in Figs. 12–28 and 12–29.

(4 in.) in diameter and conveniently close to the ground.
3. Retain nurse branches for broad-leaved evergreen trees and for deciduous trees where the winters are severe.
4. Cut off the branches properly so that the bark is not torn down the trunk.

Grafting on a cool, overcast day with no wind blowing offers the most protection from drying of the cut surfaces of the scion and stock until they can be covered with grafting wax. Grafting on hot, sunny, and windy days should be avoided. During grafting, the scion wood must not dry out by being exposed to the sun. It should be kept moist and cool in some container or be wrapped in moist burlap.

Immediately after each stub is grafted it should be thoroughly covered with grafting wax. The need for prompt and thorough coverage of all cuts, including the tip end of the scions, cannot be stressed too strongly. The wax should be worked into the bark of the stock, sealing all small cuts or cracks where air could penetrate in and around the cut surfaces where healing tissue is expected to develop. Waxes containing beeswax will attract bees, which may remove the wax, necessitating rewaxing.

Subsequent Care of Topworked Trees

After the actual top-grafting (or topbudding) operation is finished, much important work needs to be done before the topworking is successfully completed. A good grafting job can be ruined by improper care of the grafted trees.

If the grafting has been done in late spring when growth is active, trees of some species, such as the walnut, will ''bleed'' to a considerable extent from the grafted stub, even though it has been covered with grafting wax. This flow of sap around the scions can be so heavy as to interfere with the normal healing processes at the graft union. If this condition appears, it can often be corrected by making several slanting cuts around the base of the tree with a knife or saw through the bark, into the water conducting tissues in the trunk of the tree several feet below the grafted stubs. The bleeding will then take place at these cuts rather than around the graft union. This extensive sap flow is not particularly harmful to the tree and will usually stop within a few days. Boring a series of random holes into the trunk around the tree will accomplish the same purpose.

If certain bacterial diseases, such as crown gall, are present, this practice may spread the bacteria. In such cases, dipping the knife, saw, or bit into a disinfectant, such as 10 percent Clorox, is advisable.

In three to five days after grafting, the trees should be carefully inspected and the graft unions rewaxed if cracks or holes appear in the wax.

It is essential to prevent sunburn on the trunk and large branches exposed to the sun by the removal of the protecting top foliage. Protection is especially important if the grafting has been done late in the season when hot weather can be expected, and when no protecting nurse branches have been retained. The energy from the sun absorbed by the dark-colored bark can raise the temperature of the living cells below the bark to a lethal level.

It is advisable to whitewash the trunk, branches, and scions of the grafted trees. Various cold-water paints, some made especially for this purpose, are available. The white color reflects a considerable portion of the sun's radiant energy, thus keeping the temperature of the living tissues within safe limits. Interior, water-base white house paints (both latex and acrylic) mixed with water, 1:1, prevent sunburn for one season. Exterior paints give longer protection but are more likely to cause injury to the tree (*20, 34*).

Another help in preventing sunburn is to retain some of the watersprouts which soon start growth along the trunk and branches of the grafted tree. However, they must be kept under control or they will quickly shade out the developing scions. Rather than removing the watersprouts completely, they can be headed back to 20 to 25 cm (8 to 10 in.) in length, and they will shade the bark underneath. An additional benefit is that the food manufactured by this leaf area will help sustain the tree until the new scions develop sufficient foliage to do so.

Trees just grafted should be amply supplied with water so that the tissues are in a high state of turgidity. This is necessary in order to obtain good callus production, which is essential for healing of the graft union.

During the first summer after topworking, new shoots arising from below the grafted branches must be kept pruned back so as not to interfere with the growth of the scions. Nitrogen fertilizers should be withheld for a year or two after grafting because no stimulation of growth is usually needed. The water requirement of the trees will be less, owing to the removal of a considerable amount of leaf area when the tops were cut back for grafting.

Two to four scions are generally inserted in each stub. If all the scions grow, they should all be retained during the first year, because they will help heal over the stub. However, just one branch, from the best placed and strongest growing scion, should be retained permanently. Growth from the remaining scions should be retarded by rather severe pruning, keeping them alive to help heal the branch, but allowing the permanent scion to become the dominant one. To keep two or more scions for permanent branches at one point will undoubtedly result in a weak crotch, which will eventually break. The first year, then, the best practice is to retain all scions to help heal the stub but, by pruning, to retard the growth of all but the best one. Such a procedure is illustrated in Fig. 12–32. If two shoots arise from the scion to be retained permanently, select the best of these and remove the other. After the stub has healed over, in the second or third year, the temporary scions can be removed completely.

FIGURE 12–32 Problems encountered in handling limbs that have been topgrafted. *Left:* Where both scions "take," one is cut back heavily to allow the other to become predominant so as not to form a weak crotch. It is retained for two or three years, however, to assist in healing the grafting wound, but will eventually be removed. *Right:* Where only one scion in a fairly large branch "takes," difficulty may occur in rapid healing of the wound. The wood under the dead scion should be cut off at an angle, then waxed; this will give a minimum cut surface for healing.

If the permanent scion grows rather vigorously, the danger arises of its becoming topheavy and breaking off during winds. This may be handled in two ways—either by retarding the growth of the permanent branch by pruning it back or by nailing a lath or other type of stick onto the tree and tying the new branch to this stick. *When tying a cord around a branch, always make a loop so that there is no chance of the branch being girdled as it grows.*

Should only one scion grow at each stub, the problem of securing adequate healing of the stub on the side opposite the living scion may prove to be serious. Healing may be helped by sawing off the stub at an angle away from the surviving scion. The cut surface of the stub can be covered with grafting wax to retard wood decay. It may take a number of years for the single scion to heal over the stub, and it is possible that wood decay may start before it can do so.

When none of the scions grow in a stub, there are still some possibilities for getting it topworked. One is by allowing several well-placed watersprouts arising just below the cut surface to grow, then topbudding them during the summer. Or the watersprouts may be allowed to grow so as to keep the branch alive and healthy, then the grafting operation repeated the following year, making a fresh cut a foot or so below the original cut.

If nurse branches have been used, they should be cut back and away from the scions at intervals throughout the summer so that scion shoots are always fully exposed to the sun and have sufficient space to grow. Nurse branches should be removed entirely or grafted also by the beginning of the second or third year.

When the top of the tree has been finally worked over to the new cultivar, it will grow vigorously for a few years. Good pruning practices are needed to prevent badly placed branches from developing. Usually by the third year the tree will again be in production.

HERBACEOUS GRAFTING

Grafting herbaceous types of plants is used for various purposes, such as studying virus transmission, stock–scion physiology, and grafting compatibility, as well as for the commercial greenhouse production of certain cucurbitaceous crops, particularly in Europe. Usually, such grafts are made while the plants are quite small, the stock being grafted shortly after seed germination. Such material is generally very soft, succulent, and susceptible to injury. In one technique (see Fig. 12-33), a simple splice graft is used with a diagonal cut made through the seedling stock, just above the cotyledons (*23*). A piece of thin-walled polyethylene tubing of the proper size to give a snug fit is slipped over the cut end of the stock. The basal end of the scion receives a diagonal cut similar in length and angle to that given the stock. The scion is then slipped down into the plastic

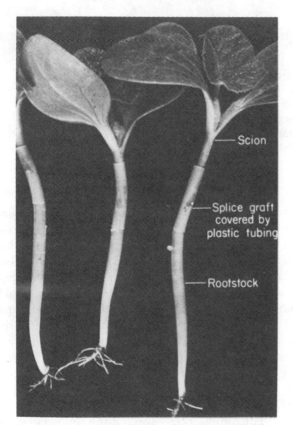

FIGURE 12-33 Grafting young, herbaceous plant material together by using a splice graft. A single, slanting cut is made in the upper part of the stock. A piece of transparent plastic tubing is slipped over this to give a snug fit. A similar slanting cut is made in the lower portion of the scion, which is then slipped into the plastic tubing so that the cut surfaces match. After the graft has healed, the tubing is cut away with a razor blade.

tubing so that the two cut surfaces make intimate contact. The tubing holds the graft in place until healing occurs—about 12 days after grafting. Then the tubing may be slipped off over the scion if there are no leaves and if the bud has not expanded; otherwise, it may be cut off with a razor blade.

In another procedure (*11*) used in grafting older herbaceous stock plants, several leaves are retained below the graft union. The cleft graft is used (but with only one scion), and the graft union is bound with raffia, budding rubbers, or adhesive latex tape. To prevent drying out, the entire plant—following grafting—is covered with a supported polyethylene bag. The grafted plant is then set in the shade until the graft has healed; then the plastic cover can be removed.

A method of establishing *Vitis vinifera* grapes on phylloxera or nematode-resistant rootstocks is by "greenwood" grafting, in which the grafting is done in the spring, placing the scion—taken from new growth—on a new green shoot developed from the rootstock plant. A simple splice graft is made with the sloping cuts 2.5 to 4 cm (1 to 1½ in.) long. The stock and scion pieces must be the same diameter. The scion has only one bud. The cuts are matched as closely as possible, and the graft union is completely covered by wrapping with a budding rubber. The graft union should be healed and the bud growing by two weeks after grafting. Since these grafts are quite fragile, they must be tied to a stake (*9, 21*).

CUTTING-GRAFTS

In the cutting-graft a leafy scion is grafted onto a leafy, unrooted stem piece (which is to become the rootstock), and the combination is then placed in a rooting medium under intermittent mist for simultaneous healing of the graft union and rooting of the stock. Leaves must be retained on the rootstock piece in order for it to root. This procedure was utilized many years ago in studying stock–scion physiology in citrus (*19*). It has been used in commercial propagation of various types of citrus on clonal dwarfing rootstocks (*12*). It is also of value in propagating certain difficult-to-root conifers (*40*), rhododendrons (*13, 33*), and macadamias (*1*), as well as a number of apple, plum, and pear cultivars (*35*). It is used in the Nether-

FIGURE 12–34   Roses in the Netherlands being propagated by simultaneous rooting and grafting. (A) *Left:* Shoot cut apart to be used for rootstocks. Only internodes are used. *Right:* Sections cut for scions. One leaf is used per scion. (B) Graft made with omega grafting machine. (C) Graft wrapped with tape for healing. (D) Completed graft with union healed and stock well rooted, ready for planting. In the Netherlands this process is called ''stenting,'' being a contraction of the Dutch words *stekken* (to strike a cutting) and *enten* (to graft). From P. A. Van de Pol et al. (*44, 45*).

lands in propagating roses (*44, 45*) (see Fig. 12–34).

For citrus a simple splice graft is used. The slope of the cut is at a 30-degree angle $1\frac{1}{3}$ to 2 cm ($\frac{1}{2}$ to $\frac{3}{4}$ in.) long; the union is tied with a rubber band. The base of the stock is dipped into a root-promoting material, such as indolebutyric acid, and then the grafts are placed under mist, or in a closed case, in flats of the rooting medium over bottom heat. After healing of the union and rooting of the stock, the grafts are allowed to harden by discontinuing the mist and bottom heat for about two weeks. Then the grafts are ready for planting in 4-liter (1-gal) cans or other containers.

MICROGRAFTING

Grafting of tiny plant parts can be done aseptically using the techniques described in Chaps. 16 and 17, in which the small grafts are grown in closed containers until they are large enough to be transferred to open conditions. Micrografting has been used mostly with citrus, apple, and some *Prunus* species, particularly in developing virus-free plants, where a virus-free shoot tip can be obtained but cannot be rooted. The shoot tip is grafted aseptically onto a virus-free seedling thus providing a complete virus-free plant from which other ''clean'' plants can then be propagated (*8, 25, 26, 37*). Tests conducted some years after the shoot-tip grafts were made and the resultant trees were fruiting showed that the fruits were normal for the cultivar, disease-free, and with no variations appearing (*36*). The procedures for micrografting are described in Chap. 17.

The nonaseptic propagation of very small nursery trees by grafting tiny seedlings with match-like scions, then growing the minute grafts long enough to have a viable plant, is a promising procedure (*43*). This is useful particularly in the tropics, where nursery plants sometimes must be shipped long distances, often by air, into inaccessible regions. Quantities of such tiny plants can be transported much more readily than full-sized nursery trees.

REFERENCES

1. Ahlswede, J. 1985. Twig grafting of macadamia. *Proc. Inter. Plant Prop. Soc.* 34:211–14.

2. Alley, C. J. 1957. Mechanized grape grafting. *Calif. Agr.* 11(6):3, 12.

3. ———. 1970. Can grafting be mechanized? *Proc. Inter. Plant Prop. Soc.* 20:244–48.

4. Anonymous. 1968. Pre-cut scions. *Agr. Res.* 17(6):11.

5. Beineke, W. F. 1978. Parafilm: A new way to wrap grafts. *HortScience* 13(3):284.

6. Bhar, D. S., R. J. Hilton, and G. C. Ashton. 1966. Effect of time of cutting and storage treatment on growth and vigor of scions of *Malus pumila* cv. McIntosh. *Can. Jour. Plant Sci.* 46:69–72.

7. Brase, K. D. 1951. The nurse-root graft, an aid in rootstock research. *Farm Res.* 17(1):16.

8. Burger, D. W. 1985. Micrografting: A tool for the plant propagator. *Proc. Inter. Plant Prop. Soc.* 34:244–48.

9. Carlson, V. 1963. How to green graft grapes. *Calif. Agr. Ext. Publ. AXT. 115.*

10. Cummins, J. N. 1973. Systems for producing multiple-stock fruit trees in the nursery. *Plant Prop.* 19(4):7–9.

11. Denna, D. W. 1962. A simple grafting technique for cucurbits. *Proc. Amer. Soc. Hort. Sci.* 81:369–70.

12. Dillon, D. 1967. Simultaneous grafting and rooting of citrus under mist. *Proc. Inter. Plant Prop. Soc.* 17:114–17.

13. Eichelser, J. 1967. Simultaneous grafting and rooting techniques as applied to rhododendrons. *Proc. Inter. Plant Prop. Soc.* 17:112.

14. Elliott, F. A. 1988. Top working (winter field grafting). *Proc. Inter. Plant Prop. Soc.* 37:52–55.

15. Fisher, E. 1977. The pre-budded interstem: A new technique. *Fruit Var. Jour.* 31(1):14–15.

16. Flemer, W., III. 1986. New advances in bench grafting. *Proc. Inter. Plant Prop. Soc.* 36:545–49.

17. Gaggini, J. B. 1985. Bench grafting of trees under polythene. *Proc. Inter. Plant Prop. Soc.* 34:646–47.

18. Garner, R. J. 1940. Studies in nursery technique: The production of double worked pear trees. *Ann. Rpt. E. Malling Res. Sta. for 1939,* pp. 84–86.

19. Halma, F. F., and E. R. Eggers. 1936. Propagating citrus by twig-grafting. *Proc. Amer. Soc. Hort. Sci.* 34:289–90.

20. Hansen, C. J., and H. T. Hartmann. 1951. Influence of various treatments given to walnut grafts on the percentage of scions growing. *Proc. Amer. Soc. Hort. Sci.* 57:193–97.

21. Harmon, F. N., and E. Snyder. 1948. Some factors affecting the success of greenwood grafting of grapes. *Proc. Amer. Soc. Hort. Sci.* 52:294–98.

22. Hearman, J., A. B. Beakbane, R. G. Hatton, and W. A. Roach. 1936. The reinvigoration of apple trees by the inarching of vigorous rootstocks. *Jour. Pom. and Hort. Sci.* 14:376–90.

23. Holt, J. 1958. A simple way of grafting herbaceous plants. *Gard. Chron.* 143:332.

24. Howard, G. S., and A. C. Hildreth. 1963. Induction of callus tissue on apple grafts prior to field planting and its growth effects. *Proc. Amer. Soc. Hort. Sci.* 82:11–15.

25. Huang, S., and D. F. Millikan. 1980. *In vitro* micrografting of apple shoot tips. *Hort-Science* 15(6):741–43.

26. Jonard, R., J. Hugard, J. Macheix, J. Martinez, L. Mosella-Chancel, J. Luc Poessel, and P. Villemur. 1983. In vitro micrografting and its application to fruit science. *Scient. Hort.* 20:147–59.

27. Jones, F. D. 1950. Hormone on root graft. *Amer. Nurs.* 72(11):6–7.

28. Kerr, W. L. 1936. A simple method of obtaining fruit trees on their own roots. *Proc. Amer. Soc. Hort. Sci.* 33:355–57.

29. Lagerstedt, H. B. 1982. A device for hot callusing graft unions of fruit and nut trees. *Proc. Inter. Plant Prop. Soc.* 31:151–59.

30. Lagerstedt, H. B. 1984. Hot callusing pipe speeds up grafting. *Amer. Nurs.* 160(8):113–17.

31. Leiss, J. 1987. Modified side graft for nursery trees. *Proc. Inter. Plant Prop. Soc.* 36:543–44.

32. Lincoln, F. B. 1938. Layering of root grafts—a ready method for obtaining self-rooted apple trees. *Proc. Amer. Soc. Hort. Sci.* 35:419–22.

33. McGuire, J. J., W. Johnson, and C. Dawson, 1987. Leaf-bud or side graft nurse grafts for difficult-to-root rhododendron cultivars. *Proc. Inter. Plant Prop. Soc.* 37:447–49.

34. Micke, W. C., J. A. Beutel, and J. A. Yeager. 1966. Water base paints for sunburn protection of young fruit trees. *Calif. Agr.* 20(7):7.

35. Morini, S. 1984. The propagation of fruit trees by grafted cuttings. *Jour. Hort. Sci.* 59(3):287–94.

36. Nauer, E. M., C. N. Roistacher, T. L. Carson,

and T. Murashige. 1983. In vitro shoot-tip graft-ing to eliminate citrus viruses and virus-like path-ogens produces uniform bud-lines. *HortScience* 18(3):308–9.

37. Navarro, L., C. N. Roistacher, and T. Mura-shige. 1975. Improvement of shoot tip grafting *in vitro* for virus-free citrus. *Jour. Amer. Soc. Hort. Sci.* 100:471–79.

38. Shippy, W. B. 1930. Influence of environment on the callusing of apple cuttings and grafts. *Amer. Jour. Bot.* 17:290–327.

39. Sitton, B. G., and E. P. Akin. 1940. Grafting wax melter. *USDA Leaflet 202.*

40. Teuscher, H. 1962. Speeding production of hard-to-root conifers. *Amer. Nurs.* 116(7):16.

41. Thompson, L. A., and C. O. Hesse. 1950. Some factors which may affect the choice of grafting compounds for top-working trees. *Proc. Amer. Soc. Hort. Sci.* 56:213–16.

42. Upshall, W. H. 1946. The stub graft as a supple-ment to budding in nursery practice. *Proc. Amer. Soc. Hort. Sci.* 47:187–89.

43. Verhey, E. W. M. 1982. Minute nursery trees, a breakthrough for the tropics? *Chronica Hort.* 22(1):1–2.

44. Van de Pol, P. A., and A. Breukelaar. 1982. Stenting of roses: A method for quick propaga-tion by simultaneously cutting and grafting. *Scient. Hort.* 7:187–96.

45. Van de Pol, M. H., A. J. Joosten, and H. Keizer, 1986. Stenting of roses, starch depletion, and accumulation during the early development. *Acta Hort.* 189:51–59.

46. Vermeulen, J. P. 1984. Side veneer grafting. *Proc. Inter. Plant Prop. Soc.* 33:422–25.

SUPPLEMENTARY READING

BALTET, C. 1910. *The art of grafting and budding* (6th ed.). London: Crosby, Lockwood.

BANTA, E. S. 1967. Fruit tree propagation. *Ohio Agr. Ext. Bul. 481.*

CHANDLER, W. H. 1957. Chapter 13 in *Deciduous or-chards* (3rd ed.). Philadelphia: Lea & Febiger.

———. 1958. *Evergreen orchards* (2nd ed.). Philadelphia: Lea & Febiger.

GARNER, R. J. 1979. *The grafter's handbook* (4th ed.). New York: Oxford Univ. Press.

HARTMANN, H. T., and J. A. BEUTEL. 1979. Propa-gation of temperate zone fruit plants. *Calif. Agr. Exp. Sta. Leaflet 21103.*

INTERNATIONAL PLANT PROPAGATORS' SOCIETY. Proceedings of annual meetings.

MINISTRY OF AGRICULTURE, FISHERIES, AND FOOD. 1965. Grafting fruit trees. *Advisory Leaflet 326,* London.

SNYDER, J. C., and R. D. BARTRAM. 1965. Grafting fruit trees. *Pacific Northwest Ext. Publ.* (Idaho, Oregon, and Washington).

SPANGELO, L. P. S., R. WATKINS, and E. J. DAVIES. 1968. Fruit tree propagation. *Can. Dept. Agr. Publ. 1289,* Ottawa.

WAY, R. D., F. G. DENNIS, and R. M. GILMER. 1967. Propagating fruit trees in New York. *N.Y. Agr. Exp. Sta. Bul. 817.*

13

Techniques of Budding

In contrast to grafting, in which the scion consists of a short detached piece of stem tissue with several buds, budding utilizes only one bud and a small section of bark, with or without wood. Budding is often termed "bud grafting," since the physiological processes involved are the same as in grafting.

The commonly used budding methods depend upon the bark's "slipping." This term indicates the condition in which the bark (periderm, cortex, phloem, and cambium) can be easily separated from the wood (xylem). It denotes the period of year when the plant is in active growth, when the cambium cells are actively dividing, and newly formed tissues are easily torn as the bark is lifted from the wood. Beginning with new growth in the spring, this period should last until the plant ceases growth in the fall. However, adverse growing conditions, such as lack of water, insect or disease problems, defoliation, or low temperatures, may reduce growth and lead to a tightening of the bark and can seriously interfere with

the budding operation. Getting the stock plants in the proper condition for budding is an important consideration. Of the methods described here, only one—the chip bud—can be done when the bark is not slipping.

The budding operation, particularly T-budding, can be performed more rapidly than the simplest method of grafting. Some rose budders insert as many as 2000 to 3000 or more T-buds a day with the tying done by helpers. If performed under the proper conditions, the percentage of successful unions in T-budding is very high—90 to 100 percent. Budding is widely used in producing nursery stock of rose and fruit tree cultivars, where hundreds of thousands of individual plants are propagated each year. Therefore, for propagation operations involving large numbers of plants, where speed and low mortality are essential, budding upon selected rootstocks is the method likely to be chosen.

The use of budding is confined generally to young plants or the smaller branches of large plants where the buds can be inserted into shoots from 6 to 26 mm (¼ to 1 in.) in diameter. Topworking young trees by topbudding (frameworking) is quite successful. Here the buds are inserted in small, vigorously growing branches in the upper portion of the tree.

Budding may result in a stronger union, particularly during the first few years, than is obtained by some of the grafting methods, and thus the shoots are not as likely to blow out in strong winds. Budding makes more economical use of propagating wood than grafting, each bud potentially being capable of producing a new plant. This may be quite important if propagating wood is scarce. In addition, the techniques involved in budding are simple and can be performed easily by the amateur.

ROOTSTOCKS FOR BUDDING

In propagating nursery stock of the various fruit and ornamental species by budding, a rootstock plant is used. It should have the desired characteristics of vigor, growth habit, and resistance to soilborne pests, as well as being easily propagated. This rootstock plant may be a rooted cutting, a rooted layer, or, more commonly, a seedling (see Chap. 11). Usually, one year's growth in the nursery row before the budding is to be done is sufficient to produce a rootstock plant large enough to be budded, but seedlings of slow-growing species, such as pecan, and those grown under unfavorable conditions, may require two or three seasons.

To produce nursery trees free of harmful pathogens (such as viruses, fungi, or bacteria) it is essential that the rootstock plant, as well as the budwood, be free of such organisms.

TIME OF BUDDING— FALL, SPRING, OR JUNE

Most budding methods are used at seasons of the year when the stock plant is in active growth and the cambial cells are actively dividing so that the bark separates readily from the wood (although chip budding can be done when the bark is not slipping). It is also necessary that well-developed buds of the desired cultivar be available at the

same time. These conditions exist for most plant species at three different times during the year. In the northern hemisphere, these periods are late July to early September (**fall budding**), March and April (**spring budding**), and late May and early June (**June budding**). In the southern hemisphere, similar periods would be late January to early March (*fall budding*), September and October (*spring budding*), and late November and early December (*"June" budding*).

Fall Budding

Late summer and early fall is the most important time of budding in the propagation of fruit tree nursery stock. Most of the budding is done in late summer. The rootstock plants are usually large enough by late summer to accommodate the bud, and the plants are still actively growing, with the bark slipping easily. Once growth has stopped and the bark adheres tightly to the wood, T- or patch budding can no longer be done.

In fall budding, the budsticks, consisting of the current season's shoots, are obtained at or near the time of budding. They should be vigorous and should contain healthy vegetative, or leaf, buds (Fig. 13–1). Do not select shoots whose buds have been broken off. Short, slowly growing shoots on the outer portion of the tree should be avoided, because they may have chiefly flower buds rather than vegetative buds. Flower buds are usually round and plump, whereas leaf buds are smaller and pointed. Some species have mixed buds, the node containing both vegetative and flower buds. These are satisfactory for use in budding.

Every effort should be made to make sure that the trees from which the budsticks are obtained are free of any bacterial, fungus, or virus diseases. *Using infected budsticks can infect every budded nursery tree with the disease.*

As the budsticks are selected, the leaves should be removed immediately, leaving only a short piece of the leaf stalk or petiole attached to the bud; this will aid in handling the bud later on. The budsticks should be kept from drying by wrapping in some material such as clean, moist burlap and keeping them in a cool, shady location until they are needed. The budsticks should be used promptly after cutting, although they can be

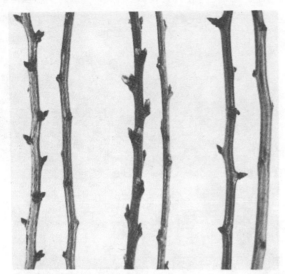

FIGURE 13-1 In budding it is important to use vegetative rather than flower buds. Vegetative buds are usually small and pointed, while flower buds are larger and more plump. Differences between vegetative and flower buds in three fruit species are illustrated here. *Left:* Almond. The shoot on the left has primarily flower buds and should not be selected for budding. The shoot on the right has vegetative buds, which are more suitable. *Center:* Peach. The shoot on the right has excellent vegetative buds while those on the left shoot are mostly flower buds. *Right:* Pear. All the buds on the shoot at the left are flower buds. Buds on the shoot at the right are good vegetative buds, suitable for budding.

Healing of the bud piece to the stock is greatly facilitated by the normal movement of water and nutrients up and down the stem of the rootstock. This would, of course, be stopped if the top of the rootstock were cut off above the bud.

If the budding operation is done properly, the bud piece should unite with the stock in two to three weeks, depending upon the growing conditions. If the leaf stalk or petiole drops off cleanly next to the bud, this is a good indication that the bud has united, especially if the bark piece retains its normal light brown or green color and the bud stays plump. On the other hand, if the leaf petiole does not drop off cleanly but adheres tightly and starts to shrivel and darken, while the bark piece also commences to turn black, it is likely that the operation has failed. If the bark of the rootstock is still slipping easily and budwood is still available, there may be time for the budding to be repeated.

Even though the bud union has healed, in most deciduous species the bud usually does not grow or "push out" in the fall, since it is in a physiological rest period or is inhibited by "apical dominance" (presence of other buds terminal to it). It remains just as it is until spring, at which time the chilling winter temperatures have overcome the rest influence and the bud is ready to grow. There are some exceptions to this; for example, in fall budding of maples, roses, honey locust, and certain other plants, some of the buds may start growth in the fall. In northern areas, if such fall-forced buds do not start early enough for the shoots to mature before cold weather starts, they are likely to be winter-killed.

In the spring, just before new growth begins, the rootstock is cut off immediately above the bud. It is desirable to make a sloping cut, slanting away from the bud. Although this cut may be waxed, it is usually not essential unless the stock is large in diameter. Cutting back the rootstock breaks apical dominance of the upper axillary buds, and forces the inserted bud into growth. In citrus budding, it is a common practice to cut the stock partially above the bud and to lop or bend it over away from the bud. The leaves of the stock plant still supply the roots with some nutrients, but the partial cutting forces the bud into growth. After the new shoot from the bud has started growth, the top is completely removed.

In colder northern regions, fall-inserted

stored for a short time if kept cool and moist. It is best, if possible, when a considerable amount of budding is being done, to collect the budsticks as they are being used, a day's supply at a time.

The best buds to use on the stick are usually those in the middle and basal portions. Buds on the succulent terminal portion of the shoot should be discarded. However, in certain species, such as the sweet cherry, buds on the basal portion of the shoots are flower buds which, of course, should not be used.

In fall budding, after the buds have been inserted and tied, there is nothing more to be done until the following spring. *Although eventually the rootstock is to be cut off above the bud, in no case should this be done immediately after the bud has been inserted.*

buds are sometimes covered with soil during the winter until danger of frost has passed and are then uncovered and topped-back in late spring.

Where strong winds occur, and in species in which the new shoots grow vigorously, support for the newly developing shoot may be necessary. One practice sometimes followed is to cut off the rootstock several inches above the bud, using this projecting stub as a support on which to tie the tender young shoot arising from the bud. This stub is removed after the shoot has become well established. In another procedure, stakes may be driven into the ground next to the stock to which the developing shoot is tied at intervals during its growth.

Cutting back to force the main bud to grow also forces many latent buds on the rootstock into growth. These must be rubbed off as soon as they appear, or they will soon choke out the inserted bud. It may be necessary to go over the budded plants several times before these "sprouts" stop appearing. Nursery people refer to this procedure as "suckering."

As shown in Fig. 13–2, the shoot arising from the inserted bud becomes the top portion of the plant. After one season's growth in the nursery with favorable conditions of soil, water and nutrients, temperature, and insect and disease control, this shoot will have developed sufficiently to enable the plant to be dug and moved to its permanent location during the following dormant season. Such a tree would have a one-year-old top and a two- or perhaps three-year-old root, but it is still considered a "yearling" tree. If the top makes insufficient growth the first year, it can be allowed to grow a second year, and is then known as a two-year-old tree. However, abnormal, slow-growing, stunted trees should be discarded.

Spring Budding

Spring budding is similar to fall budding except that insertion of the bud takes place the following spring as soon as active growth of the rootstock begins and the bark separates easily from the wood. The period for successful spring budding is limited, and budding should be completed before the rootstocks have made much new growth.

Budsticks are chosen from the same type of shoots—in regard to vigor of growth and type of

FIGURE 13–2   Row of fall-budded nursery trees about one year after budding. The inserted buds have grown through the spring and summer. The roots have grown through two seasons. The tops of the rootstocks were cut off above the inserted buds the spring following budding.

buds (Fig. 13–1)—that would have been used in fall budding, except that they are not collected until the dormant season the following winter. The leaves would, of course, have fallen by this time, and the buds would have experienced sufficient chilling to overcome their rest period. Budwood must be collected while it is still dormant—before there is any evidence of the buds swelling. Since the buds must be dormant when they are inserted, and since the rootstocks must be in active growth, it is necessary in spring budding that the budsticks be gathered some time in advance of the time of budding and stored at about −2 to 0°C (29 to 32°F) to hold the buds dormant. The budsticks should be wrapped in bundles with damp peat moss, moistened newspapers, or similar material to prevent drying out.

In spring budding, the actual budding operation should be done just as soon as the bark on the rootstock slips easily. Then, about two weeks after budding, when the bud unions have healed, the top of the stock must be cut off above the bud to force the inserted bud into active growth. At the same time, latent buds on the rootstock begin to grow and should be removed. Sometimes it is helpful to permit such shoots from the rootstock to develop to some extent to prevent sunburn and help nourish the plant. They must be held in check, however, and eventually removed.

Although the new shoot from the inserted bud gets a later start in spring budding than in fall budding, spring buds will usually develop rapidly enough, if growing conditions are favorable, to make a satisfactory top by fall. Fall budding, however, is to be preferred for several reasons: the higher temperatures at that time promote more certain healing of the union, the budding season is longer, there is no necessity to store the budsticks, the inserted buds start growth earlier in the spring, and the pressure of other work is usually not so great for propagators in the late summer as in the spring. Spring budding is used sometimes on rootstocks that were fall-budded but on which the buds failed to take.

June Budding

June budding is used to obtain a "one-year-old" budded tree in a single growing season. Both roots and tops of the budded tree develop during only one growing season. Budding is done in the early part of the growing season and the inserted bud forced into growth immediately. As a method of nursery propagation, June budding is confined to regions that have a relatively long growing season—in the United States this region includes California and some of the southern states. In Texas, June budding is not successful due to the excessive heat during this time. In the propagation of fruit trees, June budding is used mostly in producing such stone fruits as peaches, nectarines, apricots, almonds, and plums. Peach seedlings are generally used as the rootstock, but almond seedlings and plum hardwood cuttings also can be used. Budding is done by the T-bud method. If seeds are planted in the fall, or stratified seeds as early as possible in the spring, the seedlings usually attain sufficient size [30 cm (12 in.) high and at least 3 mm (⅛ in.) in diameter] to be budded by mid-May or early June (in the northern hemisphere). Preferably, June budding should not be done much after late June, or a nursery tree of satisfactory size will not be obtained

by fall. June-budded trees are not as large by the end of the growing season as those propagated by fall or spring budding, but they are of sufficient size—10 to 16 mm (⅜ to ⅝ in.) caliper and 90 to 150 cm (3 to 5 ft) tall—to produce entirely satisfactory trees (*11*).

Budwood used in June budding consists of current season's growth, that is, of new shoots which have developed since growth started in the spring. By late May or early June, these shoots will usually have grown sufficiently to have a well-developed bud in the axil of each leaf. At this time of year these buds will not have entered the rest period, so when they are used in budding they continue their growth on through the summer, producing the top portion of the budded seedling.

For June-budded trees, handling subsequent to the actual operation of budding is somewhat more exacting than for fall- or spring-budded trees. The rootstocks are smaller and have less stored food than those used in fall or spring budding. The object behind the following procedures—shown in Fig. 13–3—is to keep the rootstock (and later the budded top) actively and continuously growing so as to allow no check in growth, while at the same time changing the seedling shoot to a budded top. The bud should be inserted high enough (about 14 cm; 5½ in.) on the stem so that a number of leaves—at least three or four—can be retained below the bud. The method of T-budding with the "wood out" should be used. Healing of the inserted bud should be very rapid at this time of year, since temperatures are relatively high, and rapidly growing, succulent plant parts are used. By four days after budding, healing should have started, and the top of the rootstock can be cut back somewhat—about 9 cm (3½ in.) above the bud—leaving at least one leaf above the bud and several below it. This operation will force the inserted bud into growth and will check terminal growth of the rootstock. It will also stimulate shoot growth from basal buds of the rootstock, which will produce additional leaf area. This continuous leaf area is necessary so that there always will be enough leaves to keep manufacturing food for the small plant. Ten days to two weeks after budding, the rootstock can be cut back to the bud, which should be starting to grow. If the budding rubber has not broken, it should be cut at this time.

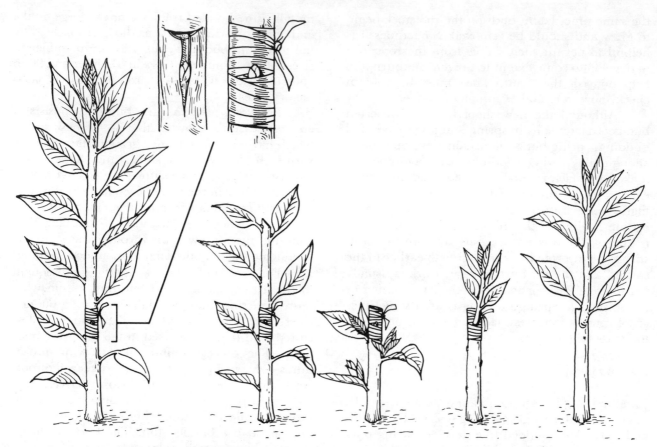

FIGURE 13–3 It is important in June budding that the stock be cut back to the bud properly. *Far left:* The bud is inserted high enough on the stock so that there are several leaves below the bud. *Center left:* Three or four days after budding, the stock is partially cut back about 9 cm (3½ in.) above the bud. *Center:* Ten days to two weeks after budding, the stock is completely removed just above the bud. *Center right:* This forces the bud and other buds on the stock into growth; the latter must subsequently be removed. *Far right:* Appearance of the budded tree after the new shoot has made considerable growth.

Other shoots arising from the rootstock should be headed back to retard their growth. After the inserted bud grows and develops a substantial leaf area, it can supply the plant with the necessary nutrients. By the time the shoot from the inserted bud has grown about 25 cm (10 in.) high, it should have enough leaves so that all other shoots and leaves can be removed. Later inspections should be made to remove any shoots arising from the rootstock below the budded shoot. The steps in fall, spring, and June budding are compared in Fig. 13–4.

METHODS OF BUDDING

T-Budding (Shield Budding)

This method of budding is known by both names, the "T-bud" designation arising from the T-like appearance of the cut in the stock, whereas the "shield bud" name is derived from the shield-like appearance of the bud piece when it is ready for insertion in the stock.

T-budding is widely used by nurserymen in propagating nursery stock of most fruit tree spe-

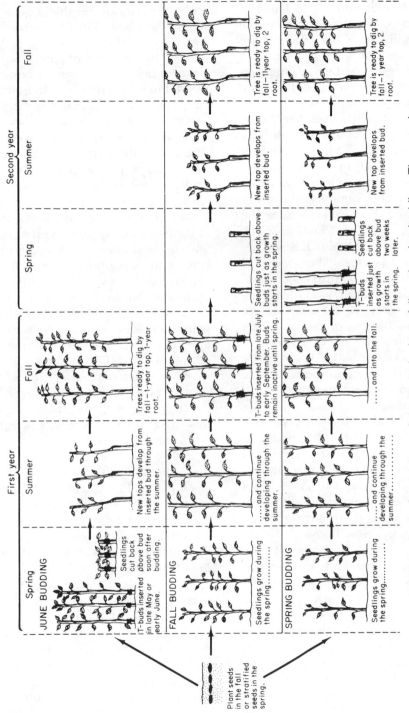

FIGURE 13-4 Comparison of the steps in June, fall, and spring budding. The actual techniques in budding are not difficult, but it is very important that the various operations be done at the proper time.

cies, many ornamental trees, roses, and some ornamental shrubs. Its use is generally limited to stocks that are about 6 to 25 mm (¼ to 1 in.) in diameter, and are actively growing so that the bark will separate readily from the wood. If the bark is so tight on the wood that it has to be pried loose forcibly, the chances of the bud healing successfully are poor. The operation should then be delayed until the bark is slipping easily.

The bud is inserted into the stock 5 to 25 cm (2 to 10 in.) above the soil level in a smooth bark surface. There are different opinions as to which is the proper side of the stock in which to insert the bud. If extreme weather conditions are likely to occur during the critical healing period just following budding, it may be desirable to place the bud on the side of the stock on which as much protection as possible may be obtained. Some believe that if the bud is placed on the windward side, there is less chance of the young shoot breaking off. Otherwise, it probably makes little difference where the bud is inserted, the convenience of the operator and the location of the smoothest bark being the controlling factors. When rows of closely planted rootstocks are budded, it is more convenient to have all the buds on the same side for later inspection and manipulations.

The cuts to be made in the stock plant are illustrated in Fig. 13–5. There are various modifications of this technique; most budders prefer to make the vertical cut first, then the horizontal crosscut at the top of the T. As the horizontal cut is made, the knife is given a twist to throw open the flaps of bark for insertion of the bud. It is important that neither the vertical nor horizontal cut be made longer than necessary, because this requires additional tying later to close the cuts.

After the proper cuts are made in the stock and the incision is ready to receive the bud, the shield piece is cut out of the budstick.

To remove the shield of bark containing the bud, a slicing cut is started at a point on the stem about 13 mm (½ in.) below the bud, continuing under and about 25 mm (1 in.) above the bud. The shield piece should be as thin as possible but still thick enough to have some rigidity. A second horizontal cut is then made 13 to 19 mm (½ to ¾ in.) above the bud, thus permitting the removal of the shield piece.

There are two methods of preparing the shield—with the "wood in" or with the "wood out." This refers to the little sliver of wood just under the bark of the shield piece and which will remain attached to it if the second, or horizontal, cut is deep and goes through the bark and wood, joining the first slicing cut. Some professional budders believe it is best to remove this sliver of wood, but others retain it. In budding certain species, however, such as maples and walnuts, much better success is usually obtained with "de-wooded" buds. If it is desired to prepare the shield with the wood out, the second horizontal cut should be just deep enough to go through the bark and not through the wood. Then if the bark is slipping easily, the bark shield can be snapped loose from the wood (which still remains attached to the budstick) by pressing it against the budstick and sliding it sideways. A small core of wood comprising the vascular tissues supplying the bud is present, and this should remain in the bud, rather than adhering to the wood and leaving a hole in the bud. If the shield is pulled outward rather than being slid sideways from the wood, this core usually pulls out of the bud, eliminating the chances of success. In June budding of fruit trees, the shield piece is usually prepared with the wood out. In most other instances, however, the wood is left in. In spring budding, using dormant budwood, this sliver of wood is tightly attached to the bark and cannot be removed.

The next step is the insertion of the shield piece containing the bud into the incision in the stock plant. The shield is pushed under the two raised flaps of bark until its upper, or horizontal, cut matches the same cut on the stock. The shield should fit snugly in place, well covered by the two flaps of bark, but with the bud itself exposed (see Fig. 11–11).

No waxing is necessary, but the bud union must be wrapped, using tape, budding rubbers, or raffia to hold the two components firmly together until healing is completed. Rubber budding strips, especially made for wrapping, are widely used for this purpose (Fig. 13–6). Their elasticity provides sufficient pressure to hold the bud securely in place. The rubber, being exposed to the sun and air, usually deteriorates, breaks, and drops off after several weeks, at which time the bud should be healed in place. If the budding rubber is covered with soil, the rate of deteriora-

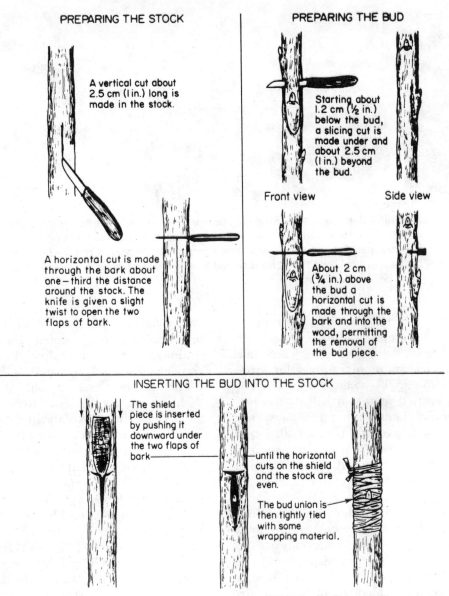

PREPARING THE STOCK

A vertical cut about 2.5 cm (1 in.) long is made in the stock.

A horizontal cut is made through the bark about one-third the distance around the stock. The knife is given a slight twist to open the two flaps of bark.

PREPARING THE BUD

Starting about 1.2 cm (½ in.) below the bud, a slicing cut is made under and about 2.5 cm (1 in.) beyond the bud.

Front view   Side view

About 2 cm (¾ in.) above the bud a horizontal cut is made through the bark and into the wood, permitting the removal of the bud piece.

INSERTING THE BUD INTO THE STOCK

The shield piece is inserted by pushing it downward under the two flaps of bark

until the horizontal cuts on the shield and the stock are even.

The bud union is then tightly tied with some wrapping material.

FIGURE 13-5 Basic steps in making the T-bud (shield bud).

tion will be much slower. This material has the advantage of eliminating cutting the wrapping ties, which can be a costly operation if many thousands of plants have been budded. The rubber will expand as the rootstock grows, and thus there is little danger of constriction.

In tying the bud, the ends of the budding rubbers are held in place by inserting them under the adjacent turn. The bud itself should not be covered. The amount of tension given the budding rubber is quite important. It should not be too loose, or there will be too little pressure holding the bud in place. On the other hand, if the rubber is stretched extremely tight, it may be so thin that it will deteriorate rapidly and break too soon—before the bud union has taken place. Often the tying is done from the top down to avoid forcing the bud out through the horizontal cut. Proper wrapping of the buds is very important.

Parafilm tape, which is a waterproof, flexible, stretchable, thermoplastic film with a paper

FIGURE 13–6   Steps in the development of a T-bud. *Left:* Bud after being inserted and wrapped. *Center:* Bud has healed in place, the budding rubber has dropped off, and the stock has been cut back above the bud. *Right:* Shoot development from the inserted bud. All buds arising from the stock have been rubbed off.

backing, has been used successfully (*4*) for covering the bud in bench chip budding of roses.

Raffia (fiber-like leaf segments of certain *Raphia* species) is used in some countries for wrapping buds. This material is soaked in water overnight before it is used so that it will be flexible. Raffia must be cut later—about ten days after budding—to prevent constriction at the bud union as the plant grows. If such nonelastic ties are not promptly cut, the resultant constriction can have a very adverse affect on subsequent growth of the bud (*14*).

Plastic ties of polyvinyl chloride (PVC) film, 10 mm (⅜ in.) wide, are quite useful for budding. Such material is moisture-proof and elastic, and since it is transparent, it permits inspection of the buds after covering (*1*). It is normally removed or cut after the bud has healed in, to prevent girdling.

Inverted T-Budding

In rainy localities, water running down the stem of the rootstock may enter the T-cut, soak under the bark, and prevent the shield piece from healing into place. Under such conditions an inverted T-bud may give better results, since it is more likely to shed excess water. In citrus budding, the inverted T method is widely used, even though the conventional method also gives good results.

In species that bleed badly during budding, such as chestnuts, the inverted T-bud allows better drainage and better healing. Proponents of both conventional and inverted T-budding can be found, and in a given locality the usage of either with a given species tends to become traditional.

The techniques of the inverted T-bud method are the same as those already described, except that the incision in the stock has the transverse cut at the bottom rather than at the top of the vertical cut, and in removing the shield piece from the budstick the knife starts above the bud and cuts downward below it. The shield is removed by making the transverse cut 13 to 19 mm (½ to ¾ in.) below the bud. The shield piece containing the bud is inserted into the lower part of the incision and pushed upward until the transverse cut of the shield meets that made in the stock.

It is important in using this inverted T-bud method that a normally oriented shield bud piece should not be inserted into an inverted incision in the stock. The bud would then have a reversed polarity. Although such upside down buds do live and grow, at least in some species, their use may not result in the expected shoot development.

Patch Budding

The distinguishing feature of patch budding and related methods is that a rectangular patch of bark is removed completely from the stock and replaced with a patch of bark of the same size containing a bud of the cultivar to be propagated.

Patch budding is somewhat slower and more difficult to perform than T-budding, but it is widely and successfully used on thick-barked species, such as walnuts and pecans, in which T-budding sometimes gives poor results, presumably owing to the poor fit around the margins of the bud. Patch budding, or one of its modifications, is also extensively used in propagating various tropical species, such as the rubber tree (*Hevea brasiliensis*).

Patch budding requires that the bark of both the stock and budstick be slipping easily. It is usually done in late summer or early fall, but can be done in the spring also. In propagating nursery stock, the diameter of the rootstock and the budstick should preferably be about the same, about 13 to 26 mm (½ to 1 in.). Although the budstick should not be much larger than about 26 mm (1 in.) in diameter, the patch can be inserted successfully into stocks as large as 10 cm (4 in.) in diameter, although healing of such large stubs may be inadequate (*15*).

Special knives (see Fig. 13–7) have been devised to remove the bark pieces from the stock and the budstick. Some type of double-bladed knife that will make two transverse parallel cuts 25 to 35 mm (1 to 1⅜ in.) apart is necessary. These cuts, about 25 mm (1 in.) in length, are made through the bark to the wood in a smooth area of

the rootstock about 10 cm (4 in.) above the ground. Then the two transverse cuts are connected at each side by vertical cuts made with a single-bladed knife.

The patch of bark containing the bud is cut from the budstick in the same manner in which the bark patch is removed from the stock. Using the same two-bladed knife, the budder makes two transverse cuts through the bark, one above and one below the bud. Then two vertical cuts are made on each side of the bud so that the bark piece will be about 25 mm (1 in.) wide. The bark piece containing the bud is now ready to be removed. It is important that it be slid off sideways rather than being lifted or pulled off. There is a small core of wood, the *bud trace*, which must remain inside the bud if a successful "take" is to be obtained. By sliding the bark patch to one side, this core is broken off, and it stays in the bud. If the bud patch is lifted off, this core of wood is likely to remain attached to the wood of the budstick, leaving a hole in the bud, as shown in Fig. 13–8.

After the bud patch is removed from the budstick, it must be inserted immediately on the stock, which should already be prepared, needing only to have the bark piece removed. The patch from the budstick should fit snugly at the top and bottom into the opening in the stock, since both transverse cuts were made with the same knife. It is more important that the bark piece fit tightly at top and bottom than that it fit along the sides.

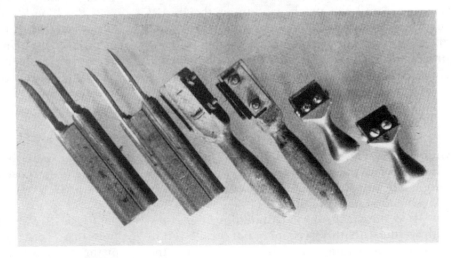

FIGURE 13–7 Tools used in patch budding and related methods. *Left:* Manufactured double-bladed knives. *Center:* Double-bladed knives made with two razor blades. *Right:* Manufactured patch-budding tools consisting of four rectangular cutting blades.

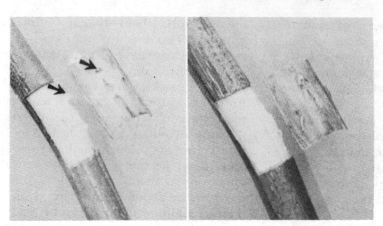

FIGURE 13–8 Removing the bud patch from the budstick in patch budding. *Left:* Incorrect. The core of wood in the bud, comprising the vascular tissues, has broken off, leaving a hole in the bud. Such a bud is not likely to grow. *Right:* Correct. The patch was pushed sideways, and the core of wood has remained inside the bud.

These procedures are illustrated in Figs. 13–9 to 13–11.

The inserted patch is now ready to be wrapped. Often the bark of the stock will be thicker than the bark of the inserted bud patch so that, upon wrapping, it is impossible for the wrapping material to hold the bud patch tightly against the stock. In this case it is necessary that the bark of the stock be pared down around the bud patch so that it will be of the same thickness, or preferably slightly thinner, than the bark of the bud patch. Then the wrapping material will hold the bud patch tightly in place.

In cases in which difficulty is experienced in obtaining successful unions with the patch bud method, it may help to make the four cuts of the rectangle in the stock one to three weeks ahead of the time when the actual budding is to be done. This bark patch is not removed from the stock, however, until the patch containing the bud is ready to be inserted. When the cuts are made ahead of time, the wounding causes the callusing process to start, so that when the new bark patch is inserted, it heals very rapidly.

In wrapping the patch bud, a material should be used that not only will hold the bark tightly in place but will cover all the cut surfaces to prevent the entrance of air under the patch, with subsequent drying and death of the tissues. The bud itself must not be covered during wrapping. The most satisfactory material is nursery adhesive tape or PVC film.

It is important in patch budding, especially with walnuts, that the wrapping not be allowed to cause a constriction at the bud union. When the

stock is growing rapidly, it is necessary to cut the tape about ten days after budding. A single vertical knife cut on the side opposite the bud is sufficient, but care should be taken not to cut into the bark. Sometimes an extra strip of tape is placed vertically before wrapping to protect the bark when the tape is cut. The cut tape should not be pulled off.

In California and other areas with hot summers patch budding is best performed in late summer when both the seedling stock and the source of budwood are growing rapidly and their bark slipping easily. The budsticks for patch budding done at this time should have the leaf blades cut off two to three weeks before the budsticks are taken from the tree. The petiole or leaf stalk is left attached to the base of the bud, but by the time the budstick is removed, this petiole has dropped off or is easily pulled off.

Patch budding can also be done in the spring after new growth has appeared on the stocks and it has been determined that the bark is slipping. There is a problem, however, in obtaining satisfactory buds to use at this time of year, since it is necessary that the bark of the budstick separate readily from the wood. At the same time, the buds should not be starting to swell. There are two methods by which satisfactory buds can be obtained for patch budding in the spring.

In one, as used in Texas for pecans, the budsticks are selected during the dormant winter period and stored at low temperatures (about 2°C; 36°F) and wrapped in moist sphagnum or peat moss to prevent their drying out. Then, about two or three weeks before the spring budding is to be

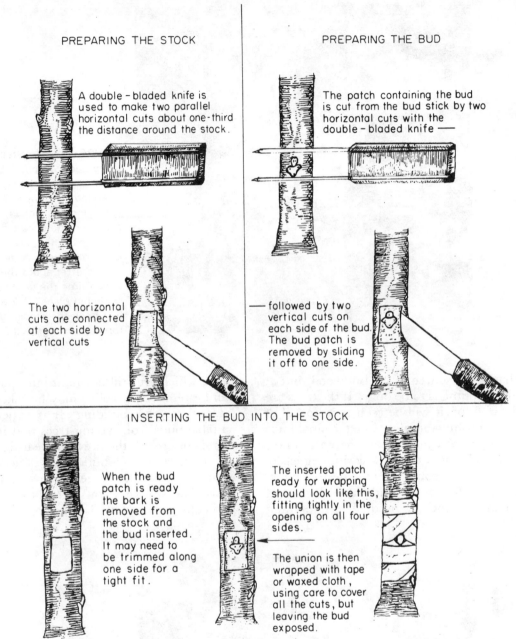

PREPARING THE STOCK

A double-bladed knife is used to make two parallel horizontal cuts about one-third the distance around the stock.

The two horizontal cuts are connected at each side by vertical cuts

PREPARING THE BUD

The patch containing the bud is cut from the bud stick by two horizontal cuts with the double-bladed knife —

— followed by two vertical cuts on each side of the bud. The bud patch is removed by sliding it off to one side.

INSERTING THE BUD INTO THE STOCK

When the bud patch is ready the bark is removed from the stock and the bud inserted. It may need to be trimmed along one side for a tight fit.

The inserted patch ready for wrapping should look like this, fitting tightly in the opening on all four sides.

The union is then wrapped with tape or waxed cloth, using care to cover all the cuts, but leaving the bud exposed.

FIGURE 13-9 Steps in making the patch bud. This method is widely used for propagating thick-barked plants.

done, they are brought into a warm room. The budsticks may be left in the damp peat moss or set with their bases in a container of water. The increased temperature will cause the cambium layer to become active, and soon the bark will slip sufficiently for the buds to be used. Although a few of the more terminal buds on each stick may start swelling in this time and cannot be used, there should be a number of buds in a satisfactory condition.

The second method of obtaining buds for spring patch budding is to take them directly from

FIGURE 13–10   Steps in the development of a patch bud. *Top left:* Patch bud after being inserted and wrapped with tape. *Top right:* After about 10 days, the tape is slit along the back side to release any constricting pressure. *Below left:* The tape is completely removed after about three weeks, at which time the patch should be healed in place. *Below right:* Cutting back the stock above the bud forces it into growth.

the tree that is the source of the budwood, but at the time the budding is to be done. If the trees are inspected carefully, it will be seen that not all of the buds start pushing at once. The terminal buds are usually more advanced than the basal ones. There is a period when the bark is slipping easily throughout the shoot containing the desired buds, but when only a few of the buds have developed so far that they cannot be used. The remaining buds, which are still dormant but upon bark that can be removed readily, may be taken and used immediately for budding. It is easier to obtain suitable buds from young trees that made vigorous shoot growth the previous year than it is from old trees. When the budding is done will be governed then by the stage of development of the buds. This will vary considerably with the species and cultivar being used. Stage of development of

FIGURE 13–11   *Left:* Shoot development from patch budding operation shown in Fig. 13–10 after two years' growth. *Right:* Strong, smooth union developed after 14 years. *Juglans regia* 'Hartley' on Paradox rootstock (*Juglans regia* × *J. hindsii*).

the rootstock is not as critical. Its bark must be slipping well, and the budding should be done before the stock plant has made much new growth.

I-Budding

In I-budding the bud patch is cut just as for patch budding, that is, in the form of a rectangle or square. Then, with the same parallel-bladed knife, two transverse cuts are made through the bark of the stock. These are joined at their centers by a single vertical cut to produce the figure I. The two flaps of bark can then be raised for insertion of the bud patch beneath them. It may make a better fit to slant the side edges of the bud patch. In tying the I-bud, care should be taken to see that the bud patch does not buckle upward and fail to touch the stock (see Fig. 13–12).

I-budding should be considered for use when the bark of the stock is much thicker than that of the budstick. In such cases, if the patch bud were used, considerable paring down of the bark of the

stock around the patch would be necessary. This operation is not necessary in the I-bud method (Fig. 13–12).

Flute Budding and Ring Budding

See details in Fig. 13–12.

Chip Budding

Chip budding, as in all budding, is actually a form of grafting where the scion is reduced to a small size, containing only one bud. It seems to be more useful in areas with cool, short growing seasons than in places that have long, hot growing seasons, where T-budding has been quite satisfactory.

Chip budding can be used at times when the bark is not slipping, that is, early in the spring before growth starts or during the summer when active growth has stopped prematurely owing to lack of water or some other cause. Chip budding

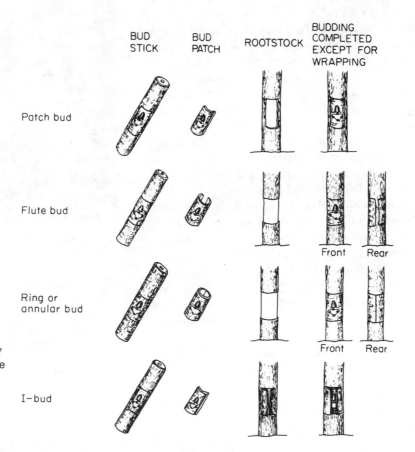

FIGURE 13–12 There are many variations of the patch bud, some of which are shown here. The naming of these types is somewhat confused; the most generally accepted names are given here.

is generally used with small material, 13 to 25 mm (½ to 1 in.) in diameter. Chip budding ordinarily is not as fast or as simple as T-budding. For many years, however, chip budding in the fall has given excellent results in budding grape cultivars on phylloxera or nematode-resistant rootstocks (6, 10). Chip budding is also used on a large scale on fruit and ornamental trees in the United States (9), Canada (12), and Great Britain (7, 8), where more vigorous initial growth has been reported with chip buds than with T-buds (Figure 13–13). Studies (3, 4) have shown that chip budding is also quite satisfactory for propagating field roses by indoor bench grafting in winter on dormant unrooted rootstock cuttings. Chip budding has been used, too, for propagating magnolias either onto container-grown rootstocks in the greenhouse in winter, or as outdoor field budding in the spring or summer (16).

As illustrated in Fig. 13–14, a chip of bark is removed from a smooth place between nodes near the base of the stock and replaced by another chip of the same size and shape from the budstick which contains a bud of the desired cultivar. The chips in both stock and budstick are cut out in the same manner. In the budstick the first cut is made just below the bud and down into the wood at an angle of 30 to 45 degrees. The second cut is started about 25 mm (1 in.) above the bud and goes inward and downward behind the bud until it intersects the first cut. (The order of making these two cuts may be reversed.) The chip is removed from the stock and replaced by the one from the budstick. To obtain a good fit, both should be cut to the same size and shape. The cambium layer of the bud piece must be placed to coincide with that of the stock, preferably on both sides of the stem, but at least on one side. Studies (13) in England have shown that a better union is obtained with chip budding than with T-budding (see p. 319).

In chip budding there are no protective flaps of bark to prevent the bud piece from drying out, as there are in T-budding. It is very important, then, that the chip bud be wrapped to seal the cut edges as well as to hold the bud piece tightly into the stock. Nursery adhesive tape works well for this purpose, although white or transparent plastic tape is more often used, covering the bud. After the bud has been inserted, it must be

FIGURE 13–13 *Above:* A block of clonal apple rootstocks that have been budded to the desired cultivar by chip budding in England. Arrows point to insertion of bud. *Below:* Close-up of two stems in each of which two chip buds have been inserted. Top buds, in both cases, have been covered by transparent plastic tape. Lower buds have been covered by opaque white tape; buds are completely covered by tape, which will be removed after healing takes place.

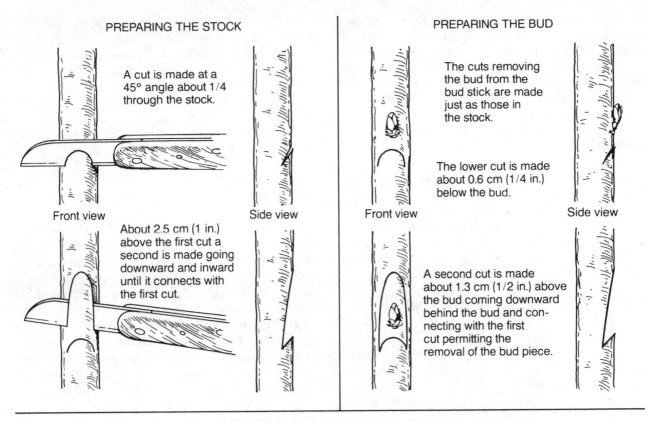

PREPARING THE STOCK

A cut is made at a 45° angle about 1/4 through the stock.

Front view   Side view

About 2.5 cm (1 in.) above the first cut a second is made going downward and inward until it connects with the first cut.

PREPARING THE BUD

The cuts removing the bud from the bud stick are made just as those in the stock.

The lower cut is made about 0.6 cm (1/4 in.) below the bud.

Front view   Side view

A second cut is made about 1.3 cm (1/2 in.) above the bud coming downward behind the bud and connecting with the first cut permitting the removal of the bud piece.

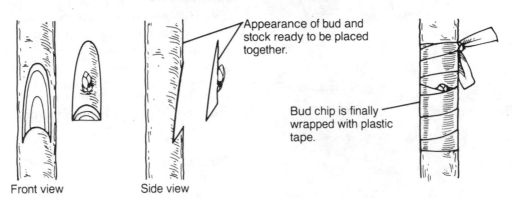

INSERTING THE BUD INTO THE STOCK

Appearance of bud and stock ready to be placed together.

Bud chip is finally wrapped with plastic tape.

Front view   Side view

FIGURE 13–14 Chip budding has been widely used in propagating vinifera grapes on nematode- and phylloxera-resistant rootstocks. It is also used successfully for propagating fruit trees and woody ornamentals, with the bud piece cut as shown here and covered completely with wide plastic tape.

wrapped immediately so that it will not dry out (*1*). When the bud starts growth, the tape is cut.

In grape propagation, if the bud is inserted into the stock close to the ground level, drying can be prevented by wrapping the bud with budding rubber and covering the whole bud union imme-

diately with several inches of finely pulverized moist soil, which can be removed after the bud has united. Since budding rubbers are slow to disintegrate in the soil, they should be cut or removed before constriction occurs.

In chip budding, as in the other methods,

the stock is not cut back above the bud until the union is complete. If the chip bud is inserted in the fall, the stock is cut back just as growth starts the next spring. If the budding is done in the spring, the stock is cut back about ten days after the bud has been inserted.

TOPBUDDING

In young trees with an ample supply of vigorous shoots at a height of 1.2 to 1.8 m (4 to 6 ft), top-budding is a fast and certain method of topwork-ing (Fig. 13–15). It can be used in older trees, too, if they are cut back rather severely the year before to provide a quantity of vigorous watersprout shoots fairly close to the ground.

FIGURE 13–15   Topworking a young tree by topbudding. T-buds were inserted in the positions shown by arrows and have grown for one season. All other shoots have been removed. From L. H. Day, ''Apple, Quince, and Pear Rootstocks in California,'' *Calif. Agr. Exp. Sta. Bul. 700.*

Depending upon the size of the tree, 10 to 15 buds are placed in vigorously growing branches 6 to 19 mm (¼ to ¾ in.) in diameter in the upper portion of the tree—about shoulder height. Al-though a number of buds could be placed in a sin-gle branch, usually only one will be saved to de-velop into secondary branches, which will then form the permanent new top of the tree. The T-bud method is used on thin-barked species, and the patch bud on those with thick bark.

Topbudding is usually done in midsummer, as soon as well-matured budwood can be obtained and while the stock tree is still in active growth with the bark slipping easily. Orchard trees gener-ally stop growth earlier in the season than young nursery trees; therefore the budding must be done earlier. When topbudding is done at this time of year, the buds usually remain inactive until the following spring. At that time, just as vegetative growth starts, the stock branches are cut back just above the buds. This forces the buds into active growth, and they should develop into good-sized branches by the end of the summer. At the time the shoots are cut back to the buds, all other un-budded branches should be removed at the trunk. It is important that the trees be inspected carefully through the summer and all shoots removed that are arising from any but the inserted buds.

Topbudding can also be done in the spring just as the tree to be topworked is starting active growth and the bark is slipping easily. The tech-niques for spring budding of nursery stock are used. Although not commonly done, June bud-ding could be used for topbudding.

DOUBLE-WORKING BY BUDDING

In propagating nursery trees, some budding methods can be used in developing double-worked trees. The intermediate stock (interstock) can be budded on the rootstocks; then the follow-ing year the desired cultivar is budded on the in-terstock. Although quite effective, this is a rather lengthy process, taking three years. It is possible to develop a double-worked tree in one operation in one year by the double-shield bud method (*2, 5, 7*). A T-bud is used, but just under and below it a thin budless shield piece of the desired inter-stock is inserted.

MICROBUDDING

This type of budding is used successfully in propagating citrus trees and probably could be utilized also for other tree and shrub species. It has been of commercial importance in the citrus districts of southeastern Australia (*17*). As described here, it is not done under aseptic conditions. Microbudding is similar to ordinary T-budding, except that the bud piece is reduced to a very small size. The leaf petiole is cut off just above the bud, and then the bud is removed from the budstick by a flat cut just underneath the bud, with a razor-sharp knife. Only the bud itself and a small piece of wood under it are used. In the stock an inverted T-cut is made, and the micro-bud is slipped into this, right side up. The entire T-cut, including the bud, is covered with thin plastic budding tape. The tape is allowed to remain for 10 to 14 days for spring budding and three weeks for fall budding, after which it is removed by cutting with a knife. By this time the buds should have healed in place; subsequent handling is the same as for conventional T-budding.

REFERENCES

1. Bremer, A. H. 1977. Chip budding on a commercial scale. *Proc. Inter. Plant Prop. Soc.* 27:366–67.

2. Bryden, J. D. 1957. Use of plastic ties in fruit tree budding. *Agr. Gaz. New S. Wales (Austral.)* 68:87–88.

3. Davies, F. T., Jr., Y. Fann, and J. E. Lazante. 1980. Bench chip budding of field roses. *HortScience.* 15(6):817–18.

4. Fann, Y., F. T. Davies, Jr., and D. R. Paterson. 1983. Correlative effects of bench chip-budded 'Mirandy' roses. *Jour. Amer. Soc. Hort. Sci.* 108(2):180–83.

5. Garner, R. J. 1953. Double-working pears at budding time. *Ann. Rpt. E. Malling Res. Sta. for 1952,* pp. 174–75.

6. Harmon, F. N., and J. H. Weinberger. 1969. The chip-bud method of propagating vinifera grape varieties on rootstocks. *USDA Leaflet 513.*

7. Howard, B. H. 1977. Chip budding fruit and ornamental trees. *Proc. Inter. Plant Prop. Soc.* 27:357–64.

8. Howard, B. H., D. S. Skene, and J. S. Coles. 1974. The effect of different grafting methods upon the development of one-year-old nursery apple trees. *Jour. Hort. Sci.* 49:287–95.

9. Kidd, E. L., Jr., 1987. Asexual propagation of fruit and nut trees at Stark Brothers Nurseries. *Proc. Inter. Plant Prop. Soc.* 36:427–30.

10. Lider, L. A. 1963. Field budding and the care of the budded grapevine. *Calif. Agr. Ext. Ser. Leaflet 153.*

11. Mertz, W. 1964. Deciduous June-bud fruit trees. *Proc. Inter. Plant Prop. Soc.* 14:255–59.

12. Osborne, R. H. 1987. Chip budding techniques in the nursery. *Proc. Inter. Plant Prop. Soc.* 36:550–55.

13. Skene, D. S., H. R. Shepard, and B. H. Howard. 1983. Characteristic anatomy of union formation in T- and chip-budded fruit and ornamental trees. *Jour. Hort. Sci.* 58(3):295–99.

14. Smith, N. G., R. J. Garner, and W. S. Rogers. 1962. Delayed growth of apple scions in relation to early budding, bud constriction, and some other factors. *Ann. Rpt. E. Malling Res. Sta. for 1961,* pp. 51–56.

15. Taylor, R. M. 1972. Influence of gibberellic acid on early patch budding of pecan seedlings. *Jour. Amer. Soc. Hort. Sci.* 97(5):677–79.

16. Tubesing, C. E. 1988. Chip budding of magnolias. *Proc. Inter. Plant Prop. Soc.* 87:377–79.

17. Wishart, R. D. A. 1961. Microbudding of citrus. *S. Austral. Dept. Agr. Leaflet 3660.*

SUPPLEMENTARY READING

BALTET, C. 1910. *The art of grafting and budding* (6th ed.). London: Crosby, Lockwood.

BRASE, K. D. 1952. Propagation of fruit trees by budding. *Farm Res. N.Y. Agr. Exp. Sta.* 18:(3).

CARLSON, R. F., and A. E. MITCHELL. 1971. Budding and grafting fruit trees. *Mich. Agr. Ext. Bul. 508.*

GARNER, R. J. 1979. *The grafter's handbook* (4th ed.). New York: Oxford Univ. Press.

HARTMANN, H. T., and J. A. BEUTEL. 1979. Propagation of temperate zone fruit plants. *Calif. Agr. Exp. Sta. Leaflet 21103.*

INTERNATIONAL PLANT PROPAGATORS' SOCIETY. Proceedings of annual meetings.

PLATT, R. G., and K. W. OPITZ. 1973. Propagation of citrus. In *The citrus industry,* Vol. 3, W. Reuther, ed. Berkeley: Univ. of Calif. Press.

14
Layering and Its Natural Modifications

Layering is a propagation method by which adventitious roots are caused to form on a stem while it is still attached to the parent plant. The rooted, or layered, stem is detached to become a new plant growing on its own roots. Layering may be a natural means of reproduction, as in black raspberries and trailing blackberries, or it may be induced artificially in many kinds of plants.

FACTORS AFFECTING REGENERATION OF PLANTS BY LAYERING

Nutrition

The stem remains attached to the plant during rooting and is continually supplied with water and minerals through the intact xylem. Since the phloem is generally disrupted by girdling, incision, or bending of the attached propagule, the base of the layer accumulates carbohydrates (*12*). Induction of adventitious roots on the intact stem is affected by several significant factors:

Stress Avoidance

Since the propagule remains attached to the stock plant and continues to perform as a shoot, water stress associated with stem cutting propagation is avoided. Not only is there continuous access to nutrients and carbohydrates from the stock plant, but leaching of nutrients and metabolites from intermittent mist is avoided. This is particularly important for difficult-to-root species that are normally maintained under mist for long periods. Greater leaf senescence and petiole abscission occur with recalcitrant leafy cuttings under mist, and, for example, is a criterion for predicting rooting ability of avocado cultivars (*42*). However, with layering, the problem of leaf drop or shedding under mist is avoided.

Stem Treatments

Adventitious roots are induced to form on the attached stems by various manipulations of the stems (as shown in Figs. 14–1 and 14–2) that cause an interruption in the downward translocation of organic materials—carbohydrates, auxin, and other growth factors—from the leaves and growing shoot tips. These materials accumulate above the point of treatment, and rooting occurs as it would on a stem cutting. Conversely, the upward flow of possible inhibitors to rooting may be checked (see Chap. 9).

Light Exclusion

Elimination of light from the part of the stem where roots are to develop is a feature common to all methods of layering. A distinction should be made between **blanching,** the covering of an intact stem after it has already formed, and **etiolation,** the effect produced as the shoot develops and elongates in the absence of light (*15*). (See Chaps. 9 and 10 for discussions on etiolation, blanching, **banding,** and **shading.**) Intact stems of some plants are able to produce roots after blanching only, but in others, phloem interruption also may be required. However, the greatest stimulus to root induction results when the ini-

tially developing shoots are continuously covered by the rooting medium—as in trench layering—so that approximately 2.5 cm (1 in.) of the base of the layered shoots is never exposed to light (*15, 43*). A large measure of the success with which shy-rooting plants are rooted by layering apparently results from etiolation and blanching.

Physiological Conditioning

Root initiation and development during layering may be associated with particular physiological conditions in the stem associated with the time of year. For many types of layering, the timing is associated with the movement of carbohydrates and other substances toward the roots at the end of a seasonal cycle of growth.

Rejuvenation

Cutting back shoots in mound and trench layering and regenerating new shoots from the base annually has a parallel in the hedging methods used to "rejuvenate" stock plants for improved rooting of the cuttings that are taken from them. Modified nonearthed stool techniques by ultra-severely pruning stockplants and using PVC black tape to cover base of shoots have enchanced rooting of apple (*25*).

BEFORE ROOTING   AFTER ROOTING

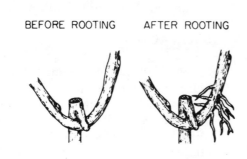

Shoot bent to a sharp v, or it may be cut, notched, girdled, or wired.

Shoot cut or broken on lower side

FIGURE 14–1 Treatments used to stimulate rooting during layering.

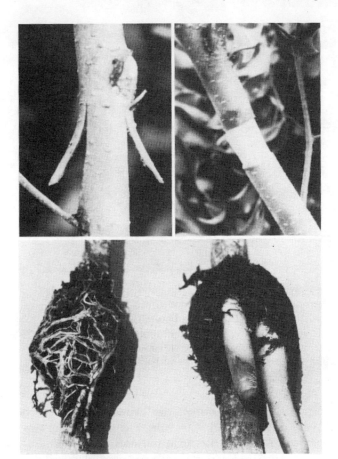

FIGURE 14–2 Air layering of *Dracena marginata*. Fine roots are induced by a double-slit method (*upper and lower left*), while coarse roots are produced when shoots are girdled during the air-layering technique (*upper and lower right*). (Courtesy T. K. Broschat and H. Donselman (*6*).)

Procedures that improve rooting of cuttings also apply to layering. For example, the use of root-promoting substances, such as indolebutyric acid (IBA), during layering is often beneficial, as it is with cuttings, although the methods of application may be somewhat different (*17, 44, 50*). Applying the material to girdling cuts as a powder, in lanolin, or as a solution can be utilized effectively. Root formation on layers depends upon supplying continuous moisture, good aeration, and avoiding high solar heat buildup at the rooting zone (*33, 46, 48*).

Uses of Layering

There are four principal uses of layering:

1. Propagation of plants of species such as black raspberries and trailing blackberries that reproduce naturally by this method.
2. Propagation of plants of clones whose cuttings do not root easily, yet are sufficiently valuable to justify the cost and labor required in layering. Filberts have been commercially propagated by simple layering, for example, and Muscadine grapes (*Vitis rotundifolia*) by compound layering. Specific size-controlling clonal rootstocks of apple, pear, and some other fruit trees are produced by mound or trench layering. Certain tropical fruits, for example, mango and litchi, are propagated by air layering.
3. Layering is useful for producing a large-sized plant in a short time (*26*). Air layering is used in greenhouses to propagate relatively large specimen plants of rubber plant (*Ficus elastica*) (Fig. 14–8) and croton. Dieffenbachia and similar plants may be propagated by simple layering (Fig. 14–6).
4. Layering is valuable for producing relatively small numbers of plants of good size with minimum propagation facilities, particularly when outdoor stock plant space is not a limiting factor.

As a commercial propagation method, layering requires considerable labor and is cumbersome and expensive. Special layering beds, or (more commonly) **stool beds,** are established but are somewhat difficult to manage. Nevertheless, for those plants where stooling is cost-effective, nurseries have developed appropriate management and mechanized techniques (*9, 24*) (see Fig. 14–3). Trends have been to shift from layering to other propagation methods as they are developed. Improved procedures for hardwood cutting propagation can replace layering procedures for all but the most difficult apple rootstocks. More significantly, micropropagation techniques (see Chap. 17) can replace many of the layering procedures now in use. It is important to note that microprop-

FIGURE 14-3 Mounding up of friable soil with a ridge plow in England. *Above:* Soil is mounded up to and between new shoots until 15 cm (6 in.) of the proximal shoot portion of apple stool is covered. During the harvest, soil is plowed away from the sides of the ridges. *Below:* The rooted apple rootstock are cut with a tractor-mounted rotary saw. (Courtesy B. H. Howard. 1987. Propagation. In *Rootstocks for fruit crops,* R. C. Rom and R. F. Carlson, eds. © John Wiley.)

agation is not widely used since micropropagules tend to be small, often multiple branched, and different from rootstock plants produced from stool beds.

PROCEDURES IN LAYERING

Tip Layering

Tip layering is a natural method of reproduction characteristic of trailing blackberries, dewberries, and black and purple raspberries. Stems of these plants are biennial in that the canes are vegetative during the first year, fruitful the second, and pruned out after fruiting. In the nursery it is advisable to set aside stock plants solely for propagation. Healthy young plants are set 3.6 m (12 ft) apart to give room for subsequent layering. The plants are cut down to within 23 cm (9 in.) of the ground as soon as they are planted. Vigorous new canes are "summer topped" by pinching off 7.6 to 9.2 cm (3 to 4 in.) of the tip after growth of 45 to 76 cm (18 to 30 in.). This encourages lateral shoot production, and will increase the number of potential tip layers, and also next year's fruit

crop. By late summer, the canes begin to arch over, and their tips assume a characteristic appearance in that the terminal ends become elongated and the leaves small and curled to give a "rat-tail" appearance. The best time for layering is when only part of the lateral tips has attained this appearance. If the operation is done too soon, the shoots may continue to grow instead of forming a terminal bud. If it is done too late, the root system will be small.

Rooting takes place near the tip of the current season's shoot. The shoot tip recurves upward to produce a sharp bend in the stem from which roots develop.

Tips are layered by hand, a spade or trowel being used to make a hole with one side vertical and one sloping slightly toward the parent plant. The tip is placed in the hole with the shoot lying along the sloping side and the returned soil is pressed firmly against it. Placed thus, the tip cannot continue to grow in length and becomes "telescoped," soon forming an abundant root system and developing a vigorous young vertical shoot.

The plants are ready for digging by the end of the same season. The rooted tip consists of a terminal bud, a large mass of roots, and 15 to 20

FIGURE 14-4 Tip layer of boysenberry.

cm (6 to 8 in.) of the old cane to serve as a "handle" and to mark the location of the new plant (Fig. 14-4). Since the tip layers are tender, easily injured, and subject to drying out, digging should be done preferably just before replanting. The remainder of the layered shoot attached to the parent plant is cut back to 23 cm (9 in.) as in the first year. Economical quantities of shoots are produced annually for as long as ten years. Rooted tip layers are planted in the late fall or early spring. New canes develop rapidly during the first season.

Simple Layering

Simple layering is illustrated in Figs. 14-5 to 14-7. The usual time for layering is in the early spring, and dormant, one-year-old shoots are used. Low, flexible branches of the plant, which can be bent easily to the ground, are chosen.

Shoots layered in the spring will usually be rooted adequately by the end of the first growing season and can be removed either in the fall or in the next spring before growth starts. Mature shoots layered in summer should be left through

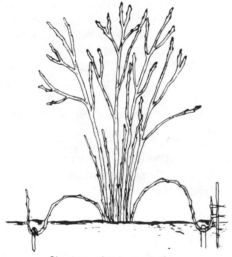

Shoots are bent over to the ground in early spring or fall. A second bend is made in branch a short distance from tip, which is covered with soil and held in place with wire or wood stakes. The stem is sometimes injured at the underground section which stimulates rooting. Includes notching, bending, wiring, or girdling.

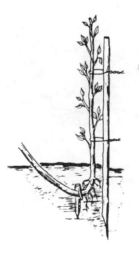

Roots form on the buried part of the shoot near the bend.

The rooted layer is removed from the parent plant.

FIGURE 14-5 Steps in propagation by simple layering.

FIGURE 14-6 Propagation of two *Dieffenbachia* plants by simple layering. Leggy stems were curved and placed into containers of soil. After several months, strong root systems forced at the curved portion of stems (*right*); new plants are then severed from the mother plant for independent growth.

the winter and either removed the next spring before growth begins or left until the end of the second growing season. When the rooted layer is removed from the parent plant, it is treated essentially as a rooted cutting of the same plant (*10*).

A supply of rooted layers can be produced over a period of years by establishing a layering bed composed of stock plants far enough apart to allow room for all shoots to be layered. This pro-

cedure has been used commercially to propagate certain hard-to-root shrubs (*27*) as well as filberts (Fig. 14-7).

Compound or Serpentine Layering

Compound layering is essentially the same as simple layering, except that the branch is alternately covered and exposed along its length. Several new plants are thus possible from a single branch. This method is used for propagating plants that have long, flexible shoots, such as the Muscadine grape. Ornamental vines, such as *Wisteria* and *Clematis* also can be propagated this way, although this method is used more often by amateurs than by commercial propagators.

Air Layering (Chinese Layering, Pot Layerage, Circumposition, Marcottage, Gootee)

Air layering is used to propagate a number of tropical and subtropical trees and shrubs (*6, 26, 38*), including the litchi (*16*) and the Persian lime (*Citrus aurantifolia*) (*45*). It is used in the greenhouse for *Ficus* species, croton, *Monstera,* and philodendron to produce large plants quickly (*26, 36, 41*). Air layering is effective also for rooting mature pines (*2, 5, 18, 31*). With polyethylene film and aluminum foil for wrapping the layers, it is possible to air-layer plants out-of-doors (Fig. 14-8) (*16, 53*).

Air layers are made in the spring on stems

FIGURE 14-7 Propagation of filberts in Oregon by simple layering. Center row consists of mother plants. Two outer rows are layered plants, which will be dug at end of growing season. Arrows point to shoots from mother plants, which have been bent over and placed under the soil for rooting in the manner shown in Fig. 14-5.

FIGURE 14-8 Steps in making an air layer on a *Ficus elastica* plant using polyethylene film. *Right (top):* The stem should be girdled for a distance of about 2.5 cm (1 in.) to induce adventitious root formation above the cut. A ball of slightly damp sphagnum moss is placed around the girdled section (*middle*). A wrapping of polyethylene film is placed around the sphagnum moss and tied at each end (*below*). *Left:* Roots on a *Ficus elastica* air layer showing through the clear plastic covering. At this stage the layer is ready to be removed from the parent plant and potted.

of the previous season's growth or, in some cases, in the late summer with partially hardened shoots. Stems older than one year can be used in some cases, but rooting is less satisfactory and the larger plants produced are somewhat more difficult to handle after rooting. The presence of numerous active leaves on the layered shoot speeds root formation. With tropical greenhouse plants, layering should be done after several leaves have developed during a period of growth.

The first step in air layering is to girdle or cut the bark of the stem (see Fig. 14–8). A strip of bark 1.8 to 2.5 cm (½ to 1 in.) wide, depending upon the kind of plant, is completely removed from around the stem. Scraping the exposed surface to insure complete removal of the phloem and cambium is desirable to retard healing. Another procedure is to make slanting upward cuts on one or both sides of the stem about 3 cm (1½ in.) long, keeping the two surfaces apart by sphagnum or a piece of wood. Wounding by girdling the stem reduces water conductivity more than a double-split technique, but does not impede rooting (Fig. 14–2). This was reported for *Ficus elastica, F. benjamina,* and *Schefflera arboricola,* but not observed with *Dracaena marginata* (*6*). Growing plants in 50 percent shade to reduce water stress is also effective (*5*).

Application of IBA to the exposed wound is beneficial. One method used in the commercial air layering of *Mahonia aquifolium* 'Compacta' is to apply a small amount of sphagnum moss soaked with 60 ppm IBA under the wounded flap of tissue (*52*). Increasing concentrations of up to 4 percent IBA in talc has increased rooting and survival in pecan air layers (*44*). About two handfuls of moistened sphagnum moss with the excess moisture squeezed out are placed around the stem to enclose the cut surfaces. If the moisture content of the sphagnum moss is too high, the stem will decay.

A piece of polyethylene film, 20 to 25 cm (8 to 10 in.) square, is wrapped carefully about the branch so that the sphagnum moss is completely covered. The ends of the sheet should be folded (as in wrapping meat) with the fold placed on the lower side. The two ends must be twisted to make sure that no water can seep inside. Aluminum foil is also useful for this purpose and in helping maintain moderate temperatures by reducing the heat load from sunlight (*7, 51*). Foil is either wrapped around the polyethylene or used as the sole wrapping material in climates with high relative humidity. Adhesive tape, such as electricians' waterproof tape, serves well to wrap the ends; the winding should be started well above the edge of the cover to enclose the ends, particularly the upper one, securely. Budding rubbers, twist ties, and florist's ties are other materials that can be used for this purpose. To avoid breakage during air layering, some operators will attach short canes as a "splint" across the girdled or incised section.

The layer is removed from the parent plant when roots are observed through the transparent film (Fig. 14–8). In some plants, rooting occurs in two to three months or less. Layers made in spring or early summer are best left until the shoots become dormant in the fall, and are removed at that time. Holly, lilac, azalea, and magnolia should be left for two seasons (*10*). In general, it is desirable to remove the layer for transplanting when it is not actively growing.

Pruning is usually advisable to reduce the top in proportion to the roots. Pot the rooted layer into a suitable container and place it in mist or under cool, humid conditions, such as in an enclosed frame, where the plants can be syringed frequently. If potted in the fall, a sufficiently large root system usually develops by spring to permit successful growth in the open. Placing the rooted layers under mist for several weeks, followed by gradual hardening-off, is probably the most satisfactory procedure (*37*).

Mound (Stool) Layering or Stooling

Stooling is a form of layering, except that there is no physical bending over of the attached propagule, which is generally done with most layering techniques. Stooling produces "stoolshoots," whereas other layering techniques produce a "layer." Mound or stool layering is used primarily commercially to produce various apple and pear clonal rootstocks, quince, currants, and gooseberries (see Figs. 14–3, 14–9, and 14–10) (*4, 8, 13, 15*). A stool bed is established by planting healthy mother plants of suitable size (8 to 10 mm diameter) in loose, fertile, well-drained soil one year before the propagation is to begin. The mother plants should be set 30 to 38 cm (12 to 15 in.) apart in the row, but the spacing between rows varies with different conditions and types of nursery equipment used. Width between rows should be sufficient to allow for cultivation and hilling operations during spring and summer. In England and in New York a minimum of 2.5 m (8 ft) row spacing is required for using tractors (*4*).

Before new growth starts the following spring, all plants are cut back to 2.5 cm (1 in.) above the ground level. Two to five new shoots usually develop from the crown the second year, more in later years. When these shoots have grown 7.6 to 12.7 cm (3 to 5 in.), loose soil, sawdust, or a soil-sawdust mixture is drawn up around each shoot to one-half its height (Fig. 14–3). When the shoots have grown to a total height

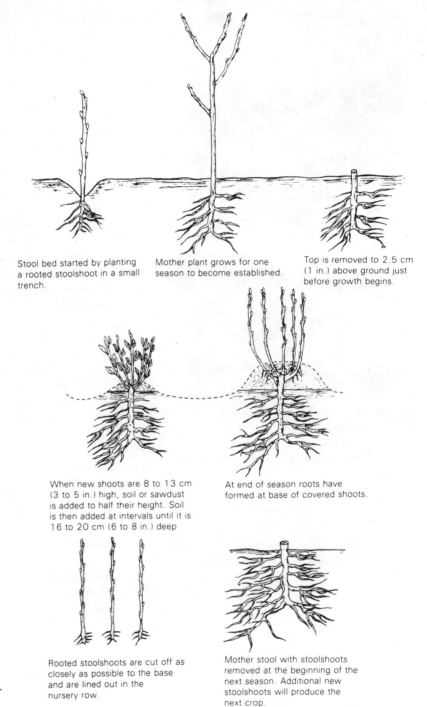

Stool bed started by planting a rooted stoolshoot in a small trench.

Mother plant grows for one season to become established.

Top is removed to 2.5 cm (1 in.) above ground just before growth begins.

When new shoots are 8 to 13 cm (3 to 5 in.) high, soil or sawdust is added to half their height. Soil is then added at intervals until it is 16 to 20 cm (6 to 8 in.) deep

At end of season roots have formed at base of covered shoots.

Rooted stoolshoots are cut off as closely as possible to the base and are lined out in the nursery row.

Mother stool with stoolshoots removed at the beginning of the next season. Additional new stoolshoots will produce the next crop.

FIGURE 14–9 Steps in propagation by mound (stool) layering.

of 19 to 25 cm (8 to 10 in.), a second hilling operation takes place. Additional rooting medium is added, to be mounded around the bases of the shoots but not to more than half their total height. A third and final hilling operation is made in midsummer when the shoots have developed to a total length of approximately 45 cm (18 in.). The base of the shoots will then have been covered with soil to a depth of 15 to 20 cm (6 to 8 in.).

Stoolshoots of easily propagated plants should have rooted sufficiently by the end of the growing season to be separated from the parent

FIGURE 14-10 *Above:* Rooting medium of wood shavings and soil pulled away showing root production by end of summer. *Below:* Stool bed used in propagating clonal apple rootstocks.

stool for lining out in the nursery row. The rooted layers are cut close to their base to keep the height of the stool plant low. These rooted layers are handled as liners and transplanted directly into the nursery row.

After the shoots have been cut away, the mother stool remains exposed until new shoots have grown 7.6 to 12.7 (3 to 5 in.), when the hilling-up begins for the next year.

Cutting back whole plants, then mound layering the vigorous juvenile shoots, has been described as a method of rooting six- to seven-year-old seedling cashew (*35*), seedling pecan (*30*), and other difficult-to-root plants. See Fig. 14–11 for mounding (stooling) pecans. A stool bed can be used for 15 to 20 years with proper handling, providing it is maintained in a vigorous condition, with disease, insect, and weed control.

Selective and biennial harvesting has been used for invigorating declining stoolbeds and pro-

ducing large shoots for high budding, but has led to increased apple mildew infection (*49*). Unlike the less vigorous M.9a apple rootstock, the vigor of MM.111 makes it capable of withstanding severe winter pruning during stooling and with sufficient reserves for the development of the mother stool (*22*). NAA-based Tipoff sprays have been used to eliminate small nonproductive shoots from apple stoolbeds (*23*).

Girdling the bases of the shoots by wiring about six weeks after they begin to grow will stimulate rooting in many plants (*28, 40*). The size of the root system in apple layers has been increased on shoots growing through the spaces of a galvanized screen 0.5 cm (³⁄₁₆ in.) square laid in a 45-cm (18-in.) strip down the row over the top of the cut-back stumps. New shoots growing through this screen gradually become girdled as the season progresses (*21*). Roots form on the stem above the screen. However, with M.9 apple rootstock, roots formed below the screen (England).

Budded plants of apple and citrus (*13, 29*) have been produced by budding the layer in place in the stool bed. This may be done in the middle of the growing season, whereupon the budded layer is transplanted to the nursery in the fall for an additional season's growth. However, budding rootstocks **in situ** is generally not recommended because the stock accumulates viruses from scions budded onto it over time. A containerized stooling system for limited quantities of clonal apple rootstock is illustrated in Fig. 14–12.

Trench Layering

Trench layering (etiolation method, as opposed to blanching during stooling) consists of growing a plant or a branch of a plant which is pegged flat in a horizontal position in the base of a shallow trench and filling in soil around the new shoots as they develop, so that the shoot bases are etiolated. Roots develop from the base of these new shoots (see Fig. 14–13). Trench layering is used primarily for woody species difficult to propagate by stooling (*43*).

The first step in this procedure involves the establishment of the mother bed which, as in mound layering, can be used over a period of years. Rooted layers or one-year-old nursery-budded or grafted trees are planted 45 to 76 cm (18 to 30 in.) apart at an angle of 30 to 45 degrees

FIGURE 14–11   Clonal propagation of pecan by mound (stool) layering. *Upper left:* Five-year-old 'Stuart' pecan trees in the background and five-year-old pecan stooling beds in the foreground. *Upper right:* Stool plants prior to mounding with soil. *Lower left:* Larger rooted propagules (*lower right*) can then be lined out in the nursery.

down a row. The rows should be 2.5 m (8 ft) apart—wide enough to allow for cultivation and to draw soil up around the plants to a height of 15 cm (6 in.). The plants are then cut back to a uniform length—45 to 60 cm (18 to 24 in.)—and left to grow one season. In some cases (in layering walnuts, for instance) plants can be placed horizontally in the trench and the developing shoots layered the first year. Further details are given in Fig. 14–13.

Trench layering could be practiced on established shrubs or trees by bending long, flexible shoots or vines to the ground as is done in simple layering, but laying them flat in trenches. The shoot is covered along its entire length, but the tip is left exposed. New shoots which develop from the buds along the stem grow upward through the soil, with roots forming at their base. The latter procedure is sometimes known as **continuous layering**.

PLANT MODIFICATIONS RESULTING IN NATURAL LAYERING

Some plants exhibit modifications of their vegetative structure or method of growth that lead to their natural vegetative increase. Those listed be-

CONTAINERIZED LAYERING

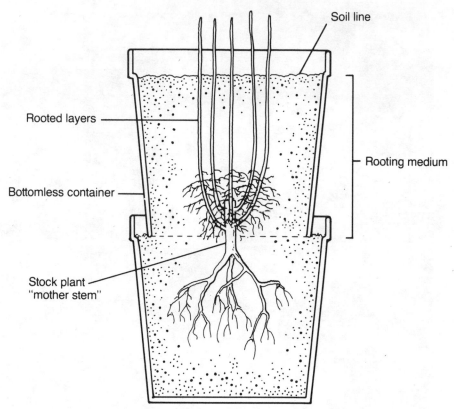

FIGURE 14–12   For limited quantities of clonal rootstock this containerized layering system can be used. It also makes a good classroom demonstration of mounding or stooling. (Redrawn from R. H. Munson (*34*).)

low could be considered natural forms of layering and are often utilized for propagation.

Runners

A **runner** is a specialized stem that develops from the axil of a leaf at the crown of a plant, grows horizontally along the ground, and forms a new plant at one of the nodes. The strawberry is a typical plant propagated in this way (Fig. 14–14). Other plants propagated by runners include bugle (*Ajuga*), the strawberry geranium (*Saxifraga sarmentosa*), and the ground cover *Duchesnea indica*. Plants of these species grow as a rosette or crown. Some ferns, such as Boston fern (*Nephrolepsis*), produce runner-like branches, as do certain orchid species such as *Dendrobium,* which form small plants known as *''keikies.''* Runners of the spider plant (*Chlorophytum comosum*) are shown in Fig. 14–15.

In most strawberry cultivars, runner formation is related to the length of day and temperature. Runners are produced in long days of 12 to 14 hours or more with high midsummer temperatures. New plants are produced at alternate nodes. These take root but remain attached to the mother plant. New runners are in turn produced by daughter plants. The connecting stems die in the late fall and winter, and each daughter plant becomes separate from the others.

In propagating by runners, daughter plants are dug when they have become well rooted, and then transplanted to the desired locations. Strawberry plants are also propagated extensively by micropropagation (*3, 11*).

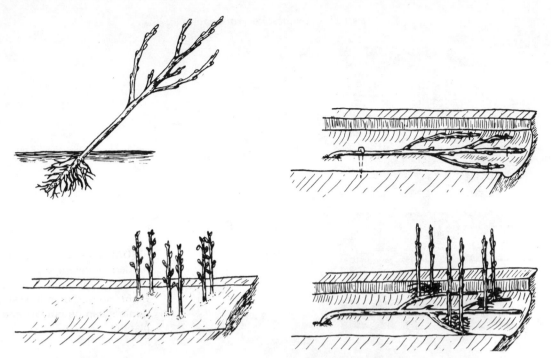

FIGURE 14-13 Steps in propagation by trench layering. *Upper left:* Mother plant after one year's growth in nursery. The trees were planted in the row at an angle of 30° to 45°. The trees are 46 to 78 cm (18 to 30 in.) apart down the row. *Upper right:* Just before growth begins, the plant is laid flat on the bottom of a trench about 5 cm (2 in.) deep. Shoots are cut back slightly and weak branches removed. Tree must be kept completely flat with wooden pegs or wire fasteners. *Lower left:* Rooting medium, as fine soil, peat moss, or sawdust, is added at intervals to produce etiolation on 5 to 7.5 cm (2 to 3 in.) of the base of the developing shoots. Apply first 2.5 to 5 cm (1 to 2 in.) layer before the buds swell. Repeat as shoots emerge and before they expand. Later coverings are less frequent and only cover half of shoot. Final medium depth is 15 to 19 cm (6 to 8 in.) *Lower right:* At the end of the season the medium is removed and the rooted layers are cut off close to the parent plant. Shoots left at internodes can be layered the following season.

FIGURE 14-14 Runners arising from the crown of a strawberry plant. New plants are produced at every second node. The daughter plants, in turn, produce additional runners and runner plants.

FIGURE 14-15   *Left:* Masses of runners developing from spider plants (*Chlorophytum comosum*). *Right:* Rooted plantlets at end of runners can be cut off and planted.

Stolons

Stolons, produced by some plants, are modified stems that grow horizontal to the ground. These may be prostrate or sprawling stems growing above ground as found in some woody species, such as *Cornus stolonifera*. The term also describes the horizontal stem structure occurring in Bermuda grass (*Cynodon dactylis*), *Ajuga*, mint (*Mentha*), and *Stachys* (*39*). Stolonlike underground stems are involved in tuberization; they are the stems that develop into a potato tuber.

The stolon can be treated as a naturally occurring rooted layer and can be cut from the parent plant and planted.

Offsets

An **offset** is a characteristic type of lateral shoot or branch that develops from the base of the main stem in certain plants. This term is applied generally to a shortened, thickened stem of rosettelike appearance. Many bulbs reproduce by producing typical offset bulblets from their base (see Chap. 15 for details). The term *offset* (or **offshoot**) also applies to lateral branches arising on stems of monocotyledons as in date palm, pineapple, or banana (see Chap. 18) (Fig. 14–16).

Offsets are removed by cutting them close to the main stem with a sharp knife. If it is well rooted, the offset can be potted and established like a rooted cutting. If insufficient roots are present, the shoot is placed in a favorable rooting medium and treated as a leafy stem cutting. Slips and suckers, which are two types of pineapple offsets, readily form aerial rootlets which facilitate commercial field propagation (Fig. 14–17).

In cases in which offset development is meager, cutting back the main rosette may stimulate the development of more offsets from the old stem just as removing the terminal bud stimulates lateral shoots in any other type of plant.

Natural increase by offsets tends to be slow and not suited for commercial propagation. However, the natural rate in many of these monocotyledonous plants can be stimulated greatly by the multiplicative techniques used in micropropagation in aseptic culture (see Chaps. 16 and 17). With the development of such procedures, commercial propagation of the desirable plants of this group has been greatly enhanced. Shoot increase takes place either by stimulation of axillary branching or by induction of adventitious shoots.

Offshoots of the date palm do not root readily if separated from the parent plant. They are usually layered for a year prior to removing. Al-

FIGURE 14–16 Removing a date offshoot with chisel and sledge hammer. (From R. W. Nixon, Date culture in the United States, *USDA Cir. 728.*)

ternate methods are available utilizing micro-propagation of shoot tips or lateral buds to multiply specific clones vegetatively on aseptic culture. Plantlets can be produced by somatic embryos from callus or from proliferation and rooting of shoot tips and lateral buds, although rates of production may be relatively low (*47*).

Suckers

A **sucker** is a shoot that arises on a plant from below ground, as shown in Fig. 14–18. The most precise use of this term is to designate a shoot that arises from an adventitious bud on a root. However, in practice, shoots that arise from the vicinity of the crown are called *suckers* even though originating from stem tissue. Propagators generally designate any shoot produced from the rootstock below the bud union of a budded tree as a *sucker* and refer to the operation of removing them as "*suckering.*" In contrast, a shoot arising from a latent bud of a stem several years old, as, for instance, on the trunk or main branches, should be termed a **watersprout**.

Suckers are dug out and cut from the parent plant. In some cases part of the old root may be retained, although most new roots arise from the base of the sucker. It is important to dig the sucker out rather than pull it, to avoid injury to its base. Suckers are treated essentially as a rooted layer or as a cutting, in case few or no roots have formed. They are usually dug during the dormant season.

Crown Division

The term **crown** as generally used in horticulture designates that part of a plant at the surface of the ground from which new shoots are produced. In trees or shrubs with a single trunk, the crown is principally a point of location near the ground surface marking the general transition zone between stem and root. In herbaceous perennials, the crown is the part of the plant from which new shoots arise annually. The crown of herbaceous perennials consists of many branches, each being the base of the current season's stem, which originated from the base of the preceding year's branch. These lateral shoots are stimulated to grow from the base of the old stem as it dies back after blooming. Adventitious roots develop along the base of the new shoots. These new shoots eventually flower either the same year they are produced or the following year. As a result of the annual production of new shoots and the dying back of old shoots, the crown may become extensive within a period of a relatively few years and may need to be divided every few years to prevent overcrowding.

Multibranched woody shrubs may develop

FIGURE 14–17 *Upper left:* Cross section of a pineapple showing the crown atop the fruit and slip attached to the peduncle supporting the fruit. *Upper right:* Slips (*left*) from the peduncle are more commonly used propagules than suckers (*right*). *Lower left:* Preformed root primordia enable slips to root readily when mass propagated in black polyethylene mulched field beds (*lower right*). Slips, suckers, and crowns are three types of pineapple offsets that can be used in pineapple vegetative propagation.

FIGURE 14-18 Suckers arising as adventitious shoots from the roots of a red raspberry plant. After they are well rooted the suckers may be cut from the parent plant and transplanted to their permanent location.

extensive crowns. Although an individual woody stem may persist for a number of years, new, vigorous shoots are continuously produced from the crown, and eventually crowd out the older shoots. Crowns of such shrubs can be divided in the dormant season and treated as a large rooted cutting.

Crown division is an important method of propagation for herbaceous perennials, and to some extent for woody shrubs, because of its simplicity and reliability. Such characteristics make this method particularly useful to the amateur or professional gardener who is generally interested in only a modest increase of a particular plant.

Crowns of outdoor herbaceous perennials are usually divided in the spring just before growth begins or in late summer or autumn at the end of the growing season. As a general rule, plants that bloom in the spring and summer and produce new growth after blooming should be divided in the fall. Those that bloom in summer and fall and make little or no new growth until spring should be divided in early spring. Potted plants are divided when they become too large for the particular container in which they are growing. Division is necessary to maintain the variegated form in some plants usually propagated by leaf cuttings, such as *Sanseviera* (*19*).

In crown division, plants are dug and cut into sections with a knife, hand-ax, handsaw, or other sharp instrument (Fig. 14–19). In herbaceous perennials, as the Shasta daisy (*Chrysanthemum superbum*) or day lily (*Hemerocallis*) (see Fig. 14–20), where an abundance of new rooted off-

FIGURE 14-19 Crown division of herbaceous perennials by (*top*) knife with blue fescue (*Festuca ovina* 'Glauca'), (*middle*) hand-ax with feather reed grass (*Calamagrostis acutiflora* 'Stricta'), and (*bottom*) handsaw with giant reed (*Arundo donax*). (Courtesy R. A. Simon.)

FIGURE 14–20   Propagation by crown division illustrated by division of day lily (*Hemerocallis*) clump.

shoots are produced from the crown, each may be broken from the old crown and planted separately, the older part of the plant clump being discarded.

For commercial production of improved cultivars, multiplication is very slow. Increased production of shoots for rooting can be produced by cutting back to the crown in the early spring after new growth starts and treating with a cytokinin (*1*). More rapid multiplication is achieved in

tissue culture propagation, regenerating plants from callus produced on flower petals and sepals (*20, 32*) (see Chaps. 16 and 17).

Herbaceous perennials generally propagated by division can be readily propagated also by micropropagation in aseptic culture systems (*54*). Axillary branching is greatly enhanced and multiplication can be made at very high rates with minimum stock plants.

REFERENCES

1. Apps, D. A., and C. W. Heuser. 1975. Vegetative propagation of *Hemerocallis*—including tissue culture. *Proc. Inter. Plant Prop. Soc.* 25:362–67.

2. Barnes, R. D. 1974. Air-layering of grafts to overcome incompatibility problems in propagating old pine trees. *New Zealand Jour. For. Sci.* 4(2):120–26.

3. Boxus, P., M. Quoirin, and J. M. Laine. 1977. Large scale propagation of strawberry plants from tissue culture. In *Applied and fundamental aspects of plant cell, tissue and organ culture,* J. Reinert and Y. P. S. Bajaj, eds. New York: Springer-Verlag, pp. 130–43.

4. Brase, K. D., and R. D.Way. 1959. Rootstocks and methods used for dwarfing fruit trees. *N. Y. Agr. Exp. Sta. Bul. 783.*

5. Broschat, T. K., and H. M. Donselman. 1981. Effects of light intensity, air layering, and water stress on leaf diffusive resistance and incidence of

leaf spotting in *Ficus elastica*. *HortScience* 16(2): 211–12.

6. ———. 1983. Effect of wounding method on rooting and water conductivity in four woody species of air-layered foliage plants. *HortScience* 18:445–47.

7. Cameron, R. J. 1968. The leaching of auxin from air layers. *New Zealand Jour. Bot.* 6(2):237–39.

8. Carlson, R. F., and H. B. Tukey. 1955. Cultural practices in propagating dwarfing rootstocks in Michigan. *Mich. Agr. Exp. Sta. Quart. Bul.* 37: 492–97.

9. Chase, H. H. 1964. Propagation of oriental magnolias by layering. *Proc. Inter. Plant Prop. Soc.* 14:67–69.

10. Creech, J. L. 1954. Layering. *Nat. Hort. Mag.* 33:37–43.

11. Damiano, C. 1980. Strawberry micropropaga-

tion. In *Proc. conf. on nursery production of fruit plants through tissue culture; applications and feasibility,* R. H. Zimmerman, ed. U.S. Dept. of Agr. Sci. and Education Administration ARR-NE-11, pp. 11–22.

12. Doud, S. L., and R. F. Carlson. 1977. Effects of etiolation, stem anatomy, and starch reserves on root initiation of layered *Malus* clones. *Jour. Amer. Soc. Hort. Sci.* 102(4):487–91.

13. Duarte, O., and C. Medina. 1971. Propagation of citrus by improved mound layering. *HortScience* 6:567.

14. Dunn, N. D. 1979. Commercial propagation of fruit tree rootstocks. *Proc. Inter. Plant Prop. Soc.* 29:187–90.

15. Garner, R. J. 1979. *The grafter's handbook* (4th ed.). New York: Oxford Univ. Press.

16. Grove, W. R. 1947. Wrapping air layers with rubber plastic. *Proc. Fla. State Hort. Sci.* 60:184–89.

17. Hanger, F. E. W., and A. Ravenscroft. 1954. Air layering experiments at Wisley. *Jour. Roy. Hort. Soc.* 79:111–16.

18. Hare, R. C. 1979. Modular air-layering and chemical treatments inprove rooting of loblolly pine. *Proc. Inter. Plant Prop. Soc.* 29:446–54.

19. Henley, R. W. 1979. Tropical foliage plants for propagation. *Proc. Inter. Plant Prop. Soc.* 29:454–67.

20. Heuser, C. W., and J. Harker. 1976. Tissue culture propagation of daylilies. *Proc. Inter. Plant Prop. Soc.* 25:269–72.

21. Hogue, E. J., and R. L. Granger. 1969. A new method of stool bed layering. *HortScience* 4:29–30.

22. Howard, B. H. 1977. Effects of initial establishment practice on the subsequent productivity of apple stoolbeds. *Jour. Hort. Sci.* 52:437–46.

23. ———. 1984. The effects of NAA-based Tipoff sprays on apple shoot production in MM.106 stoolbeds. *Jour. Hort. Sci.* 59:303–11.

24. ———. 1987. Propagation. In *Rootstocks for fruit crops,* R. C. Rom and R. F. Carlson, eds. New York: John Wiley.

25. Howard, B. H., R. S. Harrison-Murray, and S. B. Arjyal. 1985. Responses of apple summer cuttings to severity of stockplant pruning and to stem blanching. *Jour. Hort. Sci.* 60:145–52.

26. Joiner, J. N., ed. 1981. *Foliage plant production.* Englewood Cliffs, N.J.: Prentice-Hall.

27. Knight, F. P. 1945. The vegetative propagation of flowering trees and shrubs. *Jour. Roy. Hort. Soc.* 70:319–30.

28. Maurer, K. J. 1950. Möglichkeiten der vegetativen Vermehrung der Walnuss. *Schweiz. Z. Obst. V. Weinb.* 59:136–37.

29. Medina, C., and O. Duarte. 1971. Propagating apples in Peru by an improved mound layering method. *Jour. Amer. Soc. Hort. Sci.* 96:150–51

30. Medina, J. P. 1981. Studies of clonal propagation on pecans at Ica, Peru. *The Plant Propagator* 27(2):10–11.

31. Mergen, F. 1955. Air layering of slash pine. *Jour. For.* 53:265–70.

32. Meyer, M. M. 1976. Propagation of daylilies by tissue culture. *HortScience* 11:485–87.

33. Modlibowska, I., and C. P. Field. 1942. Winter injury to fruit trees by frost in England (1939–1940). *Jour. Pom. Hort. Sci.* 19:197–207.

34. Munson, R. H. 1982. Containerized layering of *Malus* rootstocks. *The Plant Propagator* 28(2):12–14.

35. Nagabhushanam, S., and M. A. Menon. 1980. Propagation of cashew (*Anacardium occidentale* L.) by etiolation, girdling and stooling. *The Plant Propagator* 26:11–13.

36. Neel, P. L. 1979. Macropropagation of tropical plants as practiced in Florida. *Proc. Inter. Plant Prop. Soc.* 29:468–80.

37. Nelson, R. 1953. High humidity treatment for air layers of lychee. *Proc. Fla. State Hort. Soc.* 66:198–99.

38. Nelson, W. L. 1987. Innovations in air layering. *Proc. Inter. Plant Prop. Soc.* 37:88–89.

39. Nitsch, J. P. 1971. Perennation through seeds and other structures. In *Plant physiology,* Vol. 6A, F. C. Steward, ed. New York: Academic Press, pp. 413–79.

40. Oppenheim, J. E. 1932. A new system of citrus layers. *Hadar* 5:2–4.

41. Poole, R. T., and C. A. Conover. 1988. Vegetative propagation of foliage plants. *Proc. Inter. Plant Prop. Soc.* 37:503–7.

42. Raviv, M., and O. Reuveni. 1984. Mode of leaf shedding from avocado cuttings and the effect of its delay on rooting. *HortScience* 19:529–31.

43. Serr, E. F. 1954. Rooting paradox walnut hybrids. *Calif. Agr.* 8(5):7.

44. Sparks, D., and J. W. Chapman. 1970. The effect of indole-3-butyric acid on rooting and survival of air-layered branches of the pecan, *Carya illinoensis* Koch, cv. 'Stuart'. *HortScience* 5(5):445–46.

45. Sutton, N. E. 1954. Marcotting of Persian limes. *Proc. Fla. State Hort. Soc.* 67:219–20.

46. Thomas, L. A. 1938. Stock and scion investigations. II. The propagation of own-rooted apple trees. *Jour. Counc. Sci. Industr. Res. Org., Austral.* 11:175–79.

47. Tisserat, B. 1981. Date palm tissue culture. *USDA, Agri. Res. Serv., Adv. in Agr. Tech., Western Ser. 17.* pp. 1–50.

48. Tukey, H. B., and K. Brase. 1930. Granulated peat moss in field propagation of apple and quince stocks. *Proc. Amer. Soc. Hort. Sci.* 27:106–13.

49. Vasek, J., and B. H. Howard. 1984. Effects of selective and biennial harvesting on the production of apple stoolbeds. *Jour. Hort. Sci.* 59:477–85.

50. Vieitez, E. 1974. Vegetative propagation of chestnut. *New Zealand Jour. For. Sci.* 4(2):242–52.

51. Vinzant, K. L. 1978. Propagation of *Schefflera arboricola* by air layering and factors affecting long term storage. *The Plant Propagator* 24(1):5–6.

52. Wells, R. 1986. Air layering: an alternative method for the propagation of *Mahonia aquifolia* 'Compacta'. *Proc. Inter. Plant Prop. Soc.* 36:97–99.

53. Wyman, D. 1952. Air layering with polyethylene films. *Jour. Roy. Hort. Soc.* 77:135–40.

54. Zilis, M., D. Zwagerman, D. Lamberts, and L. Kurtz. 1979. Commercial propagation of herbaceous perennials by tissue culture. *Proc. Inter. Plant Prop. Soc.* 29:404–13.

SUPPLEMENTARY READING

GARNER, R. J. 1979. *The grafter's handbook* (4th ed.). New York: Oxford Univ. Press.

GARNER, R. J., and S. A. CHAUDHRI. 1976. *The propagation of tropical fruit trees.* Hort. Rev. 4. East Malling, Maidstone, Kent: Commonwealth Bureau of Hort. and Plant. Crops.

MCMILLAN-BROWSE, P. D. A. 1979. *Plant propagation.* New York: Simon and Schuster.

MINISTRY OF AGRICULTURE, FISHERIES AND FOOD 1969. *Fruit tree raising—rootstocks and propagation* (5th ed.). Bul. 135. London.

ROM, R. C., and R. F. CARLSON, eds. 1987. *Rootstocks for fruit crops.* New York: John Wiley.

STILL, S. M. 1988. *Manual of herbaceous ornamental plants.* Champaign, Ill.: Stipes Publ. Co.

TUKEY, H. B. 1964. *Dwarfed fruit trees.* New York: Macmillan.

15

Propagation by Specialized Stems and Roots

Bulbs, corms, tubers, tuberous roots and stems, rhizomes, and pseudobulbs are specialized vegetative structures that function primarily in the storage of food for the plant's survival during adversity. Plants possessing these modified plant parts are generally herbaceous perennials in which the shoots die down at the end of a growing season, and the plant survives in the ground as a dormant, fleshy organ that bears buds to produce new shoots the next season. Such plants are well suited to withstand periods of adverse growing conditions in their yearly growth cycle. The two principal climatic cycles for which such performance is adapted are the warm–cold cycle of the temperate zones and the wet–dry cycle of tropical and subtropical regions (*4, 15*).

These specialized organs also function in vegetative reproduction. The propagation procedure that utilizes the production of naturally detachable structures, such as the bulb and corm, is generally spoken of as **separation.** In cases in which the plant is cut into sections, as is done with the rhizome, stem tuber, and tuberous root, the process is spoken of as **division.**

BULBS

Definition and Structure

A **bulb** is a specialized underground organ consisting of a short, fleshy, usually vertical stem axis (**basal plate**) bearing at its apex a growing point or a flower primordium enclosed by thick, fleshy scales (see Fig. 15–1). Bulbs are produced by monocotyledonous plants in which the usual plant structure is modified for storage and reproduction.

Most of the bulb consists of **bulb scales,** which morphologically on a tunicate bulb are the continuous, sheathing leaf bases (see Fig. 15–2). The outer bulb scales are generally fleshy and contain reserve food materials, whereas the bulb scales toward the center function less as storage

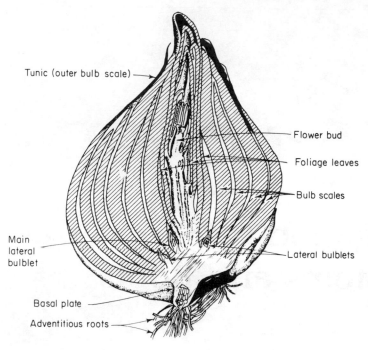

FIGURE 15-1 The structure of a tulip bulb—an example of a **tunicated laminate** bulb. Longitudinal section representing stage of development shortly after the bulb is planted in the fall. (Redrawn from Mulder and Luyten (*44*).)

organs and are more leaflike. In the center of the bulb, there is either a vegetative meristem or an unexpanded flowering shoot. Meristems develop in the axil of these scales to produce miniature bulbs, known as **bulblets,** which when grown to full size are known as **offsets.** In various species of lilies, bulblets may form in the leaf axils either on the underground portion or on the aerial portion of the stem. The aerial bulblets are called **bulbils,** while the underground organs are called **stem bulblets.** There are two types of bulbs:

Tunicate Bulbs

Tunicate (laminate) bulbs are represented by the onion, garlic, daffodil, and tulip. These bulbs have outer bulb scales that are dry and membranous. This covering, or tunic, provides protection from drying and mechanical injury to the bulb. The fleshy scales are in continuous, concentric layers, or **lamina,** so that the structure is more or less solid.

There are three basic bulb structures, de-

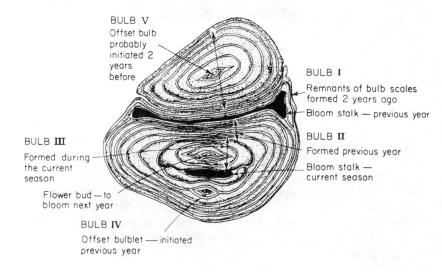

FIGURE 15-2 Cross section of a daffodil bulb. The continuous, concentric leaf scales found in the laminate bulb are shown. Also shown is the perennial nature of the daffodil bulb, which continues to grow by producing a new bulb annually at the main meristem. Lateral **offset** bulbs are also produced; parts of five individual, differently aged bulbs are shown here. (Redrawn from Huisman and Hartsema (*27*).)

fined by type of scales and growth pattern. The amaryllis (*Hippeastrum*) is an example of one type, in which the scales are expanded bases of leaves (Fig. 15–3). A second type, which includes the *Narcissus,* has both expanded leaf bases and true scales. Tulip is an example of the third type of bulb, which has only true scales, with leaves produced on the flowering or vegetative shoot (*50*).

Adventitious root primordia are present on the dormant stored bulb. They do no elongate until planted under proper conditions and at the proper time. They occur in a narrow band around the outside edge on the bottom of the basal plate.

Nontunicate Bulbs

Nontunicate (scaly) bulbs are represented by the lily (Figs. 15–4 and 15–5). These bulbs do not possess the enveloping dry covering. The scales are separate and attached to the basal plate. In general, the nontunicate bulbs are easily damaged and must be handled more carefully than the tunicate bulbs; they must be kept continuously moist because they are injured by drying. In the nontunicate lily bulb new roots are produced in

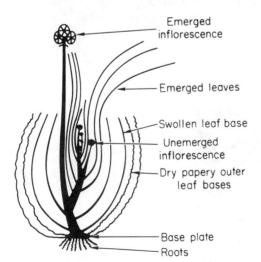

Emerged inflorescence

Emerged leaves

Swollen leaf base

Unemerged inflorescence

Dry papery outer leaf bases

Base plate

Roots

FIGURE 15–3 Diagram showing morphology and growth cycle of *Hippeastrum* bulb. New bulbs continually develop from the center in a cycle of four leaves and an inflorescence. Bases of these leaves enlarge to become the scales that contain stored food. Oldest scales disintegrate. (From A. R. Rees, *The growth of bulbs,* Academic Press, New York, 1972.)

midsummer or later, and persist through the following year (*15, 54*). In most lily species, roots also form on the stem above the bulb.

In many species thickened **contractile** roots shorten and pull the bulb to a given level in the ground (*50, 69*). Tulips do not produce contractile roots but produce **droppers** (*50*), stolon-like structures that grow from the bulb and produce a bulb at the tip.

Growth Pattern

An individual bulb goes through a characteristic cycle of development, beginning with its initiation as a meristem and terminating in flowering and seed production. This general developmental cycle is composed of two stages: (a) *vegetative* and (b) *reproductive.* In the vegetative stage the bulblet grows to flowering size and attains its maximum weight. The subsequent reproductive stage includes the *induction* and *initiation* of flowering, *differentiation* of the floral parts, *elongation* of the flowering shoot, and finally *anthesis* (flowering). Sometimes seed production occurs. Various bulb species have specific environmental requirements for the individual phases of this cycle that determine their seasonal behavior, environmental adaptations, and methods of handling. They can be grouped into classes according to their time of bloom and method of handling.

Spring-Flowering Bulbs

Important commercial crops included in the spring-flowering group are the tulip, daffodil, hyacinth, and bulbous iris, although other kinds are grown in gardens.

Bulb formation. The vegetative stage begins with the initiation of the bulblet on the basal plate in the axil of a bulb scale. In this initial period, which usually occupies a single growing season, the bulblet is insignificant in size, since it is present within another growing bulb and can be observed only if the bulb is dissected. Its subsequent pattern of development and the time required for the bulblet to attain flowering size differ for different species. The bulbs of the tulip and the bulbous iris, for instance, disintegrate upon flowering and are replaced by a cluster of new bulbs and bulblets initiated the previous season.

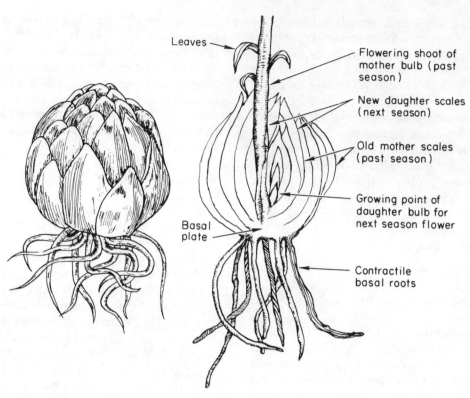

FIGURE 15–4 *Left:* Outer appearance of a **scaly** bulb of lily (*Lilium hollandicum*). *Right:* Longitudinal section of a bulb of *L. longiflorum 'Ace',* after flowering stage, showing old mother bulb scales and new daughter bulb scales. Bulb obtained in fall near digging time (*11*).

FIGURE 15–5 In the production of a lily crop, producers count leaf emergence from potted bulbs (*left*) (pot and soil removed) and use this as a gauge to speed up or slow down the production cycle, in timing flowering plants (*right*) for certain market periods. (Left photo courtesy A. E. Nightingale.)

The largest of these may have attained flowering size at this time, but smaller ones require additional years of growth (see Figs. 15–1 and 15–6). The flowering bulb of the daffodil, on the other hand, continues to grow from the center year by year, producing new offsets, which may remain attached for several years (see Figs. 15–2 and 15–6). The hyacinth bulb also continues to grow year by year, but because the number of offsets produced is limited, artificial methods of propagation are usually used.

The size and quality of the flower are directly related to the size of the bulb. A bulb must reach a certain minimum size to be capable of initiating flower primordia. Commercial value is largely based on bulb size (*2*), although the condition of the bulb and freedom from disease are also important quality factors.

Increase in size and weight of the developing bulb takes place in the period during and (mostly) after flowering, as long as the foliage remains in good condition (*9*). Cultural operations that include irrigation; weed, disease, and insect control; and fertilization encourage vegetative growth. The benefit, however, is to the next year's flower production, because larger bulbs are produced. Conversely, adverse growing conditions, removal of foliage, and premature digging of the bulb result in smaller bulbs and reduced flower production.

Moderately low temperatures tend to prolong the vegetative period, whereas higher temperatures may cause the vegetative stage to cease and the reproductive stage to begin. Thus a shift from cool to warm conditions early in the spring, as occurs in mild climates, will shorten the vegetative period, result in smaller bulbs, and consequently produce inferior blooms the following year (*50*). Commercial bulb-producing areas for hardy spring-flowering bulbs are largely in regions of cool springs and summers, such as the Netherlands and the Pacific northwest of the United States.

The relative length of photoperiod apparently is not an important factor affecting bulb formation in most species. It has been shown, however, to be significant in some *Allium* species, such as onion and garlic (*26, 36*).

Flower bud formation and flowering. The beginning of the reproductive stage and the end of the vegetative stage is indicated by the drying of the foliage and the maturation of the bulb. From then on, no additional increase in size or weight of the bulb takes place. The roots disintegrate, and the bulb enters a seemingly "dormant" period. However, important internal changes take place, and in some species the vegetative growing point undergoes transition to a flowering shoot. In nature, all bulb activity takes place un-

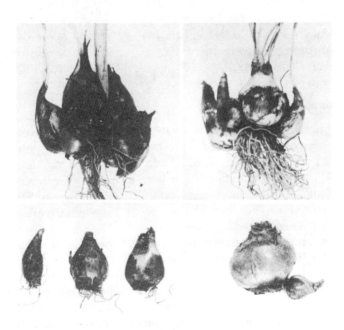

FIGURE 15–6 Propagation by offset bulbs. *Top left:* Bulbous iris. The old bulb disintegrates, leaving a cluster of bulbs. *Top right:* Daffodil. Bulb continues to grow from the inside each year, but continuously produces lateral bulbs, which eventually split away. *Below left:* Daffodil. Three types of bulbs: "split," or "slab," bulb; "round" bulb; and "double-nose" bulb. *Below right:* Hyacinth. Lateral bulblets produced, but old bulb continues to develop from inside.

derground during this period; in horticultural practice, the bulbs are dug, stored, and distributed during this three- to four-month period.

Temperature controls the progression from the vegetative stage to flowering (9, 25, 26, 66). Differentiation of flower primordia for the spring flowering group occurs at moderately warm temperatures in late summer or early fall either in the ground or in storage. Subsequent exposure to lower but above-freezing temperatures is required to promote flower stalk elongation. As temperatures increase in the spring, the flower stalk elongates and the bulb plant subsequently flowers. Bulbous iris is an exception in that flower induction is induced by low temperatures in fall and early winter. For species in groups II (i.e., *Narcissus*) and III (i.e., tulip), induction occurs in spring after the storage period.

The optimum temperature—determined by the shortest time in which the bulb would flower—has been established for these developmental phases for the important bulb species, such as tulip (9), hyacinth (8), lily (66), and daffodil, depending also upon cultivar. Such information is important in establishing forcing schedules for retail florist sales (7) (see Fig. 15-5). Holding the bulbs continuously at high temperatures (30 to 32°C; 86 to 90°F or more), or at temperatures near freezing, will inhibit or retard floral development and can be used to lengthen the period required for flowering. With a shift to favorable temperatures, flower bud development will continue. This treatment (below or above optimum) can be used when shipping bulbs from northern- to southern-hemisphere countries.

Summer Flowering Bulbs

Lilies are important plants with a growth cycle geared to the seasonal pattern of the summer–winter cycles of the temperate zone. Their nontunicate bulbs do not go "dormant" in late summer and fall as does the tunicate type of bulb, but have unique characteristics that must be understood for proper handling. Although different lily species have somewhat different methods of reproduction (54, 70) the pattern, as determined for the Easter lily (*Lilium longiflorum*), can serve as a model (4, 10, 11, 66).

Lilies flower in the late spring or early summer at the apex of the leaf-bearing stem axis. The flower-producing bulb is known as the **mother bulb** made up of the basal plate, fleshy scales, and the flowering axis (Figs. 15-3 and 15-5). Prior to flowering, a new daughter bulb(s) is developing within the mother bulb. It had been initiated the previous fall and winter from a growing point in the axil of a scale at the base of the stem axis. During spring the daughter bulb initiates new scales and leaf primordia at the growing point. Natural chemical inhibitors in the daughter scales prevent elongation of the daughter axis, which remains dormant—but it can be promoted to grow by exposure to high temperature (37.5°C; 100°F), to low temperature (4.5°C; 40°F), or by treatment with gibberellic acid (65).

After flowering of the mother bulb, no more scales are produced by the daughter bulb, but it increases in size (circumference) and weight until it equals the weight of the mother bulb surrounding it. Inhibitory effects of the daughter scales decrease, as does the response to dormancy-breaking treatments (65). Fleshy basal roots persist on the mother bulb through the fall and winter, and new adventitious roots develop in late summer or fall from the basal plate of the new bulb formed above the mother bulb. Warm temperatures promote root formation (10). Bulbs should be dug for transplanting after they "mature" in the fall. The top may or may not have died down. Bulbs should be handled carefully to avoid injury and to prevent drying. The commercial value of the bulb depends on size (transverse circumference) and weight at the time of digging (34); and on the condition of the fleshy roots and the freedom of the bulb from disease.

Transition of the meristem to a flowering shoot does not take place until the stem axis has protruded through the "nose" of the bulb and the shoot is *induced* with chilling temperatures (32). The critical temperatures are 15.5 to 18.5°C (60 to 65°F) or less, and the chilling effect becomes most effective at 2 to 4.5°C (35 to 40°F). The cold period requirement to induce flowering is known as **vernalization.** The shoot emerges 10 to 15 cm (4 to 6 in.) above ground after chilling and flower primordia starts to *develop*. Storing bulbs at warm (21°C; 70°F or more) or low (−0.5°C; 31°F) temperatures will keep bulbs dormant and delay blooming (61). Moisture content of the storage medium is important; if too dry, the bulb will deteriorate, and if too wet, it will decay.

Following the flower induction stage, and with the onset of higher temperatures, the stem elongates, initiating first leaves and then flowers. The outer scales of the "old" mother bulb rapidly disintegrate early in the spring as the new mother bulb produces the flower.

Stem bulblets may develop in the axils of the leaves underground or, in some species, bulbils may develop above ground. These appear about the time of flowering.

Flowering time and the size and quality of the Easter lily bloom can be closely regulated by manipulating the temperature at various stages following the digging of bulbs in the fall (*11*). Daylength also influences development, but to a lesser extent (*67*).

Tender, Winter-Flowering Bulbs

There are a number of flowering bulbs from tropical areas whose growth cycle is related to a wet–dry, rather than a cold–warm climatic cycle. The amaryllis (*Hippeastrum vittata*) is an example (Fig. 15–3) (*25*). This bulb is a perennial, growing continuously from the center with the outer scales disintegrating. New leaves are produced continuously from the center during the vegetative period extending from late winter to the following summer. In the axil of every fourth leaf (or scale) that develops, a meristem is initiated. Thus throughout the vegetative period a series of vegetative offsets is produced. By fall the leaves mature and the bulb becomes dormant, during which time the bulb should be dry. In this period, the *fourth* growing point from the center and any external to it differentiate into flower buds, and the shoot begins to elongate slowly. After two or three months of dry storage, the bulbs can be watered, causing the flowering shoots to elongate rapidly, with flowering taking place in midwinter. Maximum foliage development and bulb growth are essential to produce a bulb large enough to form a flowering shoot.

Propagation

Offsets

Offsets are used to propagate many kinds of bulbs. This method is sufficiently rapid for the commercial production of tulip, daffodil, bulbous iris, and grape hyacinth but, in general, is too slow for the lily, hyacinth, and amaryllis.

If undisturbed, the offsets may remain attached to the mother bulb for several years. They can also be removed at the time the bulbs are dug and replanted into beds or nursery rows to grow into flowering-sized bulbs. This may require several growing seasons, depending upon the kind of bulb and size of the offset.

Tulip. Tulip bulb planting takes place in the fall (*8, 9*). Two systems of planting are used: the bed system, used extensively in Holland, and the row or field system, used mostly in the United States and England. Beds are usually 1 m (3 ft) wide and separated by 31- to 45-cm (12- to 18-in.) paths. The soil is removed to a depth of 9 cm (4 in.), the bulbs set in rows 15 cm (6 in.) apart, and the soil replaced. In the other system, single or double rows are placed wide enough apart to permit the use of machines. To improve drainage, two or three adjoining rows may be planted on a ridge. Bulbs are spaced one to two diameters apart, with small bulbs scattered along the row. A mulch may be applied after planting but removed the following spring before growth.

Planting stock consists principally of those bulbs of the minimum size for flowering, 9 to 10 cm (3.6 to 4 in.) or smaller in circumference. Since the time required to produce flowering size varies with the size of the bulb, the planting stock is graded so that all those of one size can be planted together. For instance, an 8-cm (3.2 in.) or larger bulb normally requires a single season to become flowering size; a 5- to 7-cm (2- to 3-in.) bulb, two seasons; and those 5 cm or less, three years (*9*).

During the flowering and subsequent bulb growing period of the next spring, good growing conditions should be provided so that the size and weight of the new bulbs will be at a maximum. Foliage should not be removed until it dries or matures. Important cultural operations include removal of competing weed growth, irrigation, fungicidal sprays to control *Botrytis* blight (*16*), and fertilization. Beds should be inspected for disease early in the season and for trueness-to-cultivar at the time of blossoming. All diseased or off-type plants should be rogued out (*23*). It is desirable to remove the flower heads at blooming time, because they may serve as a source of *Botrytis* infection and can lower bulb weight.

Bulbs are dug in early to midsummer when the leaves have turned yellow or the outer tunic of the bulb has become dark brown in color. In the Pacific northwest of the United States, where summer temperatures are cool and the leaves remain green for a longer period, digging may take place before the leaves dry. If the bulbs are dug too early or if warm weather causes early maturation, the bulbs may be small in size. The bulbs are dug by machine or by hand with a short-handled spade. After the loose soil is shaken from the bulbs, they are placed in trays in well-ventilated storage houses for drying, cleaning, sorting, and grading. General storage temperatures are 18 to 20°C (65 to 68°F). To force early flowering, the bulbs should be held at 20°C for three to five weeks and then placed at 9°C (48°F) for eight weeks. Later flowering can be produced by holding bulbs at 22°C (72°F) for 10 weeks. For shipment from northern- to southern-hemisphere countries, the bulbs can be held at −1°C (31°F) until late December, when they are shifted to a higher temperature (25.5°C; 78°F) (8).

Daffodil.   Daffodil bulbs are perennial and produce a new meristem growing point at the center every year (18). **Offsets** are produced that grow in size for several years until they break away from the original bulb, although they are still attached at the basal plate. An offset bulb, when it first separates from the mother bulb, is known as a "split," "spoon," or "slab," and can be separated from the mother bulb and planted. Within a year it becomes a "round," or "single-nose," bulb containing a single flower bud. One year later a new offset should be visible, enclosed within the scales of the original bulb, indicating the presence of two flower buds. At this stage the bulb is known as a "double-nose." By the next year the offsets split away, then the bulb is known as a "mother bulb." Grading of daffodil bulbs is principally by age, that is, as splits, round, double-nose, and mother bulbs. The grades marketed commercially are the round and the double-nose bulb. The mother bulbs are used as planting stock to produce additional offsets, and only the surplus is marketed. Offsets, or splits, are replanted for additional growth.

Storage should be at 13 to 16°C (55 to 60°F) with a relative humidity of 75 percent. To force earlier flowering, they can be stored at 9°C (48°F) for eight weeks. To delay flowering, store at 22°C (72°F) for 13 to 15 weeks. For shipment from the northern to the southern hemisphere, the bulbs can be held at 30°C (86°F) until October, then stored at −1°C (31°F) until late December, and then at 25°C (77°F) (8).

Hot water treatment plus a fungicide for stem and bulb disease and nematode control is important (23). A three- or four-hour treatment at 43°C (110°F) is used, but that temperature must be carefully maintained or the bulbs may be damaged.

Lilies.   Lilies increase naturally, but except for a few species this increase is slow and of limited propagation value except in home gardens (54, 70). Several methods of bulb increase are found among the different species. For instance, *Lilium concolor, L. hansonii, L. henryi,* and *L. regale* increase by bulb splitting. Two to four lateral bulblets are initiated about the base of the mother bulb, which disintegrates during the process, leaving a tight cluster of new bulbs. *Lilium bulbiferum, L. canadense, L. pardalinum, L. parryi, L. superbum,* and *L. tigrinum* multiply from lateral bulblets produced from the rhizome-like bulb. This process is sometimes called "budding-off."

Bulblet Formation on Stems

Underground stem bulblets are used to propagate the Easter lily (*Lilium longiflorum*) and some other lily species (Fig. 15–7). In the field, flowering of the Easter lily occurs in early summer. Bulblets form and increase in size from spring throughout summer (53). Between mid-August and mid-September in the northern hemisphere the stems are pulled from the bulbs and stacked upright in the field. Periodic sprinkling keeps the stems and bulblets from drying out. Similarly, the base of the stem can be "heeled in" the ground at an angle of 30 to 45 degrees or laid horizontally in trays at high humidity.

About mid-October the bulblets are planted in the field 10 cm (4 in.) deep and an inch apart in double rows spaced 91 cm (36 in.) apart. Here they remain for the following season. They are dug in September as yearling bulbs and again re-

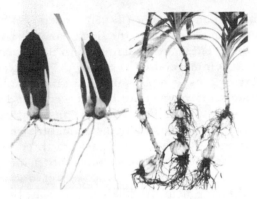

FIGURE 15–7 Two methods of lily propagation. *Left:* Bulblets are produced at the base of individual bulb scales; this method of propagation is possible for nearly all lily species. *Right:* Underground stem bulblets; these are produced only by some lily species.

planted, this time 12.5 cm (6 in.) deep and 10 to 12.5 cm (4 to 6 in.) apart in single rows. At the end of the second year, they are dug and sold as commercial bulbs.

Digging is done in September after the stem is pulled. The bulbs are graded, packed in peat moss, and shipped. Commercial bulbs range in size from 17.5 to 20 cm (7 to 10 in.) in circumference. Lily bulbs must be handled carefully so they will not be injured and must be kept from drying out. The fleshy roots should also be kept in good condition. For long-term storage that will prevent flowering, the bulbs should be packed in polyethylene-lined cases with peat moss at 30 to 50 percent moisture and stored at −1°C (31°F) (*61*).

Control of viruses, fungus diseases, and nematodes during propagation is important in bulb production. Methods of control include using pathogen-free stocks for propagation (*3, 53*), growing plants in pathogen-free locations with good sanitary procedures, and treating the bulbs with fungicides.

Aerial stem bulblets, commonly known as *bulbils,* are formed in the axil of the leaves of some species, such as *Lilium bulbiferum, L. sargentiae, L. sulphureum,* and *L. tigrinum.* Bulbils develop in the early part of the season and fall to the ground several weeks after the plant flowers. They are harvested shortly before they fall naturally and are then handled in essentially the same manner as

underground stem bulblets. Increased bulbil production can be induced by disbudding as soon as the flower buds form. Likewise, some lily species that do not form bulbils naturally can be induced to do so by pinching out the flower buds and a week later cutting off the upper half of the stems. Species that respond to the later procedure include *Lilium candidum, L. chalcedonicum, L. hollandicum, L. maculatum,* and *L. testaceum.* (*54*).

Bulbil formation rarely occurs in Easter lily (*L. longiflorum*). Exogenous cytokinin application to foliage induced large numbers of bulbils in leaf axils on aboveground stems (*46*), which might be used for Easter lily propagation.

Stem Cuttings

Lilies may be propagated by stem cuttings. The cutting is made shortly after flowering. Instead of roots and shoots forming on the cutting, as would occur in other plants, bulblets form at the axils of the leaves and then produce roots and small shoots while still on the cutting.

Leaf-bud cuttings, made with a single leaf and a small heel of the old stem, may be used to propagate a number of lily species. A small bulblet will develop in the axil of the leaf. It is handled in the same manner as for the other methods described here.

Bulblet Formation on Scales (Scaling)

In **scaling,** individual bulb scales are separated from the mother bulb and placed in growing conditions so that adventitious bulblets form at the base of each scale (Fig. 15-7). Three to five bulblets will develop from each scale. This method is particularly useful for rapidly building up stocks of a new cultivar or to establish pathogen-free stocks. Almost any lily species can be propagated by scaling (*38, 64*).

In Japan and the United States, commercial-sized Easter lily bulbs are produced after two to three growing seasons from scaling. The normal development of **de novo** bulblet formation during scale propagation is (a) callus formation followed by organized meristem formation, (b) leaf primordia formation, (c) formation of a scale bulblet which enlarges, and (d) leaf emergence from the primordia (*37*).

In the Netherlands there is a program to produce forcible commercial bulbs after only one growing season from scaling with **preformed** bulblets attached, that is, taking scales which had been induced to form a small bulblet. Basically, there are four types of plant development during detached scale propagation of lilies with preformed bulblets attached (*64*). These types include:

ETP: epigeous-type plant which produces a direct bolting bulblet with foliage leaves and no foliage scales. This is the most desirable development form.

HETP: hypoepigeous-type plant which first forms a rosette with foliage scales and bolts at a later stage of growth.

HTP: hypogeous-type plant which forms only a rosette with foliage scales—both HTP and HETP are susceptible to diseases and spray damage since leaves remain close to the soil.

NLB: non-greenleaf bulblet which does not produce foliage since the bulbs remain dormant (Fig. 15–8). Bulblet formation is influenced by temperature (*64*).

Increasing the duration of bulb storage prior to propagation by bulb scales will decrease the harvest weight of newly generated bulbs (*38*). Outer or middle scales increased bulb weight and number of forcible commercial bulbs, while innermost scales resulted in low weights and few forcible commercial bulbs.

Scaling is done soon after flowering in midsummer, although it might be done in late fall or even in midwinter. The bulbs are dug, the outer two layers of scales are removed and the mother bulb is replanted for continued growth. It is possible to remove the scales down to the core, but this will reduce subsequent growth of the mother bulb. The scales should be kept from drying and handled so as to avoid injury. Scales with evidence of decay should be discarded and the remaining ones dusted with or dipped into a fungicide. Naphthaleneacetic acid (1 ppm) will stimulate bulblet formation.

Scales may be handled by several methods. (a) They may be field planted in beds or frames no more than 6.25 cm (2½ in.) deep. Bulblets form on the scale during the first year to produce **yearlings.** These are replanted for a third year to produce **commercials.** (b) Scales may be placed in trays or flats of moist sand, peat moss, sphagnum moss, or vermiculite for six weeks at 18 to 21°C (65 to 70°F). The scales are inserted vertically to about half their length. Small bulblets and roots should form at the base within three to six weeks. The scales are transplanted either into the open ground or into pots or flats of soil, and then planted in the field the following spring. Subsequent treatment is the same as described for underground bulblets. (c) Scales may be packed in layers in moist vermiculite in plastic lined boxes. These are incubated for 6 to 12 weeks at 15 to 26°C (60 to 80°F). The lower temperature will encourage rooting. The boxes are then placed in cold temperature over winter and planted in rows in the spring. Two years is required to reach planting size (*40*).

A simple method of propagating lilies by scales, after removing them from the bulb, is to dust them with a fungicide then place the scales so they are not touching in damp vermiculite in a polyethylene bag. The bag is closed and tied, then set for six to eight weeks where the temperature is fairly constant at about 21°C (70°F). After bulblets are well developed at the base of the scales, the bag with the scales still inside should be refrigerated at 2 to 4.5°C (35 to 40°F) for at least eight weeks to overcome dormancy. The small bulblets can then be potted and placed in the greenhouse or out-of-doors for further growth.

Tissue culture utilizing aseptic techniques has been used as a means of scaling, particularly to obtain, maintain, and multiply virus-free stock (see Chaps. 16 and 17 and Fig. 15–9).

Keep in mind that a bulb is a compressed shoot system where apical dominance limits axillary bud growth. If new bulblets are to be formed then apical dominance must be removed using the same principle as in pruning an aboveground shoot structure. An exception to this is daughter bulb formation in the bulbous *Iris* cultivar 'Ideal'. In contrast to certain other bulbous plants, apical dominance does not directly limit daughter bulb number in *Iris,* but does prevent lateral bud sprouting (*12*).

JUNE   OCTOBER

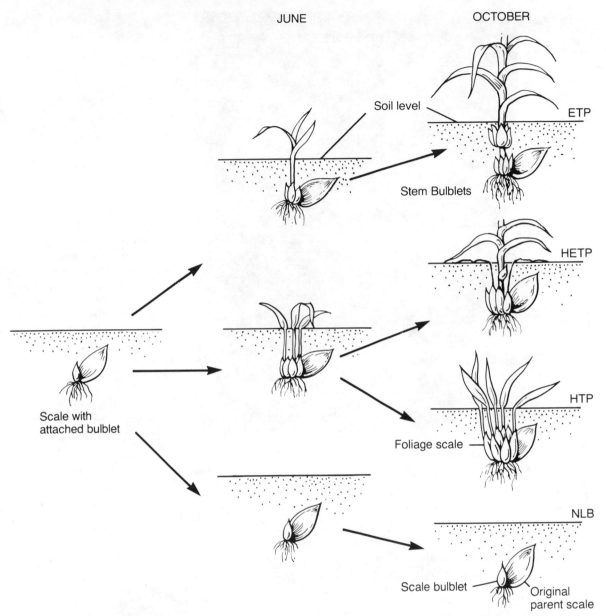

FIGURE 15-8 Types of development during scale propagation with attached bulblets. ETP, epigeous type plant (bulblet bolts and produces the most desirable form); HETP, hypoepigeous-type plant (rosettes first, then bolts); HTP, hypogeous-type plant (only a rosette with foliage leaves forms—not commercially desirable); NLB, non-green-leaf bulblet (bulblet remains dormant and does not bolt—not desirable). (Redrawn from Van Tuyl (*64*).)

FIGURE 15–9   *Above:* Micropropagation of *Iris hollandica* 'Prof Blaauw' from single- and double-scale explants. Shoots formed on bulb-scale explants (*upper left*) are sliced longitudinally and subcultured. New shoots and bulblets are then regenerated from these shoots (*upper right*). *Hyacinthus orientalis* 'Pink Pearl', form bulblets (*lower left*). The leaves from the bulblets are subcultured and many plantlets (*lower right*) are regenerated. (Courtesy P. C. G. van der Linde and J. van Aartrijk.)

Basal Cuttage

The hyacinth is the principal plant propagated by this method, although others such as *Scilla* can be handled in this way. Specific methods include "scooping" and "scoring" to remove apical dominance and encourage bulblet formation (*8, 19*). Mature bulbs which have been dug after the foliage has died down and are 17 to 18 cm or more in circumference are used. In **scooping,** the entire basal plate is scooped out with a special curve-bladed scalpel, a round-bowled spoon, or a small-bladed knife. Adventitious bulblets develop from the base of the exposed bulb scales. Depth of cutting should be enough to destroy the main shoot. In **scoring,** three straight knife cuts are made across the base of the bulb, as shown in Fig. 15–10, each deep enough to go through the basal plate and the growing point. Growing points in the axils of the bulb scales grow into bulblets.

To combat decay that may develop during the later incubation period, infected bulbs should be discarded; the tools disinfected frequently with alcohol, Physan, formalin, or mild carbolic acid

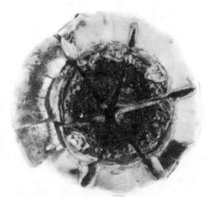

FIGURE 15–10   Basal cuttage. Hyacinth bulb which has been scored. Note the bulblets starting to appear.

solution; and the cut bulbs dusted with a fungicide. Most important is to callus the bulbs at about 21°C (70°F) for a few days to a few weeks in dry sand or soil or in open trays, cut side down. After callusing, the bulbs are incubated in trays or flats, in dark or diffuse light, at 21°C (70°F), which is increased to 29.5 to 32°C (85 to 90°F) over a two-week period and held at high humidity (85 percent) for 2½ to 3 months.

The mother bulbs are planted about 10 cm (4 in.) deep in nursery beds in the fall. The next spring bulblets produce leaves profusely. Normally the mother bulb disintegrates during the first summer. Annual digging and replanting of the graded bulblets is required until they reach flowering sizes. Bulbs for greenhouse forcing should be 17 cm (6¾ in.) or more in circumference; bulbs for bedding should be 14 to 17 cm (5½ to 6¾ in.) in circumference (2). On the average, a scooped bulb will produce 60 bulblets, but four to five years will be required to produce flowering sizes; and a scored bulb will produce 24, requiring three to four years (8).

Hot water treatment of hyacinth bulbs used for controlling *Xanthomonas hyacinthi* has been reported to induce bulblet formation and could substitute for basal cuttage (1). Bulbs must be at least 1.5 cm in circumference and treated from mid-July to early September. Treatment is 43°C (110°F) for four days or 38°C (100°F) for 30 days at relative humidity of 60 to 70 percent.

Leaf Cuttings

This method is successful for blood lily (*Haemanthus*), grape hyacinth (*Muscari*), hyacinth, and Cape cowslip (*Lachenalia*) (13), although the range of species is probably wider.

Leaves are taken at a time when they are well developed and green. An entire leaf is cut from the top of the bulb and may in turn be cut into two or three pieces. Each section is placed in a rooting medium with the basal end several inches below the surface, as described for rooting cuttings. The leaves should not be allowed to dry out, and bottom heat is desirable. Within two to four weeks small bulblets form on the base of the leaf and roots develop. At this stage the bulblets are planted in soil.

Bulb Cuttings

Among plants that respond to the bulb-cutting method of propagation are the *Albuca, Chasmanthe, Cooperia, Haemanthus, Hippeastrum, Hymenocallis, Lycoris, Narcissus, Nerine, Pancratium, Scilla, Sprekelia,* and *Urceolina* (13).

A mature bulb is cut into a series of eight to 10 vertical sections, each containing a part of the basal plate. These sections are further divided by sliding a knife down between each third or fourth pair of concentric scale rings and cutting through the basal plate. Each of these fractions makes up a bulb cutting consisting of a piece of basal plate and segments of three or four scales. This technique is also referred to as **bulb chipping** or **fractional scale-stem cuttage** (7).

The bulb cuttings are planted vertically in a rooting medium, such as peat moss and sand, with just their tips showing above the surface. The subsequent technique of handling is the same as for ordinary leaf cuttings. A moderately warm temperature, slightly higher than for mature bulbs of that kind, is required. New bulblets develop from the basal plate between the bulb scales within a few weeks, along with new roots. At this time they are transferred to flats of soil to continue development.

A variation of this method, called **twin scaling** (51, 58), involves dividing bulbs into portions, each containing a pair of bulb scales and piece of basal plate. These are kept in plastic bags and incubated with damp vermiculite three to four weeks at 21°C (70°F) or planted in compost. Bulbils then develop at the edge of the basal plate (Fig. 15–11) (22).

Twin-scale types most likely to produce a bulbil consist of a third-year leaf base plus a second-year bulb scale, or scales and (or) leaf bases. Types with first-year organs or a flower stalk are less productive but can still form bulbils (21). Twin scaling and tissue culture remain the two most important multiplication programs for *Narcissus*.

Micropropagation by Tissue Culture

Many bulb species are highly adapted to micropropagation techniques, utilizing enhanced axil-

FIGURE 15-11   Twin scale propagation of narcissus cut from whole bulbs and kept at 23°C in moist vermiculite for two months. (From A. R. Rees, *The growth of bulbs*, Academic Press, New York, 1972.)

TABLE 15-1

Partial list of bulbous crops capable of in vitro propagation and their explant source

Family	Genus	Bulbscale	Leaf	Stem	Bud	Flower Petal
				Explant source		
Liliaceae	Lilium	×	×	×	×	×
	Tulipa	×		×	×	×
	Hyacinthus	×	×	×	×	×
	Ornithogalum					
	Muscari					
	Fritillaria					
	Lachenalia					
	Alstroemeria					
Iridaceae	Iris	×		×	×	
	Gladiolus			×	×	×
	Freesia					
	Crocus					
Amaryllidaceae	Narcissus	×	×	×		
	Hippeastrum	×		×		
	Nerine	×		×		
	Haemanthis					
	Zephyranthes					
	Eucharis					

Source: Compiled from J. Van Aartrijk and P. C. C. Van der Linde (*62*).

lary shoot formation, adventitious shoot formation, bulblet induction on scales, or very often, flower scapes (Table 15–1 and Fig. 15–9). These methods are especially valuable to multiply new cultivars rapidly and to develop, maintain, and produce the initial stock of specific virus-tested propagation material. These methods are described in Chaps. 16 and 17.

CORMS

Definition and Structure

A **corm** is the swollen base of a stem axis enclosed by the dry, scale-like leaves. In contrast to the bulb, which is predominantly leaf scales, a corm is a solid stem structure with distinct nodes and internodes. The bulk of the corm consists of storage tissue composed of parenchyma cells. In the mature corm, the dry leaf bases persist at each of these nodes and enclose the corm. This covering, known as the **tunic,** protects it against injury and water loss. At the apex of the corm is a terminal shoot that will develop into the leaves and the flowering shoot. Axillary buds are produced at each of the nodes. In a large corm, several of the upper buds may develop into flowering shoots,

but those nearer the base of the corm are generally inhibited from growing. However, should something prevent the main buds from growing, these lateral buds would be capable of producing a shoot (see Figs. 15–12 and 15–13).

Two type of roots are produced from the corm: a *fibrous* root system developing from the base of the mother corm, and enlarged, fleshy *contractile* roots developing from the base of the new corm. The latter roots apparently develop in response to the fluctuating temperatures near the soil surface and with exposure of the leaves to light. At lower soil depths temperature fluctuations decrease (29), and contraction ceases once the corm is at a given depth.

Growth Pattern

Gladiolus and crocus are typical cormous plants. The gladiolus is semihardy to tender and, in areas with severe winters, the corm must be stored over winter and replanted in the spring. At the time of planting, the corm is a vegetative structure (24, 48). New roots develop from its base, and one or more of the buds begin to develop leaves. Floral initiation takes place within a few weeks after the shoot begins to grow. At the same time the base of the shoot axis thickens, and a new corm for the

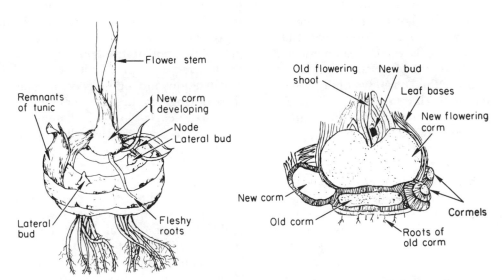

FIGURE 15–12 Gladiolus corm. *Left:* External appearance. *Right:* Longitudinal section showing solid stem structure.

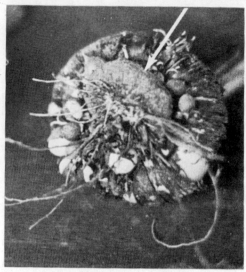

FIGURE 15–13   Stage of gladiolus corm development during the latter part of the growing season. The remnants of the originally planted corm (see arrows) are evident just below the newly formed corm. Many small white cormels also have been produced.

succeeding year begins to form above the old corm. Stolonlike structures bearing miniature corms or **cormels** on their tip develop from the base of the new corm.

In gladiolus there can be competition for assimilates between flower and corm development, which is controlled by photoperiod (*56*). Short-day (SD) conditions stimulate corm development, while corms are checked under long-days (LD), when competition with flower development and anthesis occurs. However, plant size is related to size of the corm, since final corm weight is greater under LD conditions (larger plants) than SD (smaller plants), which apparently is due to more total assimilates being available in the larger LD plants for corm production (*56*).

As new corms continue to enlarge, the old corm begins to shrivel and disintegrate as its contents are utilized in flower production. After flowering, the foliage continues to manufacture food materials, which are stored in the new corm. At the end of the summer, when the foliage dries, there are one or more new corms and perhaps a great number of little cormels. The corms are dug and stored over winter until they are planted the following spring.

Propagation

New Corms

Propagation of cormous plants is principally by the natural increase of new corms. Flower production in corms, as in bulbs, depends upon food materials stored in the corm the previous season, particularly during the period following bloom. In gladiolus, cool nights and long growing periods are favorable for production of very large corms. Fertilization and other good management practices during bloom have their greatest effect on the next year's flowers. Plants are left in the ground for two months following blooming, or until frost kills the tops. After digging, the plants are placed in trays with a screen or slat bottom arranged to allow air to circulate between them, and cured at about 32°C (90°F) at 80 to 85 percent relative humidity. A few hours at 35°C (95°F) may be helpful. Then the new corms, old corms, cormels, and tops can be easily separated. The corms are graded according to size, sorted to remove the diseased ones, treated with a fungicide, and returned to a 35°C (95°F) temperature for an additional week. This curing process suberizes the wounds and helps combat *Fusarium* infection. The corms are then stored at 5°C (40°F) with a relative humidity of 70 to 80 percent in well-aerated rooms to prevent excessive drying. It may also be desirable to treat them with a suitable fungicide (*35*) immediately before planting.

Cormels

Cormels are miniature corms that develop between the old and the new corms. One or two years' growth is required for them to reach flowering size. Shallow planting of the corms, only a few inches deep, results in greater production of cormels; increasing the depth of planting reduces cormel production.

Cormels are separated from the mother corms and stored over winter for planting in the spring. Dry cormels become very hard and may be slow to start growth the following spring, but if they are stored at about 5°C (40°F) in slightly moist peat moss, they will stay plump and in good condition. Soaking dry cormels in cool running water for one to two days and holding them moist until planting at first sign of root development will hasten the onset of growth.

Disease-free cormels can be obtained by hot-water treatments, which should be done between two and four months after digging. Holding cormels at room temperatures to keep them dormant will increase tolerance to this treatment. The cormels are soaked in water at air temperature for two days, then placed in a 1:200 dilution of commercial 37 percent formaldehyde for four hours, and then immersed in a water bath at 57°C (135°F) for 30 minutes. At the end of the treatment, the cormels are cooled quickly, dried immediately, and stored at 5°C (40°F) in a clean area with good air circulation.

The cormels are planted in the field in furrows about 5 cm (2 in.) deep in the manner of planting large seeds. Only grass-like foliage is produced the first season. The cormel does not increase in size but produces a new corm from the base of the stem axis, in the manner described for full-sized corms. At the end of the first growing season, the beds are dug and the corms separated by size. A few of the corms may attain flowering size, but most require an additional year of growth. Size grades in gladiolus are determined by diameter. There are seven grades, the smallest 0.9 to 1.2 cm (⅜ to ½ in.) in diameter, the largest 5 cm (2 in.) or more (*2*).

Division of the Corm

Large corms can be cut into sections, retaining a bud with each section. Each of these should then develop a new corm. Segments should be dusted with a fungicide because of the great likelihood of decay of the exposed surfaces.

TUBERS

Definition and Structure

A **tuber** is a special kind of swollen, modified stem structure that functions as an underground storage organ (Fig. 15–14). The potato (*Solanum tuberosum*) is a notable example of a tuber-producing plant, as is the *Caladium*, grown for its striking foliage, and the Jerusalem artichoke (*Helianthus tuberosus*).

A tuber has all the parts of a typical stem but is very much swollen. Externally the **eyes**, present in regular order over the surface, represent nodes, each consisting of one or more small buds subtended by a leaf scar. The arrangement of the nodes is a spiral, beginning with the terminal bud on the end opposite the scar resulting from the attachment to the stolon. The terminal bud is at the apical end of the tuber, oriented farthest (distally) from the crown of the plant. Consequently, tubers show the same apical dominance as any stem. Apical dominance also occurs with tuber pieces of yam (*Dioscorea alta* L.). Vine production is much greater with proximal than with distal tuber sections (*49*).

Internally, a potato tuber is composed of enlarged parenchyma-type cells containing large amounts of starch. It has the same internal structure as any stem with pith, vascular areas, and cortex.

Growth Patterns

The tuber is a storage structure that is produced in one growing season, remains dormant during the winter, and then functions to regenerate new

FIGURE 15–14   Tubers of white (Irish) potato showing their development from stolons arising from stem tissue. Note adventitious root system originating from main
plant stem. Tuber is attached to stolon at
the tuber's morphological basal (proximal)
end.

shoots the following spring. After a new seasonal
cycle begins, the shoots utilize the stored food in
the old tuber, which then disintegrates (5, 17, 59).
As the main shoot develops, adventitious roots are
initiated at the base, and lateral buds grow out
horizontally into the soil to produce elongated,
etiolated stems (*stolons*) as shown in Fig. 15–14.
Continued elongation of the stolon takes place
during long photoperiods and is associated with
the presence of auxin and a high gibberellin level.
Tuberization begins with inhibition of terminal
growth and the initiation of cell enlargement and
division in the subapical region of the stolons.
This process is associated with short or intermediate daylengths, reduced temperatures (particularly at night), high light intensity, low mineral
content, increased cytokinins and inhibitors
(ABA), and a reduction in gibberellin levels in the
plant (42).

Tuberization is caused by the production of
a tuber-inducing substance which is linked to a
tuberization regulatory protein that is produced
in the leaves and the mother tuber (14, 47). It
seems to be necessary for the stolon tip to have
attained a particular physiological age. Continued
tuber enlargement is dependent on a continuing
adequate supply of photosynthate. Conditions
that favor rapid and luxurious plant growth above
ground, such as an abundance of nitrogen, or
high temperatures, are not conducive to tuber
production. In the fall, the tops of the plants die
down and the tubers are dug. At this time, the
buds of potato tubers are dormant for six to eight
weeks. This condition must disappear before
sprouting will take place.

Propagation

Division

Propagation by tubers can be done either by
planting the tubers whole or by cutting them into
sections, each containing one or more axillary
buds or ''eyes.'' These small pieces of tuber to be
used for propagation are commonly referred to as
''seed'' potatoes (Fig. 15–15). The weight of the
tuber piece should be 28 to 56 g (1 to 2 oz) to
provide sufficient stored food for the new plant to
become well established, although in some areas
only the ''eyes'' with a very small piece of tuber
are sold and planted.

Division of tubers is done by machine or
with a sharp knife shortly before planting. The cut
pieces should be stored at warm (20°C; 68°F)
temperatures and relatively high humidities (90
percent) for two to three days prior to planting.
During this time the cut surfaces heal and become
suberized, which protects the ''seed'' piece
against drying and decay. Treatment of potato tubers prior to cutting for the control of *Rhizoctonia*
and scab may be desirable.

Caladium tubers are produced commercially
in Florida (55). The tubers are cut into sections,
usually two buds per piece. These are planted 7.6
to 9 cm (3 to 4 in.) deep, 9 to 15 cm (4 to 6 in.)
apart in rows 45 to 60 cm (18 to 24 in.) apart.
Harvest begins in November. After harvest, the
tubers are dried in open sheds for six weeks or
artificially dried for 48 hours. Further storage
should be at temperatures above 16°C (60°F).

FIGURE 15-15 "Seed" potato, which in fact is a vegetative propagule made of a diced tuber. The shoots form from eyes (axillary buds) on the tuber and the root system occurs from the newly elongating shoot. Potatoes can be commercially produced from true seed, but the production period is longer and the clone can be lost.

TUBERCLES

Begonia evansiana and the cinnamon vine (*Dioscorea batatas*) produce small aerial tubers, known as **tubercles,** in the axils of the leaves. These tubercles may be removed in the fall, stored over winter, and planted in the spring (*13*). Short days induce tuberization (*47*).

TUBEROUS ROOTS AND STEMS

Definition and Structure

The tuberous roots and stems class includes several types of structures with thickened tuberous growth that functions as storage organs. Botani-

cally, these differ from true tubers, although common horticultural usage sometimes utilizes the term "tuber" for all of them.

Fleshy and Tuberous Roots

Various herbaceous perennial species show massive enlargement of secondary roots. Typical examples are sweet potato (*Ipomoea batatus*) (Figs. 15-16 and 15-17), cassava (*Manihot esculenta*), and *Dahlia* (Figs. 15-16 and 15-18). Sweet potato has a **fleshy root** from which both adventitious buds and roots are produced, while dahlia has a **tuberous root** with a section of the attached crown containing a preformed bud for shoot development. With dahlia, fibrous roots are commonly produced on the opposite (*distal*) end, and tuberous roots closer to the crown or stem (proximal) end.

Tuberous Stems

Tuberous stems are produced by the enlargement of the hypocotyl section of the seedling plant, but may include the first nodes of the epicotyl and the upper section of the primary root (*20, 28, 52*). Typical plants with this structure are the tuberous begonia (*Begonia* × *tuberhybrida*) and cyclamen (*Cyclamen persicum*). These structures have a vertical orientation with one or more vegetative buds produced on the upper end or crown. Fibrous roots are produced on the basal part of the structure.

Growth Pattern

Tuberous roots are biennial. They are produced in one season, after which they go dormant as the herbaceous shoots die. These function as storage organs to allow the plant to survive the dormant period. In the following spring, buds from the crown produce new shoots, which utilize the food materials from the old root during their initial growth. The old root then disintegrates, and new tuberous roots are produced, which in turn maintain the plant through the following dormant period (*17, 33, 43*).

Photoperiod, not temperature, is the dominant controlling factor of tuberization in dahlia. Tuberized roots are formed under short-day conditions (five inductive cycles of an 11- to 12-hour critical photoperiod) or when a growth retardant

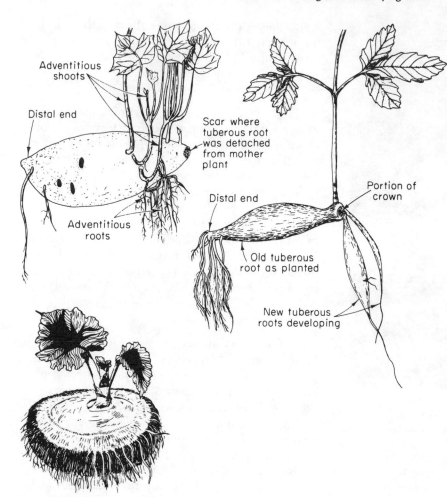

FIGURE 15-16 Types of fleshy and tuberous roots and shoots. *Top left:* Sweet potato fleshy root showing adventitious shoots. *Top right:* Dahlia during early stages of growth. The old tuberous root piece will disintegrate in the production of the new plant; the new roots can be used for propagation. *Below left:* A tuberous begonia stem, showing its vertical orientation. This type continues to enlarge each year.

is applied (*43*). Fibrous roots form under long-days or when gibberellic acid is applied. Apparently, conditions favoring tuberous root growth are antagonistic to vegetative (shoot) growth (*43*). Tuberous root formation in cassava is attributed to cytokinin control of meristematic activity (*41*).

Tuberous stems of tuberous begonia and cyclamen, on the other hand, are perennial and continue to enlarge laterally every year (*20*). Normally, these species are propagated by seed but the "tuber" can be dug, stored, and used for annual propagation over a period of years.

Propagation

Division

The usual method for propagating tuberous roots is by dividing the crown so that each section bears a shoot bud. Dahlia, for example, is dug with its cluster of roots intact, dried for a few days, and stored at 4 to 10°C (40 to 50°F) in sawdust or vermiculite. Open storage may result in shriveling. The root cluster is divided in the late winter or spring shortly before planting. In warm, moist conditions the buds begin to grow, and the tubers can be divided with assurance that each section will have a bud.

The tuberous stem of the tuberous begonia can be divided shortly after growth starts in the spring as long as each section has a bud. To combat decay, the cut surface should be dusted with a fungicide and each section dried for several days after cutting and before placing in a moist medium.

Adventitious Shoots

The fleshy roots of a few species of plants such as sweet potato have the capacity to produce

FIGURE 15–17   Propagation by sweet potato fleshy roots. Adventitious shoots ("slips") develop when the mother root is placed under warm, moist conditions. *Left:* Root on the left has been subjected to 34.5°C (100°F) for 26 hours to overcome the proximal dominance shown on the unheated root on the right. (Courtesy Welch and Little (*68*).) *Right:* After slips are well rooted, they are removed and planted.

FIGURE 15–18   Propagation of dahlia. To produce a new plant each separate tuberous root must have a section of the crown bearing a shoot bud, as shown by the detached root on the left.

adventitious shoots if subjected to the proper conditions. The roots are laid in sand so that they do not touch one another and are covered to a depth of about 5 cm (2 in.). The bed is kept moist. The temperature should be about 27°C (80°F) at the beginning and about 21 to 24°C (70 to 75°F) after sprouting has started. As the new shoots, or **slips,** come through the covering, more sand is added so that eventually the stems will be covered for 10 to 12.5 cm (4 to 5 in.). Adventitious roots develop from the base of these adventitious shoots. After the slips are well rooted, they are pulled from the parent plant and transplanted into the field. If sweet potato roots are cut in half and the pieces are subjected to 43°C (110°F) for about 26 hours, slip production increases. This treatment overcomes the apical dominance and also controls nematodes and fungus diseases (*68*) (see Fig. 15–17). This procedure can be modified in certain cultivars of the sweet potato by dividing the fleshy root into 20- to 25-g pieces, treating with a fungicide, then giving a presprouting treatment for four weeks of 26.5°C (80°F) and 90 percent relative humidity before planting (*6*).

Cyclamen can be multiplied vegetatively by cutting off the upper one-third of the tuberous stem and notching the surface into 1-cm squares. Adventitious shoots develop (12 to 13 per tuber) and can be used for propagation (45).

Leafy Cuttings

Vegetative propagation in plants of this group, such as dahlia or tuberous begonia, is often more satisfactory with stem, leaf, or leaf-bud cuttings. The cuttings will develop tuberous roots at their base. This process can be stimulated if the stem cutting initially includes a small piece of the fleshy root or stem. Vine cuttings from established beds also can be used in sweet potato propagation.

RHIZOMES

Definition and Structure

A **rhizome** is a specialized stem structure in which the main axis of the plant grows horizontally at or just below the ground surface. A number of economically important plants, such as bamboo, sugar cane, banana, and many grasses, as well as a number of ornamentals, such as rhizomatous *Iris* and lily-of-the-valley, have rhizome structures. Most are monocotyledons, although a few dicotyledons—for example, low-bush blueberry (*Vaccinium angustifolium*)—have analogous underground stems classed as rhizomes. Many ferns and lower plant groups have rhizomes or rhizome-like structures.

Figure 15–19 shows structural features of a rhizome (71). The stem appears segmented because it is composed of nodes and internodes. A leaf-like *sheath* is attached at each node; it encloses the stem and, in an expanded form, becomes the foliage leaves. When the leaves and sheaths disintegrate, a scar is left at the point of attachment identifying the node and giving a segmented appearance. Adventitious roots and lateral growing points develop in the vicinity of the node. Upright-growing, aboveground shoots and flowering stems (*culms*) are produced either terminally from the rhizome tip or from lateral branches.

Two general types of rhizomes are found (39). The first (the **pachymorph**) is illustrated by rhizomatous *Iris* in Fig. 15–20 and by ginger in Fig. 15–21. The rhizome is thick, fleshy, and shortened in relation to length. It appears as a many-branched clump made up of short, individual sections. It is *determinate;* that is, each clump terminates in a flowering stalk, growth continuing only from lateral branches. The rhizome tends to be oriented horizontally with roots arising from the lower side.

The second type (the **leptomorph**) is illustrated by the lily-of-the-valley in Fig. 15–19. The rhizome is slender with long internodes. It is *indeterminate;* that is, it grows continuously in length from the terminal apex and from lateral branch rhizomes. The stem is symmetrical and has lateral buds at most nodes, nearly all remaining dormant. This type does not produce a clump but spreads extensively over an area.

Intermediate forms between these two types also exist. These are called **mesomorphs** (39).

Growth Pattern

Rhizomes grow by elongation of the growing points produced at the terminal end and on lateral branches. Length also increases by growth in the intercalary meristems in the lower part of the internodes. As the plant continues to grow and the older part dies, the several branches arising from one plant may eventually become separated to form individual plants of a single clone.

Rhizomes exhibit consecutive vegetative and reproductive stages, but the growth cycle differs somewhat in the two types described. In the pachymorph rhizome of *Iris* (Fig. 15–20), a growth cycle begins with the initiaton and growth of a lateral branch on a flowering section. The flowering stalk dies, but these new lateral branches produce leaves and grow vegetatively during the remainder of that season. Continued growth of the underground stem, storage of food, and the production of a flower bud at the conclusion of the vegetative period are dependent upon photosynthesis. Consequently, foliage should not be removed during this period. A flowering stalk is produced the following spring and no further terminal growth can take place. In general, plants with this structure flower in the spring and grow vegetatively during the summer and fall.

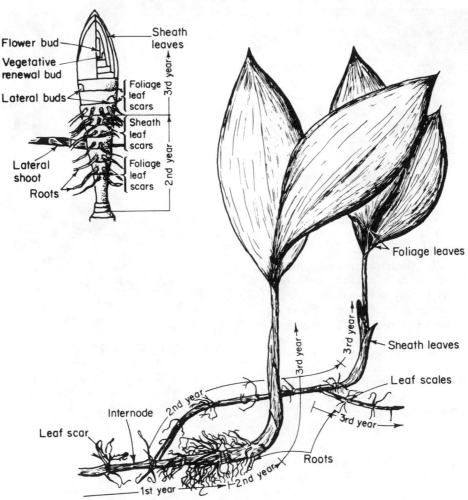

FIGURE 15–19 Structure growth cycle of lily-of-the-valley (*Convallaria majalis*). *Right:* Section of rhizome as it appears in late spring or early summer with one-, two-, or three-year-old branches. A new rhizome branch begins to elongate in early spring and terminates in a vegetative shoot bud by the fall. The following spring leaves of the bud unfold; food materials manufactured in the leaves by photosynthesis are accumulated in the rhizome. Growth the second season is again vegetative. Early in the third season a flower bud begins to form, and at the same time a vegetative growing point forms in the axil of the last leaf. *Top left:* Section of the three-year-old branch showing terminal flower bud and lateral shoot bud enclosed in leaf sheaths. Such a section is sometimes known as **pip** or **crown** and is forced for spring bloom. In the early spring the flowering shoot expands, blooms, and then dies down, the shoot bud beginning a new cycle of development. (Redrawn from Zweede (*71*).)

FIGURE 15-20 Structure of an iris (rhizomous type) plant as it appears about the time of flowering. A two-year-old section that had flowered the previous year and is now dying back is shown at D. The lateral branch arising from it consists of the one-year-old vegetative section (B); the current-season, lateral vegetative branches (A); and the current-season, terminal flowering shoot (C). The vegetative shoot (A) will flower the following year.

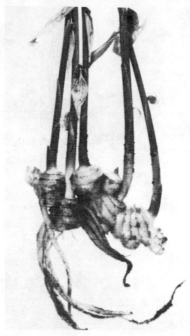

FIGURE 15-21 Rhizome of the tropical ginger plant (*Zingiber officinale*). This is easily propagated by division of the thickened rhizome, which is the source of commercial ginger.

Plants with a leptomorph habit as a general rule (with exceptions) grow vegetatively during the beginning of the growth period and flower later in the same period. The length of time during which an individual rhizome section remains vegetative varies with different kinds of plants. An individual branch in the lily-of-the-valley in Fig. 15-19, for instance, is vegetative three years before a flower bud forms.

Bamboo is divided into *clump* growers (pachymorphs), which have constricted rhizomes, and *running* bamboos (leptomorphs), which spread rapidly by vigorous rhizomes that grow several feet or more (*57*). Generally speaking, the pachymorphs are more desirable ornamentally. Some bamboo species remain vegetative for many years, but then they change abruptly and the entire plant produces flowers.

In some rhizomatous plants, such as blueberry, rhizome development is increased by higher temperatures and a long photoperiod, and is correlated with vigorous aboveground growth and high-nitrogen status (*30, 60*).

Propagation

Division of Clumps and Rhizomes

Division is the usual procedure for propagating plants with a rhizome structure, but the procedure may vary somewhat with the two types. In pachymorph rhizomes, individual sections (or *culms*) are cut off at the point of attachment to the rhizome, the top is cut back, and the piece is transplanted to the new location. Leptomorph rhizomes can be handled in essentially the same way by removing a single lateral "offshoot" from the rhizome and transplanting it. The tip of the lily-of-the-valley rhizome bearing a flower bud (see Fig. 15-19) called a "pip" is removed along with the rooted section below and transplanted.

The bird of paradise (*Strelitzia reginae*) has a slow rate of multiplication when propagated by rhizome division. Mechanical induction of branching to eliminate apical dominance of a branch leaf sheath (fan) attached to the rhizome encourages lateral shoot formation and rapid multiplication (Fig. 15-22) (*63*). Division is usually

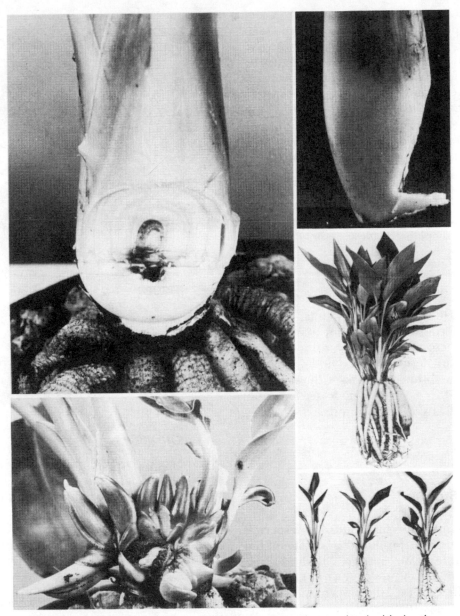

FIGURE 15-22 Propagation of *Strelitzia reginae* by mechanical induction of branching to eliminate apical dominance and increase multiplication rate. To remove the apex, a transverse lateral incision is made above the basal plate through the basal leaf sheath of a branch, keeping the leaves in contact with the roots (*upper left and right*). Lateral shoot formation occurs 4.5 months after excision of the apex (*lower left*). After one year, multiple clusters with roots (*middle right*) can be separated into individual plantlets (*lower right*). (Courtesy P. A. Van de Pol (*63*).)

carried out at the beginning of a growth period (as in early spring) or at or near the end of a growth period (i.e., in late summer or fall).

Propagation is carried out by cutting the rhizome into sections, being sure that each piece has at least one lateral bud, or "eye"; it is essentially a stem cutting. Bananas, for instance, are propagated in this way. This general method works well for the leptomorph rhizomes, in which a dormant lateral growing point is present at most nodes. The rhizomes are cut or broken into pieces, and adventitious roots and new shoots develop from the nodes. Rhizome-producing turf grasses, for instance, are cut up into sections and the individual "sprigs" transplanted. New plants can be established readily by this method.

Culm Cuttings

In large rhizome-bearing plants, such as bamboos, the aerial shoot, or culm, may be used as a cutting. These may be whole culm cuttings, in which the entire aerial shoot is laid horizontally in a trench. New branches arise at the nodes. A stem cutting of three- or four-node sections may be planted vertically in the ground.

PSEUDOBULBS

Definition and Structure

A **pseudobulb** (literally "false bulb") is a specialized storage structure, produced by many orchid species, consisting of an enlarged, fleshy section of the stem made up of one to several nodes (see Fig. 15–23). In general, the appearance of the pseudobulb varies with the orchid species. The differences can be used to identify species.

Growth Pattern

These pseudobulbs arise during the growing season on upright growths that develop laterally or terminally from the horizontal rhizome. Leaves and flowers form either at the terminal end or at the base of the pseudobulb, depending upon the species. During the growth period, they accumulate stored food materials and water and assist the

FIGURE 15–23 *Cattleya* orchid showing rhizome structure and upright elongated pseudobulbs as basal part of shoots.

plants in surviving the subsequent dormant period.

Propagation

Offshoots

In a few orchids, such as the *Dendrobium* species, the pseudobulb is long and jointed, being made up of many nodes. Offshoots develop at these nodes. From the base of these offshoots roots develop. The rooted offshoots are then cut from the parent plant and potted.

Division

The most important commercial species of orchids, including the *Cattleya, Laelia, Miltonia,* and *Odontoglossum,* may be propagated by dividing the rhizome into sections, the exact procedure used depending upon the particular kind of orchid. Division is done during the dormant season and preferably just before the beginning of a new period of growth. The rhizome is cut with a sharp knife back far enough from the terminal end to include four to five pseudobulbs in the new section, leaving the old rhizome section with a number of old pseudobulbs, or "back bulbs," from which the leaves have dehisced. The section is

FIGURE 15-24 A "back bulb" of a *Cymbidium* orchid was removed from the parent plant and placed in a rooting medium; the offshoot shown above then developed. This offshoot is now ready for removal and potting. When this is done, a second offshoot, or "break," should appear. (Courtesy A. Kofranek.)

then potted, whereupon growth begins from the bases of the pseudobulbs and at the nodes. The removal of the new section of the rhizome from the old part stimulates new growth, or "back breaks," to occur from the old parts of the rhizome. These new growths grow for a season and can be removed the following year.

An alternate procedure is to cut partly through the rhizome and leave it for one year. New back breaks will develop, which can be removed and potted.

Back Bulbs and Green Bulbs

Back bulbs (i.e., those without foliage) are commonly used to propagate clones of *Cymbidium.* These are removed from the plant, the cut surface is painted with a grafting compound, and they are placed in a rooting medium for new shoots to develop. When the stage shown in Fig. 15–24 is reached, the shoot can be removed from the bulb and potted. This back bulb can be repropagated and a second shoot developed from it.

Green bulbs (i.e., those with leaves) also can be used in *Cymbidium* propagation. Treatment with indolebutyric acid, either by soaking or by painting with a paste, has been shown to be beneficial (*31*).

Other Modified Stem Structures

Other forms of specialized structures include runners, stolons, crowns, offsets, and suckers, which are modified stem structures. These structures are discussed in Chap. 14 under the section on plant modifications for natural layering.

REFERENCES

1. Amano, N., and K. Tsutsui. 1980. Propagation of hyacinth by hot water treatment. *Acta Hort.* 109:279–87.

2. Amer. Assoc. Nurserymen, Inc., Comm. on Hort. Stand. 1986. *American standard for nursery stock.* Washington, D.C.: Amer. Assoc. Nurs.

3. Baker, K. F., and P. A. Chandler. 1957. Development and maintenance of healthy planting stock. Sec. 13 in *Calif. Agr. Exp. Sta. Man. 23.*

4. Ball, V., ed. 1985. *The Ball red book: Greenhouse growing* (14th ed.). Reston, VA.: Reston Publ. Co.

5. Booth, A. 1963. The role of growth substances in the development of stolons. In *The growth of the potato,* J. D. Ivins and F. L. Milthorpe, eds. London: Butterworth, pp. 99–113.

6. Bouwkamp, J. C., and L. D. Scott. 1972. Production of sweet potatoes from root pieces. *HortScience* 7(3):271–72.

7. Christie, C. B. 1985. Propagation of amaryllids: A brief review. *Proc. Inter. Plant Prop. Soc.* 35:351–57.

8. Crossley, J. H. 1957. Hyacinth culture; narcissus culture; tulip culture. *Handbook on bulb growing and forcing.* Northwest Bulb Growers Assoc., pp. 79–84, 99–104, 139–44.

9. De Hertogh, A. A., L. H. Aung, and M. Benschop. 1983. The tulip: Botany, usage, growth and development. *Hort. Rev.* 5:45–125.

10. De Hertogh, A. A., and N. Blakely. 1972. The influence of temperature and storage time on growth of basal roots of nonprecooled and precooled bulbs of *Lilium longiflorum* Thunb. cv. 'Ace'. *HortScience* 74:409–10.

11. De Hertogh, A. A., A. N. Roberts, N. W. Stuart, R. W. Langhans, R. G. Linderman, R. H. Lawson, H. F. Wilkins, and D. C. Kiplinger. 1971. A guide to terminology for the Easter lily. *HortScience* 6:121–23.

12. Doss, R. P. 1979. Some aspects of daughter bulb growth and development and apical dominance in bulbous iris. *Plant & Cell Phys.* 20:387–94.

13. Everett, T. H. 1954. *The American gardener's book of bulbs.* New York: Random House.

14. Ewing, E. E. 1985. Cuttings as simplified models of the potato plant. In *Potato physiology*, P. H. Li, ed. New York: Academic Press.

15. Genders, R. 1973. *Bulbs.* New York: Bobbs-Merrill.

16. Gould, C. J. 1953. Blights of lilies and tulips. *Plant diseases: USDA yearbook of agriculture.* Washington, D.C.: U.S. Govt. Printing Office, pp. 611–16.

17. Gregory, L. E. 1965. Physiology of tuberization in plants (tubers and tuberous roots). In *Encyclopedia of Plant Physiology,* Vol. 15. Berlin: Springer-Verlag, pp. 1328–54.

18. Griffiths, D. 1930. Daffodils. *USDA Cir. 122.*

19. ———. 1930. The production of hyacinth bulbs. *USDA Circ. 112.*

20. Haegeman, J. 1979. *Tuberous begonias.* Vaduz: A. R. Gantner Verlag KG.

21. Hanks, G. R. 1985. Factors affecting yields of adventitious bulbils during propagation of *Narcissus* by the twin-scaling technique. *Jour. Hort. Sci.* 60(4):531–43.

22. Hanks, G. R., and A. R. Rees. 1979. Twin-scale propagation of *Narcissus*: a review. *Scient. Hort.* 10:1–14.

23. Harrison, A. D. 1964. *Bulb and corm production.* Bul. 62. London: Minist. Agr., Fish. and Foods, pp. 1–84.

24. Hartsema, A. M. 1937. Periodieke ontwikkeling van *Gladious hybridum* var. Vesuvius. *Verh. Koninkl. Ned. Akad. van Wet.* 36(3):1–34.

25. ———. 1961. Influence of temperatures on flower formation and flowering of bulbous and tuberous plants. In *Encyclopedia of Plant Physiology,* Vol. 16. Berlin: Springer-Verlag, pp. 123–67.

26. Heath, O. V., and M. Holdsworth. 1948. Morphologenic factors as exemplified by the onion plant. *Symposia for the Soc. Exp. Biol.* II:326–50.

27. Huisman, E., and A. M. Hartsema. 1933. De periodieke ontwikkeling van *Narcissus pseudonarcissus* L. *Meded. Landbouwhoogesch., Wageningen,* DL. 37(Meded. No. 38, Lab. v. Plantenphys. onderz., Wageningen).

28. Jacobi, E. F. 1950. *Plantikunde voor tuinbouwscholen.* Zwolle, The Netherlands: W.E.J. Tjeenk Willink.

29. Jacoby, B., and A. H. Halevy. 1970. Participation of light and temperature fluctuations in the induction of contractile roots of gladiolus. *Bot. Gaz.* 131(1):74–77.

30. Kender, W. J. 1967. Rhizome development in the lowbush blueberry as influenced by temperature and photoperiod. *Proc. Amer. Soc. Hort. Sci.* 90:144–48.

31. Kofranek, A. M., and G. Barstow. 1955. The use of rooting substances in *Cymbidium* green bulb propagation. *Amer. Orch. Soc. Bul.* 24(11):751–53.

32. Langhans, R. W., and T. C. Weiler. 1968. Vernalization in Easter lilies. *HortScience* 3:280–82.

33. Lewis, C. A. 1951. Some effects of daylength on tuberization, flowering, and vegetative growth of tuberous-rooted begonias. *Proc. Amer. Soc. Hort. Sci.* 57:376–78.

34. Lin, P. C., and A. N. Roberts. 1970. Scale function in growth and flowering of *Lilium longiflorum,* Thunb. 'Nellie White'. *Jour. Amer. Soc. Hort. Sci.* 95(5):559–61.

35. Magie, R. O. 1953. Some fungi that attack gladioli. In *Plant Diseases: USDA yearbook of agriculture.* Washington, D.C.: U.S. Govt. Printing Office, pp. 601–7.

36. Mann, L. K. 1952. Anatomy of the garlic bulb and factors affecting bulb development. *Hilgardia* 21:195–251.

37. Matsuo, E., T. Ohkurano, K. Arisumi, and Y. Sakata. 1986. Scale bulblet malformations seen in *Lilium longiflorum* during scale propagation. *HortScience* 21:150.

38. Matsuo, E., and J. M. van Tuyl. 1986. Early scale propagation results in forcible bulbs of easter lily. *HortScience* 21:1006–7.

39. McClure, F. A. 1966. *The bamboos: A fresh perspective.* Cambridge, Mass.: Havard Univ. Press.

40. McRae, E. A. 1978. Commercial propagation of lilies. *Proc. Inter. Plant Prop. Soc.* 28:166–69.

41. Melis, R. J. M., and J. van Staden. 1985. Tuberization in cassava (*Manihot esculenta*): Cytokinin and abscisic acid activity in tuberous roots. *Jour. Plant Physiol.* 118:357–66.

42. Menzel, C. M. 1985. Tuberization in potato at high temperatures: Responses to exogenous gibberellin, cytokinin and ethylene. *Potato Res.* 28:263–66.

43. Moser, B. C., and C. E. Hess. 1968. The physiology of tuberous root development in dahlia. *Proc. Amer. Soc. Hort. Sci.* 93:595–603.

44. Mulder, R., and I. Luyten. 1928. De periodieke ontwikkeling van der Darwin tulip. *Verh. Koninkl. Ned. Akad. van Wet.* 26:1–64.

45. Nakayama, M. 1980. Vegetative propagation of cyclamen by notching of tuber. II. Effect of scooping size and notching size on the regeneration of cyclamen tuber. *Jour. Japan. Soc. Hort. Sci.* 49(2):228–34.

46. Nightingale, A. E. 1979. Bulbil formation on *Lilium longiflorum* Thunb. cv. Nellie White by foliar applications of PBA. *HortScience* 14(1):67–68.

47. Nitsch, J. P. 1971. Perennation through seeds and other structures. In *Plant physiology,* Vol. 6A, F. C. Steward, ed. New York: Academic Press, pp. 413–79.

48. Pfieffer, N. E. 1931. A morphological study of *Gladiolus. Contrib. Boyce Thomp. Inst.* 3:173–95.

49. Quamina, J. E., B. R. Phills, and W. A. Hill. 1982. Vine production from tuber pieces of various sizes and sections of yam (*Dioscorea alata* L.). *HortScience* 17: 73.

50. Rees, A. R. 1972. *The growth of bulbs.* New York: Academic Press.

51. Rees, A. R., and G. R. Hanks. 1980. The twin-scaling technique for narcissus propagation. *Acta Hort.* 109:211–16.

52. Reinders, E., and R. Prakken. 1964. *Leerboek der Plantkunde.* Amsterdam: Scheltema & Holkema N.V.

53. Roberts, A. N., and L. T. Blaney. 1957. Easter lilies: Culture. In *Handbook on bulb growing and forcing.* Northwest Bulb Growers Assoc., pp. 35–43.

54. Rockwell, F. F., E. C. Grayson, and J. de Graaf. 1961. *The complete book of lilies.* Garden City, N.Y.: Doubleday.

55. Sheehan, T. J. 1955. Caladium production in Florida. *Fla. Agr. Ext. Circ. 128.*

56. Shillo, R., and A. H. Halevy. 1981. Flower and corm development in gladiolus as affected by photoperiod. *Scient. Hort.* 15:187–96.

57. Simon, R. A. 1986. A survey of bamboos: Their care, culture and propagation. *Proc. Inter. Plant Prop. Soc.* 36:528–31.

58. Skelmersdale, L. 1978. Propagation of bulbous and bulbous-iike plants. *Proc. Inter. Plant Prop. Soc.* 28:209–15.

59. Slater, J. W. 1963. Mechanisms of tuber initiation. In *The growth of the potato,* J. D. Ivins and F. L. Milthorpe, eds. London: Butterworth, pp. 114–20.

60. Smagula, J. M., and P. R. Hepler. 1980. Effect of nitrogen status of dormant rooted lowbush blueberry cuttings on rhizome production. *Jour. Amer. Soc. Hort. Sci.* 105(2):283–85.

61. Stuart, N. W. 1954. Moisture content of packing medium, temperature and duration of storage as factors in forcing lily bulbs. *Proc. Amer. Soc. Hort. Sci.* 63:488–94.

62. Van Aartrijk, J., and P. C. G. Van der Linde. 1986. In vitro propagation of flower-bulb crops. In *Tissue culture as a plant production system for horticultural crops,* R. H. Zimmerman et. al., eds. Dordrecht: Martinus Nijhoff Publishers.

63. Van de Pol, P. A., and T. F. Van Hell. 1988. Vegetative propagation of *Strelitzia reginae. Acta Hort.* 226:581–86.

64. Van Tuyl, J. M. 1983. Effect of temperature treatments on the scale propagation of *Lilium longiflorum* 'White Europe' and *Lilium* × 'Enchantment'. *HortScience* 18:754–56.

65. Wang, S. Y., and A. N. Roberts. 1970. Physiology of dormancy in *Lilium longiflorum* Thunb. 'Ace'. *Jour. Amer. Soc. Hort. Sci.* 95(5):554–58.

66. Wang, Y. T., and A. N. Roberts. 1983. Influence of air and soil temperatures on the growth and development of *Lilium longiflorum* Thunb. during different growth phases. *Jour. Amer. Soc. Hort. Sci.* 108(5):810–15.

67. Weiler, T. C., and R. W. Langhans. 1972. Growth and flowering responses of *Lilium longiflorum* Thunb. 'Ace' to different day lengths. *Jour. Amer. Soc. Hort. Sci.* 97(2):176–77.

68. Welch, N. C., and T. M. Little. 1967. Heat treatment and cutting for increased sweet potato slip production. *Calif. Agr.* 21(5):4–5.

69. Wilson, K., and J. N. Honey. 1966. Root contraction in *Hyacinthus orientalis. Ann. Bot.* 30:47–61.

70. Woodcock, H. B. D., and H. T. Stearn. 1950. *Lilies of the world.* New York: Scribner's.

71. Zweede, A. K. 1930. De periodieke ontwikkeling van *Convallaria majalis. Verh. Koninkl. Ned. Akad. van Wet.* 27:1–72.

SUPPLEMENTARY READING

CROCKETT, J. V. 1971. *Bulbs*. New York: Time-Life Books.

DE HERTOGH, A. A., L. H. AUNG, and M. BENSCHOP. 1983. The tulip: Botany, usage, growth and development. *Hort. Rev.* 5:45–125.

GENDERS, R. 1973. *Bulbs, a complete handbook*. New York: Bobbs-Merrill.

GILES, F. A., R. MCINTOSH, and D. C. SAUPE. 1980. *Herbaceous perennials*. Reston, VA.: Reston Publ. Co.

HARRISON, A. D. 1964. *Bulb and corm production*. Bul. 62. London: Ministry of Agriculture, Fisheries, and Food, pp. 1–84.

HARTSEMA, A. M. 1961. Influence of temperatures on flower formation and flowering of bulbous and tuberous plants. In *Encyclopedia of plant physiology*, Vol. 16, V. Ruhland, ed. Berlin: Springer-Verlag, pp. 123–67.

KOHLEIN, F. 1988. *Iris*. Portland, Oreg.: Timber Press.

LI, P. H., ed. 1985. *Potato physiology*. New York: Academic Press.

N. A. GLADIOLUS COUNCIL. 1972. *The world of the gladiolus*. Edgewood, Md.: Edgewood Press.

REES, A. R. 1972. *The growth of bulbs*. London: Academic Press.

———. 1966. The physiology of ornamental bulbous plants. *Bot. Rev.* 32:1–22.

STILL, S. M. 1988. *Herbaceous ornamental plants* (3rd ed.). Champaign, Ill.: Stipes Publ. Co.

Third international symposium on flower bulbs. 1980. *Acta Hort.* 109:1–533.

VAN AARTRIJK, J., and P. C. G. VAN DER LINDE. 1986. In vitro propagation of flower-bulb crops. In *Tissue culture as a plant production system for horticultural crops,* R. H. Zimmerman, et. al., eds. Dordrecht: Martinus Nijhoff Publishers.

16

Principles of Tissue Culture for Micropropagation

The ability to generate and grow plant tissues (**callus, cells, protoplasts**), isolated organs (**stems, flowers, roots**), and **embryos** in aseptic culture has been the outgrowth of basic and applied research that has gone on in scientific laboratories (botany, plant pathology, and genetics) since before the turn of the century (*49, 89, 186*). Application to propagation began in the 1960s and early 1970s (*3, 15, 30, 60, 62, 74, 102, 117, 121, 122, 148*), extended to the development of commercial **laboratory-nurseries** in the 1970s in both the United States and Europe (*15, 65, 69, 128, 129, 162, 191*), and expanded in the 1980s (*32, 47, 51, 91, 100, 192*). Future growth depends upon progress in adapting the procedures to industrial conditions, reduction in costs (*118, 167*), and coordinating production to markets (*78*).

Parallel to these efforts, genetic engineers and plant breeders have utilized in vitro culture techniques to grow callus, cells, protoplasts, embryos, and other plant parts in applying molecular biology and other breeding strategies to the ge-

netic improvement of plant materials (*5, 44, 54, 88*). These efforts, however, must eventually be integrated with propagation industries to make the products available for practical use.

TERMINOLOGY

Tissue culture is a term used to indicate the aseptic culture in vitro of a wide range of excised plant parts. This technique is used for both propagation and genotype modification (i.e., plant breeding), biomass production of biochemical products, plant pathology, preservation and storage, scientific investigations, and others. These activities are included in the term **biotechnology.**

The term **micropropagation** is used specifically to refer to the application of tissue culture techniques *to the propagation of plants starting with very small plant parts grown aseptically in a test tube or other container.*

In practice, many propagators use the terms

459

micropropagation and *tissue culture* interchangeably to mean any plant propagation procedure utilizing aseptic culture. A synonymous term is *in vitro culture.*

Micropropagation and tissue culture begin with the excision of a small piece of plant, freeing it from microorganisms, and placing it into aseptic culture. The term used for this propagule to start the process is **explant** and corresponds to such other propagules as cutting, layer, scion, or seed.

Five other terms have been used to designate fundamental kinds of vegetative (somatic) regeneration used in micropropagation and tissue culture. These are based on explant selection in relation to life cycle.

Meristem-tip culture. Propagation utilizing the excision and subsequent elongation of a very small part of a shoot tip, including a single apical meristem and subtending rudimentary leaves. This type of culture is used primarily to produce a "virus-free" plant. If the meristem tip is incapable of surviving and producing roots independently, a **micrograft** might be used as an alternate procedure.

Axillary shoot proliferation. This type includes an expanded shoot of terminal and lateral growing points where elongation of the terminal shoot is suppressed and axillary shoot prolifera-tion promoted. This control allows for multiplication of **microshoots,** which can be excised and rooted in vitro to make **microplants,** or which can be cut into separate **microcuttings** to be rooted outside the in vitro system.

Adventitious shoot induction. This type involves the initiation of adventitious shoots either directly on the explant or indirectly in the callus that is produced on the explant as a result of wounding and growth regulator treatment after it is placed into culture. This method involves *de novo* production of new shoots.

Organogenesis. This term is used to describe the process by which adventitious shoots and/or roots develop from within masses of callus cells. This process occurs after an intervening period of callus growth has occurred between the time that the explant is planted and induction occurs.

Somatic embryogenesis. This term refers to the development of a complete embryo from vegetative cells produced from various sources of explants grown in culture. Parallel terms for embryo development within the plant are **zygotic embryogenesis** and **apomictic embryogenesis** (see Chap. 3). The different types of somatic embryogenesis are described later in this chapter.

BASIC PRINCIPLES OF TISSUE CULTURE

Tissue culture systems differ from traditional methods of propagation by the separation of the biological components of the system and the high degree of control that is possible in each aspect of the regeneration and developmental process. Each step of the sequence can be manipulated (or programmed) by selection of plant material and control of the in vitro environment (including disease) to maximize the production outcome in terms of number, size, and quality of propagules.

Tissue culture is based, first, on the principle of **totipotency,** i.e., the concept that every living cell has the genetic potential to reproduce the entire organism. The elabora-tion of this biological principle has been credited to the German plant physiologist Haberlandt, who first enunciated it in 1902 (*59*) and predicted that tissues, cells, and organs could be maintained indefinitely in culture. This feat was not attained however until 1934, when P. R. White (*186*) was able to grow tomato roots continuously in vitro by supplying them with yeast extract. The essential ingredients turned out to be certain B vitamins, notably B_1 (thiamine). In 1939, three investigators—Nobecourt and Gautheret in France and White in the United States—reported independently the indefinite culture of plant callus tissue in a synthetic basal medium supplemented by auxin (*186*).

A second major principle is the hormonal regulation of shoot and root regeneration made possible by the discoveries of auxin and cytokinin. The cytokinin–auxin interaction (see Fig. 9–13) was discovered by Skoog (*154*) and his coworkers utilizing tobacco pith callus. This principle provides the basis upon which all micropropagation procedures depend. In addition, a more or less universally used culture medium was developed by optimizing the inorganic concentrations by Murashige and Skoog (*123*) and the organic supplements by Linsmaier and Skoog (*95*).

When trying to free *Cymbidium* orchid shoot tips of viruses, Morel (*116*) first observed the practical application of shoot-tip culture for rapid vegetative propagation (*117*) through the extensive proliferation into rooted plantlets. Murashige (*122*) expanded this work and developed a third major concept of developmentally different consecutive stages of the culture process, leading to plantlet establishment.

A further extension of tissue culture principles was the development of cell suspension culture (*166*). The discovery that such cells could develop into embryo-like structures, which were termed *embryoids* (*159*), indicated that not only could the phenomenon of organogenesis be brought under control but also the phenomenon of embryogenesis.

Control of these developmental processes is related to a fourth major concept, e.g., **competency** and **determination** (*38, 39, 170*). Competency is a term used to describe the endogenous potential of a given cell or tissue to develop in a particular way through a kind of internal cellular programming or "memory" (*38*). The actual development into the specific kind of cell, tissue, or organ may require a specific hormonal or environmental signal. Such cells are then said to be determined and at some point of development the direction becomes irreversible.

In propagation, competency and determination determine whether a cell or tissue will undergo embryogenesis or organogenesis and whether roots, shoots, or flowers are initiated. For example, embryogenesis depends upon either the selection of embryogenic competent tissues or controlled induction of competency in nonreproductive tissues or cells. The concept similarly applies to induction of roots and/or shoots in organogenesis. Certain cells in stems become competent to develop root initials under a process of dedifferentiation and initiation of root initials (see Chap. 9).

Tissue culture and micropropagation systems are based on the control and manipulation of these processes, which are epigenetic (see p. 168) in nature. Control involves the concept of gene expression, which is the transfer of genetic information from the genes by means of RNA molecules and protein synthesis (see Chap. 8).

MICROPROPAGATION AND TISSUE CULTURE SYSTEMS

Class I—Regeneration of New Plants from Vegetative Structures

Meristem-Tip Cultures

This procedure utilizes the smallest part of the shoot tip as the explant (see Fig. 16–1). It includes the meristem dome and a few subtending leaf primordia. The amount of additional structures depends upon the length of the stem that is excised for the explant. The objective of this procedure is to produce a rooted microplant through isolation of a meristem which is free of systemic viruses, virus-like organisms, and superficial fungi and bacteria (*71, 79, 80, 135, 155, 164*). The smaller the explant, the more effective is the elimination of pathogens (*79, 89*). For example, meristem tips of 0.10 to 0.15 mm have given 100 percent virus elimination, but the percentage decreased as the size increased to 1 mm. On the other hand, the smaller the explant, the more dif-

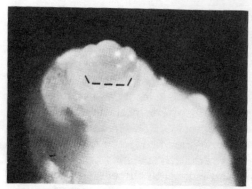

FIGURE 16-1 Shoot tip of carnation stem with outer leaves removed, showing the apical and lateral meristems (growing points). Part of shoot tip to be excised for culturing is indicated by lines. (Courtesy W. P. Hackett.)

ficult to establish and the lower the survival rate. A general compromise is to use explants 0.25 to 1.0 mm long (*108*). Virus status must be confirmed by indexing (*93*). Culturing and rooting procedures are described in Chap. 7.

Heat treatment (Chap. 8) of the test plant prior to meristem excision can increase the proba-

bility of virus removal. Some antiviral agents, such as Ribovirin (*51, 131*), added to the medium have increased the recovery of virus-free plants.

Once a rooted plant is produced, it can be used as the start of a nuclear stock ("clean stock") program for that particular species and cultivar (*136*) (see Chap. 8).

Meristem-tip culture has been successful with many important herbaceous crops. Material from this kind of program has become an essential production aspect for such commercial crops as carnation (see Fig. 16-2), chrysanthemum, orchid, geranium, potato, sweet potato, cassava, banana, and others. The procedure is more complex with woody plants, and micrografting is an alternative procedure. Improvements in the micropropagation of woody plants have been able to extend its usefulness (*14*).

Micrografting

Grafting very small meristem tips comparable to those described in the preceding section can be used as an alternative method to produce virus-free materials for various woody plants, such as citrus (*124*), apple (*73*) (see Fig. 16-3), and

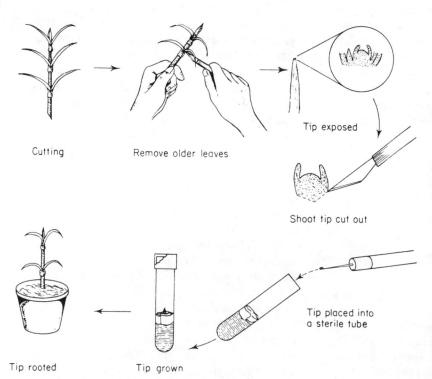

Cutting

Remove older leaves

Tip exposed

Shoot tip cut out

Tip placed into a sterile tube

Tip rooted

Tip grown

FIGURE 16-2 Meristem-tip culture of carnation. Following the arrows: the carnation cutting is obtained, the larger leaves are stripped away, and then the small, enclosing leaves at the extreme tip are removed to expose the growing point. The tip and the next subtending leaf primordia are removed with a scalpel and placed on the surface of a paper wick in a test tube with nutrient media. The shoot tip grows in the tube until large enough to be transplanted to a container. A full-size plant is shown at bottom left. (Redrawn from Holley and Baker (*70*).)

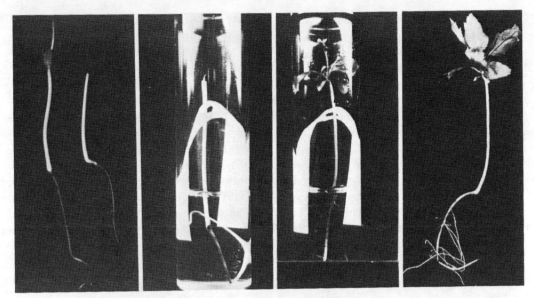

FIGURE 16–3 Micrografting of apple meristems. *Left:* Seedling hypocotyl sections before and after decapitation. *Left center:* Grafted plant one week after grafting. *Right center:* Grafted plant six weeks after grafting with scion growing. *Right:* Eight weeks after grafting, ready for transplanting. (Courtesy S. C. Huang and D. F. Millikan (*73*).)

Prunus (*14, 125*). For example, this procedure is important in *Citrus* not only because it is successful but also because explants can be used from ontogenetically mature trees (see p. 174) and avoid the juvenile phenotype of nucellar seedlings also used in virus cleanup in citrus (see p. 537). Micrografting also shows promise for early detection of graft incompatibility relationships (*77*).

SHOOT-TIP GRAFTING IN VITRO

Citrus embryos are excised from rootstock seeds, surface sterilized, and planted in standard inorganic salt medium with 1 percent agar (*124*). Embryos germinate in the dark in two weeks. Seedlings are then removed and decapitated to a 1 to 1.5 cm length; cotyledons and lateral buds are excised with a mounted razor blade. A 0.14- to 0.18-mm tip with three leaf primordia is used as the scion. This gives reasonable success and the shoot tip eliminates viruses.

An inverted T-bud cut is made in the seedling rootstock, cutting 1 mm down the stem, followed by a horizontal cut on the bottom. The excised shoot-tip is placed inside the flap next to the cambium. Grafted plants are placed in a liquid medium. A filter paper bridge with a center hole supports the stem. Cultures are kept in the light for three to five weeks to heal. When two expanded leaves appear on the scion, the grafted plant is transplanted.

A similar procedure has been used for apple (*73*) and plum (*125*). Rootstocks are either seedling plants or rooted stems. Shoot-tip scions taken from cultured plants reduce contamination problems.

Shoot-Tip Culture: Axillary Branching

Shoot-tip culture utilizing axillary shoot proliferation for multiplication is the most important commercial technique for micropropagating plants (Fig. 16–4). A shoot-tip explant is used that is larger than the meristem tip. It can vary in size from a single small shoot tip, usually at least 2 mm in length, up to 1 to 2 cm or longer. The explant is essentially a miniaturized cutting. Shoots of this size can be cleaned of major fungal and bacterial contaminants but will not be freed of systemic viruses, virus-like organisms, or even some kinds of systemic bacteria. Therefore, it is important that source plants be used that are al-

ready free of any internal pathogen significant for that particular crop (see Chap. 8).

The production system involves four more or less distinct stages of production (*30, 31, 51, 65, 122*):

1. **Establishment.** In this stage a sterile explant is produced and planted on a sterile basal medium which includes: (a) a support material consisting of a semisolid medium—agar or other commercial products, or liquid; (b) a mineral salt mixture with essential major and minor elements; (c) an energy source, usually sucrose; and (d) certain vitamin supplements. Sometimes hormones

FIGURE 16–4 Shoot tip propagation by axillary shoot formation on poplar (*Populus*). *Upper left:* Vegetative bud explant after two weeks in culture (3 mm in diameter). *Upper right:* Axillary buds on explant after four to six weeks (15 mm in diameter). *Lower left:* Advanced proliferation of microshoots (5 cm in diameter). *Lower right:* Rooted microplant. (Courtesy C. B. Christie (*24*).)

and other substances described later are added. Growth and development in the various stages is controlled by the addition (amount and ratio) of specific plant growth regulators (PGR), primarily cytokinin, auxin, and sometimes gibberellin.

The purpose of stage 1 is to establish the explant in culture, screen for contaminated explants, and allow the excised plant unit to be stabilized (*103*) to the culture environment. The explant initially grows by elongation of the main terminal shoot, with limited proliferation of axillary shoots to produce a mass of microshoots, the number depending upon the apical dominance of the particular kind of plant (*31, 131*). Depending upon the plant, the culture mass is divided after two to four weeks and subcultured on fresh medium. Division and subculturing is repeated at intervals. Stabilization may require repeated subculturing to produce a uniform, well-growing culture (see p. 482).

2. **Multiplication.** As the cluster of shoots is divided and recultured, the number of microshoots multiply at an exponential rate. Multiplication of axillary shoots is promoted by controlled inhibition of shoot elongation and stimulation of lateral shoot proliferation, primarily by controlling the concentration and ratio of cytokinin to auxin. The pattern of multiplication and subculturing continues until a population of microshoots is obtained to meet a production schedule of finished product.

The stabilized, uniformly proliferating mass of microshoots growing in culture has characteristics that differ from those of plants growing in an open environment:

a. The physiological state of the microshoot mass is one that favors high axillary shoot proliferation, reduced terminal shoot elongation, and low rooting potential. Multiplication rate can be measured by the number of new shoots per culture.

b. Shoots growing in vitro are largely **heterotrophic;** that is, they obtain their energy from sucrose in the medium and not from photosynthesis. Stomates may not be capable of functioning normally.

c. Shoots are very succulent, highly sensitive to moisture and temperature stress, and sus-

ceptible to pathogens outside the environment of the test tube. The leaves may lack a cuticle and have disorganized mesophyll.

d. In many plants, the leaf phenotype may be atypical, smaller in size, and more elongated than comparable plants outside the test tube. These characteristics tend to resemble those that are shown by seedling plants.

3. **Pretransplant.** The function of this stage is to prepare the microshoots for transplanting from the aseptic protected environment of the test tube to the outdoor conditions of the greenhouse or transplant area (*16, 35, 131, 194*). This may include rooting of individual microshoots in a medium in which the auxin level has been increased and the cytokinin concentration decreased or omitted altogether. It may involve exposure to gibberellin to stimulate shoot elongation and reduce the cytokinin effect. It may include a hardening-off process in which the concentration of sucrose is increased (*34*).

Alternate sequences for rooting include the following protocols:

a. Subculture individual microshoots in the last subculture by changing the hormone balance of the medium to lower (or no) cytokinin and higher auxin, as well as making other adjustments.

b. Transfer individual microshoots into a root-inducing (liquid?) medium for a few days and then transfer to an auxin-free rooting medium.

c. Excise individual microshoots and treat as a microcutting in a rooting medium in a greenhouse under mist or fog as a standard cutting (see Chap. 10). This is known as *ex vitro* rooting.

4. **Transplanting and acclimation.** This stage involves the shift from the heterotrophic to an **autotrophic** (free-living) condition and the **acclimation** of the microplant to the outdoor environment. Immediately upon transplanting, microplants should be kept in very high humidity, gradually exposing them to outdoor conditions similar to other transplanting operations. Keep-

ing the shoot actively growing is important because the acclimation and development of autotrophic conditions is dependent on the new growth after transfer from the test tube environment.

Adventitious Shoot Culture

This system has essentially the same culture requirements as those of shoot culture. It differs in the source of explant and in the adventitious origin of the new shoots.

New adventitious shoot apices can develop either (a) directly on the explant or (b) indirectly from the callus that develops on the cut surfaces of the explant (*68*). The choice of explant and the hormone regime in culture determine the success of adventitious shoot initiation. Some of the kinds of explants are as follows:

Pieces of leaves are used in such plants as *Saintpaulia* (*10*), *Salpiglossis* (*94*), and horseradish (*115*) that naturally regenerate in this manner. Experimentally, whole plants have been regenerated from epidermis strips or other, deeper layers (*87, 177*). An interesting variation is the "fragmented shoot apex culture" with grape (*6, 7*), in which shoot apices 0.1 mm in length have been cut into two to four segments and cultured in drops of the medium. These small segments proliferate into small leafy structures that can develop into whole plants.

Shoot-tip explants are used in a number of species, such as orchids or ferns, in which proliferations of callus-like masses develop and regenerate large numbers of shoots. Plants started from shoot-tips initially for axillary shoot formation may revert to higher percentages of adventitious shoots with time (*3*).

Cotyledons, hypocotyls, and other seedling structures have been particularly useful as starting points (Fig. 16–5) for conifers in which excised cotyledon segments can be used to regenerate new shoots in the presence of a cytokinin (*12, 22*).

Young needle fascicles have been used to regenerate shoots from older conifer trees (*12*). Explants from plant tissues that have

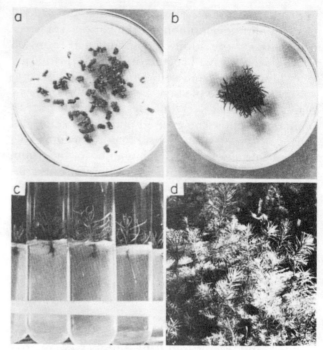

FIGURE 16–5   Adventitious bud formation on Douglas fir (*Pseudotsuga menziesii*) cotyledons. (a): cotyledons are excised from germinated seed, cut into segments, and placed on medium with 1 ppm BAP plus 0.001 ppm NAA to initiate adventitious buds. (b): medium is changed by omitting hormones; mass of shoots develop. (c): individual shoots are excised and rooted in culture with 0.001 ppm NAA, 0.5 percent sucrose, and temperature reduced to 19°C. (d): plantlets grown in greenhouse. (Courtesy Dr. Tsai-Ying Cheng.)

been rejuvenated to a juvenile condition are particularly responsive (*11*).

Segments of immature inflorescences of flower scapes are highly regenerative in some species, particularly some monocots, such as *Gladiolus* (*195*), *Hemerocallis* (*67*), *Iris* (*114*), *Hosta* (*113*), and *Freesia* (*133*), as well as some dicotyledons such as *Gerbera* (*132*) or *Chrysanthemum* (*146*).

Bulb scales of various monocotyledonous plants (*74–76*) characteristically have rings of meristematic tissue at the base near the basal plate (see Chap. 15). When scale explants are excised and put into culture, adventitious shoots develop directly from the explants.

Direct initiation begins with parenchyma cells that are located either in the epidermis or just below the surface of the stem; some of these cells become meristematic and pockets of small, densely staining cells termed **meristemoids** develop (68). These apparently originate from single cells.

Response of explants, however, depends on hormone levels (Fig. 16–6). In a study of shoot initiation on cotyledons of Douglas fir (22), cytokinin (BAP 5 μM) was necessary to induce adventitious buds, but three different response patterns resulted depending on the level of auxin also supplied. With low auxin (NAA 5 μM) shoots only developed. With a higher auxin concentration (NAA 5 μM), the cotyledon produced both callus and many shoots. When only auxin (NAA 5 μM) was added, only callus resulted.

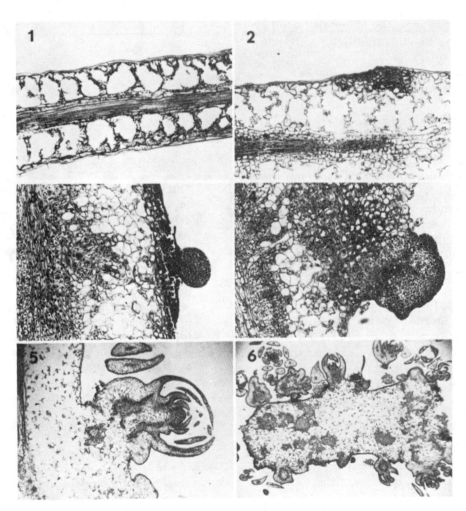

FIGURE 16–6 Adventitious shoot initiation on Douglas fir (*Pseudotsuga menziesii*) cotyledons. *Top left:* Cross section of cotyledon before culture. *Top right:* Initiation of a *meristemoid* on surface of cotyledon. *Center left:* Development of meristemoid into a tiny globular mass of tissue. *Center right:* Differentiation into a shoot primordium. *Lower left:* A completely developed shoot tip. *Lower right:* Cotyledon showing masses of adventitious shoot tips. (Courtesy Dr. Tsai-Ying Cheng.)

Indirect development of adventitious shoots first involves the initiation of basal callus from excised shoots in culture (*3*). Shoots arise from the periphery of the callus and are not initially connected to the vascular tissue of the explant.

Adventitious shoot formation can result in very high rates of multiplication, in general, higher than rates resulting from axillary shoots. On the other hand, adventitious shoots can increase rates of aberrant plant production, resulting from the breakdown of chimeras with resultant loss of variegation in some cultivars and reversions to a more juvenile condition (*99*). When this system is used, the plants that are produced must be evaluated carefully for variation.

Class II—Reproduction of Seedling Plants by Excision of Reproductive Structures

Anther and Pollen (Microspore) Culture

The discovery that pollen grains could develop into embryos was made by accident with *Datura* by Guha and Masheshwari (*55*), but their development into plants was later realized in tobacco by Nitsch and coworkers (*13, 127*) (Fig. 16–7). Since then whole plants have been produced in many species either by direct differentiation of the immature pollen grain into a small embryo (*127, 169*) or by callus development that undergoes organogenesis (*152, 169*). Anther culture is used for plant breeding to produce haploid plants. Subsequent doubling of the chromosomes results in an isogenic, homozygous line (*17*).

Two basic procedures have been used for tobacco (*169*). Figure 16–7 describes a simple technique effective with this plant. A second technique utilizes a "stress" period of the excised buds prior to culture. Buds are placed into a sealed container and stored in the dark for a period of time based on a time–temperature pattern, such as 7°C for two to three weeks. The temperature may be higher, but the time must be shorter. After this treatment the anthers are floated on liquid medium in a petri dish. Embryos become discernible in about 14 days. It is necessary to transfer the embryos to new medium, or new medium must be added. Procedures have also been described (*131*) for potato (*182*), *Brassica* spp. (*81*), cereals, and grasses (*183*).

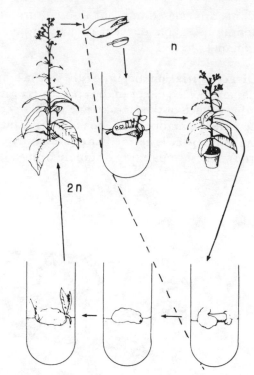

FIGURE 16–7 Anther culture. Procedure for obtaining haploid plants from *Nicotiana tabacum,* then diploid plants from the haploid ones. A flower bud is excised from a flowering plant just as corolla is emerging from the calyx (upper left). The immature anther is removed aseptically and planted on agar with a standard nutrient medium without hormones. Pollen should be at the uninucleated, microspore stage for development into haploid embryos, which germinate and form plantlets. Incubate in the dark but transfer to light as soon as embryos appear.

Ovary and Ovule (Ovulo) Culture

This system includes the aseptic culture of the excised ovary, ovule (fertilized or unfertilized) and attached placenta within the ovule. Although first utilized as a technique to investigate problems of fruit and seed development (*9, 139*), adaptations of the procedure have uses in propagation, particularly as applied to genetic improvement.

For instance, unfertilized ovules have been excised, grown in culture, then supplied with pollen, with subsequent fertilization taking place in vitro (*50, 141*). Such procedures have been used most successfully with plants having multiple fruits. Pollen can be placed directly on the placenta inside the ovule, where pollen tubes can develop and grow immediately into the ovules without passing down the style.

Cultured ovules are useful to rescue embryos that abort at a very young stage if not separated from the plant (*36, 52, 137, 140, 157*). The dissection of such small embryos is difficult at this stage and the medium required is more complex than in ovulo culture. Ovules can easily be sterilized and generally will grow in a basal medium of inorganic salts, sucrose, and sometimes hormones.

Ovary (fruit) cultures vary in nutritional requirements (*139*). For most a basic medium of inorganic salts and sucrose is essential, but auxin may be highly stimulatory.

Embryo Culture

Embryo culture is the excision of embryos from individual ovules and their eventual germination in aseptic culture (*27, 72, 131, 190*). This technique is used to rescue embryos that would have aborted within the seed before maturity. This is a common consequence of many interspecific hybrid crosses. Early-ripening cultivars of stone fruits mature their fruits before the embryos are fully developed (*66, 83, 137*). A second use is to induce prompt germination of mature dormant seeds and thus shorten breeding cycles (*92*). Excising the embryo to eliminate dormancy-inducing restraints in the surrounding endosperm has been described for *Iris* (*138, 163*), *Maranta* (*64*), peach (*83, 92*), and others. Appropriate environmental sequences may be needed for embryos with internal epicotyl dormancy, as in *Paeonia* (*112*) (Fig. 16–8).

Protocols for culture are based on the stage of embryogenesis (see Chap. 3), the culture system used, and the genotype (species and cultivar). Three broad categories of embryos can be considered: (a) *mature* embryos whose cotyledons have reached nearly full size and are desiccation resistant (see p. 64), (b) *immature embryos* that are less than one-half full size down to approximately the early heart stage, and (c) *rudimentary* embryos from proembryo stage to approximately the heart stage.

Fruits and seeds can be easily sterilized, opened, and the embryo extracted aseptically (see Chap. 17). Agar is the preferred support, but the ingredients of the basal medium vary greatly and individual species may require experimentation. Mature embryos can be grown on a basal medium of inorganic salts and sucrose (1 or 2 percent or none), but embryonic tendencies may persist with the development of abnormal seedlings. A shift from embryonic to germination potential may be promoted by exposing the embryos to controlled desiccation (*82*) or by chilling at 2°C (36°F) (*66*).

Immature embryos separated from the ovule cease embryogenesis and show precocious abnormal germination. A major breakthrough came with the discovery by van Overbeek (*179*) in 1942 that 1 to 15 percent coconut milk (a liquid endosperm) would prevent precocious germination and allow embryogenesis to proceed. Since then it has been found that success with embryos excised at this stage required high osmotic pressure and high nitrogen, either ammonium or nitrate, depending upon the species. Important media constituents include sucrose (8 to 18 percent), casein hydrolysate (5 to 30 percent), a moderate or low level of auxin and cytokinin, and/or gibberellin, depending upon the species. GA may, however, induce precocious germination, which can be overcome by abscisic acid. ABA has been shown to promote embryogenesis, but the relationship to embryo age is not clear (*45, 46*).

One protocol for clover (*Trifolium*) (*27*) calls for the following sequence of stages:

1. High sucrose, moderate auxin, and low cytokinin for one to two weeks. At this time the embryo stops growing and must be transferred.

2. Normal sucrose, low auxin, and moderate cytokinin, which allows for resumption of embryo growth and, sometimes, shoot development.

3. For embryos with shoots, shift to a medium with low auxin and high cytokinin, to stimulate shoot proliferation. Shoots are rooted and transplanted.

4. Embryos with renewed growth, but disorganized, are transferred to a somatic embryo-

FIGURE 16–8   Embryo culture of peony (*Paeonia*) seed with epicotyl dormancy. *Left top:* Seed (left). Seed coat removed (right) to show massive endosperm. *Left center:* Very small rudimentary embryo is located inside base on endosperm. Shows cutting away sections of endosperm to extract the embryo on probe. *Right top:* First stage of embryo germination after two months in test tube on agar media of mineral salts (MS medium; see Chap. 17) and 4 percent sucrose. Only radicle grows at warm temperature. *Lower right:* a subsequent cold treatment stimulates the epicotyl to grow. Shows seedling after seven weeks at 27°C (80°F), five weeks at 3°C (37°F), and four weeks at 27°C (80°F) in light. *Lower left:* Seedling transplanted to growing medium. (Courtesy M. M. Meyer, Jr. Reprinted from *Amer. Peony Soc. Bul. 217,* 1976 (*112*).)

genesis induction medium (see p. 478 for method of handling).

Rudimentary embryos of heart or less-developed stages are very difficult to handle physically and very sensitive to the medium used. Apparently, it is essential that the suspensor tissue be retained with the proembryo excised at this particular stage. Embryos at this stage could best be handled by ovule culture.

Class III—Regeneration of Plants from Tissue and Cell Culture

Callus provides an important tissue culture system because it can be subcultured and maintained more or less indefinitely. Several important pathways of development may follow.

1. Organogenesis may occur within the callus mass, to produce new plantlets.
2. Specific treatments may cause the cells to disassociate and develop a cell suspension culture.
3. Cells may be treated to produce a protoplast culture.
4. The regenerative potential may be shifted toward somatic embryogenesis, as described in a subsequent section. These relationships are shown as follows:

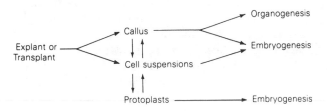

Callus and cell cultures are not directly important as a propagation system, however, because they tend to generate a significant amount of genetic and/or epigenetic aberrations (*111*). Consequently, this class of culture systems is most important for the production of new and novel genotypes. The phenomenon, known as **somaclonal variation**, when used in plant breeding and genetic engineering requires the regeneration of new plantlets and their subsequent clonal propagation (*44*). Cell suspensions are also used for the industrial production of secondary compounds, such as pharmaceuticals (*8*).

Callus Culture

Callus is produced on explants in vitro as a result of wounding and in response to hormones, either endogenous or supplied in the medium. Explants from almost any plant structure or part—seeds, stems, roots, leaves, storage organs, or fruits—can be excised, disinfested, and placed on the surface of a culture medium (*166, 173*). Continued subculture at three- to four-week intervals of small inocula taken from these callus masses can maintain the callus culture for long periods (Fig. 16–9).

The most common culture medium is the Murashige-Skoog (MS) (*123*), which established optimum rates for inorganic compounds, plus the Linsmaier and Skoog (*95*), which optimized organic supplements. This medium is rich in macroelements, particularly nitrogen, including both nitrate (NO_3) and ammonium ions (NH_4), sucrose, and certain vitamins. Initiation of cell division and subsequent callus production requires that both a cytokinin and an auxin be supplied in the proper proportion (*154*). Auxin at a moderate to high concentration is the primary hormone used to produce callus. The principal auxins include indoleacetic acid (IAA), naphthaleneacetic acid (NAA), and 2,4-dichlorophenoxyacetic acid (2,4–D), in increasing order of effectiveness. Cytokinin, as kinetin, or benzyladenine (BA) is supplied in a lesser amount if not adequate within the explant.

Although callus tissue cultures may appear outwardly to be uniform masses of cells, in reality their structure is relatively complex with considerable morphological, physiological, and genetic variation within the callus. Growth follows a typical logarithmic pattern. There is (a) a slow initial cell division *induction* period requiring auxin, (b) a rapid cell *division* phase involving active synthesis of DNA, RNA, and protein, followed by (c) a gradual *cessation* of cell division along with (d) *differentiation* into larger parenchyma and vascular-type cells. Cell division does not take place throughout the culture mass but is located primarily in a meristematic layer on the outer periphery

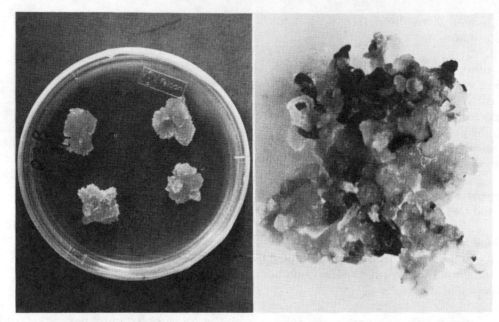

FIGURE 16-9 Tissue cultures. *Left:* Undifferentiated callus of tomato (*Lycopersicon esculentum*) growing in vitro. *Right:* Callus of corn (*Zea mays*) showing dark spots that are green meristemoid-producing areas that will initiate shoots. (Courtesy Dr. Carole Meredith.)

of cells. The inner parts of the callus remain as an undividing mass of older tissue and, in time, may differ physiologically and genetically from cells of the outer layer. Division in the exterior layer decreases and the appearance of the callus may become "knobby" as cell division becomes restricted to specific islands of cells. Thus, variations in cell age and type may occur within the tissue culture mass. The inner cells are older and the exterior cells are younger as a meristematic region persists around the periphery of the callus mass.

The induction of new plants begins with **dedifferentiation** of parenchyma cells to produce centers of meristematic activity (**meristemoids**) (*43, 172, 176*). In early studies by Skoog, tobacco callus produced shoots (*154*) if a relatively high cytokinin/auxin ratio was supplied. If the ratio was reversed, roots tended to form. An intermediate ratio produced both (Fig. 16–10). Adenine was synergistic with cytokinin and increased inorganic phosphate (PO$_4$) was useful. Although the same basic pattern tends to follow with most other plants, an exact formula for optimizing conditions

for regeneration is needed for each species or cultivar (see also p. 481).

Cell Suspensions

A suspension culture is started by placing a piece of friable callus or homogenized tissue in liquid medium so that the cells disassociate from each other. These cells are grown in various types of devices (Fig. 16–11). In one (**batch culture**), cells are grown in a flask placed on a shaking device that allows air and liquid to mix. Rotating devices that continuously bathe the tissue are available. Another device, called a **chemostat** or **turbidostat,** continuously cycles the media through the cell culture essentially the same as in microorganism culture. In a third method, cells on a **filter paper layer** are placed on a shallow liquid medium in a petri dish with no agitation.

Growth of cells follows a typical pattern based on changes in rates of cell division. Cells first divide slowly (*lag phase*), then more rapidly (*exponential*), increasing to a steady state (*linear*), followed by a declining rate (*deceleration*) until a

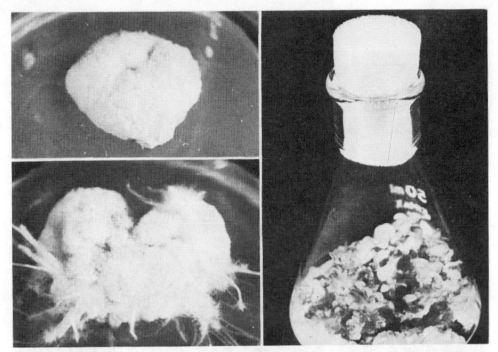

FIGURE 16-10 Organogenesis in tobacco callus. *Left top:* Undifferentiated callus tissue developed from a single pith cell produced by nurse-culture technique. *Left below:* Adventitious roots develop on the callus culture when it is grown on a medium high in auxin (IAA) and low in kinin (kinetin). *Right:* Stems and leaves are produced when the callus is grown on a medium low in IAA and high in kinetin. With proper balance both roots and stems develop and new plants are produced. (Courtesy T. Murashige.)

FIGURE 16-11 Suspension culture of cells of celery (*Apium graveolens* L.). (Courtesy Dr. Lawrence Rappaport.)

stationary state is reached. If a small inocula of liquid with new cells is transferred to new liquid medium, the process will be repeated. Under proper environmental conditions with media control, the process can go on indefinitely.

The ability to grow cell suspensions on a large commercial scale may require the use of devices known as **bioreactors** (*168*) (Fig. 16–12). These were originally developed to grow microorganisms or other living cells for fermentation or to produce various secondary products for industrial use. These devices include provision for the introduction of fresh medium and removal of spent medium, along with proper environmental controls.

Culture media are usually similar to those for callus tissue culture and include a complete range of ingredients: inorganic salts, sucrose, vitamins, and a proper balance of hormones.

Temperature control   Bioreactor (airlift)   Bioreactor (impeller)   Power source

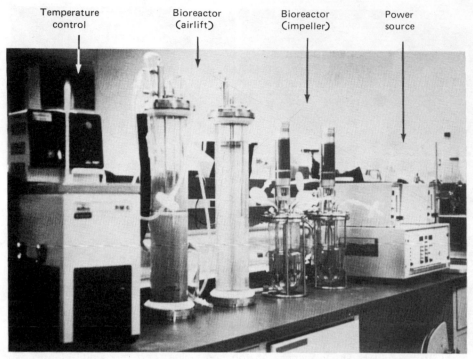

FIGURE 16-12   Laboratory version of two kinds of bioreactors. These are used to propagate plants from somatic embryos. (Courtesy Steven Strickland, Plant Genetics, Inc., Davis, Calif.)

Protoplast Culture

Protoplasts are the living parts of plant cells, containing the nucleus, cytoplasm, vacuole, and various cellular structures surrounded by a semipermeable membrane. The plant cell, in contrast to the animal cell, is surrounded by a firm nonliving cell wall composed of cellulose and hemicellulose and held together by pectin materials. Protoplasts can be obtained from cells in suspension or can be derived directly from mesophyll leaf cells.

The major advance that permitted protoplast cultures to be made (*26, 166*) was the discovery that the walls of plant cells could be removed by enzymes that digest the pectin and allow the protoplast surrounded by the cellular membrane to survive (Fig. 16-13). Commercial enzyme preparations involving cellulase complexes are available for this purpose. Maintaining an adequate osmotic pressure to prevent disruption of membranes is necessary; mannitol (0.45 to 0.8 *M*) has been used for this purpose.

Protoplasts are cultured in media similar to those for cells except for the presence of the enzymes and osmoticum. Once the enzyme is removed, regeneration of walls takes place rapidly within several days. When new cell walls are produced, the regenerated cells can be used to start tissue cultures or cell suspensions, from which it is then necessary to regenerate new plants.

Protoplasts are significant because many manipulations with the cells are possible (*151, 166, 173*) once freed of the enclosing cell walls. For example, viruses can be more easily incorporated into protoplasts. In plant breeding the fusion of protoplasts of two different genotypes, such as two species, which combine two nuclei and two cytoplasms, has been accomplished in a process called **somatic** (or **parasexual**) **hybridization** (*21*). Protoplast culture is particularly important in genetic engineering since protoplasts can absorb DNA, proteins, and other large macromolecules. New genetic material can be incorporated directly into cells of an organism. This procedure is called **transformation.** Isolated pro-

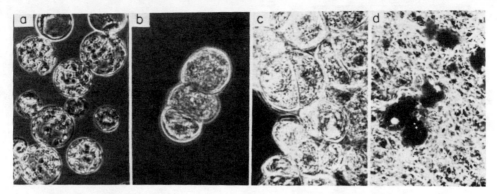

FIGURE 16–13 Protoplasts of Douglas fir (*Pseudotsuga menziesii*) cotyle-dons. (a): freshly isolated protoplasts in which cell walls have been re-moved. (b): four-cell stage after protoplast had resynthesized a cell wall and divided twice. (c): cells from a colony. (d): callus formation. (Courtesy Dr. Tsai-Ying Cheng.)

toplasts are also capable of taking up nuclei and chloroplasts (**organelle transfer**).

Class IV—Somatic Embryogenesis

Embryogenesis is described in Chap. 3 as the development of the embryo from the zygote (**zygotic embryogenesis**) or from some other cells of the reproductive apparatus, such as an unreduced gamete or a cell in the nucellus (**apomixis**). The first of these involved sexual reproduction, the second asexual (somatic) reproduction.

Early in the study of cell suspension systems, Steward et al. (*160*) discovered that when treated with coconut milk, carrot cells did not continue to multiply but differentiated into miniature embryo-like structures which were called **embryoids**. At the same time, Reinert (*145*) independently discovered the same phenomenon in carrot using high auxin concentrations in agar as the inducing material. Since then many species have been found to have a capacity for somatic embryogenesis in culture systems either through the *selection* of embryogenically competent cells or tissues or by their *induction* from treatment of the culture.

There are three basic kinds of in vitro embryogenesis:

1. **Somatic adventitious embryogenesis.** Somatic embryos may develop from cells or callus

associated with the reproduction apparatus. These have been referred to as **preembryonically determined cells** (PEDC).

Embryogenic callus and cells can originate from nucellus tissue of either **polyembryonic** or **monoembryonic** citrus species and (or) cultivars (*19, 96, 175*). Similarly, embryogenic callus production of somatic embryos can develop from very young ovule tissues with other tropical species (*96*), grasses (*180*), grape (*119*), coffee (*156*), and other species. Since these explant tissues have the same genotype as the source plant, this procedure is a potential method of clonal propagation and duplicates apomixis.

Adventitious embryos can also develop *directly* from single cells on the surface of immature embryos in culture or *indirectly* in the callus produced from them. (*53, 84, 130, 150, 181*). Although this proliferation involves somatic tissue, the original explant is zygotic in origin and would not duplicate the genotype of the source plant. Consequently, this type of culture would not be a method of clonal propagation of the original source plant but of the new sexual generation. This procedure would probably be used in a program of genetic improvement, such as to "rescue" embryos from abortive genetic combinations of hybrids (*178*).

2. **Somatic polyembryogenesis** (Fig. 16–14). This type involves the transplanting of highly embryogenic **embryonal-suspensor mass** (**ESM**),

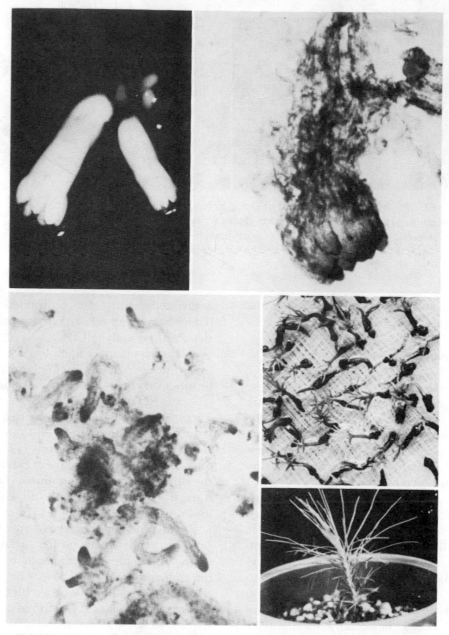

FIGURE 16–14 Somatic polyembryogenesis in conifer cell suspension cultures. *Upper left:* Transplanted embryonal-suspensor mass (ESM) of sugar pine with rescued zygotic embryos and secondary embryos developing from the ESM on agar plate. *Upper right:* Cell suspension cultures of ESM of Douglas fir showing cleavage polyembryony. Proembryonal cells in the cultures ESM divide longitudinally to produce a polyembryonic mass. *Lower left:* Cell suspension cultures of ESM of loblolly pine showing multiple embryos with suspensors developing mainly by budding polyembryogenesis. *Middle right:* Recovered plantlets from Norway spruce cell suspension cultures. Plantlets have cotyledons, new emerging shoots, and a root covered by coleorhizal-like cells and remnants of the old ESM. *Lower right:* Loblolly pine plantlet in soil derived by somatic polyembryonesis. (Courtesy D. J. Durzan.)

which precedes formation of the embryo (*39, 56*). The proliferation of these cells within very immature gymnosperm ovules gives rise to the phenomenon of polyembryogenesis (see p. 000). When the ESM tissue is transplanted to a culture medium two to four weeks after fertilization, its cells proliferate profusely and can develop directly into embryos.

This explant tissue is unique and can be distinguished from nonembryonic tissue in culture by its white mucilaginous appearance and ability to stain red with acetocarmine (0.10 percent w/v). Under UV light, embryonic cells exhibit a green influorescence, moribund cells fluoresce yellow, and suspensor cells fluoresce weakly. It is a form of "embryo rescue."

Somatic polyembryogenesis is particularly significant in gymnosperms but also occurs in angiosperms. Embryogenesis does not involve callus but results from "cleavage" and "budding" of the original ESM tissue mass. The embryogenic process in vitro follows the general protocols described below.

3. **Induced somatic embryogenesis.** This type of embryogenesis results from somatic callus and cell suspensions after the tissue is subjected to specific treatments that result in the induction of embryogenic competence. It appears to be a general biological phenomenon (*89, 90, 131, 187*). Systematic analysis of the embryogenic potential of different explant sources and the required culture conditions have been carried out with a number of crops, such as carrot (*98*) (Fig. 16-15), alfalfa (*168*), grasses (*180*), coffee (*156*), palm species (*174, 175*), and soybean (*23*).

There are two potential uses of the procedure:

a. *Mass propagation of "synthetic" somatic seeds.* Somatic embryogenesis can be applied directly to mass clonal propagation of many normally seed propagated crops, including grains, vegetables, plantation crops (oil and date palm, coffee), and forest trees. For most crops, technical problems of engineering the entire process of embryogenesis to seedling production and planting need to be solved. A certain amount of variability may be inherent from the system. Consequently, careful field testing and monitoring for variability would be required before large-scale production. The concerns of monoculture may also need to be addressed. Nevertheless, this procedure has consid-

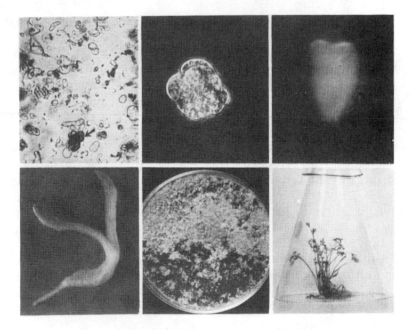

FIGURE 16-15 Embryogenesis in carrot (*159*). *Top left:* Highly magnified view of suspended cells and cell clumps derived from tissue cultures of carrot growing in a liquid medium. Arrow points to a clump of cells—the beginning stage for an embryoid. *Top center:* Single-cell clump (higher magnification) illustrating the globular stage of development. *Top right:* A more advanced heart-shaped stage of embryoid development. *Below left:* Mature embryoid that developed continuously from the globular stage, through the heart and torpedo (not shown) stages. *Below center:* Thousands of carrot plantlets that developed as embryoids. *Below right:* Single carrot plant that grew from a single embryoid transplanted from previous stage. (Courtesy F. C. Steward and M. O. Mapes.)

erable promise to have a major application to various species.

b. *Genetic improvement programs of plant cultivars.* Somatic embryogenesis could be useful to identify genetic variation within populations of cells. It could be important because of the single-cell origin of somatic embryos. Somatic variability can develop inherently within the source plant, result from the callusing system, be induced by mutagenic agents, or result from genetic engineering treatments. Somatic embryos can be grown from callus, cells, or protoplasts.

Embryogenesis from zygotic (embryo) tissue is not a direct method of clonal propagation unless the embryos are apomictic. In tree improvement programs 7 to 10 years or more might be required to evaluate individual plants derived from culture. Solution of this problem has been suggested by establishing long-term storage procedures to preserve the original embryogenic cell masses by cryopreservation (*58*) as a future propagation source.

Protocols for Somatic Embryogenesis

Protocols for the induction of embryogenically determined cells or the exploitation of pre-embryogenically determined cells need to be established for each genotype. Nevertheless, certain generalizations can be made and the system will have the following stages.

1. **Selection and culture of appropriate explant material.** The selection of explant source material is the most critical decision and may require a systematic analysis of the embryogenic potential of different explant sources within the plant. The first step is the production of callus, cell suspension, or protoplast material by methods that have already been described.

2. **Induction of embryogenic potential (conditioning) in the cell explants.** Induction is necessary for nonembryogenically determined cells and explants. Induction is achieved through the transfer of cells to a basal medium with a high concentration of auxin. The most effective auxins are (a) 2,4-dichlorophenoxyacetic acid (2,4-D) or (b) coconut milk plus a low concentration of naphthaleneacetic acid (NAA). After one to two weeks some small proembryos may appear. The larger clumps and proembryos may be separated by different-sized screens for transferring to the differentiation medium. Smaller cells can be subcultured for continued production of somatic embryos.

3. **Differentiation and maturation of somatic embryos.** Proembryo masses are shifted to an auxin-free basal medium, high in ammonium nitrogen. Somatic embryos arise from single cells in clumps or small masses, develop polarity, and follow a pattern mimicking normal zygotic embryogenesis (*1, 2, 185*). Development may be variable in rate and abnormal in appearance, with secondary embryos forming on primary embryos. Synchrony and normalized development may be obtained by separating into different sizes by screening or density gradients. Adding abscisic acid (ABA) to the medium has improved uniformity and promoted normal development. Dehydration may also induce embryo maturity similar to that produced in in vivo embryo development (see p. 64) (*45, 46, 53, 82*).

4. **Plantlet formation.** Matured somatic embryos that have reached a "normal" size can be plated onto an agar medium devoid of any auxin but containing a low level of cytokinin.

5. **Transplanting.** After leaves and roots have formed, the plantlet can be transplanted to a medium in a container and handled as with any seedling plant or plantlet. Mass propagation of somatic plants by embryogenesis to produce "synthetic seeds" requires the development of procedures: first, to generate the required "true-to-type" embryos; second, to control the development of the proembryos through "normal" patterns of embryo development and maturation; and third, to provide a delivery system to make it possible to plant the seeds under greenhouse or field conditions to produce "true-to-type" plants. Considerable progress has been made in the development and utilization of large reactor vessels (*134, 168*) to generate embryos similar to those used in fermenters or mass propagation of microbes and cells for industrial production of pharmaceuticals and other products. Fluid drilling (*53*) has some potential as a planting procedure (Chap. 7). Development of synthetic seed coverings to encapsulate the embryos in an "artificial" seed coat is another possible procedure (*143, 144*).

FACTORS AFFECTING SUCCESS IN MICROPROPAGATION AND TISSUE CULTURE SYSTEMS

Success in micropropagation depends on the optimization of many variables, each of which may differ for the particular tissue culture system utilized and the purposes for which it is done.

Genotype

Much variation exists in the ability of different species and cultivars to regenerate and the cultural conditions required to bring this about. Systematic research and development for each kind of plant have been required to provide the needed knowledge and experience. To a varying degree, a period of research and development may be needed by the propagator with any unfamiliar species or cultivar, although generalizations can be made and much information is available in the published literature.

In general, those plants that have been easiest to propagate conventionally have also been the easiest to micropropagate. For example, most herbaceous plants can usually be vegetatively propagated relatively easily by either conventional propagation or micropropagation (*51, 122, 131, 161*). Woody perennials have been much more difficult to regenerate and new principles of culture have had to be developed. Success in most woody perennial species is related to their juvenility status (*11, 37*). Shrubs are generally more easily handled than trees (*105*). Different conifer species vary in ease of propagation based on seasonal growth dynamics and axillary bud characteristics (*103, 105*).

As principles and procedures of micropropagation become refined, it is possible to extend the procedures to a wider and wider range of plant materials. Even within a species, some cultivars respond better than others and are apt to be chosen for micropropagation. As procedures become adopted and standardized, pressure accumulates to develop new cultivars that are better adapted to the procedures that have evolved. Thus a coevolution of improved genotypes and improved (and more cost-effective) procedures may be expected to develop (*34, 78, 89, 147*).

Disease Effects

External Pathogens

External contaminants include fungi, molds, bacteria, yeasts, and other microorganisms. These are present literally everywhere—in the air and on the surface of plants, tables, hands, and so on. Spores move by air currents on dust particles. One must disinfest the explants, the tools, and the working area to remove such contaminants from the surface. All work must be done in special transfer areas where contaminants have been eliminated and precautions taken to prevent recontamination.

Reduction in surface contaminants begins with control of the stock plants used as the source of explants. Usually, contaminants are present only on the surface of plant parts, although they may be lodged in cracks, between bud scales, and elsewhere. Internal structures, such as growing points of buds and inside seeds and fruits, tend to be relatively free of pathogens. However, if the plant is growing in a humid atmosphere, mycelia may invade plant interiors and become a persistent problem.

Reduction in the surface contaminants on stock plants aids in later disinfection attempts (*64*). In general, stock plants growing in containers in protected environments in a greenhouse are "cleaner" sources of explants than are those growing out-of-doors. Overhead watering, sprinkling, or any activity that increases humidity around the plant should be avoided. Likewise, keep plant parts off the ground and avoid the use of roots or underground portions as explant sources, if possible. Insect and mite populations should be controlled. In one case, withholding water and keeping plants cool and dry for three weeks was necessary to reduce contaminants on *Dieffenbachia,* even inside a greenhouse (*85*). If plants are growing out-of-doors, covering shoots with plastic bags or spraying with fungicides may be useful (*30*).

Disinfestation requires use of chemicals that are toxic to the microorganism but relatively nontoxic to plant material. Tissue culture became possible with the development of convenient and effective disinfestants, such as **calcium hypochlorite** (*89*) and **sodium hypochlorite** (now widely available as commercial household bleaches under various trade names).

Placing antibiotics such as streptomycin in the culture medium has been tried (*106*) for inhibiting microorganisms but with varying degrees of success.

Internal Pathogens

Organisms such as viruses and virus-like organisms (see Chap. 8) may be present inside the tissue of the explant at the time of culture. One should not take for granted that the explants are free of pathogens even if the extreme meristem tip of the shoot is utilized. Specific indexing or other detection procedures should be part of the system.

Some bacterial contaminants, such as *Bacillus subtilis* (*162*), *Erwinia* (*85, 86*), and *Pseudomonas* (*85*), sometimes are present inside the plantlet but do not grow readily on the medium, either because the medium has become more acid (*65*) or because cytokinin in the medium depresses their growth (*69*). These contaminants can inhibit growth and rooting potential (*65*) or remain suppressed without effect until the explant is transferred to a new culture medium (*65*).

Culture indexing can be used to eliminate those explants showing positive results for the presence of pathogens. Indexing at the initial explant stage, however, does not always identify all potential contaminants (*65*).

Juvenility (Epigenetic) Effects

Success in micropropagation of woody plant species is to a large extent a function of the juvenility status of the source plant (see Chap. 8). Juvenile seedling plants are invariably more amenable to micropropagation, whereas vegetatively propagated mature cultivars of the same species are more difficult and may require special procedures. Typically, the limiting factor has not been the ability to proliferate shoots during an establishment period but to generate roots.

Various manipulations may be useful to obtain juvenile explants from particular source plants, such as selection from the basal juvenile parts of the seedling plant, induction of adventitious shoots, such as on roots, consecutive grafting to seedling plants, etc., as described in Chap. 8.

The micropropagation sequence of subculturing itself appears to be a rejuvenating process which ultimately leads to increased competence for rooting (*192*). Changes in the leaf morphology and other juvenility markers have been observed by many individuals, and the rooted products of micropropagation invariably show some juvenile characteristics (p. 172). A progressive increase in the ability to root, paralleled by shifts in vigor and morphology of the shoot with consecutive subcultures, has been found in several woody species. This process may be the function of the stabilization period of shoot production described in a subsequent section.

Selection of Explant

Proper choice of explant is a key requirement in micropropagation, and starting a new program with a new species or cultivar may require systematic analysis of the explant potential from types of tissue (*57*). In previous sections we described various micropropagation and tissue culture systems and listed the basic kinds of explants that are used to initiate them.

For shoot-tip culture, actively growing shoots obtained early in a growth phase are usually the most successful. Less active explants may require some modification of the hormone levels during culture. Some variables include size of explant, lateral versus terminal shoots, health of the shoot where taken, and dormancy status.

Hormones and Media Effects

A **basal medium** must be selected or established through systematic analysis. Some standard media are described in Chap. 17. Probably the most widely used is the **Murashige and Skoog** (MS) medium (see p. 500). Although originally developed for tobacco pith callus, it has also proven excellent for most herbaceous plants, due possibly to the high level of nitrogen, both ammonium and nitrate. On the other hand, a reduction of overall concentration has usually been desirable for woody plants, and many variations have been published (see Chap. 17 and the general references at the end of the chapter).

Intensive commercial micropropagation may require the optimization of not only the inorganic elements but also of other factors of the medium. This can be done by systematically testing a range of concentrations of individual elements combined in various ways with other elements. Suggested methods have been a broad-spectrum approach using all possible combinations (*30, 31*), or a factorial approach (*4, 34, 161*), where a range of each ingredient (others held constant) is tested in consecutive subcultures. One subculture should occur between each test to equalize the test material.

Ingredients

Agar is the usual support, but different brands produce some variation and agar substitutes are available (*131*). Sucrose is the preferred energy source because of its availability and low cost.

Growth hormones are used to support a basic level of growth but are equally important to direct the developmental response of the propagule. Shoot growth is strongly supported by **cytokinin** concentration (Fig. 16-16). In general, a minimum level is necessary to support growth with an optimum for maximum elongation. Further increased levels inhibit apical dominance and promote lateral shoot proliferation. Requirements

may vary at different stages of culture such that variable growth responses may occur during consecutive subcultures. Adjustment of cytokinin concentrations may be necessary (*161*). Cytokinin suppresses rooting and is reduced or omitted at the third stage.

Auxin is used for root initiation, although a minimum level may be necessary to support growth. Increased concentrations of auxin are used to initiate rooting. However, the excised shoot is sensitive to auxin only during a short period of a few days, such that dipping into auxin or transferring to a new medium might be preferable to continued exposure.

Gibberellins are usually not utilized but may be important in certain instances to induce rejuvenation (or invigoration) of particular recalcitrant genotypes (*142, 161*).

Manipulations of Shoot Tips in Culture

Shoot-tip micropropagation systems have two basic periods: **culture establishment** (stage I) and **propagule production** (stage II). Stage I has three basic phases: (a) isolation, (b) stabilization, and (c) shoot production. **Isolation** involves decontamination (as described elsewhere). Woody plants may have specific problems, including difficulty in removing particularly recalcitrant bacte-

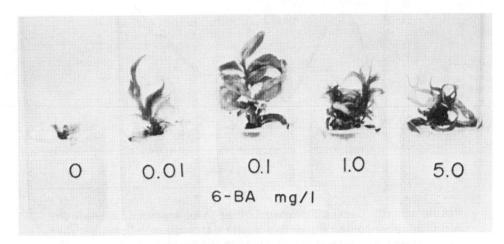

FIGURE 16-16 Cytokinin (6-benzyl-adenine) concentration changes the kind of shoot development in multiplication stage. At 0.1 mg/l the terminal shoot elongates; at 1.0 mg/l lateral shoots develop and the terminal shoot is inhibited. Propagules are almond-peach hybrid.

ria that may persist in culture, be difficult to identify, or have delayed expression in subcultures. A second problem is the production of phenolic substances that cause browning of the culture medium. This problem may be overcome by using antioxidants such as citric or ascorbic acid, very frequent subculturing, or utilizing activated charcoal.

Stabilization involves a change in growth characteristics from unpredictable, often abnormal shoot development at first, to a uniform predictable growth pattern (*103, 105*) (Fig. 16–17). Most annuals and herbaceous plants stabilize quickly within a few subcultures. Woody plants invariably require much longer and in some may never be achieved. The difficulty in stabilization appears to be associated with the phase state of the explant. The more juvenile, the more easily stabilized. Also, the characteristic seasonal growth pattern of the species being cultured is associated with ease of stabilization.

Stabilization may take a few weeks to several years. Usually, about four to six months is required. Subculture time may vary from two or three weeks to four or five weeks. Continued maintenance of shoot proliferation depends on frequent and regular subcultures to produce optimum shoot multiplication rates. Variation in routine may adversely affect multiplication rate, yield and quality of shoots, rooting, and subsequent growth. If subculturing is delayed too long, leaf yellowing and necrosis can develop.

Another variable that needs to be optimized is the subculture technique, including size of new explant, method of cutting, and sometimes orientation of planting.

Seasonal rhythmic patterns in proliferation sometimes occur, as in rose (*161*), even though subculturing is done under uniformly controlled conditions. In general, proliferation is greater in summer than in winter.

Special Characteristics in Culture

Vitrification

This phenomenon is characterized by a translucent, watersoaked, succulent appearance that can result in deterioration and failure to proliferate (*29, 48, 97*). Physiologically, expression involves excess water uptake and inhibition of lignin and cellulose synthesis. The condition appears to be a consequence of the matric potential of the medium as well as a low nitrate–ammonium relationship. Vitrification is more prevalent if plants are grown in liquid media or with low agar con-

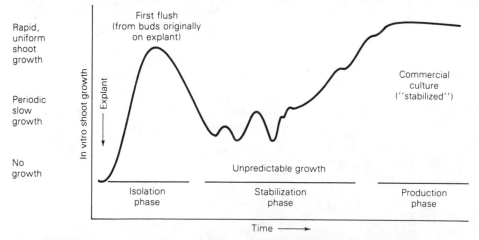

FIGURE 16–17 Generalized scheme of the three important phases in the microculture period through which a shoot must progress to be successfully microcultured. The second period is rooting and acclimation of shoots. (Reprinted by permission from D. D. McCown and B. H. McCown. In *Cell and tissue culture in forestry.* Vol 3. J. M. Bonga and D. J. Durzan, eds. Dordrecht: Martinus Nijhoff.)

centrations, high humidity, and high ammonium concentrations (as in the MS medium). Control may require manipulating all of these factors, including increase in agar concentration, change in brands of agar, inorganic ingredients in the medium, and modification of BA concentration (*131*).

Internal Pathogens

Some explants are found to have internal pathogens in the medium that are not apparent until after the culture has been grown for some time.

Shoot-Tip Necrosis and Internal Browning

Sometimes actively growing shoot tips develop "tip dieback" and subsequently die. This condition is caused by calcium-deficiency in the medium and correction requires refining the basic medium (*103*).

Habituation

This phenomenon is a characteristic of some callus and cell systems, which develop a competency to proliferate without the addition of specific hormones. It develops after long exposure to the particular hormone. It is believed to be an epigenetic phenomenon (*109, 110, 161, 165*). Shoots may become habituated to cytokinin after high cytokinin levels and continue to proliferate after cytokinin is removed.

Environmental Factors

The culture environment is important, although detailed research has not been done for many plants. Complete darkness is necessary to induce adventitious shoots in *Brunnera* and *Hosta* (*161*). On the contrary, high light intensity improves shoot intensity, increases multiplication in some plants, and decreases it in others. Increased light intensity appears to increase in importance during the acclimation and pretransplant period, as described below.

Photoperiodic response of flowering can be produced in vitro. For example, carnation shoots can be induced to flower by treatment with 16 hours of light but remain vegetative if kept at 12 hours (*161*). It is likely other plants can respond similarly.

Temperature can be a key factor in development and is highly specific or cultivar dependent. Temperate-zone plants require moderate temperature; tropical plants, high temperature.

Gaseous exchange of the flask with the outside can be important and involves the extent of the closure to the flasks.

Rooting, Acclimation, and Transplanting

The production of microplants and their transfer from the culture environment is a key step in micropropagation. The characteristics of the heterotrophic microplant in culture are described elsewhere in this chapter.

There are three key aspects: (a) induction of rooting, (b) shoot and root acclimation, and (c) physiological hardening of the tissues.

With many herbaceous plants, rooting takes place readily in culture and transfer can be accomplished with a minimum of difficulty. Shoot and root acclimation involves the formation of new roots and shoots following transfer from the culture medium. Success requires that the plantlets be kept under high-humidity conditions. It is also necessary to maintain vigorous shoot growth during the final period in culture.

Some woody perennial species are particularly recalcitrant in transplanting. For these plants, rooting microshoots from cultured plants by conventional means under high humidity or mist (*104, 193*) can reduce the problems.

A patented process (*33*) of **physiological conditioning** of the microplant in culture for direct field planting has been described for recalcitrant species, such as walnut (*34, 107*). This conditioning is accomplished by (a) optimizing the growth medium for inorganic elements and increasing sucrose concentration and time of exposure to encourage lignification, (b) reduction of temperature from 29°C to 19°C (87°F to 54°F), (c) reduction in nitrogen, and (d) increase in light intensity. After conditioning, microshoots are dipped in rooting hormone and rooted in a high-density flat, covered with plastic in a culture room. When roots become visible in one to two weeks, the unit is planted directly into the field and covered with a plastic top.

VARIATION IN MICROPROPAGATED PLANT PRODUCTS

Early results from micropropagated production showed that plants from some plant sources and micropropagation systems were true-to-type and uniform in phenotypic appearance. Other cases resulted in a great deal of variability in trueness-to-type (see Chap. 8). Much field testing and commercial experience has been obtained with various crops (*63, 161, 171, 192*). Variability can be categorized as in Chap. 8, although it is not always possible to assign a specific variant into a given category. Furthermore, much variation is species or genotype specific. The basic principle is that trueness-to-type must be tested through phenotypic and genotypic selection procedures, as described in Chap. 8.

Genetic and Chimeral Effects

Like conventional propagation, micropropagation systems that maintain and multiply preorganized shoot tips tend to remain stable. For this reason, multiple-shoot-tip culture (Fig. 16–18) is used mostly for commercial propagation. On the other hand, those systems that utilize adventitious shoots or the multiplication of callus, cells, and protoplasts result in considerably more variation in the end product.

Sources of variability are the following:

Chimeral Breakdown

This can occur in systems having adventitious regeneration because new (*de novo*) growing points develop primarily from single-cell origins. The new shoot may arise from any one of several layers of different genotypes that make up the chimera (*18, 153*). Many important horticultural crops are chimeras, where reversion would be un-

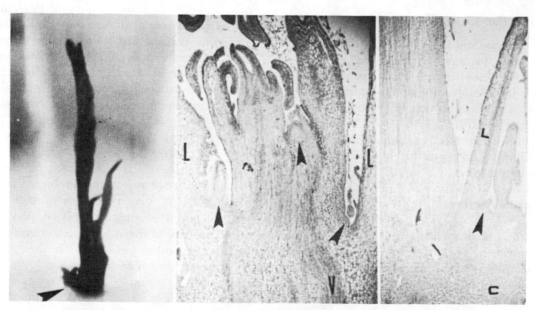

FIGURE 16–18 Shoot multiplication on shoot-tip culture of EMLA 26 apple. *Left:* After 10 days propagule consists of apical, axillary, and basal adventitious shoots. *Center:* Apical portion of the shoot-tip culture on left after six weeks. *Axillary* shoots shown by arrows. *Right:* Basal lateral shoot shown on left after six weeks. Shows that the shoot is *adventitious* and arises from a mass of basal callus (C). Leaf primordia shown by (L). Vascular area identified by (V). (Courtesy Neil Miles. From F. R. Nasir and N. Miles, Histological origin of EMLA 26 apple shoots generated during micropropagation. *HortScience* 16:417 (abstr.), 1981.

desirable (*40*). Culture from shoot tips and axillary shoots is necessary for this type of cultivar, as is true for conventional propagation methods (see Chaps. 8 and 9).

On the other hand, reversion may be desirable to convert a variable chimera into a stable form (solid mutant). For instance, a genetically uniform thornless blackberry (see p. 533) has been produced from a variable chimera thornless clone by regeneration. This concept of single cell origin is important in mutation breeding, where new variants propagated conventionally are apt to be chimeral (see p. 178).

Aberrant Cell Division in Callus or Cell Suspensions

These may appear after organogenesis (see Fig. 16–10) and result in polyploid or ancuploid plants (*28*) (see Fig. 16–19). Sometimes shoots being cultured develop a callus at the nodes and the base from which adventitious shoots may arise along with the nodal shoots. This is the case with Boston fern (*122*). Production of aberrant plants is common but in low frequency. This problem can be modified by reducing the number of subcultures from a single explant source and lowering the auxin or BA levels in the medium to lessen or eliminate the amount of callus and branching.

Preexisting Genetic Variation

This can result from variation among cells, which would not be expressed in the source plant (see Chap. 8 for an explanation). Apparently, many plants are **chimeral mosaics.** This kind of variation has been called **somaclonal variation** (*149*) and has been exploited to develop significant crop improvement programs in potato (*151*), tomato (*42*), and other crops (*149*).

Genetic and Chlorophyll Mutants

These may be characteristic of individual genotypes (see Fig. 16–20). For example, some strawberry cultivars may produce sectors (0.1 to

FIGURE 16–19 Genetic variation shown between two tobacco plants originating from single cells. Callus was developed and organogenesis occurred as described in Figure 16–10. *Left:* Normal plant. *Right:* Variant plant. (Courtesy T. Murashige.)

FIGURE 16-20 Albino plantlet appearing on an African violet (*Saintpaulia*) after micropropagation. Origin is likely to have been from a single cell; hence a solid mutant resulted.

7 percent), white streak (1 to 8 percent), and dwarfs (1 percent) (*171*). Such plants can be removed by roguing during multiplication sequences, lowering BA levels, or limiting subcultures. Visual inspection of plant propagules and careful progeny source records should be a regular aspect of production, as with other forms of propagation.

Transient Phenotypic Variation

Sometimes, growth characteristics and patterns are shown by the initial post-culture products of micropropagation that occur only temporarily and disappear with continued growth and/or repropagation. These variants may or may not be desirable, depending upon the trueness-to-type standards for the particular cultivar.

Branching

Some herbaceous ornamentals propagated by micropropagation (*161*) show desirable enhanced branching as compared to conventionally propagated plants [e.g., *Hosta, Begonia,* aster, chrysanthemum, rose (*161*), *Syngonium,* ferns, and

other foliage plants (*25*)]. These characteristics are undesirable for *Gypsophila, Kalanchoe, Gerbera,* and *Saintpaulia* (*161*). This characteristic has also occurred in strawberry, blackberry, and apple (*192*). This effect may be due to carryover effects of BA from the multiplication medium or may be epigenetic in nature (see below).

Vigor

Some kinds of plants show extra vigor and may be in a specific reproductive phase. For example, (*20, 171*) strawberry plants tend to be vigorous and produce vegetative runners (see Chap. 14). Micropropagated plants directly planted for fruit production produce somewhat smaller fruit and yield less than do conventionally propagated plants. However, these same plants are useful for nursery plant production. Consequently, meristem-tip culture to control disease and shoot-tip multiplication is a useful combination to obtain "mother plants" for conventional nursery production.

The reason for the enhanced vigor of some micropropagated plants is not completely clear. It may be due to the removal of unknown viruses or other pathogens (see Chap. 8), or it may result from epigenetic effects (see below).

Epigenetic (Rejuvenation) Effects

Variations of this category reflect primarily the juvenile–mature cycles of the plant that affect the reproductive capacity, the trueness-to-type morphology, and the capacity for further regeneration (Chap. 8).

Rejuvenation

Many, if not all woody plants that are consecutively recultured in shoot-tip culture show some degree of rejuvenation in rooting potential which is also expressed in leaf morphology and/or performance. This phenomenon has been observed in strawberry, blackberry, gooseberry, grape, and apple (*120, 192*). Enhanced rooting potential may persist after micropropagation such that the plants produced may be useful later as stock plants in conventional cutting propagation.

Age of Flowering

The effect on age of flowering varies with genotype, method of handling, and source of explant. Fruit and nut cultivars propagated from shoot-tip culture have not demonstrated significant increase in the age of bearing as compared to conventionally grafted trees (*192*), as might be expected if the plants are juvenile. In fact, examples have occurred where particularly precocious cultivars of micropropagated walnut have flowered and fruited at an unusually early age, even in the nursery row. This pattern would be particularly favorable for the production of fruit and nut cultivars and for forest species grown for seed orchards.

True-to-Type Juvenile Phenotypes

The production of a true-to-type juvenile phenotype as compared to an adult is important in the propagation of forest species to be used for lumber, biomass production, and other cases where the vegetative structure is the product (see Chap. 8). In general, production of the mature phenotype persists from vegetative propagation, whether by conventional procedures or by those micropropagation procedures that have been successful. Selecting a superior (''elite'') genotype of conifers for forest uses must be done at a mature age, but this creates problems not only in propagation but in the production of the true-to-type phenotype (*11, 37*). It is possible that this problem can be solved by explant selection from juvenile material, reversion of mature tissues, or development of suitable micropropagation techniques that will capture the genetic gains of clonal propagation, retain a juvenile phenotype, and provide an efficient economical mass-propagation system (*11, 15*).

Uneven Flowering and Reluctance to Flower

This phenomenon occurs in some herbaceous crops, such as chrysanthemum, dianthus, *Kalanchoe,* and gypsophila (*161*). Uneven performance may be a function of variation in juvenility among individual micropropagated plants. Scheduling becomes a problem and if excessive can limit use of the material. Cultivar differences exist in these responses and selection may be the long-term solution. Because of the variability in reproductive responses among different genotypes, future developments may include selection of cultivars specifically adapted to micropropagation procedures.

REFERENCES

1. Ammirato, P. V. 1985. Patterns of development in culture. In *Tissue culture in forestry and agriculture,* R. R. Henke, K. W. Hughes, M. J. Constantin, and A. Hollaender, eds. New York: Plenum Press, pp. 9–29.

2. Ammirato, P. V. 1987. Organizational events during somatic embryogenesis. In *Plant tissue and cell culture,* C. E. Green, D. A. Somers, W. P. Hackett, and D. D. Biesboer, eds. New York: Alan R. Liss, pp. 57–81.

3. Anderson, W. C. 1975. Propagation of rhododendrons by tissue culture. 1. Development of a culture medium for multiplication of shoots. *Proc. Inter. Plant Prop. Soc.* 25:129–35.

4. Anderson, W. C. 1980. Mass propagation by tissue cultures: Principles and techniques. In *Proc. conf. on nursery production of fruit plants: Applications and feasibility,* R. H. Zimmerman, ed. USDA ARR-NE-11, pp. 1–14.

5. Bajaj, Y. P. S., ed. 1986. *Biotechnology in agriculture and forestry,* Vol. 1, *Trees.* Berlin: Springer-Verlag.

6. Barlass, M., and K. G. M. Skene. 1980. Studies on the fragmented shoot apex of grapevine. I. The regenerative capacity of leaf primordial fragments in vitro. *Jour. Exp. Bot.* 31(121):483–88.

7. ———. 1980. Studies on the fragmented shoot apex of grapevine. II. Factors affecting growth and differentiation *in vitro. Jour. Exp. Bot.* 31(121):489–95.

8. Barz, W., E. Reinhard, and M. H. Zenk, eds. 1977. *Plant tissue culture and its biotechnological applications.* Berlin: Springer-Verlag.

9. Beasley, C. A. 1984. Culture of cotton ovules. In *Cell culture and somatic cell genetics of plants,* Vol. 1, *Laboratory procedures and their applications,* I. K. Vasil, ed. New York: Academic Press, pp. 232–40.

10. Bilkey, P. C., B. H. McCown, and A. C. Hildebrandt. 1978. Micropropagation of African violet from petiole cross-sections. *HortScience* 13(1):37–38.

11. Bonga, J. M. 1986. Vegetative propagation in relation to juvenility, maturity, and rejuvenation. In *Cell and tissue culture in forestry* (2nd ed.), J. M. Bonga and D. J. Durzan, eds. Dordrecht: Martinus Nijhoff Publishers, pp. 387–411.

12 Boulay, M. 1987. In vitro propagation of tree species. In *Plant tissue and cell culture,* C. E. Green, D. A. Somers, W. P. Hackett, and D. D. Biesboer, eds. New York: Alan R. Liss, pp. 367–82.

13. Bourgin, J. P., and J. P. Nitsch. 1967. Obtention de Nicotiana haploides a'partir d'elamines cultivees *in vitro. Ann. Phys. Veg.* 9:377–82.

14. Boxus, P., and P. Druart. 1986. Virus-free trees through tissue culture. In *Biotechnology in agriculture and forestry,* Vol. 1, *Trees,* Y. P. S. Bajaj, ed. Berlin: Springer-Verlag, pp. 24–30.

15. Boxus, P. H., M. Quoirin, and J. M. Laine. 1977. Large scale propagation of strawberry plants from tissue culture. In *Plant cell, tissue and organ culture,* J. Reinert and Y. P. S. Bajaj, eds. Berlin: Springer-Verlag, pp. 131–43.

16. Brainerd, K. E., L. H. Fuchigami, S. Kwiatkowski, and C. S. Clark. 1981. Leaf anatomy and water stress of aseptically cultured 'Pixy' plum grown under different environments. *HortScience* 16(2):173–75.

17. Burk, L. G., G. R. Gwynn, and J. R. Chaplin. 1972. Diploidized haploids from aseptically cultured anthers of *Nicotiana tabacum. Jour. Hered.* 63(6):355–60.

18. Bush, S. R., E. D. Earle, and R. W. Langhans. 1976. Plantlets from petal segments, petal epidermis, and shoot tips of the periclinal chimera, *Chrysanthemum morifolium 'Indianapolis'. Amer. Jour. Bot.* 63(6):729–37.

19. Button, J., and J. Kochba. 1977. Tissue culture in the citrus industry. In *Plant cell, tissue and organ culture,* J. Reinert and Y. P. S. Bajaj, eds. Berlin: Springer-Verlag, pp. 70–92.

20. Cameron, J. S., and J. F. Hancock. 1986. Enhanced vigor in vegetative progeny of micropropagated strawberry plants. *HortScience* 21(5):1225–26.

21. Carlson, P. S., H. H. Smith, and R. D. Dearing. 1972. Parasexual interspecific plant hybridization. *Proc. Nat. Acad. Sci. USA* 69(8):2292–94.

22. Cheah, Kheng-Tuan, and Tsai-Ying Cheng. 1978. Histological analysis of adventitious bud formation in cultured Douglas fir cotyledon. *Amer. Jour. Bot.* 65(8):845–49.

23. Christianson, M. L. 1985. An embryogenic culture of soybean: towards a general theory of somatic embryogenesis. In *Tissue culture in forestry and agriculture,* R. R. Henke, K. W. Hughes, M. J. Constantin, and A. Hollaender, eds. New York: Plenum Press, pp. 83–103.

24. Christie, C. B. 1978. Rapid propagation of aspens and silver poplars using tissue culture techniques. *Proc. Inter. Plant Prop. Soc.* 28:255–60.

25. Chu, I. Y. E. 1986. The application of tissue culture to plant improvement and propagation in the ornamental horticulture industry. In *Tissue culture as a plant production system for horticultural crops,* R. H. Zimmerman, R. J. Griesbach, F. A. Hammerschlag, and R. H. Lawson, eds. Dordrecht: Martinus Nijhoff Publishers, pp. 15–33.

26. Cocking, E. C. 1986. The tissue culture revolution. In *Plant tissue culture and its agricultural applications,* L. A. Withers, and P. G. Alderson, eds. London: Butterworth, pp. 3–20.

27. Collins, G. B., and J. W. Grosser. 1984. Culture of embryos. In *Cell culture and somatic cell genetics of plants,* Vol. 1, I. K. Vasil, ed. New York: Academic Presss, pp. 241–57.

28. D'Amato, F. 1977. Cytogenetics of differentiation in tissue and cell cultures. In *Plant cell, tissue and organ cultures,* J. Reinert and Y. P. S. Bajaj, eds. Berlin: Springer-Verlag, pp. 343–57.

29. Debergh, P. C. A. 1987. Recent trends in the application of tissue culture to ornamentals. In *Plant tissue and cell culture,* C. E. Green, D. A. Somers, W. P. Hackett, and D. D. Biesboer, eds. New York: Alan R. Liss, pp. 383–93.

30. de Fossard, R. A. 1976. *Tissue culture for plant propagators.* Armidale, N. S. W., Australia: Univ. of New England Printery.

31. ———. 1986. Principles of plant tissue culture. In *Tissue culture as a plant production system for horticultural crops,* R. H. Zimmerman, R. J. Griesbach, F. A. Hammerschlag, and R. H. Lawson, eds. Dordrecht: Martinus Nijhoff Publishers, pp. 1–13.

32. Dirr, M. A., and C. W. Heuser, Jr. 1987. *The*

reference manual of woody plant propagation: From seed to tissue culture. Athens, Ga.: Varsity Press.

33. Driver, J.A., 1986. Method for acclimatizing and propagating plant tissue culture shoots. *Patent No. 4,612,725.* Washington, D.C.: U.S. Patent Office.

34. Driver, J. A., and G. R. L. Suttle. 1986. Nursery handling of propagules. In *Cell and tissue culture in forestry.* (2nd ed.), J. M. Bonga and D. J. Durzan, eds. Dordrecht: Martinus Nijhoff Publishers, pp. 320–31.

35. Dunstan, D. I., and K. E. Turner. 1984. The acclimatization of micropropagated plants. In *Cell culture and somatic cell genetics of plants,* Vol. 1, *Laboratory procedures and their applications,* I. K. Vasil, ed. New York: Academic Press, pp. 123–29.

36. Dunwell, J. M. 1986. Pollen, ovule and embryo culture as tools in plant breeding. In *Plant tissue culture and its agricultural applications,* L. A. Withers, and P. G. Alderson. London: Butterworth, pp. 375–404.

37. Durzan, D. J. 1986. Special problems: adult vs. juvenile explants. In *Handbook of plant cell culture,* Vol. 4, D. A. Evans, W. R. Sharp, and P. V. Ammirato, eds. New York: Macmillan, pp. 471–503.

38. Durzan, D. J. 1988. Process control in somatic polyembryogenesis. In *Molecular genetics of forest trees,* J. Hallgren, ed. Umea: Swedish Agri. Univ., in press.

39. Durzan, D. J., and P. K. Gupta. 1988. Somatic embryogenesis and polyembryogenesis in conifers. In *Biotechnology in agriculture,* Mizrahl, A., ed. New York: Alan R. Liss, pp. 53–81.

40. Earle, E. D., and Y. Demarly. 1984. Variability in plants regenerated from tissue culture. *HortScience* 19(1):146–47.

41. Emershad, R. L., and D. W. Ramming. 1984. *In ovulo* embryo culture of *Vitis vinifera L.* cv. Thompson Seedless. *Amer. Jour. Bot.* 71:873–77.

42. Evans, D. A., and J. E. Bravo. 1986. Phenotypic and genotypic stability of tissue cultured plants. In *Tissue culture as a plant production system for horticultural crops,* R. H. Zimmerman, R. J. Griesbach, F. A. Hammerschlag, and R. H. Lawson, eds. Dordrecht: Martinus Nijhoff Publishers, pp. 73–94.

43. Evans, D. A., W. R. Sharp, and C. E. Flick. 1981. Plant regeneration from cell cultures. *Hort. Rev.* 3:214–314.

44. Evans, D. A., W. R. Sharp, and H. P. Medina-

Filho. 1984. Somaclonal and gametoclonal variation. *Amer. Jour. Bot.* 71(6):759–74.

45. Finkelstein, R. R., and M. L. Crouch. 1984. Precociously germinating rapeseed embryos retain characteristics of embryogeny. *Planta* 162:125–31.

46. Finkelstein, R. R., and M. L. Crouch. 1987. Hormonal and osmotic effects on developmental potential of maturing rapeseed. *HortScience* 22(5):797–800.

47. Fiorino, P., and F. Loreti. 1987. Propagation of fruit trees by tissue culture in Italy. *HortScience* 22(3):353–58.

48. Gaspar, T., C. Kevers, P. Debergh, L. Maene, M. Paques, and P. Boxus. 1986. Vitrification: Morphological, physiological, and ecological aspects. In *Cell and tissue culture in forestry,* Vol. 1, *General principles and biotechnology,* J. M. Bonga and D. J. Durzan, eds. Dordrecht: Martinus Nijhoff Publishers, pp. 152–66.

49. Gautheret, R. J. 1985. History of plant tissue and cell culture: A personal account. In *Cell culture and somatic cell genetics of plants,* Vol. 2, I. K. Vasil, ed. New York: Academic Press, pp. 1–59.

50. Gengenbach, B. G. 1984. In vitro pollination, fertilization, and development of maize kernels. In *Cell culture and somatic cell genetics of plants,* Vol. 1, *Laboratory procedures and their applications,* I. K. Vasil, ed. New York: Academic Press, pp. 276–82.

51. George, E. F., and P. D. Sherrington. 1984. *Plant propagation by tissue culture: Handbook and directory of commercial laboratories.* Eversley, England: Exegetics Ltd.

52. Goldy, R. G., and U. Amborn. 1987. In vitro culturability of ovules from 10 seedless grape clones. *HortScience* 22(5):952.

53. Gray, D. J. 1987. Quiescence in monocotyledonous and dicotyledonous somatic embryos induced by dehydration. *HortScience* 22(5):810–14.

54. Green, C. E., D. A. Somers, W. P. Hackett, and D. D. Biesboer, eds. 1987. *Plant tissue and cell culture.* Proc. 6th international congress plant tissue and cell culture. New York: Alan R. Liss.

55. Guha, S., and S. C. Maheshwari. 1964. In vitro production of embryos from anthers of *Datura. Nature (London)* 204:497.

56. Gupta, P. K., and D. J. Durzan. 1986. Somatic polyembryogenesis from callus of mature sugar pine embryos. *Biotechnology* 4:643–45.

57. Gupta, P. K., and D. J. Durzan. 1987. Micro-

propagation and phase specificity in mature, elite Douglas fir. *Jour. Amer. Soc. Hort. Sci.* 112(6):969–71.

58. Gupta, P. K., D. J. Durzan, and B. J. Finkle. 1987. Somatic polyembryogenesis in embryogenic cell masses of *Picea abies* (Norway spruce) and *Pinus taeda* (loblolly pine) after freezing in liquid nitrogen. *Can. Jour. For. Res.* 17:1130–34.

59. Haberlandt, G. 1902. Kulturversuche mit isolierten Pflanzenzellsn. *Sitzungsber. Akad. der Wiss. Wien, Math.-Naturwiss.* K1. 111:69–92.

60. Hackett, W. P. 1966. Applications of tissue culture to plant propagation. *Proc. Inter. Plant Prop. Soc.* 16:88–92.

61. ———. 1985. Juvenility, maturation, and rejuvenation in woody plants. *Hort. Rev.* 7:109–55.

62. Hackett, W. P., and J. M. Anderson. 1967. Aseptic multiplication and maintenance of differentiated carnation shoot tissue derived from shoot apices. *Proc. Amer. Soc. Hort. Sci.* 90:365–69.

63. Hammerschlag, F. A. 1986. Temperate fruits and nuts. In *Tissue culture as a plant production system for horticultural crops,* R. H. Zimmerman, R. J. Griesbach, F. A. Hammerschlag, and R. H. Lawson, eds. Dordrecht: Martinus Nijhoff Publishers, pp. 221–36.

64. Henny, R. J. 1980. *In vitro* germination of *Maranta leuconeura* embryos. *HortScience* 15(2):198–200.

65. Henny, R. J., J. F. Knauss, and A. Donnan, Jr. 1981. Foliage plant tissue culture. In *Foliage plant production,* J. Joiner, ed. Englewood Cliffs, N.J.: Prentice-Hall.

66. Hesse, C., and D. E. Kester. 1955. Germination of embryos of *Prunus* related to degree of embryo development and method of handling. *Proc. Amer. Soc. Hort. Sci.* 65:251–64.

67. Heuser, C. W., and J. Horker. 1976. Tissue culture propagation of daylilies. *Proc. Inter. Plant Prop. Soc.* 26:269–72.

68. Hicks, G. S. 1980. Patterns of organ development in plant tissue culture and the problem of organ determination. *Bot. Rev.* 46:1–23.

69. Holdgate, D. P., and J. S. Aynsley. 1977. The development and establishment of a commercial tissue culture laboratory. *Acta Hort.* 78:31–36.

70. Holley, W. D., and R. Baker. 1963. *Carnation production.* Dubuque, Iowa: William C. Brown.

71. Hollings, M. 1965. Disease control through virus-free stock. *Ann. Rev. Phytopath.* 3:367–96.

72. Hu, C., and P. Wang. 1986. Embryo culture: Technique and applications. In *Handbook of plant cell culture,* Vol. 4, D. A. Evans, W. R. Sharp, and P. V. Ammirato, eds. New York: Macmillan, pp. 43–96.

73. Huang, S., and D. F. Millikan. 1980. *In vitro* micrografting of apple shoot-tips. *HortScience* 15:741–43.

74. Hussey, G. 1976. *In vitro* release of axillary shoots from apical dominance in monocotyledonous plantlets. *Ann. Bot.* 40:1323–25.

75. ———. 1977. *In vitro* propagation of some members of the Liliaceae, Iridaceae and Amaryllidaceae. *Acta Hort.* 78:303–9.

76. Hussey, G., and A. Falavigna. 1980. Origin and production of *in vitro* adventitious shoots in the onion, *Allium cepa* L. *Jour. Exp. Bot.* 31(125):1675–86.

77. Jonard, R. 1986. Micrografting and its applications to tree improvement. In *Biotechnology in agriculture and forestry,* Vol. 1, *Trees,* Y. P. S. Bajaj, ed. Berlin: Springer-Verlag, pp. 31–48.

78. Jones, J. B. 1986. Determining markets and market potential. In *Tissue culture as a plant production system for horticultural crops,* R. H. Zimmerman, R. J. Griesbach, F. A. Hammerschlag, and R. H. Lawson, eds. Dordrecht: Martinus Nijhoff Publishers, pp. 175–82.

79. Kartha, K. K. 1984. Elimination of viruses. In *Cell culture and somatic cell genetics of plants,* Vol. 1, I. K. Vasil, ed. New York: Academic Press, pp. 577–85.

80. Kartha, K. K. 1986. Production and indexing of disease-free plants. In *Plant tissue culture and its agricultural applications,* L. A. Withers, and P. G. Alderson, eds. London: Butterworth, pp. 219–38.

81. Keller, W. A. 1984. Anther culture of Brassica. In *Cell culture and somatic cell genetics of plants,* Vol. 1. I. K. Vasil, ed. New York: Academic Press, pp. 302–10.

82. Kermode, A. R., J. D. Bewley, J. Dasgupta, and S. Misra. 1986. The transition from seed development to germination: A key role for desiccation? *HortScience* 21(5):1113–18.

83. Kester, D. E., and C. O. Hesse. 1955. Embryo culture of peach varieties in relation to season of ripening. *Proc. Amer. Soc. Hort. Sci.* 65:265–73.

84. Kitto, S. L., and J. Janick. 1985. A citrus embryo assay to screen water-soluble resins as synthetic seed coats. *HortScience* 20(1):98–100.

85. Knauss, J. F. 1976. A tissue culture method of producing *Dieffenbachia picta* cv. Perfection free

of fungi and bacteria. *Proc. Fla. State Hort. Soc.* 89:293–96.

86. Knauss, J.F., and J. W. Miller. 1978. A contaminant, *Erwinia carotovera*, affecting commercial plant tissue culture. *In Vitro* 14:754–56.

87. Kohlenback, H. W. 1984. Culture of isolated mesophyll cells. In *Cell culture and somatic cell genetics of plants*, Vol. 1, *Laboratory procedures and their applications*, I. K. Vasil, ed. New York: Academic Press, pp. 204–12.

88. Kosuge, T., C. P. Meredith, and A. Hollaender, eds. 1983. *Genetic engineering of plants: An agricultural perspective*. New York: Plenum Press.

89. Krikorian, A. D. 1982. Cloning higher plants from aseptically cultured tissues and cells. *Biol. Rev.* 57:151–218.

90. Krikorian, A. D., R. P. Kann, S. A. O'Connor, and M. S. Fitter. 1986. Totipotent suspensions as a means of multiplication. In *Tissue culture as a plant production system for horticultural crops*, R. H. Zimmerman, R. J. Griesbach, F. A. Hammerschlag, and R. H. Lawson, eds. Dordrecht: Martinus Nijhoff Publishers, pp. 61–72.

91. Kyte, L. 1987. *Plants from test tubes: An introduction to micropropagation* (rev. ed.). Portland, Oreg.: Timber Press.

92. Lammerts, W. E. 1942. Embryo culture, an effective technique for shortening the breeding cycle of deciduous trees and increasing germination of hybrid seed. *Amer. Jour. Bot.* 29:166–171.

93. Langhans, R. W., R. K. Horst, and E. D. Earle. 1977. Disease-free plants via tissue culture propagation. *HortScience* 12:149–50.

94. Lee, C. W., R. M. Skirvin, A. I. Soltero, and J. Janick. 1977. Tissue culture of *Salpiglossis sinuata* L. from leaf discs. *HortScience* 12(6):547–49.

95. Linsmaier, E. M., and F. Skoog. 1965. Organic growth factor requirements of tobacco tissue cultures. *Physiol. Plant* 18:101–27.

96. Litz, R. E. 1987. Application of tissue culture to tropical fruits. In *Plant tissue and cell culture*, C. C. Green, D. A. Somers, W. P. Hackett, and D. D. Biesboer, eds. New York: Alan R. Liss, pp. 407–18.

97. Loreti, F., and P. L. Pasqualetto. 1986. Vitrification of plants cultured in vitro. *Proc. Inter. Plant Prop. Soc.* 36:66–71.

98. Lutz, J. D., J. R. Wong, and J. Rowe. 1985. Somatic embryogenesis for mass cloning of crop plants. In *Tissue culture in forestry and agriculture*, R. R. Henke, K. W. Hughes, M. J. Constan-

tin, and A. Hollaender, eds. New York: Plenum Press, pp. 105–16.

99. Lyrene, P. M. 1981. Juvenility and production of fast-rooting cuttings from blueberry shoot cultures. *Jour. Amer. Soc. Hort. Sci.* 106:396–98.

100. Macdonald, B. 1986. *Practical woody plant propagation for nursery growers*. Portland, Oreg.: Timber Press.

101. Maheshwari, P., and N. S. Rangaswamy. 1963. *Plant tissue and organ culture*. Delhi: Inter. Soc. of Plant Morphologists, Univ. of Delhi.

102. Mapes, M. O. 1973. Tissue culture of bromeliads. *Proc. Inter. Plant Prop. Soc.* 23:47–55.

103. McCown, B. H. 1986. Woody ornamentals, shade trees, and conifers. In *Tissue culture as a plant production system for horticultural crops*, R. H. Zimmerman, R. J. Griesbach, F. A. Hammerschlag, and R. H. Lawson, eds. Dordrecht: Martinus Nijhoff Publishers, pp. 333–42.

104. McCown, D. D. 1986. Plug systems for micropropagules. In *Tissue culture as a plant production system for horticultural crops*, R. H. Zimmerman, R. J. Griesbach, F. A. Hammerschlag, and R. H. Lawson, eds. Dordrecht: Martinus Nijhoff Publishers, pp. 53–60.

105. McCown, D. D., and B. H. McCown. 1986. North American hardwoods. In *Cell and tissue culture in forestry*, Vol. 3, *Case histories: Gymnosperms, angiosperms and palms* (2nd ed.), J. M. Bonga and D. J. Durzan, eds. Dordrecht: Martinus Nijhoff, pp. 247–60.

106. McGarrity, G. L. 1975. Control of microbiological contamination. *Tissue Culture Assn. Man.* 1:181–85.

107. McGranahan, G. H., J. A. Driver, and W. Tulecke. 1986. Tissue culture of Juglans. In *Cell and tissue culture in forestry*, Vol. 3, *Case histories: Gymnosperms, angiosperms, and palms* (2nd ed.), J. M. Bonga, and D. J. Durzan, eds. Dordrecht: Martinus Nijhoff Publishers, pp. 261–71.

108. McGrew, J. R. 1980. Meristem culture for production of virus-free strawberries. In *Proc. conf. on nursery production of fruit plants through tissue culture: Applications and feasibility*, R. H. Zimmerman, ed. USDA. ARR-NE-11, pp. 80–85.

109. Meins, F., Jr., and A. N. Binns. 1978. Epigenetic clonal variation in the requirements of plant cells for cytokinins. In *The clonal basis of development*, S. Subtelny and I. M. Sussex, eds. New York: Academic Press, pp. 185–201.

110.   ———. 1979. Cell determination in plant development. *BioScience* 29:221–25.

111.   Meredith, C., and P. S. Carlson. 1978. Genetic variation in cultured plant cells. In *Propagation of higher plants through tissue culture,* K. W. Hughes, R. Henke, and M. Constantin, eds. Springfield, Va.: U.S. Dept. of Energy, Tech. Inform. Center, pp. 166–76.

112.   Meyer, M. M., Jr. 1976. Culture of Paeonia embryos by *in vitro* techniques. *Amer. Peony Soc. Bul.* 217:32–35.

113.   ———. 1980. *In vitro* propagation of *Hosta sieboldiana. HortScience* 15:737–38.

114.   Meyer, M. M., Jr., L. H. Fuchigami, and A. N. Roberts. 1975. Propagation of tall bearded irises by tissue culture. *HortScience* 10:479–80.

115.   Meyer, M. M., and G. M. Milbrath. 1977. *In vitro* propagation of horseradish with leaf pieces. *HortScience* 12(6):544–45.

116.   Morel, G. M. 1960. Producing virus-free cymbidiums. *Amer. Orch. Soc. Bul.* 29:495–97.

117.   ———. 1964. Tissue culture: A new means of clonal propagation of orchids. *Amer. Orch. Soc. Bul.* 33:473–78.

118.   Mudge, K. W., C. A. Borgman, J. C. Neal, and H. A. Weller. 1986. Present limitations and future prospects for commercial micropropagation of small fruits. *Proc. Inter. Plant Prop. Soc.* 36:538–43.

119.   Mullins, M. G. 1987. Propagation and genetic improvement of temperate fruits, the role of tissue culture. In *Plant tissue and cell culture,* C. E. Green, D. A. Somers, W. P. Hackett, and D. D. Biesboer, eds. New York, Alan R. Liss, pp. 395–406.

120.   Mullins, M. G., G. Y. Nair, and P. Sampet. 1979. Rejuvenation *in vitro:* Induction of juvenile characters in an adult clone of *Vitis vinifera* L. *Ann. Bot.* 44:623–27.

121.   Murashige, T. 1966. Principles of in vitro culture. *Proc. Inter. Plant Prop. Soc.* 16:80–88.

122.   Murashige, T. 1974. Plant propagation through tissue cultures. *Ann. Rev. Plant Phys.* 25:135–66.

123.   Murashige, T., and F. Skoog. 1962. A revised medium for rapid growth and bioassays with tobacco tissue cultures. *Phys. Plant.* 15:473–97.

124.   Navarro, L., and J. Juarez. 1977. Tissue culture techniques used in Spain to recover virus-free citrus plants. *Acta Hort.* 78:425–35.

125.   Negueroles, J., and O. P. Jones. 1979. Produc-

tion *in vitro* of rootstock/scion combinations of *Prunus* cultivars. *Jour. Hort. Sci.* 54(4):279–81.

126.   Ng, T. J. 1986. Use of tissue culture for micropropagation of vegetable crops. In *Tissue culture as a plant production system for horticultural crops,* R. H. Zimmerman, R. J. Griesbach, F. A. Hammerschlag, and R. H. Lawson, eds. Dordrecht: Martinus Nijhoff Publishers, pp. 259–70.

127.   Nitsch, J. P., and C. Nitsch. 1969. Haploid plants from pollen grains. *Science* 163:85–87.

128.   Oglesby, R. P., and J. L. Griffis, Jr. 1986. Commercial in vitro propagation and plantation crops. In *Tissue culture as a plant production system for horticultural crops,* R. H. Zimmerman, R. J. Griesbach, F. A. Hammerschlag, and R. H. Lawson, eds. Dordrecht: Martinus Nijhoff Publishers, pp. 253–57.

129.   Oglesby, R. P., and R. E. Strode. 1979. Commercial tissue culturing at Oglesby Nursery. *Proc. Inter. Plant Prop. Soc.* 29:341–44.

130.   Pence, V. C., P. M. Hasegawa, and J. Janick. 1979. Asexual embryogenesis in *Theobroma cacao* L. *Jour. Amer. Soc. Hort. Sci.* 104(2):145–48.

131.   Pierik, R. L. M. 1987. *In vitro culture of higher plants.* Dordrecht: Martinus Nijhoff Publishers.

132.   Pierik, R. L. M., J. L. M. Jansen, A. Maasdam, and C. M. Binnendijk. 1975. Optimalization of *Gerbera* plantlet production from excised capitulum explants. *Sci. Hort.* 3:351–57.

133.   Pierik, R. L. M., and H. H. M. Steegmans. 1976. Vegetative propagation of Freesia through the isolation of shoots *in vitro. Neth. Jour. Agr. Sci.* 24:274–77.

134.   Preil, W., P. Florek, U. Wix, and A. Beck. 1988. Toward mass propagation by use of bioreactors. *Acta Hort.,* in press.

135.   Quak, F. 1977. Meristem culture and virus-free plants. In *Plant cell, tissue and organ culture,* J. Reinert and Y. P. S. Bajaj, eds. Berlin: Springer-Verlag, pp. 598–615.

136.   Raju, B. C., and J. C. Trolinger. 1986. Pathogen indexing in large-scale propagation of florist crops. In *Tissue culture as a plant production system for horticultural crops,* R. H. Zimmerman, R. J. Griesbach, F. A. Hammerschlag, and R. H. Lawson, eds. Dordrecht: Martinus Nijhoff Publishers, p. 138.

137.   Ramming, D. W. 1985. In ovulo embryo culture of early-maturing *Prunus. HortScience* 20(3):419–20.

138.   Randolph, L. F., and L. G. Cox. 1943. Factors

influencing the germination of *Iris* seed and the relation of inhibiting substances to embryo dormancy. *Proc. Amer. Soc. Hort. Sci.* 43:284–300.

139. Rangan, T. S. 1984. Culture of ovaries. In *Cell culture and somatic cell genetics of plants,* Vol. 1, I. K. Vasil, ed. New York: Academic Press, pp. 221–26.

140. Rangan, T. S. 1984. Culture of ovules. In *Cell culture and somatic cell genetics of plants,* Vol. 1, I. K. Vasil, ed. New York: Academic Press, pp. 227–31.

141. Rangaswamy, N. S. 1977. Application of *in vitro* pollination and *in vitro* fertilization. In *Plant cell, tissue and organ culture,* J. Reinert and Y. P. S. Bajaj, eds. Berlin: Springer-Verlag, pp. 412–25.

142. Ranjit, M., and D. E. Kester. 1988. Micropropagation of cherry rootstocks: II. Invigoration and enhanced rooting of '46–1 Mazzard' by co-culture with "Colt." *Jour. Amer. Soc. Hort. Sci.* 113(1):150–54.

143. Redenbaugh, K., D. Slade, P. Viss, and J. A. Fujii. 1987. Encapsulation of somatic embryos in synthetic seed coats. *HortScience* 22(5):803–9.

144. Redenbaugh, K., P. Viss, D. Slade, and J. A. Fujii. 1987. Scale-up: artificial seeds. In *Plant tissue and cell culture,* C. E. Green, D. A. Somers, W. P. Hackett, and D. D. Biesboer, eds. New York: Alan R. Liss, pp. 473–93.

145. Reinert, J. 1959. Uber die Kontrolle die Morphogenese und die Induktion von Adventiveembryonen an gewebekulturen aus karotten. *Planta* 53:318–33.

146. Roest, S., and G. S. Bokelmann. 1973. Vegetative propagation of *Chrysanthemum cinerariaefolium in vitro. Sci. Hort.* 1:120–22.

147. Rowe, W. J. 1986. New technologies in plant tissue culture. In *Tissue culture as a plant production system for horticultural crops,* R. H. Zimmerman, R. J. Griesbach, F. A. Hammerschlag, and R. H. Lawson, eds. Dordrecht: Martinus Nijhoff Publishers, pp. 35–51.

148. Sagawa, Y., T. Shoji, and T. Shoji. 1966. Clonal propagation of cymbidium through shoot meristem culture. *Amer. Orchid Soc. Bul.* 35:118–22.

149. Scowcroft, W. R., R. I. S. Brettell, S. A. Ryan, P. A. Davies, and M. S. Pallotta. 1987. Somaclonal variation and genomic flux. In *Plant tissue and cell culture,* C. E. Green, D. A. Somers, W. P. Hackett, and D. D. Biesboer, eds. New York: Alan R. Liss, pp. 275–88.

150. Sharp, W. R., M. R. Sandahl, L. S. Caldas, and S. B. Maraffa. 1980. The physiology of in vitro asexual embryogenesis. *Hort. Rev.* 2:268–309.

151. Shepard, J. F., D. Bidney, and E. Shakin. 1980. Potato protoplasts in crop improvement. *Science* 208:17–24.

152. Sinha, S., R. P. Roy, and K. K. Jha. 1979. Callus formation and shoot bud differentiation in anther culture of *Solanum surattense. Can. Jour. Bot.* 57:2524–27.

153. Skirvin, R. M., and J. Janick. 1976. Tissue culture-induced variation in scented *Pelargonium* spp. *Jour. Amer. Soc. Hort. Sci.* 101(3):281–90.

154. Skoog, F., and C. O. Miller. 1957. Chemical regulation of growth and organ formation in plant tissues cultured *in vitro. Symp. Soc. for Exp. Biol.* 11:118–31.

155. Slack, S. A. 1980. Pathogen-free plants: Meristem-tip culture. *Plant Disease* 64(1):15–17.

156. Sondahl, M. R., and L. C. Monaco. 1981. In vitro methods applied to coffee. In *Plant tissue culture: Methods and applications in agriculture,* T. A. Thorpe, ed. New York: Academic Press, pp. 325–47.

157. Spiegel-Roy, P., N. Sahar, J. Baron, and U. Lavi. 1985. In vitro culture and plant formation from grape cultivars with abortive ovules and seeds. *Jour. Amer. Soc. Hort. Sci.* 110(1):109–12.

158. Sriskandarajal, C., and M. G. Mullins. 1981. Micropropagation of Granny Smith apple: Factors affecting root formation *in vitro. Jour. Hort. Sci.* 56:71–76.

159. Steward, F. C., M. O. Mapes, A. E. Kent, and R. D. Holstein. 1964. Growth and development of cultured plant cells. *Science* 143:20–27.

160. Steward, F. C., M. O. Mapes, and K. Mears. 1958. Growth and organized development of cultured carrots. II. Organization in cultures from freely suspended cells. *Amer. Jour. Bot.* 454:705–8.

161. Stimart, D. P. 1986. Commercial micropropagation of florist flower crops. In *Tissue culture as a plant production system for horticultural crops,* R. H. Zimmerman, R. J. Griesbach, F. A. Hammerschlag, and R. H. Lawson, eds. Dordrecht: Martinus Nijhoff Publishers, pp. 301–15.

162. Stokes, M. J. 1980. Current aspects of commercial micropropagation. *Proc. Inter. Plant Prop. Soc.* 30:255–67.

163. Stoltz, L. P. 1971. Agar restriction of the

growth of excised mature iris embryos. *Jour. Amer. Soc. Hort. Sci.* 96:611–84.

164. Stone, O. M. 1978. The production and propagation of disease-free plants. In *Propagation of higher plants through tissue culture,* K. M. Hughes, R. Henke, and M. Constantin, eds. Springfield, Va.: U.S. Dept of Energy, Tech. Inform. Center, pp. 25–34.

165. Stoutemyer, V. T., and O. K. Britt. 1969. Growth and habituation in tissue cultures of English ivy, *Hedera helix. Amer. Jour. Bot.* 56(2):222–26.

166. Street, H.E., ed. 1977. *Plant tissue and cell culture.* Berkeley, Calif.: Univ. of Calif. Press.

167. Strode, R. A., and G. Abner. 1986. Large scale tissue culture production for horticultural crops. In *Tissue culture as a plant production system for horticultural crops,* R. H. Zimmerman, R. J. Griesbach, F. A. Hammerschlag, and R. H. Lawson, eds. Dordrecht: Martinus Nijhoff Publishers, pp. 367–71.

168. Stuart, D. A., S. G. Strickland, and K. A. Walker. 1987. Bioreactor production of alfalfa somatic embryos. *HortScience* 22(5):800–803.

169. Sunderland, N. 1984. Anther culture of *Nicotiana tabacum.* In *Cell culture and somatic cell genetics of plants,* Vol. 1, I. K. Vasil, ed. New York: Academic Press, pp. 283–92.

170. Sussex, I. M. 1983. Determination of plant organs and cells. In *Genetic engineering of plants: an agricultural perspective,* T. Kosuge, C. P. Meredith, and A. Hollaender, eds. New York: Plenum Press, pp. 443–51.

171. Swartz, H. J., and J. T. Lindstrom. 1986. Small fruit and grape tissue culture from 1980 to 1985: Commercialization of the technique. In *Tissue culture as a plant production system for horticultural crops,* R. H. Zimmerman, R. J. Griesbach, F. A. Hammerschlag, and R. H. Lawson, eds. Dordrecht: Martinus Nijhoff Publishers, pp. 201–20.

172. Thorpe, T. A. 1979. Regulation of organogenesis *in vitro.* In *Propagation of higher plants through tissue culture,* K. W. Hughes, R. Henke, and M. Constantin, eds. Springfield, Va.: Natl. Tech. Inf. Ser. U.S. Dept. of Commerce, pp. 87–101.

173. ———, ed. 1981. *Plant tissue culture: Methods and applications in agriculture.* New York: Academic Press.

174. Tisserat, B. 1981. Date palm tissue culture. *USDA Agr. Res. Ser., Adv. in Agr. Tech., Western Ser. 17,* pp. 1–50.

175. Tisserat, B., E. B. Esan, and T. Murashige. 1979. Somatic embryogenesis in angiosperms. *Hort. Rev.* 1:1–78.

176. Torrey, J. G. 1977. Cytodifferentiation in cultured cells and tissues. *HortScience* 12(2):14–15.

177. Tran Thanh Van, K., and T. H. Trinh. 1986. Fundamental and applied aspects of differentiation in vitro and in vivo. In *Handbook of plant cell culture,* Vol. 4, *Techniques and applications,* D. A. Evans, W. R. Sharp, and P. V. Ammirato, eds. New York: Macmillan, pp. 316–35.

178. Tulecke, W., and G. H. McGranahan. 1985. Somatic embryogenesis and plantlet regeneration from cotyledons of walnut, *Juglans regia* L. *Plant Sci.* 40:57–63.

179. van Overbeek, J. 1942. Hormonal control of embryo and seedling. *Cold Spring Harbor Symp. Quantitative Biol.* 10:126–34.

180. Vasil, I. K. 1985. Somatic embryogenesis and its consequences in the Gramineae. In *Tissue culture in forestry and agriculture,* R. R. Henke, K. W. Hughes, M. J. Constantin, and A. Hollaender, eds. New York: Plenum Press, pp. 31–47.

181. Wang, Y., and J. Janick. 1986. Somatic embryogenesis in jojoba. *Jour. Amer. Soc. Hort. Sci.* 111(2):281–87.

182. Wenzel, G., and B. Foroughi-Wehr. 1984. Anther culture of tuberosum. In *Cell culture and somatic cell genetics of plants,* Vol. 1, I. K. Vasil, ed. New York: Academic Press, pp. 293–301.

183. Wenzel, G., and B. Foroughi-Wehr. 1984. Anther culture of cereals and grasses. In *Cell culture and somatic cell genetics of plants,* Vol. 1, I. K. Vasil, ed. New York: Academic Press, pp. 311–27.

184. Westcott, R. J., G. G. Henshaw, B. W. W. Grout, and W. M. Roca. 1977. Tissue culture methods and germ plasm storage in potato. *Acta Hort.* 78:45–49.

185. Wetherell, D. F. 1978. *In vitro* embryoid formation of cells derived from somatic plant tissues. In *Propagation of higher plants through tissue culture,* K. M. Hughes, R. Henke, and H. Constantin, eds. Springfield, Va.: U.S. Dept. of Commerce, Natl. Tech. Inform. Ser., pp. 102–24.

186. White, P. R. 1963. *The cultivation of animal and plant cells* (2nd ed.). New York: Ronald Press.

187. Williams, E. G., and G. Maheswaran. 1986. Somatic embryogenesis: factors influencing coordinated behaviour of cells as an embryonic group. *Ann. Bot.* 57:443–62.

188. Wimber, D. E. 1963. Clonal multiplication of

cymbidiums through tissue culture of the shoot meristem. *Amer. Orch. Soc. Bul.,* pp. 105–7.

189.   Withers, L. A. 1985. Cryopreservation of cultured cells and meristems. In *cell culture and somatic cell genetics of plants,* Vol. 2., I. K. Vasil, ed. New York: Academic Press, pp. 253–316.

190.   Yeung, E. C., T. A. Thorpe, and C. J. Jensen. 1981. In vitro fertilization and embryo culture. In *Plant tissue culture: Methods and applications in agriculture,* T. A. Thorpe, ed. New York: Academic Press, pp. 253–71.

191.   Zilis, M., D. Zwagerman, D. Lamberts, and L. Kurtz. 1979. Commercial propagation of herbaceous perennials by tissue culture. *Proc. Inter. Plant Prop. Soc.* 29:404–13.

192.   Zimmerman, R. H. 1986. Propagation of fruit, nut and vegetable crops—overview. In *Tissue culture as a plant production system for horticultural crops,* R. H. Zimmerman, R. J. Griesbach, F. A. Hammerschlag, and R. H. Lawson, eds. Dordrecht: Martinus Nijhoff Publishers, pp. 183–200.

193.   Zimmerman, R. H., and I. Fordham. 1985. Simplified method for rooting apple cultivars in vitro. *Jour. Amer. Soc. Hort. Sci.* 110(1):34–38.

194.   Ziv, M. 1986. *In vitro* hardening and acclimatization of tissue culture plants. In *Plant tissue culture and its agricultural applications,* L. A. Withers, and P. G. Alderson, eds. London: Butterworth, pp. 187–96.

195.   Ziv, M., A. H. Halevy, and R. Shilo. 1970. Organs and plantlets regeneration of *Gladiolus* through tissue culture. *Ann. Bot.* 34:671–76.

SUPPLEMENTARY READING

BAJAJ, Y. P. S., ed. 1986. *Biotechnology in agriculture and forestry,* Vol. 1, *Trees.* Berlin: Springer-Verlag.

BARZ, W., E. REINHARD, and M. H. ZENK, eds. 1977. *Plant tissue culture and its biotechnological applications.* Berlin: Springer-Verlag.

BONGA, J. M., and D. J. DURZAN. 1986. *Cell and tissue culture in forestry,* 4 volumes (2nd ed.). Dordrecht: Martinus Nijhoff Publishers.

CHALEFF, R. S. 1981. *Genetics of higher plants: Applications of cell culture.* Cambridge: Cambridge Univ. Press.

COCKING, E. C. 1986. The tissue culture revolution. In *Plant tissue culture and its agricultural applications,* L. A. Withers, and P. G. Alderson, eds. London: Butterworth, pp. 3–20.

DONNELLY, D. J., and W. E. VIDAVER. 1988. *Glossary of plant tissue culture.* Portland, Oreg.: Dioscorides Press.

EVANS, D. A., W. R. SHARP, and P. V. AMMIRATO, eds. 1986. *Handbook of plant cell culture,* Vol. 4. New York: Macmillan.

GAUTHERET, R. J. 1985. History of plant tissue and cell culture: A personal account. In *Cell culture and somatic cell genetics of plants,* Vol. 2, I. K. Vasil, ed. New York: Academic Press, pp. 1–59.

HENKE, R. R., K. W. HUGHES, M. J. CONSTANTIN, and A. HOLLAENDER, eds. 1985. *Tissue culture in forestry and agriculture.* New York: Plenum Press.

HUGHES, K. W., R. HENKE, and M. CONSTANTIN, eds. 1977. *Propagation of higher plants through tissue culture: A bridge between research and application.* Springfield, Va.: U.S. Dept. of Energy, Tech. Inform. Center.

KOSUGE, T., C. MEREDITH, and A. HOLLAENDER, eds. 1983. *Genetic engineering in plants.* New York: Plenum Press.

KRIKORIAN, A. D. 1982. Cloning higher plants from aseptically cultured tissues and cells. *Biol. Rev.* 57:151–218.

SALA, F., B. PARISI, R. CELLA, and O. CIFERRI, eds. 1980. *Plant cell cultures: Results and perspectives.* Amsterdam: Elsevier North Holland Biomedical Press.

THORPE, T. A., ed. 1978. *Frontiers of plant tissue culture.* Calgary: Univ. of Calgary Press.

TISSUE CULTURE ASSOCIATION. *In Vitro* (Series).

VASIL, I. K., ed. 1984. *Cell culture and somatic cell genetics of plants,* Vol. 1, *Laboratory procedures and their applications.* New York: Academic Press.

VASIL, I. K., ed. 1985. *Cell culture and somatic cell genetics of plants,* Vol. 2, *Cell growth, nutrition, cytodifferentiation, and cryopreservation.* New York: Academic Press.

WITHERS, L. A., and P. G. ALDERSON, eds. 1986. *Plant tissue culture and its agricultural applications.* London: Butterworth.

17

Techniques of In Vitro Culture for Micropropagation

Micropropagation utilizes techniques of aseptic tissue culture for the multiplication of important cultivars. It has become an aspect of commercial nursery propagation of many plants (*12, 21, 22, 77*) with the promise of increasing applications. The following advantages and uses can be cited for the wide interest in commercial nursery operations (*9, 10, 24, 40, 42, 46, 50, 57, 64*).

1. **Mass propagation of specific clones.** Multiplication rates for production can be very high since plants in culture can theoretically be multiplied at an exponential rate of consecutive subculturing (e.g., one month apart). Although such theoretical rates are not maintained in practice, the actual rates that can be attained under proper management are impressive. Many commercial laboratory nurseries in operation have the capacity to produce millions of micropropagated plants a year (*17, 55*). Commercial micropropa-

gation is particularly useful under the following conditions:

a. Plants whose natural rate of increase is relatively slow—orchids, bulbous species, Boston fern, gerbera, many foliage plants, palm species.

b. New cultivars where commercial demands require getting a cultivar on the market in as short a time as possible.

c. Cultivars that cannot readily or economically be clonally propagated by standard methods [e.g., peach–almond hybrid rootstocks (*32*), walnut rootstocks (*15, 47*)].

2. **Production of pathogen-free plants.** In Chapter 8 we described the significance of pathogen control systems for clonal propagation both for fungal and bacterial pathogens and for systemic viruses and virus-like organisms.

Micropropagation provides a method of, first, ridding the clone of the pathogen and second, providing a mass propagation system in which the plants are kept free of re-infection until delivered to users. Other related benefits include:

a. Maintain germplasm and source material in a pathogen-free condition.
b. Provide the initial step in nuclear crop production systems (see Chap. 8), which includes a micropropagation step of the nuclear plants, although the subsequent generations may be propagated conventionally. This sequence is used for geraniums (*56*), chrysanthemums, carnations (*58*), and strawberries (*68*), for example.
c. Allow movement of germplasm materials across quarantine barriers (*55*).
d. Facilitate the distribution of commercial material in international trade, which would not be feasible with conventional propagation. For example, a commercial tissue culture laboratory in California has produced minitubers of potato which are shipped throughout the world (*72*). In another example, a Florida firm was able to provide 15,000 pathogen-free propagules of a new banana cultivar to establish an industry in the West Indies through easy air shipment and without the barrier of inspection and quarantine (*57*).

3. **Clonal propagation of parental stocks for hybrid seed production.** This is done, for instance, with asparagus, tomato, cucurbits, and broccoli (*4, 14, 54*).

4. **Provide year-round nursery production.** Most nursery operations are seasonal and are limited by the ability to carry through peak periods. Micropropagation has the potential for continuous year-round operation with production scheduled according to market demands. High-volume production requires high-volume distribution and the need to stockpile items. Combining production with cold storage facilities could make it possible to hold material for peak marketing periods (*55, 67*).

PROBLEMS AND DISADVANTAGES

Micropropagation on a commercial scale has particular characteristics that may create problems which could limit use (*49, 57*).

1. Requirement for expensive and sophisticated facilities, trained personnel, and specialized techniques.
2. High cost of production resulting from facilities needed and high labor input. For example, shoot-tip propagation requires much hand labor to transfer individual propagules.
3. Requirement for a high-volume, more or less continuous distribution system or storage facilities to stockpile products.
4. Contamination that could cause very high losses in a short time.
5. Risk of variability and production of off-type individuals in the products emerging from micropropagation. Careful roguing and prior field testing of new products is essential to decrease this risk.

FACILITIES AND EQUIPMENT

Tissue culture facilities may be placed into three categories as determined by their scope, size, sophistication, and cost. These categories are: (a) research facilities where precise work requiring highly sophisticated equipment is needed (*69*), (b) large commercial propagation facilities where several millions of plants can be mass produced annually (*5, 6, 7, 8, 21, 29, 30, 55, 73, 74*), and (c) limited facilities for small research laboratories, individual nurseries, or hobbyists where a relatively small volume of material is handled (*27, 43, 66*). There is much variation in kinds of facilities and opportunities for cost cutting, but the basic principles of in vitro culture must be followed (*11, 21, 27, 37, 66*). In any case, careful consideration must be given to costs and labor required (*4, 5, 13, 43, 55, 66*).

No matter what the size, the facility should include three basic components: (a) preparation area, (b) transfer area, and (c) growing area. In addition there may be need for service areas, office, and cold storage. It is advisable to separate this facility from the regular nursery and greenhouse production area and maintain restricted entry to avoid introducing contaminants into the culture rooms. Floors, benches, and tabletops should be kept scrupulously clean. Technicians should use clean lab coats, foot coverings, and hairnets.

Preparation Area

The preparation area is essentially a kitchen with three basic functions: (a) cleaning glassware, (b) preparation and sterilization of media, and (c) storage of glassware and supplies.

An efficient method of washing is required, either by hand or by machine. Normal washing is followed by rinsing in distilled or deionized water. A sink, running water, and electrical or gas outlets for heating are necessary; air or vacuum outlets are often useful. Table surfaces should be made of a material that can be cleaned easily.

The following equipment items are used in the preparation area:

1. Refrigerator to store chemicals, stock solutions, and small batches of media.
2. Scales or analytical balance, preferably top loading.
3. Autoclave capable of reaching 120°C (250°F). A household pressure cooker can be used to sterilize small batches of media.
4. A pH meter. Indicator papers can be used as a substitute in less precise work.
5. Gas or heating plate.
6. Filters for sterilizing nonautoclavable ingredients (particularly in research laboratories). Filters are used to sterilize chemicals that are modified by autoclaving. These consist of special funnel-like units fitted with a bacteria-proof membrane through which a solution is pulled by vacuum or passed under positive pressure. Various commercial types are available.
7. Equipment to purify water, a glass still, an ion exchanger, reverse osmosis system, or a combination. In smaller operations, distilled water may be purchased and stored in plastic containers.
8. A vacuum pump or an ultrasonic cleaner.

These specific items have been used to decontaminate explants.

9. A media dispenser, valuable as a labor saving device.
10. Storage for flasks, bottles, and other supplies.

Transfer Area

The transfer area is the place where explants are inserted into culture and transfers and subcultures to fresh media take place. The key requirement is that the environment of the transfer area be sterile and free from any contaminating organisms. Transfer is most conveniently done in an open-sided **laminar airflow hood** (Fig. 17–1) where filtered air is passed from the rear of the hood outward on a positive pressure gradient. With such equipment the transfer may be carried out in a room with other activities. Less expensive alternatives, such as an enclosed walk-in room or various enclosed transfer boxes can be used if they are carefully sterilized before use.

Ultraviolet (UV) germicidal lamps can be used to sterilize the interior of the transfer chamber. These are turned on about two hours prior to using the chamber but must be turned off during operation. The light should be directed inward, not toward the worker's face because UV may adversely affect the eyes.

FIGURE 17–1 Laminar flow transfer hood within which sterile operations are carried out. Operator is sterilizing end of test tube by flaming.

Growing Area

Cultures should be grown in a separate, lighted facility where both daylength and light intensity can be controlled and where specific temperature regimes can be provided (Fig. 17–2). If various kinds of plants are to be propagated, it is useful to have several rooms, each programmed to meet the temperature and light needs of specific kinds of plants. Light requirements vary from about 3 to 30 W/m^2 PAR. Cool-white or Gro-lux fluorescent lights are generally used. A 16/8 (light/dark) photoperiod is most common. Temperatures of 21 to 30°C (70 to 86°F) are generally adequate, although some kinds of plants may need lower temperatures. The relative humidity at the temperature given is about 30 to 50 percent. Dehydration of the medium and increase in salt concentration might occur during long culture periods if the relative humidity is too low. If too high, contamination may develop.

Various kinds of rolling drum culture devices or shakers are sometimes used. Their advantage is to provide aeration in liquid culture systems.

Containers for Growing Cultures

Test tubes, Erlenmeyer flasks, petri dishes and shell vials of various sizes are routinely used in research laboratories. These may be Pyrex glass but are often less expensive soda glass and barisilicate, which can be sterilized and flamed during transfer operations. Various kinds of plastic containers are also available that are less expensive, less likely to break, and often disposable.

Petri dishes are glass or plastic, reusable or disposable. Less expensive larger glass containers commonly used in the past for home canning or commercial canning are widely used for the later stages of development in large-scale propagation. Most of these supplies are available from commercial supply houses or from food industry supply houses.

Various kinds of closures are necessary for the containers. Nonabsorbent cotton plugs have been used in research laboratories but are inconvenient in large commercial operations. Metal or plastic covers or polyurethane or polypropylene plugs are more convenient. A second cover of aluminum foil or various plastic materials is desirable to hold moisture and reduce infection but allow exchange of CO_2, O_2, and ethylene (*57*). Covering wide-mouthed jars with petri dish halves that fit works well. Parafilm (paraffined paper in a sterile roll) is used to seal petri dishes and other containers. Various kinds of specifically designed containers for in vitro culture are available from commercial dealers.

Equipment and Supplies

The following supplies are needed:

1. Beakers and Erlenmeyer flasks of various sizes in which to mix media.

FIGURE 17–2 Growing area where racks of tubes are placed for growing plants in vitro. Equipped with lights and controlled temperatures.

2. Graduated cylinders, flasks, and pipettes of various sizes to measure and dispense solutions.
3. Bottles to store reagents and stock solutions.
4. Magnetic stirring devices to mix media, or a large motordriven stirrer.
5. Dispensing pipettes and large funnels or burettes mounted in rings for dispensing media.

MEDIA PREPARATION

Ingredients

Ingredients of the culture medium vary with the kind of plant and the propagation stage at which one is working. In general, certain standard mixtures are used for most plants, but exact formulations may need to be established by testing. Media can be made from the pure chemicals or purchased as commercial premixed culture media. Empirical trials may be necessary to test available combinations of ingredients when dealing with a new kind of plant. These ingredients can be grouped into categories of (a) inorganic salts, (b) organic compounds, (c) complex natural ingredients, and (d) inert supports.

Inorganic Salts
(Groups A to E, Table 17–1)

Inorganic salts provide the macroelements (nitrogen, phosphorus, potassium, calcium, and magnesium) and microelements (boron, cobalt, copper, manganese, iodine, iron, and zinc). These salts can be made up in stock solutions by group and stored in a refrigerator. The reasons for preparing different stock solutions is that certain kinds of chemicals when mixed together will precipitate and not remain in solution.

There are many different formulations reported in the literature with detailed descriptions

TABLE 17–1

Stock solutions (grams/liter) used in preparing various media for micropropagation[a]

Group	Compound	Murashige and Skoog (MS)	Woody Plant Medium (WPM)	Anderson (AND)	Gamborg B5
A	NH_4NO_3	165.00	40.00	40.00	—
	KNO_3	190.00	—	48.00	250.00
	$Ca(NO_34H_2O)$	—	55.6	—	—
B	K_2SO_4	—	99.00	—	—
	$MgSO_4 \cdot 7H_2O$	37.00	37.00	37.00	25.00
	$MnSO_4 \cdot H_2O$	1.69	2.23	1.69	1.00
	$ZnSO_4 \cdot 7H_2O$	0.86	0.86	0.86	0.2
	$CuSO_4 \cdot 5H_2O$	0.0025	0.0025	—	0.0025
	NH_4SO_4	—	—	—	13.4
C	$CaCl_2 \cdot 2H_2O$	44.00	9.6	44.00	15.00
	KI	0.083	—	0.083	0.075
	$CoCl_2 \cdot 6H_2O$	0.0025	—	0.083	0.0025
D	KH_2PO_4	17.00	17.00	—	—
	H_3BO_3	0.62	0.62	0.62	0.30
	$Na_2MoO_4 \cdot 2H_2O$	0.025	0.025	0.025	0.025
	$NaH_2PO_4 \cdot H_2O$	[b]	—	38.00	15.00
E	$FeSO_4 \cdot 7H_2O$	2.784	2.78	5.57	2.78
	$Na_2 \cdot EDTA$	3.724	3.73	7.45	3.725
F	Thiamin·HCl	0.10	0.10	0.04	1.00
	Nicotinic acid	0.05	0.05	—	0.10
	Pyridoxine·HCl	0.05	0.05	—	0.10
	Glycine	0.20	0.20	—	—
G	Myo-inositol	10.00	10.00	10.00[c]	10.00

[a]These are 100 × the final solution. Use 10 ml of each stock solution for preparing 1 liter of the culture medium.
[b]Commonly added as a supplement directly to the culture medium at 85 to 255 mg/liter.
[c]Adenine sulfate added, 80 mg/liter.

(*21, 37, 57,*). In this chapter, we describe four representative media. The Murashige-Skoog (MS) and Linsmaier-Skoog media (*39, 53*) have been used extensively for a range of culture types and species, particularly herbaceous plants. For other types, such as woody plants, a dilution of 3 to 10 times in inorganic salts or a shift to other inorganic media is desirable. The Woody Plant Medium (WPM) (*41*) was developed for woody plants and the Anderson (AND) medium was developed for rhododendrons (*2, 37*). The Gamborg B5 medium has been used widely for cell and tissue cultures (*19*).

Organic Compounds (Group F and G, Table 17–1)

Carbohydrates. Sucrose at 2 to 4 percent is used for most cultures, but concentrations as high as 12 percent might be used in some cases, as for young embryos. Glucose has sometimes been used for monocots. Fructose and starch have been used occasionally. These materials are added at the time of making the cultures.

Vitamins. Thiamin (0.1 to 0.5 mg/liter) is almost always considered essential, and nicotinic acid (0.5 mg/liter) and pyridoxine (0.5 mg/liter) are usually added. Inositol at 100 mg/liter is beneficial in many cultures and is usually added routinely. Other vitamins sometimes beneficial include pantothenic acid (0.1 mg/liter) and biotin (0.1 mg/liter). All of these substances are soluble in water and should be prepared as stock solutions, ready for dilution at 100 times the final concentration. Stock solutions should be stored in a refrigerator.

Hormones and growth regulators. The two most important classes that are used to control organ and tissue development are the auxins and cytokinins. Gibberellins have sometimes been used to promote shoot elongation.

Auxins. The natural auxin, indoleacetic acid (IAA), is usually used at concentrations of 1 to 50 mg/liter. Synthetic auxins are usually more stable and include naphthaleneacetic acid (NAA) (0.1 to 10 mg/liter), 3-indolebutyric acid (IBA) (0.1 to 10 mg/liter), and 2,4 dichlorophenoxyacetic acid (2,4-D) (0.05 to 0.5 mg/liter).

Cytokinins. The compounds include N^6-benzyladenine (BA), kinetin, N^6-isopentenyl-adenine (2iP), and zeatin. They are used at a range of 0.01 to 10 mg/liter. Thidiazuron (*16,31*) and N-2-chloro-4-pyridyl N-phenylurea (CPPU) (*16, 31, 48*) have been used and have apparent cytokinin activity. Adenine sulfate benefits growth and development in some cultures and is often added at 40 to 120 mg/liter. Gibberellic acid (GA_3) is sometimes used but must be filter sterilized.

All of these materials should be prepared in advance and maintained as stock solutions in a refrigerator at near freezing temperature. Stock solutions of auxins and other organic materials may deteriorate with time.

Miscellaneous. Citric acid (150 mg/liter) or ascorbic acid (10 mg/liter) can be used as an anti-browning agent (antioxidant). Malic acid (100 mg/liter) is added in some embryo culture media. These should be filter sterilized since autoclaving causes decomposition. These organic acids may be incorporated in the medium or used in prewashing steps. Fine-grade activated charcoal, preferably prewashed, is used in some cases (0.1 to 1 percent) to adsorb and counteract inhibiting substances released by some tissues.

Complex natural ingredients. Various materials of unknown composition have been used to establish cultures when known substances fail to do so. Protein hydrolysates from casein or other proteins (30 to 3000 mg/liter) are sometimes helpful, primarily to provide organic nitrogen and amino acids. Coconut milk (endosperm of either green or ripe coconuts, 10 to 20 percent by volume) has been used but must be filter sterilized for some uses. Malt (500 mg/liter), yeast extract (50 to 5000 mg/liter) and such substances as tomato juice (30 percent v/v) and orange juice (30 percent v/v) also have been used. These substances are dissolved in water and added during preparation of the medium.

Preparation of Stock Solutions

Inorganic substances and many of the organic materials can be maintained under refrigeration in stock solutions of 100 times the required concentration. An aliquot of 10 ml of the stock solution can be added to each liter of medium to provide the required concentration. From the list in Table 17–1, the following materials can be combined without forming precipitates: (a) nitrates, (b) sulfates, (c) halides (i.e., chlorides and iodides), (d) phosphates, borates, and molybdates. Iron is prepared separately by combining the substances in group E, or supplied as inorganic iron—$FeCl_3 \cdot 6H_2O$ (1 mg/liter), $FeSO_4$ (2.5 mg/liter) or Fe tartrate (10 mg/liter). Iron solutions should be stored in a dark bottle.

Some organic compounds are relatively insoluble in water. A small quantity of dimethylsulfoxide (DMSO) (not more than 0.5 percent of the final medium) is effective in dissolving most organic substances. Cytokinins are weak bases and can be dissolved in a dilute acid, then diluted to a final volume with water. Auxins are weak acids and can be dissolved in a dilute base or in alcohol. Use 0.3 ml of 1 N HCl for cytokinin and NaOH (for auxin) for each 10 mg of the compound. Stock solutions and organic substances should be refrigerated.

Supports

Agar. Agar is a powdered product obtained from certain species of red algae. It owes its value in culture systems to (a) its ability to melt when heated, (b) its change to a semisolid gel at room temperatures, and (c) its relative biological inertness. It may contain certain impurities, notably salts, in its natural state. Most agars are prewashed or purified (USP grade). Quality may vary among commercial producers. Agar performance may be affected by concentration and by pH. Usually a pH of 5.0 to 6.0 is obtained, but the pH of the medium may need to be adjusted through the addition of acids or bases. As the pH is lowered (4.5 or less), the agar may fail to solidify.

The lowest concentration of agar that will support the explants is best, which is about 0.5 to 0.6 percent. This allows good contact between explant and medium and adequate uptake of nutrients. At lower concentrations the explant may sink into the agar and aeration is impaired. A minimum concentration is about 0.6 percent. As the concentration increases (up to 1.0 to 1.2

grams per liter), growth may be depressed because of higher osmotic pressure.

Agar substitutes. One commercial product is Gelrite, a polysaccharide derived from a *Pseudomonas* bacteria (*21, 23, 37, 57*). It requires heating but makes a very clear gel upon heating in combination with cations, e.g., magnesium, calcium, potassium, and sodium for gelation. Its required concentration is less than agar but is often used in combination with agar, such as in a ratio of 3 Gelrite/1 agar, with the most effective combination varying with the brand of agar (*37*).

Liquid. Nutrient solutions without agar have an advantage for some plants. Some type of support is required to keep the explants and cultures from sinking, or a shaker or rotating drum must be used to provide aeration. Filter paper bridges may be inserted into the liquid to support the culture with the paper acting as a wick. Plastic fabric supports made of 100 percent polyester fleece are available commercially. Liquid media are useful for some plant species whose explants exude toxic substances from the cut surfaces. The liquid medium can be exchanged without reculturing.

Preparation of the Medium

High water quality (i.e., low salts and/or organic or chemical impurities) is very important for growing cultures, and tests of the water supply used for growing cultures should be determined. Single distilled water may be adequate in commercial micropropagation, but it may be desirable to double distill the water or to pass it through deionizers and equipment to further purify it.

To prepare media, use a large Pyrex flask or beaker, add a portion (one-half to two-thirds) of the final volume of water. Sugar may be added at this time or not until the end. Stir. Add by pipettes or a graduated cylinder proper amounts of each ingredient from stock bottles. Agar is added next, and then the solution is made up to the prescribed volume by adding water. Adjust to the desired pH (usually about 5.7) with drop-by-drop additions of 1 N HCl or 1 N NaOH using a pH meter (or pH indicator strips) to measure.

The solution is heated to melt the agar. This

may be done in an autoclave at 121°C (218°F) for three to seven minutes or on a hot plate, stirring continuously to prevent the agar from settling and burning. The hot solution is then dispensed into containers, which are closed with stoppers immediately and sterilized in an autoclave for 20 minutes at 121°C. The greater the volume of media to be sterilized, the longer the sterilization time.

Substances unstable during autoclaving must be filter-sterilized;[1] an aliquot of such material is added to the medium after autoclaving when the medium has cooled to 35 to 40°C (95 to 104°F). The medium should be used within about two to four weeks and refrigerated if kept longer.

GENERAL METHODS OF MICROPROPAGATION

Preparation for Establishment of the Culture

Choice of Explant

The kinds of explant and where they are collected is critical for success of the micropropagation procedure, as described in Chap. 16. They may vary with the purpose of the culture, the species, and sometimes the cultivar. In the absence of definite information on a given species or cultivar, one can use available information on a closely related species or cultivar or carry out experimentation with the particular plant. When collecting for shoot tips, plants in relatively active growth are most desirable. The size of the explant may vary from as small as a 1 to 5 mm of meristem tip for meristem culture to a piece of shoot several centimeters long. A single node bearing a lateral bud also can be used. For woody plants, a shoot tip from a dormant (but not resting) bud may be utilized but is often difficult to disinfest. Removal of the bud scales, plus cutting away the leaf scar, may give sterile tissue.

Pieces of leaves with veins present, bulb scales, flower scapes, and cotyledons also may be used for obtaining explants, as described in Chap. 16 for adventitious shoot initiation.

Disinfestation

The process is called disinfestation because it removes contaminants from the surface of the organ rather than from within the organ. Primary disinfestants include alcohol (ethyl, methyl, or isopropyl, usually about 80 percent) and calcium or sodium hypochlorite, usually with 5.25 percent active ingredient. A few drops of a surfactant or detergent should be added to the bleach to improve surface coverage. The effectiveness of the bleach treatment is a time-dosage response; the disinfestation effectiveness increases with an increase in both time and concentration but the damage to living tissues also increases. Thus a compromise must be developed for the kind of explant through testing.

A typical procedure for shoot-tip propagation is to cut the shoot in short pieces several centimeters long, and wash in tap water with a detergent. For woody pieces, a quick dip in alcohol may be helpful. Wrapping the shoots with small squares of sterile gauze will hold them intact during the treatment. In a hood, the plant parts are placed into the disinfesting solution (with a surfactant or detergent added), using a solution of 1 to 10 percent of the prepared material, for 5 to 15 minutes. The solution is then poured off and the material rinsed two or three times with sterile water. Modifications of this basic procedure that may improve effectiveness include prewashing, mechanical agitation, vacuum, and consecutive treatments.

Microbial contaminants, consisting of yeasts and various species of fungi and bacteria, usually appear on the agar surface within a few days to a week as white or opaque slime or as variously colored colonies, sometimes with black spores and mycelium. When these contaminants can be identified visually, the culture vessels containing them should be discarded quickly after autoclaving. Some bacterial contaminants, such as *Bacillus subtilis, Erwinia* (*28*), or *Pseudomonas* (*3*), sometimes remain inside the plant material without being detected for several months after the initial culture or incubation. Procedures such as culture indexing to detect or to identify these contaminants

---

[1]Commercial device equipped with a cellulose acetate or cellulose nitrate membrane filter that is porous enough to allow liquid to pass but nonporous enough to prevent the movement of organisms such as bacteria. Requires a vacuum and/or positive pressure.

may be included. Antibiotics are sometimes used in the medium to improve control of bacterial growth (*19, 57*).

CULTURE INDEXING

Culture indexing tests for contamination by specific pathogens in containers are as follows: Cut off a piece of plant tissue, dice it, and place the pieces into 5 ml of nutrient broth optimum for bacterial and fungal growth. Positive results are shown in cloudiness of the medium and should appear within four to five days. (Also note Fig. 8-15.) If made at the initial explant stage, however, this procedure does not always identify all potential contaminants (*28*). Additional tests may be necessary because of latent contamination.

Aseptic Procedures

Microorganisms occur primarily as spores resting on surfaces of tables, hands, arms, clothing, and various objects and settling out of the air or blown in the dust on air currents. Aseptic transfer procedures are done in some kind of transfer hood or box to minimize the movement of dust and contaminants. These should be equipped with a high-efficiency particulate air (HEPA) filter with a positive air pressure blowing outward from the rear of the hood to prevent movement of airborne spores into the chamber. The sides and surfaces of the chamber should be washed with a disinfestant such as sodium or calcium hypochlorite or 95 percent alcohol. Hands and arms should be washed thoroughly, and gloves and a clean lab coat should be worn.

Supplies needed for a transfer operation are as follows:

1. Burner for sterilizing dissecting tools and containers. This can be an automatic gas burner or an inexpensive glass alcohol lamp with a wick. Certain types of burner-incinerators are available commercially and can be used to avoid the dangers of alcohol or gas fumes with possible attendant fires.

2. Forceps to hold and manipulate tissue. Three useful sizes (*66*) include 300 mm, 200 mm straight-tipped, and 115 mm curve-tipped.

3. Dissecting needles with wood handles, or a metal needle holder with replaceable tips.

4. Dissecting scalpels. Metal scalpel holders with replaceable knife blades are most convenient. A piece of broken razor blade glued to a holder can be used for fine work, or various kinds of microtools can be made.

5. Sterile petri dishes for holding explants and sectioning them. Small squares of sterile moist paper toweling inside each petri dish are convenient to assist cleaning later (*37*). The moist surface keeps the small explants from drying out.

6. A dissecting microscope and a top-loading balance. These may be important for specific operations. These should be wiped with a disinfectant. The dissecting microscope should have a plastic or glass tunnel over the viewing area to protect against contamination.

Scalpels, needles, and other dissecting tools are moved into position, usually on holders. Sterile petri dishes are placed into the chamber to use as working trays. Sterile water containers are included, as are containers of medium, as needed. These containers are generally kept to the front of the working area so that reaching beyond a particular "sterile" area to the rear of the hood can be avoided. Outside surfaces of containers should be wiped with a disinfectant.

In preparing explants for transfer, the implements must be sterilized. Although this can be done by dipping them into alcohol and flaming them before use, many propagators avoid this because of the fire hazard. Furthermore, simply dipping and flaming the scalpels may not kill the most resistant pathogens or spores. Longer exposure to the flame, or use of a chemical sterilant, may be required (*60, 65*). A Bacti-incinerator is a commercial device that heats tweezers and tools to a very high temperature (about 1600°F) when inserted into the cone (*37*). Needles, scalpels, and tweezers should be allowed to cool before using to prevent damage to tender tissue. When working in an open area, the propagator usually flames the mouth of the container before opening it and after inserting the material to avoid entrance of contaminants.

Stages in Growing the Cultures

A sequence of consecutive stages of culture is followed in the micropropagation procedure and is described in Chap. 16. Although general patterns of treatments and handling procedures are well established, many of the requirements of individual stages must be developed for different species and even cultivars. The process can be described as optimizing the variables that affect each stage of the process. These include subculture interval, concentration of ingredients in the medium, relative cytokinin/auxin concentrations, size of propagules, light, temperature, etc. Certain of these variables are discussed in the following section.

Micropropagation may include a preliminary pretreatment culture step in which the explants are initially placed in a petri dish on the surface of an agar medium of basal salts and sugar, but without hormones. If the explants exude phenolics or other substances, activated charcoal, ascorbic acid, or citric acid can be added to the medium. Using the antioxidants ascorbic acid and citric acid in the preliminary washing may be useful. Using a liquid medium in the initial stages or frequent transfers in agar medium for several days may leach out the toxic material.

Stage I—Establishment and Stabilization

The function of this stage is to establish the explant in culture and to stabilize the culture for multiple shoot development. This may involve (a) the stimulation of axillary shoots; (b) the initiation of adventitious shoots on excised shoots, leaves, bulb scales, flower scapes, cotyledons, and other organs, as described in Chap. 16; or (c) the initiation of callus from the cut surfaces. The medium selected may vary with the species, cultivar, and kind of explant to be used. A basic medium (BM) includes the ingredients listed in Table 17–1, plus sucrose and sometimes other supplements. Control of development is most often achieved by manipulating the levels of auxin and cytokinin. If ax-

illary shoots are desired, a moderate level of cytokinin (BA, kinetin, or 2iP) is used (0.5 to 1 mg/liter). The auxin level is kept very low (0.01 to 0.1 mg/liter) or omitted altogether.

To develop adventitious shoots on the explants (excised shoots, leaves, bulb scales, cotyledons, flower scapes, etc.), a somewhat higher cytokinin level is needed. For callus formation, increased auxin levels are needed but must be adjusted to an appropriate level of cytokinin. Usually four to six weeks are required to complete stage I and produce explants ready to be transplanted to stage II. Some woody plant species can take up to a year for this to occur.

Stage II—Shoot Multiplication

In the multiplication stage each explant has expanded into a cluster of microshoots arising from a basal mass of callus-like tissue. This structure is divided into separate microshoots, which are transplanted into new culture medium (see Fig. 17–3).

The kind of medium used depends on the species, cultivar, and type of culture. Usually it is essentially the same as that used in stage I, but often the cytokinin and mineral supplement level is increased. Adjustments may need to be determined by experimentation (*3, 11, 52*).

Different species vary in the optimum size of the microshoot and method of cutting apart. There is generally a minimum mass of tissue that is necessary to produce uniform rapid multiplication in the next transfer. Division may be made by vertically cutting the tissue mass into sections, keeping some of the base with each piece. Sometimes one dominant shoot develops and inhibits elongation of others. If this happens, the shoot may be cut off and the base recultured. Elongated shoots laid horizontally on the agar surface perform as a type of layering and stimulate lateral shoots to develop.

The number of microshoots produced ranges from 5 to 25 or more depending upon the species and conditions of culture. Multiplication may be repeated several times to increase the supply of material to a predetermined level for subsequent rooting and transplanting. Sometimes microshoots deteriorate with time, lose leaves, fail to grow, develop tip-burn, go dormant, or lose po-

FIGURE 17–3 Culture of apple rootstocks and cultivars by shoot tips in vitro. Shoots are rooted in agar in glass jars under lights in a controlled growth chamber.

tentiality to regenerate (*3*). (See p. 482 for a discussion of stabilization.)

During multiplication, off-type propagules sometimes appear, depending on the kind of plant and method of regeneration. In some plants, such as Boston fern, restricting the multiplication phase to only three propagations is recommended to avoid development of off-type microshoots. Other procedures that can reduce the potential for variability include reducing hormone concentration and avoiding basal callus which may give rise to adventitious shoots in subculturing (see Fig. 16–17).

Stage III—Pretransplant, In Vitro Rooting

The purpose of the third stage is to prepare the microplants for transplanting from the artificial heterotrophic environment of the test tube to an autotrophic free-living existence in the greenhouse and on to their ultimate location. This preparation may involve rooting, but it also may involve conditioning of the microplant to increase its potential for acclimation and survival during transplanting. For example, the agar and/or sucrose concentration might be increased. Light intensity is sometimes increased during this period.

For in vitro rooting, individual microshoots are subcultured into new containers in a sterile medium with reduced or omitted cytokinin, an increased auxin concentration, and often a reduced inorganic salt content. For some plants, rooting is best if the microcutting is kept in the auxin medium for only one or two days and then transferred to an auxin-free medium. Or the microcutting may simply be dipped into a rooting (auxin) solution immediately and inserted directly into an auxin-free medium. Some plants root best if placed in the dark during the auxin treatment period.

Some plants respond to an "elongation" phase between stages II and III by placing the microshoot into an agar medium for two to four weeks without cytokinins (or at very low levels) and, in some cases, adding (or increasing) gibberellic acid (*63*). This reduces the influence of cytokinin. The microshoots are then treated for rooting in a cytokinin- and GA-free medium.

Selection for uniformity and roguing of obviously abnormal, aberrant, or diseased microshoots should be made prior to stage III (*3*). Plants of species having an inherent dormancy or rest period (see p. 110) may need to be chilled to stimulate new growth and elongation (*3*).

HOW TO DETERMINE
CONCENTRATION IN MEDIA

Establishing the optimum concentration and range of individual chemicals in culture media is as important as establishing the proper choice of chemical, whether inorganic salt, sugar, hormone, or other substance. Concentration is established by two main systems. Concentrations are often given as **parts per million** (ppm) or **parts per billion** (ppb), or expressed as milligrams per liter (mg/liter) or grams per liter (g/liter) of water or solution weight/weight (w/w) or weight/volume (w/v). In the case of liquids, relative volumes (v/v) are used. In either case a dilution range can be established as 1:1, 1:5, 1:10, 1:100, and so on. A progression is chosen to discover the range of concentrations within which a particular substance is effective; that is, its minimum threshold value and the maximum value where injury may occur or further increase in response does not occur. With hormones the kind of response changes with the concentration.

Research laboratories and considerable literature on tissue culture utilize methods of **chemical equivalents** as a better way to express effective concentrations and compare one chemical substance with another more realistically.

Chemical equivalents are calculated from the grammolecular weight (*mole*) of a substance by adding together the atomic weights of all the atoms in the molecule plus any attached water. Such weights are available in chemistry textbooks. One molar (1 *M*) concentration means the weight of one mole in grams dissolved in enough water to make 1 liter of solution. Usually, physiological concentrations are in much smaller amounts, such as

1 molar = 1 mole/liter = 1 *M* = 1 mol liter^{-1}

1 millimolar = 0.001 *M*/liter = 1 m*M* = mmol liter^{-1}

1 micromolar = 0.000001 *M*/liter = 1 μM = μmol liter^{-1}

The significance of this system is that 1 mole of any substance has the same *number* of molecules per volume as any other substance at that molar concentration.

Stage IV—Transplanting, Acclimation, and Ex Vitro Rooting

Rooted or unrooted microshoots are removed from the culture vessel, agar is washed away completely to remove a potential source of contamination, and the microplants are transplanted into a standard pasteurized rooting or soil mix in small pots or cells in a more or less conventional manner. Initially, microplants should be protected from desiccation in a shaded, high-humidity tent or under mist or fog. Several days may be required for new functional roots to form (*62*) (Fig. 17–4).

For in vivo rooting, stage III is essentially eliminated. Individual microcuttings are made and inserted directly into a rooting medium under mist or high humidity with or without rooting hormones. All remnants of the agar medium should be washed off the individual cuttings to remove a potential source of contamination. Plugs of the type described for seeds are very useful at this stage (*45, 76*) (Fig. 17–5).

Once the microplants are established in the rooting medium, they should be gradually exposed to a lower relative humidity and a higher light intensity. Any dormancy or resting condition that develops may need to be overcome as part of the establishment process.

FIGURE 17-4 Culture of apple rootstocks and cultivars by shoot tips in vitro. *Left:* Rooted shoots are transplanted into a porous medium and placed under mist. *Right:* Rooted plantlets are hardened-off in greenhouse for transplanting into nursery.

SPORE CULTURE

Spores of various ferns can be sterilized and sown on nutrient agar. Spore germination itself may be favored by using a nutrient-free medium, but growth of the prothallus is improved by the addition of inorganic salts and sucrose (*33*).

A complete aseptic culture system is described as follows (*28*): Fern spores are placed in 2 percent Clorox® solution and centrifuged. The pellet is retrieved, resuspended, and recentrifuged several times. Resuspended spores are streaked on a culture medium (MS salts, 3 percent sucrose, thiamine, and 0.8 percent agar). After 20 days' incubation in light at 27°C (81°F), gametophyte tissue is divided, recultured, and grown for 30 days. Then 1 g of gametophyte tissue in 200 ml of ½ MS salts may be ground in a kitchen blender for five seconds. Approximately 50 ml of solution is deposited over sterile soil. Subsequent handling is by conventional methods.

A modified method has also been described (*38, 70*). One part spores to 5 parts of 5 percent Clorox® solution is shaken for 1½ minutes and then poured through a Buchner funnel filter with vacuum, followed by rinsing, followed by an additional five minutes' air vacuum. Spores are allowed to air dry in a sterile hood. The medium (2 percent agar and inorganic nutrients) is melted, cooled to approximately 49°C (120°F), and poured into a sterile pyrex oven dish or petri dishes. Spores are sown over the surface using a fine mesh screen (270 mesh). As the medium cools, the surface tension facilitates uniform distribution of spores. These are placed in light at room temperature. Germination occurs in two to three weeks and transplanting to soil is done in two to three months.

FIGURE 17–6 Orchid seedlings growing aseptically on nutrient agar medium. Seedlings are ready for transplanting.

FIGURE 17–5 Ex vitro rooting of microshoots of apple. For rooting, microcuttings were made from proliferating microshoots, and placed (10 per vial) with bases in a liquid rooting basal medium of 0.43.8 m*M* sucrose and 1.5 μM IBA in dark for three to seven days. *Top:* Individual microshoots were inserted into nonsterile preformed peat plugs (see p. 154). Trays were enclosed in transparent plastic boxes and placed in a lighted growth room. *Below:* Rooted microplants in peat plugs after three weeks. Microplants were readily acclimated and hardened-off for transplanting. (Courtesy R. H. Zimmerman (77).)

FIGURE 17–7 Plantlet of Boston fern (*Nephrolepis exaltata* 'Bostoniensis') in culture. Ready for transplanting into stage III.

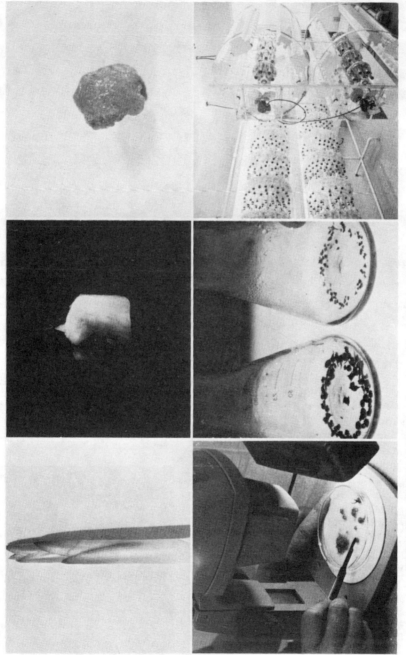

FIGURE 17–8 Steps in the micropropation of *Cymbidium* orchids (*61*). *Top row* (left to right): Growing shoot—source of meristem; excised shoot apical meristem; protocorm-like bodies, after enlarging in nutrient medium and before dividing. *Second row* (left to right): Dividing each enlarged protocorm-like structure into several pieces; before and after dividing; rotating wheels with divided mericlones in nutrient solutions for enlargement and prevention of root-shoot polarity.

FIGURE 17–8 *(cont.) Third row* (left to right): Protocorm-like material placed on nutrient agar for polarizing growth and development of roots and shoots; greenhouse filled with flasks for growing on into plants; well-developed plants large enough for flatting. *Bottom row* (left to right): Well-developed orchid plants, ready for potting; potted *Cymbidium* orchid, originating from a divided protocorm-like body; orchid house with plants coming into flowering. (Courtesy Richard Smith, Rod McLellan Co., South San Francisco.)

FIGURE 17–9 Apple rootstock propagation by shoot-tip proliferation. (A) Proliferated shoots in stage II. (B) Rooted plant ready for transplanting. (C) Apple plant after transplanting. (Courtesy Dr. Iona Snir and Dr. A. Erez.)

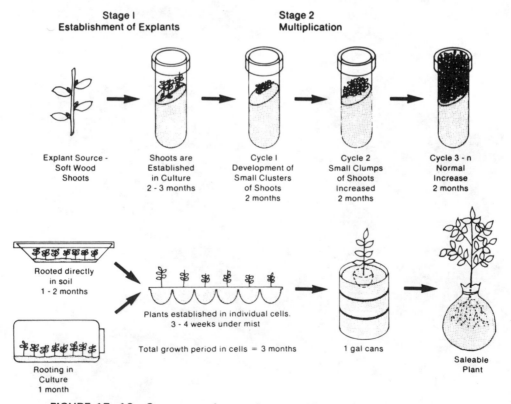

Stage I
Establishment of Explants

Stage 2
Multiplication

Explant Source -
Soft Wood
Shoots

Shoots are
Established
in Culture
2 - 3 months

Cycle I
Development of
Small Clusters
of Shoots
2 months

Cycle 2
Small Clumps
of Shoots
Increased
2 months

Cycle 3 - n
Normal
Increase
2 months

Rooted directly
in soil
1 - 2 months

Plants established in individual cells.
3 - 4 weeks under mist

Total growth period in cells = 3 months

Rooting in
Culture
1 month

1 gal cans

Saleable
Plant

FIGURE 17–10 Sequence of operations used in micropropagating rhododendrons. (Courtesy W. C. Anderson (1).)

METHODS FOR MICROPROPAGATING REPRESENTATIVE HORTICULTURAL CROPS

Many horticultural crops are being micropropagated on a regular basis. Tables 17–2 to 17–5 provide information about key steps in the propagation of representative species being grown in commercial propagation. Except where noted, methods are compiled from published information in Dirr and Heuser (*12*), George (*21*), Kyte (*37*), and Zimmerman (*77*) and are not comprehensive. Further details and additional crops need to come from surveys of manuals and texts listed at the end of this chapter or Chap. 16. Abbreviations and explanations used in the tables are listed below:

AC	activated charcoal
AD	adenine
AND	Anderson's rhododendron medium
AS	adventitious shoots
ASP	axillary shoot proliferation
BA	benzyladenine
BL	"bleach" such as sodium hypochlorite
CH	casein hydrolysate
CM	coconut milk (fresh coconuts)
DT	detergent such as Tween 20
GA	gibberellic acid
GP	growing point

IAA	indoleacetic acid
IBA	indolebutyric acid
KIN	kinetin
LS	lateral shoot
mod	modified
MS	standard Murashige-Skoog; includes standard supplements
MT	meristem tip—0.1 to 0.5 mm; primarily for virus control
NAA	Naphthaleneacetic acid
ST	Shoot tip; usually 1 cm or more long
TS	terminal shoot
2iP	iso-pentenyl adenine
WPM	Woody plant medium

rinse: sterile distilled water

dip: momentary immersion

wash: longer period in tap water; nonsterile, usually with a detergent

standard: general method described in this chapter

I. Establishment

II. Multiplication

III. Preparation for transplant; in vitro rooting

IV. Transplant and acclimation; may be ex vitro rooting

Light requirements: converted to watts per sq meter PAR

TABLE 17–2

Micropropagation methods for horticultural crops—foliage plants

Species	Method	Explant	Sterilization
1. *Anthurium*	AS, ASP	Single node with bud; excise 2 mm GP; leaf	20 min 1:10 BL + DT; GP 1:10 BL 30–45 min; rinse 5 min
2. *Dieffenbachia* 'Perfection'	ASP	Active growing shoot tips	Standard
3. *Ficus benjamina*	ASP	3–5 mm ST	Standard
4. *Nephrolepsis* (Boston fern)	ASP, AS	5–10 mm tip of actively growing runner	1/10 BL; alcohol; rinse 3×
5. *Peperomia*	AS	Leaf sections	Standard
6. *Saintpaulia* (African violet)	AS	1 cm² leaf; 2-mm cross section of petiole	Wash leaf or petiole; agitate 10 min; 1:10 BL; rinse 3×; cut
7. *Spathiphyllum*	ASP	Terminal or lateral ST from crown	Standard
8. *Syngonium*	ASP	1–2 cm young plant stems; 1–2 mm MT	15 min 1:10 BL; rinse 3×

Remarks

1. Procedure is particularly useful to avoid serious contamination problems with species. Also can use leaf explants to get adventitious shoots for later ASP culture.

2. Be sure to use source plants free of fungi and bacteria as disease control is a major benefit. Most commercial plants are grown from micropropagation.

3. Procedures also work with *F. decora* and *F. pandurata*.

4. Limit to no more than three subcultures to reduce variation; potential to produce one million plants per year. Most commercial plants are grown from micropropagation (See Fig. 17–6).

5. Propagates readily from leaf disks and follows classical hormone responses. Cultures in spring or summer yield more than in winter.

6. One of easiest to grow. Propagates from leaves or petioles as in in vivo propagation.

7. Roots develop in four weeks. Most commercial plants are grown from micropropagation. May need to restart from new explants after one year to maintain quality.

8. Propagates readily from tissue culture. Most commercial plants are grown from micropropagation.

.Propagation	Rooting; acclimation	Environmental conditions
I. MS + vitamins 15% coconut milk, 20% sucrose; rotating liquid 6 weeks	III. None	Continuous light at 16 W/m²; 25 to 28°C
II. Two-node section, same − CM, + BA (0.2)	IV. Standard	
I. MS; IAA (2.0); 2iP (16.0) separate base from tip in 30 days; repeat	III. None	27°C; 16 hrs; 3 W/m²
II. Use base; 30–60 days; liquid MS + KIN (2.0)	IV. Root in sterilized soil	
I. MS; IAA (0.3); 2iP (30.0)	III. MS − hormones or supplement; AC	I, II. 33 W/m²
II. same; subculture for 4–8 wks		III, IV. increase 21 to 229 W/m²
	IV. Standard	
I. 1/2 or 3/4 MS; KIN (1.0) NAA (0.1)	III. MS − hormones	3–6 W/m²;
II. same as I	IV. Standard	6/8; 25°C
I. MS; KIN (4.0); NAA (2.0)	III. Same	25°C; continuous light
II. Same	IV. Standard	
I. MS; BA (0.1); NAA (0.1) for petiole; mod MS + IAA (2.0) + BA (0.08) for leaf	III. Same	25° C; 10 W/m² for I, II;
	IV. Standard	1.5–3 W/m² for III
II. same		
I. 2/3 MS, 3% sucrose	III. Same but add IBA (15.0)	15 W/m²
II. Same	IV. Standard	
I. Liquid, MS + 2iP (3.0) + IAA (1.0)	III. MS salts only	25° C; 10 W/m² for I, II;
II. same	IV. Standard	25 W/m² for III

(table continues at bottom of page 514)

TABLE 17–3

Micropropagation methods for horticultural crops—florists' crops

Species	Method	Explant	Sterilization
1. *Cattleya* orchid	ASP	2 mm lateral buds or MT from 2- to 5-cm shoots	Wash + DT; alcohol dip 10 sec; 1:10 BL 15 min; rinse dry; dissect 2-mm ST in sterile antioxidant
2. *Chrysanthemum*	ASP	MT or ST	See procedure of *Dianthus*
3. *Cymbidium* orchid (see fig. 17–8)	ASP	Meristem: apical dome, 2 leaves, cube of tissue 0.5 mm	Remove outer leaves; dip in alcohol 2 sec; rinse in 1:10 BL 15 min; rinse; remove inner leaves
4. *Dianthus* (carnation)	ASP	1 mm ST + 1–2 leaves, or use larger ST	Remove shoot of ca. 12 nodes; remove leaves over 3 mm; wash with DT; add 1:10 BL 5 min; rinse 2× for 3 min each
5. *Gerbera*	ASP	3–10 mm ST from crown; emerging bud with 1-mm base	Wash running water; add water + DT 3× 10 min; rinse 3× 2 min each; 5 min 1:20 DL; excise; dip 1 min 1:20 BL and plant
6. *Kalanchoe*	ASP, AS	Leaf blade 1.5 cm; stem 2 cm; ST 2 mm	Wash 3–5 cm cuttings; rinse; add 1:20 BL 15 min; rinse 3× 1 min each; trim
7. *Hemerocallis* (daylily)	AS	2 mm sections; young (10-cm) infloresence scapes; petals and bracts from 1-mm flower buds	Discard bracts; wash; rinse; 1:10 BL 20 min; rinse 3×; slice into 2-mm sections with sterile antioxidant; place 2-mm segment upside down
8. *Lilium* (lily)	AS (scale) AS (leaf) AS (fl bud)	4–5-mm cubes from scale base; trim inner to include basal plate	Wash; rinse running water 15 min; rinse in sterile water; 1:10 DT 10–20 min; rinse 3× in sterile water

Remarks

1. Liquid medium better than agar in initial culture; same as *Cymbidium* but more complex medium. Growing in liquid on a shaker continues proliferation and overcomes polarity. Cutting into sections has same effect. Placing on stationary medium or stopping cutting allows shoot and root formation.

2. Used primarily as a disease-controlling procedure in nuclear stock program. If culture prolonged, cytokinin can stunt growth; avoid by removing for several cultures or use gibberellic acid. High light intensity improves branching. May show uneven flowering in post-culture period.

3. Initially forms protocorm-like structures; keeping in liquid on rotating wheel offsets polarity and maintains proliferation (*71*); transplant to agar for plantlet formation, or grow on agar and keep cutting into sections (Fig. 17–7). Seeds are also propagated in aseptic culture (Fig. 17–8).

4. Used primarily as a disease-controlling procedure in nuclear stock program; makes excellent material for teaching demonstration. Same comments as *Chrysanthemum* on quality of product.

5. Contamination is a persistent problem; extensively used to propagate selected forms and specific colors. Higher cytokinin may be required initially with lower concentration later. Requires renewal of stock cultures after one year.

6. Meristems hard to clean; uses principle of callus production + differentiation of plantlets, then ASP. May have uneven flowering in post-culture period particularly in some cultivars.

7. Utilizes flower buds from which adventitious shoots develop followed by ASP. Highly successful plant in culture. Relatively stable in long-term continuous culture.

8. Used mostly for Asiatic hybrids and oriental species. Easter lily for disease protection; similar methods can be used on a wide range of bulbous species. Also can use slices of young elongated stems and axillary buds for explants.

Propagation	Rooting; acclimation	Environmental conditions
Use *Cymbidium* method + coconut milk (100 mg/l)	Same as *Cymbidium*	Same as *Cymbidium*
I. Mod MS + IAA (0.5) + KIN (10) II. Same	III. omit KIN and adenine; IAA (10.0); 2 weeks IV. Standard	I, II. 27°C; 3 W/m² III. 30 W/m²
I. Liquid or agar (Knudson's 2% sucrose) II. Same; keep dividing	III. Shift to agar for shoot and root development	Continuous light; 6 W/m²; 22°C
I. Mod MS + KIN (0.2) + NAA (0.2); II. Mod MS + KIN (0.5) + NAA (0.1)	III. Omit KIN and NAA IV. Standard or root in vivo	22°C; 6 W/m² continuous
I. Mod MS + KIN (1.0) + IAA (1.0) II. Same	III. Omit KIN and IAA; IBA (1.0) IV. Standard	3–9 W/m²
I. Mod MS, i.e., KH₂ PO₄, CH, malt extract; AD, NAA (10.0), KIN (0.1) II. Same but less	III. NAA (0.5) for plantlet to form IV. Standard	Dark 4–8 weeks for callus; continuous light for plants
I. Mod MS, KIN (1.0); IAA (1.0); II. Same I. Mod MS (1/2) NAA (0.03); use 2iP for Easter lily; buds, MS + BA and IAA at 0.1 or 1.0 each	III. Omit BA and IAA; IBA (1.0) IV. Standard III. Same IV. Transfer to vermiculite	3 W/m²; 16/8 hours; 22°C Continuous dark better for bulbs; give in vitro chilling

(table continues at bottom of page 516)

TABLE 17-4

Micropropagation methods for horticultural crops—fruit and nut plants

Species	Method	Explant	Sterilization
1. *Actinidia* (kiwifruit)	MT; ASP	Store shoots 1°C; use 0.2–0.5 mm MT (dormant)	Wash 20 min; rinse; dip in alcohol; rinse; 1:10 BL; 15 min rinse 3×; dip in alcohol; rinse; dissect 0.6% BL 20 min; rinse 3×; remove 1 cm from base
	ASP	Three-cm shoot tip (spring)	
2. *Fragaria* (strawberry)	MT	0.5 mm tip from apical and lateral buds of young runner	3-cm runner before leafing; wash DT 5 min; 1:10 BL 10 min; wash 2× 10 sec each; excise under dissecting microscope
	ASP		
3. *Malus* (apple)	ASP	I. One- to two-node active shoot	I. Same as cherry
		II. Use preemergent buds	II. Use swollen, green axillary bud; cut like "chip" bud; wash vigorously DT 10 min; stir 1:2 or 1:10 BL 10–20 min; rinse 3×; remove scales; ST
4. *Prunus* (cherry)	ASP	One- to two-node active shoot	Collect 7- to 10-cm first growth in spring; wash vigorously BL 10 min; running water 2 min; 1:5 to 1:10 BL 20 min; rinse 3×; cut above and below node; plant upright
5. *Prunus* (peach)	ASP	1.5-cm active shoot	15% BL 15 min; rinse 4×; trim to 0.5 cm
6. *Rubus* (blackberry)	ASP	5 cm tip; retain young leaves	Agitate-wash in DT 5 min 4×; 1:20 BL 5 min; rinse 3× 5 min each
7. *Vaccinium* (blueberry)	ASP	One- to four-cm shoot; use node with 5 mm each side of bud	2–5 cm active shoot; trim; wash; rinse 5 min; place in fresh Ca hypochlorite 10 min; rinse 2× 10 sec; dip antioxidant; rinse 2 min
8. *Vitis* (grape)	ASP	Three-node shoot tip; 2-mm MT for virus	5-cm shoot tip; remove leaves; stir in 70% alcohol 1 min or BL 20 min; rinse 4×

Remarks

1. Appears to lend itself to rapid multiplication. Increasingly important fruit crop around the world.

2. Extensively used for nuclear stock production system for disease control programs. Microplants tend to be vigorous, runner extensively and make excellent nursery bed plants. Micropropagated plants number in the hundreds of thousands in the world.

3. Own-rooted cultivars, clonal rootstocks and ornamental crabs are being produced with varying degrees of success. Stabilization to an easy-to-root juvenile-like form appears to develop after repeated subculturing. However, field tests show early productivity and promising performance (see Fig. 17-9).

4. Rootstock clones and own-rooted cultivars can be produced with varying degrees of ease. Also, various rootstock plants have promise of production.

5. Peach cultivars can be micropropagated with varying degrees of ease. Preliminary field tests show comparable performance to conventional trees. Best success is with specific peach × almond hybrid rootstock clones of which millions have now been propagated.

6. Subculture microshoots with short piece of original explant; highly successful procedure, particularly useful in virus control programs. Field performance is favorable. Can multiply at very high rates at reasonable price.

7. Microshoots from mature clones undergo reversion to juvenile-like form prior to rooting. Variation in requirements among species and cultivars. Micropropagation comparable in time and cost to conventional propagation but growing to plant takes more time (*68*). Micropropagated plants in field show higher yield because of greater branching (*59*).

8. Inconsistent results requiring frequent transfers. Useful to multiply plants rapidly after meristem culturing. Difficult to compete with conventional propagation methods.

Propagation	Rooting; acclimation	Environmental conditions
I. 3/4 MS, BA (2.0) IBA (0.02) II. Rotating liquid works very well	III. Omit IV. Direct rooting after IBA (2.0) dip	6–9 W/m² 16/8 hr; 23°C
I. MS + BA (1.0) II. Same	III. Replace BA with IBA (1) + charcoal IV. Standard or use in vivo rooting	25°C; 3–9 W/m²
I. 1/2 MS, full Fe II. full MS but vary BA/IBA (0.1 to 1.0)	III. 1/2 MS; BA (0.2) or omit IV. Standard	
I. 1/2 MS, BA (0.5 to 1.0) II. Same, but BA (0.5 to 2.5) + IBA (0.01)	III. Omit BA; IBA (0.2), or omit IV. Standard	3–9 W/m² 16/8 hr; 25–27°C
I. Mod MS + BA (1.0) + GA (0.5); liquid or agar II. Same	III. omit BA; IBA (1.0) IV. Standard; or in vivo rooting	25°C; 6–12 W/m²
I. MS minor element but reduce major; 2 × Fe; IAA (4.0), 2iP (15) II. Same	III. Omit hormones; IBA (5.0), or omit IV. In vivo rooting with IBA dip	6–12 W/m² 16/8 hr
I. Mod MS, 2 × Fe; 2iP (15) + IAA (4) II. Same	III. Omit 2iP IV. In vivo rooting	25°C; 6–12 W/m² hr
I. 3/4 MS, BA (0.1) II. BA (0.2), NaH₂ PO₄ (170) AD (80.0)	III. Reduce additives; IAA (0.1) or omit IV. In vivo rooting slow to establish	9 W/m² 16/8 hr; 23°C; 30°C for rooting

(table continues at bottom of page 518)

TABLE 17–5

Micropropagation methods for horticultural crops—woody perennials

Species	Method	Explant	Sterilization
1. *Eucalyptus*	ASP	Seedlings; use nodes, basal shoots, shoot tips	Remove leaves; retain 1 cm petiole; wash running water 1 hr; wash in DT 5 min; agitate in fresh, filtered Ca hypochlorite 20 min; rinse 3×; trim to 1 cm; incubate on agar 24 hr
2. *Kalmia*	ASP	2–3 cm shoots	Remove leaves >1 cm; dip 70% alcohol; 1:10 BL 15 min; 1 min rinse 3×; remove any damaged tissue
3. *Nandina*	ASP	Lateral and terminal buds	5-cm shoots, remove leaves; wash DT 10 min; rinse; agitate 1:10 BL 20 min; excise under dissecting microscope
4. *Pinus radiata* (Monterey pine)	AS	Cotyledons of germinating embryos	Seeds in 1:1 BL 15–20 min; rinse 24 hr running water; 6% H_2O 5–10 min; rinse 2×; refrigerate 2 days; resterilize; rinse; excise embryo and place on agar medium
5. *Populus* (poplar)	ASP	Chilled buds; active shoots	Rinse in alcohol; sterilize in BL; rinse 3×
6. *Rhododendron* (rhododendron)	ASP	Two- to four-cm actively growing shoots	Wash 5-cm tip in BL; remove leaves and terminal bud; dip 70% alcohol 10 sec; rinse; 1:10 BL 20 min; rinse 3× 1 min each; trim off 1-cm base and lay on agar slant
7. *Rosa* (rose)	ASP	One- to two-cm active shoot	Remove lateral leaves; wash; 1:10 BL 20 min; rinse 3× 2 min each
8. *Sequoia* (redwood)	ASP	One- to three-cm ST	Wash; rinse; alcohol rinse 1 min, 0.1% fungicide 20 min; rinse 1:10 BL 20 min; rinse 1% BL; alcohol dip 10 sec; rinse 3×

Remarks

1. Difficult subject; exudate and explant selection problems; hard to establish after rooting.

2. Highly successful methods have been developed. Commercial.

3. Easy culture should produce large numbers of plants; solves virus problems with this procedure.

4. Commercial production occurs in New Zealand; similar technique being used with various other pine species; Reference: Aiken-Christie and T. A. Thorpe. 1984. In *Cell culture and somatic cell genetics of plants*, I. K. Vasil, ed. Orlando: Academic Press.

5. Utilized for biomass production. Poplars are relatively easy to propagate vegetatively. Relatively easy to "stabilize" in culture.

6. Widely propagated in commercial production. One of most successful. Unique because of specific use of 2iP as a cytokinin. Proper medium is important, particularly lower inorganic content; AC important for rooting. Essentially involves rapid subculturing that may produce reversion to more juvenile form (see Fig. 17–10).

7. Commercial production by tissue culture is a viable alternative to grafting. Requires some time to achieve "stabilization." May show seasonal patterns in proliferation. May require stock renewal at frequent intervals or will slow down in multiplication rate.

8. The first mass produced conifer. Relatively easy to micropropagate.

Propagation	Rooting; acclimation	Environmental conditions
I. 1/2 MS; 2% sucrose; BA (0.2); IBA (0.16) II. Same	III. IBA (2.0) hard to acclimate; deep shade, gradually increase light	Start in dark in 24 hr; shift to 3–6 W/m²
I. Agitated liquid WPM 1 week; change daily; stationary liquid; 2iP (1.0) II. After 2 months change to agar, 4-week interval	III. Optional IV. Root directly in 100% peat; no 2iP; high humidity	28–30°C Continuous light 3–6 W/m²
I. Mod Gamborg B5, MS Fe, NAA (0.1); BA (1.0); AC II. Same; omit BA	III. 1/3 MS; AC; NAA (3.0)	19 W/m²; 26°C
I. Low salts, 2% sucrose; BA (5.0); NAA (2.0); 3 weeks II. Same but BA omitted; transfer each 4–6 weeks	III. Single shoots, water (0.8%), agar, NAA (0.5), IBA (1.0) 5 days; rooted in trays in rooting medium	25°C/20°C with 16/8 hr light; high RH; mist in rooting
I. Mod MS, BA (0.2–0.5) II. Same; BA (0.5)	III. Same, IBA (0.2–0.5); rooting in 7–14 days	25°C; 3–9 W/m²
I, II. AND; 2iP (5.0), IAA (1.0); adenine (80); shift every 2 weeks until growth starts, then every 5 weeks; gradually remove explant	III. omit 2iP; NAA (5.0); AC IV. Can root treated shoots in 1:1 peat-perlite	3–9 W/m², 25°C
I. Mod WPM; long explants vertical in liquid; short explants on agar II. Agar; BA (1.0)	III. Omit IV. Root directly; aided by being in dark.	3–9 W/m²; 25°C
I. MS, KIN (2.0); IAA (0.5) II. Same	III. Omit KIN; IAA (2.0), IBA (3.0); AC IV. Root 4-cm cuttings after treatment with hormone and fungicide	25°C

(table continues at bottom of page 520)

REFERENCES

1. Anderson, W. C. 1975. Propagation of rhododendrons by tissue culture. 1. Development of a culture medium for multiplication of shoots. *Proc. Inter. Plant Prop. Soc.* 25:129–35.

2. ———. 1978. Rooting of tissue cultured rhododendrons. *Proc. Inter. Plant Prop. Soc.* 28:135–39.

3. ———. 1980. Mass propagation by tissue culture: Principles and practice. In *Proc. of the conference on nursery production of fruit plants through tissue culture: Applications and feasibility,* R. H. Zimmerman, ed.

U.S. Dept. of Agr. Sci and Education Administration ARR-NE-11, pp. 1–10.

4. Anderson, W. C., and J. B. Carstens, 1977. Tissue culture propagation of broccoli, *Brassica oleracea* (italica group), for use in F₁, hybrid seed production. *Jour. Amer. Soc. Hort. Sci.* 102(1):69–73.

5. Boxus, P., and P. Druart. 1980. Micropropagation, an industrial propagation method of quality plants true to type and at a reasonable price. In *Plant cell cultures: Results and perspectives,* F. Sala et

al., eds. Amsterdam: Elsevier/North-Holland Biomedical Press.

6. Bridgen, M. P., and J. W. Bartok, Jr. 1987. Designing a plant micropropagation laboratory. *Proc. Inter. Plant Prop. Soc.* 37:462–67.

7. Broome, O. C. 1986. Laboratory design. In *Tissue culture as a plant production system for horticultural crops*. R. H. Zimmerman, R. J. Griesbach, F. A. Hammerschlag, and R. H. Lawson, eds. Dordrecht: Martinus Nijhoff Publishers, pp. 351–65.

8. Brown, D. C. W., and T. Thorpe. 1984. Organization of a plant tissue culture laboratory. In *Cell culture and somatic cell genetics of plants*. Vol. 1, I. K. Vasil, ed. New York: Academic Press, pp. 1–12.

9. Chu, I. Y. E. 1986. The application of tissue culture to plant improvement and propagation in the ornamental horticulture industry. In *Tissue culture as a plant production system for horticultural crops*. R. H. Zimmerman, R. J. Griesbach, F. A. Hammerschlag, and R. H. Lawson, eds. Dordrecht: Martinus Nijhoff Publishers, pp. 15–33.

10. Debergh, P. C. A. 1987. Recent trends in the application of tissue culture to ornamentals. In *Plant tissue and cell culture,* C. E. Green, D. A. Somers, W. P. Hackett, and D. D. Biesboer, eds. New York: Alan R. Liss, pp. 383–93.

11. de Fossard, R. A. 1976. *Tissue culture for plant propagators.* Armidale, Australia: Univ. of New England Printery.

12. Dirr, M. A., and C. W. Heuser, Jr. 1987. *The reference manual of woody plant propagation: From seed to tissue culture.* Athens, Ga.: Varsity Press.

13. Donnan, A., Jr. 1986. Determining and minimizing production costs. In *Tissue culture as a plant production system for horticultural crops,* R. H. Zimmerman, R. J. Griesbach, F. A. Hammerschlag, and R. W. Lawson, eds. Dordrecht: Martinus Nijhoff Publishers, pp. 167–73.

14. Dore, C. 1987. Application of tissue culture to vegetable crop improvement. *In Plant tissue and cell culture,* C. E. Green, D. A. Somers, W. P. Hackett, and D. D. Biesboer, eds. New York: Alan R. Liss, pp. 419–32.

15. Driver, J. A., and G. R. L. Suttle. 1986. Nursery handling of propagules. *In Cell and tissue culture forestry* (2nd ed.), J. M. Bonga and D. J. Durzan, eds. Dordrecht: Martinus Nijhoff Publishers, pp. 320–31.

16. Fellman, C. D., P. E. Reed, and M. A. Hosier. 1987. Effects of Thidiazuron and CPPU on meristem formation and shoot proliferation. *HortScience* 22(6):1197–1200.

17. Fiorino, P., and F. Loreti. 1987. Propagation of fruit trees by tissue culture in Italy. *HortScience* 22(3):353–58.

18. Fogh, J., ed. 1973. *Contamination in tissue culture.* New York: Academic Press.

19. Gamborg, O. L., and L. R. Wetter, eds. 1975. *Plant tissue culture methods.* Saskatoon, Canada: Prairie Regional Laboratory.

20. Ganzer, J. 1979. Commercial application of tissue culture in fruit production. *Proc. Inter. Plant Prop. Soc.* 29:401–3.

21. George, E. F., and P. D. Sherrington. 1984. *Plant propagation of tissue culture; Handbook and directory of commercial laboratories.* Eversley, England: Exegetics Ltd.

22. Griesbach, R. J. 1986. Orchid tissue culture. In *Tissue culture as a plant production system for horticultural crops.* R. H. Zimmerman, R. J. Griesbach, F. A. Hammerschlag, and R. H. Lawson, eds. Dordrecht: Martinus Nijhoff Publishers, pp. 343–51.

23. Halquist, J. L., M. A. Hosier, and P. E. Read. 1983. A comparison of several gelling agents and concentrations on callus growth and organogenesis in vitro. *In Vitro* 19:248.

24. Hammerschlag, F. A. 1986. Temperate fruits and nuts. In *Tissue culture as a plant production system for horticultural crops.* R. H. Zimmerman, R. J. Griesbach, F. H. Hammerschlag, and R. H. Lawson, eds. Dordrecht: Martinus Nijhoff Publishers, pp. 221–36.

25. Harney, P. M., and A. Knap. 1979. A technique for the *in vitro* propagation of African violets using petioles. *Can. Jour. Plant Sci.* 59:263–66.

26. Hartman, R. D., and F. W. Zettler. 1986. Tissue culture as a plant production system for foliage plants. In *Tissue culture as a plant production system for horticultural crops,* R. H. Zimmerman, R. J. Griesbach, F. A. Hammerschlag, and R. H. Lawson, eds. Dordrecht: Martinus Nijhoff Publishers, pp. 293–99.

27. Hartmann, H. T., and J. Whisler. 1977. Micropropagation exercises in teaching plant propagation. *Proc Inter. Plant. Prop. Soc.* 27:407–13.

28. Henny, R. J., J. F. Knauss, and A. Donnan, Jr. 1981. Foliage plant tissue culture. In *Foliage plant production,* J. Joiner, ed. Englewood Cliffs, N.J.: Prentice-Hall.

29. Holdgate, D. P., and J. S. Aynsley. 1977. The development and establishment of a commercial tissue culture laboratory. *Acta Hort.* 78:31–36.

30. Jones, J. B. 1986. Determining markets and

market potential. In *Tissue culture as a plant production system for horticultural crops.* R. H. Zimmerman, R. J. Griesbach, F. A. Hammerschlag, and R. H. Lawson, eds. Dordrecht: Martinus Nijhoff Publishers, pp. 175–82.

31. Kerns, H. R., and M. M. Meyer, Jr. 1986. Tissue culture propagation of *Acer* x *freemanii* using thidiazuron to stimualte shoot tip proliferation. *HortScience* 21(5):1209–10.

32. Kester, D. E., and C. Grasselly. 1987. Almond rootstocks. In *Rootstocks for fruit crops,* R. C. Rom and R. F. Carlson, eds. New York: John Wiley, pp. 265–93.

33. Khoo, S. I., and M. B. Thomas. 1980. Studies on the germination of fern spores. *The Plant Propagator* 26:11–15.

34. Kiernan, J. M., et al. 1984. Characterization of strawberry plants produced by tissue culture and infected with specific mycorrhizal fungi. *HortScience* 19(6):883–85.

35. Knudson, L. 1922. Nonsymbiotic germination of orchid seeds. *Bot. Gaz.* 73:1–15.

36. ———. 1951. Nutrient solution for orchids. *Bot. Gaz.* 112:528–32.

37. Kyte, L. 1987. *Plants from test tubes; an introduction to micropropagation.* (rev. ed.) Portland, Oreg.: Timber Press.

38. Lane, B. C. 1980. A procedure for propagating ferns from spores using a nutrient-agar solution. *Proc. Inter. Plant Prop. Soc.* 30:94–97.

39. Linsmaier, E. M., and F. Skoog. 1965. Organic growth factor requirements of tobacco tissue cultures. *Physiol. Plant* 18:101–27.

40. Litz, R. E. 1987. Application of tissue culture to tropical fruits. In *Plant tissue and cell culture,* C. E. Green, D. A. Somers, W. P. Hackett, and D. D. Biesboer, eds. New York: Alan R. Liss, pp. 407–18.

41. Lloyd, G., and B. McCown. 1980. Commercially feasible micropropagation of mountain laurel, *Kalmia latifolia,* by use of shoot-tip culture. *Proc. Inter. Plant Prop. Soc.* 30:421–27.

42. Macdonald, B. 1986. *Practical woody plant propagation for nursery growers.* Portland, Oreg.: Timber Press.

43. Matsuyama, J. 1980. Overview of tissue culture at K. M. Nursery. *Proc. Inter. Plant Prop. Soc.* 30:40–42.

44. McCown, B. H. 1986. Woody ornamentals, shade trees, and conifers. In *Tissue culture as a plant production system for horticultural crops,* R. H. Zimmerman, R. J. Griesbach, F. A. Hammer-

schlag, and R. H. Lawson, eds. Dordrecht: Martinus Nijhoff Publishers, pp. 333–42.

45. McCown, D. D. 1986. Plug systems for micropropagules. In *Tissue culture as a plant production system for horticultural crops,* R. H. Zimmerman, R. J. Griesbach, F. A. Hammerschlag, and R. H. Lawson, eds. Dordrecht: Martinus Nijhoff Publishers, pp. 53–60.

46. McCown, D. D., and B. H. McCown. 1986. North American hardwoods. In *cell and tissue culture in forestry,* vol. 3, *Case histories: Gymnosperms, angiosperms and palms* (2nd ed.). J. M. Bonga and D. J. Durzan, eds. Dordrecht: Martinus Nijhoff Publishers, pp. 247–60.

47. McGranahan, G. H., and P. B. Catlin. 1987. *Juglans* rootstocks. In *Rootstocks for fruit crops,* R. C. Rom and R. F. Carlson, eds. New York: John Wiley, pp. 411–50.

48. Mok, M. C., D. W. S. Mok, J. E. Turner, and C. V. Mufer. 1987. Biological and biochemical effects of cytokinin-active phenylurea derivatives in tissue culture systems. *HortScience* 22(6):1194–97.

49. Mudge, K. W., C. A. Borgman, J. C. Neal, and H. A. Weller. 1986. Present limitations and future prospects for commercial micropropagation of small fruits. *Proc. Inter. Plant Prop. Soc.* 36:538–43.

50. Mullins, M. G. 1987. Propagation and genetic improvement of temperate fruits, the role of tissue culture. In *Plant tissue and cell culture,* C. E. Green, D. A. Somers, W. P. Hackett, and D. D. Biesboer, eds. New York: Alan R. Liss, pp. 395–406.

51. Murashige, T. 1974. Plant propagation through tissue cultures. *Ann. Rev. Plant Phys.* 25:135–66.

52. ———. 1977. Plant cell and organ cultures as horticultural practices. *Acta Hort.* 78:17–30.

53. Murashige, T., and F. Skoog. 1962. A revised medium for rapid growth and bioassays with tobacco tissue cultures. *Phys. Plant.* 15:473–97.

54. Ng, T. J. 1986. Use of tissue culture for micropropagation of vegetable crops. In *Tissue culture as a plant production system for horticultural crops,* R. H. Zimmerman, R. J. Griesbach, F. A. Hammerschlag, and R. H. Lawson, eds. Dordrecht: Martinus Nijhoff Publishers, pp. 259–70.

55. Oglesby, R. P., and J. L. Griffis, Jr. 1986. Commercial in vitro propagation and plantation crops. In *Tissue culture as a plant production system for horticultural crops,* R. H. Zimmerman, R. J. Griesbach, F. A. Hammerschlag, and R. H.

Lawson, eds. Dordrecht: Martinus Nijhoff Publishers, pp. 253–57.

56. Oglevee-O'Donovan, W. 1986. Production of culture virus-indexed geranium. In *Tissue culture as a plant production system for horticultural crops,* R. H. Zimmerman, R. J. Griesbach, F. A. Hammerschlag, and R. H. Lawson, eds. Dordrecht: Martinus Nijhoff Publishers, pp. 119–24.

57. Pierik, R. L. M. 1987. *In vitro culture of higher plants.* Dordrecht: Martinus Nijhoff Publishers.

58. Raju, B. C., and J. C. Trolinger, 1986. Pathogen indexing in large-scale propagation of florist crops. In *Tissue culture as a plant production system for horticultural crops,* R. H. Zimmerman, R. J. Griesbach, F. A. Hammerschlag, and R. H. Lawson, eds. Dordrecht: Martinus Nijhoff Publishers, p. 138.

59. Read, P. E., C. A. Hartley, J. R. Sandahl, and D. K. Wildung. 1987. Field performance of in vitro propagated blueberries. *Proc. Inter. Plant Prop. Soc.* 37:450–52.

60. Singa, S., G. K. Bissonnette, and M. L. Double. 1987. Methods for sterilizing instruments contaminated with *Bacillus* spp. from plant tissue cultures. *HortScience* 22(4):659.

61. Smith, R. J. 1972. Orchid propagation by in vitro culture techniques. *Proc. Inter. Plant Prop. Soc.* 22:174–77.

62. Smith, W. A. 1981. The aftermath of the test tube. *Proc. Inter. Plant Prop. Soc.* 31:47–49.

63. Snir, I., and A. Erez. 1980. In vitro propagation of Malling Merton apple rootstocks. *HortScience* 15(5):597–98.

64. Stimart, D. P. 1986. Commercial micropropagation of florist flower crops. In *Tissue culture as a plant production system for horticultural crops,* R. H. Zimmerman, R. J. Griesbach, F. A. Hammerschlag, and R. H. Lawson, eds. Dordrecht: Martinus Nijhoff Publishers, pp. 301–15.

65. Stokes, M. J. 1980. Current aspects of commercial micropropagation. *Proc. Inter. Plant Prop. Soc.* 30:249–54.

66. Stoltz, L. P. 1979. Getting started in tissue culture: Equipment and costs. *Proc. Inter. Plant Prop. Soc.* 29:375–81.

67. Strode, R. A., and G. Abner. 1986. Large scale tissue culture production for horticultural crops. In *Tissue culture as a plant production system for horticultural crops,* R. H. Zimmerman, R. J. Griesbach, F. A. Hammerschlag, and R. H. Lawson, eds. Dordrecht: Martinus Nijhoff Publishers, pp. 367–71.

68. Swartz, H. J., and J. T. Lindstrom. 1986. Small fruit and grape tissue culture from 1980 to 1985: Commercialization of the technique. In *Tissue culture as a plant production system for horticultural crops,* R. H. Zimmerman, R. J. Griesbach, F. A. Hammerschlag, and R. H. Lawson, eds. Dordrecht: Martinus Nijhoff Publishers, pp. 201–20.

69. Thorpe, T. A., ed. 1981. *Plant tissue culture: Methods and applications in agriculture.* New York: Academic Press.

70. Tjosvold, S., and A. Teasdale. 1980. Uniform fern spore dispersal on warm nutrient-agar solution. *The Plant Propagator* 26:11.

71. Wimber, D. E. 1963. Clonal multiplication of *Cymbidium* through tissue culture. *Amer. Orch. Soc. Bul.* 32:105–7.

72. Wochok, Z. S. 1986. Status of crop improvement through tissue culture. *Prop. Inter. Plant Prop. Soc.* 36:72–77.

73. Zilis, M., D. Zwagerman, D. Lamberts, and L. Kurtz. 1979. Commercial propagation of herbaceous perennials by tissue culture. *Proc. Inter. Plant Prop. Soc.* 29:404–13.

74. Zimmerman, R. H. 1979. The laboratory of micropropagation at Cesena, Italy. *Proc. Inter. Plant Prop. Soc.* 29:398–400.

75. Zimmerman, R. H., O. C. Broome. 1980. Apple cultivar micropropagation. In *Proc. of the conference on nursery production of fruit plants through tissue culture: Applications and feasibility,* R. H. Zimmerman, ed. USDA. ARR-NE-11, pp. 54–58.

76. Zimmerman, R. H., and I. Fordham. 1985. Simplified method for rooting apple cultivars in vitro. *Jour. Amer. Soc. Hort. Sci.* 110(1):34–38.

77. Zimmerman, R. H., R. J. Griesbach, F. A. Hammerschlag, and R. H. Lawson, eds. 1986. *Tissue culture as a plant production system for horticultural crops.* Dordrecht: Martinus Nijhoff Publishers.

SUPPLEMENTARY READING

BONGA, J. M., and D. J. DURZAN. 1986. *Cell and tissue culture in foresty,* 4 volumes (2nd ed.). Dordrecht: Martinus Nijhoff Publishers.

BOXUS, P. 1978. *"In vitro" multiplication of woody species; round-table conference, Gembloux,(Belgium), June 6–8, 1978.* Gembloux: Service des Relations publiques.

BRIDGEN, M. P. and J. W. BARTOK, JR. 1988. Designing a plant micropropagation laboratory. *Proc. Inter. Plant Prop. Soc.* 37:462–67.

CONGER, B. V., ed. 1981. *Cloning agricultural plants via in vitro techniques.* Boca Raton, Fla.: CRC Press.

DEBERGH, I. P., ed. 1977. Symposium on tissue culture for horticultural purposes, Ghent, Belgium, Sept. 6–9, 1977. *Acta Hort.* 78:1–459.

DE FOSSARD, R. A. 1976. *Tissue culture for plant propagators.* Armidale, Australia: Univ. of New England Printery.

GAMBORG, O. L., and L. R. WETTER, eds. 1975. *Plant tissue culture methods.* Saskatoon, Saskatchewan: National Research Council of Canada Prairie Regional Laboratory.

HENKE, R. R., K. W. HUGHES, M. J. CONSTANTIN, and A. HOLLAENDER, eds. 1985. *Tissue culture in forestry and agriculture.* New York:Plenum Press.

HUTCHINSON, J. F., and R. H. ZIMMERMAN. 1989. Tissue culture of temperate fruit and nut trees. *Hort. Rev.* 9:273–349.

INGRAM, D. S., and J. P. HELGESON. 1981. *Tissue culture methods for plant pathologists.* Oxford: Blackwell Scientific Publ.

INTERNATIONAL PLANT PROPAGATORS SOCIETY. Proceedings of annual meetings.

INTERNATIONAL PLANT TISSUE CULTURE SOCIETY. Newsletters.

KYTE, L. 1987. *Plants from test tubes: An introduction to micropropagation* (Rev. ed.). Portland, Oreg.: Timber Press.

REINERT, J., and Y. P. S. BAJAJ, eds. 1977. *Plant cell, tissue and organ culture.* Berlin: Springer-Verlag.

STREET, H. E., ed. 1977. *Plant tissue and cell culture* (2nd ed.). Berkeley: Univ. of Calif. Press.

THORPE, T. A., ed., 1981. *Plant tissue culture: Methods and applications in agriculture.* New York: Academic Press.

TORRES, K. C. 1989. *Tissue culture techniques for horticultural crops.* New York: Van Nostrand Reinhold.

WITHERS, L. A., and P. G. ALDERSON. 1986. *Plant tissue culture and its agricultural applications.* London: Butterworth.

ZIMMERMAN, R. H., ed. 1980. *Proceedings of conference on nursery production of fruit plants through tissue culture: Applications and feasibility.* USDA Sci. and Education Administration ARR-NE-11.

18

Propagation Methods and Rootstocks for the Important Fruit and Nut Species

Few fruit or nut crop cultivars reproduce true-to-type when propagated by seed. It is necessary, therefore, that they be propagated by some asexual method. Most tree fruit and nut species are propagated by budding or grafting on rootstocks—seedlings, rooted cuttings, or layered plants. In some cases, cuttings would be the simplest and easiest method to use, but for other than a few species such as the grape, fig, olive, quince, currant, gooseberry, and pomegranate, fruits and nuts are difficult to propagate by cuttings, so other methods are used. Some, such as the filbert, are propagated by layering.

Some large nurseries propagate many hundreds of thousands of fruit and nut trees each year by various vegetative methods (*133*). Large-scale propagation of some fruit tree cultivars, as well as clonal rootstocks, by tissue culture methods is becoming commonplace in some countries. (*30, 62*)

Almond (*Prunus dulcis* [Mill.] D. A. Webb. Syn. *P. amygdalus* Batsch). The almond is propagated by T-budding on seedling rootstocks by fall, spring, or June budding (*83*). Stem cuttings have given but slight success.

Rootstocks for Almond. Historically, almond seedlings have been used. Peach seedlings are favored for irrigated orchards or where nematodes are a problem. The clonal rootstock, 'Marianna 2624' plum is of value in certain situations. Almond × peach hybrids is a class of rootstocks for almond developed in recent years.

Almond (P. dulcis). Almond seedlings are quite satisfactory in deep and well-drained soils. Seeds of the bitter types, or certain commercial cultivars such as 'Texas' ('Mission'), are commonly used. Almond seeds require stratification for three or four weeks before planting. In poorly drained soils almond roots are often unsatisfactory, owing to their susceptibility to infection by crown rot (*Phytophthora* spp.) and crown gall (*Bacterium tumefasciens*). Their deep-rooting tendency is an advantage in orchards grown on unirrigated soils or where drought conditions occur. Almond seedlings are tolerant of high-lime soils and, of the rootstocks available for almonds, are the least affected by excess boron salts. Almond seedlings are susceptible to root-knot and root-lesion nematodes as well as to oak root fungus

(*Armillarea mellea*). Nematode-resistant almond selections have been developed in Israel.

Peach (P. persica).   Peach seedlings are widely used as rootstocks for almond where irrigation is practiced and where nematodes are present. Any of the peach stocks are satisfactory, but 'Lovell', 'Halford', and 'Nemaguard' (for nematode resistance) are most commonly used. Peach roots are not as susceptible to crown gall or crown rot as almond roots. Almond on peach roots in irrigated soils will grow faster for the first several years and bear heavier crops during the first 15 to 20 years than those on almond roots, but trees on almond roots tend to live longer and may eventually outgrow those on peach. Peach seeds need about three months of stratification for good germination.

'Marianna 2624' plum.   This clonal rootstock is used in heavy, wet soils or where oak root fungus or root-knot nematodes (to which it is immune) are present (*48*). Almond trees on this stock are about one-third smaller than those on the other available rootstocks. Not all almond cultivars are compatible with this stock (*132*); those that cannot be used include 'Nonpareil', 'Milow', and 'Kapareil'. Fall planting of hardwood cuttings of this rootstock (in mild winter climates), after treatment with IBA, gives good rooting.

Peach–almond hybrids (130).   These are first-generation offspring of peach and almond. Seedling populations can be produced by natural crossing between adjoining trees of the two species. Hybrid plants are identified in the nursery row by their high vigor and intermediate appearance between the parents. A hybrid, 'GF 677', resistant to alkaline soils is widely used in Europe for both almonds and peaches. 'Hansen 536' and 'Hansen 2168' are two hybrid rootstock clones released by the University of California. Rooting of selected clones is possible by hardwood cuttings treated with IBA plus a fungicide, then planted directly in the nursery. Peach–almond hybrids can also be rooted by semihardwood cuttings under mist if given IBA treatments (*162*), but micropropagation is the most important commercial method (*131*). These stocks are noted for their vigor and excellent compatibility with scion cultivars (*273*).

Apple (*Malus* × *domestica* Borkh.).   T-budding is used, either as fall budding or spring budding on seedling or clonal rootstocks. Chip budding also is successful. Root grafting is used in some places, usually as whole-root grafts. Propagation of cultivars by hardwood cuttings, with the exception of certain clonal rootstocks, is seldom used. Softwood cuttings can be rooted under mist, but this method is not used commercially. Micropropagation methods have been used to produce apple nursery trees (*127, 228, 271*).

All apple rootstocks, either seedling or clonal, are in the genus *Malus,* although the apple will grow for a time and even come into bearing grafted on pear (*Pyrus communis*) roots. Apple seedling rootstocks are widely used in the western and southeastern United States. The clonal dwarfing rootstocks are more favored in the midwestern and northeastern United States and in Europe.

Rootstocks for Apple (61, 230)

Seedling Stocks.   Apples on seedling roots produce a larger tree than those grown on most of the clonal stocks. Seedlings of 'Delicious', 'Golden Delicious,' 'McIntosh', 'Winesap', 'Yellow Newtown', and 'Rome Beauty' (but particularly 'Delicious') are widely and successfully used as rootstocks. Such seedlings are quite uniform, and no incompatibility problems have arisen. However, in purchasing seed it is often difficult to determine the seed source. In the colder portions of the United States—the Dakotas and Minnesota—the hardier Siberian crabapple (*Malus baccata*) and seedlings of such cultivars as Antonovka, containing some *M. baccata* parentage, are used. In Poland 'Antonovka' seedlings are the chief apple rootstock. Some nurseries in British Columbia, Canada, use 'McIntosh' seedlings for their winter-hardiness, upright growth in the nursery, and early fall shedding of leaves, although they tend to be somewhat susceptible to *hairy root,* a form of *crown gall,* and to *powdery mildew.* The resulting trees tend to be variable in size and performance.

Apples with the triploid number of chromosomes, such as 'Gravenstein', 'Baldwin', 'Stayman Winesap', 'Arkansas', 'Rhode Island Greening', 'Bramley's Seedling', and 'Tompkins King', produce seeds that are of low viability and are not recommended as a seed source. Seeds of 'Wealthy', 'Jonathan', or 'Hibernal' have given unsatisfactory results.

The principal source of seeds is the pomace from processed apples. For spring sowing, seeds require stratification for 60 to 90 days at 2 to 7°C (35 to 45°F) to germinate. Some nurseries fall-plant seeds to receive the natural winter chilling. To avoid seedlings that become crooked in breaking through the crust, soil must be raked over in the spring. To obtain a branched root system, the seedlings may be undercut while small to prevent the development of a taproot; a straight root may be preferred, however, when bench grafting is to be done. Seedlings that do not grow to a satisfactory size in one year should be culled out.

Various Asiatic species of *Malus* have been tried

as dwarfing or semidwarfing rootstocks or interstocks for apple cultivars (*204, 205*). Some of these, *Malus hupehensis, M. toringoides, M. sargentii,* and *M. sikkimensis,* are apomictic. These stocks are moderately hardy and resistant to crown gall. *Malus sikkimensis* seedlings are uniform, and scion cultivars worked on this stock are restricted in growth and start bearing early. However, such trees are relatively unproductive compared to those on the dwarfing Malling stocks.

Apple roots are resistant to root-knot and root-lesion nematodes, moderately resistant to oak root fungus, and highly resistant to verticillium wilt.

Clonal Stocks. Numerous clonal, asexually propagated apple rootstocks have been developed. Most are in the species, *Malus* × *domestica.* It is important in using these clonal stocks to obtain only virus-tested material. All of the clonal apple rootstocks listed below have some deficiencies. In various apple producing countries rootstock breeding and testing programs are underway to develop new dwarfing and semidwarfing rootstocks to replace them (*44*).

'Alnarp 2'. This rootstock was developed at the Alnarp Fruit Tree Station in Sweden and is widely used there for its winter-hardy properties. The roots are well anchored and give about 20 percent size reduction as compared to trees on seedling roots. It is susceptible to woolly apple aphid and to fire blight.

'Robusta No. 5' [M. robusta (M. baccata × *M. prunifolia)].* This vigorous, very hardy clonal apple rootstock, propagated by stooling or stem cuttings, originated in 1928 at the Central Experimental Farm, Ottawa, Canada. It is apparently resistant to fire blight and crown rot and seems to be compatible with most apple cultivars. It is extensively used as an apple rootstock in eastern Ontario and Quebec, as well as in the New England states. It is the best stock for use where extreme winter-hardiness is required, but it has a low chilling requirement and has the unfavorable habit of starting growth too early in the spring following three or four warm days in late winter. It is not a dwarfing stock.

'Mark'. Developed at Michigan State University as an open-pollinated seedling of 'Malling 9'. Trees on this stock are similar in size to those on 'Malling 9' or slightly larger and about the same in productivity. This rootstock does not sucker and is well anchored. It propagates easily in stool beds. Trees on this stock have an open, spreading growth habit.

Malling Series. Beginning in 1912, the East Malling Research Station in England selected and classified a series of vegetatively propagated apple rootstocks that ranged from very dwarfing to very invigorating in their effect on the scion cultivar (*268*). This size-controlling influence is modified by the scion cultivar used. The dwarfing influence of these various rootstocks does not extend to the fruit, however; fruit size, especially on young dwarfed trees, is often larger than on standard-size trees.

The Malling stocks apparently give completely compatible graft unions with most apple cultivars. Trees on this series of rootstocks have been planted in varying degrees in many parts of the world. These stocks have proven to be hardy except in regions with extremely severe winters, such as the northern United States and Canada. They have done well on both heavy and light soils, but none are resistant to woolly aphid.

Malling-Merton series (192). The John Innes Horticultural Institution in England and the East Malling Research Station began work jointly in 1928 on breeding a new series of apple rootstocks to provide resistance to woolly aphids and to give a range in tree vigor. Of this group only 'MM 111' and 'MM 106' are now widely used. Both show resistance to woolly aphids. However, these stocks have been severely attacked by this insect in South Africa, presumably because of the presence there of a different strain of woolly aphid (*73*).

Other improved cultivar characteristics associated with these stocks include high yield, induced precocity of flowering (with some stocks), well-anchored trees (with some stocks), freedom from suckering, and good propagation qualities.

All the Malling and Malling–Merton stocks are readily propagated, mostly by stoolbed layering. A number of rootstock clones, notably 'Malling 26', 'MM 106', and 'MM 111', respond very well to propagation by the hardwood cutting method. Certain of these clonal apple rootstocks have been produced by micropropagation methods (*122, 225*). Virus-tested material of both of these groups of rootstocks has been developed by a joint effort on the *E*ast *M*alling and *L*ong *A*shton Research Stations in England and is distributed under the designations EMLA 7, EMLA 9, EMLA 26, EMLA 27, EMLA 106, and EMLA 111. These stocks give trees with 10 to 15 percent more vigor than the older virus-infected stocks of the same clone.

A summary of the characteristics of the most useful of these two groups of clonal rootstocks follows, with the rootstocks grouped according to their effect on the vigor of the scion cultivar. However, the particular scion cultivar used has a definite influence on the size of the composite tree. The most popular of these rootstocks in North America are Malling 7, 9, 26 and Malling-Merton 106 and 111.

Dwarfing Stocks

'Malling 27' (193).   This is, by far, the most dwarfing of all the Malling stocks, producing trees only about four feet tall, about half the size of those on 'Malling 9'. It is a cross between 'Malling 13' and 'Malling 9' made in 1929. It may have use in high-density plantings. Virus-tested propagating material was released by the East Malling Research Station in 1970. It can also be used as an interstock to give a dwarfing effect.

'Malling 9' ('Jaune de Metz').   This originated as a chance seedling in France in 1879 and has been used widely in Europe for many years as an apple rootstock. It is a dwarfed tree itself and a valuable dwarfing rootstock much in demand for producing small trees for the home garden or for commercial high-density plantings. Recommended tree spacing is 1.8 × 3.6 m (6 × 12 ft). Such trees are seldom over nine feet tall when mature and usually start bearing in the first year or two after planting. Virus-tested propagating material is available.

'Malling 9' has numerous thick, fleshy, brittle roots and requires a fertile soil; the trees require staking or trellising for support. It is resistant to collar rot (*Phytophthora cactorum*) but susceptible to mildew, crown gall, fire blight, and woolly apple aphid. Roots are sensitive to low winter temperatures. It is propagated by stooling.

When used as an intermediate stock in double-working it will cause dwarfing of the scion cultivar but less than when it is used as the rootstock (*205, 244*).

'Malling 26'.   This stock was introduced in 1959 at the East Malling Research Station, originating from a cross between 'Malling 16' and 'Malling 9'. It is better anchored than 'M. 9' and it produces a tree somewhat larger and sturdier than 'Malling 9'—less so than 'Malling 7' or 'MM 106', but still a tree that requires staking. Suggested tree spacing is 3 × 4.2 m (10 × 14 ft). It is propagated by softwood cuttings under mist or by hardwood cuttings (*101*), but is a poor producer in stool beds. It is quite winter-hardy but does not tolerate heavy or poorly drained soils and is very susceptible to fire blight and collar rot.

Semidwarfing Stocks

'Malling 7a'.   Malling 7 was selected at East Malling from a group of French traditional rootstocks known as Doucin. This stock produces trees somewhat larger than those on 'Malling 26' roots. 'Malling 7a' has a stronger, deeper, root system than 'Malling 9' and produces an early-bearing, semidwarf tree. It is tolerant of excessive soil moisture but susceptible to crown gall and not very well anchored, requiring staking for the first few years. It seems to make good growth over a wide soil temperature range (*159*). Suggested tree spacing is 4.2 × 4.8 m (14 × 16 ft). This stock has the undesirable characteristic of suckering badly, and the trees are not very winter-hardy. 'Malling 7a' is easily propagated by stooling or by leafy cuttings under mist. (The "a" designation indicates a clone free of certain viruses present in the original 'Malling 7' but removed by heat therapy). A virus-indexed EMLA clone was introduced in 1974.

'Malling-Merton 106'.   This originated as a cross between 'Northern Spy' and 'Malling 1'. A virus-indexed clone was introduced in 1969. This stock produces trees about half the size of those on seedling rootstocks and about the same as those on 'Malling 7a', but is more productive, with earlier cropping. It is the most extensively used clonal apple rootstock in the United States and, in England, it is the most popular apple rootstock for bush trees and hedgerows. On good soils, with some scion cultivars, it can produce a standard-size tree. The roots are well anchored and do not sucker. Suggested planting distances are 4.2 × 5.4 m (14 × 18 ft). In some areas 'MM 106' is susceptible to mildew as well as collar rot, which may be its chief weakness. It has not been affected by fire blight. It grows well in the nursery but drops its leaves later in the fall than most other understocks and is not resistant to early fall freezes. Hardwood and softwood cuttings root easily and stool beds are quite productive.

Vigorous Stocks

'Malling 2' ('Doucin').   This was the most commonly used apple understock in England but has been replaced by 'MM 111'. Trees on 'Malling 2' tend to be vigorous and fruitful, come into bearing early, and are smaller than trees on seedling roots. It is moderately susceptible to crown gall but resistant to collar rot. It is somewhat difficult to propagate by cuttings and the layers on young stool plants root rather sparsely. It is no longer planted in the United States.

'Malling-Merton 111'.   This stock originated as a cross between 'Northern Spy' and 'Merton 793' ('Northern Spy' × 'Malling 2'). A virus indexed EMLA clone was introduced in 1969. Trees on this stock are comparable in vigor to those on 'Malling 2', but grow better when young, start bearing earlier, and survive periods of drought better. It does well on a wide range of soil types. It is susceptible to mildew but not to collar rot or woolly aphid. Stool beds are highly productive with heavy root systems developing. Hardwood cuttings root well with proper treatment (*101*) and softwood cuttings under mist root well. 'MM 111' is more winter-hardy than 'Malling 7' or 'MM 106'. Suggested planting distances are 4.8 × 6 m (16 × 20 ft). It shows excessive vigor in some situations.

Very Vigorous Stocks

'Malling 16' ('Ketziner Ideal').   Trees on this rootstock are large, well anchored, and come into bearing

about the same time as those on seedling roots (or later). 'Malling 16' seems to do well over a wide range of soil temperatures (*172*). It is usually propagated by stooling, but can also be started by root cuttings. It is very susceptible to woolly aphid and is no longer used in the United States.

'Malling-Merton 109'.   Trees on this stock are about the same size as those on seedling stocks. Its chief advantage is its resistance to woolly aphid and its stimulation of early bearing. Trees on this stock sometimes tend to lean badly and are not tolerant of waterlogged soils. It is not used in the United States.

'Malling 25'.   Although not as resistant to woolly aphid as the Malling-Merton stocks, it is more resistant than 'Malling 16'. It produces large, well-anchored, vigorous trees, induces early fruit-bud formation, and gives excellent fruit set and yields but has not gained acceptance by U.S. apple growers.

Apricot (*Prunus armeniaca* L.).   Apricot cultivars are propagated commercially by T-budding or chip budding on various seedling rootstocks in the genus *Prunus*. Fall budding is the usual practice, but spring and June budding may be used. Bench grafting also has been successful (*47*).

Rootstocks for Apricot (*43, 176, 227*).   Three stocks are commercially suitable—apricot seedlings, peach seedlings, and in some cases, myrobalan plum seedlings. Seeds of all these species require low-temperature 5°C (41°F) stratification before planting in the spring—three to four weeks for the apricot and about three months for the peach and myrobalan plum. On good, well-drained soils, apricot seedlings are the recommended rootstock for apricot cultivars.

Apricot (P. armeniaca).   Apricot seeds can be obtained from drying yards and canneries. Seeds of 'Royal' or 'Blenheim' produce excellent rootstock seedlings in California. In the eastern United States seedlings of 'Manchurian', 'Goldcot', and 'Curtis' are recommended. Since the apricot root is almost immune to the root-knot nematode (*Meloidogyne* spp.), it should be used where this pest is present. In addition, it is somewhat resistant to the root-lesion nematode. It is susceptible to crown rot (*Phytophthora* spp.) and intolerant of poor soil-drainage conditions. Apricot roots are not as susceptible to crown gall (*Agrobacterium tumefaciens*) as are peach and plum roots. Apricot seedling roots are susceptible to oak root fungus and highly susceptible to verticillium wilt. Apricot trees on apricot roots live longer and produce heavier crops than trees on either peach or plum roots, if grown on well-drained loam soils.

Peach (P. persica).   In California peach seedlings are satisfactory as a rootstock for apricot cultivars. Although the peach itself is short-lived, apricot trees 85 years old growing satisfactorily on peach are known. In unirrigated orchards or where drought conditions prevail, apricots on peach seedling roots make better growth than those on apricot roots. Peach roots are not tolerant of wet soils, growing better on light or well-drained soils. For trees to be planted in a location formerly occupied by peach roots, some stock other than peach should be used, because new peach roots often grow poorly on these soils. In the eastern United States and Canada apricot cultivars show definite incompatibility on peach seedlings, so it cannot be assumed that all cultivars will do well on peach roots (*24, 138*).

Myrobalan plum (P. cerasifera).   Although there are successful high-yielding apricot orchards grown on this rootstock, it cannot be recommended without qualification. In a few instances the trees have broken off at the graft union in heavy winds, and die-back conditions have been noted. Nurseries often have trouble starting apricots on myrobalan roots, some of the trees failing to grow rapidly and upright or else having weak or rough unions. After these weaker trees are culled out, the remaining ones seem to grow satisfactorily. This stock is useful for apricot when the trees are to be planted in heavy soils or under excessive soil moisture conditions, which the myrobalan root will tolerate. An alternative to myrobalan plum seedlings is the related vegetatively propagated 'Marianna 2624' plum, on which apricots seem to do well.

Avocado (*Persea americana* Mill.) (*17, 185*).   Nursery trees of the avocado are propagated commercially in California by T-budding, tip grafting (whip or splice graft) (*236*), or wedge or cleft grafting on avocado seedlings or clonal rootstocks (*18*). In Florida and some of the Caribbean countries T-budding is used occasionally, but the usual nursery practice is to graft mature tip scions, either as side grafts—or sometimes the side-veneer—or cleft grafts on young succulent seedlings. To avoid sunblotch virus, the seeds (and budwood) must be taken from source trees that have been registered by state certifying agencies as free of the disease. Clonal rootstocks resistant to *Phytophthora* can be produced in large numbers by a rather complicated procedure using a nurse-seed method, which has a U.S. patent (*18*).

Seeds are generally planted shortly after removal from fruit taken from the tree (not picked from the ground), care being taken not to allow them to dry out, although they can be stored six to eight months if packed in dry peat moss and held at 5°C (41°F) with 90 percent relative humidity. To eliminate infection from *Phytophthora cinnamomi* (avocado root rot), seeds should be dipped in hot water at 49 to 52°C (120 to

125°F) for 30 minutes before planting (*269*). Germination of the seed is hastened by removing the brown seed coats and cutting a thin slice from the apical and basal end of each seed before planting. The seed coats can be removed by wetting the seeds and allowing them to dry in the sun. Seeds of the Mexican race, which ripen its fruit in the fall, may be planted in beds in late fall or early winter. They should be placed with the large, basal end down, just deep enough to cover the tips. If grown in a warm area, the sprouted seeds are ready to line out in the nursery row the following spring. By summer or fall the seedlings are usually large enough to permit T-budding; if not, they can be budded the following spring.

It is important to select the budwood properly. The best buds are usually near the terminal ends of completed growth cycles with fully matured, leathery leaves. To prevent drying, the leaves should be removed when the budwood is taken.

Four to six weeks after budding, the seedling rootstocks should be cut off 20 to 25 cm (8 to 10 in.) above the bud, or bent over a few inches above the bud. The remaining portion of the seedling above the bud is not cut off until the bud shoots have completed a cycle of growth. The new shoots are usually staked and tied. In digging from the nursery, the trees are "balled and burlapped" for removal to their permanent location following the first or second growth cycle or just as the first flush starts.

In California, avocado propagation has changed in recent years to a container operation. Seeds are planted in polyethylene bags and the resulting seedlings are cleft or wedge grafted about 9 cm above the seed two to four weeks after germination. Scions are taken from freshly cut terminal shoots showing strong plump buds. The plastic containers, with the grafts, are placed on raised benches in plastic-covered houses, which facilitates sanitation procedures to avoid *P. cinnamomi.* After four to six weeks the grafts are moved to a 50 percent shade house to harden, then transplanted into large containers for moving to an outdoor area (*184*).

In Florida, side grafting on young, succulent West Indian seedlings is used (*151*). The seedlings, grown in gallon containers, are grafted when they are 15 to 25 cm (6 to 10 in.) high and 6 to 10 mm (¼ to ⅜ in.) in diameter. The scions are shoot terminals, 5 to 7.5 cm (2 to 3 in.) long, with a plump terminal bud, taken just as it resumes growth.

Semihardwood stem cuttings from mature trees of the Mexican race can be rooted under intermittent mist using bottom heat at 25°C (77°F), provided seven or more leaves are retained on a 20 cm (8 in.) cutting. If the leaves drop, rooting ceases (*197*). It is difficult to root cuttings taken from mature trees of the Guatemalan or West Indian race, but they can be rooted if the basal portion of the leafy shoot to be made into the cutting is etiolated, that is, allowed to grow only in complete darkness. The terminal portion of such a shoot develops in the light until three to five leaves have formed. Those shoots with etiolated bases can then be detached and rooted in a propagating case (*67, 68*). Use of rooting hormones has not been beneficial. Avocados started as rooted cuttings eventually make satisfactory trees but, in general, grow poorly in the initial stages.

Rootstocks for Avocado

Mexican race (P. americana var. *drymifolia).* Seedlings of this race are preferred in California for their cold-hardiness and their partial resistance to *Phytophthora cinnamomi,* lime-induced chlorosis, and *Dothiorella* and *Verticillium.* They are, however, susceptible to injury from high salinity. In Florida, where seedlings of large diameter are preferred for grafting, the Mexican types are little used, owing to their thin shoots.

Guatemalan race (P. americana). These are occasionally used in California when there is a scarcity of Mexican seeds. Guatemalan seedlings are often initially more vigorous than Mexican seedlings but are more susceptible to diseases and to injury from cold.

West Indian race (P. americana). These seedlings are too liable to frost injury to be used commercially under California conditions but are widely used in Florida. The large seed produces a pencil-size shoot suitable for side grafting in two to four weeks after germination.

Banana (*Musa* spp.**) (***212, 216***).** The banana "tree" is a large perennial herb. The "stem" consists of compressed, curved leaf stalk bases arranged spirally in strips. The bases of the leaf stalks are attached to the true stem, a rhizome (a horizontal, underground stem that develops into a so-called "corm" structure). New "suckers" grow from buds on the corm and soon develop their own roots and a base as large as the parent plant, which dies and deteriorates shortly after the fruit bunch is harvested. A banana plant may live for a considerable time but it is really a succession of new plants, each arising as a sucker from a rhizomatous bud; any given sucker fruits only once, then dies.

Since the edible types rarely produce seeds, commercial propagation of the banana is asexual, consisting of division of the rhizome and replanting of the pieces or the suckers. A large rhizome is cut into pieces weighing 3 to 4.5 kg (7 to 10 lb), depending on the cultivar, which are termed "heads." Each head should contain at least two buds capable of growing into suckers. Each sucker produces two branches in the first crop. Fairly large "sword" suckers 0.9 to 1.8 m (3 to 6 ft) high with well-developed roots also are used, but the leaves must be shortened considerably to reduce water loss after the sucker is cut from the parent plant. These suckers are removed with a sharp cutting tool inserted vertically about half way between the parent stalk and stem of the sucker. These sword suckers produce only one bunch of fruit in the first crop, but they are often preferred, because of the large size of the bunch.

To propagate planting material from sources free of fusarium wilt, in vitro meristem cultures can be used starting with explants obtained from a decapitated shoot apex from "clean" banana suckers. In one year about 1 million pathogen-free plantlets can be produced for commercial plantings (*118*).

Blackberry (*Rubus* subgenus *Eubatus*) (*125*).   The upright type of blackberry produces suckers readily. Its propagation consists of digging up suckers with attached root pieces during early spring and replanting them in a new location. The suckers are often grown an additional year in the nursery to develop stronger plants before being set out in a new planting.

The trailing type of blackberry, such as the youngberry, boysenberry, loganberry, or dewberry, does not produce many suckers. This type is usually propagated by tip layering.

All blackberries can be propagated by root cuttings, but some thornless forms, such as the thornless youngberry and 'Evergreen Thornless', revert to the thorny type if propagated in this manner.

Both the upright and the trailing types of blackberries may be started easily by conventional stem cuttings, by one-node stem cuttings (*272*), or by leaf-bud cuttings taken from leafy shoots and rooted under high humidity, particularly in mist-propagating beds. Treatment of such cuttings with root-promoting chemicals is beneficial. Micropropagation methods also can be used for rapid, large-scale production of new plants (*19, 21*).

Blueberry, Highbush (*Vaccinium corymbosum* L. and *V. australe* Small) (*38, 54, 55, 209*).   Dormant hardwood stem cuttings are used. The blueberry can also be started by leaf-bud cuttings. Most plants are grown in the nursery for one year after the year of rooting and then sold as two-year plants for setting in their permanent location. Micropropagation of some highbush blueberry cultivars has been successful (*34*).

Blueberry cultivars can be propagated by T-budding in midsummer to seedling plants or rooted cuttings. The highbush blueberry is successfully worked onto the rabbiteye blueberry (*V. ashei*) as a rootstock to take advantage of the latter's wide soil adaptability and vigor (*70*). In propagating blueberries one should obtain their propagating material from virus-tested plants.

Hardwood Cuttings.   The blueberry is not easy to propagate by hardwood cuttings, but good results can be obtained. Special stock blocks with plants free of viruses and mycoplasma-like diseases should be used for the cutting material rather than production fields, which may be highly diseased (*55*). Cutting material—consisting of vigorous, firm, pencil-size, unbranched shoots of the previous season's growth—should be taken from dormant plants in winter and placed in cold storage until early spring when the cuttings are made and planted. Only vegetative wood, without fruit buds, should be used. The cuttings should contain three or four buds and be 10 to 13 cm (4 to 5 in.) long. A polyethylene-covered frame in a lathhouse is a suitable rooting structure. Bottom heat is helpful, but root-promoting chemicals have generally failed to improve rooting. A mixture of half sand and half ground sphagnum peat moss is a satisfactory rooting medium.

Cuttings should be spaced about 5 cm (2 in.) apart in the rooting bed and set with the top bud just showing. When leaves appear, the frame should be raised slightly to allow for ventilation. Either mist or frequent watering to maintain a high humidity is required. Roots start to form in about two months.

Blueberry, Lowbush (*Vaccinium angustifolium* Ait.) (*82*).   This is probably best propagated by leafy softwood cuttings under intermittent mist with bottom heat, using sand and peat moss (1:1) as a rooting medium (*129*). Cuttings taken in late spring and early summer from actively growing shoots root well, some clones giving almost 100 percent rooting. Transfer rooted cuttings to peat pots for further growth and overwintering. Cuttings made from rhizomes also can be rooted. These are best taken in early spring or late summer and fall, avoiding the midsummer rest period of the rhizome buds.

Seed propagation is sometimes used. Seeds are removed from ripe berries, then spread over a well-drained, acid-type soil mix containing one-third peat moss. Cover with a layer of finely ground sphagnum moss and keep it moist until the seeds germinate, usually in three to four weeks. Seedlings 2 cm (¾ in.) tall are transferred to peat pots.

Blueberry, Rabbiteye (*Vaccinium ashei* Reade).   This can be propagated by hardwood cuttings as well as by leafy softwood cuttings taken in midsummer and rooted under mist with a 1 percent indolebutyric acid in talc treatment. Added lights to give a 16-hour daylength may improve root production (*40*). Rabbiteye blueberry plants also can be produced by micropropagation (*152*).

Butternut (*Juglans cinerea* L.).   There are several butternut cultivars and these are propagated by grafting onto *Juglans nigra* seedlings, using a form of the bark graft or any method successful with the other nut species. Excessive sap ''bleeding'' is a problem in graft union healing and necessary steps must be taken to overcome it.

Cacao (*Theobroma cacao* L.) (*245*).   Almost all commercial cacao plantings consist of seedling trees, which are highly variable. Cacao can be propagated vegetatively, however, and with the development of superior clones and the use of modern techniques for rooting cuttings, it is probable that more vegetatively propagated plantings will be made (*58*).

In the large cacao-producing areas of West Africa and South America, emphasis is on seedling propagation with seed taken from selected clones. Vegetative propagation is used to produce seed parents. Cross-pollination between high-yielding clones and use of the resultant seeds to produce bearing trees is an important propagation method.

Seedling Propagation.   Freshly harvested mature seeds should be planted immediately. Cacao seeds quickly deteriorate after harvesting, normally losing all capacity to germinate within a week after removal from the pod. Prevention of drying plus storage at 24 to 29°C (75 to 85°F) prolongs seed life (*10*).

A common practice is to plant three or four seeds in a hole. If they all germinate they are allowed to grow, the plants being treated as branches of one tree. Another method is to start the seedlings in a nursery bed and transplant the small trees to their permanent location. Alternatively, the seedlings may be started in polyethylene bags, baskets, bamboo or paper cylinders, or clay pots, from which they are removed later and planted. A germination temperature of about 26°C (80°F) should be used.

Asexual Propagation.   After young seedling trees have attained sufficient size, they can be used as rootstocks on which superior clones are budded. The patch bud is most successful, but T-budding may be used (*57*). In areas where unfavorable soil conditions occur, superior clones are sometimes grafted on resistant clonal rootstocks. In the world industry, however, budding and grafting are done on a negligible scale. Air layering is quite successful.

Carob (*Ceratonia siliqua* L.)   This subtropical evergreen tree is usually propagated by seeds, which germinate without difficulty when freshly harvested. If seeds

dry out and the seed coats are hard, they should be softened by hot water or sulfuric acid treatment (see Chap. 7). Transplanting of bare root seedlings gives poor results so the seeds are best planted in their permanent location or started in containers for later transplanting. Seedlings are best budded to selected cultivars; this is most successful in late spring. Cuttings can be rooted if taken in mid-spring and treated with indolebutyric acid at about 7500 ppm. Air layering in late summer is successful.

Micropropagation using explants from both seedling and mature trees has been done successfully (*211*).

Cashew (*Anacardium occidentale* L.) (*6, 166*).   This tender tropical evergreen tree, grown principally in India, is usually propagated by seed; two or three are planted directly in place in the orchard, since seedlings transplant with difficulty. They are later thinned to one tree per location. There is no seed dormancy but seeds should be tested for the presence of embryos by placing in water; those that float should be discarded. Germination takes place in 15 to 20 days. For transplanting, the seeds are started in some type of container which will disintegrate, and the container—with the seedling—is set in the ground. Cashew seedlings show great variability in growth habit, yield, and nut quality. There are few named cultivars but efforts are being made to select superior, high-yielding types and propagate them by asexual methods, particularly by stooling (*171*), approach-grafting (*194*), and air layering (*175*). Other methods, such as T-budding, patch budding, and rooting leafy cuttings, also have been used successfully in limited trials.

Cherimoya (*Annona cherimola* Mill.) (*72*).   This is propagated by cleft grafting or T-budding selected cultivars on seedlings of cherimoya or the related sugar apple (*Annona squamosa*), which gives a dwarf plant, or on custard apple (*Annona* hybrids). The latter two species should not be used as rootstocks in cold areas or in poorly drained soils where they are subject to root rot. Cuttings or air layers have been difficult to root.

Some seedling forms developed in Mexico and South America come nearly true from seed, and in many regions seed propagation is used exclusively. The seeds retain their viability for many years if kept dry, and germinate in a few weeks after planting. After the young seedlings are 7.5 to 10 cm (3 to 4 in.) high they should be transferred from flats to small pots, and when about 20 cm (8 in.) high, to larger pots or to open ground.

Cherry (*Prunus avium* L., *P. cerasus* L.).   Cherry trees are propagated by T-budding or chip budding the desired cultivar on a seedling rootstock. Rootstock seedlings are often grown closely planted in a seed bed for one year, then are lined-out about 10 cm (4 in.) apart in the nursery and grown a second year before budding. If growing conditions are good, seeds may be planted directly in the nursery in early spring, and the seedlings will be large enough for budding by late summer or early fall.

Rootstocks for Cherry (183). The two most common stocks are Mazzard (*Prunus avium*) and Mahaleb (*P. mahaleb*) seedlings. The vegetatively propagated 'F 12/1' and the dwarfing 'Colt' have also been used. All these rootstocks are used for sweet cherry (*P. avium*) cultivars. Sour cherry (*P. cerasus*) and Duke cherry (hybrids between sweet and sour cherries) cultivars are propagated on either Mahaleb or Mazzard roots. All these cherry rootstocks are susceptible to verticillium wilt. *P. avium* is moderately resistant to oak root fungus, but *P. mahaleb* is susceptible.

Mazzard (P. avium). Mazzard seedlings used in the United States are available from sources in Oregon and Washington. There is considerable variation among the sources of Mazzard seeds, some undoubtedly better than others. Seeds from trees indexed for freedom from ring spot virus should be used if possible. In germinating Mazzard seeds it is beneficial to soak them in water changed daily for about eight days prior to stratification. A warm, moist stratification period at 21°C (70°F) for four to six weeks prior to cold stratification should improve germination. Finally a cold (4°C; 40°F) stratification period of 150 days is used. When a fair percentage of the seeds show cracking of the endocarp with root tips emerging, they should be removed and planted (*139*).

The clonal Mazzard stock, 'Malling 12/1', developed at the East Malling Research Station, produces trees that are vigorous, uniform, and resistant to bacterial canker. A virus-free clone is available. Propagation is by trench or mound layering. It is widely used in the United Kingdom, Australia, New Zealand, and South Africa. In Oregon this stock has worked well for the root, trunk, and primary scaffold system onto which the scion cultivar is worked.

Sweet cherry cultivars make an excellent graft union with Mazzard roots, giving trees that are vigorous and long-lived, but may be too large for economical harvest. Mazzard roots are not particularly suitable for heavy, poorly aerated, wet soils, but will tolerate such conditions better than Mahaleb. Under dry, unirrigated, drought conditions Mahaleb is more likely to survive than Mazzard, presumably because of the deep, vertical rooting habit of Mahaleb in contrast to Mazzard's shallow, horizontal root system.

Mazzard roots are immune to one species of root-knot nematode (*Meloidogyne incognita*) and resistant to *M. javanica,* but susceptible to the root-lesion nematode (*Pratylenchus vulnus*).

Mahaleb (P. mahaleb). This is a very heterogeneous species with considerable variability in many characteristics. The seeds should be soaked in water for 24 hours, then stratified for about 100 days at 4°C (40°F). Leafy Mahaleb cuttings are easily rooted under intermittent mist if treated with indolebutyric acid, thus permitting the establishment of clonal Mahaleb stocks. This is the principal rootstock in the United States for 'Montmorency', the leading sour cherry cultivar. Mahaleb roots may produce a somewhat dwarfed tree, particularly if the budding is done high, e.g., 38 to 50 cm (15 to 20 in.) on the rootstock; there is evidence, however, that in good soils the trees will become as large as those on Mazzard roots. Trees on Mahaleb roots are resistant to the buckskin virus. Mahaleb-rooted trees do not grow satisfactorily in heavy, wet soils with high water tables, being susceptible to *Phytophthora* root rot. This rootstock should be used for nonirrigation or drought conditions. Trees on Mahaleb roots are more cold-hardy than those on Mazzard. Sweet cherries often grow faster for the first few years on Mahaleb than on Mazzard roots, and some cultivars start heavy bearing rather early, which may result in some dwarfing of the tree. There is evidence, especially in England, that trees on Mahaleb roots are relatively short-lived, but in the United States good, productive trees known to be more than 50 years old are grown on this stock. Mahaleb roots are more resistant to root-lesion nematode (*Pratylenchus vulnus*) than Mazzard. They are also resistant to one species of the root-knot nematode (*Meloidogyne incognita*) but susceptible to *M. javanica.* Mahaleb roots are more resistant to bacterial canker than Mazzard roots.

'Colt' (*Prunus avium* × *P. pseudocerasus*) is a dwarfing clonal cherry rootstock developed in 1958 at the East Malling Research Station. It can be used for both sweet and sour cherries and is easily propagated by cuttings. A virus-tested strain, EMLA Colt, was released in 1977.

Chestnut spp. (*Castanea mollissima Blume*). (*123, 124*). The Chinese chestnut is resistant to the blight fungus, *Endothia parasitica*, which killed most American chestnut (*C. dentata*) trees in eastern United States upon its introduction from China in the late 1800s.

C. mollissima is usually propagated by seed, many plantings consisting of seedling trees. Clonal selections have been made, however, and named cultivars are available. Seeds should be prevented from drying. The nuts are gathered as soon as they drop and either planted in the fall or kept in moist storage one or two months at 0 to 2°C (32 to 40°F) for spring planting. Seed nuts are satisfactorily stored in tight tin cans, with one or two very small holes for ventilation, at 0°C (32°F) or slightly higher; this storage temperature also aids in overcoming embryo dormancy (*158*). Weevils in the nuts, which will destroy the embryo, can be killed by hot-water treatment at 49°C (120°F) for 30 minutes. After one year's growth, the seedlings should be large enough to transplant to their permanent location or be grafted to the desired cultivar. Although the chestnut is difficult to graft or bud, splice or bark grafting and inverted T-budding have given good results. In regular T-budding, the buds tend to "drown," owing to excessive bleeding. Only *C. mollissima* seedlings should be used as rootstocks. The Chinese chestnut can be propagated by rooting leafy cuttings under mist if treated with IBA (*117*).

The well-known edible European (Spanish) sweet chestnut (*C. sativa* Mill.) is imported in large quantities into the United States from Europe. It is usually propagated by nursery grafting but can just as easily and faster be propagated by chip budding (*111*).

A chestnut hybrid (*C. sativa* × *C. castanea*) has been propagated in vitro, taking explants from mature trees (*248*).

Citrus (*Citrus* spp.) (*186, 195, 198, 201, 270*). Propagation methods are the same for all species of citrus. Members of this genus are readily intergrafted and can also be grafted to other closely related genera such as *Fortunella* (kumquat) and *Poncirus* (trifoliate orange). Citrus cultivars are propagated commercially by T-budding on seedling rootstocks. In all types of citrus propagation it is very important to use true-to-type material free of transmissible pathogens (*134*). Citrus propagation has changed but little in the past century. There are some new practices, however, such as container production, more fumigation of nursery soils, use of plastic tape in budding, and more attention paid to controlling viruses during propagation (*267*).

Many citrus species can be propagated by rooting leafy cuttings, or by leaf-bud cuttings, although nursery trees are not commonly propagated in this manner (*46*). Except for the psorosis virus, transmissible diseases do not appear in seedlings unless they become infected from insect vectors or by budding.

In Florida the Persian lime (*C. aurantifolia*) is propagated to some extent by air layering, as is the pummelo (*C. grande*) in southeast Asian countries.

Probably the most rapid method of obtaining a citrus cultivar worked on a given rootstock is the use of "cutting-grafts" to produce dwarf plants. (*49*).

Growing Citrus Nursery Stock—Field Production. It is important to avoid using soils infested with citrus nematodes (*Tylenchulus semipenetrans*) or burrowing nematodes (*Radopholus similis*), or soilborne diseases, although citrus is resistant to verticillium wilt. For nurseries it is preferable to use virgin soil or at least a soil that has not been formerly planted to citrus. For small operations, raised seedbeds enclosed by 12-in. boards can be used. A soil mixture of ¾ sandy loam and ¼ peat moss is satisfactory. Treating the soil with a fumigant such as DD (dichloropropane–dichloropropene) at the rate of 700 to 1000 lb per acre will minimize the chances of nematode infestations. To reduce the hazard of fungus infection, methyl bromide is often used to fumigate the seedbed and nursery site. Following treatment, planting should be delayed for six to ten weeks to allow the fumigant to dissipate. Some soils in California, however, have remained toxic to citrus for a year following such treatment. The seedbed should be in a lathhouse, or some other provision should be made for screening the young seedlings from the full sun.

Since there is considerable variation in the performance of seedlings taken from different trees, it is best to select the seeds from healthy, virus-tested old trees, known to produce vigorous, uniform seedlings that develop into satisfactory orchard trees after being budded to the desired cultivar.

Citrus seeds generally have no dormancy but are injured by being allowed to dry; they may be planted immediately after being extracted from the ripe fruit. Certain species, such as the trifoliate orange or its hybrids, mature their fruits in the fall. If the seeds are to be planted at that time, they should be held in moist storage at −1 to 4°C (30 to 40°F) for at least four weeks before planting. Trifoliate orange seedlings, if grown during times of the year having short days, will respond strongly with increased growth when supplementary light is provided to lengthen the day. The same is true for seedlings of certain citrange and sweet orange cultivars (*250*).

Seeds may be stored in polyethylene bags at a low temperature of (4°C; 40°F); before storage they should be soaked for ten minutes in water at 49°C (120°F) to aid in eliminating seed-borne diseases. Treatment of the seed with a fungicide, such as thiram, also is beneficial.

The best time to plant the seeds is in the spring after the soil has warmed (above 15°C; 60°F). Seeds should be planted in rows 5 to 7.5 cm (2 to 3 in.) apart, and 2.5 cm (1 in.) apart in the row. They are pressed lightly in the soil and covered with a 2 cm (¾ in.) layer of clean, sharp river sand. The sand prevents crusting and aids in the control of "damping off" fungi. The soil should be kept moist at all times until the seedlings emerge. Either extreme, allowing the soil to become dry and baked or overly wet, should be avoided. Electric soil-heating cables placed below the seedbed to maintain a temperature of 27 to 29°C (80 to 85°F) will hasten germination. By this method, seeds may be planted in the winter months, and the seedlings will be large enough to line-out in the nursery in the spring. Many can be budded by fall or the following spring. This often shortens the propagation time by 6 to 12 months.

After the seedlings are 20 to 30 cm (8 to 12 in.) tall, they are ready to be transplanted from the seedbed to the nursery row, preferably in the spring after danger of frost has passed. The seedlings are dug with a spading fork after the soil has been wet thoroughly to a depth of 45 cm (18 in.). They can then be loosened and removed with little danger of root injury. All stunted or off-type seedlings or those with crooked, misshapen roots should be discarded.

The nursery site should be in a frost-free, weed-free location on a medium-textured, well-drained soil at least 61 cm (24 in.) deep and with irrigation water available. Old citrus soils should be avoided unless heavily fumigated with DD or methyl bromide before planting. The seedlings should be planted at the same depth as in the seed bed and spaced 25 to 30 cm (10 to 12 in.) apart in 0.9 to 1.2 m (3 to 4 ft) rows.

Citrus seedlings are usually budded in the fall in Florida and California, starting mid-September, early enough so that warm weather will insure a good bud union, yet late enough so that bud growth does not start and the wound callus does not grow over the bud itself.

Growing Citrus Nursery Stock—Container Production (*1*). Citrus seeds are mechanically extracted and washed. They are then germinated in special elongated square plastic pots in a greenhouse. After three or four months the seedling rootstock plants are large enough for inverted T-budding; finished budded trees can be developed in about 12 months grown in plastic containers (*26*). Sometimes cleft grafting is used on the young seedlings after they are growing in their final containers (*161*). Citrus nursery trees apparently grow well in containers and develop into good orchard trees upon field planting providing they are not held in the container so long that root binding occurs (*167*).

Most commercial citrus-producing areas of the world have programs to determine the presence of virus and virus-like diseases in trees used as budwood sources (*134*). There are about 12 known viruses, a viroid causing "exocortis," and a mycoplasmalike organism causing "stubborn" disease, which can infect citrus. Good indexing procedures have been worked out for most of them (*20, 198*). Budwood should be taken only from known high-producing, disease-free trees. It is desirable to select the budsticks from a single tree, avoiding any off-type "sporting" branches. The

best type of budwood is that next to the last flush of growth, or the last flush after the growth hardens. A round budstick gives more good buds than an angular one. The best buds are those in the axils of large leaves. The budsticks are usually cut at the time of budding, the leaves removed, and the budsticks protected against drying. Budsticks may be stored for several weeks if kept moist and held under refrigeration at 4 to 13°C (40 to 55°F).

The T-bud method is satisfactory for citrus. The bud piece is cut to include a sliver of wood beneath the bud. Fall buds are unwrapped in six to eight weeks after budding, spring buds in about three weeks. In California and Texas the buds are usually inserted at a height of 30 to 45 cm (12 to 18 in.) but in Florida the buds are inserted very low on the stock—2.5 to 5 cm (1 to 2 in.) above the soil. Such low budding is often necessitated by profuse branching in rough lemon and sour orange seedlings, which is caused by partial defoliation due to scab and anthracnose spot.

Micro-budding is used to some extent for citrus, particularly in Australia. (See Chap. 13.)

Buds inserted in the fall are forced into growth in the spring by "lopping" the top of the seedlings 5 to 7.5 cm (2 to 3 in.) above the bud. This is done just before spring growth starts, and consists of partly severing the top, allowing it to fall over on the ground. The top thus continues to nourish the seedling roots, but the bud is forced into growth. Lopping of spring buds is done when the bud wraps are removed—about three weeks after budding. If possible, the "lops" should be left until late summer, at which time they are cut off just above the bud union. Although lopping is satisfactory, it may make irrigation and cultivation difficult. An alternative practice is to first cut the seedling completely off 30.5 to 35.5 cm (12 to 14 in.) above the bud, then later cut it back immediately above the bud. However, this practice does not force the bud as well as lopping or cutting the seedling just above the bud.

Young citrus nursery trees may be dug "balled and burlapped" or bare-root. Bare-rooted trees should have the tops pruned back severely before digging. Transplanting of such trees is best done in early spring, but balled trees can be moved any time during spring before hot weather starts.

Polyembryony is high in seeds of most citrus species used as rootstocks. The sexual seedlings are often weak and make a poor rootstock so are usually rogued out. The other seedlings arising from the nucellus are apomictic and have the same characteristics as the seed-bearing plant. Consequently they are uniform, and make good rootstocks if the parent tree is desirable. If used as orchard trees, nucellar seedlings are vigorous, thorny, and upright-growing, but slow to start bearing. These undesirable qualities (thorniness and delayed bearing) are less pronounced in nursery trees propagated from budwood taken from the upper part of old nucellar seedling trees. Nucellar cultivars have been developed for all the commercial citrus cultivars, and are used because of their increased vigor, tree size, and yields, and their freedom from viruses (23) (see Chap. 8).

Rootstocks for Citrus (25, 46, 105, 252, 267)

Sweet orange (C. sinensis). This is a good rootstock for all citrus cultivars, producing large, vigorous trees resistant to tristeza. Sweet orange is adapted to well-drained, light- to medium-loam soils, but owing to its susceptibility to gummosis (*Phytophthora* spp.), it is not suited to poorly drained, heavy soils. It produces standard-size fruits that are thin-skinned, juicy, and of fairly high quality. The seeds germinate readily, but the seedlings, which are 70 to 90 percent nucellar, are relatively slow growers and tend to produce low-branched, bushy trunks in the nursery. It is not widely used today principally because of its susceptibility to *Phytophthora.*

Sour orange (C. aurantium). This is an excellent stock for most citrus species, owing to its vigor, hardiness, deep root system, resistance to gummosis diseases, and to the high quality, smooth, thin-skinned, and juicy fruit produced by cultivars worked on it. However, it lacks tolerance to "tristeza" (quick decline), due to a virus disease that is transmitted by an insect vector or from infected budwood (13, 84). The scion top itself may be quite tolerant to the virus, but in combination with sour orange root the stock is affected, owing to death of the phloem tissues in the bud union area, which then results in starvation of the root. Grapefruit, mandarin, and orange (but not lemon) cultivars on sour orange are subject to tristeza. In California sour orange is no longer recommended as a rootstock for orange or grapefruit cultivars. In Florida, too, the use of sour orange stock has been declining, due to tristeza. In Texas, Mexico, Cuba, Venezuela, Honduras, Sicily, and Israel sour orange rootstock is widely used. In Australia strains of sour orange—originating in Israel—have been tested that are tolerant to tristeza (233).

Rough lemon (C. jambhiri). This stock is well adapted to warm, humid areas with deep sandy soils. It has an extensive root system, which makes it very drought tolerant. Trees on this rootstock outyield those on other stocks, although the fruit produced is of lower quality. Rough lemon formerly was widely used in Florida, but due to blight, has lost its popularity there. Its use in California is restricted to the lighter soil areas, particularly in the desert regions. This is the ma-

jor citrus rootstock in Arizona. Its susceptibility to *Phytophthora* precludes its use on heavy soils. Both trees and fruit are more susceptible to cold injury on this stock than on most other commercial stocks (71). Fruits from trees on rough lemon roots are early maturing, but have a thick rind, are low in both sugar and acid, and are often coarse-textured compared with fruit from trees on other rootstocks.

Rough lemon produces numerous seeds that germinate well. Ninety to 100 percent of the seedlings are nucellar. They grow upright with single, unbranched trunks that are easy to bud and handle in the nursery. Rough lemon also can be easily propagated by cuttings. Sweet orange cultivars are resistant to quick decline when worked on rough lemon roots. The chief advantages of this stock are its vigor and its ability to produce a bearing tree quickly, particularly on light, sandy soils. Some orange cultivars on rough lemon develop an incompatible bud union.

Trifoliate orange (Poncirus trifoliata). This dwarfing citrus rootstock has been used to some extent for many years. It is the main citrus rootstock in Japan and it is used in China as a stock for Satsumas and kumquats. In northern Florida and along the Gulf coast to Louisiana, it has long been used as a stock for Satsuma oranges and kumquats, for which it is excellent. *Poncirus trifoliata,* as a rootstock, does best on medium-textured soils. It will tolerate heavy soil but grows quite slowly. On light, sandy soil its growth may be so poor that trees on this stock are worthless. Trifoliate orange is commonly used as a stock for ornamental citrus and in home orchards for dwarfed trees. Trees on this stock yield heavily and produce high-quality fruits. Trifoliate orange is a deciduous species noted for its winter-hardiness and its *Phytophthora* and nematode resistance. Both the tree and its fruits, when worked on this stock, are more resistant to cold than other rootstock-top combinations, making it particularly adaptable to the colder citrus-growing regions. A citrus cultivar worked on trifoliate orange is one of the few examples of an evergreen top on a deciduous rootstock.

Trees on trifoliate roots are often affected by exocortis, or "scaly butt." This may be avoided by using uncontaminated nucellar buds or by taking buds from trees on trifoliate stock that are free of the disease. This stock is quite resistant to gummosis, but it is very susceptible to citrus canker (*Phytomonas citri*) and moderately so to scab. It is susceptible to lime-induced chlorosis but tolerant of excess boron.

Trifoliate orange fruits produce large numbers of plump seeds that germinate easily. The upright-growing, thorny seedlings, about 60 percent of which are nucellar, are easy to bud and handle in the nursery, but their slow growth often necessitates an extra year in the nursery before salable trees are produced.

'Cleopatra' mandarin (C. reshni). This is a good rootstock, particularly in Florida, as a stock for oranges, grapefruit, and other mandarin types, and has come into use in California and Texas as a replacement for sour orange rootstock. Its resistance to gummosis, comparative salt tolerance, and resistance to tristeza seemingly justify its greater use. In addition, trees on this stock show good yields of high-quality fruit, although fruit size is somewhat smaller than average. Its chief disadvantages are the slow growth of the seedlings, slowness in coming into bearing, and susceptibility to *Phytophthora parasitica* root rot. About 80 percent of the seedlings are nucellar. Some cases of incompatibility, especially with 'Eureka' lemon, have appeared.

Citranges (Trifoliate orange × sweet orange hybrids). There are several named cultivars—'Savage', 'Morton', 'Troyer', 'Carrizo', and so on—some of which are useful as rootstocks both in California and Florida.

'Savage' is especially suitable as a dwarfing stock for grapefruit and is also satisfactory for the mandarins. It is resistant to gummosis, and trees on this stock are more cold-hardy than those on many of the other rootstocks (*71*).

'Morton' citrange is not particularly dwarfing. Trees worked on it are heavy producers of excellent-quality fruit. It produces so few seeds, however, that to get quantities of nursery trees started is difficult. In addition, orange cultivars on this stock appear to be susceptible to quick decline.

Sweet orange on 'Troyer' citrange is vigorous, cold-hardy, and resistant to gummosis, with high-quality fruit. This stock is more widely used than any other for oranges in California. In replanting citrus on old citrus soils, trees on the 'Troyer' citrange have shown outstanding vigor. 'Troyer' itself is relatively fruitful and produces 15 to 20 plump seeds per fruit, facilitating the propagation of nursery trees. 'Troyer' and 'Carrizo' citrange have proven best for 'Valencia' oranges in Australia on sandy and sandy loam soils (*240*).

Seedlings of citrange cultivars are mostly nucellar and develop strong, single trunks, easily handled in the nursery. As with trifoliate orange, only exocortis-free buds should be used on citrange rootstocks; otherwise, dwarfing—and eventual low production—will result. 'Troyer' and 'Carrizo' are the two commercially most important citrus rootstocks in California and Arizona. 'Carrizo' is an important rootstock in Florida.

Rangpur lime (C. aurantifolia × C. reticulata). This is the most widely used citrus rootstock in Brazil. It produces vigorous, fruitful trees, which are resistant to tristeza but highly susceptible to exocortis. In Texas it has been more salt-tolerant than other citrus rootstocks. Some strains of Rangpur lime are susceptible to *Phytophthora*.

Alemow (Citrus macrophylla). This is used in California as a rootstock for lemons in high-boron areas, owing to its boron tolerance. It is susceptible to tristeza when sweet orange scion cultivars are used and to a rootstock necrosis first detected about 1960.

Coconut (*Cocos nucifera* L.) (*31, 189*). Trees are propagated only by seed, and there are some named cultivars that maintain their characteristics quite dependably by seed propagation. It is important to select seeds from trees that produce large crops of high-quality nuts.

The nuts are usually germinated in seedbeds. Seedlings that develop rapidly and have strong, vigorous shoots are selected. The nuts, still in the husk, are set at least 12 in. apart in the bed and laid on their sides with the stem end containing the "eyes" slightly raised. The sprout emerges through the eye on the side that has the longest part of the triangular hull. As soon as this occurs (about a month after planting), the sprout sends roots downward through the hull and into the soil. In 6 to 18 months the seedlings are large enough to transplant to their permanent location.

Coffee (*Coffea arabica* L.) (*259*). The most common method of propagation is by seed, preferably obtained from selected superior trees. Coffee seeds lose viability quickly and are subject to drying through the seed coverings. Seeds held at a moisture content of 40 to 50 percent and at 4 to 10°C (40 to 50°F) will keep for several months. There are no dormancy problems. Seeds are usually planted in seedbeds under shade (*234*). Sometimes the seedlings are started in soil in containers formed from leaves, or in polyethylene bags, to facilitate transplanting. Germination takes place in four to six weeks. When the first pair of true leaves develops, the seedlings are transplanted to the nursery and set 12 in. apart. After 12 to 18 months in the nursery, by which time they have formed six to eight pairs of laterals, the young trees are ready to set out in the plantation.

Coffee can be propagated asexually by almost all methods, but leafy cuttings probably hold the most promise for commercial use. Cutting material should be taken only from upright shoot terminals in order to produce the desired upright-growing tree. Leafy cuttings of partially hardened wood can be rooted fairly easily, especially if treated with a root-promoting chemical. High-humidity conditions, such as in intermittent mist, must be maintained, and the cuttings should be partially shaded during rooting (*219*). Coffee plants have been propagated by the formation of adventitious embryos in aseptic callus cultures (*108*).

Crabapple.   Siberian crabapple (*Malus baccata* Borkh.), Western crabapple (*M. ioensis* Britt.), and other *Malus* species. The usual propagation method is to bud or graft the desired cultivars on seedling rootstocks, either one of the crabapple species or the common apple, *Malus* × *domestica*. In areas where winter-hardiness is important, *M. baccata* seedlings should be used.

Crabapples can be propagated, although with difficulty, by softwood or hardwood cuttings, especially if root-promoting chemicals are used.

Cranberry (*Vaccinium macrocarpon* Ait.) (*42, 81*). This vine type of evergreen plant produces trailing runners upon which are numerous short upright branches. Propagation is by cuttings made from either runners or upright branches. Cutting material is obtained by mowing the vines in early spring before new growth has started. The cuttings are then set directly in place in their permanent location without previous rooting at distances of 6 to 18 in. apart each way. Two to four cuttings are set in sand in each "hill." The cuttings are 13 to 25 cm (5 to 10 in.) long and set deep enough so that only an inch is above ground. A more rapid method of starting a cranberry bog is to scatter the cuttings over the ground and work them into the soil with a special disk-type planter. This is justifiable when there is an abundance of cutting material and a scarcity of labor for setting the cuttings by hand. Water is applied to the bog immediately after planting. The cuttings root during the first year and make some top growth, but the plants do not start bearing until three or four years later.

Currant (*Ribes* spp.).   Currants are readily propagated by hardwood cuttings prepared from well-matured wood of the previous summer's growth. Cuttings 20 to 25 cm (8 to 10 in.) long are made in late fall, stored in moist sand, sawdust, or peat moss at about 2°C (35°F), then planted in early spring. The plants can be transplanted to their permanent location in one or two years, depending upon their growth. Currants can also be propagated by mound layering.

Date (*Phoenix dactylifera* L.) (*174, 202*).   Propagation of the date palm is either by seed or by offshoots. This is a monocotyledonous plant having no continuous cambial cylinder, so it cannot be propagated by budding or grafting. In commercial plantings, most of the

trees are female, but a few male trees are necessary for pollination. The date palm is capable of regeneration from various tissues in in vitro culture systems (*241*).

In seed propagation about half of the trees produced are males, whereas the seedling female trees produce fruits of variable and generally inferior types. The commercial grower is therefore little interested in seedling trees, preferring the superior named cultivars, even though the latter must be propagated by the vegetative offshoot method. Date seeds germinate readily.

Offshoots arise from axillary buds near the base of the tree. If they are near the ground level, they will develop roots in the soil after three to five years on the parent tree. Large, well-rooted offshoots, weighing 18 to 45 kg (40 to 100 lb), are more likely to grow than smaller ones. Offshoots higher on the trunk can be induced to root if moist rooting medium is held against the base of the offshoot by means of a box or a polyethylene tube. Unrooted offshoots arising higher on the stem can be cut off and rooted in the nursery, but in relatively low percentages.

Considerable skill is required to cut off a date offshoot properly. Soil is dug away from the rooted offshoot, but with a ball of moist earth as thick as possible attached to the roots. The connection with the parent tree should be exposed on each side by removing loose fiber and old leaf bases. A special chisel, having a blade flat on one side and beveled on the other, is used to sever the offshoot. The first cut is made to the side of the base of the offshoot close to the main trunk. The beveled side of the chisel is toward the parent tree, which gives a smooth cut on the offshoot. A single cut may be sufficient, but usually one or more cuts from each side are necessary to remove the offshoot. The offshoot should never be pried loose; it should be cut off cleanly. After removal, it should be handled carefully and replanted as soon as possible, care being taken to prevent the roots from drying out (see Fig. 14–16).

Feijoa (*Feijoa sellowiana* Berg.) (*120*).   Propagation can be done by seeds, which germinate without difficulty. They should be started in flats of soil and later transplanted to the nursery row. Cultivars can be propagated by layering or by grafting onto feijoa seedlings. A successful method has been by bench grafting using the cleft graft (*59*). Leafy softwood cuttings treated with root-promoting substances and started under closed frames can be rooted.

Fig (*Ficus carica* L.) (*36, 135*).   The fig is easily propagated by hardwood cuttings. Two- or three-year-old wood or basal parts of vigorous one-year shoots with a minimum of pith are suitable for cuttings. For best results cuttings should be prepared in early spring, well before bud break, with the bases allowed to callus for about 10 days at about 24°C (75°F) in damp wood shavings or sawdust before planting. The cuttings are grown for one or two seasons in the nursery, then transplanted to their permanent location. A common method in European countries is to plant long cuttings, 0.9 to 1.2 meters (3 to 4 ft), their full length in the ground where the tree is to be located permanently; sometimes two cuttings are set in one location to increase the chance that one will survive. The fig can be budded, using either T-buds inserted in vigorous one-year-old shoots on heavily pruned trees, or patch buds on older shoots.

Fig roots are resistant to oak root fungus (*Armillaria mellea*) and verticillium wilt but quite susceptible to both root-knot (*Meloidogyne* spp.) and lesion nematodes (*Pratylenchus vulnus*) (*36*).

Seed propagation is used only for breeding new cultivars. The small seeds can be germinated easily in flats of well-prepared soil. Fertile seeds should first be separated from the sterile ones by placing all of them in water; fertile seeds sink, but sterile ones float.

Figs can be air-layered. One-year-old branches, if layered in early spring, are usually well-rooted by midsummer. Figs can also be propagated by tissue culture methods (*188*).

Filbert (*Corylus avellana* L. and *C. maxima* Mill.) (*137, 160*). Simple layering has been the usual method of commercial filbert propagation. The suckers arising from the base of vigorous young trees four to eight years old are layered in early spring, or special stool mother plants are maintained in a bush form just for layering purposes (see Fig. 14–7). After one season's growth, a well-rooted tree 0.6 to 1.8 m (2 to 6 ft) tall may be obtained, ready to set out in the orchard. Old orchard trees are not suitable to use for layering. Sometimes suckers arising from the roots are dug and grown in the nursery row for a year or, if well-rooted, planted directly in place in the orchard.

Filberts are difficult to propagate by cuttings. Budding or grafting filbert cultivars on seedling stocks was rarely practiced, because of the difficulty in obtaining successful unions, but grafting of filberts, using the "hot grafting" procedure, as shown in Fig. 12–24, has now revolutionized filbert propagation (*113, 232*).

Filbert seeds are easily germinated but require a stratification period of several months at 0 to 4°C (32 to 40°F).

Filberts can readily be propagated by tissue culture methods, giving 90 percent or better survival of the micropropagated plantlets (*5*).

Gooseberry (*Grossularia* or *Ribes* spp.). Mound layering is used commercially. Layered shoots of American cultivars usually root well after one season. They are then cut off and transferred to the nursery row for a second season's growth before they are set out in their permanent location. The slower-rooting layers of European cultivars may have to remain attached to the parent plant for two seasons before they develop enough roots to be detached. Some cultivars, such as the 'Houghton', 'Poorman', and 'Van Fleet', can be started fairly easily by hardwood cuttings. Micropropagation of gooseberries can be easily done and the rooted shoots stored under refrigeration for as long as 130 days with 100 percent survival (*257*).

Grape (*Vitis* spp.) (*2, 251, 265*). Grapevines are propagated by seeds, cuttings, layering, budding, or grafting. New plants have been produced by several in vitro techniques, including embryoid formation and fragmented shoot tip cultures (*136*). Seeds are used only in breeding programs to produce new cultivars. Grape propagation methods have been modernized by the use of virus-indexed, "clean" planting stock, mist-propagation techniques for leafy cuttings, and rapid machine-grafting procedures. Most commercial propagation is by dormant hardwood cuttings. For types difficult to root, such as the Muscadine (*Vitis rotundifolia*), layering or the use of leafy cuttings under mist is necessary (*79*). Budding or grafting on rootstocks is used occasionally to increase vine life, plant vigor, and yield. Where noxious soil organisms, such as phylloxera (*Dactylosphaera vitifoliae*) or root-knot nematodes (*Meloidogyne* spp.), are present, and cultivars of susceptible species such as *V. vinifera* are to be grown, it is necessary to graft or bud onto a resistant rootstock. Root-knot nematodes can be eradicated from grapevine rootings by dipping them in hot water (51.5 to 54.5°C; 125 to 130°F) for five to three minutes, respectively (*142*).

Seeds. Grape seeds are not difficult to germinate. Best results with vinifera grape seeds are obtained after a moist stratification period at 0.5 to 4°C (33 to 40°F) for about 12 weeks before planting (*92*).

Dormant Cuttings. Most grape cultivars are traditionally propagated by dormant hardwood cuttings, which root readily. Cutting material should be collected during the winter from healthy, vigorous, mature vines. Well-developed current season's canes should be used; they should be of medium size and have moderately short internodes. Cuttings 8 to 13 mm (⅓ to ½ in.) in diameter and 36 to 46 cm (14 to 18 in.) long are generally used, planted in the spring deep enough to cover all but one bud. One season's growth in the nursery should produce plants large enough to transplant to the vineyard. Root-promoting auxin-type chemicals have not been particularly helpful in rooting hardwood grape cuttings.

Leafy Cuttings. Leafy greenwood grape cuttings root profusely under mist in about 10 days if given relatively high (26.5 to 29.5°C; 80 to 85°F) bottom heat and if treated with indolebutyric acid. Scarce planting stock (such as virus-indexed material) can be increased very rapidly using one-budded stem cuttings (see Fig. 18–1), then consecutively taking additional cuttings from the shoot arising from the bud on the rooted cutting, and so on (*255*).

Layers. Grape cultivars difficult to start by cuttings can be propagated by simple, trench, or mound layering (see Chap. 14).

Grafting. Bench grafting (*93*) is widely used; scions are grafted on either rooted or unrooted disbudded rootstock cuttings by the whip graft or, better, by machine grafting (see Figs. 12–22 to 12–24). The grafts are made in late winter or early spring from completely dormant scion and stock material. The stocks are cut to 30.5 to 35.5 cm (12 to 14 in.) with the lower cut just below a node and the top cut an inch or more above a node. All buds are removed from the stock to prevent subsequent suckering. Scion wood should have the same diameter as the stock.

After grafting, using a one-bud scion, the union is stapled together or wrapped with budding rubber. The grafts should be held for three to four weeks in well-aerated, moist wood shavings or peat moss at about 26.5°C (80°F) for callusing.

The callused grafts are removed from the callusing boxes or plastic bags and any roots and the scion shoot are carefully trimmed back to an 18 mm (½ in.) stub. The scion is dipped quickly into a temperature controlled container of melted (low melting point) paraffin to a depth of 2.5 cm (1 in.) below the graft union and then quickly into cool water. The paraffined bench

graft is then planted into a 5.0 × 5.0 × 25 cm (2 × 2 × 10 in.) milk carton or planting tube, which contains a mixture of perlite and pumice, or perlite and peat moss (see Fig. 18–2). The cartons or tubes are placed upright in flats. The flats are set on pallets and moved into a heated greenhouse for six to eight weeks. The growing bench grafts are then transferred to a 50 percent shade screen house for about two weeks for hardening off prior to planting in the nursery or vineyard.

The bench grafts are planted deeply so that the top of the carton or planting tube is at least 2 to 3 in. below the soil level to ensure that water will get inside the carton or tube. At no time, however, should bench grafts be planted in the vineyard with the graft union at or below ground level.

Greenwood Grafting. Greenwood grafting is a simple and rapid procedure for propagating vinifera grapes on resistant rootstocks (*91*). A one-budded greenwood scion is splice-grafted during the active growing season on new growth arising either from a one-year-old rooted cutting or from a cutting during midseason of the second year's growth.

Budding (145). A satisfactory method of establishing grape cultivars on resistant rootstocks is by field budding on rapidly growing, well-rooted cuttings that were planted in their permanent vineyard location the previous winter or spring. T-budding in late spring can be done using dormant budwood held under refrigeration. Shortly after budding, the trunks should be cut with diagonal slashes at the base to allow the ''bleeding'' to take place there rather than where the bud was inserted (*3*). In another method chip budding is performed in late summer or early fall just as soon as fresh mature buds from wood with light brown bark can be obtained and before the stock begins to go dormant. In areas where mature buds cannot be obtained early in

FIGURE 18–1 Leafy, one-node grape stem cutting rooted under mist. In the grape, roots arise readily from the internodes.

FIGURE 18–2 Steps in grafting a grape cultivar onto a resistant rootstock. *Left:* A one-budded scion grafted on an unrooted rootstock cutting using a French grafting device. *Center:* After callusing, followed by paraffining scion and graft union, the stock is planted in a tube of felt building paper (or a plastic-coated cardboard cylinder) and then set in a protected place, such as a greenhouse or lathhouse. *Right:* Several weeks later the scion bud starts growth. Graft union has healed and the rootstock cutting is well rooted, so that after seven to ten days hardening-off, the plant can be set out (still in the tube) directly in the vineyard.

the fall, growers may store under refrigeration budsticks collected in the winter and bud them in late spring or early summer.

The bud is inserted in the stock 5 to 10 cm (2 to 4 in.) above the soil level, preferably on the side adjacent to the supporting stake. It is tied in place with budding rubber but is not waxed. The bud is then covered with 13 to 25 cm (5 to 10 in.) of well-pulverized, *moist* soil to prevent drying. In areas of extremely hot summers, or in soils of low moisture, variable results are likely to be obtained, and bench or nursery grafted vines should be used. If the buds are tied with white 13

mm (½ in.) plastic tape, it is unnecessary to mound them over with soil.

Top-Grafting Grapevines. Two methods can be used for changing cultivars of mature grapevines: By one method the tops of the vines are cut off in early spring 30 to 53 cm (12 to 21 in.) below the lower wire and side whip-grafted, using a two-bud scion of the desired cultivar. The scions are wrapped with 1-in. white plastic tape and the cut surfaces are covered with grafting wax.

After the bark "slips" in late spring, the vines may be T-budded with the inverted T wrapped with white plastic tape. No grafting compound is needed. Both methods give highly successful takes.

Rootstocks for American Grapes (*115, 146, 150, 173, 187*)

'Salt Creek' (Ramsay), 'Dog Ridge', 'Champanel', 'Lukfata' (V. champini). These stocks have been very effective in the southern coastal region of the United States in increasing yields and prolonging vine life of the "bunch" type of grapes.

V. rupestris 'Constantia', 'Couderc 3309', 'V. cordifolia × V. riparia 125-1', 'Cynthiana', 'Wine King', 'Lenoir'. These stocks have been useful in the inland states of the United States in increasing yields, plant vigor, and evenness of fruit ripening.

'Couderc 3309', 'SO-4 ('Oppenheim #4'), 'Teleki 5-A', 'Couderc 1616', 'Mal gue 44-53', and 'Castel 188-15'. These stocks are used in New York State for their resistance to root parasites, especially in replant situations.

Rootstocks for Vinifera Grapes (*115, 143, 144, 173, 187*)

'St. George' (V. rupestris). This vigorous, phylloxera-resistant stock, noted for its drought tolerance, is especially suitable for shallow, nonirrigated soils; it is not resistant to root-knot nematodes. It is readily propagated by cuttings and easily grafted. Because of high vigor and low vine productiveness with this stock, it is planted less frequently than in the past. Its best use is in plant situations where phylloxera has built up in the soil.

'Ganzin No. 1' ('A × R. #1'). This phylloxera-resistant rootstock is highly recommended for fertile, irrigated soils. Vines on this stock under such conditions are generally more productive than those on *'St. George'.* It is susceptible to root-knot nematodes and does not do well on dry, hillside soils. Cuttings root easily but it does not bench graft well, as it calluses poorly.

'Couderc 1613'. This stock is the one most widely used in the interior valley grape growing region of California. It is resistant to root-knot nematodes. Although suitable for fertile, irrigated, sandy loam soils,

it produces weak, unproductive vines in nonirrigated or sandy soils of low fertility. Cuttings are easily rooted.

'Harmony'. This rootstock, developed by the USDA, is a cross between a 'Dog Ridge' seedling and a '1613' seedling. It has more vigor and more phylloxera and nematode resistance than '1613'. Cuttings root readily and it buds and grafts easily. It is recommended as a replacement for '1613', especially for table grape vineyards.

'Freedom'. This stock originated as a sister seedling of 'Harmony'. It is very vigorous, equal to 'Salt Creek'. It has the same resistance to phylloxera and nematodes as 'Harmony', but cuttings root more easily and grow more uniformly and vigorously. It buds and grafts easily. It should be used as a replacement for 'Couderc 1613' where high vigor is desired.

'Dog Ridge and 'Salt Creek' ('Ramsay') (V. champini). These closely related stocks are resistant to phylloxera and to root-knot nematodes and are extremely vigorous. They should be used only in low-fertility and sandy soils. In fertile soils, the vines are often so vigorous that they are unproductive. Cuttings of these stocks are difficult to root, especially 'Salt Creek'. On better soils, 'Salt Creek' is usually preferred to 'Dog Ridge' because the latter's extreme vigor causes poor fruit set. These two stocks have performed best for raisin and wine grape vineyards in sandy soils.

Poorly growing vinifera grapevines on infertile soil can be invigorated by inarching with rooted cuttings of one of these cultivars—'Dog Ridge' for very sandy soils or 'Salt Creek' for sandy soils.

Grapefruit. *See* Citrus.

Guava (*Psidium cattleianum* Sabine—the Cattley or strawberry guava—and *Psidium guajava* L.—the common, tropical, or lemon guava) (*151*). The Cattley guava has no named cultivars, and most nursery plants are propagated by seed. This species comes nearly true from seed, large-fruited, superior trees being used as the seed source. It is difficult to propagate by vegetative methods.

The common guava has several cultivars, but they are not grown extensively, because of difficulties in asexual propagation. Most trees of this species are propagated by seeds, which germinate easily and in high percentages, but do not come true to type. Seeds should be taken from the best type of tree available, and the flowers to produce the seeds should be self-pollinated, which will reduce seedling variability.

Seedlings are somewhat susceptible to damping-off organisms, and should be started in sterilized soil or otherwise protected by fungicides. When about 4 cm

(1½ in.) high, the seedlings should be transplanted into individual containers. In six months the plants should be about 30 cm (12 in.) high and can be transplanted to their permanent location.

For large-scale propagation of guava cultivars, grafting or budding is necessary. Chip budding has been successful, done any time during the summer using greenwood buds from selected cultivars inserted into seedling stocks about 5 mm in thickness. Plastic wrapping tape is used to cover the buds. The stock is cut off above the inserted bud after about three weeks (*121*).

Cuttings are difficult to root, best results being obtained by rooting succulent shoots under mist after treating them with indolebutyric acid (*182*).

Air layering gives good results. Simple and mound layering are effective methods of starting new plants; the layers may be tightly wrapped with wire just below the point where roots are wanted, or can be ringed, with indolebutyric acid (500 ppm) in lanolin rubbed into the cuts to promote rooting (*156*).

Hickory (*Carya ovata* [Mill] C. Koch—Shagbark hickory). Hickory cultivars are propagated by grafting or budding on seedlings of the Shagbark hickory (*C. ovata*), or of the pecan (*C. illinoensis*). However, owing to grafting and transplanting difficulties and slow growth of the trees, there is little planting of grafted trees. Most trees grown are seedlings.

To avoid transplanting failure caused by the long taproot, several stratified nuts are planted in the spring where the trees are to be located permanently. The best tree is saved and topworked, usually by bark grafting, to the desired cultivar, although standards for cultivars have not been well established. If grafted nursery trees are to be used, the taproot should be cut a year previous to digging or the tree transplanted in the nursery once or twice to force out lateral roots.

Hickory seeds will germinate when planted in the spring without previous stratification but should be kept in moist, cool storage until planting. Fall planting is successful if the soil in cold climates is well mulched to prevent excessive freezing and thawing.

Patch budding is used by nurseries in the commercial propagation of hickories, usually in late summer. The seedling stocks are grown for two years or more before they are large enough to bud.

Jujube (*Zizyphus jujuba Mill.*). Chinese date. The leading cultivars—'Lang' and 'Li'—are propagated by budding or grafting on jujube seedlings. The seeds should be stratified at about 4°C (40°F) for several months before planting. Jujubes may be propagated also by root cuttings or hardwood stem cuttings.

Kiwifruit (*Actinidia deliciosa* (*A. chev.*) C. F. Liang et A. R. Ferguson). (*22, 60, 64*).   Chinese gooseberry. Both male and female vines of this dioecious subtropical fruit must be planted to ensure fruiting. It is propagated by budding or grafting cultivars on seedling rootstocks, but it is also propagated by leafy, semihardwood cuttings under mist. Hardwood cuttings will root.

Seed.   Seed propagation can be used but the sex of the vine cannot be determined until fruiting at seven years or more. Select seed from soft, well-ripened fruit; dry and store at 5°C (41°F). After at least two weeks at this temperature, subject seed to fluctuating temperatures—10°C (50°F) night and 20°C (68°F) day—for two or three weeks before planting.

Cuttings.   Leafy semihardwood cuttings taken from apical and central parts of current season's growth in late spring to midsummer may be rooted under mist in coarse vermiculite when treated with 0.6 percent indolebutyric acid. Dipping the bases of the cuttings in captan is helpful. Hardwood cuttings, taken in midwinter and planted in a greenhouse, can be rooted if treated with a strong concentration of a rooting hormone. Plants can be started also by root cuttings.

Grafting and Budding.   Seedlings can be grafted successfully by the whip graft in late winter using dormant scion wood. T-budding in late summer is also successful. Plants grafted on seedlings are believed to be more vigorous than those started as rooted cuttings and to come into bearing sooner.

Tissue Culture.   Kiwifruit can easily be propagated by tissue culture methods, using either short meristem tips or longer apical shoots as the initial explants (*89, 229*).

Kola (*Cola nitida* Vent.).   Propagation of the kola, the tropical caffein-containing nut used to make refreshing beverages, is largely by seed planted directly in the field or in nurseries for later field planting. Seeds for planting should be harvested only when completely mature. A pronounced juvenility pattern occurs in kola; seven to nine years are required for the seedlings to flower so that asexual propagation using mature growth is essential for early production.

Rooting of terminal leafy cuttings in polyethylene-covered frames can be done but root-promoting hormones have not been successful (*246.*). Air layering is quite successful in obtaining early bearing, commercial production taking place 12 to 18 months after planting the rooted layers. IBA treatments aid in rooting the layers. Patch budding is also a successful method of vegetative propagation (*8*).

Lemon.   *See* Citrus.

Lime.   *See* Citrus.

Litchi (*Litchi chinensis* Sonn.).   Seed-produced litchi plants are inferior, so the many cultivars of this species are propagated by asexual methods, principally air layering. This can be done at any time of the year, but with best results in spring or summer (*151*). Large limbs air layer easier than small ones.

Tip cuttings from a flush of growth in the spring have been rooted in fairly high percentages under mist in the full sun. Root-promoting chemicals are beneficial. Hardwood cuttings from an active flush of new growth have rooted more readily than those from dormant hardwood cuttings.

Litchi seeds germinate in two to three weeks if planted immediately upon removal from the fruit, but they lose their viability in a few days if not planted. Seedling trees are rarely grown for their fruit; they require 10 to 15 years to start bearing, and the fruit is likely to be of low quality (*33*).

Loganberry.   *See* Blackberry.

Loquat (*Eriobotrya japonica* [Thunb.] Lindl.).   This is propagated by T-budding or side grafting superior cultivars on loquat seedlings or by top-grafting older, established trees, using the cleft graft. The quince can be used as a rootstock, producing a dwarfed tree. Air layering is successful; indolebutyric acid in lanolin at 250 ppm rubbed into the layered surface will increase rooting (*218*). For ornamental use, seedling trees are satisfactory.

Macadamia (*Macadamia integrifolia* Maiden & Betche, and *M. tetraphylla* L. Johnson) (*85, 200*).   Trees of this subtropical evergreen nut tree may be grown as seedlings, but vegetative propagation from a few superior selections by grafting is usually practiced. Macadamia is highly resistant to *Phytophthora cinnamomi* and will tolerate heavy clay soil.

Fresh seeds should be planted in the fall as soon as they mature, either directly in the nursery or in sand boxes in a lathhouse, then transplanted to the nursery or into polyethylene containers after the seedlings are 10 to 15 cm (4 to 6 in.) tall. It is important not to crack the seeds, because they are readily attacked by fungi. Only seeds that sink when placed in water should be used (*85*). Seeds retain viability for about 12 months at 4°C (40°F), but at room temperature viability starts decreasing after about four months. Scarifying or soaking the seed in hot water hastens germination.

Selected clones are propagated by one of the side graft methods. Leaves should be retained for a time on the rootstock. Rapid healing of the union is promoted if the rootstock is checked in growth prior to grafting by water or nitrogen deficiency to permit carbohydrate accumulation. Also, ringing the branches that are to be the source of the scions several weeks before they are taken increases their carbohydrate content and promotes healing of the union. Budding has generally been unsuccessful (*51*). *M. tetraphylla* is the preferred rootstock for *M. integrifolia* cultivars (*85, 231*). Macadamia can also be propagated by the cutting-graft method (see p. 385).

Macadamia can be propagated by rooting leafy, semihardwood cuttings of mature, current season's growth; 3- to 4-in. tip cuttings are best. Treatment with indolebutyric acid at 8000 to 10,000 ppm is beneficial. Cuttings should be placed in a closed propagating frame or under intermittent mist for rooting. Bottom heat at 24°C (75°F) is beneficial. There are pronounced cultivar differences in ease of rooting. *M. tetraphylla* cuttings root more readily than those of *M. integrifolia.* Rooted cuttings are not widely used in commercial plantings because of their shallow root system.

Mandarin.   *See* Citrus.

Mango (*Mangifera indica* L.) (*29, 217*).   Most plantings of this evergreen tropical fruit are seedlings, although many superior selections are maintained vegetatively. Polyembryonic cultivars commonly occur in mango. The seedlings may be sexual or nucellar in origin, either condition occurring in the seed, or both conditions occurring simultaneously. However, growth of several shoots from one seed does not necessarily indicate the presence of nucellar embryos, since in certain cultivars shoots develop from below ground, arising in the axils of the cotyledons of one embryo, which may or may not be of zygotic origin (*7*). Monoembryonic cultivars should not be propagated by seed, as they do not come true.

Mango seeds are used either to produce a true-to-type nucellar seedling of some superior clone or a rootstock on which the desired clone is budded or grafted. The seeds should be planted as soon as they are mature, although they can be stored in the fruits or in polyethylene bags at about 21°C (70°F) for at least two months. Low-temperature (below 10°C; 50°F) storage and excessive drying should be avoided. Removing the tough endocarp that surrounds the seed, followed by planting in a sterilized medium, should result in good germination in two to three weeks. Soon after the seedlings start to grow, they should be transplanted to pots or the nursery.

Mangos are commonly propagated in Florida by a type of veneer grafting or by chip budding. A week after budding, the stock is cut off two to three nodes above the bud, with a final removal to the bud when the bud shoot is 7.5 to 10 cm (3 to 4 in.) long. The best budwood is prepared from hardened terminal growth 6 to 10 mm (¼ to ⅜ in.) in diameter. The leaves are removed, with the exception of two or three terminal ones. The buds swell in two to three weeks, and are then ready to use. If the buds are to be used on stocks older than three weeks, ringing the base of the shoots from which the buds are to be taken about 10 days before they are used increases their carbohydrate supply and seems to promote healing. Budding is best done when the rootstock seedlings are two to three weeks old—in the succulent red stage. Four to six weeks after budding, the inserted bud should start growth (*151*). T-budding also has been successful.

Approach grafting, termed "inarching" in India, has been used since ancient times in propagating the mango. Veneer grafting also is successful (*169*), as is saddle grafting (*239*). Rootstocks are mostly from assorted seedlings, although selected monoembryonic rootstocks can be multiplied by air layering or by cuttings, and subsequently clonally propagated by stooling (*168*).

Air layering is successful, especially when etiolated shoots are treated with indolebutyric acid at 10,000 ppm; such treatments have given 100 percent rooting with 90 percent survival (*170*). The mango is difficult to propagate by cuttings, but it can be done by using leafy cuttings under mist along with indolebutyric acid treatments.

Mulberry (*Morus* spp.).   The mulberry is readily started by hardwood cuttings 20 to 30 cm (8 to 12 in.) long, made from wood of the previous season's growth and planted in early spring. It can also be propagated by leafy softwood or semihardwood cuttings under mist. Micropropagation permits faster, large-scale propagation (*119*).

Nectarine (*Prunus persica*).   *See* Peach.

Olive (*Olea europaea* L.) (*103*).   Olives are propagated by budding or grafting on seedling or clonal rootstocks, by hardwood or semihardwood cuttings, or by suckers from old trees. Seeds of small-fruited cultivars germinate more easily than those of large-fruited ones. Often germination is prolonged over one or two years. The usual practice is to plant many more seeds than will be needed as seedlings to offset the low germination percentage. The true olive seed is enclosed in a hard, stony pit, the endocarp. It has long been known that clipping the end of the endocarp will hasten germination. It has been shown (*41*) that softening the endocarp with either sulfuric acid or sodium hydroxide, then stratifying at 15°C (59°F), gives high seed germination percentages. In other tests, stratifying the seeds (still in the endocarp) at 10°C (50°F), then germinating at 20°C (68°F), gave good germination, provided that the endocarp was removed before planting (*249*). Removing the seed from the pit and germinating the seed (without the seed coats) in petri dishes on moist filter paper will give high germination percentages if the fruits are collected shortly after pit hardening.

The seedlings grow slowly and may take a year or two to become large enough to be grafted or budded. Several methods have been used successfully, including T-budding, patch budding, whip grafting, and side-tongue grafting. In Italy, a widely used method is to bark graft small seedlings in the nursery row in the spring. The stocks are cut off several inches above ground, and one small scion is inserted in each seedling, followed by tying and waxing. After grafting or budding, one or two more years are required before a tree large enough for transplanting to the orchard is produced.

Hardwood cuttings may be made from two- to three-year-old wood about an inch in diameter and 20 to 30 cm (8 to 12 in.) long. All leaves are removed. It is often helpful to soak the basal ends of the cuttings in a solution of indolebutyric acid (15 ppm for 24 hours), followed by storage in moist sawdust at 15 to 21°C (60 to 70°F) for a month preceding spring planting in the nursery (*95*).

Semihardwood cuttings from vigorous, one-year-old wood about 6 mm (¼ in.) in diameter can be successfully rooted. They should be 10 to 15 cm (4 to 6 in.) long with two to six leaves retained on the upper portion of the cutting. It is best to take the cuttings in early summer or midsummer and root them under high humidity (*102*). Olive cuttings root well under intermittent mist and respond markedly to treatment with indolebutyric acid at about 4000 ppm for five seconds. There is considerable variability among cultivars in ease of rooting.

Rootstocks for Olives. *Olea europaea* seedlings are often used with budding or grafting, although considerable variation in tree vigor and size may result from this practice (*97*). A more suitable rootstock, giving uniform trees, is rooted cuttings of a strong-growing cultivar, such as 'Mission'. Olive trees are very susceptible to *Verticillium* wilt; where this occurs resistant clonal rootstocks should be used such as 'Oblonga' or 'Swan Hill' (*104*). Other *Olea* species do not make satisfactory rootstocks for the edible olive.

Orange. *See* Citrus.

Papaya (*Carica papaya* L.) (*90, 238*). Propagation by seed is the most practical method. They can be sown in flats of soil or in seedbeds in the open with germination in two to three weeks. The seeds do equally well if taken from stored or fresh fruit. Seeds can be stored for up to six years at 5°C (41°F) in sealed, moisture-proof bags without losing viability (*11*). The best seeds are from controlled pollinations between superior trees. When the seedlings are about 10 cm (4 in.) tall they are transplanted. This is generally done once or twice before they are put in their permanent location.

A method of starting seedlings without disturbing the roots is to plant four to eight seeds in a container and then thin to two to four of the strongest when they are about 10 cm (4 in.) tall. They can then be set in the field without disturbing the root system. Young papaya seedlings are very susceptible to damping-off organisms, so the soil in which they are started should be pasteurized if possible.

In Florida, the usual practice is to plant seeds in midwinter and set the young plants in the field by early spring. They grow during spring and summer, and usually mature their first fruits by fall, the plants bearing all winter and the following season.

Papaya trees can also be produced by budding (*226*), by in vitro micropropagation (*147*), and by cuttings. Papaya is sometimes called pawpaw, but it is not the same as the native American pawpaw.

Passion Fruit (*Passiflora edulis* Sims.) (*52*). This subtropical tender evergreen fruit is propagated chiefly by grafting onto seedling roots. Seeds germinate two to three weeks after planting. To avoid attacks of fusarium wilt, which is a serious problem, it is necessary to cleft-graft the purple-fruited hybrid fruiting cultivars onto seedlings of golden passion fruit, *P. edulis* forma *flavicarpa* (*237*), which are resistant to this disease and to nematodes.

Peach and Nectarine (*Prunus persica* Batsch) (*32, 65, 254*). Nursery trees of these fruits are propagated by T-budding or chip budding on seedling rootstocks. Fall budding is most common, but spring budding or—in regions with long growing seasons—June budding also is done.

Some peach cultivars can be propagated by leafy, succulent softwood cuttings taken in spring or summer, treated with a root-promoting material, and rooted in a mist propagating bed (*39, 100*). Also, in areas with mild winters, some peach cultivars can be started from hardwood cuttings if they are treated with indolebutyric acid (4000 ppm for five seconds), then set out in the nursery in the fall (*87*). Propagation of peaches by cuttings is being done commercially where high density tree production systems, requiring a large number of inexpensive nursery trees, are used (*56*).

Rootstocks for Peach (*140, 141, 177, 215*). Most peach cultivars are propagated on peach seedlings. Apricot and almond seedlings are sometimes used, as well as (in Europe) various plum clones. Peach seeds must be stratified at about 4°C (40°F) for three to four months, and almond and apricot for four weeks, before they are planted. Plum rather than peach is usually used as peach rootstocks in the United Kingdom.

Peach (P. persica). Peach seedlings are the most satisfactory rootstock for peach and should be used unless certain special conditions warrant other stocks. Seeds of 'Halford', 'Lovell', 'Elberta', 'Rutgers Red Leaf', or 'Nemared' are usually used, since they germinate well and produce vigorous seedlings. Seedlings of these cultivars are not resistant to root-knot nematodes, however, and where this is a problem, resistant peach stocks should be considered. Seeds from peach cultivars whose fruits mature early in the season should not be used because their germination percentage is usually low. It is best to obtain seeds from the current season's crops, since viability decreases with each year of storage. Both 'Lovell' and 'Nemaguard' seeds from trees of known viruses are available and should be used where possible. Seeds are sometimes obtained from wild types as, in the United States, the Tennessee Naturals or Indian peaches.

Peach roots are quite susceptible to the root-knot nematode (*Meloidogyne* spp.), especially in sandy soils. They are also susceptible to the root-lesion nematode, *Pratylenchus vulnus* (*48*). 'Nemaguard', a *Prunus persica* × *P. davidiana* hybrid rootstock introduced by the USDA in 1959, produces seedlings that are uniform and resistant to both *M. incognita* and *M. javanica,* but not to the ring nematode, *Criconemella xenoplax*. Nemaguard roots give strong, well-anchored, high-yielding trees, although certain peach and plum cultivars have not done well on it and it has shown susceptibility to bacterial canker (*215*) and to crown rot. 'Nemaguard' is not hardy in the colder peach growing areas.

A very winter-hardy peach rootstock, 'Siberian C', withstanding −11°C (12°F) soil temperatures, was introduced in 1967 by the Canadian Department of Agriculture Research Station at Harrow, Ontario. It does well on light, sandy soils and will give about 15 percent dwarfing to the scion cultivar. It is, however, susceptible to root-knot and root-lesion nematodes (*140*) and resulted in high tree mortality in tests in the southeastern United States (*53*).

Nursery trees on peach roots often make unsatisfactory growth when planted on soils previously planted to peach trees. Peach roots are susceptible to oak root fungus, crown rot, crown gall, and verticillium wilt.

Plum. After peach, plums are the most commonly used rootstock for peach cultivars. These are: *Prunus insititia* ('St. Julien d'Orleans', 'St. Julien Hybrid No. 1', 'St. Julien GF 655.2'); *P. cerasifera* (Myrobalan plums); *P. domestica* ('GF 43'); and *P. domestica* × *P. spinosa* ('Damas GF 1869'). In the United Kingdom, rootstocks such as these have been compatible with all peach and nectarine cultivars worked on them, and produce medium-sized to large trees. Plum rootstocks are especially adapted to wet, waterlogged soils.

Apricot (P. armeniaca). Apricot seedlings are occasionally used as a rootstock for the peach. The graft union is not always successful, but numerous trees and commercial orchards of this combination have produced fairly well for many years. Seedlings of the 'Blenheim' apricot seem to make better rootstocks for peaches than those of 'Tilton'. The apricot root is highly resistant to root-knot but not to root-lesion nematodes.

Almond (P. amygdalus). Almond seedlings have been used with limited success as a rootstock for peaches. There are trees on this combination growing well, but in general this is not a satisfactory combination. The trees are often dwarfed and tend to be shortlived.

Peach/almond hybrids. These are clonal rootstocks, such as 'GF556' and 'GF667', that are important in France and other Mediterranean countries. They are excellent rootstocks for peach cultivars and can be propagated by rooting semihardwood cuttings under mist (*162*). Micropropagation is used commercially for 'GF667'.

Western sand cherry (P. besseyi). When this stock was used in limited experiments as a dwarfing rootstock for several peach cultivars, it was found that although the bud unions were excellent, about 40 percent of the nursery trees failed to survive. The remainder grew well, however, and developed into typical dwarf trees with healthy, dark green foliage. The trees bore normal-sized fruit in the second or third year after transplanting to the orchard (*16*).

Nanking cherry (P. tomentosa). This may be suitable for some peach cultivars as a dwarfing rootstock.

Pear (*Pyrus communis* L.). Pears are propagated by fall budding, using T-budding or chip budding on either seedling pear rootstocks or rooted quince cuttings. Pear trees are also started by whole-root grafting, using the whip or tongue method. Pear seeds must be stratified for 60 to 100 days at about 4°C (40°F). They are then planted thickly about 13 mm (½ in.)

deep in a seedbed, where they are allowed to grow during one season. The following spring they are dug, the roots and top are cut back, and then they are transplanted to the nursery row, where they are grown a second season, ready for budding in the fall.

Some pear cultivars, such as 'Old Home' and 'Bartlett', can be propagated by hardwood cuttings or by leafy cuttings under mist if treated with indolebutyric acid (*98, 99, 260*). In pear rootstock test plantings, own-rooted 'Bartlett' trees have shown excellent production with large, well-shaped fruit; they are resistant to pear decline (*75, 98*) and with age become partially dwarfed, a desirable attribute for high-density plantings.

Rootstocks for Pear (*75, 77, 94, 149, 261*). In using rootstocks from seedlings of one of the several *Pyrus* species available, the seed source is very important if a rootstock of a certain species is expected. The various *Pyrus* species hybridize freely, some bloom at the same time, and cross-pollination is necessary for seeds to develop. It is best to use as the seed source an isolated group of trees of a given known species and to avoid collecting seeds from a planting containing trees of several different species, seeds that are likely to produce hybrid seedlings.

"Pear decline," caused by mycoplasma-like organisms (*110*) known to be spread by pear psylla (*Psylla pyricola*), devastated pear orchards, first in Italy (*214*) and later in western North America. Pear decline is associated with the rootstock used (*12, 14, 206*), so it is an important factor to be considered in rootstock selection. For plantings in decline-prevalent areas only rootstocks known to be resistant to decline should be used.

Seedlings from cultivars that may have resulted from pollination by a decline-susceptible species, such as *Pyrus pyrifolia* or *P. ussuriensis,* should not be used. Where pear psylla exist, and where they may be carrying the mycoplasmalike bodies causing pear decline, seedlings of the Oriental pears—*P. pyrifolia* or *P. ussuriensis*—should *not* be used, as trees with these as rootstocks are highly susceptible to decline.

French pear (P. communis). French pear seedlings are generally grown from seeds of 'Winter Nelis' or 'Bartlett'. This rootstock is moderately vigorous and winter-hardy, produces moderately productive, uniform trees with a strong, well-anchored root system, and is resistant to pear decline. It forms an excellent graft union with all pear cultivars and will tolerate relatively wet (but not waterlogged) and heavy soils. French pear roots are resistant to verticillium wilt, oak root fungus (*Armillaria mellea*), root-knot and root-lesion nematodes, and crown gall.

The two serious defects of French pear roots are their susceptibility to pear root aphid, *Eriosoma pyricola,* and to fire blight, *Erwinia amylovora.* Millions of pear trees on French pear roots have died from fire blight, owing to the high susceptibility of this stock.

Seedlings of the 'Kieffer' pear—a hybrid between *P. communis* and *P. pyrifolia*—have been satisfactorily used for many years in Australia as a pear rootstock.

Blight-resistant French pear rootstocks. Pear cultivars resistant to blight, such as 'Old Home' (*76, 98*) and 'Farmingdale' (*196*), have been used as an intermediate stock grafted on seedling rootstocks. Topworking with the desired cultivar takes place after the trunk and primary scaffold branches develop. If a blight attack occurs in the top of the tree, it will stop at the resistant body stock, which can subsequently be regrafted after blight has been cut out. 'Old Home' × 'Farmingdale' crosses have been made in Oregon and clonal blight-resistant pear rootstocks have been developed, some of which give dwarfed trees.

Japanese Pear [P. pyrifolia (P. serotina)]. This is the primary rootstock in Japan for cultivars of the same species. This was widely used in the United States as a pear rootstock from about 1900 to 1925, but is no longer considered of value, owing to its high susceptibility to pear decline and to the physiological defect, "black-end" or "hard-end," which may occur in 'Bartlett', 'Anjou', 'Winter Nelis', and other cultivars when they are propagated on it (*107*). Fruits affected with black-end are unusable because of hardening and cracking of the flesh at the blossom end, often with the development of blackened areas.

P. calleryana. This stock is blight resistant and produces vigorous trees with a strong graft union; fruit quality is good, with no black-end. 'Bartlett' orchards with this rootstock have produced well in California, and it is popular in the southern part of the United States for 'Keiffer' and other hybrid pears. In areas with cold winters it lacks winter-hardiness. Where good pear psylla control has been practiced, trees on *P. calleryana* roots are resistant to pear decline. However, trees with *P. calleryana* roots show less resistance to oak root fungus than those with *P. communis* roots. Seedlings of a selection of this species, known as 'D-6', are widely used as a pear rootstock in Australia.

P. ussuriensis. This Oriental stock has been used in the past to some extent; many pear cultivars on this root develop black-end, although not to the extent found with *P. pyrifolia* roots. It also produces small trees that are susceptible to pear decline, and should not be used in areas where this may occur.

P. betulaefolia. This species has vigorous seedlings, resistance to leaf spot and pear root aphid, tolerance to alkali soils, an adaptability to a wide range of climatic conditions, good resistance to pear decline, and produces large high-yielding trees.

Quince (Cydonia oblonga) (262). This species has been used for centuries as a dwarfing stock for pear and is now widely used as a pear rootstock in western Europe. Some cultivars, however, fail to make a strong union directly on the quince, hence double-working, with an intermediate stock such as 'Hardy' or 'Old Home' is necessary. Cultivars that require such a compatible interstock when worked on quince roots include 'Bartlett', 'Bosc', 'Winter Nelis', 'Seckel', 'Easter', 'Claireau', 'Conference', 'Guyot', 'Clapp's Favorite', 'Farmingdale', and 'El Dorado'. A selection of 'Bartlett' ('Swiss Bartlett'), compatible with 'Quince A' rootstock, originated in Switzerland, presumably as a bud mutation from an incompatible form (*191*). The following pears also appear to be compatible when worked directly on quince: 'Anjou', 'Old Home', 'Hardy', 'Packham's Triumph', 'Gorham', 'Comice', 'Flemish Beauty', 'Duchess', and 'Maxine'.

Quince roots are resistant to pear root aphids and nematodes, but are susceptible to oak root fungus, fire blight, and excess lime, and are not winter-hardy in areas where extremely low temperatures occur. In some areas trees on quince roots have developed pear decline, but in others they have not, possibly because different quince stocks were used. The black-end trouble has not developed with pears on quince roots. There are a number of quince cultivars, most of which are easily propagated by hardwood cuttings or layering. 'Angers' quince is commonly used as a pear rootstock because its cuttings root readily, it grows vigorously in the nursery, and does well in the orchard. 'Provence' quince roots produce larger pear trees than 'Angers' and are more winter hardy.

The East Malling Research Station has selected several clones of quince suitable as pear rootstocks and designated them as 'Quince A', 'B', and 'C'. 'Quince A' ('Angers') has proved to be the most satisfactory stock. 'Quince B' (Common quince) is somewhat dwarfing, whereas 'Quince C' produces very dwarfing but highly productive trees. It is important to use only virus-tested quince stock.

Provence quince (LePage Series C and BA 29C) originated in France. These are winter hardy and, when used as rootstocks for pears, give trees one-half to two-thirds the size of standard pear trees. The BA-29C series is a virus-free selection of LePage Series C.

Rootstocks recommended for Bartlett pear orchards in California are: *P. betulaefolia* seedlings, own-rooted 'Old Home', *P. communis* 'Winter Nelis' seedlings, as well as own-rooted 'Bartlett' (no rootstock) (*75*).

Pecan (*Carya illinoensis* Wangenh. [K. Koch] (*86, 148, 154*). Pecans are propagated by budding or grafting selected cultivars on pecan seedling rootstocks. Hardwood cuttings can be rooted. They should be taken in the fall and treated with about 10,000 ppm IBA. Long (50 cm), thick (20 mm) cuttings root best (*221*). Plants have been started also by root cuttings and by trench and air layering, the latter with the aid of indolebutyric acid treatments. Shoots from pecan seedlings have been rooted and developed into plants using micropropagation methods (*88*).

Pecan seeds sometimes start growing in the hulls even before the nuts are harvested (vivipary). The seeds lose their viability, however, in warm, dry storage. To prevent this loss, they should be stored at 0°C (32°F) immediately after harvest and until planted. Pecan seeds should be given a stratification treatment for 12 to 16 weeks at about 1 to 5°C (34 to 41°F) to ensure good, rapid germination, or else germination should take place at high temperatures—30 to 35°C (86 to 95°F)—to overcome growth restrictions of the shell (*50, 247*). Midwinter planting, with seedling emergence in the spring, is a successful procedure.

Only deep, well-drained, shady soil should be used for growing pecan nursery trees. Young seedlings are tender and should be shaded against sunburn. In the summer, toward the end of the second growing season, the seedlings are large enough to bud to the desired cultivar. Patch budding, or sometimes ring budding, is the usual method employed. After the new top grows for one or two seasons, the nursery tree is ready to transplant to its permanent location. Young pecan trees have a long taproot, which must be handled carefully in digging and replanting.

Sometimes the seedlings are changed to the desired cultivar by crown grafting in late winter or early spring, using the whip graft method. Pecan seedling trees of various sizes, growing in place in the orchard, can be topworked to the desired cultivar by bark grafting limbs 4 to 9 cm (1 ½ to 3 ½ in.) in diameter.

Rootstocks for Pecans (*86*). Commercially, pecan cultivars are propagated on pecan (*C. illinoensis*) seedlings. Seeds from the 'Apache', 'Riverside', 'Elliott', and 'Curtis' cultivars are reported to produce excellent seedlings. As a possible rootstock for wet soils, seedlings of one of the hickory species, *C. aquatica,* have been used experimentally. Although pecan scions will grow on hickory species, the nuts generally do not attain normal size. Pecan roots are susceptible to verticillium wilt.

Persimmon (*Diospyros* spp.) (*74, 190, 222*). **American Persimmon** (*D. virginiana* L.) (*63*). The named cultivars of this species are commonly propagated by budding or grafting on *D. virginiana* seedlings. Root cuttings are also successful.

Seed germination is rather slow, owing to the seed's slow rate of water absorption. Fall sowing, or stratification at about 10°C (50°F) for 60 to 90 days, is advisable. In transplanting any but small seedlings, the taproot should be cut 30 cm (12 in.) below ground a year before moving to force out lateral roots.

Oriental Persimmon (*D. kaki* L.) (*203*).   The cultivars of this fruit are propagated by grafting on seedling rootstocks. Crown grafting by the whip method in early spring when both scionwood and stock are still dormant is a common practice, although bench grafting also is used. Budding can be done but it is less successful than grafting (*190*).

The usual practice in germinating seeds of both *D. lotus* and *D. kaki* is to stratify the seeds for 120 days at about 10°C (50°F). If the seeds have dried out, they should be soaked in warm water for two days before stratification. Excessive drying of the seed is harmful, especially for *D. kaki*. The seeds are planted either in flats or in the nursery row. Young persimmon seedlings require shading. Oriental persimmon cultivars can be propagated by in vitro tissue culture methods (*235*).

Rootstocks for Oriental Persimmon (*112*)

Diospyros lotus.   This stock has been widely used in California; it is very vigorous and drought-resistant, and produces a rather fibrous type of root system, which transplants easily. This stock is quite susceptible to crown gall and verticillium and will not tolerate poorly drained soils but is highly resistant to oak root fungus. 'Hachiya' does not produce well on *D. lotus* stock, owing to the excessive shedding of fruit in all stages (*207*). 'Fuyu' scions usually do not form a good union with *D. lotus,* although 'Fuyu' topworked on a *D. kaki* cultivar established on *D. lotus* roots makes a satisfactory tree.

D. kaki.   This stock is the one most favored in Japan and is probably best for general use, because it develops a good union with all cultivars and is resistant to crown gall and oak root fungus but susceptible to verticillium. Trees worked on it grow well and yield satisfactory commercial crops. The seedlings have a long taproot with few lateral roots, making transplanting somewhat difficult.

D. virginiana.   Seedlings of this species are utilized in the southern part of the United States and seem to be adapted to a wide range of soil conditions. It has not proven satisfactory in some localities, however. In California, 'Hachiya' on this stock is distinctly dwarfed and yields poorly, owing to the sparse bloom (*207*). Most oriental cultivars make a good union with this stock, but there are diseases carried by *D. kaki* scions

that will move into the *D. virginiana* roots, causing them to die. This stock seems to be quite tolerant either of drought or of excess soil moisture conditions. It produces a fibrous type of root system easy to transplant, but tending to sucker badly.

Pineapple (*Ananas comosus* Merr.) (*35*).   All commercial propagation of the pineapple is by asexual methods. As shown in Fig. 14–17, there are three main types of planting material—*suckers, slips,* and *crowns.* Suckers coming from below ground are rarely used as planting material. Those above ground are cut from the mother plant about one month after the peak fruit harvest. Slips are taken from the plant two to three months following the peak harvest. When used in propagation, crowns are taken either before or at the time of harvest.

Slips are the most popular type of planting material for commercial use. More of these are produced by the plant than either suckers or crowns. After removal from the mother plant, slips can be stored for a relatively long time and still retain sufficient vigor for replanting. If the slips are all about the same size and age when planted, they will flower and fruit at approximately the same time.

Slips will produce fruit 18 to 20 months after planting, suckers at about 15 months, and crowns at about 22 months. Because of these differences, the three types of planting material are never mixed in one field.

Before planting, all types of planting material must be cured or dried for one to several weeks after they are cut from the mother plant. This allows a callus layer to develop over the cut surface, reducing losses from decay organisms after they are planted.

Pistachio (*Pistacia vera* L.) (*83, 126*).   The pistachio nut is usually propagated by T-budding on *Pistacia atlantica* or *P. terebinthus* seedlings as rootstocks. However, these stocks are highly susceptible to verticillium. Seedlings of *P. intergerrima* show resistance to this soil-borne disease and are being used in areas having verticillium problems. Budding is possible over a considerable period of time, but if done before midspring, when sap flow may be excessive, the percentage of takes is apt to be low. A marked improvement in bud union occurs as budding is extended through the summer and fall. Best results are usually obtained by planting seedling rootstocks in their permanent orchard location, then T-budding them 2 ft or more above ground after the seedlings are well established. Seedlings are grown in long tubular, biodegradable pots ready for planting in the orchard.

To collect seeds for rootstocks, obtain fruits when

the hulls turn blue-green. The hulls should be removed, as they apparently contain a germination inhibitor. *P. terebinthus* seed should be held in moist sand at 5 to 10°C (41 to 50°F) for six weeks before sowing, then germinated at 20°C (68°F) or slightly below. With early planting of seed and adequate irrigation and fertilization through the summer, the more vigorous seedlings may reach budding size by fall. Buds of *P. vera* cultivars are quite large, requiring a fairly large seedling to accommodate them. Because of their long taproot, bare-root nursery trees should be transplanted to their permanent planting site as early as possible. Commercial cultivars of *Pistacia vera* have been propagated by in vitro methods (*9*).

Plum (*Prunus* spp.).   Plums are propagated by T-budding or chip budding in the fall on seedling rootstocks or, with certain stocks, on rooted cuttings or layers. Budding can also be done in the spring. Some plums can be propagated by hardwood cuttings (*99*) and some by leafy, softwood cuttings under intermittent mist (*100*). Bench grafting in winter and planting the grafts in the nursery in spring gives good results using the whip graft (*47*). Japanese plum (*P. salicina*) nursery trees have been produced by micropropagation methods (*199*).

Rootstocks for Plum (*106, 178, 179*)

Myrobalan plum (P. cerasifera).   This is the most widely used plum rootstock, being particularly desirable for the European plums, *P. domestica* (which includes the commercially important prune cultivars). It is also a satisfactory stock for the Japanese plums, *P. salicina*. Some kinds of plums—'President', 'Kelsey', 'Stanley', and 'Robe de Sergeant'—are not entirely compatible with this stock, however. Cultivars that are Japanese-American hybrids (*P. salicina* × *P. americana*) are best worked on American plum seedlings (*P. americana*).

Myrobalan roots are adapted to a wide range of soil and climatic conditions. They will endure fairly heavy soils and excess moisture, and are resistant to crown rot but susceptible to root-knot nematodes and oak root fungus. They grow well on light sandy soils.

Myrobalan seeds require stratification for about three months at 2 to 4°C (36 to 40°F), They may then be planted thickly in a seedbed for one season, then transplanted to the nursery and grown for a second season before budding, or the seeds may be planted directly in the nursery row and grown there for one season, the seedlings being budded in late summer or fall.

Certain very vigorous myrobalan selections are propagated by hardwood cuttings. One of these, 'Myro C', is immune to root-knot nematodes. In England a selection, 'Myrobalan B', was developed at the East

Malling Research Station and is propagated by hardwood cuttings. It is particularly valuable in producing vigorous trees, although there are cultivars not completely compatible with it. Most plum cultivars in England belong to the *P. domestica* species.

Marianna plum (P. cerasifera × *P. munsoniana?).* This is a clonal rootstock that originated in Texas as an open-pollinated seedling of the myrobalan plum and, supposedly, *P. munsoniana*. It is propagated by hardwood cuttings (*96, 99*). Some plums have grown well on it; others have not. An exceptionally vigorous seedling selection of the parent 'Marianna' plum, made by the California Agricultural Experiment Station in 1926, is widely used under the identifying name, 'Marianna 2624'. It is adaptable to heavy, wet soils and is immune to root-knot nematodes, resistant to crown rot, crown gall, oak root fungus, and verticillium wilt, but is susceptible to bacterial canker. Since it is propagated as rooted cuttings, it has a shallow-root system for the first few years, but as the trees grow older they develop a deeper root system. Hardwood cuttings root well if fall-planted (in mild winter climates) after IBA treatments.

Peach (P. persica).   Many plum and prune orchards in California are on peach seedling rootstocks. This stock has proved satisfactory for light, well-drained soils. However, peach rootstocks should be avoided if the trees are to be planted on a site formerly occupied by a peach orchard. In some areas plum on peach roots tends to overbear and develop a die-back condition. Peach is not satisfactory as a stock for some plum cultivars, including 'Sugar' prune and 'Robe de Sergeant'.

Apricot (P. armeniaca).   Apricot seedlings can be used as a plum stock in nematode-infested sandy soils for those cultivars that are compatible with apricot. Japanese plums tend to do better than European plums on apricot roots.

Almond (P. dulcis).   Some plum cultivars can be grown successfully on almond seedlings. The 'French' prune does very well on this stock, the trees growing faster and bearing larger fruit than when myrobalan roots are used. Plum cultivars on almond roots tend to overbear, sometimes to the detriment of the tree. Except where plantings are to be made on well-drained, sandy soils, high in lime or boron, this stock probably should not be used for plums.

'Brompton' and 'Common' plum (P. domestica). These two clonal plum stocks are used chiefly in England. 'Brompton' seems to be compatible with all plum cultivars and tends to produce medium to large trees. 'Common' plum, which produces small to medium trees, has shown incompatibility with some cultivars (*165*). Propagation is by hardwood cuttings.

'St. Julien', 'Common Mussel', 'Damas', and 'Damson' (*P. insititia*). The first three stocks are used mostly in England. 'Damas C' produces medium to large trees, whereas 'Common Mussel' generally produces small to medium trees. The latter stock seems to be compatible with all plums. 'St. Julien A' produces small to medium trees for all compatible scion cultivars. Results have been variable with these stocks, since they vary widely in type. At the East Malling Research Station, 'St. Julien', 'Mussel', and 'Damas' clones have been selected and listed as A, B, C, D, and so forth. 'St. Julien' has been used to some extent as a plum stock for the *P. domestica* cultivars in the United States. 'Pixy' has been released by the East Malling Research Station as a dwarfing plum rootstock (*253*).

In England the following clonal stocks are recommended for plums: for *vigorous* trees, 'Myrobalan B'; for *semivigorous* trees, 'Brompton'; for *intermediate* trees, 'Marianna', 'Pershore'; for *semidwarf* trees, 'Common' plum, 'St. Julien A'; for *dwarf* trees, 'St. Julien K'.

Florida sand plum (*P. angustifolia*). This species may be useful as a dwarfing stock for compatible plum cultivars. In California, after 16 years, 'Giant', 'Burbank', and 'Beauty' on this stock were healthy and very productive, with a dwarf type of growth.

Japanese plum (*P. salicina*). Seedlings of this species are used as plum stocks in Japan but apparently not elsewhere. European plums (*P. domestica*), when topworked on Japanese plum stocks, result in very short-lived trees; the reverse combination, though, produces compatible unions.

Western sand cherry (*P. besseyi*). This stock has produced satisfactory dwarfed plum trees of the Japanese and European types, but poor bud unions and shoot growth developed when it was used as a stock for cultivars of *P. insititia* (*16*).

Pomegranate (*Punica granatum* L.)   The pomegranate is easily propagated by hardwood cuttings. After one season's growth, the plants are usually large enough to be moved to their permanent location. Softwood cuttings taken during the summer are also easily rooted if maintained under high-humidity conditions. The pomegranate forms suckers readily, and these may be dug up during the dormant season with a piece of root attached.

Quince (*Cydonia oblonga* Mill.).   The quince is usually propagated by hardwood cuttings, which root readily, "heel" cuttings more so than those made entirely from one-year-old wood. Cuttings from two- or three-year-old wood also root easily. The burrs or knots found on this older wood are masses of adventitious root initials. Cuttings made from the basal portion of one-year-old wood root better than those made from the terminal end. The cuttings usually make sufficient growth in one season to transplant to their permanent location. Quince can also be propagated by mound layering. In vitro micropropagation of 'Provence' quince provides a method of large-scale clonal production of a dwarfing pear rootstock (*4*). Occasionally, commercial quince cultivars are T-budded on rooted cuttings (such as 'Angers') or sometimes on quince seedlings.

Raspberry, Black (*Rubus occidentalis* L.) **and Purple** (*R. occidentalis* × *R. idaeus*) (*125*).   Tip layering is the usual method of propagating these species, the black rooting more easily than the purple. The black raspberry can be propagated also by leaf-bud cuttings, roots forming in about three weeks. Cuttings should be taken in early summer and rooted under high-humidity conditions or in a mist-propagating frame.

Raspberry, Red (*Rubus strigosus* Michx., *R. idaeus* L.) (*125*).   The red raspberry is easily propagated by removing suckers of one-year-old growth. These are dug in early spring, care being taken to leave a piece of the old root attached. Also, young, green suckers of new wood may be dug in the spring shortly after they appear above ground. When these plants are dug, they should have some new roots starting. A piece of the old root is taken with them. Such suckers should be dug and transplanted during cool, cloudy weather and irrigated well after resetting; otherwise, they will not survive. Sucker production can be stimulated by inserting a spade deeply at intervals in the vicinity of old plants to cut off roots, each root piece then sending up a shoot. Production of suckers can also be stimulated by mulching with straw or sawdust. Viruses and crown gall can be problems with red raspberries so suckers should be taken only from clean plants obtained from nurseries specializing in certified plants.

The red raspberry can be propagated by root cuttings (*116, 243*). For thick root pieces 15 cm (6 in.) root lengths are best; for thin roots use 5 cm (2 in.) pieces. Shallow 13 mm (½ in.) planting is advised (*109*). If given good care, strong nursery plants will be produced in one year.

Leafy softwood cuttings made in early spring from young sucker shoots just emerging from the soil, with a 2.5 to 5 cm (1- to 2-in.) etiolated section from below the surface, will give almost 100 percent rooting when placed in a propagating frame (*264*). Shoot cuttings taken directly from the canes are difficult to root, although it can be done by using pre-etiolated shoots, treating with IBA, and rooting under ventilated high-humidity fog (*114*).

R. idaeus has been easily propagated in vitro, giving high survival of several cultivars tested, thus providing a method for mass propagation. (*256*).

Strawberry (*Fragaria* × *ananassa* Duch.) (*180, 208, 242, 258, 263*). Strawberries are propagated by runners, or in certain everbearing types—which produce few runners—by crown division. Plants can be easily mass-produced by micropropagation using in vitro culture of shoot tips (*45, 128*). Seed propagation is used only in breeding programs for the development of new cultivars. Treatment of seed with sulfuric acid for 15 to 20 minutes before planting gives good germination (*210*).

Commercial strawberry nursery plant production involves some form of registration and certification to make available to growers pathogen-free, true-to-name plants. Viruses have been eliminated from strawberry clones by meristem-tip culture, with or without additional heat treatment (*37, 223*).

Most strawberry cultivars are short-day plants that produce runners in summer, initiate flower buds in autumn, are chilled during the winter to enable them to produce flowers and fruits in the spring.

Nursery fields are planted in the spring and the new runner plants are dug the following fall and winter after they are mature and dormant. Such planting stock comes from special "mother blocks," maintained in isolation to prevent infection with viruses and other pathogens.

In California plants are dug from the nursery row mostly in the fall in order to produce optimum plant response at particular berry production areas. The digging should be completed in the spring before the plants start to grow, particularly if they are to be held in cold storage. Special machine diggers handle great quantities of plants (Fig. 18–3). Immediately after digging, the plants are hauled to packing sheds, where leaves and petioles are removed. Plants are packed in vented cardboard or wood boxes for storing or shipping. A polyethylene liner (0.75 to 1.5 mil) is used to keep the plants from drying out (*266*). They are usually handled in 1000- or 2000-plant units, depending upon plant size. The plants are placed firmly in the boxes with roots to the center. Such plants can be stored and kept in good condition for one year if held between −2 and −1°C (28 to 30°F) to enable

FIGURE 18–3 Equipment used for large-scale digging of strawberry nursery plants. The entire row is scooped up and elevated to a revolving trommel, which separates the plants from the soil. The plants are then packed in wet burlap bags and taken to sheds for trimming and packing. (Courtesy Wheeler's Nursery, Los Molinos, Calif.)

growers in certain commercial producing areas to plant in the summer and fall (*80*).

Micropropagation has become an important commercial alternative for strawberry plant propagation, and laboratories capable of producing several million plants annually have been built (*15, 45*). However, actual commercial production may involve a combination of mother block laboratory procedures and field multiplication operations (*128*).

Everbearing cultivars that produce few runners are propagated by crown division. Certain cultivars, such as the 'Rockhill', may produce 10 to 15 strong crowns per plant by the end of the growing season. In the spring, such plants are dug and carefully cut apart; each crown may then be used as a new plant. True day-neutral cultivars have been bred which flower and fruit in continuous cycles throughout the year in areas with mild winter climates.

Tangerine. *See* Citrus.

Walnut, Black (*Juglans nigra* L.) (*69*). The several named black walnut cultivars are propagated by patch or ring budding or side grafting on *J. nigra* rootstocks (*28, 220*).

This method is suitable for seedling stocks up to about 2.5 cm (1 in.) in diameter. The scion is cut to a wedge, slightly thicker on one side than on the other. The stock is cut diagonally rather than split. The scion is inserted into this cleft, the thicker part out, with the cambium of stock and scion matching.

In growing Eastern black walnut for timber production, seedlings are started in nurseries and only the strongest, most vigorous trees are set out in the plantation. Careful planting is necessary to obtain the essential rapid, early growth. Cuttings are very difficult to root.

Walnut, Carpathian (*Juglans regia* L.) (*78*). The Carpathian strain of *J. regia* originated in cold winter areas of the Carpathian Mountains of Poland and the Kiev and Poltava regions of the Ukraine, whereas the Persian walnuts originated in the area of present-day Iran. Carpathian walnut trees are much more winter hardy than trees of the Persian walnut strain commonly grown in California. Carpathian cultivars have been developed that are winter hardy in the eastern United States. These are propagated by chip budding or bench grafting (using the whip graft) on Persian (*J. regia*) or black (*J. nigra*) walnut seedling rootstocks.

Walnut, Persian, English (*Juglans regia* L.). Walnut cultivars are propagated by patch budding (or one of its variations), T-budding, or whip grafting one-year-

old seedling rootstocks, or by topworking seedling trees, one to four years old or more, planted in place in the orchard. Bark grafting in late spring or patch budding in the spring or summer works well for the latter method. Persian walnut cultivars can be propagated by in vitro tissue culture methods (*157*).

Rootstocks for Persian Walnut (*66, 157, 213*). Nuts of most of the *Juglans* species used as seedling rootstocks should be either fall-planted, or stratified for about three months at 2 to 4°C (36 to 40°F) before they are planted in the spring, to obtain good germination. It is better to plant the seeds before they start sprouting in the stratification boxes but, with care, sprouted seeds can be planted successfully. At the end of one season, the seedlings should be large enough to bud. Seedlings of *Juglans* species are very sensitive to waterlogged soil conditions under which they will not survive (*27*).

Northern California black walnut (J. hindsii). This is the stock most commonly used in California. The seedlings are vigorous and make a strong graft union. They are resistant to verticillium, oak root fungus (*Armillaria mellea*), and root-knot nematode (*Meloidogyne* spp.), but are susceptible to crown rot (*Phytophthora* spp.), crown gall, and the root-lesion nematode (*Pratylenchus vulnus*). Persian walnut trees grafted on this rootstock are subject to a serious defect known as "blackline." The latter has appeared in California, Oregon, England, Italy, and France. It is characterized by a breakdown of the tissues in the cambial region at the graft union, leading to a lethal girdling of the trees. The cause of this problem is a pollen-transmissible virus, the cherry leaf roll virus (*163, 164*). Symptoms are yellow, drooping leaves, premature leaf drop, poor terminal shoot growth, reduction in crop, black lesions at the graft union, and a heavy production of suckers from the black walnut stock below the graft union. The Persian walnut top above the graft union eventually dies from the virus, leaving the *J. hindsii* stock alive.

Persian walnut (J. regia). Worldwide, this is probably the most common rootstock for Persian walnut cultivars due to availability of seeds. Seedlings of this species as a rootstock produce good trees with an excellent graft union and are highly resistant to crown rot. The roots are susceptible to crown gall, oak root fungus, and salt accumulation in the soil, and are not as resistant to root-knot nematodes as *J. hindsii*. Nurserymen object to the slow initial growth of the seedlings. Trees on this stock should be quite satisfactory in soils free of oak root fungus and salt accumulation. Its use is recommended in some localities as a means of avoiding the black-line virus (*163*). In Oregon, where blackline is very serious, seedlings of the 'Manregian' clone (of the Manchurian race of *J. regia*), imported by the

USDA as P.I. No. 18256, are used as rootstocks. They are vigorous, cold-hardy, and—in Oregon—no more susceptible to oak root fungus than *J. hindsii* seedlings. (*181*).

   Paradox walnut (J. hindsii × J. regia).   This is a first-generation (F₁) hybrid stock; the seedlings are obtained from seed taken from *J. hindsii* trees, whose pistillate (female) flowers have been wind pollinated with pollen from nearby *J. regia* trees. When seeds from such a *J. hindsii* tree are planted, some of the seedlings may be the hybrid progeny. The amount varies widely, from none to almost 100 percent, depending upon the individual mother tree. The hybrids are easily distinguished by their large leaves in comparison with the smaller-leaved, self-pollinated *J. hindsii* seedlings (see Fig. 4-4). Seedlings from Paradox trees themselves should not be used for producing rootstocks, because of their great variability in all characteristics. Although first-generation (F₁) seedlings are variable in some characteristics, most of them exhibit hybrid vigor and make excellent vigorous rootstocks for the Persian walnut. They are resistant to root-lesion nematodes and crown rot, and are tolerant of saline and heavy, wet soils. Paradox seedlings are very susceptible to oak root fungus and crown gall. Persian walnut on Paradox roots are just as susceptible to the black-line virus as *J. hindsii* seedlings. Trees on Paradox rootstocks grow and yield as well as, or better than, those on *J. hindsii* roots, and may produce large-size nuts with better kernel color (*213*). In very heavy or low-fertility soils, trees on Paradox roots grow faster than those on *J. hindsii*.

   Since it is difficult to secure Paradox seeds in quantity, vegetative propagation methods would be very desirable in order to establish superior Paradox clones from which large numbers of rootstock plants could be obtained. Propagation by leafy cuttings under mist, hardwood cuttings, and trench layering can be done but with difficulty.

   Eastern black walnut (J. nigra L).   This has been recommended in some cases as a rootstock for Persian walnut cultivars in Europe and the Soviet Union—but not in California. As compared to *J. regia* roots, it is reported to have greater tolerance to nematodes, oak root fungus, *Phytophthora,* and crown gall, but in California rootstock trials, trees on this stock showed poor yields.

REFERENCES

1. Adams, W. G. 1985. Improved techniques used in producing budded citrus nursery trees for commercial fruit production. *Proc. Inter. Plant Prop. Soc.* 34:496–99.

2. Alley, C. J. 1980. Propagation of grapevines. *Calif. Agr.* 34(7):29–30.

3. Alley, C. J., and A. T. Koyama. 1978. Vine bleeding delays growth of T-budded grapevines. *Calif. Agr.* 32(8):6.

4. Al Maarri, K., Y. Arnaud, and E. Miginiac. 1986. *In vitro* micropropagation of quince (*Cydonia oblonga* Mill.). *Scient. Hort.*. 28:315–21.

5. Anderson, W. C. 1984. Micropropagation of filberts, *Corylus avellana. Proc. Inter. Plant Prop. Soc.* 33:132–37.

6. Argles, G. K. 1976. *Anacardium occidentale* (cashew). In *The propagation of tropical fruit trees,* R. J. Garner and S. A. Chaudri, eds. Hort. Rev. 4. East Malling, England: FAO and Commonwealth Agricultural Bureaux.

7. Arndt, C. H. 1935. Mango polyembryony and other multiple shoots. *Amer. Jour. Bot.* 22:26.

8. Ashiru, G. A., and T. Quarcoo. 1971. Vegetative propagation of kola (*Cola nitida*). *Trop. Agr.* 48(1):85–92.

9. Barghchi, M., and P. G. Alderson. 1985. In vitro propagation of *Pistacia vera* L. and the commercial cultivars, Ohadi and Kalleghochi. *Jour. Hort. Sci.* 60(3):423–30.

10. Barton, L. V. 1965. Viability of seeds of *Theobroma cacao* L. *Contr. Boyce Thomp. Inst.* 23(4):109–22.

11. Bass, L. N. 1975. Seed storage of *Carica papaya* L. *HortScience* 10(3):232–33.

12. Batjer, L. P., and H. Schneider. 1960. Relation of pear decline to rootstocks and sievetube necrosis. *Proc. Amer. Soc. Hort. Sci.* 76:85–97.

13. Bitters, W. P., and E. R. Parker. 1953. Quick decline of citrus as influenced by top-root relationships. *Calif. Agr. Exp. Sta. Bul. 733.*

14. Blodgett, E. C., H. Schneider, and M. D. Aichele. 1962. Behavior of pear decline disease on different stock-scion combinations. *Phytopathology* 52:679–84.

15. Boxus, P. 1974. The production of strawberry plants by in vitro micropropagation. *Jour. Hort. Sci.* 49:209–10.

16. Brase, K., and R. D. Way. 1959. Rootstocks and methods used for dwarfing fruit trees. *N.Y. Agr. Exp. Sta. Bul. 783.*

17. Brokaw, W. H. 1977. Subtropical fruit tree production; avocado as a case study. *Proc. Inter. Plant Prop. Soc.* 27:113–21.

18. Brokaw, W. H. 1988. Avocado clonal rootstock propagation. *Proc. Inter. Plant Prop. Soc.* 87:97–102.

19. Broome, O. C., and R. H. Zimmerman. 1978. In vitro propagation of blackberry. *HortScience* 13(2):151–53.

20. Calavan, E. C., E. F. Frolich, J. B. Carpenter, C. N. Roistacher, and D. W. Christiansen. 1964. Rapid indexing of exocortis of citrus. *Phytopathology* 54:1359–62.

21. Caldwell, J. D. 1984. Blackberry propagation. *HortScience* 19(2):193–95.

22. Caldwell, J. D., D. C. Coston, and K. H. Brock. 1988. Rooting of semi-hardwood 'Hayward' kiwifruit cuttings. *HortScience* 23(4):714–17.

23. Cameron, J. W., and R. K. Soost. 1953. Nucellar lines of citrus. *Calif. Agr.* 7(1):8, 15, 16.

24. Carlson, R. F. 1965. Growth and incompatibility factors associated with apricot scion/rootstock in Michigan. *Mich. Quart. Bul.* 48(1):23–29.

25. Castle, W. S. 1987. Citrus rootstocks. In *Rootstocks for fruit crops*, R. C. Rom and R. F. Carlson, eds. New York: John Wiley.

26. Castle, W. S., W. G. Adams, and R. L. Dilley. 1979. An indoor container system for producing citrus nursery trees in one year from seed. *Proc. Fla. State Hort. Sci.* 92:3–7.

27. Catlin, P. B., and E. A. Olsson, 1986. Response of eastern black walnut and northern California black walnut to waterlogging. *HortScience* 21(6):1379–80.

28. Chase, S. B. 1947. Budding and grafting eastern black walnut. *Proc. Amer. Soc. Hort. Sci.* 49:175–80.

29. Chaudri, S. A. 1976. *Mangifera indica—Mango*. In *The propagation of tropical fruit trees*, R. J. Garner and S. A. Chaudri, eds. Hort. Rev. 4. East Malling, England: FAO and Commonwealth Agricultural Bureaux.

30. Cheng, T. 1978. Clonal propagation of woody plant species through tissue culture techniques. *Proc. Inter. Plant Prop. Soc.* 28:139–55.

31. Child, R. 1964. *Coconuts.* London: Longmans, Green.

32. Childers, N., ed. 1988. *The peach: World cultivars to marketing.* Gainsville, Fla.: Horticultural Publications.

33. Cobin, M. 1954. The lychee in Florida. *Fla. Agr. Exp. Sta. Bul. 546.*

34. Cohen, D., and D. Elliott. 1979. Micropropagation methods for blueberries and tamarillos. *Proc. Inter. Plant Prop. Soc.* 29:177–79.

35. Collins, J. L. 1960. *The pineapple—history, cultivation, utilization.* London: Leonard-Hill.

36. Condit, I. J. 1969. *Ficus, the exotic species.* Berkeley: Univ. Calif. Div. Agr. Sci.

37. Converse, R. A. 1972. The propagation of virus-tested strawberry stocks. *Proc. Inter Plant Prop. Soc.* 22:73–76.

38. Coorts, G. D., and J. W. Hull. 1972. Propagation of highbush blueberry (*Vaccinium australe* Small) by hard and softwood cuttings. *The Plant Propagator* 18(2):9–11.

39. Couvillon, G. A., and A. Erez. 1980. Rooting, survival, and development of several peach cultivars propagated from semi-hardwood cuttings. *HortScience* 15(1):41–43.

40. Couvillon, G. A., and F. A. Pokorny. 1968. Photoperiod, indolebutyric acid, and type of cutting wood as factors in rooting of rabbiteye blueberry (*Vaccinium ashei Reade*), cv. Woodward. *HortScience* 3(2):74–75.

41. Crisosto, C., and E. Sutter. 1985. Improving 'Manzanillo' seed germination. *HortScience* 20(1)100–102.

42. Cross, C. E., I. E. Demoranville, K. H. Deubert, R. M. Devlin, J. S. Norton, W. E. Tomlinson, and B. M. Zuckerman. 1969. Modern cultural practice in cranberry growing. *Mass. Agr. Exp. Sta. Ext. Ser. Bul. 39.*

43. Crossa-Raynaud, P., and J. M. Anderson. 1987. Apricot rootstocks. In *Rootstocks for fruit crops*, R. C. Rom and R. F. Carlson, eds. New York: John Wiley.

44. Cummins, J. N., and H. S. Aldwinckle. 1974. Breeding apple rootstocks. *HortScience* 9(4):367–72.

45. Damiano, C. 1980. Strawberry micropropagation. In *Proc. Conf. Nurs. Prod. Fruit Plants through Tissue Cult.*, R. H. Zimmerman, ed. USDA, SEA, ARR-NE-11.

46. Davies, F. S. 1986. The navel orange. In *Horticultural reviews*, Vol. 8, J. Janick, ed. Westport, Conn.: AVI Publ. Co., pp. 129–80.

47. Davis, B., II. 1976. Bench grafting plum and apricot as compared to T-budding. *Proc. Inter. Plant Prop. Soc.* 26:253–55.

48. Day, L. H., and E. R. Serr. 1951. Comparative resistance of rootstocks of fruit and nut trees to

attack by root lesion or meadow nematode. *Proc. Amer. Soc. Hort. Sci.* 57:150–54.

49. Dillon, D. 1967. Simultaneous grafting and rooting of citrus under mist. *Proc. Inter. Plant Prop. Soc.* 17:114–18.

50. Dimalla, G. G., and J. Van Staden. 1978. Pecan nut germination: A review for the nursery industry. *Scient. Hort.* 8:1–9.

51. Doggrell, L. M. 1976. Commercial propagation of macadamias. *Proc. Inter. Plant Prop. Soc.* 26:391–93.

52. Doncaster, T. 1981. The whys and hows of passion fruit growing in Queensland. *Proc. Inter. Plant Prop. Soc.* 30:616–17.

53. Dozier, W. A., J. W. Knowles, C. C. Carlton, R. C. Rom, E. H. Arrington, E. J. Wehunt, U. L. Yadava, S. L. Doud, D. F. Ritchie, C. N. Clayton, E. I. Zehr, C. E. Gambrell, J. A. Britton, and D. W. Lockwood. 1984. Survival, growth, and yield of peach trees as affected by rootstocks. *HortScience* 19(1):26–30.

54. Eck, P. 1988. *Blueberry science.* New Brunswick, N.J.: Rutgers Univ. Press.

55. Eck, P., and A. W. Stretch. 1979. Cutting wood production from highbush blueberry mother block plants as affected by nitrogen nutrition and fungicide treatment. *HortScience* 14(5):599–600.

56. Erez, A., and Z. Yablowitz. 1981. Rooting of peach hardwood cuttings for the meadow orchard. *Scient. Hort.* 15:137–44.

57. Evans, H. 1951. Investigations on the propagation of cacao. *Trop. Agr.* 28:147–203.

58. ———. 1953. Physiological aspects of the propagation of cacao from cuttings. *Rpt. 13th Inter. Hort. Cong.* 2:1179–90.

59. Fankhauser, I. 1985. Propagating feijoa by bench grafting. *Proc. Inter. Plant Prop. Soc.* 34:401–3.

60. Ferguson, A. R. 1984. Kiwifruit: a botanical review. In *Horticultural reviews,* Vol. 6, J. Janick, ed. Westport, Conn., AVI Publ. Co., pp. 1–53.

61. Ferree, D. C., and R. F. Carlson. 1987. Apple rootstocks. In *Rootstocks for fruit crops,* R. C. Rom and R. F. Carlson, eds. New York: John Wiley.

62. Fiorino, P., and F. Loreti. 1987. Propagation of fruit trees by tissue culture in Italy. *HortScience* 22(3):353–58.

63. Fletcher, W. F. 1942. The native persimmon. *USDA Farmers' Bul. 685.*

64. Fletcher, W. A. 1971. Growing Chinese gooseberries. *New Zealand Dept. Agr. Bul. 349.*

65. Fogle, H. W., H. L. Keil, W. L. Smith, S. M. Mircetich, L. C. Cochran, and H. Baker. 1974. *Peach production.* USDA, ARS Agr. Handbook 463. Washington, D.C.: U.S. Govt. Printing Office.

66. Forde, H. I. 1979. Persian walnut in the western United States. In *Nut tree culture in North America,* R. A. Jaynes, ed. Hamden, Conn.: Northern Nut Growers Assn.

67. Frolich, E. F. 1966. Rooting citrus and avocado cuttings. *Proc. Inter. Plant Prop. Soc.* 16:51–54.

68. Frolich, E. F., and R. G. Platt. 1972. Use of the etiolation technique in rooting avocado cuttings. *Calif. Avoc. Soc. Yearbook, 1971–1972,* pp. 97–109.

69. Funk, D. T. 1979. Black walnuts for nuts and timber. In *Nut tree culture in North America,* R. A. Jaynes, ed. Hamden, Conn.: North American Nut Growers Assn.

70. Galletta, G. J., and A. S. Fish, Jr. 1971. Interspecific blueberry grafting, a way to extend *Vaccinium* culture to different soils. *Jour. Amer. Soc. Hort. Sci.* 96(3):294–98.

71. Gardner, F. E., and G. E. Horanic. 1963. Cold tolerance and vigor of young citrus trees. *Proc. Fla. State Hort. Soc.* 76:105–10.

72. George, A. P., and R. J. Nissen. 1987. Propagation of *Annona* species: A review. *Scient. Hort.* 33:75–85.

73. Giliomee, J. H., D. K. Strydom, and H. J. Van Zyl. 1968. Northern Spy, Merton, and Malling-Merton rootstocks susceptible to woolly aphid, *Eriosoma lanigerum,* in the Western Cape. *S. Afr. Jour. Agr. Sci.* 11:183–86.

74. Gould, H. P. 1940. The Oriental persimmon. *USDA Leaflet 194.*

75. Griggs, W. H., J. A. Beutel, W. O. Reil, and B. T. Iwakiri. 1980. Combination of pear rootstocks recommended for new Bartlett plantings. *Calif. Agr.* 34(10):20–24.

76. Griggs, W. H., and H. T. Hartmann. 1960. Old Home pears show resistance to decline when on own roots. *Calif. Agr.* 14:8, 10.

77. Griggs, W. H., D. D. Jensen, and B. T. Iwakiri. 1968. Development of young pear trees with different rootstocks in relation to psylla infestation, pear decline, and leaf curl. *Hilgardia* 39:153–204.

78. Grimo, E. 1979. Carpathian (Persian) walnuts. In *Nut tree culture in North America,* R. A. Jaynes, ed. Hamden, Conn.: North American Nut Growers Assn.

79. Goode, D. Z., Jr., and R. P. Lane. 1983. Rooting leafy muscadine grape cuttings. *HortScience* 18(6):944–46.

80. Guttridge, C. G., D. T. Mason, and E. G. Ing. 1965. Cold storage of strawberry runner plants at different temperatures. *Exp. Hort.* 12:38–41.

81. Hall, I. V. 1969. Growing cranberries. *Can. Dept. Agr. Publ. 1282* (rev.).

82. Hall, I. V., L. E. Aalders, L. P. Jackson, G. W. Wood, and C. L. Lockhart. 1975. Lowbush blueberry production. *Can. Dept. Agr. Publ. 1477.*

83. Hall, W. J. 1975. Propagation of walnuts, almonds, and pistachios in California. *Proc. Inter. Plant Prop. Soc.* 25:53–57.

84. Halma, F. F., K. M. Smoyer, and H. W. Schwalm. 1944. Quick decline associated with sour rootstocks. *Calif. Citrogr.* 29:245.

85. Hamilton, R. Z., and W. Yee. 1974. Macadamia: Hawaii's dessert nut. *Univ. Hawaii Cir. 485.*

86. Hanna, J. D. 1987. Pecan rootstocks. In *Rootstocks for fruit crops*, R. C. Rom and R. F. Carlson, eds. New York: John Wiley.

87. Hansen, C. J., and H. T. Hartmann. 1968. The use of indolebutyric acid and captan in the propagation of clonal peach and peach-almond hybrid rootstock. *Proc. Amer. Soc. Hort. Sci.* 92:135–40.

88. Hansen, K. C., and J. E. Lazarte. 1984. In vitro propagation of pecan seedlings. *HortScience* 19(2):237–39.

89. Harada, H. 1975. In vitro organ culture of *Actinidia chinensis* Pl. as a technique for vegetative multiplication. *Jour. Hort. Sci.* 50:81–83.

90. Harkness, R. W. 1967. Papaya growing in Florida. *Fla. Agr. Exp. Sta. Cir. S-180.*

91. Harmon, F. N. 1954. A modified procedure for greenwood grafting of *Vinifera* grapes. *Proc. Amer. Soc. Hort. Sci.* 64:255–58.

92. Harmon, F. N., and J. H. Weinberger. 1959. Effects of storage and stratification on germination of vinifera grape seeds. *Proc. Amer. Soc. Hort. Sci.* 73:147–50.

93. ———. 1963. Bench grafting trials with Thompson seedless grapes on various rootstocks. *Proc. Amer. Soc. Hort. Sci.* 83:379–83.

94. Hartman, Henry. 1961. Historical facts pertaining to root and trunkstocks for pear trees. *Oreg. Agr. Exp. Sta. Misc. Paper 109.*

95. Hartmann, H. T. 1952. Further studies on the propagation of the olive by cuttings. *Proc. Amer. Soc. Hort. Sci.* 59:155–60.

96. ———. 1955. Auxins for hardwood cuttings. *Calif. Agr.* 9(4):7, 12, 13.

97. ———. 1958. Rootstock effects in the olive. *Proc. Amer. Soc. Hort. Sci.* 72:242–51.

98. Hartmann, H. T., W. H. Griggs, and C. J. Hansen. 1963. Propagation of own-rooted Old Home and Bartlett pears to produce trees resistant to pear decline. *Proc. Amer. Soc. Hort. Sci.* 82:92–102.

99. Hartmann, H. T., and C. J. Hansen. 1958. Rooting pear, plum rootstocks. *Calif. Agr.* 12(10):4, 14–15.

100. ———. 1955. Rooting of softwood cuttings of several fruit species under mist. *Proc. Amer. Soc. Hort. Sci.* 66:157–67.

101. Hartmann, H. T., C. J. Hansen, and F. Loreti. 1965. Propagation of apple rootstocks by hardwood cuttings. *Calif. Agr.* 19(6):4–5.

102. Hartmann, H. T., and F. Loreti. 1965. Seasonal variation in rooting leafy olive cuttings under mist. *Proc. Amer. Soc. Hort. Sci.* 87:194–98.

103. Hartmann, H. T., and K. Opitz. 1977. Olive production in California. *Univ. Calif. Div. Agr. Sci. Leaflet 2474.*

104. Hartmann, H. T., W. C. Schnathorst, and J. Whisler. 1971. Oblonga, a clonal olive rootstock resistant to verticillium wilt. *Calif. Agr.* 25(6):12–15.

105. Hearn, C. J., D. J. Hutchinson, and H. C. Barrett. 1974. Breeding citrus rootstocks. *HortScience* 9(4):357–58.

106. Hedrick, U. P. 1923. Stocks for plums. *N.Y. Agr. Exp. Sta. Bul. 498.*

107. Heppner, M. J. 1927. Pear black-end and its relation to different rootstocks. *Proc. Amer. Soc. Hort. Sci.* 24:139–42.

108. Herman, E. B., and G. J. Hass. 1975. Clonal propagation of *Coffea arabica* L. from callus culture. *HortScience* 10(6):588–89.

109. Heydecker, W., and Margaret Marston. 1968. Quantitative studies on the regeneration of raspberries from root cuttings. *Hort. Res.* 8(2):142–46.

110. Hibino, H., G. H. Kaloostian, and H. Schneider. 1971. Mycoplasmalike bodies in the pear psylla vector of pear decline. *Virology* 43:34–40.

111. Hilton, H. 1986. Budding of European (Spanish) chestnut (*Castanea sativa* Mill.) *Proc. Inter. Plant Prop. Soc.* 35:72–76.

112. Hodgson, R. W. 1940. Rootstocks for the Oriental persimmon. *Proc. Amer. Soc. Hort. Sci.* 37:338–39.

113. Holden, V. L. 1984. Propagation of filbert trees by layering and grafting. *Proc. Inter. Plant Prop. Soc.* 33:51–52.

114. Howard, B. H., E. Tal, and S. K. Miles. 1987. Red raspberry propagation from leafy summer cuttings. *Jour. Hort. Sci.* 62(4):485–92.

115. Howell, G. S. 1987. *Vitis* rootstocks. In *Rootstocks for fruit crops,* R. C. Rom and R. F. Carlson, eds. New York: John Wiley.

116. Hudson, J. P. 1954. Propagation of plants by root cuttings I. Regeneration of raspberry root cuttings. *Jour. Hort. Sci.* 29:27–43.

117. Hunter, A. G., and J. D. Norton. 1985. Rooting stem cuttings of Chinese chestnut. *Scient. Hort.* 26:43–45.

118. Hwang, S. C., C. L. Chen, J. C. Lin, and H. L. Lin. 1984. Cultivation of banana using plantlets from meristem culture. *HortScience* 19(2):231–33.

119. Ivanička, J. 1987. In vitro micropropagation of mulberry, *Morus nigra* L. *Scient. Hort.* 32:33–39.

120. Ivey, I. D. 1979. Feijoas: Selection and propagation. *Proc. Inter. Plant Prop. Soc.* 29:161–68.

121. Jaffee, A. 1970. Chip grafting guava cultivars. *The Plant Propagator* 16(2):6.

122. James, D. J., and I. J. Thurbon. 1981. Shoot and root initiation *in vitro* in the apple rootstock M.9 and the promotive effects of phloroglucinol. *Jour. Hort. Sci.* 56(1):15–20.

123. Jaynes, R. A. 1961. Buried-inarch technique for rooting chestnut cuttings. *North. Nut Grow. Assoc. 52nd Ann. Rpt.,* pp. 37–39.

124. ———. 1979. Chestnuts. In *Nut tree culture in North America,* R. A. Jaynes, ed. Hamden, Conn.: Northern Nut Growers Assn.

125. Jennings, D. L. 1988. *Raspberries, blackberries; their breeding, diseases, and growth.* London: Academic Press.

126. Joley, L. E. 1979. Pistachios. In *Nut tree culture in North America,* R. A. Jaynes, ed. Hamden, Conn.: Northern Nut Growers Assn.

127. Jones, O. P., C. A. Pontikis, and M. E. Hopgood. 1979. Propagation *in vitro* of five apple scion cultivars. *Jour. Hort. Sci.* 54(2):155–58.

128. Kartha, K. K., N. L. Leung, and K. Pahl. 1980. Cryopreservation of strawberry meristems and mass propagation of plantlets. *Jour. Amer. Soc. Hort. Sci.* 105(4):481–84.

129. Kender, W. J. 1965. Some factors affecting the propagation of lowbush blueberries by softwood cuttings. *Proc. Amer. Soc. Hort. Sci.* 86:301–6.

130. Kester, D. E., and R. A. Asay. 1975. Almond breeding. In *Advances in fruit breeding,* J. M. Moore and J. Janick, eds. West Lafayette, Ind.: Purdue Univ. Press.

131. Kester, D. E., and C. Grasselly. 1987. Almond rootstocks. In *Rootstocks for fruit crops,* R. C. Rom and R. F. Carlson, eds. New York: John Wiley.

132. Kester, D. E., C. J. Hansen, and C. Panetsos. 1965. Effect of scion and interstock variety on incompatibility of almond on Marianna 2624 rootstock. *Proc. Amer. Soc. Hort. Sci.* 86:169–77.

133. Kidd, E. L., Jr. 1987. Asexual propagation of fruit and nut trees at Stark Brothers Nurseries. *Proc. Inter. Plant Prop. Soc.* 36:427–30.

134. Klotz, L. J., E. C. Calavan, and L. G. Weathers. 1972. Virus and virus-like diseases of citrus. *Calif. Agr. Exp. Sta. Circ. 559.*

135. Krezdorn, A. H., and G. W. Adriance. 1961. *Fig growing in the South.* U.S. Dept. Agr. Handbook 196. Washington, D.C.: U.S. Govt. Printing Office.

136. Krul, W. R., and J. Meyerson. 1980. In vitro propagation of grape. In *Proc. Conf. Nurs. Prod. Fruit Plants through Tissue Culture,* R. H. Zimmerman, ed. USDA, SEA, ARR-NE-11.

137. Lagerstedt, H. B. 1979. Filberts. In *Nut tree culture in North America,* R. A. Jaynes, ed. Hamden, Conn.: Northern Nut Growers Assn.

138. Lapins, K. 1959. Some symptoms of stock-scion incompatibility of apricot varieties on peach seedling rootstock. *Can. Jour. Plant Sci.* 39:194–203.

139. Lawyer, E. M. 1978. Seed germination of stone fruits. *Proc. Inter. Plant Prop. Soc.* 28:106–9.

140. Layne, R. E. C. 1974. Breeding peach rootstocks for Canada and the northern United States. *HortScience* 9(4):364–66.

141. ———. 1987. Peach rootstocks. In *Rootstocks for fruit crops,* R. C. Rom and R. F. Carlson, eds. New York: John Wiley.

142. Lear, B., and L. Lider. 1959. Eradication of root-knot nematodes from grapevine rootings by hot water. *Plant Dis. Rpt.* 43:314–17.

143. Lider, L. 1958. Phylloxera-resistant grape rootstocks for the coastal valleys of California. *Hilgardia* 27:287–318.

144. ———. 1960. Vineyard trials in California with nematode-resistant grape rootstocks. *Hilgardia* 30:123–52.

145. ———. 1963. Field budding and the care of the budded grapevine. *Calif. Agr. Ext. Ser. Leaflet 153.*

146. Lider, L., and N. Shaulis. 1965. Resistant rootstocks for New York vineyards. *N.Y. Agr. Exp. Sta. Res. Circ. 2.*

147. Litz, R. E., and R. A. Conover. 1978. In vitro propagation of papaya. *HortScience* 13(3):241–42.

148. Livingston, R. L., and T. F. Crocker. 1975. Pecans in Georgia. *Univ. Ga. Col. Agr. Bul. 609.*

149. Lombard, P. B., and M. N. Westwood. 1987. Pear rootstocks. In *Rootstocks for fruit crops,* R. C. Rom and R. F. Carlson, eds. New York: John Wiley.

150. Loomis, N. H. 1948. Rootstocks for grapes in the south. *Proc. Amer. Soc. Hort. Sci.* 42:380–82.

151. Lynch, S. J., and R. O. Nelson. 1956. Current methods of vegetative propagation of avocado, mango, lychee, and guava in Florida. *Cieba* (Escuela Agricola Panamericana, Tegucigalpa, Honduras) 4:315–77.

152. Lyrene, P. M. 1980. Micropropagation of rabbiteye blueberries. *HortScience* 15(1):80–81.

153. ———. 1979. Hickories. In *Nut tree culture in North America,* R. A. Jaynes, ed. Hamden, Conn.: Northern Nut Growers Assn.

154. Madden, G. E. 1979. Pecans. In *Nut tree culture in North America,* R. A. Jaynes, ed. Hamden, Conn.: Northern Nut Growers Assn.

155. Madden, G. E., and H. W. Tisdale. 1975. Effects of chilling and stratification on nut germination of northern and southern pecan cultivars. *HortScience* 10(3):259–60.

156. Majumdar, P. K., and S. K. Mukherjee. 1968. Guava: A new vegetative propagation method. *Indian Hort.,* Jan.–Mar.

157. McGranahan, G. H., and P. B. Catlin. 1987. *Juglans* rootstocks. In *Rootstocks for fruit crops,* R. C. Rom and R. F. Carlson, eds. New York: John Wiley.

158. McKay, J. W., and H. L. Crane. 1953. Chinese chestnut: A promising new orchard crop. *Econ. Bot.* 7:228–42.

159. McKenzie, D. W. 1964. Apple rootstock trials: Jonathan on East Malling, Merton, and Malling-Merton rootstocks. *Jour. Hort. Sci.* 39:69–77.

160. McMillan-Browse, P. D. A. 1970. Vegetative propagation of *Corylus. Proc. Inter. Plant Prop. Soc.* 20:356–58.

161. Maxwell, N. P., and C. G. Lyons. 1979. A technique for propagating container-grown citrus on sour orange rootstock. *HortScience* 14(1):56–57.

162. Mehlenbacher, S. A. 1986. Rooting of interspecific peach hybrids by semi-hardwood cuttings. *HortScience* 21(6):1374–76.

163. Mircetich, J. S. M., J. Refsguard, and M. E. Matheron. 1980. Blackline of English walnut trees traced to graft-transmitted virus. *Calif. Agr.* 34(11, 12):8–10.

164. Mircetich, J. S. M., and A. Rowhani. 1984. The relationship between cherry leafroll virus and blackline disease of English walnut trees. *Phytopathology* 74:426–28.

165. Montgomery, H. B. S. 1963. Fruit tree raising: Rootstocks and propagation. *Minist. Agr. Fish. Foods (London) Bul. 135.*

166. Morton, J. F., and F. D. Venning. 1972. Avoid failures and losses in the cultivation of cashew. *Econ. Bot.* 26(3):245–54.

167. Moss, G. I., and R. Dalgleish. 1981. Rapid propagation of citrus in containers. *Proc. Inter. Plant Prop. Soc.* 31:262–74.

168. Mukherjee, S. K., and P. K. Majumdar. 1963. Standardization of rootstock of mango. I. Studies on the propagation of clonal rootstocks by stooling and layering. *Indian Jour. Hort.* 20:204–9.

169. ———. 1964. Effect of different factors on the success of veneer grafting in mango. *Indian Jour. Hort.* 21:46–50.

170. Mukherjee, S. K., and N. N. Bid. 1965. Propagation of mango. II. Effect of etiolation and growth regulator treatment on the success of air layering. *Indian Jour. Agr. Sci.* 35:309–14.

171. Nagabhushanam, S., and M. A. Menon. 1980. Propagation of cashew (*Anacordium occidentale* L.) by etiolation, girdling, and stooling. *The Plant Propagator* 26(1):11–13.

172. Nelson, S. H., and H. B. Tukey. 1955. Root temperature affects the performance of East Malling apple rootstocks. *Quart. Bul. Mich. Agr. Exp. Sta.* 38:46–51.

173. Nesbitt, W. B. 1974. Breeding resistant grape rootstocks. *HortScience* 9(4):359–61.

174. Nixon, R. W. 1969. Growing dates in the United States. *USDA Agr. Inf. Bul. 207.*

175. Northwood, P. J. 1964. Vegetative propagation of cashew (*Anacardium occidentale*) by the air layering method. *E. Afr. Agr. For. Jour.* 30:35–37.

176. Norton, R. A., C. J. Hansen, H. J. O'Reilly,

and W. H. Hart. 1963. Rootstocks for apricots in California. *Calif. Agr. Ext. Ser. Leaflet 156.*

177. ———. 1963. Rootstocks for peaches and nectarines in California. *Calif. Agr. Ext. Ser. Leaflet 157.*

178. ———. 1963. Rootstocks for plums and prunes in California. *Calif. Agr. Ext. Ser. Leaflet 158.*

179. Okie, W. R. 1987. Plum rootstocks. In *Rootstocks for fruit crops,* R. C. Rom and R. F. Carlson, eds. New York: John Wiley.

180. Otterbacher, A. G., and R. M. Skirvin. 1978. Derivation of the binomial *Fragaria* × *ananassa* for the cultivated strawberry. *HortScience* 13(6):637–39.

181. Painter, J. H. 1967. Producing walnuts in Oregon. *Ore. Agr. Ext. Bul. 795.*

182. Pennock, W., and G. Maldonado. 1963. The propagation of guavas from stem cuttings. *Jour. Agr. Univ. Puerto Rico* 47:280–90.

183. Perry, R. L. 1987. Cherry rootstocks. In *Rootstocks for fruit crops,* R. C. Rom and R. F. Carlson, eds. New York: John Wiley.

184. Platt, R. G. 1976. Current techniques of avocado propagation. In *Proc. 1st Inter. Trop. Short Course: The avocado,* J. W. Sauls, R. L. Phillips, and L. K. Jackson, eds. Gainesville, Fla.: Fruit Crops Dept., Univ. of Fla.

185. Platt, R. G., and E. F. Frolich. 1965. Propagation of avocados. *Calif. Agr. Exp. Sta. Cir. 531.*

186. Platt, R. G., and K. Opitz. 1973. The propagation of citrus. In *The citrus industry,* Vol. 3, W. Reuther, ed. Berkeley: Univ. Calif. Div. Agr. Sci.

187. Pongracz, D. P. 1983. *Rootstocks for grapevines.* Capetown, South Africa: D. Philips.

188. Pontikis, C. A., and P. Melas. 1986. Micropropagation of *Ficus carica* L. *HortScience* 21(1):153.

189. Popenoe, J. 1970. Coconut varieties and propagation. In *Coconuts: Production, processing, products,* J. G. Woodroof, ed. Westport, Conn.: AVI Publ. Co.

190. Porter, G. 1988. Non-astringent persimmon propagation in southeast Queensland. *Proc. Inter. Plant Prop. Soc.* 37:148–50.

191. Posnette, A., and R. Cropley. 1962. Further studies on a selection of Williams Bon Chretien pear compatible with Quince A rootstocks. *Jour. Hort. Sci.* 37:291–94.

192. Preston, A. P. 1966. Apple rootstock studies: Fifteen years' results with Malling-Merton clones. *Jour. Hort. Sci.* 41:349–60.

193. ———. 1971. Apple rootstocks 3431 (M. 27). *Ann. Rpt. E. Malling Res. Sta. for 1970,* pp. 143–47.

194. Rao, V. N. M., and I. K. S. Rao. 1957. Studies on the vegetative propagation of cashew (*Anacardium occidentale* L.). Approach grafting with and without the aid of plastic film wrappers. *Indian Jour. Agr. Sci.* 27:267–75.

195. Ray, R., and L. Walheim. 1980. *Citrus: How to select, grow and enjoy.* Tucson, Ariz.: HP Books.

196. Reimer, R. C. 1925. Blight resistance in pears and characteristics of pear species and stocks. *Oreg. Agr. Exp. Sta. Bul. 214.*

197. Reuveni, O., and M. Raviv. 1981. Importance of leaf retention to rooting of avocado cuttings. *Jour. Amer. Soc. Hort. Sci.* 106(3):127–30.

198. Reuther, W., ed. *The citrus industry,* Vol. 1, 1967; Vol. 2, 1968; Vol. 3, 1973; Vol. 4, 1978. Berkeley: Univ. Calif. Div. Agr. Sci.

199. Rosati, P., G. Marino, and C. Swierczewski. 1980. In vitro propagation of Japanese plum (*Prunus salicina* Lindl. cv. Calita.). *Jour. Amer. Soc. Hort. Sci.* 105(1):126–29.

200. Rosedale, D. O. 1963. Growing macadamia nuts in California. *Calif. Agr. Ext. Ser. Publ. AXT-103.*

201. Rouse, R. E. (presiding). 1988. Description and characteristics of major citrus cultivars of the world. Proc. of workshop sponsored by A.S.H.S. Citrus Working Group; J. B. Forsyth (Australia); M. Iwamasa (Japan); J. M. Ortiz, S. Zaragosa, and R. Bone (Spain); W. Reuther (United States). *HortScience* 23(4):679–97.

202. Rygg, G. L. 1975. *Date development, handling, and packing in the United States.* USDA, ARS Agr. Handbook 482. Washington, D.C.: U.S. Govt. Printing Office.

203. Ryugo, K., C. A. Schroeder, A. Sugiura, and K. Yonemori. 1988. Persimmons for California. *Calif. Agr.* 42(4):7–9.

204. Sax, K. 1949. The use of *Malus* species for apple rootstocks. *Proc. Amer. Soc. Hort. Sci.* 53:219–20.

205. ———. 1953. Interstock effects in dwarfing fruit trees. *Proc. Amer. Soc. Hort. Sci.* 62:201–4.

206. Schneider, H. 1970. Graft transmission and host range of the pear decline causal agent. *Phytopathology* 60:204–7.

207. Schroeder, C. A. 1947. Rootstock influence on fruit set in the Hachiya persimmon. *Proc. Amer. Soc. Hort. Sci.* 50:149–50.

208. Scott, D. H. 1977. Growing strawberries in the

southeastern and Gulf coast states. *USDA ARS Farmers' Bul. 2246.*

209. Scott, D. H., A. D. Draper, and G. M. Darrow. 1973. Commercial blueberry growing. *USDA Farmers' Bul. 2254.*

210. Scott, D. H., and D. P. Ink. 1948. Germination of strawberry seed as affected by scarification treatment with sulfuric acid. *Proc. Amer. Soc. Hort. Sci.* 51:299–300.

211. Sebastian, K. T., and J. A. McComb. 1986. A micropropagation system for carob. *Scient. Hort.* 28:127–31.

212. Seelig, R. A. 1969. Bananas. In *Fruit & vegetable facts & pointers* (3rd ed.). Washington, D.C.: United Fresh Fruit & Vegetable Assn.

213. Serr, E. F., and A. D. Rizzi. 1964. Walnut rootstocks. *Univ. Calif. Agr. Ext. Publ. AXT-120.*

214. Shalla, T., and L. Chiarappa. 1961. Pear decline in Italy. *Bul. Calif. Dept. Agr.* 50:213–17.

215. Sharpe, R. H. 1974. Breeding peach rootstocks for the southern United States. *HortScience* 9(4):362–63.

216. Simmonds, N. W. 1966. *Bananas* (2nd ed.). London: Longmans, Green.

217. Singh, L. B. 1960. *The mango: Botany, cultivation, and utilization.* New York: Interscience Publishers.

218. Singh, S., and D. V. Chugh. 1961. Marcotting with some plant regulators in loquat (*Eriobotrya japonica* Lindl.). *Indian Jour. Hort.* 18:123–29.

219. Singh-Dhaliwal, T., and A. Torres-Sepulveda. 1961. Recent experiments on rooting coffee stem cuttings in Puerto Rico. *Univ. Puerto Rico Agr. Exp. Sta. Tech. Paper 33.*

220. Sitton, B. G. 1931. Vegetative propagation of the black walnut. *Mich. Agr. Exp. Sta. Tech. Bul. 119.*

221. Smith, I. E., B. N. Wolstenholme, and P. Allan. 1974. Rooting and establishment of pecan (*Carya illinoensis* [Wang] K. Koch.) stem cuttings. *Agroplantae* 6:21–28.

222. Smith, J. C., D. J. Jordon, and F. H. Wood. 1986. Persimmon propagation: research highlights from Ruakura. *Proc. Inter. Plant Prop. Soc.* 35:323–28.

223. Smith, S. H., R. E. Hilton, and N. W. Frazier. 1970. Meristem culture for elimination of strawberry viruses. *Calif. Agr.* 24(8):8–10.

224. Smith, M. W. 1987. Grafting and cutting propagation of pecans. *Proc. Inter. Plant Prop. Soc.* 36:414–18.

225. Snir, I. and A. Erez. 1980. *In vitro* propagation of Malling–Merton apple rootstocks. *HortScience* 15(5):597–98.

226. Sookmark, S., and E. A. Tai. 1975. Vegetative propagation of papaya by budding. *Acta Hort.* 49:85–90.

227. Spiegel-Roy, P. 1962. Experience with rootstocks for apricot and stock–scion compatibilities in Israel. *Proc. 16th Inter. Hort. Cong.,* pp. 587–92.

228. Sriskandarajah, S., and M. G. Mullin. 1981. Micropropagation of Granny Smith apple: Factors affecting root formation *in vitro. Jour. Hort. Sci.* 56(1):71–76.

229. Standardi, A., and F. Catalano. 1985. Tissue culture propagation of kiwifruit. *Proc. Inter. Plant Prop. Soc.* 34:236–43.

230. Stebbins, R., and F. E. Larsen. 1981. Rootstocks for apple in the Pacific northwest. *PNW Coop. Ext. Publ. 208.*

231. Storey, W. B. 1976. *Macadamia tetraphylla*—the preferred rootstock. *Calif. Macadamia Soc. Yearbook* 22:101–7.

232. Strametz, J. R. 1984. Hot callus grafting of filbert trees. *Proc. Inter. Plant Prop. Soc.* 33:52–53.

233. Stubbs, L. L. 1963. Tristeza-tolerant strains of sour orange. *F.A.O. Plant Prot. Bul* 11:8–10.

234. Suarez de Castro, F. 1961. Semilleros o germinadores de cafe (Coffee seed beds or germinators). *Agri. Trop.* (Bogota) 17:317–24.

235. Sugiura, A., R. Tao, H. Murayama, and T. Tomana. 1986. In vitro propagation of Japanese persimmon. *HortScience* 21(5):1205–7.

236. Teague, C. P. 1966. Avocado tip-grafting. *Proc. Inter. Plant Prop. Soc.* 16:50–51.

237. Teulon, J. 1971. Propagation of passion fruit (*Passiflora edulis*) on a fusarium-resistant rootstock. *The Plant Propagator* 17(3):4–5.

238. Theakston, F. E. 1976. Carica papaya. In *The propagation of tropical fruit trees,* R. J.Garner and S. A. Chaudri, eds. Hort. Rev. 4. East Malling, England: FAO and Commonwealth Agricultural Bureaux.

239. Thomas, C. A. 1981. Mango propagation by saddle grafting. *Jour. Hort. Sci.* 56(2):173–75.

240. Thornton, I. R., and R. T. Dimsey. 1987. A comparison of rootstocks for Valencia orange in the Sunraysia district of Australia. *Jour. Hort. Sci.* 62(2):253–61.

241. Tisserat, B. 1984. Propagation of date palms by shoot tip cultures. *HortScience* 19(2):230–31.

242. Tomkins, J. P., and D. K. Ourecky. 1972. Growing strawberries in New York State. *N. Y. State Col. Agr. Inform. Bul. 15.*

243. Torre, L. C., and B. H. Barritt. 1979. Red raspberry establishment from root cuttings. *Jour. Amer. Soc. Hort. Sci.* 104(1):28–31.

244. Tukey, H. B., and K. D. Brase. 1943. The dwarfing effect of an intermediate stempiece of Malling IX. *Proc. Amer. Soc. Hort. Sci.* 42:357–64.

245. Urquhart, D. H. 1961. *Cocoa* (2nd ed.). London: Longmans, Green.

246. van Eignatten, C. L. M. 1969. Propagation of kola (*Cola nitida*) Commun. NEDERF (Amsterdam), Sept.

247. van Staden, J., B. N. Wolstenholme, and G. G. Dimala. 1976. Effect of temperature on pecan seed germination. *HortScience* 11(3):261–62.

248. Vietez, A. M., A. Ballester, M. L. Vieitez, and E. Vieitez. 1983. In vitro plantlet regeneration of mature chestnut. *Jour. Hort. Sci.* 58(9):457–63.

249. Voyiatzis, D. C., and I. C. Porlingis. 1987. Temperature requirements for the germination of olive seeds (*Olea europaea* L.) *Jour. Hort. Sci.* 62(3):405–11.

250. Warner, R. M., Z. Worku, and J. A. Silva. 1979. Effect of photoperiod on growth responses of citrus rootstocks. *Jour. Amer. Soc. Hort. Sci.* 104(2):232–35.

251. Weaver, R. J. 1976. *Grape growing.* New York: John Wiley.

252. Webber, H. J. 1948. Rootstocks: their character and reactions. In *The citrus industry,* Vol. II, H. J. Webber and L. D. Batchelor, eds. Berkeley: Univ. of Calif. Press.

253. Webster, A. D. 1974. Dwarfing plum rootstocks emerging at East Malling. *Grower* 81:382–86.

254. Weinberger, J. H. 1975. Growing nectarines. *USDA, ARS, Agr. Inf. Bul. 379.*

255. Weinberger, J. H., and N. H. Loomis. 1972. A rapid method for propagating grapevines on rootstocks. *USDA Agr. Res. Ser. ARS-W-2.*

256. Welander, M. 1985. In vitro culture of raspberry (*Rubus ideaus*) for mass propagation. *Jour. Hort. Sci.* 60(4):493–99.

257. Welander, M. 1985. Micropropagation of gooseberry, *Ribes grossularia. Scient. Hort.* 26:267–72.

258. Welch, N. C., R. Bringhurst, A. S. Greathead, V. Voth, W. S. Seyman, N. F. McCalley, and H. W. Otto. 1982. *Strawberry production in California.* Leaflet 2959. Berkeley: Univ. Calif. Div. Agr. Sci.

259. Wellman, F. L. 1961. *Coffee: Botany, cultivation, utilization.* London: Leonard-Hill.

260. Westwood, M. N., and L. A. Brooks. 1963. Propagation of hardwood pear cuttings. *Proc. Inter. Plant Prop. Soc.* 13:261–68.

261. Westwood, M. N., H. R. Cameron, P. B. Lombard, and C. B. Cordy. 1971. Effects of trunk and rootstock on decline, growth, and performance of pear. *Jour. Amer. Soc. Hort. Sci.* 96(2):147–50.

262. Westwood, M. N., and A. N. Roberts. 1965. Quince root used for dwarf pears. *Oreg. Orn. and Nurs. Dig.* 9(1):1–2.

263. Wilhelm, S. 1974. The garden strawberry: A study of its origin. *Amer. Scient.* 62(3)264–71.

264. Williams, M. W., and R. A. Norton. 1959. Propagation of red raspberries from softwood cuttings. *Proc. Amer. Soc. Hort. Sci.* 74:401–6.

265. Winkler, A. J., J. A. Cook, L. A. Lider, and W. M. Kliewer. 1974. Propagation. Chapter 9 in *General viticulture* (2nd ed.). Berkeley: Univ. of Calif. Press.

266. Worthington, T., and D. H. Scott. 1957. Strawberry plant storage using polyethylene liners. *Amer. Nurs.* 105(9):13, 56–57.

267. Wutscher, H. K. 1979. Citrus rootstocks. *Hort. Rev.* 1:237–69.

268. Zeiger, D., and H. B. Tukey. 1960. An historical review of the Malling apple rootstocks in America. *Mich. State Univ. Circ. Bul. 226.*

269. Zentmyer, G. A., A. O. Paulus, and R. M. Burns. 1962. Avocado root rot. *Calif. Agr. Exp. Sta. Circ. 511.*

270. Zeigler, L. W., and H. S. Wolfe. 1975. *Citrus growing in Florida.* Gainesville, Fla.: Univ. Presses of Florida.

271. Zimmerman, R. H., and D. C. Broome. 1980. Apple cultivar micropropagation. *USDA, Agr. Research Results ARR-NE-11,* pp. 54–63.

272. Zimmerman, R. H., G. J. Galleta, and O. C. Broome. 1980. Propagation of thornless blackberries by one-node cuttings. *Jour. Amer. Soc. Hort. Sci.* 105(3):405–7.

273. Zuccherelli, G. 1979. Moltiplicazione *in vitro* dei portainnesti clonali del pesco. *Frutticoltura* 41:15.

SUPPLEMENTARY READING

CARLSON, R. F. 1971. Fruit trees—dwarfing and propagation. *Mich. State Univ. Hort. Rpt. 1* (rev.).

CHANDLER, W. H. 1957. *Deciduous orchards* (3rd ed.). Philadelphia: Lea & Febiger.

———. 1958. *Evergreen orchards* (2nd ed.). Philadelphia: Lea & Febiger.

DAY, L. H. 1953. Rootstocks for stone fruits. *Calif. Agr. Exp. Sta. Bul. 736.*

GARNER, R. J., and S. A. CHAUDRI. 1976. *The propagation of tropical fruit trees.* Hort. Rev. 5. East Malling, England: FAO and Commonwealth Agricultural Bureaux.

HARTMANN, H. T., and J. A. BEUTEL. 1979. Propagation of temperate zone fruit plants. Leaflet 21103. Berkeley: Univ. Calif. Div. Agr. Sci.

INTERNATIONAL DWARF FRUIT TREE ASSOCIATION. *Compact tree fruit* and *Proceedings of Annual Meetings.*

JAYNES, R. A., ed. 1979. *Nut tree culture in North America.* Hamden, Conn.: Northern Nut Growers Assn.

MORTON, JULIA F. 1987. *Fruits of warm climates.* Winterville, N.C.: Creative Resource Systems.

ROACH, F. A. 1969. *Fruit tree raising: Rootstocks and propagation.* Bul. 135 (5th ed.) London: Ministry of Agriculture, Fisheries, and Food.

ROM, R. C., and R. F. CARLSON, eds. 1987. *Rootstocks for fruit crops.* New York: John Wiley.

ROM, R. C., and F. S. DAVIES, eds. 1988. Symposium: Rootstock breeding and evaluation. *HortScience* 23(1):99–118.

RYUGO, K. 1988. *Fruit culture: Its science and art.* Somerset, N.J.: John Wiley.

TUKEY, H. B. 1964. *Dwarfed fruit trees.* New York: Macmillan.

WESTWOOD, M. N. 1988. Rootstocks: Their propagation, function, and performance. Chapter 4 in *Temperate zone pomology* (rev. ed.). Portland, Oreg.: Timber Press.

19

Propagation of Ornamental Trees, Shrubs, and Woody Vines

Abelia (*Abelia* × *grandiflora*). Glossy abelia. Semi-hardwood cuttings can be rooted easily under mist in summer or fall. Talc or liquid of 1000 to 2000 ppm IBA or 1000 ppm NAA have produced superior results. Hardwood cuttings also may be rooted in fall or late winter, but less successfully than with semihardwood cuttings. Layering can also be used (*59, 119*).

Abies spp. *See* Fir.

Abutilon (*Abutilon* spp.). Flowering maple, Chinese bell flower. Started by leafy cuttings rooted under mist from summer through fall. Seeds germinate without difficulty.

Acacia (*Acacia* spp.). Acacia is usually propagated by seeds. The impervious seed coats must be softened before planting by soaking in concentrated sulfuric acid for 20 minutes to two hours or by pouring boiling water over the seeds and allowing them to soak for 12 hours in the gradually cooling water. *A. cyanophylla* (beach acacia, golden willow), *A. farnesiana* (sweet acacia, West Indian blackthorn), and *A. koa* (koa

acacia) seeds are scarified by soaking in warm water overnight. Leafy cuttings of partially matured wood can be rooted under mist if treated with 8000 ppm IBA talc. *A. redolens* and *A. subprosa* root during the fall with 3000 ppm IBA liquid. Cuttings with heels form vigorous new wood (*42*). All but the youngest plants are difficult to transplant because of a pronounced taproot. *A. koa* is difficult to root from stem cuttings, and better results have occurred with micropropagation (*119*).

Acer spp. *See* Maple.

Aesculus spp. *See* Buckeye.

Agarita (*Berberis trifoliata* [*Mahonia trifoliata*]) (*117*). A dense evergreen rounded shrub with holly-like leaves that is an early spring bloomer in Texas with yellow fragrant flowers in March. This drought-resistant plant also has edible red berries and grows best in full sun. Seeds require a two- to three-month cold stratification.

Ailanthus (*Ailanthus altissima* [Mill] Swingle). Tree of Heaven. Seed propagation is easy, and self-sowing usually occurs when both male and female trees are grown close together. Embryo dormancy apparently is present in freshly harvested seed. Stratification at about 4°C (40°F) for two months aids germination. Seed propagation produces both types of trees, but planting male trees should be avoided, since the staminate flowers produce an obnoxious odor. The more desirable female trees can be propagated by root cuttings planted in the spring.

Albizzia (*Albizia julibrissin* Durazz.). Silk tree, mimosa. This species is propagated by seeds with some treatment to overcome the impermeable seed coat, such as soaking in sulfuric acid for half an hour plus thorough washing before planting. Seed selection is important for obtaining plants with good flower color. Stem cuttings do not root, but root cuttings about 7.5 cm (3 in.) long and 12 mm (½ in.) or more in diameter taken and planted in early spring are successful (*81*). Juvenile shoots arising from root pieces can be removed rooted with 1000 to 3000 IBA liquid or talc (*57*).

Alder (*Alnus* spp.). *A. incana* (gray alder), *A. serrulata* (tag alder), *A. rhombifolia* (sierra alder), *A. rugosa* (speckled alder), *A. cordata* (Italian alder), and *A. glutinosa* (European alder). Seeds of this small deciduous tree should be thoroughly cleaned and can be planted immediately after fall harvest, but three to five months of stratification improves germination. *A. incana* stem cuttings obtain 65 percent rooting with 8000 ppm IBA talc, and can also be grafted on potted seedling understock. Softwood cuttings of *A. glutinosa* root with 3000 ppm IBA talc; cultivars can also be side-veneer grafted (*118, 119*).

Allamanda cathartica. Yellow allamanda, golden-trumpet. An evergreen shrub species commonly propagated by division (*119*). Also easily propagated using semisoftwood cuttings in summer and fall (Texas) with 1000–2000 ppm IBA liquid.

Alnus. *See* Alder.

Amelanchier spp. *See* Serviceberry.

Amorpha spp. (*57, 119*). Indigobush. A large deciduous shrub in the legume family. They have an impermeable seed coat which requires acid scarification. *A. canescens* (leadplant amphora) seed is acid scarified for 15 minutes followed by two to eight weeks of stratification; can be rooted with cuttings using 8000 ppm IBA talc. Seed of *A. fruticosa* (indigobush amorpha) can be

fall planted or given 10 minutes of acid scarification; roots readily from untreated softwood cuttings.

Anise (*Illicium* spp.). Evergreen, medium-sized shrubs which are fragrant. Seed propagation requires no pretreatment; however, most species are propagated by semihardwood cuttings which root readily. Summer cuttings treated with 3000 ppm IBA quick-dip rooted in peat–perlite media with mist (*57*).

Aralia spp. Walking stick, Angelica tree. Suckering shrubs and trees. Seeds should be sown in early fall. Scarify with sulfuric acid for 30 to 40 minutes, and cold stratify for three months to overcome the double dormancy requirement. *A. spinosa* has three-month cold stratification requirement. Asexual propagation is with 4- to 5-in.-long root cuttings, which are taken in fall and stored in cool storage until spring planting. With periclinal chimeras, avoid using root cuttings; instead, use patch budding for variegated forms (*57*).

Araucaria spp. (*119*). These large evergreen trees are seed propagated. Seeds of *A. bidwillii* (bunya-bunya), *A. columnaris* (Cook or New Caledonia pine), *A. araucana* (monkey puzzle tree), and *A. cunninghamii* (colonial pine) must be cleaned and then sown within a month after collection. There are no dormancy problems and germination time is two to four weeks. *A. heterophylla* (Norfolk Island pine) is generally seed propagated; plant seed directly in pots, placing the seed vertical and burying half the seed height. Keeping seeds moist is critical for the first 10 to 14 days after planting and can be accomplished by misting once each hour (Texas). Can also be propagated by semihardwood cuttings. Cuttings of side branches will root but produce horizontally growing plants. Plants produced from terminal cuttings grow upright.

Arborvitae. American (*Thuja occidentalis* L.), v. Oriental (*T. orientalis* L.), Korean (*T. koraiensis*), and Western (*T. plicata*).

Seeds. Germination is relatively easy, but stratification of seeds for 60 days at about 4°C (40°F) may be helpful. *T. plicata* seed generally does not require stratification (*59*). *T. occidentalis* are fall planted and *T. orientalis* are spring planted.

Cuttings

Thuja occidentalis. Hardwood cuttings can be rooted in midwinter under mist in the greenhouse. Best rooting is often found with cuttings taken from older plants no longer making rapid growth. The cuttings should be about 20 cm (6 in.) long and may be taken either from succulent, vigorously growing terminals or from more mature side growths several years old.

Wounding and treating with 3000 to 8000 ppm IBA liquid or talc are beneficial. No shading should be used.

Cuttings may also be made in midsummer and rooted out-of-doors in a shaded, closed frame. They should be several inches long and of current season's growth with somewhat matured wood at the base. They should be rooted by fall.

Thuja orientalis. Cuttings of this species are often more difficult to root than those of *T. occidentalis.* Small, soft cuttings several centimeters long, taken in late spring, can be rooted in mist beds if treated with 8000 ppm IBA talc. Cuttings 10 to 15 cm (4 to 6 in.) can be taken in summer or winter, treated with 5000 ppm liquid IBA, and rooted under mist (Texas). *T. plicata* roots easily from fall cuttings treated with 3000 ppm IBA talc (*57*). In Southern California, conifers are rooted outdoors in full sun; cuttings are taken in November and December before spring rains, since these species have cultural problems with too much water.

Grafting. The side graft is used in propagating selected clones of *T. orientalis,* with two-year-old potted *T. orientalis* seedlings as the rootstock. Grafting is done in late winter in the greenhouse. After making the grafts, the potted plants are set in open benches filled with moist peat moss just covering the union. The grafted plants should be ready to set out in the field for further growth by midspring. Selected cultivars of *T. occidentalis* can also be side-veneer grafted. In Southern California, *T. orientalis* is own rooted, rather than grafted.

Arbutus menziesii. *See* Madrone, Pacific.

Arctostaphylos. *See* Manzanita.

Ardisia. Coralberry, spiceberry (*A. crenata*), and Japanese (*A. japonica*). These medium-sized evergreen shrubs can be seed propagated without stratification. *A. crenata* can be propagated by softwood cuttings (*119*). *A. japonica* is easily rooted with 1000 ppm IBA liquid and this stoloniferous plant can be divided (*57*).

Aronia arbutifolia. *See* Chokeberry.

Ash (*Fraxinus* spp.). Seeds of most species germinate if stratified for two to four months at about 4°C (40°F). Seeds of *F. excelsior, F. nigra,* and *F. quadrangulata* should have one to three months moist storage at room temperature, followed by five to six months at about 4°C (40°F). *F. excelsior* and *F. ornus* seedlings are usually used as rootstocks for grafting or budding ash cultivars.

Asimina triloba. Common pawpaw and dwarf pawpaw (*A. parviflora*). This small tree and deciduous shrub

are propagated primarily by seed which is stratified at 4.0°C (40°F) for two to three months (*119*). A southern ecotype of *A. triloba* had 100 percent germination without stratification (*59*). Successful cutting propagation has not been reported.

Aspen. *See* Poplar.

Aspen (Quaking). *See* Poplar.

Australian Pine. *See* Casuarina.

Azalea (*Rhododendron* spp.) (*134*.). *See also* Rhododendron.

Evergreen and Semievergreen Types. Although evergreen azaleas can be propagated by seeds, grafting, and layering, almost all azaleas of these types are started by cuttings (*90*).

Seeds. The seed capsules should be gathered in the fall after they turn brown, then stored at room temperature in a container that will hold the seed when the capsules open. The seeds have no dormancy problems but will lose viability if held for extended periods in open storage; sealed, low-temperature storage is preferable. The seeds germinate satisfactorily in a thin layer of screened sphagnum moss over an acid soil mixture. Seeds are usually planted in the greenhouse from midwinter to early spring. Indoor lighting during seed propagation has been used. Optimum germinating temperatures are about 21°C (70°F) during the day and 13°C (55°F) at night. Germination usually occurs within a month. In areas having hard water, the seedlings should be watered with distilled or rain water because the alkaline salts will soon cause injury.

Cuttings. Azalea cuttings of the evergreen and semievergreen types are not difficult to propagate; roots form in three to four weeks under proper conditions. They are best taken in midsummer to fall after the current season's new growth has become somewhat hardened but before the wood has turned red or brown. Propagators may use an average of 3000 ppm IBA or no hormone treatment at all. Azaleas do very well under mist if the medium is well drained. After roots form, the mist should be reduced and finally discontinued.

Tissue Culture. Most members of the genus *Rhododendron* (rhododendrons and azaleas) can be successfully tissue cultured using shoot tips. The cytokinin 2iP is far more effective than kinetin and less toxic than BA. See Chap. 17 and the review of Dirr and Heuser (p. 189) on *Rhododendron* tissue culture (*59*).

Deciduous Types

Seeds. Deciduous azaleas are often propagated by seed, since it is inexpensive and rooting cuttings is difficult, as is inducing new growth after rooting. The same germination procedures can be used as described for the evergreen types. Growth of the seedlings is sometimes slow, but with early spring germination and early transplanting to a good growing medium, it is possible to have plants large enough to set flower buds in their second year.

In a different procedure, which will bring the seedlings into bloom sooner, seeds are sown in late summer (from seed harvested in midwinter) on top of screened peat moss in flats covered with glass. After the seeds germinate, the day length is extended by added light to keep them growing well. The seedlings are transplanted in midfall into flats of 10 percent sand and 90 percent peat moss, where they are left (in the greenhouse) until midsummer (*112*).

Cuttings (*167*). Deciduous azaleas are difficult to propagate by cuttings, but by taking soft, succulent material in the spring, particularly from stock plants brought into the greenhouse and forced, cuttings can be rooted. Timing is very important; it is much better to take the cuttings in spring than in summer.

Wounding, plus the use of root-promoting chemicals, is important in obtaining good rooting. Treatment with 75 ppm IBA liquid for 15 hours, or IBA talc at 0.8 percent is effective. Closed frames or intermittent mist with bottom heat (20°C; 68°F) and a rooting medium of sphagnum peat or a sand–peat–perlite mixture provide good rooting conditions. It is also possible to use 7- to 10-cm (3- to 4-in.)-long root cuttings.

Grafting. This is the chief propagation method for most Ghent and Mollis hybrids. *Rhododendron luteum* (*Azalea pontica*) seedlings, two or three years old, are used as the understock. This method has the disadvantages of undesirable suckering from the understock and a general lack of plant vigor, perhaps owing to an unsatisfactory graft union or to a lack of adaptability of the seedling stocks to some climates. The seedlings are potted in early spring and grafted during late summer. A side-veneer graft may be used. The grafted plants are set in closed frames in the greenhouse during healing of the union (*152*). Tree-type azaleas can be produced by grafting high in rooted cuttings of *R. concinnum*.

Layering. Azaleas are easily layered. It is a method worth using for obtaining a limited number of new plants, especially of the deciduous type difficult to start by cuttings. Simple, mound, trench, and air layering have all been used satisfactorily.

Baccharis. False willow, narrowleaf baccharis (*Baccharis angustifolia*), and eastern baccharis (*Baccharis halimifolia*). These small evergreen shrubs have excellent salt tolerance. Baccharis is seed propagated with no seed pretreatment required. Germination occurs within one to two weeks (*119*).

Bald Cypress (*Taxodium distichum* [L.] Rich.). Bald cypress is propagated by seeds, which should be fall-planted or stratified at 5°C (41°F) for 90 days. Soaking the seeds for 24 to 48 hours enhances germination.

Bamboo (*Arundinaria, Bambusa, Dendrocalamus, Phyllostachys, Pseudoasa, Sasa, Sinarundinaria, Thamnocalamus* spp.) (*184, 199*). Bamboo has about 1000 species in some 50 genera. Two classes of bamboo are commercially propagated in containers (Louisiana) to eliminate much of the hand labor and extensive field stock space needed in the old traditional method of hand digging and separating established clumps. Clump-forming bamboo (pachymorphs with constricted rhizomes) are propagated by dividing the young, peripheral rhizomes in late spring, early summer; the culms (aboveground stem with branches and leaves) are cut back and the rhizome is containerized in 5 bark:1 sand media with slow-release macro and micro elements, and watered as needed. The following spring rhizomes from the containerized stock plant are further divided and transplanted with attached culms and roots to new containers for further propagation or finishing-off as a marketable plant. The running-type bamboos (leptomorphs with vigorous rhizomes extending beyond parent plants) are more cold tolerant but less desirable in the garden. They are propagated by the division technique described above for rhizomes less than 1 cm (½ in.), or by using 1-ft rhizomes 2 cm (¾ in.) thick which are containerized and grown for two years as stock plants. Rhizomes 15 cm (6 in.) long are then containerized and grown on as a finished crop. Prevention of drying during transplanting is important. Best results are obtained from rhizomes taken and planted in late winter or early spring before the buds begin to push. *Phyllostachys* has been micropropagated, which offers potential for those bamboo species which are currently difficult to propagate or transplant.

Bambusa. *See* Bamboo.

Banksia spp. (*11*). Banksia are native Australian shrubs and small trees in the Proteaceae family. Two species with promise as a plantation-grown flower crop with export potential are *B. coccinea* (scarlet banksia) and *B. menziesii* (raspberry frost banksia). These species are propagated by seed, although they are difficult to remove from the seed-containing structures. One method is to soak several days in water and then dry quickly. There are no pregermination treatments; however, 15°C is the optimum germination (20 days) temperature for *B. coccinea*. This species can also be propagated from semihardwood cuttings using 8000 to 12,000 ppm IBA–50 percent ethanol with a five-second quick-dip. Cuttings should be wounded and inserted in a porous mix under intermittent mist with 25°C (77°F) bottom heat. It is best to root cuttings directly in liner pots and avoid root disturbance. Grafting *Banksia* spp. to understock resistant to phytohphora and tolerant to high soil phosphorus and poorly drained sites has not been successful, due to graft incompatibility.

Barberry (*Berberis* spp.).   Barberries are propagated without difficulty by fall-sowing or by spring-sowing seeds that have been stratified for two to six weeks at 4°C (40°F). Some species that require two to three months of stratification or seed can be sown in fall. It is important to remove all pulp from seeds. Semihard-wood cuttings taken from spring to fall can be rooted under mist (*224*). Some propagators have success with a hardwood cutting (leaves still attached) taken from October to November. IBA from 2000 to 8000 ppm aids rooting in some species. *B. thunbergii* 'Atropur-purea nana' is propagated with 5- to 6-in. (14-cm) semi-hardwood cuttings in May and June (Texas), which are collected when the new growth is firm at the cutting base (9°C) (*7*). The stem of the cutting should have a greenish-yellow color; avoid using brown wood. Cut-tings are quick-dipped into 1870 ppm IBA with 50 per-cent alcohol and direct stuck into 5-cm (2¼-in.) liner pots. Timing, application of mist, and the hardening-off process are critical. Greenhouse grafting of some selected types is also practiced, and layering is done occasionally. Division of crowns is useful for small quantities of plants. Micropropagation of *B. thunbergii* 'Atropurpurea' has been reported (*215*).

Basswood.   *See* Linden.

Bauhinia congesta (*117*).   Anacho orchid tree. This multistemmed deciduous. shrub has emerald leaves, which are cleft to the petiole, and produces abundant white flowers from March through summer (Texas). Seeds germinate easily without pregermination treat-ment.

Bayberry (*Myrica* spp.).   Wax myrtle. Propagated by seed or by cuttings, both with difficulty. *M. pennsylvani-cum* best started by seed given a scarification, then cold-stratification treatment, followed by kinetin applica-tions. Gibberellic acid treatments also may be helpful (*100*). *M. confera* (southern wax myrtle) is fairly easy to root under mist using 2000 ppm IBA liquid (Texas).

Beech (*Fagus* spp.).   Seeds germinate readily in the spring from fall planting or after being stratified for three months at about 4°C (40°F). Seeds should not be allowed to dry out. Selected clones are grafted by either the cleft, the whip, or the side-veneer method,

on seedlings of *F. sylvatica,* the European beech. Fre-quent transplanting or root pruning of nursery trees is necessary to prevent development of a single long tap-root, which, if present, makes subsequent transplant-ing difficult. There are clonal differences in cuttings taken from seedling plants. Overwintering survival of rooted cuttings (Denmark) is a problem (*139*). Etiola-tion of stock plants and covering shoot bases with black adhesive tape enhanced rooting 69 percent (*10*).

Berberis spp.   *See* Barberry.

Betula spp.   *See* Birch.

Birch (*Betula* spp.).   Seeds should be either fall-planted or spring-planted following stratification at about 4°C (40°F) for one to three months. *Betula nigra* matures its seeds in the spring. If planted promptly, the seed will germinate at once without pretreatment. Some of the selected, weeping forms are grafted on *Betula pubescens* or *B. pendula* seedlings (*53*). *B. papyrifera* seeds germi-nate best if held below freezing temperature for six weeks before planting. Birch is considered difficult to propagate by cuttings, but leafy, semihardwood cut-tings root under mist if taken in midsummer. Softwood cuttings of *B. alleghaniensis, B. lenta, B. nigra, B. papyri-fera,* and *B. pendula* are rooted with 8000 ppm IBA talc (*119*). Softwood one-node cuttings of 18-year-old *B. papyrifera* collected in mid-July (Vermont) treated with 4000 to 6000 ppm IBA in 50 percent ethanol had 40 to 60 percent rooting; problems occurred with survival of post-rooted cuttings (*174*). Low-growing species can be propagated by layering. Birch is commercially micro-propagated (*22, 153*).

Bittersweet (*Celastrus* spp.).   These dioecious twining vines have male flowers on one plant and female on another, and the two types must be near each other to produce attractive berries. To propagate, seeds must be removed from the berries and then fall-planted or stratified for about three months at 4°C (40°F) before planting. Clones of known sex can be propagated by softwood cuttings taken in midsummer, or by hard-wood cuttings taken in winter. Semihardwood summer cuttings root well when treated with 8000 ppm IBA talc and thiram (*59*). IBA-treated hardwood, softwood, and root cuttings can also be used.

Black Haw.   *See* Bumelia.

Bottle Brush (*Callistemon* spp.).   Although seeds ger-minate without difficulty, seedlings should be avoided because many of them prove worthless as ornamentals. The preferred method of propagation is by semihardwood cuttings taken from selected cultivars. These root under mist quite easily. Both *C. viminalis* and *C. citrinis* can be layered.

Bougainvillea (*Bougainvillea* spp.).   This showy tropical, woody, evergreen vine, grows outdoors only in mild climates, and is propagated by leafy cuttings taken at any time of the year. Rooting is aided by natural or supplemental bottom heat. Difficult to root cultivars should be treated with IBA. Micropropagation by shoot-tip culture has been successful (*39*).

Boxwood (*Buxus sempervirens* L. and *B. microphylla* Sieb. and Zucc.).   Cuttings are commonly used—either softwood taken in spring or summer, or semi-hardwood taken in the fall. In the latter method, which is ordinarily used, the cuttings are rooted in a poly-frame during the winter and spring or under mist at any time. Seeds are rarely used, because of the very slow growth of the seedlings. Young plants should always be grown in a container or transplanted with a ball of soil around their roots.

Boxwood, Oregon.   *See* Paxistima myrsinites.

Breath of Heaven.   *See* Diosma (*Ericoides*).

Broom (*Cytisus* spp.).   Seeds of many of the species germinate satisfactorily if gathered as soon as mature and treated with sulfuric acid to soften the hard seed coat before planting. The seeds are best germinated in a warm location and then transferred to a cooler place when the seedlings are several inches high. Since the various *Cytisus* species crossbreed readily, stock plants for seed sources should be isolated. Cuttings can be rooted rather easily under mist in midsummer if treated with IBA and given bottom heat; in the southern United States, natural heat is normally sufficient.

Buckeye (*Aesculus* spp.) (*59*).   Horsechestnut. This tree may be propagated by seeds, but prompt sowing or stratification after gathering in the fall is necessary. If the seeds lose their waxy appearance and become wrinkled, their viability will be reduced. For best germination, seeds of *A. arguta* (Texas buckeye), *A* × *carnea* (red horsechestnut), *A. hippocastanum* (common horsechestnut), *A. indica*, *A. octandra*, *A. sylvatica*, and *A. turbinata* should be stratified for four months at about 4°C (40°F) immediately after collecting. *A. californica* and *A. parvifolia* seed need no pretreatment. In Southern California, seeds of *A. carnea* and *A. hippocastanum* are leached and will germinate while being leached in a bag with running water within 10 days. In propagating the low-growing species, simple layering in either

spring or fall is often used. T-budding, cleft grafting, or bench grafting, using the whip graft, can be used for selected cultivars grafted on *A. hippocastanum* as the rootstock. Although normally propagated by seed, grafting is done to perpetuate superior sterile forms (*A. hippocastanum* 'Baumannii'), more quickly produce a large caliper tree, and/or avoid suckering problems. The dwarf buckeye or bottlebrush, *A. parviflora*, is propagated by underground stem pieces treated from 0 to 5000 ppm IBA liquid; it can also be propagated by root cuttings. Seeds of buckeye are poisonous if eaten.

Buckeye, Mexican.   *See* Ungnadia speciosa.

Buckthorn (*Rhamnus* spp.).   This can be propagated by planting seed out-of-doors in the fall. Macerate fruits and clean seeds. Those of some species may germinate better if given a 20-minute treatment with sulfuric acid before planting.

Buddleia.   *See* Butterfly Bush.

Bumelia (*Bumelia lanuginosa*).   Also known as the black haw or false buckthorn, this deciduous small tree is propagated by seed. Seed is fall collected and stratified at 4°C (40°F) for two months. Seed is sown in the spring in outdoor beds (*119*).

Bunya-Bunya.   *See* Araucaria spp.

Butterfly Bush (*Buddleia* spp.).   Softwood cuttings can be taken in the summer or fall and rooted in a poly-frame or under mist. Seeds started in the greenhouse in early spring will provide flowering plants by fall, although reproduction is not genetically true by seed.

Buxus spp.   *See* Boxwood.

Callistemon spp.   *See* Bottle Brush.

Calluna vulgaris.   *See* Heather.

Calocedrus decurrens.   *See* Incense Cedar.

Calycanthus spp. (*57*).   Sweetshrub, allspice. Stoloniferous shrubs. *C. floridus* is common to the eastern and southern United States. *C. floridus* 'Athens' ('Katherine') is very fragrant with yellow flowers, while *C. occidentalis* has reddish flowers. Seed should be collected when receptacles change from green to brown and immediately planted. Seed collected later should be cold stratified for three months. Good rooting success is reported with summer semihardwood cuttings using 2000 to 3000 ppm IBA liquid.

Camellia (*Camellia* spp.) (*150*). Camellias can be propagated by seed, cuttings, grafting, layering, and by micropropagation (*99, 192*). They do not come true from seed. To perpetuate cultivars, cuttings, grafts, or layers must be used. Seedlings are used in breeding new cultivars, as rootstocks for grafting, or in growing hedges where foliage is the only consideration.

Seeds. In the fall when the capsules begin to turn reddish brown and split, seeds should be gathered before the seed coats harden and the seeds become scattered. The seed should not be allowed to dry out and should be planted before the seed coat hardens. If the seeds must be stored for long periods, they will keep satisfactorily mixed with ground charcoal and stored in an airtight container placed in a cool location. After the hard seed coat develops, scarification is accomplished by pouring boiling water over the seeds and allowing them to remain in the cooling water for 24 hours. For germination, a well-drained acid soil high in organic matter should be used. It takes four to seven years to bring a camellia into flowering from seed.

Cuttings. Most *C. japonica, C. sasanqua, C. reticulata,* and *C. sinensis* cultivars and hybrids are produced commercially from cuttings, which are not difficult to root. *Camellia reticulata* cuttings do not root easily, however, and this species is generally propagated by either cleft, side-veneer, or approach grafting.

Cuttings are best taken from midsummer to midfall from the flush of growth after the wood has matured somewhat and changed from green to light brown in color. Tip cuttings are used, 3 to 6 in. long, with two or three terminal leaves. Rooting is much improved if the cuttings are treated with IBA at 20 ppm for 24 hours, quick-dips of 3000 to 5000 ppm, or 3000 to 8000 ppm IBA talc (*59*), or 6000 to 8000 ppm IBA + 2500 ppm NAA (*15*). Wounding the base of the cuttings before they are treated is also likely to improve rooting. Cuttings root best either in a polyethylene closed frame or under mist.

Camellias may be started also as leaf-bud cuttings which are handled as stem cuttings. In this case, excessive concentrations of a root-promoting substance should be avoided because this may inhibit development of the single bud.

Grafting. Camellias are frequently grafted, since some cultivars (i.e., 'Pink Pagoda') are poor rooters and grow poorly on their own roots, for multiplying new cultivars faster and to change cultivars of older established plants. Vigorous seedlings or rooted cuttings of either *C. japonica* or *C. sasanqua* can be used as stocks for grafting. Any of the side-graft methods is suitable. *C. japonica, C. sinensis,* and *C. reticulata* can be container grafted, by using a whip graft.

Both cleft and bark graft methods are used for grafting larger, established plants. This is best done in the spring, two or three weeks before new vegetative growth starts. The stock and the scion should be dormant at the time of grafting. Scions several inches long from terminal shoots containing one or two leaves and several dormant buds are used, inserted into the stock plant, which is cut off 5 to 7.5 cm (2 to 3 in.) above the soil level. The "nurse-seed" method of grafting has been used successfully for cultivars hard to root (*158*).

Layering. To obtain a few additional plants from a single mother plant, simple layering may be performed in the spring. Branches must be present close to the ground which, preferably, are young and not over 13 mm (½ in.) in diameter. One or two years may be required for enough roots to form before the layer can be removed.

Camphor Tree (*Cinnamomum camphora* (L.) J. Pressl). Best propagated by semihardwood cuttings taken in spring and rooted under mist. Seed propagation also is satisfactory.

Campsis spp. *See* Trumpet Creeper.

Cape Jasmine. *See* Gardenia.

Carob. *See* Ceratonia siliqua.

Carpinus spp. *See* Hornbeam.

Carya spp. *See* Hickory. *See also* Pecan (Chap. 18).

Castanea spp. *See* Chestnut.

Casuarina. Ironwood (*Casuarina cunninghamiana*), Australian pine (*C. equisetifolia*), and longleaf casuarina (*C. glauca*). These evergreen, large trees from Australia are easily propagated from seed. Seed require no pretreatment, and after harvesting and cleaning, seed is stored and spring sown in outdoor beds.

Catalpa (*Catalpa* spp.). Seeds germinate readily without any previous treatment. They are stored dry over winter at room temperature and planted in late spring. For ornamental purposes, *C. bignonioides* 'Nana' is often budded or grafted high on stems of *C. speciosa,* giving the "umbrella tree" effect. A strong shoot is forced from a one-year-old seedling rootstock, which is then budded with several buds in the fall at a height of 6 ft or more. Catalpa species can also be propagated in summer by softwood cuttings rooted under mist. Hardwood cuttings of *C. bignonioides* and *C. speciosa* root with 8000 ppm IBA talc and thiram (*59*); the terminal bud of cuttings should be removed.

Ceanothus (*Ceanothus* spp.). Propagation is by seed, cuttings, layering, and sometimes grafting. Seeds must be gathered shortly before the capsules open or they will be lost. Those of *C. arboreus, C. cuneatus, C. jepsoni, C. megacarpus, C. oliganthus, C. rigidis,* and *C. thyrsiflorus* have only seed coat dormancy. Placing the seeds in hot water (82 to 87°C; 180 to 190°F) and allowing them to cool for 12 to 24 hours, or even boiling in water for five minutes, will aid germination. To obtain germination in other *Ceanothus* species, which have both seed coat and embryo dormancy, the seed should be immersed in hot water (as described above), then stratified at 2 to 4°C (35 to 40°F) for two to three months.

Semihardwood cuttings can be rooted under mist at any time from spring to fall when treated with 0.5 percent IBA/NAA in alcohol and DMSO (*203*). Terminal softwood cuttings taken from vigorously growing plants in containers, and then treated with 0.1 percent IBA give good results.

Ceanothus americanus seedlings are often used as rootstocks for grafting selected clones. Layering can be done with *C. prostratus* and *C. diversifolius* when nodes touch the ground.

Cedar (*Cedrus* spp.). Seeds germinate if not permitted to dry out. No dormancy conditions occur, but soaking the seeds in water several hours before planting may be helpful. Also, a one-month cold stratification period will improve the germination rate; *C. atlantica* and *C. libani* are difficult to root. *C. deodara* can be rooted by wounding cuttings and using bottom heat; cuttings are collected in late fall to early winter and quick dipped with 5000 ppm IBA (*165*). Side-veneer grafting of selected forms on one- or two-year-old potted seedling stocks may be done in the spring (or winter in Southern California). Scions should be taken from vigorous terminal growth of current season's wood rather than from lateral shoots. In winter grafting, both lateral and terminal scions from the previous season's growth can be used, producing identical results. *Cedrus atlantica* selections and other species are grafted on *Cedrus deodara* seedling rootstock.

Cedrus spp. *See* Cedar.

Celastrus spp. *See* Bittersweet.

Celtis spp. *See* Hackberry.

Ceratonia siliqua (*119*). Carob, locust bean. This evergreen, medium-sized tree is propagated primarily by seed. Seed is fall harvested, cleaned, scarified, and sown in greenhouse beds. Taproot is easily injured, so best to sow seeds in air-pruning flats. This species can be layered and grafted.

Cercis spp. *See* Redbud.

Chaenomeles spp. *See* Quince, Flowering.

Chamaecyparis (*Chamaecyparis* spp.). Falsecypress. After collection in the fall, the seeds should be carefully dried in a warm room or in a kiln at 32 to 43°C (90 to 110°F). Stratification at about 4°C (40°F) for two to three months will aid germination. Some species such as *C. nootkatensis, C. thyoides,* and *C. praecox* have a double dormancy and a warm–cold stratification is required. Cuttings of most species are not difficult to root, particularly if juvenile forms are used. They may be taken in fall and rooted in a cold frame or in winter and rooted under mist or a polytent using bottom heat. IBA 3000 to 8000 ppm quick-dips are used (*137*).

Cherry, Flowering (*Prunus* spp.). Cultivars of *P. serrulata, P. sargentii, P. sieboldi, P. yedoensis, P. campanulata,* and *P. subhirtella* comprise the flowering cherries. Seedlings of *P. avium,* the Mazzard cherry, or *P. serrulata* are suitable as rootstocks upon which these ornamental forms may be T-budded, either in the fall or in the spring. *Prunus dropmoreana* is a suitable stock for *P. serrulata* 'Kwanzan'. If cross-pollination with other species can be avoided, *P. sargentii, P. campanulata,* and *P. yedoensis* will reproduce true from seed. Leafy cuttings of some of the flowering cherry species can be rooted under mist in high percentages if treated with IBA, but subsequent survival and overwintering are sometimes difficult. *P. tenella* semihardwood cuttings collected in July (Ireland) quick-dipped five seconds in 1250 ppm K-IBA + 1250 K-NAA had 76 percent rooting under mist; IBA alone was ineffective (*129*). *P. serrulata* 'Kwanzan' is successfully own-rooted from softwood cuttings by wounding and using 1 percent IBA and 0.5 percent NAA liquid and propagated under mist (*149*). Many of the *Prunus* species are commercially micropropagated (*22, 204*).

Chestnut. American chestnut (*Castanea dentata*), Chinese chestnut (*C. mollissima*), and sweet chestnut (*C. sativa*). These deciduous medium-sized trees are predominantly seed propagated. Seeds are harvested in early fall and stratified two to three months. Mature cuttings are very difficult to root. Rooting success has occurred with juvenile cuttings of *C. mollissima* (8000 ppm IBA talc or 7000 IBA in 95 percent alcohol), and *C. sativa* (9 percent IBA talc) (*59*). Limited success has occurred with hardwood cuttings quick-dipped with 12,000 ppm IBA and with air layering (*119*). Leafy softwood summer cuttings rooted with 1250 ppm IBA five-second quick-dip in a high-humidity ventilated fog system; pretreatment with a blanching tape also improved rooting (177). Grafting success has occurred with cleft graftage, inlay bark graft, and chip budding. Juvenile (186) and mature (220) explants of *Castanea* spp. have been successfully micropropagated.

Chionanthus spp. *See* Fringe Tree.

Chokeberry. Red (*Aronia arbutifolia*) and black (*A. melanocarpa*). Propagated by seed, which should be collected in autumn, then stratified for about three months at 5°C (41°F) before planting. Can also be started by leafy cuttings treated with 4000 ppm IBA liquid under mist, or by suckers or layers.

Cinnamomum camphora. *See* Camphor Tree.

Clematis (*Clematis* spp.) (*73, 74*). Clematis can be propagated by seed, cuttings, grafting, division of roots, or layering. Seeds of some clematis species have embryo dormancy, so stratification for one to three months at about 4°C (40°F) is likely to aid germination. Some species need a warm–cold stratification. Clematis is probably best propagated by cuttings taken from young plants; these root under mist in about five weeks. Young wood with short internodes taken in the spring gives satisfactory results, but partially matured wood taken in late spring to late summer is more commonly used. Leaf-bud cuttings taken in midsummer will also root readily under mist. Semihardwood and hardwood cuttings can be treated with 3000 to 8000 ppm IBA talc (*119*). The large-flowering hybrids are generally root-grafted by the cleft or side-veneer graft or a similar method (*191*) in the spring on *C. flammula*, *C. vitalba*, or *C. viticella* seedlings. The grafts are planted deeply with scion roots forming, the rootstock acting as a temporary nurse-root graft. When only a few plants are needed, layering the long canes gives satisfactory results.

Clethra spp. Clethra are low-growing shrubs or small trees with attractive foliage, bark, and fragrant flowers. Can be propagated by seed which is sown in flats under mist. *C. alnifolia* is rooted with softwood cuttings (June–September, Delaware) treated with Woods Rooting Compound, 1:3 Chloromone, or 1000 ppm IBA (*26*). Other species are rooted with 1000 to 5000 ppm IBA.

Coccoloba spp. (*173*). Some of these species are useful landscape plants with high salt tolerance. *C. uvifera* (sea grape) and *C. diversifolia* (pigeon plum) are primarily seed propagated. Seeds are collected from August to November (Florida), the seed coat peeled off, and no other pregermination is needed. It is important that seed not dry out before planting. Some nursery producers propagate *C. uvifera* by cuttings and air-layering.

Coralberry. *See* Ardisia.

Cornus spp. *See* Dogwood.

Cotinus coggygria. *See* Smoke Tree.

Cotoneaster (*Cotoneaster* spp.) (*86*). Seeds of most species should be soaked for about 90 minutes in concentrated sulfuric acid and then stratified for three to four months at about 4°C (40°F). As a substitute for the acid treatment, a moist, warm (15 to 24°C; 60 to 75°F) stratification treatment for three to four months may be used. This must be followed by the cold stratification treatment. Leafy cuttings of many species taken in spring or summer will root under mist without much difficulty. Evergreen/semievergreen types root better than deciduous; treat with 1000 to 3000 ppm IBA liquid. Cotoneaster can be budded high onto a pear nursery tree to produce a "tree" cotoneaster. A blight-resistant pear rootstock, such as 'Old Home', should be used. *C. bullatus* and *C. actifolius* scions are more commonly used. Simple layering can be done.

Cottonwood. *See* Poplar.

Crabapple, Flowering (*Malus* spp.). Four species of crabapples—*M. toringoides, M. hupehensis, M. sikkimensis,* and *M. florentina*—will reproduce true from seed. Selected forms of all other crabapple species, such as *M. sargentii, M. floribunda,* and *M.* 'Dolgo', should be propagated by asexual methods. *M.* 'Hopa', 'Almey', and *M.* × *eleyi* are commercially propagated with July semihardwood cuttings (Florida) which are wounded at the base, quick-dipped in 5000 ppm K-IBA salt, and rooted under mist within four to six weeks (*48*). Flowering crabapples are also micropropagated (*201*). Nursery trees are commonly propagated either by root grafting, using the whip graft, or by T-budding seedlings in the nursery row; the latter is done either as spring or as fall budding. Fall budding is considered by most nurserypeople to be the faster and most desirable method of propagating crabapples. In addition, older *Malus* trees may be topworked to the desired flowering crab cultivar. Various seedling rootstocks are used, such as *M.* × *domestica* (common apple), *M. baccata, M. ioensis,* and *M. coronaria,* as well as a number of the Malling series of apple rootstocks. The flowering 'Bechtel's Crab' is said to show delayed incompatibility, which appears in 10 to 15 years, when worked on *M.* × *domestica* seedlings. Dwarf trees of the 'Carmine' crabapple (*M. atrosanguinea*) have been produced by grafting onto *Cotoneaster divaricata*. Asexual propagation by in vitro propagation (*236*) and by cuttings is becoming more widespread. Hardwood cuttings of crabapples are difficult to root. Softwood cuttings taken in late spring root when treated with 2500 to 10,000 ppm IBA liquid in peat–perlite media under mist (*57, 59*).

Crape Myrtle (*Lagerstroemia indica* L.). This species is easily propagated from softwood or hardwood cuttings. An IBA quick dip of 1000 to 1250 ppm will aid root formation. Most cultivars are easy to transplant; however, some dwarf cultivars must be transplanted with a ball of soil (*69*). Hardwood cuttings of field-grown plants are gathered after the first hard frost, sawed into 8-in. (20-cm) cuttings, graded, bunched, stored over winter, and then planted in open fields in spring (March in Alabama) without auxin treatment and 80 percent rooting achieved (*29*). Mature explants of *L. indica* L. 'Near East' were successfully tissue cultured on woody plant medium (WPM) using the cytokinin PBA; microshoots rooted easily *ex vitro* without auxin pretreatment (*235*).

Crataegus spp. *See* Hawthorn.

Cryptomeria, Japanese (*Cryptomeria japonica* [L. f] D. Don). This can be propagated either by seeds or by cuttings. Seeds should not dry out. Cuttings, 5 to 15 cm (2 to 6 in.) long, should be taken from green wood at a stage of maturity at which it breaks with a snap when bent. Root with bottom heat; keep the cuttings shaded and cool. After roots start to form, in about two weeks, give more light; transplant to pots when roots are about 13 mm (½ in.) long. IBA promotes rooting.

× **Cupressocyparis leylandii.** *See* Leyland Cypress.

Cupressus spp. *See* Cypress.

Currant (Red Flowering). *See* Ribes sanquineum.

Cycads. Species of Encephalartos (Zamiaceae) and Kunze. Many are seed propagated, but seed viability can be a problem. Slow growth rates can limit the potential for vegetative propagation. The American *Zamia* can be regenerated from its underground tuberous root and stem tissue. The South African cycad, *Stangeria eriopus,* has been micropropagated (*168*).

Cypress (*Cupressus* spp.). Seeds have embryo dormancy, so stratification for about four weeks at 2 to 4°C (35 to 40°F) will improve germination. Cuttings

can be rooted if taken during winter months. Treatments with IBA 2000 to 8000 ppm aid rooting. Side-veneer grafting of selected forms on seedling *Cupressus* rootstocks in the spring is often practiced.

Cytisus spp. *See* Broom.

Daphne (*Daphne* spp.) (*32, 36*). Daphne can be propagated by seed, stem and leaf-bud cuttings, layering, and grafting. Seeds should be sown at once after harvest or, if dried, scarified, then given a moist-chilling period before planting. Daphne is probably best propagated by leafy cuttings in sand and peat moss (2:1) under mist; cuttings are taken in summer from partially matured current season's growth. The daphnes do not transplant easily and should be moved only when young. Berries are very poisonous if eaten.

Dawn Redwood. *See* Metasequoia.

Davidia involucrata. Dove tree. Seeds have a double dormancy requirement of five to six months warm stratification followed by three months of cold stratification (*85*). Polyethylene bags containing 1 sand:1 peat is a suggested system for the pretreatment of *Davidia* seeds. Rooting is variable. Leaf-bud cuttings treated with 3000 ppm IBA talc rooted 85 percent; using four leaves per cutting from current season's growth, wounded, and treated with 8000 ppm IBA talc, there was 50 percent rooting under mist (*59*). Disturbance of rooted cuttings and overwintering of cuttings can be a problem.

Delonix regia. *See* Royal Poinciana.

Deutzia (*Deutzia* spp.). Deutzia is easily propagated either by hardwood cuttings lined-out in the nursery row in spring or by softwood cuttings under mist in summer.

Diosma (*Diosma ericoides*). Breath of Heaven. Propagated by leafy cuttings taken in summer and rooted under mist.

Diospyros texana (*117*). Texas persimmon. A small multitrunked tree of 15 to 30 ft. It has intricate branching and smooth gray bark. Females bear black edible fruit 2.5 cm (1 in.) in diameter. Propagation is by seed and no pregermination treatment is necessary.

Disanthus cercidifolius (*57*). A multistemmed shrub of potential in the Hammamelidaceae with red fall color and purple flowers. Reported to have a double dormancy requirement requiring seed be warm stratified five months and cold stratified three months. Cuttings collected in July (Georgia) root best when treated with 10,000 ppm IBA alcohol quick-dip and propagated under mist with peat–perlite. Successfully overwintered in a polyhouse if potted up immediately after rooting, fertilized, and allowed to harden off naturally in the fall.

Dogwood (*Cornus* spp.). Seeds have various dormancy conditions; those of the popular flowering dogwood (*C. florida*) require either fall planting or a stratification period of about four months at 4°C (40°F). Best germination is obtained if the seeds are gathered as soon as the fruit starts to color, and sown or stratified immediately. If allowed to dry out it is best to remove seeds from the fruit and soak in water. Other species require, in addition, a treatment to soften the seed covering. Two months in moist sand at diurnally fluctuating temperatures (21 to 30°C; 70 to 85°F), followed by four to six months at 0 to 4°C (32 to 40°F), is effective. With some species, the warm stratification period may be replaced by mechanical scarification or soaking in sulfuric acid.

Some dogwoods can be started easily by cuttings. *C. kousa* cuttings collected in mid-June through July (Massachusetts) root well when wounded on one side and treated with either 8000 ppm IBA talc and 15 percent thiram or five-second quick-dips of 2500 ppm IBA and NAA (*84*). Those of *C. florida* are best taken in late spring or early summer from new growth after flowering, then rooted under mist. Treatments with IBA at 20,000 ppm liquid formulation have given good rooting, as has 2 percent IBA talc, removing basal leaves, and wounding cuttings (*95*). In Florida semihardwood cuttings taken in May and late August are quick-dipped in 1 percent K-IBA for 3 or 10 seconds, respectively (*48*). Hardwood cuttings taken in the spring are successful with certain species, such as *C. alba*.

C. florida 'Rubra' can be rooted successfully if cuttings are taken in early summer after the second flush growth and rooted under mist in peat moss–sand 1:3, or in perlite–peat 3:2 (*194*), after treating with 3000 ppm IBA. To ensure survival through the following winter in cold climates, the potted cuttings should be kept in heated cold frames or polyhouses to hold the temperature between 0 and 7°C (32 and 45°F). Rooted cuttings that had shoot growth in the fall, but were not given nitrogen, had the best overwinter survival in a cold frame covered with microfoam (*95*).

Selected types such as the red flowering dogwood, *C. florida* 'Rubra', and the weeping forms (difficult to start by cuttings) are often propagated by T-budding in late summer or by whip grafting in the greenhouse in winter on *C. florida* seedling rootstock. *C. kousa* and *C. florida* can be reciprocally grafted (*84*).

Douglas Fir (*Pseudotsuga menziesii* [Mirb.] Franco) (*16*). Seeds of this important lumber and Christmas tree species exhibit varying degrees of embryo dormancy. For prompt germination, it is best to sow the seeds in the fall or stratify them for two months at about 4°C (40°F). Plantlets have been micropropagated successfully using cotyledon explants (*37*).

Douglas fir cuttings are rather difficult to root, but by taking them in late winter, treating them with IBA, and rooting them in a sand–peat moss mixture, it is possible to obtain fairly good rooting. Cuttings from young trees root much more easily than do those from old trees, and cuttings from certain source trees are easier to root than those from others. In selection criteria of Douglas fir Christmas trees, both rooting of cuttings are strongly related to clones and seasonal fluctuation; NAA is more effective than IBA in stimulating rooting (*182*).

Dove tree. *See* Davidia involucrata.

Elaeagnus (*Elaeagnus* spp.). Russian olive, silverberry, silverthorn. Seeds planted in the spring germinate readily following a stratification period of three months at 4°C (40°F). Removal of the pit (endocarp) for silverberry seeds (*E. commutata*) resulted in about 90 percent germination of unstratified seeds, since a germination inhibitor is apparently present in the pit; if not fall planted, seed should be cold stratified three months. Seeds of the Russian olive, *E. angustifolia*, should be treated with sulfuric acid for 30 to 60 minutes before fall planting or stratification; this deciduous species is commercially seed propagated, and only limited rooting success (28 percent) has been obtained with 3000 ppm IBA talc (*59*). Leafy cuttings of the evergreen species root readily. Hardwood and semihardwood cuttings of *E. pungens* root well when treated with 8000 to 20,000 ppm IBA talc (*15*).

Elderberry (*Sambucus* spp.). Seed propagation is difficult because of complex dormancy conditions involving both the seed coat and embryo. Probably the best treatment is a warm (21 to 30°C; 70 to 85°F), moist stratification period for two months, followed by a cold (4°C; 40°F) stratification period for three to five months. These conditions could be obtained naturally by planting the seed in late summer, after which germination should occur the following spring. Since softwood cuttings root easily if taken in spring or summer, this method is generally practiced.

Elm (*Ulmus* spp.). Seed propagation is commonly used. Elm seed loses viability rapidly if stored at room temperature, but it can be kept for several years in sealed containers at 0 to 4°C (32 to 40°F). Seed ripening in the spring should be sown immediately, and germination usually takes place promptly. For those species that ripen their seed in the fall, either fall planting or stratification for two months at about 4°C (40°F) should be used. To obtain tree uniformity, selected clones are propagated by budding on seedling rootstocks of the same species.

Softwood cuttings of several elm species can be rooted under mist when taken in early summer. Treatments with IBA at 50 ppm for 24 hours have been beneficial (*181*). Semilignified cuttings have been rooted without mist or hormone treatments (*193*). Softwood cuttings taken from new growth arising from cutoff stumps root readily if treated with IBA and placed under mist (*195*). *U. hollandica* can be propagated by hardwood cuttings taken in late winter, treated with IBA at 1500 ppm, then placed in a bin over bottom heat for six weeks before planting (*230*).

English Ivy (*Hedera helix* L.).   English ivy is readily propagated by rooting cuttings of the juvenile (nonfruiting, lobed-leaf) form. It is also sometimes grafted onto *Fatshedera* (*Fatsia japonica* × *Hedera helix*) as a rootstock. *Acanthopanax sieboldianus* is a suitable rootstock where root rot is likely to result from excessive soil moisture.

Enkianthus (*Enkianthus* spp.) (*59, 166*).   Attractive shrubs with pronounced fall color. Blossoms vary from white to creamy yellow to shrimp color to crimson to rosy red. Seed propagation requires no pretreatment, and germination occurs within two to three weeks after sowing. Cuttings root quite easily. Leafy cuttings are taken mid-June (Massachusetts), treated with talc or liquid IBA at 5000 to 8000 ppm, and stuck in 1 sand:1 peat under mist. Rooted cuttings should be allowed to harden off and overwinter before they are disturbed.

Erica spp.   *See* Heath.

Eriobotrya japonica.   *See* Loquot.

Escallonia (*Escallonia* spp.).   Escallonia is easily started by leafy cuttings taken after a flush of growth. Cuttings root well under mist and respond markedly to treatment with IBA. It is best to direct root in liner pots since transplanting is difficult.

Eucalyptus (*Eucalyptus* spp.) (*47*).   Eucalyptus is almost entirely propagated by seeds planted in the spring. Mature capsules are obtained just before they are ready to open. No dormancy conditions occur in most species, so seeds are able to germinate immedi-

ately following ripening. Seeds of some species—for example, *E. dives, E. niphophila,* and *E. pauciflora*—however, require stratification for about two months at 4°C (40°F) for best germination. Eucalyptus seedlings are very susceptible to damping-off. Seeds are usually planted in flats of pasteurized soil placed in a shady location. From flats they are transplanted into small pots, from which they are later lined out in a nursery row. The roots of young trees will not tolerate drying, so the young plants should be handled as container-grown stock. Seeds may be sown directly into containers in which the seedlings are grown until planting in their permanent location (*98*).

Eucalyptus is difficult to start from cuttings, but good rooting can be obtained. For example, leafy cuttings of *E. camaldulensis* taken in early spring from shoots arising from the base of young trees, wounded, treated with a 4000 ppm solution of IBA plus NAA, 1:1, and placed in perlite under mist over bottom heat at 21°C (70°F) gave 65 percent rooting (*76*).

Eucalyptus ficifolia has been grafted successfully by a side-wedge method, using young vigorous *Eucalyptus* seedling rootstocks growing in containers and placed under very high humidity following grafting. Use of scions taken from shoots that had been girdled at least a month previously increased success (*189*). Eucalyptus can now be routinely micropropagated, even with mature explant tissue (*27, 54*).

Euonymus (*Euonymus* spp.) (*89*).   Stratification for three to four months at 0 to 10°C (32 to 50°F) is required for satisfactory seed germination. Remove seeds from fruit and prevent drying. Euonymus is easily started by cuttings, hardwood in early spring for the deciduous species and leafy semihardwood under mist after a flush of growth has partially matured for the evergreen types. IBA can be applied as a 2000-ppm dip or 3000 to 8000 ppm talc (*59*). Layering is also successful.

Euphorbia pulcherrima.   *See* Poinsettia.

Fagus spp.   *See* Beech.

False Cypress.   *See* Chamaecyparis.

× Fatshedera (*Fatsia* × *Hedera*).   *See* English Ivy.

Feijoa (*Feijoa sellowiana*).   See Chap. 18.

Ficus spp.   Fig, rubber plant. A wide range of ornamental trees, vines, and ornamental potted plants. Many species form aerial roots, are easily rooted by cuttings and layering, and are micropropagated commercially (*178*). Single-node propagation of *Ficus* spp. produces many more plants per stock plant than the common method of air layering (*179*).

F. benjamina. Weeping fig. Easily started by leafy semihardwood cuttings taken in spring or early summer and rooted under mist. Can be propagated by shoot-tip culture in vitro.

F. elastica (*17*). Rubber plant. Propagated as cuttings taken from 5- to 27-cm (6- to 12-in.) shoots; single buds or ''eyes'' can be removed and rooted. These cuttings are made in spring, inserted in sand or a similar medium, and held in a warm greenhouse. Shoots of trees growing outdoors in the tropics are air-layered and the rooted air layers shipped to wholesale nurseries for further development. Indoor plants that become too ''leggy'' also can be air-layered (see Fig. 14–8).

F. lyrata. Fiddleleaf fig. Can be propagated by cuttings, or by air-layering (*125*). Can also be micropropagated by adventitious shoots developing on excised leaf pieces in vitro.

F. pumila. Creeping fig. Propagated with 4 to 5 in. (10- to 13-cm)-long cuttings. The juvenile form, which is a climbing vine with aerial rootlets, roots easily all year round; 1000 to 1500 ppm IBA quick-dips enhance rooting (*52*). The mature form lacks aerial roots, but can be rooted successfully with 2000 to 3000 ppm IBA.

Fig. *See* Ficus spp.

Fir (*Abies* spp.). Seed propagation is not difficult, but fresh seed should be used, since most species lose their viability after one year in ordinary storage. Embryo dormancy is generally present; fall planting or stratification at about 4°C (40°F) for one to three months is required for good germination. Fir seedlings are very susceptible to damping-off. They should be given partial shade during the first season, since they are injured by excessive heat and sunlight.

Fir cuttings are considered difficult to root, but *Abies fraseri* cuttings can be rooted in high percentages, especially if taken from young trees, wounded, and treated with IBA (*115*). This species, along with white fir (*A. concolor*) and red fir (*A. magnifica*), the California ''silver tip,'' are important Christmas tree species.

Firethorn. *See* Pyracantha.

Forsythia (*Forsythia* spp.). Forsythia is easily propagated by hardwood cuttings set in the nursery row in early spring or by leafy softwood cuttings taken during late spring and rooted under intermittent mist.

Fraxinus spp. *See* Ash.

Fringe Tree (*Chionanthus virginicus* L. and *C. retusus* L.). Seed propagation can be used, but is very slow. Embryo dormancy, as well as some endosperm inhibition, seems to be present. Probably the best practice is stratification for 30 days or more at room temperature, followed by one or two months of stratification at about 4°C (40°F). A three-month warm stratification followed by a three-month cold stratification has also been recommended (*57*).

Cutting propagation of the Chinese fringe tree (*C. retusus*) generally has been considered almost impossible, but by taking softwood cuttings in late spring, rooting under mist, treating with 8000 to 20,000 ppm IBA talc, and using a mixture of sand and vermiculite as the rooting medium, excellent rooting percentages can be obtained. Rooting improves when stock plants are kept juvenile or when serial propagation is used.

Fuchsia (*Fuchsia* spp.) (*222*). Fuchsia is easily rooted by leafy cuttings maintained under humid conditions. Roots develop in two or three weeks.

Gardenia (*Gardenia jasminoides* Ellis). Cape jasmine. Leafy terminal cuttings can be rooted under mist in the greenhouse, full sun or shade from fall to spring. Sand and peat moss, 1:1, is a good rooting medium. Gardenias are difficult to transplant and should be moved only when small. *G. jasminoides* 'Ellis' was successfully micropropagated with 2iP; microshoots rooted easily during acclimation without auxin pretreatment (*66*).

Garrya elliptica (*185*). *G. elliptica* 'James Roof' is an ornamental tree producing catkins 300 mm long. It is generally considered difficult to propagate asexually. Tip nodal cuttings 10 cm long on side shoots taken with a heel with well-developed terminal buds are collected late summer to December (England). Wounding is optional and cuttings are treated with 8000 ppm IBA talc and propagated in 6 peat moss:4 perlite media. Cuttings are rooted under mist or plastic film. Direct rooting in small liner pots is desirable to avoid root disturbance problems. Rooted cuttings should complete their first spring flush of growth before transplanting.

Ginkgo (*Ginkgo biloba* L.) (*146*). Seed propagation may be used for this ''living fossil,'' shown by records in rocks to have existed on earth 150 million years ago during the dinosaur age. A satisfactory procedure is to collect the "fruits" in midfall, remove the pulp, and pack the cleaned seeds in layers of moist sand for 10 weeks at temperatures of 15 to 21°C (60 to 70°F) to permit the embryos to finish developing. After this the seeds require a stratification period of several months at about 4°C (40°F) for good germination. Seedlings produce either male or female trees, but the sex cannot be determined until the trees flower—after about 20 years. The plum-like "fruits" on the female trees have a very disagreeable odor, so only male trees are used for ornamental planting. The fruit contains the same chemical irritants as poison oak and can cause very bad rashes. For these reasons, seed propagation should not be used except for understock production; propagation by cuttings from male trees is advisable. Softwood cuttings taken in early summer can be rooted under mist if treated with 8000 ppm IBA talc or liquid. Commercial propagation is by T-budding or chip budding using buds from male trees inserted into ginkgo seedlings, since cutting produced liners do not grow as quickly. Can also be cleft grafted in January (Southern California).

Gleditsia triacanthos.   *See* Honeylocust, Common.

Golden Chain.   *See* Laburnum.

Goldenrain Tree (*Koelreuteria* spp.).   This tree is usually propagated by seed. The seeds have double dormancy, germinating best if the seed coats are softened by soaking for about 60 minutes in concentrated sulfuric acid, or by mechanical scarification followed by stratification for about 90 days at 2 to 4°C (35 to 45°F) to overcome embryo dormancy. In Southern California, stratification is avoided by pouring 83°C (180°F) water on seed and soaking overnight. Root cuttings can be used and softwood cuttings of new growth taken in the spring can be rooted under mist.

Gold Tree.   *See* Tabebuia argentea.

Golden Trumpet.   *See* Allamanda cathartica.

Gordonia spp. (*59, 119*).   Loblolly bay, black laurel. These medium-sized trees can be increased by seed, which germinate readily without pregermination requirements. Semihardwood cuttings are easily rooted with 2500 to 3000 ppm IBA liquid or 3000 ppm IBA talc. These species transplant easily after rooting.

Grevillea spp.   These Australian native shrubs or trees are propagated by seed or by cuttings. Cuttings of the low-growing species root readily, but larger-growing species, such as *G. robusta,* silk oak, are best propagated by seed. *G. johnsonii* is difficult to root but fall cuttings (Australia) treated with 4000 ppm IBA–ethanol five-second quick-dip have 70 percent rooting (*70*).

Hackberry (*Celtis* spp.) (*154*).   Seeds are ordinarily used, sown either in the fall or stratified for two or three months at about 4°C (40°F) and planted in the spring. Prior to stratification, treatments to soften the seed coat, such as soaking in concentrated sulfuric acid, may hasten germination. Clones of two species, *C. occidentalis* and *C. laevigata* (sugarberry), can be started by cuttings, but the rooting percentage is low. Grafting and chip budding also have been used with *C. occidentalis* and *C. laevigata* as rootstock for other species.

Halesia.   *See* Silverbell.

Hamamelis.   *See* Witch Hazel.

Hawthorn (*Crataegus* spp.) (*49*).   Hawthorns tend to reproduce true by seed. Pronounced seed dormancy is present because of a combination of an impermeable seed coat and embryo conditions. Probably the best procedure for rapid germination is stratification of freshly collected and cleaned seed in moist peat moss for three or four months at 21 to 27°C (70 to 80°F) (or treatment with sulfuric acid), followed by stratification for five months at about 4°C (40°F). Planting the seed in early summer will provide these conditions naturally, resulting in germination the following spring. Untreated seed may require two or three years to germinate. Seeds of some *Crataegus* species do not have an impermeable seed coat, so the initial high-temperature storage period is unnecessary. Since hawthorn develops a long taproot, transplanting is successful only with very young plants. Air pruning of seedling roots may be beneficial.

Selected clones may be T-budded or root-grafted on seedlings of *C. crus-galli,* or *C. coccinea* for the American (entire leaf) types and seedlings of *C. laevigata* or *C. monogyna* for the European (cut-leaf) types.

Heath (*Erica* spp.) and **Heather** (*Calluna vulgaris* Hull) (*60*).   The propagation of these two closely related genera is about the same. Seeds may be germinated in flats in the greenhouse in winter or in a shaded outdoor cold frame in spring. Leafy, partially matured cuttings taken at almost any time of year, but especially in early summer, root readily under mist in a glasshouse or polyethylene-covered cold frame. Most species root readily; however, IBA (1000 ppm quick-dip or 4000 ppm talc and 15 percent thiram) speeds up rooting.

Hebe (*Hebe* spp.).　Veronica. Propagated by seed, by leafy cuttings in summer under mist, or by layering.

Hedera helix.　*See* English Ivy.

Hemlock (*Tsuga* spp.).　Hemlocks are propagated by seed without difficulty. Seed dormancy is variable; some lots exhibit embryo dormancy, whereas others do not. To ensure good germination it is advisable to stratify the seeds for two to four months at about 4°C (40°F). Fall planting outdoors generally gives satisfactory germination in the spring. The seedlings should be given partial shade during the first season. Hemlock cuttings are somewhat difficult to root, but success has been reported with cuttings taken at all times of the year. *T. canadensis* cultivars can be rooted with either hardwood (bottom heat) or softwood cuttings treated with 1 percent IBA–50 percent ethanol quick-dip (Massachusetts). Softwood cuttings have a lower rooting percentage, but once rooted have a greater growth rate than hardwood cuttings (*55*). Layering also is successful, and grafting is also done.

Heteromeles arbutifolia (Ait.).　M. J. Roemer, Toyon, Christmas Berry. Usually propagated by seed, which requires a long (three month) stratification period, or by fall planting to obtain outdoor winter chilling of the seed. Softwood tip cuttings taken in midspring, treated with 0.8 percent IBA in talc, and placed under mist will root (*97*). It can also be started by layering.

Hibiscus (*Hibiscus syriacus* L.).　Shrub-althea, Rose of Sharon. Easily propagated, either by hardwood cuttings in the nursery row in spring, or by softwood cuttings in midsummer under mist. Lateral shoots make good cutting material. Softwood cuttings respond well to treatment with 1000 ppm IBA liquid. Hardwood cuttings have rooted well when treated with 8000 ppm IBA talc.

Hibiscus, Chinese (*Hibiscus rosa-sinensis* L.).　Seeds, cuttings, budding or grafting, division, and air layering can be used. Hibiscus is commercially micropropagated.

　Cuttings.　Softwood and hardwood cuttings are not difficult to root; however, there are cultivar differences (*127*). Softwood and semihardwood cuttings are rooted under mist with Styrofoam: peat, or 1 bark:1 sand:1 peat, or Oasis Horticubes to prevent root damage during transplanting. Mild IBA with tip cuttings and a slightly stronger solution on wounded hardwood cuttings have been used (*127*). Soft-tip cuttings 11 cm (4 in) long are rooted December to April (Australia) with 0.5 percent IBA talc, basal heat (72°F, 22°C),

and mist (*67*). A reduction in light intensity (65 percent natural light—in Florida) enhanced rooting of leafy cuttings more than did 8000 ppm IBA talc dip; both Ethephon (1000 ppm) and IBA enhanced rooting of cuttings taken from stock plants grown in 100 percent natural light (*123*).

　Grafting.　Strongly growing cultivars which are resistant to soil pests and can be started easily by cuttings, such as 'Single Scarlet', 'Dainty', 'Euterpe', or 'Apple Blossom', are used as rootstocks. Some clones develop into much better plants when grafted on these rootstocks than when they are on their own roots, propagated by cuttings. Whip grafting in the spring or cleft grafting or side grafting in late spring or early summer is successful. Scions of current season's growth, about pencil size, are grafted on rooted cuttings of about the same size (*196*).

　Budding.　T-budding, using an inverted T, is sometimes practiced, generally in the spring, although it is successful at any time during the year when the bark is slipping.

　Air Layering.　Air layering is practiced during the spring or summer, particularly for cultivars difficult to start by cuttings. Roots will usually form in six to eight weeks.

Hickory. Water hickory (*Carya aquatica*), bitternut hickory (*c. cordiformis*), pignut hickory (*C. glabra*), shellbark (*C. laciniosa*), nutmeg hickory (*C. myristicaeformis*), shagbark hickory (*C. ovata*), and Mockernut hickory (*C. tomentosa*).　Some of these tree species have edible fruits. Seeds (nuts) are fall collected and will erratically germinate without pretreatment. Most species have a three- to four-month cold stratification requirement for uniform germination. No reports exist on successful rooting or grafting of hickory. However, *C. illinoiensis* (see Chap. 18) is successfully grafted by patch budding, inlay bark graft, and four-flap (banana graft); cutting propagation is difficult; however, Medina (*157*) reports high rooting success with pecans using mounding and stooling techniques. Some of these asexual propagation techniques could be applied to other *Carya* species.

Holly (*Ilex* spp.) (*80, 106, 200*).　Holly can be propagated by seeds, cuttings, grafting, budding, layering, and division. Most hollies are dioecious. The female plants produce the very desirable decorative berries if male plants are nearby for pollination. In seed propagation, both male and female plants are produced in ratios of one female to three, or sometimes up to 10, male plants. Sex cannot be determined, however, until the seedlings start blooming, at 4 to 12 years.

Seeds. Germination of holly seed is very erratic; those of some species, *I. crenata, I. cassine, I. glabra, I. vomitoria, I. amelanchier,* and *I. myrtifolia,* germinate promptly and should be planted as soon as they are gathered. Seeds of other species, *I. aquifolium* (English holly), *I. cornuta* (Chinese holly), *I. vericilliata, I. decidua,* and most *I. opaca* (American holly), do not germinate until a year or more after planting even though stratified, probably due to rudimentary embryos at time of harvest.

Seeds of *I. aquifolium, I. opaca,* and *I. cornuta* should be collected and cleaned as soon as the fruit is ripe in the fall and then stored at about 4°C (40°F) until spring in a mixture of moist sand and peat moss. Germination in these species generally does not occur until a year later, and then growth is very slow, two seasons being required to bring seedlings of *I. opaca* to a size large enough to be used for grafting.

Cuttings. This is the method most used commercially, permitting large-scale production of choice clones. Semihardwood tip cuttings from well-matured current season's growth produce the best plants. Cuttings taken from flat, horizontal branches of *I. crenata* tend to produce plants having this type of growth (plagiotropic) and those from upright growth (orthotropic) produce upright plants.

Timing is important; best rooting is usually obtained from mid- to late summer, but cuttings may be successfully taken on into the following spring. Wounding the base of the cuttings helps induce root formation. The wounding induced by stripping off the lower leaves may be sufficient.

The use of a root-promoting chemical, particularly IBA at relatively high concentrations (8000 to 20,000 ppm), is essential in obtaining rooting of some cultivars, such as *I. opaca* 'Savannah' (10,000 ppm IBA liquid), whereas 2500 ppm IBA is sufficient for medium-difficult species such as *I. cornuta (15)*. *I. vomitoria* 'Nana' are direct stuck into 6-cm (2¼-in.) liner pots using 3000 ppm K-IBA spring–summer (Alabama) and 5000 ppm K-IBA fall–winter (62). Bottom heat at 21 to 24°C (70 to 75°F) is beneficial. The maintenance of a high relative humidity is essential. The use of intermittent mist in a greenhouse, where high temperatures can be maintained, provides good rooting conditions. A 1 perlite:1 peat moss or 3 pinebark:2 peat:2 perlite rooting medium is satisfactory.

Grafting and Budding. Hollies are easily grafted, the cleft, whip, and side grafts being used. The operation is best performed during the dormant season for field grafting. Grafting is often done on greenhouse potted stock. T-budding also is suitable, and is done in late summer or early spring. *Ilex opaca* is a satisfactory stock for its own cultivars and those of *I. aquifolium,* but probably the best stocks for English holly are *I. aquifolium* and *I. cornuta* 'Burfordii'.

Air Layering. This is successful for a number of *Ilex* species. Layers are best started in early summer; after 10 to 14 weeks, plants 30 to 60 cm (1 to 2 ft) high are produced.

Honeylocust, Common (*Gleditsia triacanthos* L.**).** This is readily propagated either by seeds planted in the spring or by cuttings. The thornless honey locust, *G. triacanthos,* var. *inermis,* and the thornless and fruitless patented 'Moraine' locust are usually propagated by grafting on seedlings of the thorny type. In seed propagation, soaking the seed in sulfuric acid for one hour, followed by stratification at 2°C (36°F) for three months, gives good germination. Hardwood cuttings planted in the spring root successfully.

Honeysuckle (*Lonicera* spp.**).** Seeds show considerable variation in their dormancy conditions, some species having both seed coat and embryo dormancy, some only embryo dormancy, and some no dormancy. This variability also occurs among different lots of seeds of the same species. In *L. tatarica* some lots have no seed dormancy. In general, however, for prompt germination, stratification for two to three months at about 4°C (40°F) is recommended. Seeds of *L. hirsuta* and *L. oblongifolia* should have two months of warm stratification (21 to 30°C; 70 to 85°F), followed by two to three months of stratification at about 4°C (40°F).

Most honeysuckle species are propagated easily by either hardwood cuttings in the spring or leafy softwood cuttings of summer growth under mist. Layering of vine types, such as 'Hall's' honeysuckle, is very easy, roots forming wherever the canes become buried under moist soil.

Hornbeam (*Carpinus* spp.**).** For seed propagation, collect seeds while the wings are still soft and pliable. Check to make sure that seeds have embryos, which may be absent after stressful growing conditions *(145)*. Do not allow seeds to dry out. Sow outdoors in autumn or stratify (three to four months) over winter and sow in spring. If the seed dries, a hard seed coat germination block develops, and this double dormancy requires some type of scarification before stratification; two months of warm stratification followed by two months of cold stratification has been used with *C. caroliniana (21)*. Cultivars may be grafted or budded on seedlings of the same species. Stem cuttings of *C. betulus* 'Fastiagata' rooted with 2 percent IBA, *C. carolina* 'Pyramidalis' with 1.6 percent IBA, *C. japonica* with 3000 ppm IBA-talc and thiram *(59)*; wounding cuttings has been recommended. Stock plant etiolation improved rooting of *C. betulus* softwood cuttings (*10*).

Horse Chestnut.　*See* Buckeye.

Huckleberry, Evergreen.　*See* Vaccinium ovatum.

Hydrangea (*Hydrangea* spp.).　This genus is generally easy to root, with certain exceptions such as *H. quercifolia,* which has overwintering problems as rooted cuttings. Softwood, semihardwood, and hardwood cuttings are used with 1000 ppm IBA talc or solution; 3000 to 5000 ppm IBA talc or solution is used with hardwood cuttings and more-difficult-to-root species (*59*). Florists' hydrangea have been rooted with 5000 to 10,000 ppm IBA talc under mist (*6*). There are benefits of micropropagating *H. macrophylla* for mass propagation, such as elimination of high-cost greenhouse stockplant space and producing virus-free stock plants for conventional cutting propagation (*5, 206*).

Hypericum spp.　*See* St. Johnswort.

Ilex spp.　*See* Holly.

Illicium.　*See* Anise.

Incense Cedar (*Calocedrus decurrens* [Torr.] Florin.). This is propagated by seed. Germination is promoted by a stratification period of about eight weeks at 0 to 4°C (32 to 40°F).

Indigo bush.　*See* Amorpha.

Ironwood.　*See* Casuarina.

Jacaranda (*Jacaranda acutifolia* Humb. & Bonpl.) (*225*).　This is easily propagated by seed taken from capsules after blooming. Vegetative propagation is normally not used; however, a white flowering form is grafted (Southern California).

Jasmine (*Jasminum* spp.).　Jasmine is propagated without difficulty by leafy semihardwood cuttings taken in late summer and rooted under mist. Layers and suckers also can be used.

Jasmine, Asiatic.　*See* Trachelospermum asiaticum.

Jasmine, Star, Confederate.　*See* Trachelospermum jasminoides.

Juniper (*Juniperus* spp.).　The junipers are generally propagated by cuttings, but in some cases difficult-to-

root species such as the upright types are grafted on seedlings or select cutting grown species. The low-growing, prostrate forms are easily layered.

Seeds.　Seedlings of the red cedar, *Juniperus virginiana,* or of *J. chinensis,* are ordinarily used as stocks for grafting ornamental clones. Seeds should be gathered in the fall as soon as the berry-like cones become ripe. For best germination, seeds should be removed from the fruits, then treated with sulfuric acid for 30 minutes before being stratified for about four months at 4°C (40°F). Rather than the acid treatment, two to three months of warm 21 to 30°C (70 to 85°F) stratification, or summer planting, could be used. As an alternative for cold stratification, the seed may be sown in the fall. Germination is delayed at temperatures above 15°C (60°F). Viability of the seed varies considerably from year to year and among different lots, but it is never much over 50 percent. Treated seed is usually planted in the spring, either in outdoor beds or in flats in the greenhouse. Two or three years are required to produce plants large enough to graft.

Cuttings.　The spreading, prostrate types of junipers are more easily rooted than upright kinds. Cuttings are made 5 to 15 cm (2 to 6 in.) long from new lateral growth tips stripped off older branches. A small piece of old wood—a heel—is thus left attached to the base of the cutting. Some propagators believe this to be advantageous. In other cases, good results are obtained when the cuttings are just clipped without the heel from the older wood. Terminal growth of current season's wood also roots well.

Cuttings to be rooted in the greenhouse can be taken at any time during the winter (*114, 141*) or rooted outdoors on heated beds (Southern California). Exposing the stock plants to several hard freezes seems to give better rooting. Optimum time for taking cuttings is when stock plants have ceased growth (i.e., late fall–winter propagation period is more successful than summer). For propagating in an outdoor cold frame, cuttings are taken in late summer or early fall. There may be advantages to using bottom heat. Lightly wounding the base of the cuttings is sometimes helpful, and the use of root-promoting chemicals, especially IBA, is beneficial. Recommendations have included 2500 IBA quick-dip (Alabama) for medium-difficult juniper species to 3000 to 8000 ppm IBA liquid to 0.3 to 4.5 percent IBA talc (*59*). For upright junipers, one large California nursery uses combinations of 3000 to 6000 ppm K-IBA, IBA in 55 percent methanol, and NAA in 55 percent methanol (*45*). A medium-coarse sand or a 10:1 (v/v) mixture of perlite and peat moss is a satisfactory rooting medium. Maintenance of a humid environment without excessive wetting of the cuttings is desirable, as is a relatively high light intensity.

A light, intermittent mist can be used. Bottom heat of 60 to 65°F (12°C) is critical the first six weeks of propagation to allow the basal wound of cuttings to callus; in Southern California, heat is withheld for six weeks to allow callusing. Bottom heat should then be raised to 70 to 75°F (23°C) to encourage rooting (45). Hardwood cuttings can be rooted in outdoor field beds.

Grafting (176). Vigorous seedling understocks with straight trunks, about pencil size, are dug in the fall from the seedling bed and potted in small pots set in peat moss in a cool, dry greenhouse. Seedlings potted earlier—in the spring—may also be used. After about 30 days, the greenhouse is heated and the plants kept well watered. This procedure stimulates growth so that after one or two weeks the plants resume root activity and are in a suitable condition for grafting.

The scions should be selected from current season's growth taken from vigorous, healthy plants and preferably of the same diameter as the stock to be grafted. Scion material can be stored at −1 to 4°C (30 to 40°F) for several weeks until used if kept in a saturated atmosphere.

Side-veneer or side-graft methods are ordinarily used. The unions are best tied with budding rubber strips. The grafted plants are set in a greenhouse bench filled with peat moss deep enough to cover the union. The temperature around the graft union should be kept as constant as possible at 24°C (75°F) with a relative humidity of 85 percent or more around the tops of the plants. A lightly shaded greenhouse should be used to avoid burning the grafts. Adequate healing will take place in two to eight weeks, after which the temperature and humidity can be lowered. The stock plant is then cut off above the graft union to allow the scion to develop.

Kalmia spp. *See* Laurel.

Koelreuteria spp. *See* Goldenrain Tree.

Laburnum (*Laburnum* spp.). Golden Chain. Propagated by seeds or layering. Cultivars are propagated by grafting or budding on laburnum seedling rootstocks. Seeds are poisonous if eaten.

Lagerstroemia indica. *See* Crape Myrtle.

Larch (*Larix* spp.) (*111*). Most of these deciduous conifers are propagated easily by fall-planted seeds. Cones should be collected before they dry and open on the tree. Several species have empty or improperly developed seed. Seeds of some species have a slight embryo dormancy, so for spring planting, stratification for one month at about 4°C (40°F) is advisable. Cut-

ting propagation is best done by rooting leafy-tipped softwood cuttings in late summer under mist. IBA promotes rooting. Cutting material should be taken from young trees only.

Larix spp. *See* Larch.

Laurel (*Kalmia* spp.) (*68, 121, 122*). The laurels can be propagated readily by seed germinated at about 20°C (68°F). Germination of *K. latifolia* seed is enhanced by cold stratification for eight weeks, or by a 12-hour soak in 200 ppm gibberellic acid. *Kalmia* cultivars can be propagated by leafy cuttings under mist or in polyethylene-covered frames (*68, 82*), by cleft or side grafting, or by layering; cuttings of some *Kalmia* selections respond to rooting hormones, whereas others do not. *K. latifolia* cultivars are micropropagated commercially (*22, 148*).

Leptospermum spp. Tree tea. Some species of this Australian native must be propagated by seed. Cultivars of other species, such as *L. scaparium,* can be readily propagated by cuttings.

Leyland Cypress (× *Cupressocyparis leylandii* (*231*). This bigeneric hybrid of *Cypressus macrocarpa* and *Chamaecyparis nootkatensis* is propagated by cuttings, rooted under mist with bottom heat. Cuttings can be taken any time from late winter to autumn and should be treated with IBA at 3000 ppm, or 8000 ppm in December–January (Georgia) (*180*).

Libocedrus decurrens. *See* Incense Cedar.

Ligustrum spp. *See* Privet.

Lilac, French Hybrid (*Syringa vulgaris* cvs.) (*43, 101, 226*). Grafting or budding on California privet (*Ligustrum ovalifolium*) or Amur privet (*L. amurense*) cuttings, or on lilac or green ash (*Fraxinus pennsylvanica*) seedlings, is used commercially in lilac propagation. *L. ovalifolium* is not reliably hardy in zone five, which should be considered when using it as a rootstock. Cuttings can be rooted, however, if attention is given to proper timing. Layering or division of old plants is quite satisfactory when only a few new plants are needed. The common purple lilac is propagated by root suckers. *S. vulgaris* can be micropropagated (*113*).

Seeds. Seedlings are used mostly as an understock for grafting or in hybridization. Lilac cultivars will not reproduce true from seed. Seeds require fall planting out-of-doors or a stratification period of 40 to 60 days at about 4°C (40°F) for good germination.

Cuttings. Ordinarily, good rooting of lilacs can be obtained only with terminal leafy cuttings taken within a narrow period shortly after growth commences in the spring. When the new, green shoots have reached a length of 10 to 15 cm (4 to 6 in.) they should be cut off and trimmed into cuttings. Since they are very succulent at this state it is difficult to prevent wilting. In a mist propagating bed rooting occurs in three to six weeks. Rooting can also be obtained in a polyethylene-covered bed in the greenhouse with bottom heat, or outdoor mist beds. Sprays with captan (2 tsp per gallon of water) and other fungicides will help avoid fungus attack. Cuttings of hybrid lilac cultivars showed improved rooting with prior etiolation of stock plants, and the period over which lilac cuttings could be propagated successfully was lengthened considerably (*10*). IBA talc of 3000 to 8000 ppm can enhance rooting

Grafting and Budding. Because of the difficulty in rooting lilac cuttings and the fact that they must be taken at a definite time in the spring, often at the peak of the nurseryman's busy season, many propagators practice nurse-root grafting. When privet or green ash is used as the understock, the lilac may show incompatibility symptoms, but if the grafts are planted deeply, scion roots rapidly develop from the lilac and soon become the predominant root system of the plant. *Syringa vulgaris* seedlings are sometimes used as the stock, but if suckers arise from this stock, there is difficulty in distinguishing them from the selected hybrid lilac top. The plant thus becomes a mixture of growth from the scion and rootstock and is a maintenance problem in the landscape. It is best in any case, to plant lilacs "on their own roots," either as rooted cuttings or with the privet or lilac "nurse root" already removed.

Grafting is done during the winter season, using rootstock plants which have been dug and brought inside. Vigorous, one-year-old scion wood is used, taken from plants that have been heavily pruned and well fertilized to induce such growth. Different grafting methods can be used, such as the side or the whip graft. Cleft grafting on pieces of privet root is also practiced and tends to eliminate subsequent suckering from the rootstock.

T-Budding. T-budding in late summer or early fall is sometimes practiced, the lilac buds being inserted below ground on one-year-old privet cuttings or seedlings. The following spring the privet stock is cut off above the bud and soil mounded around the shoot as it develops so as to encourage subsequent rooting from the lilac. Unless this is done, the plant will be short-lived.

Layering. Simple layering of one-year shoots arising from the base of plants on their own roots provides an easy propagation method where only a few plants are needed. Air-layering of one- or two-year-old branches also is successful.

Linden (*Tilia* spp.) (*79, 217*). Basswood. The seeds have a dormant embryo plus an impermeable seed coat which, in some species, is surrounded by a hard, tough pericarp. Such seeds are slow and difficult to germinate. Removing the pericarp, either mechanically or by soaking the seeds in concentrated nitric acid for one-half to two hours, rinsing thoroughly and drying, then soaking the seeds for about 15 minutes in concentrated sulfuric acid to etch the seed coat, followed by stratification for four months at 2°C (35°F), may give fairly good germination; otherwise, warm (15 to 27°C; 60 to 80°F) stratification for four to five months, followed by an equal period at 2 to 4°C (35 to 40°F) can be used. Collecting the seed from the tree just as the seed coats turn completely brown (but before the seeds drop and the seed coats become hard and dry), followed by immediate planting has given good germination.

Suckers arising around the base of trees cut back to the ground have been successfully mound layered, and softwood cuttings taken from stump sprouts have been rooted. Propagation of selected clones by cuttings or by grafting is slow and difficult but T-budding in late summer on seedling stocks of the same species gives good results (*79*). *T. cordata* has been micropropagated successfully (*35*).

Liquidambar (*Liquidambar styraciflua* L.). American sweet gum. Propagation is usually by seeds, which are collected in the fall but not allowed to dry out. Stratification for one to three months at about 4°C (40°F) is recommended to overcome seed dormancy. Selected clones are grafted or T-budded from spring through fall onto *L. styraciflua* seedlings. They can also be propagated by stem (4000 to 8000 ppm IBA) or root cuttings; overwintering may be a problem. Girdling of 10 year old *L. formosana* prior to taking cuttings, and the use of a rooting powder (IBA, PPZ, sucrose, and captan) stimulated 90 percent rooting (*103*). Sweet gum has been micropropagated successfully (*211*).

Liriodendron tulipifera. *See* Tulip Tree.

Loblolly Bay. *See* Gordonia spp.

Locust, Black (*Robinia pseudoacacia* L.). Black locust is readily propagated by seeds, which are soaked in concentrated sulfuric acid for one hour, followed by thorough rinsing in water, before planting. A hot-water scarification followed by a 24-hour soak prior to sowing can also be used. This species can be propagated by root cuttings and by grafting using a whip, side-veneer graft, cleft, or wedge graft (*171*). Micropropagation is also feasible (*35*).

Lonicera spp.   *See* Honeysuckle.

Loquot (*Eriobotrya japonica*).   These large evergreen shrubs or small trees can be propagated by seed, requiring no pretreatment. Considered difficult to root. Side grafting and budding is another form of propagation. *E. japonica* has been successfully micropropagated. (See Chap. 18.)

Maclura pomifera (*170*).   Osage-orange. A tree for difficult sites. Can be seed propagated with a 30-day stratification treatment. A two-day water soak overcomes dormancy and permits germination without stratification (*59*). Male and thornless cultivars 'Altamont', 'Park', and 'Wichita' can be asexually produced by softwood cuttings treated with 5000 to 10,000 ppm IBA or hardwood cuttings (January—Kansas). This species can also be budded. Osage-orange can also be micropropagated (132).

Madrone, Pacific (*Arbutus menziesii* Pursh.).   This Pacific coast evergreen tree is usually propagated by seeds, which are stratified for three months at 2 to 4°C (35 to 40°F). Seedlings are started in flats, then transferred to pots. They are difficult to transplant and should be set in their permanent location when not over 18 in. tall. Propagation can be done by cuttings, layering, and also grafting.

Magnolia (*Magnolia* spp.) (*3, 23, 38*).   Seeds, cuttings, grafting, and layering are utilized in propagating magnolias. Magnolia nursery trees are difficult to transplant, so they should be set out from containers or balled and burlapped, but only in early spring.

Seeds.   Magnolia seeds are gathered in the fall as soon as possible after the fruit is ripe, when the red seeds are visible all over the fruit. After cleaning, the seeds should either be sown immediately in the fall or—prior to spring planting—stratified for two to three months at about 4°C (40°F). Allowing the seeds to dry out at any time seems to be harmful. After sowing, the germination medium must not become dry. *M. grandiflora* seeds, and perhaps those of other species, lose their viability if stored through the winter at room temperature. If prolonged storage is necessary, the seeds should be held in sealed containers at 0 to 4°C (32 to 40°F). Magnolia seedlings grow rapidly, and generally are large enough to graft by the end of the first season. Transplanting should be kept at a minimum, since this retards the plants.

Cuttings.   Some species, such as *M. soulangeana* and *M. stellata,* are successfully propagated commercially by leafy softwood cuttings. These may be taken from late spring to late summer after terminal growth has stopped and the wood has become partly matured.

Excellent rooting can be obtained if cuttings are taken from very young plants, the bases wounded, treated with auxin, then rooted in sand or coarse perlite in outdoor mist beds. Under such conditions, rooting is rapid and in high percentages, and there is little trouble from diseases.

Leafy cuttings of *M. grandiflora,* taken from late spring to late summer, wounded, and treated with IBA at 5000 to 20,000 ppm have rooted well with bottom heat (24°C; 75°F) and intermittent mist. Semihardwood cuttings root well when five-second quick-dipped in 5000 to 10,000 ppm NAA in 50 percent alcohol (*57*). To obtain survival of the rooted cuttings through the following winter, they should be rooted early enough in the season so that some resumption of growth will occur before fall.

Grafting.   *Magnolia kobus* is probably the best rootstock for the oriental magnolias, whereas *M. acuminata* can be used as a stock for either oriental or American species. *M. sprengeri* 'Diva' makes excellent, comparably vigorous rootstocks for the large Asian species and their hybrids. *Magnolia grandiflora* seedlings are used for *M. grandiflora* cultivars.

Winter graftage using side or side-veneer grafts are satisfactory, with the union and scion waxed after grafting. Some propagators pot the seedlings in the fall, then bring them into the greenhouse and do the grafting in midwinter. The newly grafted plants may be set on open benches in the greenhouse or placed in closed propagating frames, where they stay for 7 to 10 days, while the union is healing. After six weeks they can be removed from the case and the understock cut off above the union. Chip budding is possible throughout the growing season (*214*).

Layering.   Simple or mound layering gives good results. One- or two-year-old shoots arising from the base of stock plants are started in spring, but often two seasons are required to produce well-rooted layers.

Mahogany.   *See* Swietenia mahagoni.

Mahonia (*Mahonia* spp.) (*46*).   Seed propagation is generally easy for most cultivars, while cuttings can be difficult. *M. bealei, M. lomarifolia,* and *M. japonica* are easy to grow from seed and do not require special treatments. *M. aquifolia* and *M. repens* seed must be separated from the fruit pulp, leached of inhibitors and stratified for either three months (Georgia) or a total of five months (California). Cuttings of *M. aquifolium*

'Compacta' are collected in fall or winter, quick-dipped with 3000 ppm IBA, and rooted under mist with 75 to 80°F (25°C) bottom heat; 8000 ppm IBA talc has also been used with *M. bealei*, *M. nervosa*, *M. pinnata*, *M. repens*, and *M. wagneri* (*59*). Air layering of *M. aquifolium* yields 80 percent rooted plants.

Malus spp.   *See* Crabapple, Flowering.

Manzanita (*Arctostaphylos* spp.) (*71*). There is a double dormancy requirement which makes seed propagation difficult. Seed scarification is done by using controlled flash fire burning or with sulfuric acid treatments of three to six hours. There may be some advantages of an initial insertion of seed in boiling water than soaking of seed in gradually cooling water for 24 hours. Cutting propagation is much more practical. Terminal cuttings are taken from November to February (California), submerged in 5 to 10 percent Clorox, given a 60-second quick-dip in Hormex (more satisfactory than Rootone) and rooted under intermittent mist with bottom heat (70°F; 21°C).

Maple (*Acer* spp.) (*219*). Various methods of propagation are used—seeds, grafting, budding, cutting, layering, and micropropagation (*22, 130*).

    Seeds. Most maples produce seed in the fall, but two species—*A. rubrum* and *A. saccharinum*—produce seed in the spring. Such spring-ripening seeds should be gathered promptly when mature and sown immediately without drying. For other species, stratification, usually for 90 days at 4°C (40°F), followed by spring planting, gives good germination. Fall planting out-of-doors may be done if the seeds are first soaked for a week, changing the water daily. Seeds of the Japanese maple, *A. palmatum,* germinate satisfactorily if they are placed in warm water (about 43°C; 110°F) and allowed to soak for two days, followed by stratification. Soaking seeds of *A. rubrum* and *A. negundo* in cold running water for five days and two weeks, respectively, before planting may increase germination. Seeds of *A. negundo*, *A. barbatum*, and *A. floridanum* are treated with a cold–warm–cold stratification (*119*). *Acer* seeds should not be allowed to dry out. Cultivars of some maples, such as *A. palmatum* 'Atropurpureum', will reproduce fairly true from seed, especially if the stock plants are isolated. The few off-type plants can be removed from the nursery row.

    Cuttings (*59, 227*). Leafy Japanese maple (*A. palmatum*) cuttings, as well as those of other Asiatic maples, will root in a sand–peat moss medium if they are made from tips of vigorous pencil-sized shoots in late spring and placed under mist. Wounding and relatively strong applications of 2 percent IBA talc and 5

percent Benlate are helpful. IBA at 8000 to 20,000 ppm in talc or quick-dips has given good results in rooting leafy cuttings of various *Acer* species under mist and over bottom heat in the greenhouse. Semihardwood cuttings of *A. floridanum* are propagated successfully at 8000 ppm IBA talc and wounding. Softwood cuttings of *A. griseum* from stock blocks are rooted successfully with 8000 ppm IBA talc (*116*) or 5000 ppm IBA liquid. *A. negundo* semihardwood cuttings are rooted with 8000 ppm IBA talc. Single-node cuttings of *A. glabrum* subsp. *Douglasii* (Douglas maple) root with 0.8 percent IBA talc or 0.25 percent IBA in 50 percent ethanol; rooted cuttings should be overwintered before transplanting (*213*). A mixture of 1:1 peat moss and perlite is a good rooting medium. It is often difficult to overwinter rooted maple cuttings. To overcome this problem, new growth should be induced on the cuttings, shortly after rooting, by using supplementary lighting and supplying fertilizers. Hardwood cuttings of *Acer palmatum* taken in midwinter have been rooted successfully in the greenhouse after wounding and treating with IBA (*33*). *A. truncatum* softwood cuttings taken from trunk sprouts of a 10-year-old tree had 79 percent rooting with a 5000-ppm IBA 10-second quick-dip (*169*). Softwood cuttings taken in August (Kansas) from three-year-old seedling stock had 85 percent rooting success with five-second 5000-ppm IBA quick-dips. Growth after rooting via long-day manipulation is important for winter survival of transplanted rooted cuttings and cuttings left in the propagation bed before lifting in spring. Maple species have been micropropagated with shoot-tip explants.

    Sugar maple (*A. saccharum*) cuttings are best taken in late spring after cessation of shoot elongation, then rooted under mist. Long, thick cuttings are preferred. IBA treatments give variable results. Overwintering is a problem but survival is best with large, vigorous, well-rooted cuttings. In cold climates it may be necessary to pot the cuttings in late summer, harden them off, then store until spring at about 1°C (34°F) (*61*).

    Grafting and Budding. *Acer palmatum* seedlings are used as the understock for Japanese maple cultivars, *A. saccharum* for the sugar maples, *A. rubrum* for red maple cultivars, *A. pseudoplatanus* for striped maple (*A. pensylvanicum*), and *A. platanoides* for such Norway maple clones as 'Crimson King', 'Schwedler', and the pyramidal forms. There is some evidence of delayed incompatibility in using *A. saccharinum* as a stock for red maple cultivars.

    Seedling rootstock plants are grown for one year in a seedbed, then in the fall or early spring are dug and transplanted into small pots in which they grow, plunged in propagating frames, through the second summer. In late winter, the stock plants are brought into the greenhouse preparatory to grafting. As soon as roots show signs of growth, the stock is ready for grafting. Dormant scions are taken from outdoor plants. The side graft or side-veneer graft is ordinarily used with *A. palmatum* 'Dissectum', *A. pensylvanicum, A. pseudoplatanus,* and *A. saccharinum*. While the union is healing, the plants are set in a grafting case with peat moss covering the union. Sometimes grafting wax is used, covering both the scion and the graft union (*94*). In Southern California, summer grafted *A. palmatum* cultivars are side grafted and wrapped with budding rubbers or plastic tape for three weeks with 90 percent grafting success.

Maples are also T-budded successfully on one-year seedlings in the nursery row, the buds being inserted from mid- to late summer. The wood is removed from the bud shield, which then consists only of the actual bud and attached bark. The seedling is cut back to the bud the following spring just as growth is starting. *A. platanoides* 'Crimson King' is more successfully chip budded (70 percent) than T-budded (7 percent).

Melaleuca spp. The seeds of these native Australian species are almost dust-like but germinate easily and can be handled like eucalyptus seed.

Metasequoia (*Metasequoia glyptostroboides* Hu and Cheng) (*41*). Dawn redwood. Seeds of this "living fossil," discovered in western China in 1945, germinate without difficulty, and both softwood and hardwood cuttings will root. The leafless hardwood cuttings may be lined-out in the nursery row in early spring. Leafy cuttings root easily under mist if taken in summer and treated with 3000 to 8000 IBA talc or a 2 percent quick-dip.

Mimosa. *See* Albizzia.

Mock Orange (*Philadelphus* spp.). The many cultivars of mock orange are best propagated by cuttings, which root easily. Hardwood cuttings can be planted in early spring or leafy softwood cuttings under mist in early summer. Removing rooted suckers arising from the base of old plants is an easy means of obtaining a few new plants.

Monkey Puzzle Tree. *See* Araucaria spp.

Morus alba. *See* Mulberry, Fruitless.

Mountain Ash. *See* Sorbus.

Mulberry, Fruitless (*Morus alba* L.). Some mulberry trees produce only male flowers and hence do not bear fruits. Propagated vegetatively, these are suitable as ornamental shade trees. Some are very rapid growers. They may be propagated by cuttings or by budding or grafting on mulberry seedling rootstocks. Leafy softwood cuttings taken in midsummer will root under mist.

Myrica. *See* Bayberry.

Myrtle (*Myrtus* spp.). Myrtle is usually propagated in summer by leafy softwood cuttings of partially matured wood rooted under glass. Treatments with IBA have been helpful in some instances.

Nandina (*Nandina domestica* Thun.). Can be propagated by seed. The embryos in the mature fruits are rudimentary, but will develop in cold storage. Seeds can be collected in late fall, held in dry storage at 4°C (40°F), then planted in late summer, germination starting in about 60 days. Germination tends to take place in autumn regardless of planting date. No low-temperature moist stratification period is necessary.

Digging suckers from old plants can be done.

Nandina, Dwarf. 'Compacta nana', 'Purpurea nana', 'Gulf stream', 'Harbour Dwarf', and 'Moon bay' are easy to root (*8*). Shoot tips with no brown wood which are 4 cm (1.5 in.) long are stripped of the bottom leaves and quick-dipped in 1250 ppm IBA and 500 ppm NAA and stuck directly into 6-cm (2.2-in.) liner pots. The propagation medium is 2 pine bark:1 peat:1 sand; 2.5 lb/yd^3 of Osmocote and Micromax are added. Can be rooted any time of the year (Texas), except during spring flush. Winter rooting requires bottom heat. 'Harbour Dwarf' can be propagated by separation of suckers at the base.

Nandinas are micropropagated commercially (*22*).

Nerium Oleander. *See* Oleander.

Norfolk Island Pine. *See* Araucaria spp.

Oak (*Quercus* spp.). Seed propagation is generally practiced. Wide variations exist in the germination requirements of oak seed, particularly between the black oak (acorns maturing the second year) and white oak (acorns maturing the first year) groups. Seeds of the white oak group have little or no dormancy and, with few exceptions, are ready to germinate as soon as they mature in the fall. Seeds of most species of the black oak group have embryo dormancy, requiring either stratification (0 to 2°C; 32 to 35°F) for one to three months of fall planting. Seeds of the following species will germinate without a low-temperature stratification period: *Quercus agrifolia, Q. alba, Q. arizonica, Q. bicolor, Q. chrysolepis, Q. douglassii, Q. garryana, Q. lobata, Q. macrocarpa, Q. montana, Q. petraea, Q. prinus, Q. robur, Q. stellata, Q. suber, Q. turbinella,* and *Q. virginiana.*

Acorns are often attacked by weevils. Soaking in water held at 49°C (120°F) for 30 minutes will rid the acorns of this damaging pest. However, research with *Q. virginiana* indicates no commercial advantage with heat treatment for weevil control. In fact, there is a loss of seed viability.

Seeds are usually floated in water, and those seeds that float are discarded. Acorns of many species tend to lose their viability rapidly when stored dry at room temperature. Seeds of some species can be stored for several years without losing viability by holding them at 1 to 3°C (34 to 37°F) in polyethylene bags. The seeds should have a 60 to 70 percent moisture level at the start of storage (*75*).

To obtain lateral root branching—which makes the seedlings more adaptable to transplanting—the acorns can be planted in a box that has a copper wire screen (mesh about 6 in.) below the seeds. The tip of the taproot, upon contacting this mesh, is air pruned and killed, forcing development of many lateral roots. A system for speeding up the production of red oak whip production in containers has been described (*209*).

Bench grafting of potted seedling stocks in the greenhouse in late winter or early spring is moderately successful. Side or whip grafting is ordinarily used, with dormant one-year-old wood for scions. Seedlings in place in the nursery row are occasionally crown grafted in the spring, after the stock plants start to leaf out. Scions are taken from wood gathered when dormant and stored under cool, moist conditions until used. Various grafting methods are satisfactory—whip, cleft, or bark. Budding generally has been unsatisfactory. In grafting, only seedlings of the black oak group should be used for scion cultivars of the same group and, in the same manner, only seedlings of the white oak group should be used as stocks for other members of the white oaks. The use of seedlings of the same species is preferable. Although some distantly related species of oak will unite satisfactorily, incompatibility symptoms usually appear later. Root grafting with a side-veneer graft on *Q. robur* has been reported to produce plants with more uniform growth and no suckering; this technique was not successful with *Q. rubra* or *Q. palustris* (*144*).

Attempts to propagate oaks by cuttings or layering usually have been unsatisfactory, although some success has been obtained in rooting leafy softwood cuttings of *Quercus robur* 'Fastigata' under outdoor mist in midsummer after treatment with IBA at 20,000 ppm (*78*). Recently, commercial propagation of *Q. virginiana* and *Q. laurifolia* has been done by taking July semi-hardwood cuttings (Florida) 2 to 3 in. (6 cm), wounding, and quick-dipped in 12,000 ppm K-IBA salt and rooting under mist; rooting takes seven to nine weeks (*48*). *Q. shumardii* cuttings taken in July (Florida—wood is not as brittle nor hard as wood of other oak species) is quick-dipped in 10,000 ppm K-IBA salt and also roots in seven to nine weeks. Some success has occurred with serial propagation of rooted *Q. virginiana* which are maintained under accelerated growth techniques in a greenhouse and used as stock plants for future propagules (*160*). Girdling of stock-plant shoots prior to collecting cuttings and treating cuttings with a rooting powder of auxins, sucrose, and fungicide increased rooting (76 percent) and survival of 19 to 57-year-old water oak (*104*).

Single-node-stem sections of *Q. shumardii* were micropropagated successfully in liquid woody plant medium (WPM) with BA; 73 percent rooting success occurred *ex vitro* with 500-ppm 15-minute basal dips of microshoots (*13*). Juvenile and mature explants of *Q. robur* have been micropropagated and rooted successfully in vitro (*221*).

Oleander (*Nerium oleander* L.).   Seedlings reproduce fairly true-to-type, although a small percentage of plants with different flower colors will appear. The seeds should be collected in late fall after a frost has caused the seed pods to open. Rubbing the seeds through a coarse mesh wire screen removes most of the fuzzy coating. The seeds are then planted immediately in the greenhouse in flats without further treatment. Germination occurs in about two weeks. Leafy cuttings root easily under mist if taken from rather mature wood during the summer. Simple layering also is successful. Plant parts are very poisonous.

Olive (*Olea europaea* L.).   A fruitless cultivar, 'Swan Hill', is available for use as a patio or street tree; it is usually grafted on *O. oblonga*. Cuttings are difficult to root. They should be placed under mist and treated with IBA at 2000 to 3000 ppm, or it can be grafted on easily rooted cultivars used as a rootstock. *O. europaea* 'Wilsonii' and 'Majestic Beauty' are also fruitless and root more easily. Olives will not survive outdoor winter temperatures below about −9°C (15°F) (*107*).

Orchid (Anacho) Tree.   *See* Bauhinia congesta.

Osage-Orange.   *See* Maclura pomifera.

Osmanthus spp.   Seeds are slow and difficult to propagate. Semihardwood cuttings or cuttings with firm wood are used and 3000 to 8000 IBA talc or liquid enhance rooting. Softwood cuttings of *Osmanthus × fortunei* treated with 2500 IBA had 92 percent rooting (*18*).

Oxydendrum arboreum L.   Sourwood is propagated commercially by seed. Fall-harvested seed needs no pretreatment and flats containing seeds are generally placed under continuous light. Stratification of two to three months hastened germination and decreased the light required for germination (*9*). Propagation by cuttings is extremely difficult.

Pachysandra terminalis (Siebold & Zucc.).   Japanese spurge. This evergreen ground cover for shady areas spreads naturally by rhizomes. Can be propagated easily by division or by leafy cuttings under mist or glass.

Paeonia suffruticosa.   *See* Peony, Tree.

Palms (*25, 34*).   There are numerous species and genera of ornamental palms. They are propagated by mature seed, which should be planted as soon as possible after harvesting and not allowed to dry out. Palm seeds remain viable for only a short period. Palm seeds are susceptible to surface molds and should be protected by dusting with a fungicide (e.g., captan). Any seeds that float in water should be discarded. A mixture of one-half peat moss and one-half perlite is a good germination medium. Seeds of most species germinate in one to three months, especially if bottom heat, 28°C (80°F), is maintained, but some may take as long as one to two years. Germination in seed of some species can be accelerated by scarification, plus soaking in gibberellic acid at 1000 ppm for 72 hours, plus germination over bottom heat at 27°C (81°F) (*164*). Some palm species have limited optimum temperatures for seed germination, and temperatures above and below the optimum level contribute to irregular and low germination; four native Florida palms had optimum germination temperatures of 95°F (35°C) (*30*).

Parthenocissus (*P. quinquefolia* Planch.).   Virginia creeper; (*P. tricuspidata* Planch.) Boston Ivy. These two ornamental vines can be propagated by seeds planted in the fall or stratified for two months at about 4°C (40°F) before planting in the spring. Leafy softwood cuttings taken in late summer root easily under mist, as do hardwood cuttings planted outdoors in early spring. Plants may be started by compound layering. Grafting of named cultivars on *P. quinquefolia* seedlings or rooted cuttings is done by some nurseries using the whip or cleft graft.

Passiflora × alatocaerulea (*Lindl.*).   Passion vine. This tender, subtropical vine is propagated by leafy cuttings under glass or mist.

Pawpaw.   *See* Asimina spp.

Paxistima myrsinites (*213*).   Oregon boxwood. An evergreen shrub native to the Pacific northwest. Readily propagated from semihardwood cuttings from midsummer until bud break in spring (Vancouver, B.C.). Can be stuck directly using 0.8 percent IBA talc using basal heat (70°F; 21°C) and propagated under mist or in fall and winter under contact polyethylene film (*213*).

Pear, 'Bradford' (*Pyrus calleryana* Dcne. 'Bradford'). This ornamental pear introduced by the USDA can be propagated by T-budding or root grafting onto *P. calleryana* seedlings. It is not compatible with *P. communis* roots. *P. calleryana* can be rooted with 8000 ppm IBA talc, 10,000 ppm IBA + 5000 ppm NAA, or 2 percent K-IBA; the cutting wood should be firm and is taken June to August and rooted under mist in bark:sand or 1 peat:1 perlite:1 vermiculite media (*1, 59*). This species is also micropropagated (*22*).

Pear, Evergreen (*Pyrus kawakamii* Hayata).   Evergreen pear is propagated by cuttings or, more commonly, by grafting on *P. calleryana* seedlings. Bench grafting is done in midwinter using the cleft graft. The grafts can be planted in containers or in the nursery row.

Peony, Tree (*Paeonia suffruticosa* Andr.).   Moutan. (*234*).   Seed propagation is complicated by "epicotyl dormancy." The seeds should be planted in a moist medium and, after roots have developed, transplanted to pots of soil placed in a cold room (4 to 10°C; 40 to 50°F) or outdoors (in winter) for 2½ months. This overcomes dormancy conditions in the shoot tip, which then grows readily in spring or upon transfer of the plants to warm temperatures. Selected cultivars are propagated by grafting in late summer on herbaceous peony (*P. lactiflora*) roots as the understock. The grafts are callused in a sand-peat medium in a greenhouse until fall, when they are potted.

Persimmon, Texas.   *See* Diospyros texana.

Philadelphus spp.   *See* Mock Orange.

Photina arbutifolia.   *See* Heteromeles arbutifolia.

Photinia (*Photinia* spp.). Large evergreen shrubs with small pome fruits. *P. × fraseri* (Fraser or red-tip photinia) can be propagated by two-month-cold-stratified seed. Semihardwood cuttings are best rooted with 10,000 ppm IBA liquid (*15*) or 8000 ppm talc (*119*). Wounding cuttings, trimming leaves of cuttings, and 3000 to 8000 ppm IBA talc or quick-dips are optional for *P. × fraseri* 'Red Robin', which is a hybrid of *P. glabra* and *P. serrulata* (*40*).

In another report on *Photinia* (*56*), best rooting response occurred with Woods rooting compound (containing 4 percent dimethylformamide solvent) and Dip 'N Grow. K-IBA and K-NAA had only 23 percent rooting while 0.2 percent IBA–50 percent alcohol and 0.1 percent NAA–50 percent alcohol had 83 percent rooting. Boron at 50 ppm greatly enhanced the rooting response of K-IBA. The solvent system (carrier) can have a pronounced effect on the rooting compound response (*57*).

Picea spp. *See* Spruce.

Pieris spp. (*83*). *P. floribunda* and *P. japonica* reproduce readily by seed with no treatment necessary. Some *Pieris* species can be started easily by cuttings, but *P. floribunda* cuttings are difficult to root. Rooting can be obtained under mist or polyethylene—covered frames, wounding, IBA talc or liquid at 5000 to 8500 ppm, and bottom heat. Growool propagating sheets are reported to reduce transplant shock (*4*). Commercial micropropagation has been successful.

Pine (*Pinus* spp.) (*212*). Pines are ordinarily propagated by seed. Considerable variability exists among the species in regard to seed dormancy conditions. Seeds of many species have no dormancy and will germinate immediately upon collection, whereas others have embryo dormancy. With the latter, stratification at 0 to 4°C (32 to 40°F) for one to three months will increase or hasten germination. Seed coat dormancy also seems to be present in *P. cembra* and *P. monticola*. With these species, concentrated sulfuric acid treatment for three to five hours and for 45 minutes, respectively, followed by stratification for three months at 2°C (36°F) will aid germination. Species whose seeds have no dormancy conditions and can be planted without treatment include *Pinus aristata*, *P. banksiana*, *P. canariensis*, *P. caribaea*, *P. clausa*, *P. contorta*, *P. coulteri*, *P. edulis*, *P. halepensis*, *P. jeffreyi*, *P. latifolia*, *P. mugo*, *P. nigra*, *P. palustris*, *P. pinaster*, *P. ponderosa*, *P. pungens*, *P. radiata*, *P. resinosa*, *P. roxburghii*, *P. sylvestris*, *P. thunbergii*, *P. virginiana*, and *P. wallichiana*. However, if seeds of the above species have been stored for any length of time, it is advisable to give them a cold stratification period before planting. Pine seeds can be stored for considerable periods of time without losing viability if held in sealed containers between −15 to 0°C (5 and 32°F). Seeds should not be allowed to dry out.

Pine cuttings are difficult to root, although those of mugo pine (*Pinus mugo*) root easily if taken in early summer (*131*), and *P. radiata* roots well if cuttings are taken from young trees. Success is more likely if cutting material is taken in winter from low-growing lateral shoots on young trees. Treatment with IBA is beneficial. Considerable study has been given to the rooting of cuttings of Monterey pine (*P. radiata*) because of its importance as a lumber tree in New Zealand and as a Christmas tree species. There are clonal variations in the commercial rooting of this species. Best rooting is from cuttings taken in early winter. Wounding, plus a concentrated-dip treatment of IBA at 4000 ppm, was beneficial. A more symmetrical root system could be induced by clipping the ends of the original roots and allowing the root system to develop from the secondary roots. Outdoor rooting under intermittent mist has been successful using IBA at 5 percent, with peat moss, redwood sawdust, and *P. radiata* litter (1:1:1) as the rooting medium (*147*). Rooting of *P. strobus* improved to 83 percent when stock plants were etiolated and shoot bases covered with black adhesive tape (*10*). Accelerated growth techniques of supplementary lighting, CO_2, temperature manipulation, foliar nutrients, and root powder formulations containing auxins, growth retardants, sucrose, and captan enhanced rooting from 83 to 100 percent in *P. echinata*, *P. thunbergiana*, and *P. elliottii* (*102*). Air layering of loblolly pine has been successful (*105*).

Pines can be propagated asexually by rooting needle fascicles (needle leaves held together by the scale leaves, containing a base and a diminutive shoot apex) (*44*). Rooting is best when the fascicles are taken from trees less than four years old. IBA treatments are helpful. By selecting certain seedlings whose cuttings root easily and by using critical timing in taking cuttings, it is possible to select clones in which cutting propagation is commercially feasible. This was shown to be true for the mugo pine (*92*). Cuttings of the Scotch pine have been rooted by a unique method of forcing out interfascicular shoots from young stock plants. These shoots, when made into cuttings and given the proper treatment, rooted in high percentages.

Side-veneer grafting is used for propagating selected clones; well-established two-year-old seedlings of the same or closely related species should be used as rootstocks. Scions should be of new growth, taken from firm, partly matured wood. Commercial systems of micropropagation of pine are being developed (*22, 233*).

Pinus spp. *See* Pine.

Pistache, Chinese (*Pistacia chinensis* Bunge). Seeds should be collected from relatively large fruits, blue-green in color, in midfall. Pulp must be removed. Soak fruits in water, then rub over a screen. Seeds that float in water have aborted embryos and should be discarded. Stratification at 4 to 10°C (40 to 50°F) for 10 weeks gives good germination. Seedlings exhibit a wide range of variability. T-budding selected clones on seedling *P. chinensis* rootstocks in late summer is used to produce uniform, superior trees. Propagation by cuttings is very difficult. Pistache nursery trees do not transplant well if the roots become exposed, so they are usually handled as container-grown nursery stock (*124*).

Pistacia chinensis. *See* Pistache, Chinese.

Pittosporum (*Pittosporum* spp.). These are started by seeds or cuttings. The seeds are not difficult to germinate; dipping a cloth bag containing the seeds for several seconds in boiling water may hasten germination. Leafy semihardwood cuttings taken after a flush of growth has partially matured will root readily, particularly under mist. IBA 1000 to 3000 ppm liquid or talc are beneficial to rooting of *P. tobira* and 6000 to 8000 ppm IBA for *P. tenuifolium*.

Plane Tree (*Platanus* spp.). Sycamore. Seeds are ordinarily used in propagation, but they should not be allowed to dry out. The best procedure is to allow the seeds to overwinter in the seed balls right on the tree. They may then be collected in late winter or early spring and planted immediately. Germination usually occurs promptly. If the seeds are collected in the fall, then stratification over winter at about 4°C (40°F) should be used. The hybrid London plane tree, *P* × *acerifolia*, can be propagated by hardwood cuttings, taken and planted in the nursery in autumn or by leafy softwood cuttings taken in midsummer and rooted under mist. Auxin treatments tend to inhibit rooting (*163*).

Platanus spp. *See* Plane Tree.

Plumbago (*Plumbago* spp.). Seeds sown in late winter usually germinate easily. Leafy cuttings taken from partially matured wood can be rooted without difficulty under mist. Root cuttings also can be used, and old plants can be divided.

Plumeria (*Plumeria* spp.). Leafy cuttings 15 to 20 cm (6 to 8 in.) long of this tender tropical shrub widely

grown in Hawaii will root readily under mist if treated with 2500 ppm IBA. Terminal, leafless shoots, about the same length, can be cut off before new leaves appear and planted. Root and shoot growth develop readily.

Podocarpus (*Podocarpus* spp.). These evergreen trees and shrubs have foliage resembling the related yews (*Taxus*) and make good container plants. They can be propagated by stem cuttings taken in late summer; however, generally they are seed propagated.

Poinsettia (*Euphorbia pulcherrima* Willd.) (*64, 140*). Propagation by leafy cuttings under mist is the usual procedure. Low-strength root-promoting chemicals (IBA, NAA) are helpful. Stock plants of self-branching cultivars should be used as a source of cuttings. It is best to root cuttings in containers so that roots will not be disturbed. Cuttings can be rooted in the greenhouse from spring to fall. Specialists or their licensed propagators will sell rooted liners to greenhouse producers for either stock plants or for shifting up to a larger pot for finishing.

Poplar (*Populus* spp.). Poplar, Cottonwood, Aspen (*175*). These trees can be propagated by seeds; they should be collected as soon as the capsules begin to open and planted at once, because they lose viability rapidly, and should not be allowed to dry out. However, if held in sealed containers near 0°C (32°F), seeds of some species can be stored for as long as three years. There are no dormancy conditions and seeds germinate within a few days after planting. The seedlings are highly susceptible to damping-off fungi and will not tolerate excessive heat or drying. Poplars are difficult to propagate in quantity by seed.

Hardwood cuttings of *Populus* (except the aspens) planted in the spring root easily. Treatment with IBA is likely to improve rooting. Leafy softwood cuttings (of some species at least) taken in midsummer also root well. Micropropagation using shoot-tip explants has been successful in poplar propagation (*232*).

P. tremuloides, quaking aspen, can be propagated by removing root pieces, root suckers, and layering; inducing adventitious shoots to form from these propagules in vermiculite; then rooting shoots as stem cuttings under mist with IBA treatments. Many *Populus* species are micropropagated (*22*).

Populus spp. *See* Poplar.

Potentilla (*Potentilla* spp.) (*87*). Cinquefoil. Propagation is usually by cuttings, but seed and division can be used. Cuttings are taken from early summer through fall. Rooting is best under light mist with bottom heat. Auxin treatments are helpful.

Privet (*Ligustrum* spp.). Seed propagation is easily done. The cleaned seed should be stratified for two to three months at 0 to 10°C (32 to 50°F) before planting. Hardwood cuttings of most species planted in the spring root easily, as do softwood cuttings in summer under mist. Japanese privet (*L. japonicum*) is somewhat difficult to start from cuttings, the best results being obtained with actively growing terminal shoots rather than more mature wood. Treat with 2500 ppm IBA liquid (*15*).

Protea spp. (*172*.). This South African native, popular for its cut flowers, is propagated by seed, cuttings, and grafting. Cuttings are difficult to root but respond to auxin treatments.

Prunus *campanulata, P. sargentii, P. serrulata, P. sieboldi, P. subhirtella, P. yedoensis.* *See* Cherry, Flowering.

Pseudotsuga menziesii. *See* Douglas Fir.

Pyracantha (*Pyracantha* spp.). Firethorn. Propagation is almost always by cuttings. Partially matured, leafy, current-season's growth is taken from late spring to late fall and rooted either in the greenhouse under mist for good results. Treatments with 2500 ppm IBA liquid are beneficial (*15*). Semihardwood cuttings collected in fall rooted well when wounded and treated with 8000 ppm IBA talc or liquid (*59*). *Pyracantha* spp. have a three-month-cold-stratification requirement when propagated sexually.

Pyrus calleryana. *See* Pear, 'Bradford'.

Pyrus kawakamii. *See* Pear, Evergreen.

Quercus spp. *See* Oak.

Quince, Flowering (*Chaenomeles* spp.). Flowering quince is easily started by seeds, which should be fall-planted or stratified for two or three months at 4°C (40°F) before sowing. Clean the seeds from the fruit. Root cuttings can be taken in late fall, cut into 5- to 10-cm (2- to 4-in.) lengths, and stored at 2 to 4°C (35 to 40°F) until spring, when they can be lined-out horizontally in the nursery row. Leafy cuttings of partially matured wood may be rooted under mist in late spring. Treatment with IBA is beneficial. Older plants tend to produce suckers freely at the base; these suckers may be removed and used if they are well rooted.

Redbud (*Cercis* spp.). Seed propagation is successful, but seed treatments are necessary because of dormancy resulting from an impervious seed coat plus a dormant embryo. Probably the most satisfactory treatment is a 30 to 60-minute soaking period in concentrated sulfuric acid, or 180°F hot-water soak; scarification treatments must be followed by stratification for three months at 2 to 4°C (35 to 40°F). Fall sowing outdoors of untreated seeds also may give good germination. Seed provenance is important since ecotypic variation affects survival, dormancy, and plant growth potential (i.e., a Florida versus Massachusetts seed source).

Leafy softwood cuttings of some *Cercis* species root readily under mist if taken in spring or early summer. Simple layering also is used successfully. T-budding in midsummer on *C. canadensis* seedlings is used commercially for *Cercis* cultivars (*223*). Both *C. canadensis* and *C. mexicana* have been micropropagated (*12, 14*).

Redwood. Coast redwood (*Sequoia sempervirens* [D. Don] Endl.); giant sequoia or Sierra redwood (*Sequoiadendron giganteum* [Lindl.] Buchh.). Both species are ordinarily propagated by seed, however, some ornamental *S. sempervirens* are propagated by cuttage. Seeds of *S. sempervirens* are mature at the end of the first season, but those of *S. giganteum* require two seasons for maturity of the embryos. Cones are collected in the fall and allowed to dry for two to four weeks, after which the seeds can be separated. Seeds may be kept for several years in sealed containers under 4°C (40°F) storage without losing viability. Stratification for 10 weeks before planting at about 4°C (40°F) promotes germination of *S. giganteum* seed. Seeds of *S. sempervirens* will germinate without a stratification treatment.

Fall planting also may be done, sowing the seed about 3 mm (⅛ in.) deep in a well-prepared seedbed. The young seedlings should be given partial shade for the first 60 days.

Introduction of vegetatively propagated ornamental cultivars of *S. sempervirens* has resulted in vast improvement over highly variable seed propagated specimens. Hardwood cuttings (February in southern California) were trimmed to 12 cm (5 in.) in length so that the outer tissue on the main stem of the cutting was brown at the base and green above. Cuttings were propagated in 9 perlite:1 peat, placed in cutting flats on outdoor heated beds [21°C (70°F)], and rooted with intermittent mist under full sun for five months. The cultivars 'Majestic Beauty' rooted best with 3000 ppm IBA and 3000 ppm NAA liquid, 'Santa Cruz' with 16,000 ppm IBA powder, and 'Soquel' with 6000 ppm IBA and 6000 ppm NAA liquid (*19*). Winter grafting on seedling understock can also be done with a whip and tongue graft. Young *S. giganteum* trees are grown for Christmas trees. *S. giganteum* cuttings root easily if taken from young trees, treated with IBA, and rooted under mist (*77*). Micropropagation is also being used.

Redwood, Dawn.   *See* Metasequoia.

Rhamnus spp.   *See* Buckthorn.

Rhododendron (*Rhododendron* spp.) (*218, 228*).   *See also* Azalea. These can be propagated by seeds, cutting, grafting, and layering. Micropropagation of *Rhododendron* P.J.M. hybrids with 2iP produced plantlets with more uniform height and more basal branching than macropropagated plants produced from stem cuttings (*72*). Rhododendron, evergreen, and deciduous azaleas are micropropagated commercially (*2, 65, 126, 207*).

Seeds.   Seedlings may be used as rootstocks for grafting or for propagation of ornamental species. *Rhododendron ponticum* is the principal rootstock for grafting. The seed should be collected just when the capsules are beginning to dehisce, and may be stored dry and planted in late winter or early spring in the greenhouse. Seed to be kept for long periods should be put in sealed bottles and held at about 4°C (40°F). A good germination medium is a layer of shredded sphagnum moss or vermiculite over a mixture of sand and acid peat. The very small seeds are sifted on the surface of the medium and watered with a fine spray. The flats should be covered with glass or propagated under mist (*63, 229*). Careful attention must be given to provide adequate moisture and ventilation as well as even heat: 15 to 21°C (60 to 70°F). The plants grow slowly, taking three months to reach transplanting size. After two or three true leaves form, they are moved to another flat and spaced 2.5 to 5 cm (1 to 2 in.) apart, where they remain through the winter in a cool greenhouse or in cold frames. In the spring the plants are set out in the field in an acid soil, and by fall they are ready to be dug and potted preparatory to grafting in the winter.

Cuttings (*96, 218*).   Rooting cuttings is the chief method of rhododendron propagation. These are best taken, midsummer into fall, from stock plants in full sun grown especially for this purpose. However, stem cuttings—or leaf-bud cuttings—of some hybrids taken in midwinter will root well. Any flower buds should be removed from the cuttings. Treatments with indolebutyric acid at relatively high concentrations are required; IBA in talc or 1 or 2 percent concentration, plus an added fungicide (benomyl) works well. Wounding the base of the cutting on both sides is a strong stimulus to rooting in rhododendrons. A rooting medium of two-thirds sphagnum peat moss and one

perlite or 1 peat:1 perlite (v/v) is suitable. Bottom heat at 24°C (75°F) should be used. Rhododendron cuttings are best rooted under mist in the greenhouse or in a closed polytent, and should be lifted soon after roots are well formed (about three months) or the roots will deteriorate. After rooting and transplanting (into peat moss, with added fertilizers) the cuttings should be held at 4°C (40°F) for about 20 days, after which the night temperature can be raised to a minimum of 18°C (65°F). Supplementary light at this stage to extend the daylength will give good growth response. Plants started from cuttings usually develop rapidly and are free of the disadvantage of suckering from the rootstock, which occurs with grafted plants.

Grafting.   A side-veneer graft is most successful. The best scionwood is taken from straight, vigorous current season's growth. After grafting, the plants are kept in closed frames under high humidity and a temperature near 21°C (70°F) until the union has healed. The plants should then be moved to cooler conditions—10 to 15°C (50 to 60°F)—and the top of the stock removed above the graft union. After the plant has hardened, it is transplanted to the nursery row in acid soil and grown for two years, after which it is ready to dig as a salable nursery plant. A modified chip bud method using leafbuds in early summer (Rhode Island) was successful for grafting difficult-to-root rhododendron cultivars (*155*).

Layering.   Rhododendrons are easily reproduced by trench and simple layering. All parts of rhododendron and azalea plants are poisonous if eaten.

Rhus spp.   *See* Sumac.

Ribes sanguineum (*213*).   Red flowering currant. An attractive flowering deciduous shrub native to the Pacific northwest. Easily propagated by softwood cuttings taken during summer to early September (Vancouver, B.C.), or by hardwood cuttings made with a "heel" of two-year-old wood. Both cutting types do best with basal heat (21°C; 70°F) and 0.8 percent IBA talc.

Robinia pseudoacacia.   *See* Locust, Black.

Rose (*Rosa* spp.) (*110, 135, 210*).   All the rose cultivars selected are propagated by asexual methods. T-budding on vigorous rootstocks is most common, although the use of softwood or hardwood cuttings, chip budding, simultaneous budding and rooting (stenting), layering, or the use of suckers is sometimes practiced. Seed propagation is used in breeding new cultivars, in producing plants in large numbers for conservation projects or mass landscaping, and in growing rootstock plants of certain species, such as *R. canina*. Some commercial rosebush producers prefer propagating rootstock by seed to avoid virus transmission through conventional asexual techniques. Selected rose cultivars are also micropropagated and "own rooted" to avoid more costly grafting procedures (*109, 110*).

Seeds (31). As soon as the rose fruits ("hips") are ripe but before the flesh starts to soften, they should be collected and the seeds extracted. It is best to stratify them immediately at 2 to 4°C (35 to 40°F). Six weeks is sufficient for *Rosa multiflora,* but others—*R. rugosa* and *R. hugonis*—require four to six months, and *R. blanda* 10 months (*197*). *Rosa canina* germinates best if the seeds are held at room temperature for two months in moist vermiculite and then transferred to 0°C (32°F) for an additional two months. Hybrid rose seeds usually respond best to a stratification temperature of 1 to 4°C (34 to 40°F) for 60 to 90 days, although some seeds may germinate with no cold stratification treatment. Germination is probably prevented in rose seeds by inhibitors occurring in the seed coverings, as well as by the mechanical restriction imposed by the massive pericarp (fruit wall). The seeds may be planted either in the spring or in the fall in seedbeds or in the nursery row. In areas of severe winters, seedlings are likely to be winter-killed if they are smaller than 10 cm (4 in.) by the onset of cold weather.

Hardwood Cuttings. Hardwood cuttings are widely used commercially in the propagation of rose rootstocks, and to some extent in propagating the strong growing polyanthas, pillars, climbers, and hybrid perpetuals. The hybrid teas and other similar everblooming roses also can be started by cuttings, but more winter-hardy and nematode-resistant plants are produced if they are budded on selected vigorous rootstocks. In mild climates (Texas, California, Spain) the cuttings are taken and field planted in the nursery in the fall. In areas with severe winters, cuttings may be made in late fall or early winter, tied in bundles, and stored in damp peat moss or sand at about 4°C (40°F) until spring, when they are planted in the nursery row. The rootstocks are ready to bud by the following spring, summer, or fall. The cuttings are made into 15- to 20-cm (6- to 8-in.) lengths from previous season's canes of 6 to 9 mm (¼ to ⅜ in.) diameter. Commercially, large bundles of canes are run through bandsaws to cut them to the correct length. Disbudding ("de-eying") is done in rootstock propagation; all buds except the top two or three are removed to prevent subsequent sucker growth in the nursery row. The use of auxins and other rooting pretreatments are of little benefit in the commercial propagation of *Rosa multiflora* hardwood cuttings (*50*).

Softwood Cuttings. Softwood cuttings are made from current season's growth, from early spring to late summer, depending upon the time the wood becomes partially mature. Rooting is fairly rapid, occurring in 10 to 14 days. At the end of the season the cuttings may be transplanted to their permanent location, potted and overwintered in a cold frame, or transferred to the nursery row for another season's growth or to be budded to the desired cultivar. Cultivars of most miniature roses are easily propagated by softwood or semihardwood cuttings under mist.

Budding. T-budding is the method ordinarily used. The buds are inserted into 5- to 10-mm (³⁄₁₆- to ⅜-in.)-diameter rootstock plants. In mild climates budding can be done during a long period, from late winter until fall, but mostly in the spring. Early buds will make some growth during the summer and produce a salable plant by fall. Some propagators break over the top of the rootstock about two weeks after budding to force the bud out. After the bud has reached a length of 10 to 20 cm (4 to 8 in.), the top of the stock is entirely removed. In areas with shorter growing seasons, budding is done during the summer. Buds inserted late in the summer either make little growth or remain dormant until the following spring. In this case the rootstock is cut off just above the bud in late winter or early spring, forcing the inserted bud into growth. Shoots from buds started in the fall are cut back to 13 mm (½ in.) in the spring. The shoot then grows through the following entire summer, producing a well-developed plant by fall. After the shoot has grown about 15 cm (6 in.), it is generally cut back to 5 to 7.5 cm (2 or 3 in.) to force out side branches.

Budwood may be obtained during the budding season from the current season's growth of the desired cultivar, only a day's supply being taken at a time. It is best collected early in the morning, clipping off the leaves immediately and leaving about 6 mm (¼ in.) of the petiole attached to the bud. Lateral buds from the stems producing the flowers are the best to use. Plump, but dormant buds three or four nodes below the flower are the most desirable. The wood should be at a stage of maturity in which the thorns are easily removed.

In California, Texas, and Spain, commercial landscape rose bush produ..ers collect budwood in late fall (prior to digging two-year-old rose bushes), store the budwood at 29°F (−2°C), and then T-bud in the following spring (*51*). The budwood is collected in late fall after the flowers are shed and the thorns become dark. The leaves are removed by hand, or "sweated-off" under high humidity, and the thorns are left intact. Sticks 25 to 38 cm (10 to 15 in.) long are put up in bundles of 30 or 40 each. The bundles are wrapped as tightly as possible in moist paper and then placed in polyethylene bags.

From spring to early summer, the budwood is

taken out of storage, the thorns are removed, and shield buds from the budwood are T-budded by inserting them into the medial, de-eyed portion of the original cuttings (rather than into new growth arising from the rootstock).

Over the last several years, there has been an interest in simultaneously grafting (budding) and rooting landscape roses through bench chip budding and direct rooting in the field (*51*), and using the stenting system in grafting and rooting greenhouse cut-flower roses (*216*). (See Chap. 13.)

Rootstocks for Rose Cultivars (24). Most rose rootstock clones have been in use for many years, propagated by cuttings. Many of the clones are virus-infected, thus infecting the cultivar top after budding. However, these clonal rootstocks are available with the viruses eliminated by heat treatments. Holding potted rose plants at a dry heat of 37 to 38°C (98 to 100°F) for four to five weeks will rid infected stocks of the virus. Again, seed propagation of rootstocks helps avoid most virus problems.

Rosa multiflora. This is a useful rootstock, especially in its thornless forms, for landscape roses. Several "strains" have been developed, some giving better bud unions and bud development than others. Cuttings of this species root readily, develop a vigorous root system, and do not sucker excessively. It is adaptable to a wide range of soil and climatic conditions. Seedlings are used in the eastern part of the United States, Great Britain, and Australia, and cuttings on the Pacific coast. The bark often becomes so thick late in the season that budding is impossible.

Rosa canina. Dog rose. Although this species has not done well under American conditions, the stock is commonly used in Europe. It is usually propagated by seed, since the cuttings do not root easily; however, the seeds are difficult to germinate. The prominent thorns make it difficult to handle. It also tends to sucker. Young plants on this stock grow slowly, but they are long-lived. *Rosa canina* is adaptable to drought and alkaline soil conditions, and is used as greenhouse rootstock for cut flower rose production in the Netherlands.

Rosa chinensis. 'Gloire de Rosomanes', 'Ragged Robin'. This old French stock is popular in California for outdoor roses, resisting heat and dry conditions well. It is also resistant to nematodes and does not sucker if the lower buds on the cuttings are removed. This stock grows steadily through the summer, permitting budding at any time. The fibrous root system is easy to transplant but requires good soil drainage. In some areas, however, it is difficult to propagate and is injured by leafspot. Owing to its susceptibility to verticillium wilt, it should not be planted on land previously planted with tomatoes or cotton.

Rosa 'Dr. Huey'. This is the principal rootstock in Arizona and the southern San Joaquin Valley, California, rose districts, replacing 'Ragged Robin' to a large extent. It has also performed well in Australia (*187*). It is useful for late season budding because of its thin bark. It is very vigorous and well adapted to irrigated conditions, and its cuttings root readily. It is very good as a stock for weak-growing cultivars. Defects are its injury from subzero temperatures and susceptibility to blackspot, mildew, and verticillium.

Rosa × noisettiana 'Manetti'. This is an old stock, very popular for greenhouse forcing roses. It is also of value for dwarf roses and for planting in sandy soils. It is easily propagated by cuttings, produces a plant of moderate vigor, and is resistant to some strains of verticillium.

Rosa odorata (Odorata 22449). Tea rose. This stock is excellent for greenhouse forcing roses. Cuttings root easily under suitable conditions, and produce a large symmetrical root system. It is adapted to both excessively dry and wet soil conditions. Since it is not cold-hardy, it should be used only in areas with mild winters. Some propagation stock of this clone is badly diseased and does not root well. The plants are not adaptable to cold-storage handling. It is more susceptible than *R. manetti* to verticillium.

Rosa dumetorum 'Laxa'. This stock is the most widely used rose rootstock in Great Britain. It produces few suckers and starts growth early in the season. It is propagated by seed, which is given a sulfuric acid treatment followed by stratification at 24°C (76°F) for 30 days and then at 5°C (42°F) for 85 days.

IXL (Tausendschon/Veilchenblau). This stock is used primarily as a trunk for tree roses. It is very vigorous and has no thorns. The canes tend to sunburn and are somewhat susceptible to low-temperature injury.

Multiflore de la Grifferaie. This stock is useful as a trunk for tree roses, producing desirable straight canes. It is vigorous, extremely hardy, and resistant to borers, but very susceptible to mite injury.

Rosa rugosa. This form, which is used as an understock, bears single, purplish-red flowers. For bush roses it is propagated by cuttings, and for tree roses by seed. The root system is shallow and fibrous and tends to sucker badly, but the plants are very long-lived. It is also used as the upright stem in producing standard (tree) roses.

Propagating Tree (Standard) Roses. A satisfactory method of producing this popular form of rose is to use *Rosa multiflora* as the rootstock, which is budded in the first summer to IXL or, preferably, the Grifferaie stock. These are trained to an upright form and kept free of suckers. In the second summer, at a height of about 0.9 m (3 ft) three or four buds of the desired flowering cultivar are inserted into the trunk stock. During the winter, the cane above the inserted buds is removed. The buds develop the following summer, as do buds from the stock, which must be removed. In the fall the plants may be dug and moved to their permanent location. Tree roses are sometimes dug as balled and burlapped plants because an extensive root system is formed during the two years in which the top is being developed.

Propagation of Miniature Roses (159). Soft or semihardwood cuttings are taken the year around and rooted under mist, after dipping in IBA talc. Miniature rose cultivars are especially bred for their ease of rooting. A good rooting mix is ⅓ peat moss:⅓ fir bark:⅓ perlite. Cuttings root in three to four weeks under warm conditions.

Rose of Sharon. *See* Hibiscus.

Rosemary (*Rosmarinus officinalis* L.). This ground cover with aromatic leaves is easily propagated by leafy cuttings under mist or glass, and also by seeds.

Rosewood, Arizona. *See* Vauquelinia spp.

Royal Poinciana (*Delonix regia* [Bajer.] Raf.). This spectacular tropical flowering tree is propagated by seed. Germination is rapid when seeds are treated to soften seed coats, as by pouring boiling water over seeds or by soaking in concentrated sulfuric acid for one hour.

Rubber Plant. *See* Ficus elastica.

Russian Olive. *See* Elaeagnus.

St. Johnswort (*Hypericum* spp.). St. Johnswort is easily started by softwood cuttings taken in late summer from the tips of current growth and rooted under high humidity or mist.

Salix spp. *See* Willow.

Sambucus spp. *See* Elderberry.

Sapindus drummondii. *See* Soapberry, Western.

Sea Grape. *See* Coccoloba spp.

Senicio cineraria. *See* Senicio spp. (Chap. 20).

Sequoiadendron giganteum. *See* Redwood.

Sequoia sempervirens. *See* Redwood.

Serviceberry (*Amelanchier* spp.) (*58, 59*). These trees and shrubs are generally seed propagated. Seeds show embryo dormancy which can be overcome by stratification at 2°C (36°F) for three to six months. Seeds should not be allowed to dry out. Seeds of *A. alnifolia* (Saskatoon) and *A. laevis* (Allegheny serviceberry) require 15 to 30 minutes of H_2SO_4 scarification prior to three months of cold stratification. Many species are readily propagated by leafy softwood cuttings taken when the new growth is several inches long and rooted under mist. Talc and liquid IBA from 1000 to 10,000 ppm is reported to enhance rooting. The newer cultivars 'Prince Charles' and 'Princess Diana', which are selections of *A. laevis* and *A.* × *grandiflora*, respectively, are micropropagated commercially.

Shrub-althea. *See* Hibiscus.

Silk Tree. *See* Albizzia.

Silverbell (*Halesia* spp.). A small, valuable landscape tree, native to the southeastern United States, that has striped bark, bell-shaped flowers, and interesting fruit (two- to four-winged drupe). Seed propagation of *Halesia* requires complex stratification regimes and success is often limited (*91*). *H. carolina* is warm stratified for two to four months, followed by two to three months of cold stratification; extracting stratified embryo and fall planting has been successful (*59*). Spring and summer cutting propagation with 1000 to 10,000 ppm IBA quick-dips gave 80 to 100 percent rooting success (*57*). *H. carolina* has been micropropagated successfully with mature explants (*20*).

Smoke Tree (*Cotinus coggygria* Scop.) (*129*). This tree can be propagated by leafy softwood cuttings under mist. Tip cuttings taken from spring growth and treated with IBA should root in about five weeks. After the mist is discontinued, the rooted cuttings are left in place to be transplanted bareroot the following spring. Smoke tree should not be propagated by seeds, since many of the seedlings are male plants, lacking the showy flowering panicles. Only vegetative methods should be used, with propagating wood taken from plants known to produce large quantities of the desirable fruiting clusters.

Snowberry (*Symphoricarpos* spp.). Seed propagation is difficult because of a hard, impermeable endocarp and a partially developed embryo at harvest. Give seeds a three- to four-month warm, moist stratification followed by cold stratification at 5°C (41°F) for six months before planting. Propagation is possible also by suckers, division, and cuttings.

Soapberry, Western (*Sapindus drummondii*) (*162*). Western soapberry is a deciduous landscape tree with a bright yellow fall color and a low water requirement, and it performs well on highly calcareous soils. For maximum germination, fall or winter collection of seeds followed by a 60-minute acid scarification and three-month cold stratification is recommended. May and June cuttings can be rooted when treated with 1.6 percent IBA (*59*).

Sophora (*Sophora* spp.). Texas mountain laurel (*S. secundiflora*) is a medium-sized shrub native to western Texas and northern Mexico. This legume has dark evergreen foliage, fragrant blue flowers, and is resistant to drought and temperature stress. *Sophora* is commercially seed propagated by scarifying with H_2SO_4 for 35 minutes or nicking the seed coat with a band saw (*208*). Asexual propagation by cuttings has not been successful; however, *S. secundiflora* can be micropropagated, which offers potential for clonal selection (*88*). *S. japonica* (Japanese pagoda tree) is generally seed propagated. Seed must be cleaned by soaking fruit for 48 hours; generally, no scarification or stratification treatments are needed (*59*). Cutting propagation is limited. Grafting is done by T-budding or with a side graft. Cultivars of *S. microphylla* are rooted successfully by taking semihardwood cuttings in winter (New Zealand), which are stripped of lower leaves, wounded, dipped in 8000 ppm IBA talc (Seradix 3), and inserted into 1 bark:1 pumice medium under mist with 20°C (68°F) bottom heat (*28*).

Sorbus (*Sorbus* spp.) (*142*). Seeds should be collected as soon as the fruits mature; fleshy parts are removed to eliminate inhibitors. Seeds require stratification for at least two months at about 5°C (41°F) for good germination. Cuttings do not root, and layering is difficult. Either fall budding or bench grafting (whip graft) is successful. Selected cultivars are best worked on seedlings of their own species, although *S. aria, S. aucuparia,* and *S. cuspidata* (European mountain ash) seedlings seem satisfactory as a rootstock for other species. *S. aucuparia* has been micropropagated (*35*).

Spiraea (*Spiraea* spp). Spiraea is usually propagated by cuttings, although some species, such as *S. thunbergi,* are more easily started by seeds, which should not be allowed to dry out. Leafy softwood cuttings taken in midsummer are generally successful. Treatments with auxin are beneficial. Some species, such as *S.* × *vanhouttei,* can be started readily by hardwood cuttings, planted in early spring.

Spruce (*Picea* spp.). Spruces are ordinarily propagated without difficulty by seed, either fall-planted or stratified over winter. Most species have embryo dormancy, requiring one to three months' stratification at about 4°C (40°F) for good germination. Seeds of *P. abies, P. engelmannii,* and *P. glauca* var. *albertiana* are among those giving good germination without stratification. Colorado blue spruce (*Picea pungens* 'Glauca') grown from seed produces trees with a slight bluish cast. Only a small percentage of the seedlings have the very desirable bright blue color. Several exceptionally fine blue seedling specimens have been selected as clones and are perpetuated by grafting. The two best known are the Koster blue spruce (*Picea pungens* 'Koster') developed at the Koster Nursery in Holland many years ago, and the compact 'Moerheim' blue spruce (*Picea pungens* 'Moerheimii'), which originated in Europe around 1930.

Selected clones of spruce are difficult to propagate by cuttings, but there are instances in which good percentages of cuttings have rooted (*93, 120*), especially from certain young source trees. Taking cuttings from vigorous containerized trees gives good results.

Cuttings taken in spring, midsummer, and mid-autumn have been rooted. As cuttings, it is best to use only shoot terminals, which should be gathered in early morning when the wood is turgid. Wounding, mist, and high light intensity during rooting are helpful. Medium, slightly acid sand or coarse perlite is a good rooting medium. Auxin treatments have given variable results. In making cuttings of upright-growing types, terminal shoots should be selected rather than lateral branches, since the latter, if rooted, tend to produce abnormal, sprawling plants rather than the desired upright form.

The Koster blue spruce is propagated commercially by grafting scions on Norway spruce (*Picea abies*) or Sitka spruce (*P. sitchensis*) seedlings (*151*).

Spurge, Japanese. *See* Pachysandra terminalis.

Stangeria eriopus. *See* Cycads.

Stewartia (*Stewartia* spp.) (*57, 59*). Deciduous or evergreen shrubs and trees with good fall color and bark characteristics. Double dormancy is overcome by a three- to five-month warm stratification followed by a three-month cold stratification. Semihardwood cuttings collected in early summer have been treated with 8000 ppm IBA + 15 percent thiram, 2500 ppm IBA + NAA, or 5000 ppm IBA–alcohol quick-dip, and propagated in sand–perlite under mist. Problems occur with winter survival, and rooted cuttings should be hardened-off and not transplanted during the fall.

Sumac (*Rhus* spp.). Sumacs are commonly propagated by seeds, which are collected in the fall; they should not be allowed to dry out. For prompt germination the seeds should be soaked in concentrated sulfuric acid for one to six hours, depending upon the species, then either fall-planted out-of-doors or stratified for two months at about 4°C (40°F) before planting. *R. virens* (evergreen sumac) is an excellent shrub or hedge for hot, well-drained locations in the southwest; seeds need to be scarified (*117*). Not all species have embryo dormancy, however, and the latter treatment may sometimes be omitted (e.g., with *R. ovata* and *R. integrifolia*). Some species, such as *R. aromatica,* need no pretreatments and will germinate with fall sowing. Some sumac plants bear only female flowers and others only male flowers, whereas still others have both flower types on the same plant. To ensure fruiting, plants of the latter type should be propagated asexually. In seed propagation many of the seedlings produced will not bear fruit.

For those species that sucker freely, such as *R. typhina* and *R. copallina,* root cuttings several inches long planted in the nursery row in early spring may be used. Leafy softwood cuttings, at least of some species, such as *R. aromatica,* taken in midsummer, root well if treated with 1 percent IBA mixed with 50 percent captan, 1:1.

Sweetshrub. *See* Calycanthus spp.

Swietenia mahagoni (*173*). Mahogany is a salt-tolerant, semideciduous landscape tree for southern Florida. Seeds are collected when mature in late winter and germinate easily without soaking or pregermination treatments.

Sycamore. *See* Plane Tree.

Symphoricarpos. *See* Snowberry.

Syringa spp. *See* Lilac, French Hybrid.

Tabebuia argentea (*173*). Gold tree or silver trumpet tree has a cork-like light-colored bark with spectacular golden yellow flowers. It is generally seed propagated without any pregermination treatment. It is important that the propagation medium be drenched with fungicides (captan, Ferbam, etc.) to control damping-off. This species can also be produced by cuttings and air layering.

Tamarisk (*Tamarix* spp.). These are easily rooted by hardwood cuttings, which are usually made about 12 in. long and planted deeply. Softwood cuttings taken in early summer also will root readily under mist.

Tamarix spp. *See* Tamarisk.

Taxodium distichum. *See* Bald Cypress.

Taxus spp. *See* Yew.

Telopea speciosissima. *See* Waratah.

Thuja spp. *See* Arborvitae.

Tilia spp. *See* Linden.

Toyon. *See* Heteromeles arbutifolia.

Trachelospermum asiaticum. Asiatic jasmine. An important ground cover that can be rooted any time of the year (Texas). Cuttings are quick-dipped in 3500 ppm IBA and rooted under mist or under heavy shade in outdoor beds. They root easily from the nodes and can be direct stuck in any suitable rooting media.

Trachelospermum jasminoides. Star or Confederate jasmine. Leafy cuttings of partially matured wood root easily, especially when placed under mist and treated with 1000 to 3000 ppm IBA liquid or talc.

Tree of Heaven. *See* Ailanthus.

Tree Tea. *See* Leptospermum spp.

Trumpet Creeper (*Campsis* spp.). This vine is usually propagated by cuttings, but seeds can also be used. With the latter method, stratification for two months at 4 to 10°C (40 to 50°F) hastens but does not increase germination. Both softwood and hardwood cuttings root readily. *C. radicans* can be started by root cuttings. Layering also is successful.

Tsuga spp. *See* Hemlock.

Tulip Tree (*Liriodendron tulipifera* L.) (*161*). Yellow poplar. Seed propagation is somewhat difficult. Artificial cross-pollination gives a higher percentage of "filled" seeds. Seeds should be stratified for about two months before planting and should not be allowed to dry out. A daily varying stratification temperature between freezing and about 10°C (50°F) has given good results, although a constant temperature around 4°C (40°F) would probably be equally satisfactory. Fall planting, with outdoor stratification through the winter, also has given good germination. Seeds of this species are often devoid of embryos, so cutting tests of each seed lot should be made. Although considered difficult to root, leafy stem cuttings taken in the summer have been rooted in fairly good percentages. Root cuttings also have been successful. Propagation by both budding and grafting is successful.

Cultivars can also be chip budded.

Young tulip trees are very difficult to transplant, so they should always be propagated into containers or dug, balled and burlapped, for transplanting from the nursery row.

Ulmus spp.   *See* Elm.

Ungnadia speciosa (*117*).   Mexican buckeye. An outstanding native of central Texas and northern Mexico which grows as a tree or a large multistemmed shrub. The deciduous foliage is similar to pecan, and rose-colored flowers resemble redbud. It is easy to grow and flowers within one to two years. Propagation is by seed, which is collected in late summer and needs no pregermination treatment. Seeds turn from green to dark brown and become mature in July (Texas). Seedlings require full sun (*205*).

Vaccinium ovatum (*213*).   Evergreen huckleberry. Similar to Japanese holly, but with reddish-colored new growth and pale pink flowers in profusion. Propagated by cuttings from fully matured shoots taken in fall and winter; cuttings made from previous year's growth taken the third week in April rooted 100 percent (Vancouver, B.C.). Basal heat (21°C; 70°F) and 0.3 to 0.4 percent IBA talc enhance rooting.

Vauquelinia spp. (*202*).   Arizona rosewood (*V. californica*) is a drought-tolerant, rosaceous evergreen shrub, which is one of the most popular landscape plants used in semiarid regions of the U.S. southwest. It can be propagated by seed, but is highly variable. Cuttings taken from 6- to 10-year-old stock plants root best when propagated May to June (Arizona) under mist and bottom heat (32°C; 90°F). IBA at 8000 ppm in 50 percent ethanol was of some benefit, but clonal variability (0 to 95 percent rooting) and cutting month were the greatest factors.

Veronica.   *See* Hebe.

Viburnum (*Viburnum* spp.) (*17, 156*).   This large group of desirable shrubs can be propagated by a number of methods, including seeds, cuttings, grafting, and layering. At least one species (*V. dentatum*) is readily started by root cuttings.

Seeds.   The viburnums have rather complicated seed dormancy conditions. Seeds of some species, such as *V. sieboldi*, will germinate after a single ordinary low-temperature (4°C; 40°F) stratification period, but for most species a period of two to nine months at high temperatures (20 to 30°C; 68 to 86°F), followed by a

two- to four-month period at low temperatures (4°C; 40°F) is required. The initial warm temperatures cause root formation, and the subsequent low temperature causes shoot development. Cold stratification alone will not result in germination. Such rather exacting treatments may best be given by planting the seeds in summer or early fall (at least 60 days before the onset of winter), thus providing the initial high-temperature requirement; the subsequent winter period fulfills the low-temperature requirement. After this, the seeds should germinate readily in the spring. Often, collecting the seeds early, before a hard seed coat has developed, will hasten germination. Viburnum seed can be kept up to 10 years if stored dry in sealed containers and held just above freezing. *V. lantana, V. opulus,* and *V. rhytidophyllum* are commonly propagated by seed.

Cuttings.   Although some viburnum species (*V. opulus, V. dentatum,* and *V. trilobum*) can be propagated by hardwood cuttings, softwood cuttings rooted in sand or perlite under mist are successful for most species. Soft, succulent cuttings taken in late spring root faster than those made from more mature tissue in midsummer, but the latter are more likely to grow on into sturdy plants that will survive through the following winter. Treatments with 8000 ppm IBA talc and thiram, and 2500 ppm IBA liquid, have been recommended (*15*). One of the chief problems with viburnum cuttings is to keep them growing after rooting. Cuttings made from succulent, rapidly growing material often die in a few weeks after being potted. This problem may be overcome by not transplanting the cuttings too soon, and allowing a secondary root system to form, which will better stand the transplanting shock. It may help to feed the rooted cuttings with a nutrient solution about 10 days before the cuttings are to be removed. Placing the rooted cuttings under artificial lights to increase daylength also is helpful. Cuttings of some species root more easily than others. *Viburnum carlesii*, for example, is difficult to root, but *V. burkwoodii, V. rhytidophylloides, V. lantana, V. sargentii,* and *V. plicatum* form a *tomentosum* root readily.

Grafting.   Selected types of viburnum are often propagated by grafting on rooted cuttings, layers, or seedlings of *V. dentatum* or *V. lantana*. Often, grafted viburnums will develop into vigorous plants more quickly than those started as cuttings. *V. opulus* 'Roseum' (Snowball) is dwarfed when grafted onto *V. opulus* 'Nanum' cuttings. It is important that all buds be removed from the rootstock so that subsequent suckering from the stock does not occur. The understocks are potted in the fall and brought into the greenhouse, where they are grafted in midwinter by the side graft method, using dormant scionwood. After grafting, the potted plants are placed in a closed, glass-covered frame with the unions buried in damp peat moss.

Grafting can be done in late summer also, using potted understock plants and scion material that has stopped growing and become hardened. The grafted plants are plunged in slightly damp peat moss in closed frames in the greenhouse until the unions heal, after which they are moved to outdoor, glass-covered cold frames for hardening-off for the winter.

Layering. Simple layering is widely used, especially in Europe, for propagating most viburnum species. Wood of the previous season's growth will produce roots in 18 to 24 months if layered in the spring. Some species are best layered in midsummer, using current season's wood.

Walking Stick.   *See* Aralia spp.

Waratah (*Telopea speciosissima*).   Seeds of this Australian native shrub with beautiful chrysanthemumlike flowers germinate easily, but the seedlings are difficult to transplant and grow outside their native environment. It requires soil of extremely low phosphorus content.

Wax Myrtle.   *See* Bayberry.

Weigela (*Weigela* spp.).   This shrub is easily propagated either by hardwood cuttings planted in early spring or by softwood tip cuttings under mist taken any time from late spring into fall. Treatment with IBA promotes rooting.

Willow, False Willow.   *See* Baccharis.

Willow (*Salix* spp.).   Willow seeds must be collected as soon as the capsules mature (when they have turned from green to yellow) and planted immediately since they retain their viability for only a few days at room temperature. Even under the most favorable conditions, maximum storage is four to six weeks. No dormancy occurs, germination taking place 12 to 24 hours after planting if the seeds are kept constantly moist. Willows are difficult to propagate in quantity by seed.

Willows root so readily by either stem or root cuttings that there is little need to use other methods. Hardwood cuttings planted in early spring root promptly.

Wisteria (*Wisteria* spp.).   It may take eight to twelve years for seed propagated wisteria to flower, which is why it is vegetatively propagated. Wisterias may be started by softwood cuttings under mist taken in midsummer. IBA often aids rooting. Some species can be started by hardwood cuttings set in the greenhouse in the spring. Simple layering of the long canes is quite

successful. Choice types are often grafted on rooted cuttings of less desirable types. Suckers arising from roots of such grafted plants should be removed promptly. Wisterias do not transplant easily, so young nursery plants are best started in containers. Seeds and pods are poisonous if eaten.

Witch Hazel (*Hamamelis* spp.) (*143*).   Witch hazel is propagated by seed, budding (*183*), grafting, or layering. Cutting propagation is difficult but possible. Seeds are gathered and planted outdoors in early fall. Prevent seeds from drying. *H. japonica* seeds should be soaked for a week with water changed daily. Cultivars of *H. mollis, H.* × *intermedia,* and *H. japonica* are propagated by budding or grafting (side-veneer graft) on *H. virginiana* seedlings; however, understocks tend to sucker and compete with scions. Leafy cuttings of *H. mollis, H. virginiana, H. japonica,* and *H. vernalis* can be rooted under mist with 8000 ppm IBA talc or 10,000 IBA liquid. Survival during winter is a problem, however (*138*). Commercial micropropagation of *Hamamelis* may become more feasible (*22*).

Xylosma congestum (Lour.) Merr. (*136*).   This is propagated by rooting leafy cuttings, taken in late summer or early fall, using the first and second subterminal cuttings on the shoot. Cuttings are dipped in IBA at 5000 ppm and rooted in a 1:3 peat moss–perlite mixture with bottom heat (21°C; 70°F). Results under mist have been contradictory. Rooted cuttings should be hardened in a cool, humid greenhouse.

Yellow Poplar.   *See* Tulip Tree.

Yew (*Taxus* spp.) (*190, 198*).   Most clonal selections of yews are propagated by cuttings, which root without much difficulty. Seedling propagation is little used, because of variation in the progeny, complicated seed dormancy conditions, and the slow growth of seedlings. Side or side-veneer grafting is practiced for those few cultivars that are especially difficult to start by cuttings, with easily rooted cuttings used as rootstock.

Seeds. This method is confined in commercial practice almost entirely to the Japanese yew, *Taxus cuspidata,* which comes fairly true from seed if isolated plants can be located as sources of seed. Seed imported from Japan is believed to produce uniform offspring.

For good germination, seeds should be given a warm 20°C (68°F) stratification period in moist peat moss or other medium for three months, followed by four months at a lower temperature (5°C; 41°F). Seedling growth is very slow. Two years in the seedbed, followed by two years in a lining-out bed, then three or four years in the nursery row are required to produce a plant of salable size of Japanese yew.

Cuttings. *Taxus* cuttings can be rooted outdoors in cold frames or in the greenhouse under mist, the latter giving much faster results.

For the cold frame, fairly large cuttings, 20 to 25 cm (8 to 10 in.) long, are made in early fall from new growth with a section of old wood at the base. Many species and cultivars respond best by stripping the needles at the base of the cutting (wounding) and treating with 8000 ppm IBA talc or an IBA quick dip of 5000 to 10,000 ppm. Cuttings may be kept in closed frames through the winter. Rooting takes place slowly during the following spring and summer.

For greenhouse propagation, cuttings should be taken in early winter, after several frosts have occurred, and rooted under mist in sand with bottom heat at about 21°C (70°F) and an air temperature of 10 to 13°C (50 to 55°F). Rooting in the greenhouse takes only about two months, but cuttings should not be dug too soon. Allow time for secondary roots to develop from the first-formed primary roots. There may be advantages of a two-month cold period after cuttings have rooted before planting out liners. There is evidence that cuttings from male plants (at least in *T. cuspidata expansa*) root more readily than cuttings from female plants (those that produce fruits). Seeds are very poisonous if eaten.

REFERENCES

1. Ackerman, W. L., and G. A. Seaton. 1969. Propagating the Bradford pear from cuttings. *Amer. Nurs.* 130(7):8.

2. Anderson, W. C. 1978. Rooting of tissue-cultured rhododendrons. *Proc. Inter. Plant Prop. Soc.* 28:135–39.

3. Argles, G. K. 1969. The propagation of magnolias. *Nurs. and Gard. Cent.* 148 (10):361–65; 148 (11):399–406.

4. Artlett, E. G. 1985. Propagation of *Pieris japonica. Proc. Inter. Plant Prop. Soc.* 35:128–29.

5. Bailey, D. A., G. R. Seckinger, and P. A. Hammer. 1986. In vitro propagation of florists' hydrangea. *HortScience* 21:525–26.

6. Bailey, D. A., and T. C. Weiler. 1984. Rapid propagation and establishment of florists' hydrangea. *HortScience* 19:850–52.

7. Barr, B. 1985. Propagation of *Berberis thunbergii* 'Atropurpurea Nana'. *Proc. Inter. Plant Prop. Soc.* 35:711–12.

8. ———. 1987. Propagation of dwarf *nandina* cultivars. *Proc. Inter. Plant Prop. Soc.* 37:507–8.

9. Barton, S. S., and V. P. Bonaminio. 1986. Influence of stratification and light on germination of sourwood [*Oxydendrum arboreum* (L.) DC]. *Jour. Environ. Hort.* 4:8–11.

10. Bassuk, N., D. Miske, and B. Maynard. 1984. Stock plant etiolation for improved rooting of cuttings. *Proc. Inter. Plant Prop. Soc.* 34:543–50.

11. Bennell, M., and G. Barth. 1986. Propagation of *Banksia coccina* by cuttings and seed. *Proc. Inter. Plant Prop. Soc.* 36:148–52.

12. Bennett, L. K. 1987. Tissue culturing redbud. *Amer. Nurs.* 166:85–91.

13. Bennett, L. K., and F. T. Davies, Jr. 1986. In vitro propagation of *Quercus shumardii* seedlings. *HortScience* 21:1045–47.

14. ———. 1985. Micropropagation of *Cercis. HortScience* 20:592.

15. Berry, J. B. 1984. Rooting hormone formulations: A chance for advancement. *Proc. Inter. Plant Prop. Soc.* 34:486–91.

16. Bhella, H. S. 1975. Some factors affecting propagation of Douglas fir. *Proc. Inter. Plant Prop. Soc.* 25:420–24.

17. ———. 1980. Vegetative propagation of viburnum, cvs. Alleghany, Mohican, and Onondaga. *The Plant Propagator* 26(3):5–9.

18. Blazich, F. A., and J. R. Acedo. 1987. Propagation of *Osmanthus* × *fortunei* by softwood cuttings. *Jour. Environ. Hort.* 5:70–71.

19. Blythe, G. 1984. Cutting propagation of *Sequoia sempervirens. Proc. Inter. Plant Prop. Soc.* 34:204–11.

20. Brand, M. H., and R. D. Lineberger. 1986. In vitro propagation of *Halesia carolina* L. and the influence of explantation timing on initial shoot proliferation. *Plant, Cell, Tissue, Organ Culture* 7:103–13.

21. Bretzloff, L. V., and N. E. Pellet. 1979. Effect of stratification and gibberellic acid on the germination of *Carpinus caroliniana* Walt. *HortScience* 14:621–22.

22. Briggs, B. A., and S. M. McCulloch. 1983.

Progress in micropropagation of woody plants in the United States and Canada. *Proc. Inter. Plant Prop. Soc.* 33:239–48.

23. Browse, P. M. 1986. Dormancy control in *Magnolia* seed germination. *Proc. Inter. Plant Prop. Soc.* 36:116–20.

24. Buck, G. J. 1951. Varieties of rose understocks. *Amer. Rose Ann.* 36:101–16.

25. Bunker, E. J. 1975. Germinating palm seed. *Proc. Inter. Plant Prop. Soc.* 25:377–78.

26. Bunting, B. A. 1985. Propagation of *Clethra alnifolia. Proc. Inter. Plant Prop. Soc.* 35:712–13.

27. Burger, D. W. 1987. In vitro micropropagation of *Eucalyptus sideroxylon. HortScience* 22:496–97.

28. Butcher, S. M., and S. M. N. Wood. 1984. Vegetative propagation and development of *Sophora microphylla* Ait. *Proc. Inter. Plant Prop. Soc.* 34:407–16.

29. Byers, D. 1983. Selection and propagation of crape myrtle. *Proc. Inter. Plant Prop. Soc.* 33:542–45.

30. Carpenter, W. J. 1988. Temperature affects seed germination of four Florida palm species. *HortScience* 23:336–37.

31. Carter, A. R. 1969. Rose rootstocks—performance and propagation from seed. *Proc. Inter. Plant Prop. Soc.* 19:172–80.

32. ———. 1979. *Daphne* propagation. *Proc. Inter. Plant Prop. Soc.* 29:248–51.

33. Carville, L. 1975. Propagation of *Acer palmatum* cultivars from hardwood cuttings. *Proc. Inter. Plant Prop. Soc.* 25:39–42.

34. Caulfield, H. W. 1976. Pointers for successful germination of palm seed. *Proc. Inter. Plant Prop. Soc.* 26:402–5.

35. Chalupa, V. 1987. Effect of benzylaminopurine and thidiazuron on in vitro shoot proliferation of *Tilia cordata* Mill., *Sorbus aucuparia* L. and *Robinia pseudoacacia* L. *Biolog. Plantar.* (*Praha*) 29:425–29.

36. Chandler, G. P. 1969. Rooting daphnes from cuttings. *Proc. Inter. Plant Prop. Soc.* 19:205–6.

37. Chang, T. Y., and T. H. Vogue. 1977. Regeneration of Douglas fir plantlets through tissue culture. *Science* 198:306–7.

38. Chase, H. H. 1964. Propagation of Oriental magnolias by layering. *Proc. Inter. Plant Prop. Soc.* 14:67–69.

39. Chaturvedi, A., K. Sharma, and P. N. Prasad. 1978. Shoot apex culture of *Bougainvillea glabra* 'Magnifica'. *HortScience* 13(1):36.

40. Christie, C. B. 1986. Factors affecting root formation on *Photinia* 'Red Robin' cuttings. *Proc. Inter. Plant Prop. Soc.* 36:490–94.

41. Chu, K., and W. S. Cooper. 1950. An ecological reconnaissance in the native home of Metasequoia glyptostroboides. *Ecology* 31:260–78.

42. Coate, B. 1983. Vegetative propagation of *Acacia iteaphylla. Proc. Inter. Plant Prop. Soc.* 33:118–20.

43. Coggeshall, R. C. 1977. Propagating French hybrid lilacs by softwood cuttings. *Proc. Inter. Plant Prop. Soc.* 27:442–44.

44. Cohen, M. A. 1975. Vegetative propagation of *Pinus strobus* by needle fascicles. *Proc. Inter. Plant Prop. Soc.* 25:413–19.

45. Connor, D. A. 1985. Propagation of upright junipers. *Proc. Inter. Plant Prop. Soc.* 35:719–21.

46. ———. 1985. Propagation of *Mahonia* species and cultivars. *Proc. Inter. Plant Prop. Soc.* 35:279–81.

47. Costin, J. J. 1977. Production of eucalyptus. *Proc. Inter. Plant Prop. Soc.* 27:44–48.

48. Covan, D. A. 1986. Softwood cutting propagation of oaks, magnolias, crabapples and dogwoods. *Proc. Inter. Plant Prop. Soc.* 36:419–21.

49. Cumming, W. A. 1964. *Crataegus* rootstock studies. *Proc. Inter. Plant Prop. Soc.* 14:146–49.

50. Davies, F. T., Jr. 1985. Adventitious root formation in *Rosa multiflora* 'Brooks 56' hardwood cuttings. *Jour. Environ. Hort.* 3:55–56.

51. Davies, F. T., Jr., Y. Fann, J. E. Lazarte, and D. R. Paterson. 1980. Bench chip budding of field roses. *HortScience* 15:817–18.

52. Davies, F. T., Jr., and J. N. Joiner. 1980. Growth regulator effects on adventitious root formation in leaf bud cuttings of juvenile and mature *Ficus pumila. Jour. Amer. Soc. Hort. Sci.* 105:91–95.

53. Deering, T. 1979. Bench grafting of *Betula* species. *The Plant Propagator* 25(3):8–9.

54. de Fossard, R. A., and H. de Fossard. 1988. Micropropagation of some members of the Myrtaceae. *Acta Hort.* 227:346–51.

55. Del Tredici, P. 1985. Propagation of *Tsuga canadensis* cultivars: Hardwood versus softwood cuttings. *Proc. Inter. Plant Prop. Soc.* 35:565–69.

56. Dirr, M. A. 1983. Comparative effects of selected rooting compounds of the rooting of *Photinia* × *Fraseri. Proc. Inter. Plant Prop. Soc.* 33:536–40.

57. ———. 1985. Ten woody plants that deserve a

longer look. *Proc. Inter. Plant Prop. Soc.* 35:728–34.

58. ———. 1987. Native amelanchiers: a sampler of northeastern species. *Amer. Nurs.* 166:64–83.

59. Dirr, M. A., and C. W. Heuser, Jr. 1987. *The reference manual of woody plant propagation.* Athens, Ga.: Univ. Press.

60. Dodge, M. H. 1986. Heath and heather propagation. *Proc. Inter. Plant Prop. Soc.* 36:563–67.

61. Donnelly, J. R., and H. W. Yawney. 1972. Some factors associated with vegetatively propagating sugar maple by stem cuttings. *Proc. Inter. Plant Prop. Soc.* 22:413–31.

62. Duck, P. H. 1985. Propagation of *Ilex vomitoria* 'Nana'. *Proc. Inter. Plant Prop. Soc.* 35:710–11.

63. Duncan, P. J., and T. E. Bilderback. 1982. Effect of irrigation systems, gibberellic acid and photoperiod on seed germination of *Kalmia latifolia* L. and *Rhododendron maximum* L. *HortScience* 17:916–17.

64. Ecke, P., and O. A. Matkin. 1976. *The poinsettia manual.* Encinitas, Calif.: Paul Ecke Poinsettias.

65. Economou, A. S., and P. E. Read. 1984. In vitro shoot proliferation of Minnesota deciduous azaleas. *HortScience* 19:60–61.

66. Economou, A. S., and K. M. J. Spanoudaki. 1985. In vitro propagation of *Gardenia. HortScience* 20:213.

67. Eden, W. S. 1985. Hibiscus propagation in cool climates. *Proc. Inter. Plant Prop. Soc.* 35:132–34.

68. Eichelser, J. E. 1978. Propagation of *Kalmia. Proc. Inter. Plant Prop. Soc.* 28:133–35.

69. Einert, A. E. 1974. Propagation of dwarf crape myrtles. *Proc. Inter. Plant Prop. Soc.* 24:370–73.

70. Ellyard, R. K., and P. J. Ollerenshaw. 1984. Effect of indolebutyric acid, medium composition and cutting type on rooting of *Grevillea johnsonii* cuttings at two basal temperatures. *Proc. Inter. Plant Prop. Soc.* 34:101–08.

71. Emery, D. E. 1985. *Arctostaphylos* propagation. *Proc. Inter. Plant Prop. Soc.* 35:281–84.

72. Ettinger, T. L., and J. E. Preece. 1985. Aseptic micropropagation of *Rhododendron* P.J.M. hybrids. *Jour. Hort. Sci.* 60:269–74.

73. Evison, R. J. 1977. Propagation of clematis. *Proc. Inter. Plant Prop. Soc.* 27:436–40.

74. ———. 1979. *Making the most of clematis.* Nottingham, England: Floraprint.

75. Farmer, R. E., Jr. 1975. Long term storage of northern red and scarlet oak seed. *The Plant Propagator* 20(4) and 21(1):11–14.

76. Fazio, S. 1964. Propagating *Eucalyptus* from cuttings. *Proc. Inter. Plant Prop. Soc.* 14:288–90.

77. Fins, L. 1980. Propagation of giant sequoia by rooting cuttings. *Proc. Inter. Plant Prop. Soc.* 30:127–32.

78. Flemer, W., III. 1962. The vegetative propagation of oaks. *Proc. Inter. Plant Prop. Soc.* 12:168–71.

79. ———. 1980. Linden propagation—a review. *Proc. Inter. Plant Prop. Soc.* 30:333–36.

80. Fleming, R. A. 1978. Propagation of holly in southern Ontario. *Proc. Inter. Plant Prop. Soc.* 28:553–57.

81. Fordham, A. J. 1972. Vegetative propagation of *Albizia. Amer. Nurs.* 128:(4):7, 63.

82. ———. 1977. Propagation of *Kalmia latifolia* by cuttings. *Proc. Inter. Plant Prop. Soc.* 27:479–83.

83. ———. 1977. *Pieris floribunda* and its propagation. *Proc. Inter. Plant Prop. Soc.* 27:495–97.

84. ———. 1984. *Cornus kousa* and its propagation. *Proc. Inter. Plant Prop. Soc.* 34:598–602.

85. ———. 1987. *Davidia involucrata* var. *vilmoriniana*—dove tree and its propagation by seeds. *Proc. Inter. Plant Prop. Soc.* 37:343–45.

86. Fox. B. S. 1972. Propagation of cotoneasters. *Proc. Inter. Plant Prop. Soc.* 22:213–18.

87. Freeland, K. S. 1977. Propagation of potentilla. *Proc. Inter. Plant Prop. Soc.* 27:441–42.

88. Froberg, C. A. 1985. Tissue culture propagation of *Sophora secundiflora. Proc. Inter. Plant Prop. Soc.* 35:750–54.

89. Fuller, C. W. 1979. Container production of *Euonymus alata* 'Compacta'. *Proc. Inter. Plant Prop. Soc.* 29:360–62.

90. Galle, F. 1988. *Azaleas.* Portland, Oreg.: Timber Press.

91. Giersbach, J., and L. V. Barton. 1932. Germination of seeds of the silverbell, *Halesia carolina. Contrib. Boyce Thomp. Inst.* 4:27–37.

92. Girouard, R. M. 1971. Vegetative propagation of pines by means of needle fascicles—a literature review. *Inform. Rpt. Dept. Environ., Can. For. Ser., Quebec.*

93. ———. 1973. Rooting, survival, shoot formation, and elongation of Norway spruce stem cuttings as affected by cutting types and auxin treatments. *The Plant Propagator* 19(2):16–17.

94. Goddard, A. N. 1970. Grafting Japanese maples. *The Plant Propagator* 16(4):6.

95. Goodman, M. A., and D. P. Stimart. 1987. Factors regulating overwinter survival of newly

propagated tip cuttings of *Acer palmatum* Thumb 'Bloodgood' and *Cornus florida* L. var. *rubra*. *HortScience* 22:1296–98.

96. Goreau, T. 1980. Rhododendron propagation. *Proc. Inter. Plant Prop. Soc.* 30:532–37.

97. Greever, P. T. 1979. Propagation of *Heteromeles arbutifolia* by softwood cuttings or by seed. *The Plant Propagator* 25(2):10–11.

98. Grossbechler, F. 1981. Mass production of eucalyptus seedlings by direct sowing method. *Proc. Inter. Plant Prop. Soc.* 31:276–79.

99. Haldemann, J. H., R. L. Thomas, and D. L. McKanny. 1987. Use of benomyl and rifampicin for in vitro shoot tip culture of *Camellia sinensis* and *C. japonica*. *HortScience* 22:306–7.

100. Hamilton, D. F., and P. L. Carpenter. 1977. Seed germination of *Myrica pennsylvanicum* L. *HortScience* 12(6):565–66.

101. Hand, N. P. 1978. Propagation of lilacs. *Proc. Inter. Plant Prop. Soc.* 28:348–50.

102. Hare, R. C. 1974. Chemical and environmental treatments promote rooting of pine cuttings. *Can. Jour. For. Res.* 4:101–6.

103. ———. 1976. Rooting of American and Formosan sweetgum cuttings taken from girdled and nongirdled cuttings. *Tree Plant. Notes* 27:6–7.

104. ———. 1977. Rooting of cuttings from mature water oak. *S. Jour. Appl. For.* 1:24–25.

105. ———. 1979. Modular air layering and chemical treatments improve rooting of loblolly pine. *Proc. Inter. Plant Prop. Soc.* 29:446–54.

106. Hartline, J. B. 1970. Holly propagation. *Amer. Hort. Mag.* 49(4):213–18.

107. Hartmann, H. T. 1967. 'Swan Hill': A new fruitless ornamental olive. *Calif. Agr.* 21(1):4–5.

108. Hasegawa, P. M. 1979. *In vitro* propagation of rose. *HortScience* 14:610–12.

109. ———. 1980. Factors affecting shoot and root initiation from cultured rose shoot tips. *Jour. Amer. Soc. Hort. Sci* 105(2):216–20.

110. Hasek, R. F. 1980. Roses. In *Introduction to floriculture*, R. A. Larson, ed. New York: Academic Press.

111. Heit, C. E. 1972. Propagation from seed: Growing larches. *Amer. Nurs.* 135(8):14–15, 99–110.

112. Henny, J. 1963. Exbury azaleas. *Proc. Inter. Plant Prop. Soc.* 13:231–33.

113. Hildebrandt, V., and P. M. Harney. 1983. In vitro propagation of *Syringa vulgaris* 'Vesper'. *HortScience* 18:432–34.

114. Hill, J. B. 1962. The propagation of *Juniperus chinensis* in greenhouse and mist bed. *Proc. Inter. Plant Prop. Soc.* 12:173–78.

115. Hinesley, L. E., and F. A. Blazich. 1980. Vegetative propagation of *Abies fraseri* by stem cuttings. *HortScience* 15(1):96–97.

116. Hoogendoorn, D. P. 1984. Propagation of *Acer griseum* from cuttings. *Proc. Inter. Plant Prop. Soc.* 34:570–73.

117. Hubbard, A. C. 1986. Native ornamentals for the U.S. southwest. *Proc. Inter. Plant Prop. Soc.* 36:347–50.

118. Huss-Danell, K., L. Eliasson, and I. Ohberg. 1980. Conditions for rooting of leafy cuttings of *Alnus incana*. *Phys. Plant.* 49:113–16.

119. Ingram, D. L. 1987. *Landscape plant propagation information retrieval system*. Circular 760. Gainesville: Univ. of Florida.

120. Iseli, J., and D. Howse. 1981. New cultivars of *Picea pungens glauca*—their attributes and propagation. *The Plant Propagator* 27(1):5–8.

121. Jaynes, R. A. 1988. *Kalmia, the laurel book II*. Portland, Oreg.: Timber Press.

122. ———. 1976. Mountain laurel selections and how to propagate them. *Proc. Inter. Plant Prop. Soc.* 26:233–36.

123. Johnson, C. R., and D. F. Hamilton. 1977. Rooting of *Hibiscus rosa-sinensis* as influenced by light intensity and ethephon. *HortScience* 12:39–40.

124. Joley, L. 1960. Experiences with propagation of the genus *Pistacia*. *Proc. Inter. Plant Prop. Soc.* 10:287–92.

125. Jona, R., and I. Gribaudo. 1987. Adventitious bud formation from leaf explants of *Ficus lyrata*. *HortScience* 22:651–53.

126. Kavanagh, J. M., S. A. Hunter, and P. J. Crossan. 1986. Micropropagation of catawba hybrid *Rhododendron* 'Nova Zembla', 'Cynthia' and 'Pink Pearl'. *Proc. Inter. Plant Prop. Soc.* 36:264–72.

127. Kelety, M. M. 1984. Container-grown hibiscus: Propagation and production. *Proc. Inter. Plant Prop. Soc.* 34:480–86.

128. Kelley, J. D., and J. E. Foret, Jr. 1977. Effect of timing and wood maturity on rooting of cuttings of *Cotinus coggygria* 'Royal Purple'. *Proc. Inter. Plant Prop. Soc.* 27:445–48.

129. Kelly, J. C. 1986. Propagation of *Prunus tenella* 'Firehill' from cuttings. *Proc. Inter. Plant Prop. Soc.* 36:279–81.

130. Kerns, H. R., and M. M. Meyer. 1986. Tissue

culture propagation of *Acer* × *freemanii* using thidiazuron to stimulate shoot tip proliferation. *HortScience* 21:1209–10.

131. Kiang, Y. T., O. M. Rogers, and R. B. Pike. 1974. Rooting mugho pine cuttings. *HortScience* 9(4):350.

132. King, S. M. 1988. Tissue culture of osage-orange. *HortScience* 23:613–15.

133. Klapis, A. J., Jr. 1964. Grafting junipers. *Proc. Inter. Plant Prop. Soc.* 15:340–41.

134. Kofranek, A. M., and R. Larson, eds. 1975. *Growing azaleas commercially*. Berkeley: Univ. Calif. Div. Agr. Sci.

135. Krussmann, G. 1981. *The complete book of roses*. Portland, Oreg.: Timber Press.

136. Kubo, E. 1965. Propagation of *Xylosma congestum*. *Proc. Inter. Plant Prop. Soc.* 15:340–41.

137. Lamb, J. G. D. 1970. Trials on propagation of *Chamaecyparis* at Kinsealy. *Proc. Inter. Plant Prop. Soc.* 20:334–38.

138. ———. 1976. The propagation of understocks for *Hamamelis*. *Proc. Inter. Plant Prop. Soc.* 26:127–30.

139. Larsen, O. N. 1985. Clonal propagation of *Fagus sylvatica* L. by cuttings. *Proc. Inter. Plant Prop. Soc.* 35:438–42.

140. Larson, R. A., J. W. Love, D. L. Strider, R. K. Jones, J. R. Baker, and K. F. Horn. 1978. *Commercial poinsettia production*. Raleigh, N.C.: N.C. State Univ. Dept. Agr. Inf.

141. Lanphear, F. O. 1963. The seasonal response in rooting of evergreen cuttings. *Proc. Inter. Plant Prop. Soc.* 20:334–38.

142. Lawyer, D. A. 1968. Propagating *Sorbus*. *Amer. Nurs.* 127(2):7, 54, 56, 58, 60.

143. Leiss, J. 1969. *Hamamelis* propagation. *Proc. Inter. Plant Prop. Soc.* 19:349–52.

144. ———. 1984. Root grafting of oaks. *Proc. Inter. Plant Prop. Soc.* 34:526–27.

145. ———. 1985. Seed treatments to enhance germination. *Proc. Inter. Plant Prop. Soc.* 35:495–99.

146. Li, Hui-Lin. 1961. Ginkgo—the maidenhair tree. *Amer. Hort. Mag.* 40:239–49.

147. Libby, W. J., and M. T. Conkle. 1966. Effects of auxin treatment, tree age, tree vigor, and cold storage on rooting young Monterey pine. *For. Sci.* 12:484–502.

148. Lloyd, G., and B. McCown. 1980. Commercially-feasible micropropagation of mountain laurel, *Kalmia latifolia*, by use of shoot-tip culture. *Proc. Inter. Plant Prop. Soc.* 30:421–27.

149. Lohnes, J. P., B. C. Van Duyne, and C. H. Case. 1988. Propagating 'Kwanzan' cherries. *Amer. Nurs.* 167:69–85.

150. MacDonald, B. 1974. Camellia propagation. *Proc. Inter. Plant Prop. Soc.* 24:152–54.

151. Mahlstede, C. 1962. A new technique in grafting blue spruce. *Proc. Inter. Plant Prop. Soc.* 12:125–26.

152. March, S. G. 1959. Propagating Ghent and Mollis azaleas. *Amer. Nurs.* 110(12):98–101.

153. McCown, B., and R. Amos. 1979. Initial trials with micropropagation of birch selections. *Proc. Inter. Plant Prop. Soc.* 29:387–93.

154. McDaniel, J. D. 1964. A look at some hackberries. *Proc. Inter. Plant Prop. Soc.* 14:143–46.

155. McGuire, J. J., W. Johnson, and C. Dawson. 1987. Leafbud or side-graft nurse root grafts for difficult to root *Rhododendron* cultivars. *Proc. Inter. Plant Prop. Soc.* 37:447–49.

156. McMillan-Browse, P. D. A. 1970. Notes on the propagation of viburnums. *Proc. Inter. Plant Prop. Soc.* 20:378–86.

157. Medina, J. P. 1981. Studies of clonal propagation of pecans at Ica, Peru. *The Plant Propagator* 27:10–11.

158. Moore, J. C. 1963. Propagation of chestnuts and camellias by nurse seed grafts. *Proc. Inter. Plant Prop. Soc.* 13:141–43.

159. Moore, R. S. 1981. Miniature rose production. *Proc. Inter. Plant Prop. Soc.* 30:54–60.

160. Morgan, D. L. 1985. Propagation of *Quercus virginiana* cuttings. *Proc. Inter. Plant Prop. Soc.* 35:716–19.

161. Morsink, W. A. G. 1974. Propagation of hardy strains of tulip trees (*Liriodendron tulipifera* L.) for southern Ontario. *The Plant Propagator* 20(3):14–15.

162. Munson, R. H. 1984. Germination of western soapberry as affected by scarification and stratification. *HortScience* 19:712–13.

163. Myers, J. R., and S. M. Still. 1979. Propagating London plane tree from cuttings. *The Plant Propagator* 25(3):8–9.

164. Nagao, M. A., K. Kanegawa, and W. S. Sakai. 1980. Accelerating palm seed germination with gibberellic acid, scarification, and bottom heat. *HortScience* 15(2):200–201.

165. Nicholson, R. 1984. Propagation notes on *Cedrus deodara* 'Shalimar' and *Calocedrus decurrens*. *The Plant Propagator* 30:5–6.

166. ———. 1987. Enkianthus—a worthy genus

waits to emerge from companion-plant status. *Amer. Nurs.* 166:91–97.

167. Nienhuys, H. C. 1981. Propagation of deciduous azaleas. *Proc. Inter. Plant Prop. Soc.* 30:457–59.

168. Osborne, R., and J. Van Staden. 1987. In vitro regeneration of *Stangeria eriopus*. *HortScience* 22:13–26.

169. Pair, J. C. 1986. Propagation of *Acer truncatum*, a new introduction to the southern great plains. *Proc. Inter. Plant Prop. Soc.* 36:403–13.

170. ———. 1986. New plant introductions with great stress tolerance to conditions in the southern great plains. *Proc. Inter. Plant Prop. Soc.* 36:351–55.

171. Parr, G. 1983. Summer grafting of golden robinia. *Proc. Inter. Plant Prop. Soc.* 33:164–67.

172. Parvin, P., R. A. Criley, and R. M. Bullock. 1973. Proteas: Developmental research for a new cut flower crop. *HortScience* 8(4):299–303.

173. Patel, S. I. 1983. Propagation of some rare tropical plants. *Proc. Inter. Plant Prop. Soc.* 33:573–80.

174. Pelleh, N. E., and K. Alpert. 1985. Rooting softwood cuttings of mature *Betula papyrifera*. *Proc. Inter. Plant Prop. Soc.* 35:519–25.

175. Phipps, H. M., D. A. Belton, and D. A. Netzer. 1977. Propagating cuttings of some *Populus* clones for tree plantations. *The Plant Propagator* 23(4):8–11.

176. Pinney, J. J. 1970. A simplified process for grafting junipers. *Amer. Nurs.* 131(10):7, 82–84.

177. Ponchia, G., and B. H. Howard. 1988. Chestnut and hazel propagation by leafy summer cuttings *Acta Hort.* 227:236–41.

178. Pontikis, C. A., and P. Melas. 1986. Micropropagation of *Ficus carica* L. *HortScience* 21:153.

179. Poole, R. T., and C. A. Conover. 1984. Propagation of ornamental *Ficus* by cuttings. *HortScience* 19:120–21.

180. Powell, J. C. 1985. Production of × *Cupressocyparis Leylandii*. *Proc. Inter. Plant Prop. Soc.* 35:722–23.

181. Pridham, A. M. S. 1964. Propagation of American elm from cuttings. *Proc. Inter. Plant Prop. Soc.* 14:86–88.

182. Proebsting, W. E. 1984. Rooting of Douglas-fir stem cuttings: Relative activity of IBA and NAA. *HortScience* 19:845–56.

183. Purcell, G. V. 1973. The budding of *Hamamelis*. *Proc. Inter. Plant Prop. Soc.* 23:129–32.

184. Richard, M. A. 1984. Propagation of bamboo by vegetative means. *Proc. Inter. Plant Prop. Soc.* 34:440–43.

185. Ridgway, D. 1984. Propagation and production of *Garrya elliptica*. *Proc. Inter. Plant Prop. Soc.* 34:261–65.

186. Rodríquez, R. 1982. In vitro propagation of *Castanea sativa* Mill. through meristem-tip culture. *HortScience* 17:888–89.

187. Ross, D. M. 1977. Rose rootstock, 'Dr. Huey', in South Australia. *Proc. Inter. Plant Prop. Soc.* 27:562–63.

188. Rouland, H., and N. E. Pellett. 1974. Propagation of Norway spruce (*Picea abies* [L.] Karst) by stem cuttings. *The Plant Propagator* 20(1):20–26.

189. Ryan, G. F. 1966. Grafting *Eucalyptus ficifolia*. *The Plant Propagator* 12(2):4–6.

190. Sabo, J. E. 1976. Propagation of *Taxus* in Northern Ohio. *Proc. Inter. Plant Prop. Soc.* 26:174–76.

191. Salter, C. E. 1970. *Clematis armandii* grafting. *Proc. Inter. Plant Prop. Soc.* 20:330–32.

192. Samartin, A. 1984. In vitro propagation of *Camellia japonica* seedlings. *HortScience* 19:225–26.

193. Saul, G. H., and L. Zsuffa. 1978. Vegetative propagation of elms by green cuttings. *Proc. Inter. Plant Prop. Soc.* 28:490–94.

194. Savella, L. 1981. Propagating pink dogwoods from rooted cuttings. *Proc. Inter. Plant Prop. Soc.* 30:405–7.

195. Schreiber, L. R., and M. Kawase. 1975. Rooting of cuttings from tops and stumps of American elm. *HortScience* 10(6):615.

196. Scott, A. 1976. A successful technique for grafting hibiscus. *Proc. Inter. Plant Prop. Soc.* 26:389–91.

197. Semeniuk, P., and R. N. Stewart. 1964. Low temperature requirements for after-ripening of seed of *Rosa blanda*. *Proc. Amer. Soc. Hort. Sci.* 85:639–41.

198. Shugert, R. 1985. Taxus production in the U.S.A. *Proc. Inter. Plant Prop. Soc.* 35:149–53.

199. Simon, R. A. 1986. A survey of hardy bamboos: Their care, culture and propagation. *Proc. Inter. Plant Prop. Soc.* 36:528–31.

200. Simpson, R. C. 1981. Propagating deciduous holly. *Proc. Inter. Plant Prop. Soc.* 30:338–42.

201. Singha, S. 1982. In vitro propagation of crabapple cultivars. *HortScience* 17:191–92.

202. Smith, E. Y., and C. W. Lee. 1983. Propaga-

tion of Arizona rosewood by stem cuttings. *HortScience* 18:764–65.

203. Smith, M. N. 1985. Propagating *Ceanothus*. *Proc. Inter. Plant Prop. Soc.* 35:301–5.

204. Snir, I. 1982. In vitro propagation of sweet cherry cultivars. *HortScience* 17:192–93.

205. Stanford, G. 1982. *Ungnadia speciosa* (Mexican buckeye). *The Plant Propagator* 28(2):5–6.

206. Stoltz, L. P. 1984. In vitro propagation and growth of *Hydrangea*. *HortScience* 19:717–19.

207. Strode, R. E., P. A. Travers, and R. P. Oglesby. 1979. Commercial micropropagation of rhododendrons. *Proc. Inter. Plant Prop. Soc.* 29:439–43.

208. Strong, M. E., and F. T. Davies, Jr. 1982. Influence of selected vesicular-arbuscular mycorrhizal fungi on seedling growth and phosphorus uptake of *Sophora secundiflora*. *HortScience* 17:620–21.

209. Struve, D. K., M. A. Arnold, and D. H. Chinery. 1987. Red oak whip production in containers. *Proc. Inter. Plant Prop. Soc.* 37:415–20.

210. Stump, D. S. 1980. Roses [*Plants & Gardens* 36(1)]. Brooklyn, N.Y.: Brooklyn Botanic Garden.

211. Sutter, E., and P. Barker. 1985. In vitro propagation of mature *Liquidambar styraciflua*. *Plant Cell, Tissue, Organ Culture* 5:13–21.

212. Ticknor, R. L. 1969. Review of the rooting of pines. *Proc. Inter. Plant Prop. Soc.* 19:132–37.

213. Tubesing, C. E. 1985. Cutting propagation of *Paxistima myrsinites, Vaccinium ovatum, Ribes sanguineum* and *Acer glabrum* v. *Douglasii*. *Proc. Inter. Plant Prop. Soc.* 35:293–96.

214. ———. 1987. Chip budding of magnolias. *Proc. Inter. Plant Prop. Soc.* 37:377–79.

215. Uno. S., and J. E. Preece. 1987. Micro- and cutting propagation of 'Crimson Pygmy' barberry. *HortScience* 22:488–91.

216. Van de Pol, P. A., and A. Breukelaar. 1982. Stenting of roses: A method for quick propagation by simultaneously cutting and grafting. *Scient. Hort.* 17:187–96.

217. Vanstone, D. E. 1978. Basswood (*Tilia americana*) seed germination. *Proc. Inter. Plant Prop. Soc.* 28:566–69.

218. Van Veen, T. 1971. The propagation and production of rhododendrons. *Amer. Nurs.* 133(8): 15–16, 52–58.

219. Vertrees, J. D. 1987. *Japanese maples*. Portland, Oreg.: Timber Press.

220. Vieitez, A. M., A. Ballester, M. L. Vieitez, and E. Vieitez. 1983. In vitro plantlet regeneration of mature chestnut. *Jour. Hort. Sci.* 58:457–63.

221. Vieitez, A. M., M. C. San-Jose, and E. Vieitez, 1985. In vitro plantlet regeneration from juvenile and mature *Quercus robur* L. *Jour. Hort. Sci.* 60:99–106.

222. Wallis, J. S. 1976. The propagation and training of standard fuchsias. *Proc. Inter. Plant Prop. Soc.* 26:346–48.

223. Warren, P. 1973. Propagation of *Cercis* cultivars by summer budding. *The Plant Propagator* 19(3):16–17.

224. Wasley, R. 1979. The propagation of *Berberis* by cuttings. *Proc. Inter. Plant Prop. Soc.* 29:215–16.

225. Watkins, J. V. 1972. Jacaranda. *Horticulture* 50(5):22–23.

226. Wedge, D. 1977. Propagation of hybrid lilacs. *Proc. Inter. Plant Prop. Soc.* 27:432–36.

227. Wells, J. S. 1980. How to propagate Japanese maples. *Amer. Nurs.* 151(9):14, 117–20.

228. ———. 1981. A history of rhododendron cutting propagation. *Amer. Nurs.* 154(9):14–15, 114–26.

229. ———. 1985. *Plant propagation practices*. Chicago: American Nurseryman Publishing Co.

230. Whalley, D. N. 1975. Propagation of Commelin elm by hardwood cuttings in heated bins. *The Plant Propagator* 21(3):4–6.

231. ———. 1979. Leyland cypress—rooting and early growth of selected clones. *Proc. Inter. Plant Prop. Soc.* 29:190–202.

232. Whitehead, H. C. M., and K. L. Giles. 1976. Rapid propagation of poplars by tissue culture methods. *Proc. Inter. Plant Prop. Soc.* 26:340–43.

233. Wisniewski, L. A., L. J. Frampton, and S. E. McKeand. 1986. Early shoot and root quality effects on nursery and field development of tissue-cultured loblolly pine. *HortScience* 21: 1185–86.

234. Wister, J. C., and G. S. Wister, eds. 1962. *The peonies*. Washington, D.C.: American Horticultural Society.

235. Zhang, Z. M., and F. T. Davies, Jr. 1986. In vitro culture of crape myrtle. *HortScience* 21:1044–45.

236. Zimmerman, R. H., R. J. Griesbach, F. A. Hammerschlag, and R. H. Lawson. 1986. *Tissue culture as a plant production system for horticultural crops*. Dordrecht: Martinus Nijhoff Publishers.

SUPPLEMENTARY READING

American Nurseryman Magazine. Chicago: American Nurseryman Publishing Co.

DIRR, M. A. 1983. *Manual of woody landscape plants.* Champaign, Ill.: Stipes Publ. Co.

DIRR, M. A., and C. W. HEUSER, JR. 1987. *The reference manual of woody plant propagation.* Athens, Ga.: Univ. Press.

DUNMIRE, J. R., ed. 1988. *Western garden book.* Menlo Park, Calif.: Lane.

FORDHAM, A. J., and L. S. SPRAKER. 1977. Propagation manual of selected gymnosperms. *Arnoldia* 37:1–88.

HENLEY, R. W. 1980. Growing trees for interior use. *Proc. Inter. Plant Prop. Soc.* 30:505–10.

INGRAM, D. L. 1987. Landscape plant propagation information retrieval system. Circular 760. Gainesville: Univ. of Florida.

INTERNATIONAL PLANT PROPAGATORS' SOCIETY. Proceedings of annual meetings.

KRUSSMAN, G. 1984. *Manual of cultivated broadleaved trees and shrubs.* Portland, Oreg.: Timber Press.

LAMB, J. G. D., J. C. KELLY, and P. BOWBRICK. 1985. *Nursery stock manual.* London: Grower Books.

MACDONALD, B. 1986. *Practical woody plant propagation for nursery growers,* Vol 1. Portland, Oreg.: Timber Press.

MCCLINTOCK, E., and A. T. LEISER. 1979. An annotated checklist of woody ornamental plants of California, Oregon and Washington. *Univ. Calif. Div. Agr. Sci. Priced Publ. 4091.*

MCMILLAN-BROWSE, P. D. A. 1979. *Hardy woody plants from seed.* London: Grower Books.

New Zealand Journal of Forestry Sciences. 1974. Special issue on vegetative propagation of conifers and hardwoods, 4(2).

SCHOPMEYER, C. S., ed. 1974. *Seeds of woody plants in the United States.* U.S. Dept. Agr. For. Ser. Handbook 450. Washington, D.C.: U.S. Govt. Printing Office.

WRIGLEY, J. W., and M. FAGG. 1979. *Australian native plants: Propagation, cultivation, and use in landscaping.* Sydney, Australia: William Collins Publishers.

20

Propagation of Selected Annuals and Herbaceous Perennials Used as Ornamentals

Herbaceous plants are classified as *annuals, biennials,* or *perennials,* although the differences among these types may not be obvious. They may also be classified as *hardy, semihardy,* or *tender.* In general, the propagation procedures for such plants depend upon their categories and the locality where they are to be grown.

In the following list of plants, seed germination data are given for some species, including suggested approximate temperatures that should give the most rapid and complete germination, along with the expected germination time. If two temperatures are given, separated by a dash, the first is the minimum night temperature, and the second the maximum day temperature. A single figure indicates a constant temperature. The propagation methods indicated will serve as a guide, but some variation from these methods may be necessary with individual cultivars (*9, 10*)

Achillea spp. Yarrow. Hardy perennial. Seeds germinate in one to two weeks at 70°F (21°C), but seeds of

A. filipendulina, which is used as a field-grown cut flower crop (*5*), sometimes produce inferior plants. Other species can be seed propagated. Fast propagation can also be achieved with cuttings taken in midsummer. Division is easy and clumps of mature plants should be divided every three to four years (*110*).

Achimenes spp. Tender perennial. Seeds germinated in a warm greenhouse can be used for propagating species. Plants grow from small scaly rhizomes, which can be divided for propagation. Softwood cuttings in spring or leaf cuttings in summer can be rooted. Partially dried leaf scales can be planted and rooted.

Aconitum spp. Monkshood. Hardy perennial. Seeds often show dormancy and must be moist-chilled below 5°C (41°F) for six weeks before planting. Considered difficult to propagate by seed. Plants have tuberous roots that can be divided, but once established they should not be transplanted. All parts of the plants are poisonous.

Adiantum. *See* Fern.

Aegopodium spp.   Goutweed. A hardy perennial that is used as a groundcover. Propagated by division.

African Violet.   *See* Saintpaulia ionantha.

Agapanthus spp.   Lily-of-the-Nile. The thick rhizomes can be divided to produce new plants. Seed propagation can be used, but time to flowering is longer.

Agave spp.   Many species of succulents, including the century plant. Perennial. Seeds should be sown in sandy soil when mature. Reproduces vegetatively by offsets from base of plant, or from the flower stalk of some species; these are removed along with roots and repotted in spring. Some species produce bulbils that can be used for propagation. Agave are commercially mass produced by tissue culture.

Ageratum houstonianum.   Ageratum. Half-hardy annual. Seeds germinate in one to two weeks at 20 to 30°C (68 to 86°F). Ageratum may be propagated by cuttings also.

Aglaonema spp.   An important foliage plant which is easily propagated by canes (long stems), shoot cuttings, division, or seeds. Canes should be treated as carefully as delicate leaf cuttings. Rooting is enhanced with IBA and bottom heat [24 to 28°C (75 to 80°F)] (*89*).

Agrostemma githago.   Corncockle. Hardy annual. Seeds germinate in two to three weeks at 20°C (68°F).

Allium spp.   Ornamental onion; also onion, chives, and garlic. Propagated by seed. Plants grow from bulbs, which produce offsets. Clumps can also be divided. Many species produce bulbils.

Aloe spp.   Succulents of the lily family. Propagated by seed in well-drained sandy soil. Germination takes place in three to four weeks at 20 to 24°C (68 to 75°F). Plants produce offshoots that can be detached and rooted. Plants with long stems can be made into cuttings, which should be exposed to air for a few hours to allow cut surfaces to suberize. These species are commercially micropropagated.

Althaea rosea.   Hollyhock. Half-hardy biennial. Seeds germinate in two to three weeks at 20° (68°F). Where winters are not too severe sow seeds in summer, transplant in fall for bloom the following year, or sow seeds in warm greenhouse in winter and transplant outdoors.

Alyssum saxatile.   Goldentuft. Hardy perennial. Seeds germinate in three to four weeks at 20 to 30°C (68 to 86°F). Sow in summer for bloom the following year. Germination may be stimulated by light or exposure of moist seeds at 15°C (50°F) for five days (*7*). Propagate by division or by softwood cuttings in spring. Double forms must be propagated by cuttings or division.

Alyxia olivaeformis.   Maile. An important foliage plant which is native to Hawaii. It is propagated almost exclusively by seeds which need to be depulped (*114*). Rooting of single-node maile stem cuttings is improved by removing one-half leaf surface area, placing greenhouse grown cuttings in water prior to treatment, and propagating in a shade cloth–covered greenhouse. Cuttings were irrigated once daily with spray stakes. A five-second quick dip in 3000 to 8000 ppm IBA was most effective (*115*).

Amaranthus caudatus.   Love-lies-bleeding. Half-hardy annual. Seeds germinate in two to three weeks at 20 to 30°C (68 to 86°F). Light may increase germination (*7*). Sow in warm greenhouse for later transplanting or sow out-of-doors when frost danger is past. *A. gangeticus tricolor,* Joseph's coat. Same as for *A. caudatus.* Sensitive to excess water.

Amaryllis belladonna.   Belladonna lily. Perennial. Grows from bulbs outdoors in mild areas or in pots in cold climates. Propagate by bulb cuttings, separation of bulbs, or tissue culture (*23*).

Anchusa capensis.   Bugloss. Hardy annual or biennial. Seeds germinate in two to three weeks at 20 to 30°C (68 to 86°F). Sow seeds in summer for bloom next year or plant in greenhouse in winter for later transplanting to garden.
      A. azurea.   Perennial. Selected clones best propagated by root cuttings or clump division (*35*).

Anemone spp.   Poppy anemone (*A. coronaria*). Tender perennials. Seeds germinate in five to six weeks at 20°C (68°F) and may be sensitive to higher temperatures (*7*). Plants develop clusters of small, claw-like tuberous roots.
      A. japonica.   Japanese anemone. Hardy perennial. Since seeds do not come true, cultivars are propagated by division or by root cuttings. Roots are dug in fall and cut into 5-cm (2-in.) pieces, which are laid in flats or in a cold frame, then covered with 2.5 cm (1 in.) of soil. Plants can be potted after shoots appear.
      A. pulsatilla.   Pasque flower. Hardy perennial. Seeds germinate in five to six weeks at 20°C (68°F),

but may be sensitive to high temperature (*5*). Plants can be divided or propagated from root cuttings taken in spring. There is an export market for *Anemone* tubers (*41*).

Anigozanthus spp. Kangaroo paw. (*67*). This native perennial Australian genus is used for cut flower production and as containerized plants. Seed supplies are often scarce and germination rates of available seed are usually low and variable for many species. Hot-water and chemical pretreatment can be used to improve germination. Some hybrids are sterile and do not set seed at all. Clumps of rhizomes can be divided, but the rate of multiplication is low and unreliable. The most effective means of commercial propagation is through micropropagation.

Anthemis spp. Golden marguerite, camomile. Hardy perennial. Seeds germinate in one to three weeks at 20°C (68°F). Plants can be divided or propagated by stem cuttings.

Anthurium andraeanum. Anthurium. Remove offshoots with attached roots from the parent plant or root two- or three-leaved terminal cuttings under mist. Anthurium can be propagated by in vitro methods using a vegetative bud explant (*63*). Seed propagation is a lengthy process requiring 1½ to 3 years for flowering, and cultivars do not come true from seed (*47*).

Antirrhinum majus. Snapdragon. Tender perennial, treated as an annual. Seeds germinate in one to two weeks at 13°C (55°F) and may respond to light (*7*). Some hybrids are best started at 16 to 18°C (60 to 65°F). Seeds germinate well in mist (*10*). Start indoors for later outdoor planting (in fall in mild climates or in spring in severe winter areas). Softwood cuttings root readily. This species is tissue cultured (*87*).

Aquilegia spp. Columbine. Hardy perennials. Seeds germinate in three to four weeks at 20 to 30°C (68 to 86°F) and may respond to light and three to four weeks moist-chilling at 5°C (41°F) (*7*). Many of the columbines come true from seed.

Arabis spp. Rockcress. Hardy perennials. Seeds germinate in three to four weeks at 20°C (68°F) and may respond to light (*5*). Softwood cuttings taken from new growth immediately after bloom root readily. Plants can be divided in spring or fall.

Arctotis stoechadifolia. African daisy. Half-hardy annual. Seeds germinate in two to three weeks at 20°C (68°F). Sow indoors for later transplanting.

Armeria spp. Thrift. Hardy evergreen perennials. Seeds germinate in three to four weeks at 20°C (68°F). Best propagated by clump division in spring or fall.

Artemisia spp. *A. ludoviciana* can be used as a foliage plant and is propagated by division or stem cuttings. *A. schmidtiana* (wormwood) is a hardy perennial used as a specimen plant and is propagated by stem cuttings, rather than by division.

Aruncus dioicus (*A. sylvester*). Goat's beard. A hardy perennial used as a specimen or border plant. Seeds have a cold stratification requirement. Can also be propagated by division.

Asclepias tuberosa. Butterfly weed. Hardy perennial. Seeds germinate in three to four weeks at 20 to 30°C (68 to 86°F). Fresh seed may need chilling. Plants should not be disturbed once established. *A. curassavica.* Bloodflower. Tropical perennial. Propagated by seed or by rooting softwood cuttings. Long taproot makes division difficult.

Asparagus asparagoides. Smilax. Tender perennial. Propagated by seeds, which germinate in three to four weeks at 20 to 30°C (68 to 86°F). Sow seeds soon after they ripen, since they are short-lived (*10*). Cuttings can be made of young side shoots taken from old plants in spring; clumps can be divided. *A. plumosus,* fern asparagus, and *A. sprengeri,* Sprenger asparagus. Seeds germinate in three to six weeks at 20 to 30°C (68 to 86°F). Crack the seed coats with a knife. This can also be propagated vegetatively as described for smilax.

Aster spp. Hardy perennials. Seeds germinate in two to three weeks at 20°C (68°F). Cultivars are propagated by lifting clumps in fall and dividing into rooted sections, discarding the older parts.

Astilbe spp. Astilbe. Hardy perennials. Propagated by division in early spring when 2.5 cm (1 in.) tall (*45*). Seed germination is slow and produces a mixed progeny. Germination takes three to four weeks at 16 to 21°C (60 to 70°F) (*110*).

Astrantia spp. Masterwort. Perennial with unusual and attractive flowers. Generally propagated by division. Seeds require a cold stratification.

Aubrieta deltoidea. Aubrieta. Hardy perennials, sometimes treated as annuals. Seeds germinate in two to three weeks at 13°C (55°F). Clumps are difficult to divide; cuttings may be taken immediately after blooming.

Aucuba japonica. *See* Gold Dust Plant.

Baptisia spp. False Indigo. Hardy perennial. Seed germination at 20°C (68°F) is slow and uneven. Gather seeds when ripe and sow outdoors to overwinter. Clumps are difficult to divide because of a long taproot.

Begonia spp. Begonia (*98*). Tropical perennials. Seeds, which are very fine and need light, germinate in two to four weeks at 20°C (68°F). Sow on moist, light medium with little or no covering. Begonia species, tuberous begonias, and wax begonias are propagated by seed, but other types are propagated vegetatively.

Tuberous Begonias. In addition to seed propagation these can be grown from tuberous stems, which are divided into sections, each bearing at least one growing point. Leaf, leaf-bud, and short stem cuttings (preferably with piece of tuberous stem attached) will root readily.

Fibrous-Rooted Begonias. Wax begonias, Christmas begonias, and others are propagated by leaf cuttings or softwood cuttings taken from young shoots in spring or summer. The cytokinin PBA was more effective in bud and shoot development from leaf cuttings than BA or kinetin (*31*).

Rhizomatous Types. Various species and cultivars, including *Rex begonia*. Plants are divided or rhizomes are cut into sections. Propagation is usually by leaf cuttings, but stem cuttings also will root. Treatment of leaf cuttings with a cytokinin increases the number of plantlets produced per leaf (*126*). *B. evansiana* produces small tubercles, which are detached and planted. Begonia can also be micropropagated using leaf petioles (*79*), petiole explants (*99, 126*) and somatic embryos (*84*).

Bellis perennis. English daisy. Hardy perennial often treated as annual or biennial. Seeds germinate in one to two weeks at 20°C (68°F) and may respond to light (*7*). Clumps should be divided every year to prevent crowding.

Boltonia spp. Boltonia. Hardy perennial. Seeds germinate in two to three weeks at 20°C (68°F). Divide plants in spring or fall.

Bouvardia ternifolia (*51*). Scarlet bouvardia. An outstanding perennial with scarlet tubular flowers which blooms from midsummer to frost in Texas and New Mexico. It is propagated by semihardwood cuttings throughout the growing season.

Bromeliads. About 2000 species of tropical herbs or subshrubs in 45 genera. The pineapple (*Ananas*) is the best known. Propagation is mainly by seeds or by asexual division of lateral shoots, but micropropagation has been used successfully with some species (*49*). Conditions vary for the successful micropropagation of *Guzmania, Tillandsia,* and *Vriesea* species found with the Bromeliaceae (*77*).

Browallia spp. Amethyst flower. Tender, blue-flowered perennial often treated as annual. Seeds germinate in two to three weeks at 20°C (68°F). Softwood cuttings can be taken in fall or spring. Can be used as flowering pot plant indoors in winter.

Cactus (*21, 85*). Large group of many genera, species, and some cultivars. Tender to semihardy perennials. Seed propagation can be used for most species, but seeds often germinate slowly. Sow fungicide-treated seed in well-drained, sterile mixture, and water sparingly, but do not allow medium to dry out. Pieces of stem can be broken off and rooted as cuttings (*120*) or small offsets, which root readily, can be removed. Allow offsets to dry for a few days to heal (suberize) cut surface before rooting. Cuttings require two to three weeks to heal. High humidity during rooting is unnecessary, but bottom heat is beneficial. Grafting is used to provide a decay-resistant stock for certain kinds and to produce unusual growth forms. For example, the pendulous *Zygocactus truncatus* is sometimes grafted on tall erect stems of *Pereskia aculeata*. Intergeneric grafts are usually successful. A type of cleft graft is used. The stem of the stock is cut off, and a wedge-shaped piece is removed. The scion is prepared by removing a thin slice from each side of the base; this is fitted into the opening in the stock. The scion is held in place with a pin or a thorn. A grafting adhesive can be used to adhere scions to stocks of transversally cut (tip grafted) cactus (*129*). The completed graft is held in a warm greenhouse until healed (*20, 38*) (see Fig. 11–4). The development of cacti shoots by micropropagation can be extremely rapid in comparison with greenhouse-germinated seedlings (*8*) and where poor branching limits propagules for traditional vegetative propagation methods (*68*).

Caladium bicolor. This tropical perennial, grown for its strikingly colorful foliage, produces tubers. Propagation is by removing the tubers from the parent plant at the end of the four- to five-month dormancy period just before planting. Sometimes the tubers are cut into pieces, each containing at least two buds ("eyes"). Caladiums do best out-of-doors when planted after the minimum night temperature is above 18°C (65°F) or as pot plants maintained with night temperatures of 18 to 21°C (65 to 70°F) and day temperatures of 24 to 29.5°C (75 to 85°F).

Calceolaria spp. Tender perennials often grown as annuals. Seeds germinate in two to three weeks at 20°C (68°F). Propagation is also by softwood cuttings.

Calendula officinalis. Pot marigold. Hardy annual; gives winter bloom in mild climates from seed sown in late summer. Seeds germinate in one to two weeks at 20 to 30°C (68 to 86°F). Thin plants 30 cm (12 in.) apart.

Calla. *See* Zantedeschia spp.

Callistephus chinensis. China aster. Half-hardy annual. Seeds germinate in two to three weeks at 20°C (68°F). Plant only wilt-resistant types.

Caltha palustris Marsh marigold. A hardy perennial used in water gardens or ponds. Propagation is by division.

Campanula carpatica. Tussock, Bellflower. Hardy perennial. Seeds germinate in two to three weeks at 20 to 30°C (68 to 86°F) and may respond to light (7). *C. lactiflora.* Bellflower. Hardy perennial. Seeds germinate in two to three weeks at 13 to 32°C (55 to 90°F).

C. medium. Canterbury bells. Hardy biennial. Seeds, which germinate in two to three weeks at 20 to 30°C (68 to 86°F), are sown in late spring or early summer for bloom the following year.

C. persicifolia. Peach bells. Hardy perennial. Seeds germinate in two to three weeks at 13 to 32°C (55 to 90°F) and may respond to light (7). Small offsets can be detached and rooted.

C. pyramidalis. Chimney bellflower. Hardy perennial, often treated as a biennial. Seeds germinate in two weeks at 20 to 30°C (68 to 86°F). Many of the named cultivars of *Campanula* spp. cannot be produced by seed, so cuttings are used. Cuttings are produced from the rhizomatous growth of stock plants and rooting occurs from the etiolated base. Cuttings are placed in peat–perlite media, and given basal heat under glass or in a tunnel (England) (101).

Canna spp. Canna. Tender perennial. Cultivars do not come true from seed. Seeds, which have hard coats and must be scarified before planting, are germinated in a warm greenhouse. Cultivars are propagated by dividing the rhizome, keeping as much stem tissue as

possible for each growing point. In mild climates this is done after the shoots die down in the fall or before growth starts in spring. In cold climates the plants are dug in fall, stored over winter, divided in spring, then started in sand or sandy soil for transplanting outdoors when frost danger is over.

Carnation. *See* Dianthus caryophyllus.

Catananche caerulea. Cupidsdart. Hardy perennial. Seeds germinate in two to four weeks at 20 to 30°C (68 to 86°F). Plants may be divided in fall.

Celosia argentea. Cockscomb. Tender annual. Seeds germinate in one to two weeks at 20 to 30°C (68 to 86°F) and may respond to light (7).

Centaurea, Cineraria, and Other Species. Dusty miller. Tender perennial. Seeds germinate in two to four weeks at 20 to 30°C (68 to 86°F). Cuttings can be rooted. *C. cyanus.* Cornflower or bachelor button. *C. moschata.* Sweet sultan. These are hardy annuals whose seeds germinate in three to four weeks at 20 to 30°C (68 to 86°F). *C. hypoleuca* (knapweed) and *C. macrocephala* (Globe centaurea) and *C. montana* are hardly perennials which are propagated by division or seed.

Cerastium tomentosum. Snow-in-summer. Hardy perennial. Seeds germinate in two to four weeks at 20°C (68°F). This is easily propagated by division in the fall or by softwood cuttings in summer.

Cheiranthus cheiri. Wallflower. Semihardy perennial often treated as a biennial. Seeds germinate in two to three weeks at 13°C (54°F) and may respond to light (7). Choice plants may be increased by cuttings taken in early summer.

Chelone spp. (110). Turtlehead. A hardy perennial used in wet areas. Propagation is by division or by cuttings. Seeds may require cold stratification for germination.

Chlorophytum comosum. Spider plant. Propagated mainly by planting miniature plants developing at ends of stolons. Stolon formation is under daylength control; short days (12 hours or less daily) promote stolon production (42). It can also be propagated by division.

Chrysanthemum carinatum, C. coronarium, and C. segetum Hybrids. Many cultivars. Hardy annuals and perennials. Seeds germinate in two to four weeks at 20°C (68°F).

C. coccincum. Painted daisy. Propagate by seed as described above or by division.

C. parthenium. Feverfew. Hardy perennial usually grown as an annual. Start from seeds, as described above. Plants easily self-seed and can be divided.

C. maximum. Shasta daisy. Hardy perennial but often treated as a biennial, since it is short-lived. Plants increase from seeds or from the stolon-like shoots, which can be lifted and divided in the fall. Individual pieces have roots and can be transplanted.

C. morifolium. Garden and greenhouse chrysanthemum; and *C. frutescens,* Marguerite. Hardy and semihardy perennials. After flowering, lateral shoots develop from the base of the flowering stems, particularly if the tops are cut back. When the new side shoots are 9 to 10 cm (3½ to 4 in.) long and firm but not woody, they are cut off and rooted as softwood cuttings under mist and with a treatment of indolebutyric acid rooting hormone. The best source of new cuttings is a mother block (or increase block) grown in an isolated area away from the producing area. Such plants are grown in programs designed to keep them pathogen- and virus-free and true-to-type (*10*). It is becoming more common for stock plants to be periodically replaced with new disease-indexed stock plants that have been regenerated through micropropagation (*18*). Softwood cuttings which are disease indexed are then propagated by conventional means. Unrooted cuttings can be held for as long as 30 days at 0.5°C (33°F). For outdoor planting, cuttings may be taken in the same way. In areas with mild winters, cuttings can be taken in late winter for rooting and later transplanting to the garden. In cold-winter areas, the plants should be dug in the fall and brought into the greenhouse or cold frame; cuttings should be made in winter. The plants may also be left in place and divided in spring or fall. If cuttings are taken from the ends of stems high above ground, they are not likely to be infected with soil-borne insects and diseases.

C. nipponicum. Nippon daisy. Propagated by spring tip cuttings. Division of clumps is difficult. Chrysanthemum is readily micropropagated by shoot-tip and petal segment explants (*95*).

Clarkia spp. Hardy annuals. Seeds germinate in one to two weeks at 13 to 32°C (54 to 90°F). Seeds of some strains need light.

Cleome spinosa. Spiderflower. Tender annual. Seeds germinate in one to two weeks at 13 to 32°C (54 to 90°F). Seeds may respond to light (*7*). Can also be propagated by division.

Codiaeum variegatum (*92*). Croton. Tropical perennial. Propagated by leafy terminal cuttings in spring or summer. Tall, "leggy" plants can be propagated by air layering. Stem cutting root number (root initiation) was unaffected by (28°C; 83°F) bottom heat or increasing light intensity; however, root length was increased (*124*). During shipping of cuttings, exposure to light results in shorter roots. Unrooted cuttings can be shipped for 5 to 10 days at 15 to 30°C (60 to 86°F) (*123*).

Colchicum autumnale. Autumn crocus, saffron. Hardy perennial that grows from corms. Seeds are sown as soon as ripe in fall but may require chilling over winter to germinate. Several years are required for plants to reach flowering size.

Coleus blumei. Tender perennials. Seeds germinate in two to three weeks at 20 to 30° C (68 to 86°F) but seedlings will be variable. Selected individuals are propagated by softwood cuttings, which root easily.

Convallaria majalis. Lily-of-the-valley. Hardy perennial that grows as a rhizome, whose end develops a large underground bud, commonly called a "pip." In fall the plants are dug, and the pip, with attached roots, is removed and used as the planting stock. Digging should take place in early autumn, with replanting completed by late autumn. Single pips may be stored in plastic bags in the refrigerator, then planted in late winter for spring bloom (see Fig. 15–19).

Cordyline spp. Ti (*C. terminalis*) is easily propagated by cuttings and by micropropagation (*62*). Other *Cordyline* species can be seed propagated. A vegetative propagation technique for *C. australis, C. kaspar,* and *C. pumillo* by division of the underground stems of stock plants has been described (*88*).

Coreopsis spp. Hardy annuals and perennials. Seeds, which germinate in two to three weeks at 20°C (68°F), may respond to light (*2*). Perennial clumps can be divided in spring or fall. *C. verticillata* can be propagated by cuttings and is hardy to zone 3 (*27*).

Cortaderia selloana. Pampas grass. Best feathery plumes are found on female plants. Propagated by clump division.

Cosmos bipinnatus and C. sulphureus. Half-hardy annual. Seeds germinate in one to two weeks at 20 to 30°C (68 to 86°F) and may respond to light (*7*).

Crassula argentea. Jade plant. Can be propagated at any time by leafbud or stem cuttings.

Crocosmia spp. Crocosmia is used for cut flowers or as a border plant. Offsets that form at the base of corms are propagated (*45*).

Crocus vernus. Dutch crocus. Also other *Crocus* species. Hardy perennial that grows from a corm. Seeds germinate as soon as ripe in summer; several years are required for plants to flower. When leaves die in fall, plants are dug and corms and cormels are separated and planted.

Cucurbita pepo var. ovifera. Ornamental gourds. Tender annuals. Seeds germinate in one to three weeks at 20 to 30°C (68 to 86°F).

Cyclamen spp. (*69, 75*). Tender perennials. Plants grow from a large tuberous underground stem. Cyclamen is propagated best by seeds, which germinate in three to four weeks in the dark at temperatures about 20°C (68°F), no higher than 22°C. Seeds are planted from midsummer to midwinter. Germination is best in a medium of peat moss to which pulverized limestone and mineral nutrients have been added (*127*). Seedlings require one to several years to flower and should be shaded in spring and summer to prevent leaf scald. The tubers can be divided for the production of a few plants identical to the parent. Short shoots of cyclamen with two to three leaves are easily rooted in three weeks when given a 10-second dip of 3000 to 5000 ppm K-IBA under intermittent mist, 21°C (70°F) basal heat and 1 perlite:1 vermiculite rooting medium (*69*).

Cymbalaria muralis. Kenilworth ivy. Semihardy perennial. Seeds germinate in one to four weeks at 12°C (54°F). Self-seeds readily. Softwood cuttings or clump division may be used.

Cynoglossom amabile. Chinese forget-me-not. Hardy biennial grown as an annual. Seeds germinate in two to three weeks at 20°C (68°F) and may respond to light (*7*).

Dahlia. Tender perennials consisting of hundreds of cultivars. Seeds germinate in two to three weeks at 20 to 30°C (68 to 86°F) when planted indoors for later transplanting outdoors. Cultivars must be propagated vegetatively. Plant grows from large tuberous roots. Clumps are dug in the fall before frost and are stored over winter at 2 to 10°C (30 to 50°F), covered with a material such as soil or vermiculite to prevent shrivel-

ing. In spring, when new sprouts begin to appear, divide the clumps so that each root section has at least one sprout. Plant outdoors when danger of frost is over. Dahlias can also be propagated by softwood or leaf-bud cuttings.

Daylily. *See* Hemerocallis spp.

Delphinium spp. Hardy perennials, usually propagated by seeds, which germinate in three to four weeks at 12°C (54°F). Seeds are short-lived and should be used fresh, or stored in containers at low temperature and reduced moisture. Seeds are usually sown outdoors in spring or summer to produce plants that flower the following year. Delphiniums can be propagated by softwood cuttings (*34*). Clumps can be divided in spring or fall, but such plants tend to be short-lived.

 D. ajacis. Larkspur. Hardy annual. Seeds germinate in three to four weeks at 12°C (54°F). Young plants and seeds can be poisonous if eaten.

Deutzia crenata. Deutzia is an excellent ground cover with white flowers. It is propagated from cuttings and hardy to zone 6 (*27*).

Dianthus caryophyllus. (*12*). Carnation. Tender to semihardy perennial grown as an annual that has many cultivars used in florist's trade. Seeds germinate readily but are used primarily for breeding. Carnations are readily propagated by softwood cuttings (*48*). With mist, and with growth regulator treatment, rooting can be done any time of the year. The best source of cuttings is a mother (or stock) block isolated from the producing area, this block originating from cuttings taken from stock plants maintained under a complex program designed to keep them pathogen- and virus-free and true-to-type (see Fig. 8–10). As with chrysanthemum, carnation stock blocks are periodically replenished with meristem-tip culture for disease-free plants. Conventional cutting propagation then proceeds with these clean stock blocks (see Chap. 10). Such rooted cuttings are produced to a large extent by specialist growers, but commercial growing benches may be another source, if careful disease control and selection is practiced in the blocks. Lateral shoots ("breaks") that arise after flowering are removed and used as cuttings. Cuttings root in two to four weeks and may be planted directly to a greenhouse bench or transplanted to peat pots or to a nursery bed. Carnations can also be micropropagated on a large scale using shoot-tip explants (*37*).

 D. chinensis, D. plumarius, and related species. Garden pinks. Hardy perennials, although some kinds are grown as annuals or biennials. Seeds germinate easily in two to three weeks at 20°C (68°F), but may not reproduce the cultivar. Softwood cuttings are taken in early summer and rooted to produce next year's plants. Layering also can be used. Division of *D. plumarius* is the propagation method used by most gardeners.

    D. barbatus.   Sweet William. Perennial but grown as a biennial. Start by seed planted outdoors in spring. In mild-winter areas, transplant to permanent location in fall. In cold-winter areas, overwinter in a cold frame and transplant in spring.

Dicentra spp. Bleedingheart. Hardy perennials. Seeds are sown in late summer or fall for overwintering at low temperatures; alternatively, seeds should be stratified for six weeks below 5°C (41°F) before planting. Seeds will germinate in three to four weeks at 10 to 13°C (50 to 55°F). Divide clumps in spring or fall. Stem cuttings can be rooted if taken in spring after flowering. Root cuttings about 7.5 cm (3 in.) long can be taken from large roots after flowering.

Dictamnus albus. Gasplant. Hardy perennial. Propagation is the same as given for *Dicentra.* Plants should not be disturbed after establishment; thus division is not done. Some people are allergic to plants of this species.

Dieffenbachia spp. Dumbcane. Tropical perennial. Cut stem into 5-cm (2-in.) segments, with one or two nodes per section, and place horizontally, half exposed in sand. New shoots and roots will develop from nodes. If plant gets tall and "leggy," the top may be cut off and rooted as a cutting, or the plant may be air-layered. Leaves and stem are poisonous and may cause rashes on skin (Fig. 10–12).

Digitalis spp. Foxglove. Seeds germinate in two to three weeks at 20 to 30°C (68 to 86°F) and may respond to light (7). Sow seeds outdoors in spring, transplant to a nursery row at 9-in. spacing, then transplant to a permanent location in fall. Perennial species increased by clump division.

Dimorphotheca spp. Cape marigold. Half-hardy annual. Seeds germinate in two to three weeks at 20 to 30°C (68 to 86°F).

Dionaea muscipula. Venus fly trap. Carnivorous plants which have unique appearance, unusual mode of life, and are in demand by plant collectors. Can be propagated by tissue culture, by leaves, adventitious buds, and peduncle explants; see the literature review on micropropagation of other insectivorous plants (80).

Doronicum spp. Leopard bane. Hardy perennial. Seeds germinate in two to three weeks at 20°C (68°F). Divide plants in spring or fall.

Dracaena spp. Variable group of tropical perennial foliage plants which are available in bush, cane, tree, and stump forms. Seeds germinate in three to four weeks at 30°C (86°F). Some species are propagated from leaf-bud cuttings which are treated with IBA and rooted under intermittent mist. *D. fragrans,* which is an important cane form used for interiorscapes, is propagated from cane stem cuttings which are cut into 30- to 183-cm (1- to 6-ft) sections, waxed on the distal (top) end; basal ends are treated with IBA and placed in a porous medium without intermittent mist under shade (indoors or field propagated). Branching of canes during field propagation is done by cutting one-third to one-half way through the cane, which results in the development of lateral buds anywhere from directly below to 15 cm (6 in.) below the cut (24).

Dusty Miller. *See* Senecio spp.

Echeveria. *See* Succulents.

Echinops exaltatus. Globe thistle. Hardy perennials. Seeds germinate in one to four weeks at 20 to 30°C (68 to 86°F). Plants may be divided in spring. Root cuttings, 5- to 7.5-cm (2- to 3-in.) long, may be made in the fall and planted in sandy soil in a cold frame.

Epiphyllum spp. Leaf-flowering cactus. Tender perennial. Seeds do not germinate well when fresh but will after 6 to 12 months' storage if planted in a warm greenhouse. Propagated readily by leaf cuttings or by grafting to *Optunia.* Can be micropropagated (68). *See* Cactus.

Epipremnum aureum. Golden pothos. Among the most important commercially produced foliage plants. Producers cut the long vines into single node, leaf-bud cuttings for propagation. Leaf-bud cutting propagation is enhanced with light intensity of about 2000 ft-c and basal heat of 28°C (83°F) (122). Stock plants should be maintained at four to five nodes (14 to 15 leaves), and a 3-cm or longer internode section below the node and a fraction of the old aerial root should be retained on the cuttings for most rapid axillary shoot development (125).

Eryngium spp. Sea-holly, eryngo. A diverse species of perennials, some of which are used as specimen and border plants. Propagation by division is possible, but a long taproot makes transplanting difficult. Root cutting propagation is the commercial method for species not coming true from seed (*110*). Seeds of *E. bourgatii* have a warm-cold stratification requirement of four weeks at 21°C (70°F), followed by six weeks of 3°C (38°F) and then a warm temperature of 18 to 23°C (65 to 75°F).

Eschscholzia californica. California poppy. Hardy annual. Sow seeds outdoors in fall in mild climates or in early spring in colder areas. Tends to self-sow. Seedlings should not be transplanted because of a long tap root.

Euphorbia spp. Euphorbia, spurge. Perennials used as border and sometimes specimen plants. Propagation is normally done by division. The thick seeds of *E. epithymoides (polychroma)* take 15 to 20 days to germinate at 18 to 21°C (65 to 70°F) (*110*).

Exacum affine. Exacum. A popular greenhouse-grown pot plant with fragrant flowers, multiple blooms, and good postharvest quality. Can be propagated from seed or cuttings. Exacum can also be micropropagated (*118*).

Fatshedera lizei. Tree ivy. Cross between *Hedera helix* and *Fatsia japonica*. Propagated by stem cuttings or by air layering.

Fern (86, 93, 113). Many genera and species. Spores are collected from the spore cases on lower sides of fronds. Examine these sporangia with a magnifying glass to be sure they are ripe but not empty. Place fronds with the spores in a manila envelope and dry for a week at 21°C (70°F). Screen them to separate spores from the chaff. Transfer to a vacuum-tight bottle and store in a dry, cool place. Sow spores evenly on top of sterilized moist soil mixture (e.g., two-thirds peat moss, one-third perlite in flats), paying particular attention to sanitation. Leave 1 in. space on top and cover with a pane of glass. Use 18 to 24°C (65 to 75°F) air temperature; bottom heat may be helpful. Keep moist, preferably using distilled water to avoid salt injury.

Spores germinate and produce moss-like growth ⅛ in. thick, composed of many prothallia. Fertilization of the archegonium on the underside of the prothallus occurs in three to six months. In a first transplanting, a small piece of prothallus is removed with tweezers and transplanted to wider spacing in a new flat of soil mixture. The prothallia expands to about ½ in. in diameter and produce tiny sporophyte plants with primary leaves and roots. The fern plant will grow from a second transplanting. Procedures have been devel-

oped for propagating ferns from spores in vitro using nutrient agar solutions (*58, 66, 117*).

Several vegetative propagation methods are possible. Ferns grow from thick rhizomes, which can be divided. Certain species (e.g., *Cystopteris bulbifera*) produce small "bulblets," about the size of a pea, on the underside of the leaf. These drop when mature, are planted, and produce a fern plant by the second year. Other species produce small vegetative buds on the upper surface or edge of the leaves; these detach and form new plants. *Polystichum setiferum* also produce bulbils in the axils of fronds which will root and produce young plants when the fronds are pegged down on propagation media (*113*).

Cultivars of the sterile (nonpropagatable by spores) Boston fern group (*Nephrolepsis*) are now largely micropropagated starting with rhizome tips (*17*) (see p. 509). This technique is also applicable to other fern genera, such as *Adiantum* (maidenhair fern), *Alsophila* (Australian tree fern), *Pteris* (brake fern), *Microlepia, Platycerium* (staghorn fern), and *Woodwardia* (chain fern).

Freesia spp. Tender perennials. Seeds planted in fall germinate in four to six weeks and will bloom the next spring. A germination temperature of about 18.5°C (65°F) is best (*40*). Plants grow from corms, which are planted in spring and dug in fall. Small cormels are removed at this time and replanted to grow for future flowering.

Fuchsia × hybrida. *Fuchsia magellanica* hybrids. Fuchsia are tender perennials treated as annuals that are utilized as hanging baskets, containers, or trained to tree form on standards. The most effective way to propagate cultivars is by cuttings taken in spring or late summer. For optimum cutting production it is best to maintain stock plants on 10-hour photoperiods (*121*).

Gaillardia spp. Blanketflower. Annual and hardy perennials. Seeds germinate in two to three weeks at 20°C (68°F) and may respond to light (*7*). Perennial kinds are planted in spring to bloom the following year. These may also be started from root cuttings or may be divided in spring or fall but are not long-lived.

Galanthus spp. Snowdrop. Hardly perennial. Bulbs are planted in the fall for bloom the following spring. Offsets are removed when bulbs are dug. Can also be propagated by clump division and micropropagation (*23*).

Gasteria. *See* Succulents.

Gazania spp.   Tender perennial often grown as an annual. Propagated by seeds sown in spring or fall, or by softwood cuttings taken in late summer that are rooted in a cold frame, then transplanted in spring. Divide clumps after three or four years.

Gentiana spp.   Gentian. Many species, mostly hardy perennials, although some are annuals and biennials. Plant fresh seed in the fall to overwinter outdoors. Seeds germinate in one to four weeks at 20°C (68°F). However, seeds need to be precooled [three weeks at 2°C (36°F)] or treated with 300 ppm GA₃ (*13*). Seedlings are very delicate and should not be transplanted until roots are established during the first month. Cutting propagation is used for some white cultivars which have poor seed germination. Micropropagated liners are now becoming available.

Geranium.   *See* Pelargonium × hortorum.

Gerbera jamesonii.   Transvaal daisy. Tender perennial. Seeds germinate in two to three weeks at 20°C (68°F); it is important to use fresh seed. Or remove basal shoots from the rhizome and use as cuttings. Micropropagation from shoot tips can be used for rapid, large-scale multiplication (*65, 82*).

Geum spp.   Avens. Hardy perennials. Seeds germinate in three to four weeks at 20 to 30°C (68 to 86°F). Propagate also by clump division in spring or fall.

Gladiolus.   Tender perennial grown from a corm. Seed propagation is used for developing new cultivars. Seeds are planted in spring either indoors for later transplanting or outdoors when danger of frost is over (see Chap. 15).

Gloxinia (*Sinningia speciosa*).   Grown from seed for commercial production. Seeds germinate rapidly if placed in an artificial medium under intermittent mist and held at about 21°C (70°F) night temperature. Flowering plants develop in six to seven months. Gloxinia can also be micropropagated using leaf explants (*55*).

Godetia spp.   Hardy annuals. Sow seeds in early spring; these germinate in two to three weeks at 20°C (68°F).

Gold Dust Plant (*Aucuba japonica* 'Variegata').   Propagated by leafy stem cuttings under mist, which root easily, or by root cuttings. Grows well in shade.

Gypsophila spp. (*G. elegans*).   Baby's breath. Annual. Seed germinates in two to three weeks at 20°C (68°F).

G. paniculata.   Hardy perennial. Started by seed as above. Plants can be divided in spring and fall. Double-flowered cultivars are grafted on seedling *G. paniculata* (single-flowering) roots. Grafting can be done in summer and fall, using outdoor-grown plants for rootstocks and placing them in a cold frame for healing of the graft; grafting is also done in winter and early spring, using greenhouse-grown stock plants. *G. paniculata* can be micropropagated using shoot-tip explants (*64*).

Haworthia.   *See* Succulents.

Helenium autumnale.   Sneezeweed. Hardy perennial. Seeds germinate in one to two weeks at 20°C (68°F). Cultivars are increased by division. Separate rooted shoots in spring, line-out in nursery, then transplant in fall and winter.

Helianthemum nummularium.   Sunrose. Half-hardy perennial. Seeds germinate in two to three weeks at 20 to 30°C (68 to 86°F). Cultivars are propagated by softwood cuttings taken from young shoots in spring. Transplant to pots and place in permanent location the following winter or spring. Division of clumps is also possible, but plants tend to be short-lived.

Helianthus annuus.   Sunflower. Hardy annual. Seeds germinate in two to three weeks at 20 to 30°C (68 to 86°F). *H. decapetalus* and other hardy perennial species are increased by division.

Heliconia spp. (*107*).   These tropical ornamental herbaceous perennials are prized for their showy inflorescences. They are easily propagated by division of the rhizomes.

Heliopsis scabra.   Heliopsis. Hardy perennial. Seeds germinate in one to two weeks at 20°C (68°F). Divide clumps in fall.

Heliotropium spp.   Heliotrope. Tender perennial usually grown as an annual. Seed germinates in three to four weeks at 20 to 30°C (68 to 86°F) and may respond to light (*7*). Start indoors for later spring planting. Take softwood cuttings of side shoots in fall or spring and root at low temperatures 10°C (50°F) in slightly moist conditions.

Helleborus spp.   Hellebore, Christmas rose. Perennials used as border plants. Propagation by seed is very slow. *H. lividus* has a double dormancy requirement of warm stratification (21°C; 70°F) for 8 to 10 weeks followed by cool stratification of 3°C (37°F) for 8 to 10 weeks; seeds still may take up to two years to germinate. Division is the most common method by carefully separating the crown (*110*). Roots and leaves are poisonous.

Hemerocallis spp. (*30*). Daylily. Hardy perennial. Seeds require about six weeks of moist-chilling for good germination. Seed propagation is used only to develop new cultivars and requires six weeks of stratification; germination takes three to seven weeks at 16 to 21°C (60 to 70°F) (*110*). Divide clumps in fall or spring, separating into rooted sections, each with about three offshoots. Clones can also be micropropagated using flower petals and sepals as explants (*1, 78*).

Heuchera spp. American alumroot, coralbells. Perennials used as border plants for their foliage and flowers. Propagated by clump division or by leaf cuttings. Seeds germinate in three weeks at 70°F (21°C).

Hippeastrum spp. Amaryllis. Tender bulbous perennial. Remove and pot bulb offsets, which will flower the second year. Make bulb cuttings in late summer. Dry membranous seeds are borne in loculicidal dehiscing capsules. Seeds germinate under warm conditions 20 to 30°C (68 to 86°F). Seedlings take two to four years to produce flowers. Can also be micropropagated (*23*).

Hosta spp. Plantain-lily. Herbaceous perennials which are used for massed plantings or as specimen plants for their foliage and flowers. Propagated by clump division in spring. One producer removes the apical dominance of the crown (terminal) bud and slices (divides) the remaining clumps into quarters which are then placed outdoors in trays (England) which are winter protected and then planted when they begin to shoot (*45*). It takes three years to produce a flowering plant from seed. Micropropagation is being used with new cultivars to speed up propagation.

Hunnemannia fumariifolia. Goldencup. Tender perennial often grown as annual. Seeds germinate in two to three weeks at 20°C (68°F). For bloom first year, sow seeds early indoors then transplant outdoors when danger of freezing is over.

Hyacinthus spp. Hyacinth. Hardy, spring-flowering perennial; bulbs are planted in the fall. Removal of offset bulbs gives small increase. For commercial propagation, new bulbs are obtained by scoring or scooping mature bulbs. Micropropagation, using segments of the bulb, leaf, inflorescence, or stem as explant, is successful (*52*). Seeds may be planted outdoors in fall, but up to six years are required to produce blooms (see Chap. 15).

Hydrangea macrophylla. Florists' hydrangea. See Chap. 19, page 582.

Iberis spp. Candytuft. Hardy annual and perennial species. Seeds germinate in one to two weeks at 20 to 30°C (68 to 86°F) but may need light. Root softwood cuttings in summer or divide clumps in fall.

Impatiens spp. Snapweed, touch-me-not, balsam. Perennials and half-hardy annuals. Seeds germinate in two to four weeks at 20°C (68°F) and may respond to light (*7*). Perennial species can be started by cuttings, and by micropropagation (*108*).

Incarvillea spp. Incarvillea. Hardy perennial. Seeds germinate in one to two weeks at 20°C (68°F). Divide in fall or, preferably, in spring.

Ipomoea spp. Morning glory. Tender perennial grown as an annual. Seeds germinate in one to three weeks at 20 to 30°C (68 to 86°F). Notch seed coats or soak seeds overnight in warm water before planting.

Iresine spp. Bloodleaf. Tender perennial. Softwood cuttings root easily. Keep stock plants over winter in greenhouse and take cuttings in late winter or spring.

Iris spp. Perennials. There are several different groups of hardy or semihardy iris, which grow either from rhizomes or from bulbs. Rhizomes are divided after bloom. Discard the older portion and use only the vigorous side shoots. Leaves are trimmed to about 15 cm (6 in.).

Bulbous species follow a typical spring-flowering, fall-planting sequence. The old bulb completely disintegrates, leaving a cluster of various-size new bulbs. These are separated and graded, the largest size being used to produce flowers, the smaller for further growth.

Seeds, which are used to propagate species and to develop new cultivars, should be planted as soon as ripe after being given a moist-chilling period; germination is often irregular and slow. Removal of embryo from the seed and growing it in artificial culture has given prompt germination in some cases. Iris can be micropropagated, which greatly hastens production of new cultivars over the customary division of rhizomes (*52*).

Ixia spp. Corn lily. Tender, summer- or fall-flowering perennials grown from corms. In cold climates these are dug in fall and stored over winter. Small cormels are removed and planted in the ground or in flats to reach flowering size, as is done with gladiolus.

Ixora spp. (90). Ixora. Several species are used in Hawaii as landscape and flowering pot plants. Many species are easy to propagate by cuttings. Difficult-to-root *I. acuminata* three-node cuttings had optimal rooting when given a five-second dip of IBA-NAA, both at 2500 ppm.

Kalanchoe spp. Tropical perennials. *See also* Succulents.

Kangaroo Paw. *See* Anigozanthus spp.

Kniphofia Hybrids (*K. tritoma*) (110). Torch lily or poker plant. Perennials used as specimen plants, borders, and cut flowers. Normally propagated by division. Seed propagation takes two to three years to produce a flowering plant.

Lantana sellowiana, L. camara. Lantana. Tender perennials. Seeds germinate in six to seven weeks at 20°C (68°F). Softwood cuttings root easily.

Lathyrus latifolius. Perennial pea vine. Hardy perennial. Seeds germinate in two to three weeks at 20 to 30°C (68 to 86°F). Clumps may be divided. *L. odoratus.* Sweet pea. Hardy annual. Seed germinates in two weeks at 20°C (68°F). Notching seed or soaking in warm water may hasten germination. Plant outdoors in fall where winters are mild, in spring where winters are severe.

Lavandula spp. Lavender. Half-hardy perennial native to Mediterranean. Seeds, which may be planted in winter, germinate in two to three weeks at 11 to 32°C (52 to 90°F). Take cuttings from side shoots in later summer or fall; plant in soil and cover or start in cold frame. Divide clumps in the fall.

Lespedeza spp. (33). Bushclover. Considered a herbaceous perennial in more northern latitudes or a semi-woody shrub in the southern United States. Excellent landscape shrub for massing and screening with blue-green foliage and purple flowers. Seeds can be direct sown after harvest or scarified with a 15-minute acid treatment if stored. Roots easily from softwood cuttings with 1000 ppm IBA in 50 percent ethanol quick dip.

Leucojum spp. (110). Spring snowflake. Plants produced from seed take four to five years to flower. Propagation is normally done by separating bulbs, which is done by digging bulbs after foliage has turned brown. Can be micropropagated (23).

Liatris spp. Gayfeather. Herbaceous perennials that are being utilized for cut-flower crops (4). Seeds will germinate in three to four weeks at 21°C (70°F), but flower development takes two years. Asexual propagation is by woody corms or rhizomes which are divided in the spring (110).

Lilium spp. (76). Lily. Hardy perennials. These are spring- and summer-flowering plants grown from scaly bulbs; most have a vertical axis, but in some species growth is horizontal with a rhizomatous structure. Lilies include many species, hybrids, and named cultivars. Seed propagation is used for species and for new cultivars. Seeds of different lily species have different germination requirements (94).

Immediate seed germinators include most commercially important species and hybrids (*L. amabile, L. concolor, L. longiflorum, L. regale, L. tigrinum,* Aurelian hybrids, Mid-Century hybrids, and others). Germination is epigeous; shoots generally emerge three to six weeks after planting at moderately high temperatures. Treat seeds with a fungicide to control *Botrytis.* Sow ¾ in. deep in flats during winter or outdoors in a seedbed in early spring. Dig the small bulblets in fall, sort for size, and replant with similar sizes together. Plants normally grow two years in a seedbed and two years in a nursery row before producing good-sized flowering bulbs.

Another group is the *slow seed germinators* of the epigeal type [(ETP) L. *candidum,* L. *henryi,* Aurelian hybrids, and others] in which seed germination is slow and erratic; the procedures used are essentially the same as described above.

The most difficult group to propagate are the *slow seed germinators* of the hypogeous type [(HTP) L. *auratum,* L. *bolanderi,* L. *canadense,* L. *martagon,* L. *parvum,* L. *speciosum,* and others]. Seeds of this group require three months under warm conditions for the root to grow and produce a small bulblet, then a cold period of about six weeks, followed by another warm period in which the leaves and stem begin to grow. This sequence can be provided by planting the seeds outdoors in summer as soon as they are ripe, or planting seeds in flats and then storing under appropriate conditions to provide the required temperature sequence. Vegetative methods of propagation include natural increase of the bulbs, such as bulblet production on underground stems (either naturally or artificially), aerial stem bulblets (bulbils), or scaling. Outer and middle scales are used for scale propagation to increase the number of forcible commercial bulbs (74). These procedures are described in Chap. 15.

Lilies can be micropropagated from bulb scales (112) and pedicels (71). L. *longiflorum* can be propagated by leaf cuttings.

Limonium spp. Statice. L. *sinuatum* is a perennial herb native to the eastern Mediterranean which is grown commercially around the world as a cut flower for both fresh and dry-flower arrangements. Plants are propagated by seed and will remain in a rosette form until sufficient cool temperature is obtained to satisfy vernalization and bolting, and subsequent anthesis is induced. Statice has recently been micropropagated (*19, 44*).

Linaria spp. Toadflax. Hardy annual and perennials. Seeds germinate in two to three weeks at 12°C (54°F). Perennial species take two years to produce bloom from seed. Clumps can be divided in spring or fall.

Linum spp. Flax. Hardy annual and perennial species. Seeds germinate in three to four weeks at 12°C (54°F). Divide clumps of perennial species in fall or spring.

Lobelia erinus. Lobelia. Tender perennial grown as an annual. Seeds germinate in two to three weeks at 20 to 30°C (68 to 86°F), but seedling growth is slow. May respond to light (*7*). Start indoors 10 to 12 weeks before transplanting outdoors after last frost. Mature plants, if potted in the fall and kept in greenhouse over winter, can be used to provide new growth for cuttings to be taken in late winter.

Lobelia spp. Hardy perennials. Seeds germinate in three to four weeks at 20 to 30°C (68 to 86°F). Species self-seeds. Divide clumps in fall or spring.

Lobularia maritima. Sweet alyssum. Perennial but grown as hardy annual. Seed germinates in one to two weeks at 20°C (68°F) and blooms appear in six weeks.

Lunaria annua. Honesty. Biennial, sometimes grown as an annual. Seeds germinate in two to three weeks at 20°C (68°F). *L. rediviva.* Hardy perennial. Propagated by seed as described above. Also increased by division.

Lupinus hartwegii, L. nanus, and Others. Hardy annuals. Seeds germinate in two to three weeks at 20°C (68°F). *L. polyphyllus.* Perennial. Seeds germinate in three to four weeks at 20°C (68°F). Seed coats may be hard and should be scarified. *L. arboreus.* Tree lupine. Hardy perennial. Start from seeds indoors and transplant to permanent location. *L.* 'Russell Hybrid'. Sow seeds in spring or summer, or propagate by cuttings taken in early spring with small piece of root or crown attached.

Lychnis spp. Campion. Mostly hardy perennials but some are grown as annuals or biennials. Seeds germinate in three to four weeks at 20°C (68°F). Clumps can be divided in spring or fall.

Lycoris spp. Spider lily. Tender and semihardy bulbous perennials. Propagation is by bulb offsets, which are removed when the dormant bulbs are dug. These are replanted to grow larger. Bulb cuttings can also be used for increase.

Malva alcea. Hollyhock mallow. Drought tolerant perennial native to Italy; utilized in border plantings. Propagated by division. Seed is another method since this species self-seeds.

Maranta leuconeura. Prayer plant. Perennial herbaceous plants that are used in terrariums or as small potted plants. They are normally propagated asexually by stem cuttings because seeds rarely germinate. Cuttings are rooted in humidity tents for four to six weeks. *M. leuconeura* 'Kerchoviana' can be micropropagated (*36*).

Marigold. *See* Tagetes spp.

Matthiola incana. Common stock. *M. longipetala bicornis.* Evening scented stock. Perennial grown as biennial or annual. Seed germinates in two weeks at 12 to 32°C (54 to 90°F) and may respond to light (*7*). Seeds are sown in summer or fall for winter bloom, in late winter indoors for spring bloom, or outdoors in spring for summer bloom.

Mesembryanthemum spp. *See* Succulents.

Mimulus spp. Monkey flower. Includes many species of tender to hardy plants. Mostly perennials, but sometimes grown as annuals. Seeds germinate in one to two weeks at 12°C (54°F). Softwood cuttings taken from young shoots can be rooted.

Miscanthus sinensis. 'Gracillimus' (maiden grass) and 'Variegatus' (variegated Japanese silver grass) are perennial grasses that make good specimens in the landscape with their showy feathery inflorescences. They are propagated by division during the cooler months (California) and are hardy to zone 5 (*27*).

Molluccella laevis. Bells of Ireland. Half-hardy annual. Seed germinates in three to five weeks at 10°C (50°F). Difficult to transplant; sow seeds in place.

Monarda didyma (*110*). Bee balm. A perennial garden plant native to eastern North America. Can be propagated by seed, which germinates in two to three weeks at 16 to 21°C (60 to 70°F). Can also be propagated by softwood cuttings or by clump division.

Monstera deliciosa.   Often misnamed cut-leaf philo-dendron. Easily propagated by rooting sections of the main stem, by stem cuttings, or by air layering.

Muscari spp.   Grape hyacinth. Hardy bulbous perennial. Plant blooms in spring; bulbs become dormant in fall when they are lifted and divided. Increase by removing bulb offsets. Seed propagation can also be used.

Myosotis sylvatica.   Forget-me-not. Hardy biennial. Seeds germinate in two to three weeks at 20°C (68°F). Sow in summer and transplant to permanent location the following spring. *M. scorpioides* is a perennial started from seed; division in spring is also used.

Narcissus, Pseudo-narcissus, and Other Species. Daffodil. Hardy perennial spring-flowering bulb. Vegetative propagation procedures are described in Chap. 15. Vegetative techniques include twin scaling, chipping (almost the same method as twin scaling, except that the bulb is cut across the root plate into 16 pieces with up to two bulbils developing per section), and micropropagation (*23, 50*).

Nasturtium.   *See* Tropaeolum majus.

Nelumbo lutea.   American lotus. An aquatic plant for water gardens in the urban landscape. Usually propagated vegetatively through rhizome division or tuber production. Rhizome cultures were established in vitro from excised embryos, and maximum rhizome growth and development occurred with medium supplemented with GA_3; similar in vitro techniques could be used to propagate commercially important flowering water garden plants (*56*).

Nemesia strumosa.   Half-hardy annual. Seeds germinate in two to three weeks at 13°C (55°F). Sow outdoors in spring in cold climates, in fall in mild-winter areas.

Nepeta mussinii.   Nepeta, catmint. Hardy perennial. Seeds germinate in two to three weeks at 20°C (68°F). Softwood cuttings of nonflowering side shoots taken in early summer root readily. Cuttings may be inserted directly into soil if protected. Plants may be divided in spring, using newest parts and discarding older portion of clumps.

Nicotiana spp.   Flowering tobacco. Half-hardy annuals. Seeds germinate in one to two weeks at 20 to 30°C (68 to 86°F) and may respond to light (*7*). Sow indoors four to six weeks before last frost then transplant out-of-doors.

Nierembergia spp.   Cupflower. Tender perennial, sometimes grown as an annual. Seeds germinate in two to three weeks at 20 to 30°C (68 to 86°F). Softwood cuttings removed from new growth in spring root readily. Clumps can be divided.

Nymphaea spp.   Water lily. Consists of numerous species and many named cultivars. Plants grow as rhizomes. Clumps are divided in spring. Seeds are used to grow natural species and to develop new cultivars. Tropical water lily hybrids and species grow from tubers. Seeds do not reproduce hybrids. Both kinds of seeds are planted 2.5 cm (1 in.) deep in sandy soil, then immersed in water 3 to 4 in. deep. Hardy species should be started at 16°C (60°F), tropical species at 21 to 27°C (70 to 80°F). Vegetative propagation is either from small tubers that can be removed from old tubers in fall or from small plantlets growing from the leaf (*100*). *N.* 'Gladstone' production can be extended by photoperiod control (*57*).

Oenothera spp.   Evening-primrose. Hardy perennials, but some kinds are biennial. Seeds germinate in one to three weeks at 20 to 30°C (68 to 86°F). Plants can also be increased by division of clumps in the fall.

Opuntia.   *See* Cactus.

Orchids (*81, 97, 109*).   Many genera, hybrids, and cultivars are cultivated, and many more are found in nature. Some, such as *Aerides, Arachnis, Phalaenopsis, Renanthera,* and *Vanda,* exhibit *monopodial* habit of growth. This means they are erect and grow continuously from the shoot apex and can be propagated by tip cuttings. Adventitious roots are produced along the stem and inflorescences are produced laterally from leaf axils. Most others, including *Brassovola, Calanthe, Cattleya, Cymbidium, Laelia, Miltonia, Odontoglossom, Oncidium,* and *Phaius* have a *sympodial* habit of growth, are procumbent, and do not grow continuously from the apex. Their main axis is a rhizome in which new growth arises from offshoots, or "breaks." Pseudobulbs are usually present on plants of this type. Many orchids are *epiphytes* (i.e., air plants), typically growing on branches of trees. Others (*Cymbidium, Cypripedium, Paphiopedilum*) are terrestrial and grow in the ground.

   Seed propagation is mainly used for hybridization. Many important cultivars are seedling hybrids, either from species or between genera, resulting from controlled crosses of carefully selected parents. Many such important crosses are between tetraploid and diploid parents to produce triploids. These offspring are sterile and are not in turn usable as parents. Seedling variation occurs since orchids are heterozygous. Five to seven years is required for a seedling plant to bloom.

Orchid flowers are hand-pollinated. A seed capsule requires 6 to 12 months to mature. A single capsule will contain many thousands of tiny seeds with relatively undeveloped embryos. In vitro culture is universally used for seed propagation. (The procedure is described in Chap. 17.) Knudson's C medium is usually used. Arditti (2) has summarized the many experiences of testing various nutritional and other factors for orchid seed germination. Orchid seed can be stored for many years if held in sealed containers over calcium chloride at about 2°C (36°F).

Vegetative methods for orchids are generally slow, difficult for many genera, and usually too low-yielding for extensive commercial use. Sympodial species are increased by division of the rhizome while it is dormant or just as new growth begins. Four or five pseudobulbs are included in each section. "Back-bulbs" and "greenbulbs" can be used for some genera.

Orchids with long cane-like stems, as *Dendroium* and *Epidendrum,* sometimes produce offshoots ('keiki") that produce roots. Offshoots can also be produced if the stem is cut off and laid horizontally in moist sphagnum or some other medium. Flower stems of *Phaius* and *Phalaenopsis* can be cut off after blooming and handled in the same way. A drastic method of inducing offshoots is to cut out or mutilate the growing point of *Phalaenopsis,* remove the small leaves, and treat the injured portion with a fungicide. Offshoots may then be produced. Monopodial species can be propagated by long (30 to 37 cm) tip cuttings with a few roots already present. Air layering also is possible.

Vegetative propagation by proliferation of shoot-tip (meristem) cultures in vitro has revolutionized orchid propagation, particularly for *Cymbidium, Cattleya,* and some other genera (3, 81). The shoot growing point is dissected from the plant and grown on a special, sterile medium; a proliferated mass of tissue and small protocorms develops, which can be divided periodically. Many thousands of separate protocorms can be developed in this way within a matter of months, each of which will eventually differentiate shoots and roots to produce an orchid plant. The procedure is described in Chap. 17. *Vanilla planifolia,* which is an orchid essential for its oil that grows as a vine, is normally propagated commercially by cuttings; nodal stem explants of this species were successfully micropropagated with BA and microshoots rooted *ex vitro* (59).

Paeonia spp. Paeonia hybrids (*P. hybryda*), fernleaf peony (*P. tenuifolia*), tree peony [*P. suffructicosa* (*P. arborea*)]. Hardy perennials native to China used for specimen plants in borders and as cut flowers. Seed propagation is difficult, taking five to seven years to produce a flowering plant from seed (110). Germina-

tion may take one to two years to meet epicotyl dormancy requirement. Seeds are sown in fall for cold stratification requirement during the winter. Roots develop during the first summer, and shoots develop the second spring. Plants developed from seed are generally not true to type. Another method is to collect seeds before they become black and completely ripe. Do not allow them to dry out; sow in pots, which should be buried in the ground for six to seven weeks. Roots will develop; dig up and plant in a protected location or under mulch over winter. Best propagation method is to divide clumps in fall; each piece should have at least one bud or "eye," preferably three to five. *P. suffruticosa* is grafted in late summer on *P. lactiflora* understock.

Pansy. *See* Viola cornuta.

Papaver nudicaule. Iceland poppy. Hardy perennial grown as biennial. Seeds germinate in one to two weeks at 12°C (54°F). Sow in permanent location in summer for bloom next year. *P. orientale.* Oriental poppy. Hardy perennial. Very fine seeds, which may respond to light, should be covered very lightly (7). Seeds germinate in one to two weeks at 12°C (54°F). Cultivars are propagated by root cuttings. Dig when leaves die down in fall, cut into 7.5- to 10-cm (3- to 4-in.) sections, lay horizontally in sandy soil in a flat covered to 2.5 cm (1 in.). Root cuttings are transplanted in spring. Or dig plants in spring, prepare root cuttings, and plant directly in a permanent location. *P. rhoeas.* Corn poppy, Shirley poppy. Hardy annual. Seed germinates in one to two weeks at 13°C (55°F). Sow in late summer for early spring bloom or in early spring for summer bloom. Do not transplant.

Pelargonium × hortorum. Geranium. Started by cuttings and by seed. Traditionally propagated by cuttings, which root easily, but *Pythium* and *Botrytis* infection can be serious problems. Pathogen-free stock, identified by culture indexing, should be used and can be supplied by specialists (116). There may be practical value in applying ABA in the shipment and storage of geranium cuttings (6).

In the mid-1970s large-scale seed propagation of geraniums began with the introduction of certain cultivars that would grow from seed to flower in 14 to 16 weeks. Seeds germinate best at about 22°C (72°F) in an artificial medium (4, 7). Chlormequat (cycocel) is utilized in greenhouse geranium production to produce compact, early-flowering, well-branched plants; there may be advantages in applying this growth retardant in seed-propagated garden geraniums (96).

Pennisetum spp. Fountain grass. Perennial grasses with feathery inflorescences that do not set seed. They are propagated by division during cooler weather (California) (*27*).

Pentstemon spp. Beardtongue. Semihardy to hardy perennials, sometimes handled as annuals. Seeds germinate in two to three weeks at 20 to 30°C (68 to 86°F) but growth is slow and uneven; seeds may respond to light (*7*). Plants started indoors in early spring and transplanted outdoors later may bloom the first year. Plants are usually short-lived. Softwood cuttings taken from nonflowering side shoots of old plants root readily. Make cuttings in fall to obtain plants for next season. Clumps may be divided.

Peperomia spp. Tender perennial. Softwood stem, leaf-bud, or leaf cuttings root readily. Plants can be divided. Peperomia can also be micropropagated from excised leaf explants (*46*).

Periwinkle. *See* Vinca.

Petunia Hybrids. Petunia. Tender perennial often grown as annual. Seeds germinate in one to two weeks at 20°C (68°F). Seeds of double-flowered cultivars and some F₁ hybrids may need light and high temperature (27 to 29°C; 80 to 85°F) for good germination. It is best to start plants indoors for later outdoor planting. Softwood cuttings taken in late summer or fall from side shoots root easily.

Philodendron spp. Tropical vines. Seeds germinate readily at about 25°C (77°F) if sown as soon as they are ripe and before they become dry. Best propagation methods are use of leaf-bud or stem cuttings, rooting sections of main stem, or by air layering.

Phlox divaricata. Sweet William. Hardy perennial. Expose seeds to cold during winter before planting. Softwood cuttings taken in spring root easily. Divide clumps in spring or fall.

Phlox drummondii. Annual phlox. Hardy. Seed germinates in two to three weeks at 20°C (68°F). Start indoors for later outdoor planting, or outdoors after frost.

Phlox paniculata. Garden phlox. Hardy perennial. Plants do not come true from seed. Sow seeds as soon as ripe in fall to germinate the next spring. They will germinate in three to four weeks at 20°C (68°F). Grow plants one season and transplant in fall. Softwood cuttings taken from young shoots in spring or summer root easily, but they are subject to damping-off if kept too wet. It is best to propagate from root cuttings. Dig

clumps in fall; remove all large roots to within 5 cm (2 in.) of crown (which is replanted). Cut roots into 2-in. lengths and place in flats of sandy soil; cover 13 mm (½ in.) deep. Transplant next spring. Divide clumps in fall or spring. Phlox can be micropropagated using shoot explants.

Physostegia virginiana. False dragonhead or lionsheart. A herbaceous perennial which is used as a field-grown cut flower crop (*5*). Seeds will germinate in two to three weeks at 23°C (75°F). This species can also be propagated by division.

Poinsettia (*Euphorbia pulcherrima*). See Chap. 19.

Polemonium spp. Jacob's ladder. Hardy perennial. Seeds germinate in three to four weeks at 20 to 30°C (68 to 86°F). Divide clumps or root stem cuttings.

Polianthes tuberosa. Tuberose. Tender bulbous perennial. Propagated by removing offsets at planting time. The small bulbs take more than one year to flower. Divide clumps every four years.

Poliomintha longiflora (*51*). Mexican oregano. Has small evergreen leaves which smell like the spice oregano and is a striking landscape plant which produces light lavender flowers throughout the summer. It is easily propagated by semihardwood cuttings.

Polygonum spp (*110*). Polygonatum, fleeceflower, Solomon's seal, bistort. Versatile perennials and annuals used as ground covers, rock gardens, planters, and hanging baskets. Generally propagated by division or seed. *P. capitatum* 'Magic Carpet', which is an excellent annual ground cover, is propagated by seed at 21 to 27°C (70 to 80°F) and germinates in three weeks.

Portulaca grandiflora. Rose moss. Half-hardy annual. Seeds germinate in two to three weeks at 20 to 30°C (68 to 86°F); they respond to light (*7*). Reseeds itself. Can be propagated by cuttings.

Primula obconica. Top primrose. Tender perennial grown as an annual. Seeds germinate well in a cool greenhouse or after three to four weeks at 12 to 32°C (54 to 90°F). The very tiny seeds respond to light and should not be covered. *Primula malacoides*. Fairy primrose. Seeds will germinate in two to three weeks at 20 to 30°C (68 to 86°F). *Primula sinensis*. Chinese primrose. Seeds will germinate in three to four weeks at 20°C (68°F). In these species double-flowering cultivars do not produce seeds but are propagated by cuttings taken in spring, or by division. Other *Primula* spp., *P.* × *polyantha* and *P. vulgaris*. Hardy perennials grown outdoors. Seeds germinate in three to six weeks at 20°C (68°F), but some species may require lower temperatures (*26, 39*). It is best to collect and sow seeds as soon as they are ripe in the fall. Cuttings taken in spring root easily. Clumps can be divided just after flowering.

Ranunculus asiaticus.   Turban or Persian ranunculus. Tender perennial. Seeds germinate in one to four weeks at 20°C (68°F). Grows from tuberous roots which can be divided. *Ranunculus* spp., Buttercup. Hardy perennial. Seeds germinate in one to four weeks at 20°C (68°F). Plant seeds as soon as they are ripe. Divide plants in spring or fall.

Reseda odorata.   Mignonette. Hardy annual. Seeds should not be covered; they germinate in two to three weeks at 12°C (54°F) and respond to light (*7*).

Ricinus communis.   Castor bean. Soak seeds in water for 24 hours or nick with file before planting. Plant each seed in individual pot and after frost danger is past, transplant to outdoor location. Seeds are poisonous.

Rodgersia pinnata.   Featherleaf, Rodgersflower. Perennial for moist border with attractive foliage and ornamental flower. Propagation is by division.

Rudbeckia spp.   Black-eyed Susan, coneflower. Hardy annual, biennial, and perennial species. Seeds germinate in two to three weeks at 20 to 30°C (68 to 86°F). Perennial kinds are propagated by clump division. Will reseed naturally.

Saintpaulia ionantha.   African violet (*53, 61, 128*). Tropical perennial. Propagation is by seed, division, or cuttings. The very fine seeds, which germinate in three to four weeks at 30°C (86°F), should not be covered. Seedlings are subject to damping-off. Vegetative methods are necessary to maintain cultivars. Plants can be divided. Leaf cuttings (blade and petiole) are easily rooted, either in a rooting medium or in water. Rapid, large-scale propagation can be accomplished using leaf petiole sections by in vitro culture techniques (*14, 29, 106*).

Salpiglossis sinuata.   Painted tongue. Semihardy annual. Seeds are difficult to germinate but some will start in one to two weeks at 20 to 30°C (68 to 86°F). Start indoors in peat pots. Can be micropropagated (*70*).

Salvia spp.   Sage. Annual, biennial, and perennial species. Salvia are used as borders or as cut flowers (*5*). Seeds germinate in two to three weeks at 20 to 30°C (68 to 86°F) and may respond to light (*7*). Soak flats thoroughly, and do not rewater until sprouted; they also grow well under mist (*10*). Softwood cuttings of young shoots 3 to 4 in. long root readily. *S. officinalis* basal cuttings root better than apical cuttings. Flowering reduces rooting and removal of flowers enhances propagation. Rooting is enhanced with basal dips of 1000 ppm K-IBA salt and 100 ppm dithiothreitol (DTT) (*91*). Rooting ability was highest in spring (Israel). *S. greggii* is easily propagated by semihardwood cuttings (*51*). Plants can be divided, but such divisions are slow to recover.

   S. splendens. Scarlet sage. Tender perennial grown as an annual. Germinate seeds at 20 to 30°C (68 to 86°F), then grow at 13°C (55°F) night temperature. Softwood cuttings taken in fall root readily.

Sansevieria trifasciata and S. 'hahnii' (Dwarf Form).   Bowstrip hemp. Snakeplant. Tropical perennial. Plants grow from a rhizome, which can be readily divided. Leaves may be cut into sections, several inches long, and inserted into a rooting medium; a new shoot and roots will develop from base of leaf cutting. The variegated form, *S. t.* 'Laurentii', is a chimera, which can be maintained only by division. *S. trifasciata* has been micropropagated (*15*).

Santolina chamaecyparissus (*incana*).   Lavender cotton. A perennial native to the Mediterranean that is used as carpet bedding or a low hedge. It needs to be trimmed to maintain compact growth. It is normally propagated by cuttings.

Sanvitalia procumbens.   Creeping zinnia. Hardy annual. Seeds germinate in one to two weeks at 20°C (68°F) and may respond to light (*7*). Sow in place in spring, or in fall in mild climates.

Saponaria officinalis.   Bouncing Bet. Hardy perennial. Seeds germinate in two to three weeks at 20°C (68°F). The plant spreads rapidly by an underground creeping stem which can be divided. *Saponaria vaccaria.* Soapwort. Hardly annual. Seeds germinate in two to three weeks at 20°C (68°F).

Saxifraga spp.   Many interesting unusual species and hybrids. Mostly hardy perennials. Seeds germinate easily; they are sown preferably when ripe. Some hybrids and cultivars are maintained only by vegetative methods and are propagated once flowering is finished in June or July (England) (*101*). Cuttings are small and slow growing—it takes one year to grow a liner. Cuttings are rooted in a cold frame or seed tray. Most plants grow as small rosettes and are easily propagated by making small cuttings involving single rosettes. Small plants from stolons root readily. Plants can be divided in spring or fall. *S. stolonifera*, strawberry geranium, is a tender perennial that reproduces by runners.

Scabiosa spp. Pincushion flower. Annual and hardy perennial species. Seeds germinate in two to three weeks at 20 to 30°C (68 to 86°F). Perennial kinds can be divided.

Scarlet Sage. *See* Salvia splendens.

Schefflera arboricola (43). Schefflera is an important foliage plant that can be propagated easily by seeds, cuttings or air layering. Basal cuttings develop more roots and longer shoots and require less time to break lateral buds than do apical cuttings. As single-node cutting length increased to 20 cm (8 in.) so did rooting, bud-break, and shoot growth.

Schizanthus spp. Butterfly flower. Tender annual. Seeds germinate in one to two weeks at 12°C (54°F) and are sensitive to high temperatures (7). Sow seeds in fall for early spring bloom indoors, or sow in early spring to be transplanted outdoors for summer bloom.

Scilla spp. Squill. Includes several kinds of bulbous hardy spring-flowering perennials. Dig plants when leaves die down in summer and remove the bulblets. *S. autumnalis* is planted in spring and blooms in fall.

Sedum spp. (101). Sedum is composed of a wide range of species including herbaceous perennials, evergreens, and monocarps. Many of the *Sedum* spp. can be raised by seed, but generally this method is limited to herbaceous perennials only; seeds germinate at 15 to 18°C (65°F). The mat-forming species are propagated by division, since the creeping shoots root into the ground as they travel and mats are easily pulled apart. Direct sticking of cuttings into containers is done since many species root so readily.

Sempervivum. *See* Succulents.

Senecio spp. Dusty miller (*S. cineraria*). These tender perennials are grown as annuals. Seeds germinate in two to three weeks at 20°C (68°F). Stem-tip cuttings root rapidly if treated with IBA and placed under mist over bottom heat. Florists' cineraria (*S. cruentus*) is a cool-season crop with true blue or lavender flowers.

Seeds can be obtained from self-pollinated plants, but size and quality of flowers deteriorate after one to two generations. Plants produced vegetatively by cuttings have reduced vigor and flower size. This species is micropropagated (25).

Sidalcea malviflora (110). Sidalcea. A native western U.S. perennial used as a border plant. Cultivars are commonly propagated by division. Can be seed propagated, but cultivars do not come true from seed.

Sinningia speciosa. Gloxinia. Tropical perennial. Commonly grown from seeds, which are very fine and require light. Sow in uncovered, well-drained, peat moss medium; they germinate in two to three weeks at 20°C (68°F). Vegetative methods are required to reproduce cultivars. Plant grows from a tuberous root on which a rosette of leaves is produced. The root can be divided as described for tuberous begonia. Softwood cuttings or leaf cuttings taken in spring from young shoots starting from the tubers root easily.

Snapdragon. *See* Antirrhinum majus.

Spider Plant. *See* Chlorophytum comosum.

Stachys spp. Lamb's ear (*S. byzantina*), big betony (*S. grandiflora*). Hardy perennials used as border plants or ground covers. Propagated by clump division or by seed.

Stock. *See* Matthiola.

Stokesia laevis. Stokes aster. Hardy perennial. Seeds germinate in four to six weeks at 20 to 30°C (68 to 86°F) and the plants bloom the first year. Make root cuttings or divide clumps in spring.

Strelitzia reginae. Bird-of paradise. Seed propagation of this tropical perennial is undesirable due to juvenility and genetic variation. Seeds are sown under warm conditions, and freshly harvested seed should be used to avoid seed coat impermeability. Bottom heat of 37°C (98°F) aids germination, which occurs in 6 to 10 weeks (22). This species grows from a rhizome, which can be divided in the spring; however, division is limited by a low rate of multiplication with 0.5 to 1.5 divisions per branch per year. A technique has been developed to overcome the strong apical dominance which inhibits branching of axillary buds into propagules. A triangular incision with a knife is made at the base of a separated branch 8 to 12 mm above the basal plate to reach and remove the apex from adult plants (119). After two to six months, 2 to 30 lateral shoots develop from each fan (separated branch). During the next six months, newly formed laterals root and can be divided into individual plants (see Chap. 10 and Fig. 15–22).

Streptocarpus × hybridus. Cape primrose is a herbaceous perennial in the Gesneriaceae. It is generally propagated by seed or leaf cuttings. Many of the new commercial cultivars bloom throughout the year with large blooms on compact plants, but are sterile hybrids and must be vegetatively propagated. Cape primrose has been successfully micropropagated using explants of leaf discs, shoot apices, stem, petiole, pedicel segments, and corolla flower parts (*83*).

Succulents (*11, 38, 85, 107*). This loosely defined horticultural group includes many genera, such as *Agave, Aloe, Crassula, Echevaria, Euphorbia, Gasteria, Haworthia, Hoya, Kalanchoe, Portulacaria, Sedum,* and *Yucca.* These are plants with fleshy stems and leaves that store water, or plants that are highly drought resistant. Most are half-hardy or tender perennials. Seed propagation is possible, although young plants are often slow to develop and to produce flowers. It is best to germinate seeds indoors at high day temperatures (29 to 35°C; 85 to 95°F). Seedlings are susceptible to damping-off.

Cuttings of most species root readily—either stem, leaf-bud, or leaf—in a 1:1 peat–perlite medium. They should be exposed to the open air or inserted into dry sand for a few days to allow callus to develop over the cut end. Some protection from drying is needed during rooting. Some species can be reproduced by removing offsets. Grafting is possible as described for cacti (see Fig. 11–4). Many of the succulents can be micropropagated (*73, 102, 104, 111*).

Sweet Alyssum. *See Lobularia maritima.*

Sweet Pea. *See* Lathyrus.

Tagetes spp. Marigold. Tender annuals. Seeds germinate readily in one week at 20 to 30°C (68 to 86°F) and sometimes respond to light (*7*). Sow in place in spring after frost. *T. erecta* (African marigold) has been micropropagated from leaf segments (*60*).

Thalictrum spp. Meadow rue. Hardy perennial. Seeds germinate in four to six weeks at 20°C (68°F). Sometimes hard seeds are present. Plants can be divided in spring or fall.

Thermopsis caroliniana. Tender or hardy perennials. Use fresh seeds, which will germinate in two to three weeks at 20 to 30°C (68 to 86°F). Plants can be divided, but it is best to leave them undisturbed.

Thunbergia spp. Clockvine. Tender perennials grown as annuals. Seeds germinate in two to three weeks at 20 to 30°C (68 to 86°F), but seedlings grow slowly. Softwood cuttings taken from new shoots root readily.

Thymus spp. Thyme. Hardy perennials. Seeds germinate in one to two weeks at 12 to 32°C (54 to 90°F). Sometimes germination is promoted by light. May be increased by division or by softwood cuttings taken in summer.

Ti. *See* Cordyline terminalis.

Tigridia pavonia. Tiger flower. Tender bulbous perennials. Plant bulbs in spring and dig in fall when leaves die. Increase by removal of small bulblets just before they are planted. Easily started by seed.

Tithonia rotundifolia. Mexican sunflower. Tender perennial grown as an annual. Seeds germinate in two to three weeks at 20 to 30°C (68 to 86°F).

Tolmiea menziesii. Piggyback plant. New plantlets form on upper surface of leaves. Such leaves are removed and the petiole is stuck in rooting medium to depth of new plantlet, which then resumes growth.

Torenia fournieri. Wishbone flower (*103*). Torenia is a tender perennial grown as an annual. Seeds germinate at 21 to 23°C (70 to 75°F) media temperature in 10 to 14 days. Seedling and plugs should be grown at 750 foot-candles on 18-hour photoperiods.

Trachymene coerulea. Laceflower. Tender perennial grown as an annual. Seeds germinate in two to three weeks at 20°C (68°F).

Trollius spp. Globeflower. Hardy perennial. Plant seeds in fall, which produce plants that flower by next spring (*16*). Also increases by clump division.

Tropaeolum majus. Nasturtium. Tender perennial grown as an annual. Plant seeds in place; they germinate in one to two weeks at 20°C (68°F), but are difficult to transplant. Double-flowering kinds must be propagated vegetatively, usually by softwood cuttings.

Tulipa spp. and Hybrid Cultivars. Tulip. Hardy bulbous perennials. Plant bulbs in fall for spring bloom. Seeds are used to reproduce species and for breeding new cultivars. They germinate readily if planted as soon as ripe. Vegetative methods include removal of offset bulbs in the fall. Different bulb sizes are planted separately, since the time required to produce flowers varies with size. For details of procedure, see Chap. 15.

Valeriana officinalis. Valerian. Hardy perennial. Seeds germinate in three to four weeks at 12 to 32°C (54 to 90°F). New plants can be obtained by clump division.

Venidium fastuosum. Venidium. Half-hardy annual. Seeds germinate in four to six weeks at 20 to 30°C (68 to 86°F); or sow outdoors at 10 to 13°C (50 to 55°F).

Verbascum spp. Mullein. Hardy perennials and biennials. Seeds are slow to germinate; best temperature is 30°C (86°F). Propagate named cultivars by root cuttings taken in early spring.

Verbena spp. Verbena [*Verbena* × *hybrida* (*V.* × *hortensis*)]. Tender perennial grown as an annual. Seeds germinate in three to four weeks at 20 to 30°C (68 to 86°F); sometimes promoted by light (*7*). May be propagated by softwood cuttings taken in summer. *V. canadensis.* Clump verbena. Hardy perennial that blooms first year from seed. Seeds germinate in two to four weeks at 12 to 32°C (54 to 90°F). Plants can be propagated by division or by softwood cuttings.

Veronica spp. Speedwell. Hardy perennials. Seeds germinate in two weeks at 12 to 32°C (54 to 90°F). Plants are increased by division in spring or fall or by softwood cuttings taken in the spring or summer. *V. spicata* has been micropropagated successfully and 80 percent rooting occurred *ex vitro* without auxin (*105*).

Vinca major. Tender perennial. Propagate by division or by softwood cuttings taken in summer. *V. minor.* Periwinkle. Hardy perennial. Seeds germinate in two to three weeks at 20°C (68°F). Easily propagated by softwood cuttings or by division. *V. minor* is also micropropagated (*105*). *V. rosea.* Madagascar periwinkle. Tender perennial grown at 20°C (68°F).

Viola cornuta. Horned violet, tufted pansy. Hardy perennial. Seeds germinate in two to three weeks at 12 to 32°C (54 to 90°F). Seeds of some cultivars need light. Vegetative propagation is by cuttings taken from new shoots obtained by heavy cutting back in the fall. Clumps may also be divided. *V. tricolor hortensis.* Pansy. Hardy or semihardy, short-lived perennial often grown as an annual. Usually propagated by seeds as described for *V. cornuta* but may also be increased by cuttings or by division. *V. odorata.* Sweet violet. Tender to semihardy perennials. Grows by rhizome-like stems, which can be separated from others on the crown and treated as a cutting with some roots present. *Viola* species. Many hardy perennial kinds. These are grown by seeds as above, but germination may be slow and seeds are best exposed to cold before planting. Many species produce seeds in inconspicuous, enclosed (cleistogamous) flowers near the ground, whereas the conspicuous, showy flowers produce few or no seeds. These plants can also be reproduced by cuttings or by division (*26*).

Yucca spp. Yucca. Tender to semihardy perennials. Seeds germinate at 20°C (68°F) but rather slowly and require four to five years to flower. Plants are monocots; some are essentially stemless and grow as a rosette, while others have either long or short stems. Offsets growing from the base of the plant can be removed and handled as cuttings; sometimes entire branches or the top of the plant can be detached a few inches below the place where leaves are borne and replanted in sandy soil. Sections of old stems can be laid on sand or other medium in a warm greenhouse, and new side shoots that develop can be detached and rooted. *Y. elephantipes* is rooted by long canes (*89*).

Zantedeschia spp. Calla. Several species, which have similar propagation requirements. Tropical perennials. Plants grow by thickened rhizomes, which produce offsets or rooted side shoots; these are removed and planted. Calla are also micropropagated (*54*).

Zebrina pendula. Wandering Jew. Easily propagated at any season by stem cuttings.

Zinnia elegans and Other Species. Zinnia. Half-hardy hot-weather annual. Seeds germinate outdoors in one week at 20 to 30°C (68 to 86°F). Sometimes seeds respond to light (*7*).

REFERENCES

1. Apps, D. A., and C. W. Heuser. 1975. Vegetative propagation of *Hemerocallis*—including tissue culture. *Proc. Inter. Plant Prop. Soc.* 25:362–67.

2. Arditti, J. 1967. Factors affecting the germination of orchid seeds. *Bot. Rev.* 33:1-97.

3. ———. 1977. *Orchid biology.* Ithaca, N.Y.: Cornell Univ. Press.

4. Armitage, A. M. 1986. *Seed propagated geraniums.* Portland, Oreg.: Timber Press.

5. Armitage, A. M. 1987. The influence of spacing on field grown perennial crops. *HortScience* 22:904-7.

6. Arteca, R. N., D. S. Tsai, and C. Schlangnhaufer. 1985. Abscisic acid effects on photosynthesis and transpiration in geranium cuttings. *HortScience* 20:370-72.

7. Assn. Off. Seed Anal. 1970. Rules for seed testing. *Proc. Assn. Off. Seed Anal.* 60(2):1-115.

8. Ault, J. R., and W. J. Blackmon. 1987. In vitro propagation of *Ferocactus acanthodes* (Cactaceae). *HortScience* 22:126-27.

9. Bailey, L. H., E. Z. Bailey, and Staff of Bailey Hortorium. 1976. *Hortus third.* New York: Macmillan.

10. Ball, V., ed. 1985. *The Ball red book: Greenhouse growing* (14th ed.). Reston, Va.: Reston Publ. Co.

11. Bayer, M. B. 1982. *The new Haworthia handbook.* Pretoria: National Botanic Gardens of South Africa.

12. Besemer, S. 1980. Carnations. In *Introduction to floriculture,* R. A. Larson, ed. New York: Academic Press.

13. Bicknell, R. A. 1984. Seed propagation of *Gentiana scabra. Proc. Inter. Plant Prop. Soc.* 34:396-401.

14. Bilkey, P. C., B. H. McCown, and A. C. Hildebrandt. 1978. Micropropagation of African violet from petiole cross-sections. *HortScience* 13(1):37-38.

15. Blazich, F. A., and R. T. Novitzky. 1984. In vitro propagation of *Sansevieria trifasciata. HortScience* 19:122-23.

16. Brumback, W. E. 1985. Propagation of wildflowers. *Proc. Inter. Plant Prop. Soc.* 35:542-48.

17. Burr, R. W. 1975. Mass production of Boston fern through tissue culture. *Proc. Inter. Plant Prop. Soc.* 25:122-24.

18. Bush, S. R., E. D. Earle, and R. W. Langhans. 1976. Plantlets from petal segments, petal epidermis, and shoot tips of the periclinal chimera, *Chrysanthemum morifolium* 'Indianapolis'. *Amer. Jour. Bot.* 63:729-37.

19. Butcher, S. M., R. A. Bicknell, J. F. Seelye,

and N. K. Borst. 1986. Propagation of *Limonium peregrinum. Proc. Inter. Plant Prop. Soc.* 36:448-50.

20. Carter, F. M. 1973. Grafting cacti. *Horticulture* 51(3):34-35.

21. Chidamian, C. 1984. *Book of cacti and other succulents.* Portland, Oreg.: Timber Press.

22. Chopping, N. 1986. Replacing the bird: Pollination in the genus *Strelitzia. Proc. Inter. Plant Prop. Soc.* 36:204-7.

23. Christie, C. B. 1985. Propagation of amaryllids: A brief review. *Proc. Inter. Plant Prop. Soc.* 35:351-57.

24. Cialone, J. 1984. Developments in *Dracena* production. *Proc. Inter. Plant Prop. Soc.* 34:491-94.

25. Cockrel, A. D., G. L. McDaniel, and E. T. Graham. 1986. In vitro propagation of florists' cineraria. *HortScience* 21:139-40.

26. Colborn, L. N. 1986. Primroses and violets: What's new. *Proc. Inter. Plant Prop. Soc.* 36:245-49.

27. Connor, D. M. 1985. Plants for the discriminating propagator. *Proc. Inter. Plant Prop. Soc.* 35:274-77.

28. Conover, C. A., and R. T. Poole. 1978. Production of *Ficus elastica* 'Decora' standards. *HortScience* 13(6):707-8.

29. Cooke, R. C. 1977. Tissue culture propagation of African violets. *HortScience* 12(6):549.

30. Darrow, G. M., and F. G. Meyer, eds. 1968. *Day lily handbook.* American Horticultural Society, Vol. 47, No. 2.

31. Davies, F. T., Jr., and B. C. Moser. 1980. Stimulation of bud and shoot development of Rieger begonia leaf cuttings with cytokinins. *Jour. Amer. Soc. Hort. Sci.* 105:27-30.

32. Deburg, P., and J. DeWall. 1977. Mass propagation of *Ficus lyrata. Acta Hort.* 78:361-64.

33. Dirr, M. A. 1985. Ten woody plants that deserve a longer look. *Proc. Inter. Plant Prop. Soc.* 35:728-34.

34. Dodge, M. H. 1978. Propagation of named delphinium cultivars. *Proc. Inter. Plant Prop. Soc.* 28:496-98.

35. Dodge, M. H. 1985. Propagation of herbaceous perennials by root cuttings. *Proc. Inter. Plant Prop. Soc.* 35:548-55.

36. Dunston, S., and E. Sutter. 1984. In vitro propagation of prayer plants. *HortScience* 19:511-12.

37. Earle, E. D., and R. W. Langhans. 1975. Car-

nation propagation from shoot tips cultured in liquid medium. *HortScience* 10(6):608–10.

38. Edinger, P., ed. 1970. *Succulents and cactus.* Menlo Park, Calif.: Lane.

39. Erickson, D. M. 1985. Propagating and growing primroses in the Pacific Northwest. *Proc. Inter. Plant Prop. Soc.* 35:219–21.

40. Gilbertson-Ferriss, T. L., and H. F. Wilkins. 1977. Factors influencing seed germination of *Freesia refracta* Klatt cv. Royal Mix. *HortScience* 12(6):572–73.

41. Gill, L. M. 1984. Anemone tuber production in southwest England. *Proc. Inter. Plant Prop. Soc.* 34:290–94.

42. Hammer, P. A. 1976. Stolon formation in *Chlorophytum. HortScience* 11(6):570–72.

43. Hansen, J. 1986. Influence of cutting position and stem length on rooting of leaf-bud cuttings of *Schefflera arboricola. Scient. Hort.* 28:177–86.

44. Harazy, A., B. Leshem, A. Cohen, and H. D. Rabinowitch. 1985. In vitro propagation of statice as an aid to breeding. *HortScience* 20:361–62.

45. Hardy, G. 1986. Herbaceous plant production at Blooms of Bressingham Ltd. *Proc. Inter. Plant Proc. Soc.* 36:314–19.

46. Henny, R. J. 1978. In vitro propagation of *Peperomia* 'Red Ripple' from leaf discs. *HortScience* 13(2):150–51.

47. Higaki, T., and D. P. Watson. 1973. Anthurium culture in Hawaii. *Univ. Hawaii Coop. Ext. Ser. Circ. 420.*

48. Holley, W. D., and R. Baker. 1963. *Carnation production.* Dubuque, Iowa: William C. Brown Co., Publishers.

49. Hosoki, T., and T. Asahira. 1980. In vitro propagation of bromeliads in liquid culture. *HortScience* 15(5):603–4.

50. Houghton, W. J. 1984. New narcissus and their propagation. *Proc. Inter. Plant Prop. Soc.* 34:294–96.

51. Hubbard, A. C. 1986. Native ornamentals for the U.S. southwest. 1986. *Proc. Inter. Plant Prop. Soc.* 36:347–50.

52. Hussey, G. 1975. Totipotency in tissue explants of some members of the Liliaceae, Iridaceae, and Amaryllidaceae. *Jour. Exp. Bot.* 26:253–62.

53. Jackson, H. C. 1975. Propagation and culture of African violets. *Proc. Inter. Plant Prop. Soc.* 25:269–71.

54. Jamieson, A. C. 1988. New Zealand callas. *Grower Talks* 51:56–60.

55. Johnson, B. B. 1978. In vitro propagation of gloxinia from leaf explants. *HortScience* 13(2): 149–50.

56. Kane, M. E., T. J. Sheehan, and F. H. Ferwerda. 1988. In vitro growth of American lotus embryos. *HortScience* 23:611–13.

57. Kelly, J. W., and J. J. Frett. 1986. Photoperiodic control of growth in water lilies. *HortScience* 21:151.

58. Knauss, J. F. 1976. A partial tissue culture method for pathogen-free propagation of selected fern from spores. *Proc. Fla. State Hort. Soc.* 89:363–65.

59. Kononowicz, H., and J. Janick. 1984. In vitro propagation of *Vanilla planifolia. HortScience* 19:58–59.

60. Kothari, S. L., and N. Chandra. 1984. In vitro propagation of African marigold. *HortScience* 19:703–5.

61. Kramer, J. 1971. *How to grow African violets* (4th ed.). Menlo Park, Calif.: Lane.

62. Kunisaki, J. T. 1975. In vitro propagation of *Cordyline terminalis* (L.) Kurth. *HortScience* 10(6): 601–2.

63. ———. 1980. In vitro propagation of *Anthurium andreanum* Lind. *HortScience* 15(4):508–9.

64. Kusey, W. E., Jr., P. A. Hammer, and T. C. Weiler. 1980. In vitro propagation of *Gypsophila paniculata* L. 'Bristol Fairy'. *HortScience* 15(5):600–601.

65. Laliberté, S., L. Chretien, and J. Vieth. 1985. In vitro plantlet production from young capitulum explants of *Gerbera jamesonii. HortScience* 20:137–39.

66. Lane, B. C. 1981. A procedure for propagating ferns from spore using a nutrient agar solution. *Proc. Inter. Plant Prop. Soc.* 30:94–97.

67. Lawson, G. M., and P. B. Goodwin. 1985. Commercial production of kangaroo paws. *Proc. Inter. Plant Prop. Soc.* 35:57–65.

68. Lazarte, J. E., M. S. Gaiser, and O. R. Brown. 1982. In vitro propagation of *Epiphyllum chrysocardium. HortScience* 17:84.

69. Lee, C. I., and H. C. Kohl. 1985. Note on vegetative reproduction of *Cyclamen indicum. Plant Prop.* 31:4.

70. Lee, C. W., R. M. Skirvin, A. I. Soltero, and J. Janick. 1977. Tissue culture of *Salpiglossis sinuata* L. from leaf discs. *HortScience* 12(6):547–49.

71. Liu, L., and D. W. Burger. 1986. In vitro prop-

agation of Easter lily from pedicels. *HortScience* 21:1437–38.

72. Lyons, R. E., and R. E. Widmer. 1980. Origin and historical aspects of *Cyclamen persicum* Mill. *HortScience* 15(2):132–35.

73. Makino, R. K., P. J. Makino, and T. Murashige. 1977. Rapid cloning of *Ficus* cultivars through application of in vitro methodology. *In Vitro* 13(3):160.

74. Matsuo, E., and J. M. van Tuyl. 1986. Early scale propagation results in forcible bulbs of Easter lily. *HortScience* 21:1006–7.

75. McMillan, R. N. 1980. Cyclamen production problems. *Proc. Inter. Plant Prop. Soc.* 29:173–76.

76. McRae, E. A. 1978. Commercial propagation of lilies. *Proc. Inter. Plant Prop. Soc.* 28:166–69.

77. Mekers, O. 1977. In vitro propagation of some Tillandsiodeae (Bromeliaceae). *Acta Hort.* 78:311–20.

78. Meyer, M. M. 1976. Propagation of daylilies by tissue culture. *HortScience* 11(5):485–87.

79. Mikkelsen, E. P., and K. C. Sink, Jr. 1978. In vitro propagation of Rieger Elatior begonias. *HortScience* 13(3):242–44.

80. Minocha, S. C. 1985. In vitro propagation of *Dionaea muscipula*. *HortScience* 20:216–17.

81. Morel, G. M. 1966. Meristem culture: Clonal propagation of orchids. *Orchid Digest* 30(2):45–49.

82. Murashige, T., M. Serpa, and J. B. Jones. 1974. Clonal multiplication of gerbera through tissue culture. *HortScience* 9(3):175–80.

83. Peck, D. E., and B. G. Cumming. 1984. In vitro vegetative propagation of cape primrose using the corolla of the flower. *HortScience* 19:399–400.

84. Peck, D. E., and B. G. Cumming. 1984. In vitro propagation of *Begonia × tuberhydra* from leaf sections. *HortScience* 19:395–97.

85. Perl, P. 1978. *Cacti and succulents.* Alexandria, Va.: Time-Life Books.

86. ———. 1977. *Ferns.* Alexandria, Va.: Time-Life Books.

87. Pfister, J. M., and J. M. Widholm. 1984. Plant regeneration from snapdragon tissue culture. *HortScience* 19:852–54.

88. Platt, G. C. 1985. Propagation of the cordylines by vegetative means. *Proc. Inter. Plant Prop. Soc.* 35:364–66.

89. Poole, R. T., and C. A. Conover. 1987. Veg-

etative propagation of foliage plants. *Proc. Inter. Plant Prop. Soc.* 37:503–7.

90. Rauch, F. D., and R. M. Yamakawa. 1980. Effects of auxin on rooting of *Ixora acuminata*. *HortScience* 15:97.

91. Raviv, M., E. Putieusky, and D. Sanderovich. 1984. Rooting stem cuttings of sage (*Salvia officinalis* L.). *The Plant Propagator* 30:8–9.

92. Raward, I. D. 1975. Propagation of *Codiaeum* (croton) by tip cuttings. *Proc. Inter. Plant Prop. Soc.* 25:386.

93. Roberts, D. J. 1965. Modern propagation of ferns. *Proc. Inter. Plant Prop. Soc.* 15:317–21.

94. Rockwell, F. F., G. C. Grayson, and J. de Graff. 1961. *The complete book of lilies.* Garden City, N.Y.: Doubleday.

95. Roest, S., and G. S. Bokelmann. 1975. Vegetative propagation of *Chrysanthemum morifolium* Ram in vitro. *Scient. Hort.* 3:317–30.

96. Schwartz, M. A., R. N. Payne, and G. Sites. 1985. Residual effects of chlormequat on garden performance in sun and shade of seed and cutting propagated cultivars of geraniums. *HortScience* 20:368–70.

97. Sheehan, T. J. 1983. Recent advances in botany, propagation and physiology of orchids. In *Horticultural reviews*, Vol. 5, J. A. Janick ed., Westport, Conn., AVI Publ. Co., pp. 279–315.

98. Sheerin, P. 1974. Propagation of various types of begonia. *Proc. Inter. Plant Prop. Soc.* 24:292–93.

99. Simmonds, J. A., and T. Werry. 1987. Liquid-shake culture for improved micropropagation of *Begonia × hiemalis*. *HortScience* 22:122–24.

100. Slocum, P. D. 1985. Propagating water lilies and aquatics. *Brooklyn Botanic Garden Record* 41:25–28.

101. Small, D. J. 1986. Propagation of choice alpines. *Proc. Inter. Plant Prop. Soc.* 36:241–44.

102. Smith, R. H., and A. E. Nightingale. 1979. In vitro propagation of *Kalanchoe*. *HortScience* 14(1):20.

103. Solem, M. 1988. Torenia. *Grower Talks* 52:18.

104. Standifer, L. C., E. N. O'Rourke, and R. Porche-Sorbet. 1984. Propagation of *Haworthia* from floral scapes. *The Plant Propagator* 30:4–6.

105. Stapfer, R. E., and C. W. Heuser. 1985. In vitro propagation of periwinkle. *HortScience* 20:141–42.

106. Start, N. D., and B. G. Cumming 1976. In

vitro propagation of *Saintpaulia ionantha* Wendl. *HortScience* 11(3):204–6.

107. Stefanis, J. P., and R. W. Langhans. 1980. Factors affecting the production and propagation of xerophytic succulent species. *HortScience* 15(4):504–5.

108. Stephens, L. C., S. L. Krell, and J. L. Weigle. 1985. In vitro propagation of Java, New Guinea and Java × New Guinea *Impatiens. HortScience* 20:362–63.

109. Stewart, J., and E. Hennessy. 1981. *Orchids of Africa.* Boston: Houghton Mifflin.

110. Still, S. M. 1988. *Herbaceous ornamental plants* (3rd ed.). Champaign, Ill.: Stipes Publ. Co.

111. Stimart, D. P. 1986. Commercial micropropagation of florist flower crops. In *Tissue culture as a plant production system for horticultural crops*, R. H. Zimmerman et al., eds. Dordrecht: Martinus Nijhoff Publishers.

112. Stimart, D. P., and P. D. Ascher. 1978. Tissue culture of bulb scale sections for asexual propagation of *Lilium longiflorum* Thumb. *Jour. Amer. Soc. Hort. Sci.* 103:182–84.

113. Stokes, P. 1984. Hardy ferns. *Proc. Inter. Plant Prop. Soc.* 34:332–33.

114. Tanabe, M. J. 1980. Effect of depulping and growth regulators on seed germination of *Alyxia olivaeformis. HortScience* 15:199–200.

115. Tanabe, M. J. 1982. Single node stem propagation of *Alyxia olivaeformis.*

116. Thorn-Horst, A., R. K. Horst, S. H. Smith, and W. A. Oglevee. 1977. A virus-indexing tissue culture system for geraniums. *Flor. Rev.* 160(4148):28–29, 72–74.

117. Tjosvold, S. 1978. Uniform spore dispersal on warm nutrient agar solution. *Univ. Calif. Nursery and Flower Rpt.*, summer issue.

118. Torres, K. C., and N. J. Natarella. 1984. In

vitro propagation of Exacum. *HortScience* 19:224–25.

119. Van de Pol, P. A., and T. F. van Hell. 1988. Vegetative propagation of *Strelitzia reginae. Acta Hort.* 226:581–86.

120. Van Dyk, M., and R. Currah. 1983. Vegetative propagation of prairie forbs native to southern Alberta, Canada. *The Plant Propagator* 28:12–14.

121. Von Hentig, W. U., M. Fisher, and K. Köhler. 1984. Influence of daylength on the production and quality of cuttings from *Fuchsia* mother plants. *Proc. Inter. Plant Prop. Soc.* 34:141–49.

122. Wang, Y. T. 1987. Effect of warm-medium, light intensity, BA and parent leaf on propagation of golden pothos. *HortScience* 22:597–99.

123. Wang, Y. T. 1987. Effect of temperature, duration and light during simulated shipping on quality and rooting of croton cuttings. *HortScience* 22:1301–2.

124. Wang, Y. T. 1987. Influence of light and heated medium on rooting and shoot growth of two foliage plant species. *HortScience* 23:346–47.

125. Wang, Y. T., and C. A. Boogher. 1988. Effect of nodal position, cutting length, and root retention on the propagation of golden pothos. *HortScience* 23:347–49.

126. Welander, T. 1978. In vitro organogenesis in explants from different cultivars of *Begonia × hiemalis. Physiol. Plant.* 41:142–45.

127. Widmer, R. E. 1980. Cyclamens. In *Introduction to floriculture*, R. A. Larson, ed. New York: Academic Press.

128. Wilson, H. V. 1980. *Saintpaulia* species. *Amer. Hort.* 58(6):35–39.

129. Zieslin, N., and A. Keren. 1980. Effects of rootstock on cactus grafted with an adhesive. *HortScience* 21:153–54.

SUPPLEMENTARY READING

ADEN, P. 1988. *The hosta book.* Portland, Oreg.: Timber Press.

ARMITAGE, A. M. 1986. *Seed propagated geraniums.* Portland, Oreg.: Timber Press.

BALL, V. 1985. *Ball red book: Greenhouse growing* (14th ed.) Reston, Va.: Reston Publ. Co.

BENSON, L. 1983. *The cacti of the U.S. and Canada.* Palo Alto, Calif.: Stanford Univ. Press.

CHIDAMIAN, C. 1984. *Book of cacti and other succulents.* Portland, Oreg.: Timber Press.

GEORGE, A. S. 1985. *The banksia book.* Portland, Oreg.: Timber Press.

GILES, F. A., R. MCINTOSH, and D. C. SAUPE. 1980. *Herbaceous perennials.* Reston, Va.: Reston.

GRAF, A. B. 1986. Exotica 4: *Pictorial encyclopedia of exotic plants from tropical and near-tropical regions* (12th ed.). East Rutherford, N.J.: Roehrs Company.

Growers Books. 1980. *New cut flower crops. Grower guide 18.* London: Grower Books.

International Plant Propagators' Society. Proceedings of annual meetings.

JOINER, J. N., ed. 1981. *Foliage plant production.* Englewood Cliffs, N.J.: Prentice-Hall.

KOHLEIN, F. 1988. *Iris.* Portland, Oreg.: Timber Press.

LARSON, R. A., ed. 1980. *Introduction to floriculture.* New York: Academic Press.

LAURIE, A., D. C. KIPLINGER, and K. S. NELSON. 1979. *Commercial flower forcing* (8th ed.). New York: McGraw-Hill.

LEVY, M. 1988. Perennial production—a grower's notebook for summer. *Grower Talks* 52:72–78.

MASTALERZ, J. W., ed. 1976. *Bedding plants* (2nd ed.). University Park, Pa.: Pennsylvania State Univ.

NEEL, P. L. 1979. Macropropagation of tropical plants as practiced in Florida. *Proc. Inter. Plant Prop. Soc.* 29:468–80.

PADILLA, V. 1986. *Bromeliads.* New York: Crown Publ. Co.

POOLE, R. T., and C. A. CONOVER. 1988. Vegetative propagation of foliage plants. *Proc. Inter. Plant Prop. Soc.* 37:503–7.

PROFESSIONAL PLANT GROWERS ASSOCIATION. Annual proceedings.

REILLY, A. 1978. *Park's success with seeds.* Greenwood, S.C.: Geo. W. Park Seed Co., Inc.

RICE, G. 1986. *A handbook of annuals and bedding plants.* Portland, Oreg: Timber Press.

STILL, S. M. 1988. *Herbaceous ornamental plants* (3rd ed.). Champaign, Ill.: Stipes Publ. Co.

STIMART, D. P. 1986. Commercial micropropagation of florist flower crops. In *Tissue culture as a plant production system for horticultural crops,* R. H. Zimmerman, et al., eds. Dordrecht: Martinus Nijhoff Publishers.

WELLS, J. S. 1988. *Modern miniature daffodils.* Portland, Oreg.: Timber Press.

YOUNG, J. A. C. 1986. *Collecting, processing and germinating seeds of wildland plants.* Portland, Oreg.: Timber Press.

ZILIS, M, D. ZWAGERMAN, D. LAMBERTS, and L. KURTZ. 1979. Commercial propagation of perennials by tissue culture. *Proc. Inter. Plant Prop. Soc.* 29:404–13.

Indexes

SUBJECT INDEX

[1]Boldface numbers indicate figures.

PLANT INDEX, SCIENTIFIC NAMES

[1]Boldface numbers indicate figures.

PLANT INDEX, COMMON NAMES

[1]Boldface numbers indicate figures.